USING MULTIVARIATE STATISTICS

SECOND EDITION

Barbara G. Tabachnick
Linda S. Fidell

California State University, Northridge

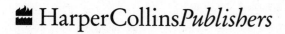
HarperCollins*Publishers*

Editor in Chief: Judith L. Rothman
Project Editor: Ellen MacElree
Cover Design: CIRCA 86, Inc.
Text Art: RDL Artset Ltd.
Production Manager: Willie Lane
Compositor: Arcata Graphics/Kingsport
Printer and Binder: R. R. Donnelley & Sons Company
Cover Printer: Phoenix Color

Using Multivariate Statistics, Second Edition

Library of Congress Cataloging in Publication Data

Tabachnick, Barbara G.,
  Using multivariate statistics / Barbara G. Tabachnick, Linda S. Fidell.—2nd ed.
    p. cm.
  Includes index.
  ISBN 0-06-046571-9
  1. Multivariate analysis.   I. Fidell, Linda S. II. Title
QA278.T3 1989
519.5'35—dc19   88-39830
                                              CIP

      93 9

The second edition is dedicated to Troupe Mosaic—because we still can't think about statistics all the time.

Contents

Preface

This second edition differs from the first in several respects. The first edition illustrated multivariate analyses through BMDP and SPSS. As a result of grumbling from our East Coast colleagues (and the thought of more potential sales), we added SAS to this edition. We also added SYSTAT to illustrate the analyses through another flexible and powerful package currently available for microcomputers. We ran the examples on the most current versions of each of the packages so the output and options shown are those available as of spring 1988.

There are two new chapters in this new edition, on multiway frequency analysis and profile analysis of repeated measures. Both of these techniques are growing in popularity and use. Their inclusion should give you even greater flexibility in your analysis of data. The chapter on screening data prior to analysis (Chapter 4) has been completely rewritten and now includes comprehensive computer output along with flow diagrams for screening both grouped and ungrouped data.

Our students continue to be our best teachers of what needs to be said about these statistical techniques. After teaching several more semesters with the first edition as text, we learned where changes needed to be made and what changes were likely to work better. As a result, the narrative has been completely rewritten, first sentence to last. Our goal, as with the first edition, is clarity of presentation. We try to tell you what your options are when you do an analysis, how to set up the analysis, and how to interpret and write up your results. We do not explain in gruesome detail the mathematical origins of the procedure, or, for the most part, differences of opinion among those who are developing the analyses. Where such differences of opinion exist, we have chosen the alternative that we think most likely to lead you to the safest and best analysis of your data. The battle over alternatives is being fought out in the journals, not here.

Because of this orientation, some have accused this book of being a cookbook, to which we plead both guilty and not guilty. "Guilty" because we do not focus on the mathematical underpinning of the procedures; "not guilty" because the fourth section of Chapters 5 through 12 has a small example worked out by hand that ties the calculations into those for univariate statistics, where possible. For those who want to understand how the analysis develops, the relevant material is in those sections plus the appendix on matrix algebra. Just as the chapters can be read in any order, or excluded, so also the material in the fourth sections and appendix can be used or omitted, at the interest and discretion of the reader.

Most people do not have to understand the intricacies of the internal-combustion engine to drive to work, and we believe that most people do not have to understand the subtleties of matrix algebra and calculus to analyze their data. In both cases it is nicer and safer if you do thoroughly understand the material, but in both cases there are also signals that something might be amiss that, if attended, help prevent disaster. Look to your dashboard, and Chapter 4 as well as Section 3 of Chapters 5 through 12, respectively, for those signals.

For those who would like to play with the setups and data for the large examples in their own computing facilities, we have developed a floppy diskette, available from the authors upon request (with a small donation for the floppy and mailing). Contact us—we'd like to hear from you anyway.

Many people have contributed to the production of the second edition: buyers of the first edition who created the demand, the staff at Harper & Row—especially Ellen MacElree—who were patient and helpful, Barbara Nicholson and Kenya Moore who typed the first edition onto floppies so that we could edit it, and the following reviewers who provided very helpful reviews of both the second edition and the new chapters: Dale Berger, Claremont Graduate School; Robert Gardiner, University of Western Ontario; Lawrence Gordon, University of Vermont; Mark Leary, Wake Forest University; Henry Morlock, SUNY-Plattsburgh; Thomas Nygren, Ohio State University; and Paul Spector, University of South Florida. We especially want to thank Curt Dommeyer, William Vincent, and Jim Dole, who provided a colleague's-eye view of the first edition, and Betty Rose and David Barber, who gave us a very helpful student's-eye view.

And very most especially we want to thank our husbands (and, in one case, children) who still haven't divorced us!

Barbara G. Tabachnick
Linda S. Fidell

Chapter 1

Introduction

1.1 MULTIVARIATE STATISTICS: WHY?

Multivariate statistics are increasingly popular statistical techniques used for analyzing complicated data. They provide analysis when there are many independent variables (IVs) and/or many dependent variables (DVs) all correlated with one another to varying degrees. Because of the difficulty of addressing complicated research questions with univariate analyses and because of the availability of canned software for performing multivariate analyses, multivariate statistics are becoming widely used. Indeed, the day is near when a standard univariate statistics course ill prepares a student to read research literature or a researcher to produce it.

But how much harder are the multivariate techniques? Compared with the multivariate methods, univariate statistics are so straightforward and neatly structured that it is hard to believe they once took so much effort to master. Yet many researchers apply and validly interpret results of intricate analysis of variance designs before the grand structure is apparent to them. While we are delighted if you gain insights into the full multivariate general linear model,[1] we have accomplished our goal if you feel comfortable selecting and setting up multivariate analyses and interpreting the computer output.

Multivariate methods are more complex than univariate by at least an order of magnitude. But for the most part the greater complexity requires few conceptual leaps. Familiar concepts such as sampling distributions and homogeneity of variance simply become more elaborate.

Multivariate models have not gained popularity by accident—or even by sinister design. Their growing popularity parallels the greater complexity of contemporary

---

[1] Chapter 13 is an attempt to foster such insights.

research. In psychology, for example, we are increasingly less enamored of the simple, clean, laboratory study where pliant, first-year college students each provide us with a single behavioral measure on cue.

1.1.1 The Domain of Multivariate Statistics: Numbers of IVs and DVs

Multivariate statistics are an extension of parametric univariate and bivariate statistics. Multivariate statistics are the "complete" or general case, while univariate and bivariate statistics are special cases of the general multivariate model. If you measure many variables, multivariate techniques let you perform a single analysis instead of a series of univariate or bivariate analyses.

Variables are roughly dichotomized into two major types—independent and dependent—and are taken on subjects (people, companies, rats, stocks, airplanes, or whatever is considered the unit of analysis). Independent variables (IVs) are the differing conditions (treatment vs. placebo) to which you expose your subjects, or characteristics (tall or short) the subjects themselves bring into the research situation. IVs are usually considered either predictor or causal variables because they predict or cause the DVs—the response or outcome variables. Note that IV and DV are defined within a research context; a DV in one research setting is often an IV in another.

Additional terms for IVs and DVs are predictor-criterion, stimulus-response, task-performance, or simply input-output. We use IV and DV throughout to identify variables that belong on one side of the equation or the other, without causal implication. That is, the terms are used for convenience rather than to indicate that the variables were derived from experiment.

Univariate statistics refers to analyses of research in which there is a single DV. There may, however, be more than one IV. For example, social behavior of undergraduates (the DV) is studied as a function of geographical area (one IV) and type of training in social skills to which students are exposed (another IV). Analysis of variance is the prime example of univariate statistics (with *t* test as a special case).

Bivariate statistics frequently refers to analysis of two variables where neither is an experimental IV and the desire is simply to study the relationship between the variables (e.g., the relationship between income and amount of education). The prototypical example of a bivariate statistic is the Pearson product-moment correlation coefficient. (Chapter 3 reviews univariate and bivariate statistics.)

With *multivariate statistics,* you simultaneously analyze multiple dependent and multiple independent variables. This capability is important in both nonexperimental (correlational or survey) and experimental research.

1.1.2 Experimental and Nonexperimental Research

A critical distinction between experimental and nonexperimental research is whether the researcher manipulates the levels of the IVs. In an experiment, the researcher has control over the levels (or conditions) of at least one IV to which a subject is exposed. Further, the experimenter randomly assigns subjects to levels of the IV, and controls all other influential factors by holding them constant or counterbalancing their influence. Scores on the DV are expected to be the same, within random variation, except for the influence of the IV (Campbell and Stanley, 1966). If there are systematic differences in the DV associated with levels of the IV, these differences are attributed to the IV.

For example, if groups of undergraduates are randomly assigned to the same material but different types of teaching techniques, and afterward some groups of undergraduates perform better than others, the difference in performance is said, with some degree of confidence, to be caused by the difference in teaching technique. In this type of research the terms *independent* and *dependent* have obvious meaning: the value of the DV *depends* on the manipulated level of the IV. The IV is independently manipulated by the experimenter and the score on the DV depends on the level of the IV.

In nonexperimental (correlational or survey) research, the levels of the IV(s) are not manipulated by the researcher. The researcher assigns labels to categories of the IV, but has no control over the assignment of subjects to categories. For example, groups of people may be categorized in terms of geographic area of residence (Northeast, Far West, etc.), but only the definition of the variable is under researcher control. Except for the military, place of residence is rarely subject to manipulation by a researcher. Nevertheless, a naturally occurring difference like this is often considered an IV and is used to predict some other nonexperimental (dependent) variable such as income. In this type of research the distinction between IVs and DVs is usually arbitrary and many researchers prefer to call IVs predictors and DVs criterion variables.

In nonexperimental research, it is very difficult to attribute causality to an IV. If there is a systematic difference in a DV associated with levels of an IV, the two variables are said to be related (again with some degree of confidence) but the cause of the relationship is unclear.

One of the few considerations *not* relevant to choice of statistical technique is whether or not data are collected experimentally. The statistics "work" whether or not the researcher manipulated the IV. But attribution of causality for results is crucially affected by the experimental-nonexperimental distinction.

1.1.2.1 Multivariate Statistics in Nonexperimental Research Nonexperimental research takes many forms, but a common example is the survey. Typically, many people are surveyed and each respondent provides answers to many questions, producing a large number of variables. Usually these variables are interrelated in highly complex ways, and univariate or bivariate statistics are simply not sensitive to this complexity. Bivariate correlations between all pairs of variables, for example, could not reveal that the 20 to 25 variables measured really represent only two to three "supervariables."

Or, if a research goal is to distinguish among subgroups in our sample (e.g., between Catholics and Protestants) on the basis of a variety of attitudinal variables, we could use several univariate *t* tests (or analyses of variance) to examine group differences on each variable separately. But if the variables are related, which is highly likely, the results of many *t* tests are misleading and statistically suspect.

All tests of statistical significance are made with a margin of error, usually 5%. If several independent tests are done on the same sample, the margin of error increases as the number of tests increases, and at a known rate. If the variables tested are not only from the same sample, but also themselves related, the probability of error increases at an unknown rate.

With the use of multivariate statistical techniques, complex interrelationships among variables are revealed and are accounted for in statistical inference. Further, it is possible to keep the overall error rate at some stated level, say 5%, no matter how many variables there are and how they are related.

1.1.2.2 Multivariate Statistics in Experimental Research   For the most part, multivariate techniques were developed for use in nonexperimental research. But they are also useful in experimental research where there are multiple DVs. The problem of inflated error rate with multiple variables generalizes directly to experimental research. With multiple IVs, the research is usually designed so that the IVs are independent of each other (see Chapter 3) and there is a straightforward correction for numerous statistical tests. With multiple DVs, however, it is highly unlikely that they are uncorrelated, and separate univariate analyses lead to unacceptably high, but unknown, levels of error.

Experimental research designs with multiple DVs were unusual at one time. Now, however, with attempts to make experimental designs more realistic, and with the availability of computer programs, experiments often have several DVs. It is simply wasteful to run an experiment with only one DV and run the risk of missing the impact of the IV because the most sensitive DV is not measured. Multivariate statistics help the experimenter design more efficient and more realistic experiments by allowing measurement of multiple DVs without violation of acceptable levels of error.

1.1.3 Computers and Multivariate Statistics

One answer to the question "Why multivariate statistics?" is that the techniques are now accessible by computer. Only the most dedicated number cruncher would consider doing real-life sized problems in multivariate statistics without a computer. Fortunately, excellent multivariate programs are available in a number of computer packages.

In this book, four packages are demonstrated. Discussion and examples are based on programs in SPSS (Statistical Package for the Social Sciences) (Nie et al., 1975 and SPSS Inc., 1986), BMDP Statistical Software (Dixon, 1985), SAS (SAS Institute Inc., 1985), and SYSTAT (Wilkinson, 1986). Each package contains a variety of specialized programs for both univariate and multivariate statistics. The manuals are excellent guides to all these packages and additional documentation appears in various newsletters, pocket guides, technical reports, and other materials that are printed as the programs are continuously corrected, revised, and upgraded. It is wise to work closely with your local computing facility to keep abreast of changes in the programs as they occur.

These four packages are selected because of their popularity and ready availability at most large- and medium-sized computing facilities. If you are affiliated with a university or with a research organization that has access to a computer network or to well-equipped microcomputers, it is likely that several packages are available to you.

If you have access to all four packages, you are indeed fortunate. Programs within the packages do not completely overlap, and some analyses are better handled through one package than through another. For example, doing several versions of the same basic analysis on the same set of data is particularly easy with SPSS. BMDP is well designed to do preliminary analysis of data and to test assumptions. SAS has the most extensive facilities for saving derived scores resulting from intermediate analyses. And SYSTAT is ideal for fast, interactive work.

These four packages are available for mainframe computers and IBM-compatible microcomputers. SYSTAT is also available for Macintosh computers. Microcomputer versions tend to have more "user-friendly" procedures for entering data and instructions but, with the exception of the SPSS-PC version of SPSS^x, setup and output match those of the mainframe versions of the programs.

The packages differ in ease of use, with SPSS among the easiest. In SPSS, all programs use the same basic setup language. Once you learn the language, it is easy to describe the data set, and you need to refer to the manual only to specify the statistical analysis desired. Further, the manuals themselves (Nie et al., 1975, SPSS Inc., 1986) are highly informative, although the original manual (Nie et al., 1975) was better than the more recent one because the statistical techniques associated with the various programs were discussed more fully. Both manuals have especially helpful indexes for locating information. Several different forms of each type of analysis are set up and annotated segments of computer output are given. Discussion of the output in the manual tends to be minimal but the output itself is beautifully formatted and easy to follow; everything is nicely labeled and can be found without much difficulty.

However, the manuals do not provide the statistical algorithms (equations) used in any detail. If a technique has several alternative algorithms, it is often impossible to tell from the manual which one is in use. SPSS publishes a supplementary manual of algorithms (SPSS Inc., 1985).

SPSS has had three major releases. The original package (Nie et al., 1975) was designed primarily for survey-type research and did not include programs for multivariate analysis of variance or repeated-measures analysis of variance. This lack was remedied in the SPSS 7–9 update (Hull and Nie, 1981), which included both SPSS MANOVA and an expanded program for multiple regression (SPSS NEW REGRESSION). Both these programs and upgraded versions of many old programs are available now in SPSS^x (SPSS, Inc., 1986; Norusis, 1985) which also features simplified but yet more flexible control language. Most computing facilities have now implemented SPSS^x (or SPSS-PC) but there are a few computers on which only the older programs can be used.

In BMDP, some, but not all, conventions for setup are standard for all the programs. The manual includes helpful discussion and many examples of setup and annotated output, but lacks a really useful index. The output itself is well formatted and labeled, and there are numerous explanatory comments for nonstandard statistics. For regular users, each chapter concludes with a checklist containing critical information for setup in abbreviated form. The BMDP programs are statistically quite sophisticated but present the information at a level that most researchers can deal with. The package has been more comprehensive than SPSS, but with additional programs regularly being added to the SPSS^x package, the gap is diminishing.

SAS is really two sets of programs in one. The basic programs provide for extensive data manipulation and some simple statistics, mostly descriptive. The statistics programs include a comprehensive, sophisticated assortment of techniques. The manual has examples of input and annotated output for many variations on the basic statistical techniques, gathered together at the end of each chapter. Output, however, is more difficult to interpret than that of the other packages. Labeling is often esoteric and unfamiliar unless your statistical training was in a mathematics/ statistics department. SAS output lacks the helpful comments of BMDP and the pleasing visual format of both SPSS and BMDP.

SYSTAT is less comprehensive than the other packages but is the easiest to use, particularly on microcomputers. The manual is written in an informal style and is full of examples of input and output, set off in color. The author has strong opinions on many of the controversial issues in statistics, and reading the manual is educational as well as pleasurable. Among the four packages, this is the only manual you would consider reading rather than just consulting.

In Chapters 5 through 12 of this book (the chapters that cover the specialized multivariate techniques) are explanations and illustrations of a variety of programs within each package and a comparison of the features of the programs. Our hope is that once the techniques are understood, you will be able to generalize to virtually any multivariate program, so that you are not limited to programs within these four packages.

1.1.3.1 Program Updates   With commercial computer packages, it is essential that you know which version of the package you are using. Programs are continually being changed, and not all changes are immediately implemented at each facility. Therefore, many versions of the various programs are simultaneously in use at different institutions and even at one institution, more than one version of a package is sometimes available.

Often program updates are corrections of errors discovered in earlier versions. Occasionally, though, there are major revisions in one or more programs or a new program is added to the package. Check with your computing facility to find out which version of each package is in use. Then be sure that the manual you are using is consistent with the version in use at your facility. Also check updates for error correction in previous releases.

Except where noted, this book reviews SPSS Release 9.0 and SPSS^x Version 2.0, the 1987 update of BMDP, SAS Version 5, and Systat 3.0.

1.1.3.2 Garbage In, Roses Out?   The trick in multivariate statistics is not in computation; that is easily done by computer. The trick is to choose the appropriate program, use it correctly, and know how to interpret the output. Output from commercial computer programs, with their beautifully formatted tables, graphs, and matrices, can make garbage look like roses. Throughout this book we try to suggest clues for when the true message in the output more closely resembles the fertilizer than the flowers.

When you use multivariate statistics, you rarely get as close to the raw data as you do when you apply univariate statistics to a relatively few cases. Errors and anomalies in the data that would be obvious if the data were processed by hand are less easy to spot when processing is entirely by computer. But the computer packages have programs to graph and describe your data in the simplest univariate terms and to display bivariate relationships among your variables. As discussed in Chapter 4, these programs provide preliminary analyses that are absolutely necessary if the results of multivariate programs are to be interpretable.

1.1.4 Why Not?

There is, as usual, "no free lunch." Certain costs are associated with the benefits of using multivariate procedures. Benefits of increased flexibility in research design, for instance, are sometimes paralleled by increased ambiguity in interpretation of results. In addition, multivariate results can be quite sensitive to which analytic strategy is chosen (cf. Section 1.2.5) and do not always provide better protection against statistical errors than their univariate counterparts. Add to this the fact that occasionally you *still* can't get a firm statistical answer to your research questions, and you may wonder if the increase in complexity and difficulty is warranted.

Frankly, we think it is. Slippery as some of the concepts and procedures are, they provide insights into relationships among variables that may more closely resemble the compexity of the "real" world. And sometimes you get at least partial answers to questions that couldn't be asked at all in the univariate framework. For a complete analysis, making sense out of your data usually requires a judicious mix of multivariate and univariate statistics.

And the addition of multivariate statistics to your repertoire makes data analysis a lot more fun. If you liked univariate statistics, you'll love multivariate statistics!

1.2 SOME USEFUL DEFINITIONS

In order to describe multivariate statistics easily, it is useful to review some common terms in research design and basic statistics. Distinctions were made between independent and dependent variables and between experimental and nonexperimental research in the preceding section. Additional terms are described in this section.

1.2.1 Continuous, Discrete, and Dichotomous Data

In applying statistics of any sort, it is important to consider the type of measurement and the nature of the correspondence between numbers and the events that they represent. The distinction made here is between continuous, discrete, and dichotomous variables; you may prefer to substitute the terms "interval" for continuous and "nominal" for discrete.

Continuous variables are measured on a scale that changes values smoothly rather than in steps. Continuous variables take on any value within the range of the scale. Precision is limited by the measuring instrument, not by the nature of the scale itself. Some examples of continuous variables are time as measured on an old-fashioned analog clock face, annual income, age, temperature, distance, and scores on the Graduate Record Exam (GRE). Normally distributed continuous variables are appropriately analyzed by multivariate techniques.

Discrete variables take on a finite number of values (usually a fairly small number) and there is no smooth transition from one value or category to the next. Examples include time as measured by a digital clock, continents, categories of religious affiliation, and type of community (rural or urban). Discrete variables may be used in multivariate analyses if there are several categories and the categories represent an attribute that changes in a quantitative way. For instance, a variable that represents age categories, where, say, 1 stands for 0 to 4 years, 2 stands for 5 to 9 years, 3 stands for 10 to 14 years, and so on up through the normal age span, is useful because there are a lot of categories and the numbers designate a quantitative attribute (increasing age). But if the same numbers are used to designate categories of religious affiliation, they are not in appropriate form for analysis because religions do not fall along a quantitative continuum.

In the latter case, the discrete variable sometimes is analyzed after it is changed into a number of dichotomous or two-level variables (e.g., Catholic vs. non-Catholic, Protestant vs. non-Protestant, Jewish vs. non-Jewish, and so on until the degrees of freedom are used). Recategorization of a discrete variable into a series of dichotomous ones is called dummy variable coding. The conversion of a discrete variable into a series of dichotomous ones is done to create linearity between pairs of variables. The discrete variable can have a relationship of any shape with another variable, and the relationship is changed arbitrarily if assignment of numbers to categories is changed. Dichotomous variables, however, have only linear relationships with other variables and are, therefore, appropriately analyzed by methods that rely on correlation.

The distinction between continuous and discrete variables is not always clear. If you add enough digits to the digital clock, for instance, it becomes for all practical purposes a continuous measurement, while time as measured by the analog device can also be read in discrete categories, say, hours. In fact, any continuous measurement may be rendered discrete (or dichotomous) by specifying cutoffs on the continuous scale.

The property of variables that is crucial to application of multivariate procedures is not type of measurement but rather shape of distribution, as discussed in Chapter 4 and elsewhere. Nonnormally distributed continuous variables and dichotomous variables with very uneven splits between the categories present problems to mulivariate analyses. This issue and its resolution are discussed at some length in Chapter 4.

Another type of measurement that is used sometimes produces a rank order (ordinal) scale. This scale assigns a number to each subject to indicate the subject's position vis-à-vis other subjects along some quantitative dimension. For instance, ranks are assigned to contestants (first place, second place, third place, etc.) to provide an indication of who was best, but not by how much. The problem with ordinal measures is that their distributions are rectangular (one frequency per number) instead of normal, unless tied ranks are permitted and they pile up in the middle of the distribution. However, as with discrete scales, rank order data can be analyzed through multivariate techniques if a reasonably large number of ranks is assigned (say 20 or more for each subsample to be analyzed) and if the variable has a linear relationship with other variables.

1.2.2 Samples and Populations

Samples are measured in order to make generalizations about populations. Ideally, samples are selected, usually by some random process, so that they represent the population of interest. In real life, however, populations are frequently best defined in terms of samples, rather than vice versa; the population is the group from which you were able to randomly sample.

Sampling has somewhat different connotations in nonexperimental and experimental research. In nonexperimental research, you investigate relationships among variables in some predefined population. Typically you take elaborate precautions to ensure that you have achieved a representative sample of that population; you define your population, then do your best to randomly sample from it.[2]

In experimental research, you attempt to *create* different populations by treating subgroups from an orginally homogeneous group differently. The sampling objective here is to assure that all subjects come from the same population *before* you treat them differently. Random sampling consists of randomly assigning subjects to treatment groups (levels of the IV) to ensure that, before differential treatment, all samples come from the same population. Statistical tests provide evidence as to whether, after treatment, all samples still come from the same population. Generalizations about treatment effectiveness are made to the type of subjects who participated in the experiment.

1.2.3 Descriptive and Inferential Statistics

Descriptive statistics describe samples of subjects in terms of variables or combinations of variables. Inferential statistics test hypotheses about differences in populations on the basis of measurements made on samples of subjects. If reliable differences are found, descriptive statistics are then used to provide estimations of central tendency, and the like, in the population. Descriptive statistics used in this way are called parameter estimates.

Use of inferential and descriptive statistics is rarely an either-or proposition. With a data set and array of research questions, we are usually interested in both describing and making inferences about the results. We describe the data, find reliable differences or relationships, and estimate population values for the reliable findings. However, there are more restrictions on inference than there are on description. Many assumptions of multivariate statistics are necessary only for inference. If simple description of the sample is the major goal, many assumptions are relaxed, as discussed in Chapters 5 through 12.

---

[2] Strategies for random sampling are discussed in detail in such texts as Kish (1965) and Moser and Kalton (1972).

1.2.4 Orthogonality

Orthogonality is perfect *non*association between variables. If two variables are orthogonal, knowing the value of one variable gives no clue as to the value of the other; the correlation between them is zero.

Orthogonality is often desirable in statistical applications. For instance, factorial designs for experiments are orthogonal when two or more IVs are completely crossed with equal sample sizes in each combination of levels. Hypotheses about main effects and interactions are tested independently of each other, except for use of a common error term. The outcome of each test gives no hint as to the outcome of the others. In orthogonal experimental designs with random assignment and good controls, causality can be unambiguously attributed to various main effects and interactions.

Similarly, in multivariate analyses, there are advantages if sets of IVs or DVs are orthogonal. If the set of IVs is orthogonal, each adds, in a simple fashion, to prediction of the DV. If, for example, 80% of the variance in income is predicted from education and occupational prestige, and education and occupational prestige are orthogonal, 35% may be predicted from education and 45% predicted from occupational prestige. Or, if the set of DVs is orthogonal, the overall effect of an IV can be partitioned into effects on each DV in an additive fashion.

Consider the Venn diagram in Figure 1.1. Venn diagrams represent variance as overlapping areas between two (or more) sources. The total variance for income is the circle identified as Y. X_1 represents education and X_2 represents occupational prestige; the circle for education overlaps that for income 35% while the circle for occupational prestige overlaps 45%. Together, they account for 80% of the variability in income because education and occupational prestige are orthogonal and do not overlap each other.

1.2.5 Standard and Hierarchical Analyses

Usually, however, the variables are correlated with each other (nonorthogonal). IVs in nonexperimental designs are often correlated naturally; in experimental designs IVs become correlated, for instance, if unequal numbers of subjects are measured in different cells. DVs are usually correlated because individual differences among subjects tend to be consistent over many attributes.

When variables are correlated, they have shared or overlapping variance. In the example, if education and occupational prestige correlate with each other and the independent contribution made by education is 35% and that by occupational prestige 45%, their joint contribution to prediction of income is not 80% but rather

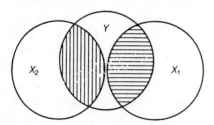

Figure 1.1   Venn diagram for Y (income), X_1 (education), and X_2 (occupational prestige).

something less because some of the prediction of income is overlapping. A major decision for the multivariate analyst is how to handle the variance that is predictable from more than one variable. Many multivariate techniques permit at least two strategies for handling it; some permit more.

In standard analysis, the overlapping variance contributes to the size of summary statistics of the overall relationship but is not assigned to either variable. Overlapping variance is disregarded in assessing the contribution of each variable to the solution. Figure 1.2(a) depicts a standard analysis as a Venn diagram where overlapping variance is shown as overlapping areas in the circles, the unique contributions of X_1 and X_2 to prediction of Y are shown as shaded areas, and the total relationship between Y and the combination of X_1 and X_2 is the area bounded by the heavy line. If X_1 is education and X_2 is occupational prestige, then in standard analysis, X_1 is "credited with" the area marked by the horizontal lines and X_2 by the area marked by vertical

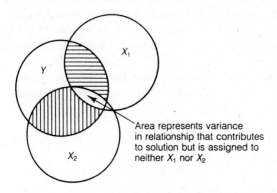

Area represents variance
in relationship that contributes
to solution but is assigned to
neither X_1 nor X_2

(a) Standard analysis

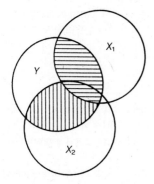

(b) Hierarchical analysis in which X_1
is given priority over X_2

Figure 1.2   Standard (a) and hierarchical (b) analyses of the relationship between Y, X_1, and X_2. Horizontal shading depicts variance assigned to X_1. Vertical shading depicts variance assigned to X_2.

lines. If X_1 and X_2 overlap substantially, very little area is assigned to either of them despite the fact that they are both related to Y. They have knocked each other out of looking important to the solution.

Hierarchical analyses differ in that the researcher assigns priority for entry of variables into equations and the first one to enter is assigned both unique variance and any overlapping variance it has with other variables. Lower-priority variables then are assigned on entry their unique and any remaining overlapping variance. Figure 1.2(b) shows a hierarchical analysis for the same case as Figure 1.2(a) where X_1 (education) is given priority over X_2 (occupational prestige). The total variance explained is the same as in Figure 1.2(a) but the relative contributions of X_1 and X_2 have changed; education now shows a stronger relationship with income than in the standard analysis, while the relation between occupational prestige and income remains the same.

The choice of strategy for dealing with overlapping variance is not trivial. If variables are correlated, the apparent importance of variables to the solution changes depending on whether a standard or a hierarchical strategy is used. If the multivariate procedures have a reputation for unreliability it is because solutions change, sometimes dramatically, when different strategies for entry of variables are chosen. However, the strategies also ask different questions of the data and it is incumbent on the researcher to determine exactly which question is asked. We try to make the choices clear in the chapters that follow.

1.3 COMBINING VARIABLES

Multivariate analyses combine variables to do useful work such as predict scores or predict group membership. The combination that is formed depends on the relationships between the variables and the goals of analysis, but in all cases the combination that is formed is a *linear* combination.[3] A linear combination is one in which variables are assigned weights, and then products of weight and variable scores are summed to produce a score on a combined variable. Equation 1.1 is a linear combination between Y', the DV, and X_1 and X_2, the IVs.

$$Y' = W_1X_1 + W_2X_2 \qquad (1.1)$$

If, for example, Y' is predicted reading ability, X_1 is age, and X_2 is sex, the best prediction of reading ability is obtained by weighting age (X_1) by W_1, and sex (X_2) by W_2 before summing. No other values of W_1 and W_2 produce as good a prediction of reading ability.

Notice that Equation 1.1 includes neither X_1 or X_2 raised to powers (exponents) nor a product of X_1 and X_2. This seems to severely restrict multivariate solutions until one realizes that X_1 could itself be a product of two different variables or a single variable raised to a power. For example, X_1 might be age squared. A multivariate solution does not produce exponents or cross products of IVs to improve a solution,

---

[3] Nonlinear analyses are available but beyond the scope of this book.

but the researcher can include $X's$ that are cross products of IVs or are IVs raised to powers.

The size of the W values (or some function of them) often reveals a great deal about the relationship between DV and IVs. If, for instance, the W value for some IV is zero, the IV is not needed in the best DV-IV relationship. Or if some IV has a large W value, then the IV is important to the relationship. Although complications (to be explained later) prevent interpretation of the multivariate solution from the sizes of the W values, they are nonetheless important in most multivariate procedures.

The combination of variables represents a supervariable, not directly measured but often interpretable. The supervariable is thought of as an underlying dimension that performs the work desired from the multivariate analysis of predicting something or optimizing relationships. Therefore the attempt to understand the meaning of the combination of IVs is critical to many multivariate procedures.

In the search for the best weights to apply in combining variables, computers do not try out all possible sets of weights. Various algorithms have been developed to arrive at the weights. The algorithms involve manipulation of a correlation matrix, a variance-covariance matrix, or a sum-of-squares and cross-products matrix. Section 1.5 describes these matrices in very simple terms and shows their development from a very small data set. Appendix A describes some terms and manipulations appropriate to matrices. In each of Chapters 5 through 12, a small hypothetical sample of data is analyzed by hand to show how the weights are derived for each analysis. Though this information is useful for a basic understanding of the multivariate statistics, it is *not* necessary for applying multivariate techniques fruitfully to your research questions.

1.4 NUMBER AND NATURE OF VARIABLES TO INCLUDE

Attention to the *number* of variables included in analysis is important. A general rule is to get the best solution with the fewest variables possible. As more and more variables are included, the solution usually improves, but only slightly. Sometimes the improvement does not compensate for the cost in degrees of freedom of including more variables, so the power of the analyses diminishes. If there are too many variables relative to sample size, the solution provides a good fit to the sample that doesn't generalize to the population, a condition termed *overfitting*. To avoid overfitting, include only a limited number of variables in each analysis.[4]

There are several important considerations such as cost, availability, meaning, and relationships regarding the *nature* of the variables to include in a multivariate analysis. Except in factor analysis, one usually wants a small number of cheaply obtained, easily available, unrelated variables. Further considerations for variable selection are mentioned as they apply to each analysis.

But one of the most important considerations in variable selection for all analyses is reliability. How stable is the position of a given score in a distribution of scores

---

[4] The exception is factor analysis, where numerous variables are measured.

when measured at different times or in different ways? Unreliable variables degrade an analysis while reliable ones enhance it. A few reliable variables give a more meaningful solution than a large number of unreliable variables. Indeed, if variables are sufficiently unreliable, the entire solution may reflect only measurement error.

1.5 DATA APPROPRIATE FOR MULTIVARIATE STATISTICS

For multivariate statistics, an appropriate data set consists of values on a number of variables for each of several subjects. For continuous variables, the values are scores on variables. For example, if the continuous variable is score on the GRE, the values for the various subjects are scores such as 500, 650, 420, and so on. For discrete variables, values are number codes for group membership or treatment. For example, if there are three teaching techniques, students who receive one technique are arbitrarily assigned a code "1" on the teaching technique variable, those receiving another technique are assigned a code "2," and so on.

1.5.1 The Data Matrix

The data matrix is an organization of values in which rows (lines) represent subjects and columns represent variables. An example of a data matrix with six subjects[5] and four variables is in Table 1.1. For example, X_1 might be type of teaching technique, X_2 score on the GRE, X_3 GPA, and X_4 gender, with women coded 1 and men coded 2.

In order to apply computer techniques to these data, they are entered into a data file with long-term storage accessible by computer. Each subject starts with a new row (line). If there are more data for each subject than can be accommodated on a single line, each subject is assigned additional lines. Information identifying the subject is typically entered first, followed by the value on each variable for that subject. Scores for each variable are entered in the same order for each subject. Any computer package manual provides information on setting up a data matrix.

In this example, there are values for every variable for each subject. This is not always the case with research in the real world. With large numbers of subjects

TABLE 1.1 A DATA MATRIX OF HYPOTHETICAL SCORES

Student	X_1	X_2	X_3	X_4
1	1	500	3.20	1
2	1	420	2.50	2
3	2	650	3.90	1
4	2	550	3.50	2
5	3	480	3.30	1
6	3	600	3.25	2

[5] Normally, of course, there are many more than six subjects.

and variables, scores frequently are missing on some variables for some subjects. For instance, respondents may refuse to answer some kinds of questions, or some students may be absent the day that one of the tests is given, and so forth. This creates blanks in the data matrix that must be dealt with. Chapter 4 covers this messy (but often unavoidable) problem.

1.5.2 The Correlation Matrix

Most readers are familiar with a correlation matrix, or **R**. **R** is a square, symmetrical matrix. Each row (and each column) represents a different variable, and the value at the intersection of each row and column is the correlation between the two variables. For instance, the value in the second row, third column, is the correlation between the second variable and the third variable. The same correlation also appears at the intersection of the third row and the second column. Thus correlation matrices are said to be symmetrical about the main diagonal, which means they are mirror images of themselves above and below the diagonal going from top left to bottom right. For this reason it is common practice to show only the bottom half or the top half of an **R** matrix. The entries in the main diagonal are often omitted as well, since they are all ones—correlations of variables with themselves.[6]

Table 1.2 shows the correlation matrix for X_2, X_3, and X_4 of Table 1.1. The .85 is the correlation between X_2 and X_3 and it appears twice in the matrix (as do others). Other correlations are as indicated in the table.

Many programs allow the researcher a choice between analysis of a correlation matrix and analysis of a variance-covariance matrix. If the correlation matrix is analyzed, a unit-free result is produced. That is, the solution reflects the relationships among the variables but not in the metric in which they are measured. If the metric of the scores is somewhat arbitrary, analysis of **R** is appropriate.

1.5.3 The Variance-Covariance Matrix

If, on the other hand, scores are measured along a meaningful scale, it is sometimes appropriate to analyze a variance-covariance matrix. A variance-covariance matrix, $\mathbf{\Sigma}$, is also square and symmetrical, but the elements in the main diagonal are the variances of each variable, while the off-diagonal elements are covariances between different variables.

TABLE 1.2 CORRELATION MATRIX
FOR PART OF HYPOTHETI-
CAL DATA FOR TABLE 1.1

	X_2	X_3	X_4
X_2	1.00	.85	−.13
R = X_3	.85	1.00	−.46
X_4	−.13	−.46	1.00

[6] Alternatively, other information such as standard deviations are inserted.

TABLE 1.3 VARIANCE-COVARIANCE
MATRIX FOR PART OF
HYPOTHETICAL DATA OF
TABLE 1.1

	X_2	X_3	X_4
X_2	7026.66	32.80	−6.00
$\Sigma = X_3$	32.80	.21	−.12
X_4	−6.00	−.12	.30

Variances, as you recall, are averaged squared deviations of each score from the mean of the scores. Because the deviations are averaged, the *number* of scores included in computation of a variance is not relevant, but the metric in which the scores are measured is. Scores measured in hundreds tend to have largish numbers as variances; scores measured in decimals tend to have very small variances.

Covariances are averaged cross products (the deviations between one variable and its mean times the deviation between a second variable and its mean). Covariances are similar to correlations except that they, like variances, retain information concerning the scales in which the variables are measured. The variance-covariance matrix for the continuous data of Table 1.1 appears in Table 1.3.

1.5.4 The Sum-of-Squares and Cross-Products Matrix

This matrix, **S,** is a precursor to the variance-covariance matrix in which deviations are not yet averaged. Thus their sizes depend on the number of cases of which the elements are composed as well as on the metric in which they were measured. The sum-of-squares and cross-products matrix for X_2, X_3, and X_4 in Table 1.1 appears in Table 1.4.

The entry in the major diagonal of the **S** matrix is the sum of squared deviations of scores from the mean for that variable, hence "sum of squares." That is, for each variable, the value in the major diagonal is

$$\text{Sum of squares } (X_j) = \sum_{i=1}^{N} (X_{ij} - \overline{X}_j)^2 \tag{1.2}$$

where $i = 1, 2, \ldots, N$. N is the number of subjects. $j = 1, 2, \ldots, P$. P is the number of variables. X_{ij} is the score on variable j by subject i, and $\overline{X}_j$ is the mean of all scores on the jth variable.

TABLE 1.4 SUM-OF-SQUARES AND
CROSS-PRODUCTS MATRIX FOR
PART OF HYPOTHETICAL DATA
OF TABLE 1.1

	X_2	X_3	X_4
X_2	35133.33	164.00	−30.00
$S = X_3$	164.00	1.05	−0.59
X_4	−30.00	−0.59	1.50

For example, for X_4, the mean is 1.5. The sum of squared deviations around the mean and the diagonal value for the variable is

$$\sum_{i=1}^{6} (X_{i4} - \overline{X}_4)^2 = (1 - 1.5)^2 + (2 - 1.5)^2 + (1 - 1.5)^2$$
$$+ (2 - 1.5)^2 + (1 - 1.5)^2 + (2 - 1.5)^2$$
$$= 1.50$$

The off-diagonal elements of the sum-of-squares and cross-products matrix are the cross products of the variables. For each pair of variables, represented by row and column labels in Table 1.4, the entry is the sum of the product of the deviation of one variable around its mean times the deviation of the other variable around its mean.

$$\text{Cross product } (X_j X_k) = \sum_{i=1}^{N} (X_{ij} - \overline{X}_j)(X_{ik} - \overline{X}_k) \qquad (1.3)$$

where $k = 1, 2, \ldots, P,$ and all other terms are as defined in Equation 1.1, and $j \neq k$. (Note that if $j = k$, Equation 1.3 becomes identical to Equation 1.2.)

For example, the cross-product term for variables X_2 and X_3 is

$$\sum_{i=1}^{N} (X_{i2} - \overline{X}_2)(X_{i3} - \overline{X}_3) = (500 - 533.33)(3.20 - 3.275)$$
$$+ (420 - 533.33)(2.50 - 3.275)$$
$$+ \cdots + (600 - 533.33)(3.25 - 3.275)$$
$$= 164.00$$

Most computations start with **S** and proceed to Σ or **R**. The progression from a sum-of-squares and cross-products matrix to a variance-covariance matrix is simple.

$$\Sigma = \frac{1}{N - 1} S \qquad (1.4)$$

The variance-covariance matrix is produced by dividing every element in the sum-of-squares and cross-products matrix by $N - 1$, where N is the number of subjects.

The correlation matrix is derived from an **S** matrix by dividing each sum of squares by itself (to produce the 1s in the main diagonal of **R**) and each cross product of the **S** matrix by the square root of the product of the sum of squared deviations around the mean for each of the variables in the pair. That is, each cross product is divided by

$$\text{Denominator } (X_j X_k) = \sqrt{\sum_{i=1}^{N} (X_{ij} - \overline{X}_j)^2 \sum_{i=1}^{N} (X_{ik} - \overline{X}_k)^2} \qquad (1.5)$$

where terms are defined as in Equation 1.3.

For some multivariate operations, it is not necessary to feed the data matrix to a computer program. Instead, a Σ or an **R** matrix is entered, with each row (representing a variable) starting a new line. Considerable computing time and expense are saved by entering one or the other of these matrices rather than raw data into many programs.

1.5.5 Residuals

Often a goal of analysis, or test of its efficiency, is to reproduce the values of a DV or the correlation matrix of a set of variables. For example, we might want to predict scores on the GRE (X_2 of Table 1.1) from knowledge of GPA (X_3) and gender (X_4). After applying the proper statistical operations, a multiple regression in this case, a predicted GRE score for each student is derived. By applying the proper weights to GPA and gender for a student, we make a prediction as to what that student's GRE is. But because we already obtained GRE scores for our sample of students, we are able to compare the prediction with the obtained GRE score. The difference between the predicted and obtained values is known as the residual and is a measure of error of prediction.

In most analyses the residuals for the entire sample sum to zero. That is, sometimes the prediction is too large and sometimes it is too small, but the average of all the errors is zero. The squared value of the residuals, however, provides a measure of how good the prediction is. When the predictions are close to the obtained values, the squared errors are small. The way that the residuals are distributed is of further interest in evaluating the degree to which the data meet the assumptions of multivariate analyses, as discussed in Chapter 4 and elsewhere.

1.6 ORGANIZATION OF THE BOOK

Chapter 2 gives a guide to the multivariate techniques that are covered in this book and places them in context with the more familiar univariate and bivariate statistics where possible. Included in Chapter 2 is a flow chart that organizes statistical techniques on the basis of the major research questions asked. Chapter 3 provides a brief review of univariate and bivariate statistics for those who are interested.

Chapter 4 deals with the assumptions and limitations of multivariate statistics. Assessment and violation of assumptions are discussed, along with alternatives for dealing with violations when they occur. Chapter 4 is meant to be referred to often, and the reader is guided back to it frequently in Chapters 5 through 12.

Chapters 5 through 12 cover specific multivariate techniques. They include descriptive, conceptual sections as well as a guided tour through a real-world data set for which the analysis is appropriate. The tour includes an example of a Results section, appropriate for submission to a professional journal, describing the outcome of the statistical analysis. Each technique chapter includes a comparison of computer programs available in the four packages described in Section 1.1.3. You may want to vary the order in which you cover these chapters.

Chapter 13 is an attempt to integrate univariate, bivariate, and multivariate statistics through the multivariate general linear model. The common elements underlying all the techniques are emphasized, rather than the distinctions between them. Chapter 13 is meant to pull together the material in the remainder of the book with a conceptual rather than pragmatic emphasis. Some may wish to consider this material earlier, for instance, immediately after Chapter 2.

Chapter 2

A Guide to Statistical Techniques: Using the Book

2.1 RESEARCH QUESTIONS AND ASSOCIATED TECHNIQUES

All parametric statistics—univariate, bivariate, and multivariate—are special applications of the general linear model. However, it is often more helpful to emphasize differences rather than similarities among the analyses when considering which technique is appropriate for what data. This chapter, then, emphasizes differences in research questions answered by the different techniques while Chapter 13, the last chapter, provides an integrated overview of the techniques.[1]

There is a decision tree at the end of this chapter that leads you to an appropriate analysis for your data. On the basis of a few characteristics of your data set, you determine which statistical technique(s) is best able to answer your research questions. The first, most important criterion for choosing a technique is the major research question to be answered by application of the statistical analysis. Research questions are categorized here into degree of relationship among variables, significance of group differences, prediction of group membership, and structure.

2.1.1 Degree of Relationship among Variables

If the major purpose of analysis is to assess the associations among two or more variables, some form of correlation or regression is appropriate. The choice among five different statistical techniques is made by determining the number of independent

[1] You may find it helpful to read Chapter 13 soon instead of waiting until the end.

and dependent variables, the nature of the variables (continuous or discrete), and whether or not any of the IVs are best conceptualized as covariates.[2]

2.1.1.1 Bivariate *r* Bivariate correlation and regression, as reviewed in Chapter 3, assess the degree of relationship between two continuous variables such as belly dancing skill and years of musical training. Bivariate correlation measures the association between the two variables with no distinction necessary between IV and DV. Bivariate regression, on the other hand, predicts a score on one variable from knowledge of the score on another variable (e.g., predicts skill in belly dancing from years of musical training). The predicted variable is considered the DV, while the predictor is considered the IV. Bivariate correlation and regression are not multivariate techniques but they are integrated into the general linear model in Chapter 13.

2.1.1.2 Multiple *R* Multiple correlation assesses the degree to which one continuous variable (the DV) is related to a set of other (usually) continuous variables (the IVs) that have been combined to create a new, composite variable. Multiple correlation is a bivariate correlation between the original DV and the composite variable created from the IVs. For example, how large is the association between belly dancing skill and a number of IVs such as years of musical training, body flexibility, and age?

Multiple regression is used to predict the score on the DV from scores on the IVs. Other examples are prediction of success in an educational program from scores on a number of achievement tests, or outcome of psychotherapy from demographic variables and scores on various personality measures, or stock market behavior from a variety of political and economic variables.

In multiple correlation and regression the IVs may or may not be correlated with each other. With some ambiguity, the techniques also allow assessment of the relative contribution of each of the IVs toward predicting the DV, as discussed in Chapter 5.

2.1.1.3 Hierarchical *R* In hierarchical multiple regression, IVs are given priorities by the researcher before their contribution toward prediction of the DV is assessed. For example, the researcher might first assess the effects of age and flexibility on belly dancing skill before looking at the contribution that years of musical training makes to that skill. Differences among dancers in age and flexibility are statistically "removed" before assessment of the effects of years of musical training.

In the example of psychotherapy, success of outcome might first be predicted from demographic variables such as age, sex, and marital status. Then scores on personality tests are added to see if prediction of outcome is enhanced after adjusting for the demographic variables.

In general, then, the effects of higher-priority IVs are assessed and removed before the effects of lower-priority IVs are assessed. For each IV in a hierarchical multiple regression, then, higher-priority IVs act as covariates for lower-priority IVs.

The degree of relationship between the DV and the IVs is reassessed at each

---

[2] If the effects of some IVs are assessed after the effects of other IVs are statistically removed, the latter are called covariates.

step of the hierarchy. That is, multiple correlation is recomputed as each new IV is added to predict the DV. Hierarchical multiple regression, then, is also useful for developing a reduced set of IVs (if that is desired) by determining when IVs no longer add to predictability. Hierarchical multiple regression is discussed in Chapter 5.

2.1.1.4 Canonical *R*   In canonical correlation, there are several continuous DVs as well as several continuous IVs and the goal is to assess the relationship between the two sets of variables. For example, we might study the relationship between a number of indices of belly dancing skill (the DVs, such as knowledge of steps, ability to play finger cymbals, responsiveness to the music) and the IVs (flexibility, musical training, and age). Or we might ask whether there is a relationship between achievement in arithmetic, reading, and spelling as measured in elementary school and a set of variables reflecting early childhood development (e.g., age at first speech, walking, toilet training).

Or, one might study the relationship between demographic variables (e.g., religious affiliation, age, sex, marital status, socioeconomic status) on the one hand and a set of personality variables on the other (e.g., introversion-extraversion, degree of neuroticism, submissiveness-dominance). Or, if several measures of therapeutic outcome (e.g., ratings by psychotherapist, client, and client's family) are available, it might be interesting to know the degree to which they are predicted by a set of demographic and/or personality variables. Such research questions are answered by canonical correlation, the subject of Chapter 6.

2.1.1.5 Multiway Frequency Analysis   The goal of multiway frequency analysis is to assess relationships among discrete variables. There may be no distinctions among the variables, or one of them may be considered a DV with the rest serving as IVs (logit analysis). For example, you might want to predict whether or not someone is a belly dancer (considered the DV) from knowledge of gender, occupational category, and preferred type of reading material (science fiction, romance, history, statistics). Or, you might simply be interested in the relationships among belly dancing, gender, occupational category, and type of reading material with none considered the DV. Chapter 7 deals with both forms of multiway frequency analysis.

2.1.2 Significance of Group Differences

When subjects are randomly assigned to groups or are in naturally occurring groups, the major research question usually is the extent to which reliable mean differences on DVs are associated with group membership. This is particularly true in experimental work. But once reliable differences are found, the researcher often then assesses the degree of relationship (strength of association) between IVs and DVs.

The choice among techniques hinges on the number of IVs and DVs, and whether some variables are conceptualized as covariates. Further distinctions are made as to whether all DVs are measured on the same scale, and how within-subjects IVs are to be treated.

2.1.2.1 One-Way ANOVA and *t* Test These two statistics, reviewed in Chapter 3, are strictly univariate in nature and are adequately covered in most standard statistical texts.

2.1.2.2 One-Way ANCOVA One-way analysis of covariance is designed to assess group differences on a single DV after the effects of one or more covariates are statistically removed. Covariates are chosen because of their known association with the DV; otherwise there is no point to their use. For example, age and degree of reading disability are usually related to outcome of a program of educational therapy (the DV). If groups are formed by randomly assigning children to different types of educational therapy (the IV), it is useful to remove differences in age and degree of reading disability before examining the relationship between outcome and type of educational therapy. Prior differences among children in age and reading disability are used as covariates. The ANCOVA question is: Are there mean differences in outcome associated with type of educational therapy after adjusting for differences in age and degree of reading disability?

ANCOVA gives a more powerful look at the IV-DV relationship by minimizing error variance (cf. Chapter 3). The stronger the relationship between the DV and the covariate(s), the greater the power of ANCOVA over ANOVA.

ANCOVA is also used to adjust for differences among groups when groups are naturally occurring and random assignment to them is not possible. Chapter 8 discusses this somewhat problematical use of ANCOVA. For example, one might ask if attitude toward abortion (the DV) varies as a function of religious affiliation. However, it is not possible to randomly assign people to religious affiliation. In this situation, there may well be other systematic differences between groups, such as level of education, that are related to attitude toward abortion. Because education is also related to abortion attitude, apparent differences between religious groups might well be due to differences in education rather than differences in religious affiliation. To get a "purer" measure of the relationship between attitude and religion, attitude scores are first adjusted for educational differences, that is, education is used as a covariate.

When there are more than two groups, planned or post hoc comparisons are available in ANCOVA just as in ANOVA. With ANCOVA, selected and/or pooled group means are adjusted for differences on covariates before differences in means on the DV are assessed. ANCOVA is covered in Chapter 8.

2.1.2.3 Factorial ANOVA Factorial ANOVA, reviewed in Chapter 3, is the subject of numerous statistics texts (e.g., Keppel, 1982; Winer, 1971; Myers, 1979) and is usually introduced in elementary texts, such as Young and Veldman (1977). Although there is only one DV in factorial ANOVA, its place within the general linear model is discussed in Chapter 13.

2.1.2.4 Factorial ANCOVA Factorial ANCOVA differs from one-way ANCOVA only in that there is more than one IV. The desirability and use of covariates are the same. For instance, in the educational therapy example of Section 2.1.2.2, another interesting IV might be gender of the child. The effects of gender, type of educational therapy and their interaction on outcome are assessed after adjusting for age and prior degree of reading disability. By adding gender as an IV, one can determine if boys and girls differ as to which type of educational therapy is more effective after adjustment for covariates.

2.1.2.5 Hotelling's T^2   Hotelling's T^2 is used when the IV has only two groups and there are several DVs. For example, there might be two DVs, such as score on an academic achievement test and attention span in the classroom, and two levels of type of educational therapy, emphasis on perceptual training vs. emphasis on academic training. It is not legitimate to use separate t tests for each DV to look for differences between groups because that inflates Type I error due to unnecessary multiple significance tests with (likely) correlated DVs. Instead, Hotelling's T^2 is used to see if groups differ on the two DVs combined. The researcher asks if there are reliable differences in the centroids (average on the combined DVs) for the two groups.

Hotelling's T^2 is a special case of multivariate analysis of variance, just as the t test is a special case of univariate analysis of variance, when the IV has only two groups. Multivariate analysis of variance is discussed in Chapter 9.

2.1.2.6 One-Way MANOVA   Multivariate analysis of variance evaluates differences among centroids for a set of DVs when there are two or more levels of an IV (groups). MANOVA is useful for the educational therapy example in the preceding section with two groups and also when there are more than two groups (e.g., if a nontreatment control group is added).

With more than two groups, planned and post hoc comparisons are available. For example, if a main effect of treatment is found in MANOVA, it might be interesting to ask post hoc if there are differences in the centroids of the two groups given different types of educational therapy, ignoring the control group, and, possibly, if the centroid of the control group differs from the centroid of the two educational therapy groups combined.

Any number of DVs may be used; the procedure deals with correlations among them and the entire analysis is accomplished within the preset level for Type I error. Once reliable differences are found, techniques are available to assess which DVs are influenced by the IV. For example, assignment to treatment group might affect the academic DV but not attention span. MANOVA is also available when there are within-subject IVs. For example, children might be measured on both DVs three times: 3, 6, and 9 months after therapy begins.

MANOVA is discussed in Chapter 9 and a special case of it (profile analysis) in Chapter 10. Profile analysis is used with one-way between-subjects designs where the DVs are all measured on the same scale. Discriminant function analysis is also available for one-way between-subjects designs, as described in Section 2.1.3.1 and Chapter 11.

2.1.2.7 One-Way MANCOVA   In addition to dealing with multiple DVs, multivariate analysis of variance can be applied to problems where there are one or more covariates. In this case, MANOVA becomes multivariate analysis of covariance—MANCOVA. In the educational therapy example of Section 2.1.2.6, it might be worthwhile to adjust the DV scores for pretreatment differences in academic achievement and attention span. Here the covariates are pretests of the DVs, a classic use of covariance analysis. After adjustment for pretreatment scores, differences in posttest scores (DVs) can more clearly be attributed to treatment (the two types of educational therapy plus control group that make up the IV).

In the one-way ANCOVA example of religious groups in Section 2.1.2.2, it might be interesting to test political liberalism vs. conservatism, and attitude toward ecology, as well as attitude toward abortion, to create three DVs. Here again, differences in attitudes might be associated with both differences in religion and differences in education (which, in turn, varies with religious affiliation). In the context of MANCOVA, education is the covariate, religious affiliation the IV, and attitudes the DVs. Differences in attitudes among groups with different religious affiliation are assessed after adjustment for each respondent's education.

If the IV has more than two levels, planned and post hoc comparisons are useful, with adjustment for covariates. Either MANCOVA (Chapter 9) or hierarchical discriminant function analysis (Chapter 11) is available for both the main analysis and comparisons.

2.1.2.8 Factorial MANOVA   Factorial MANOVA is the extension of MANOVA to designs with more than one IV and multiple DVs. For example, gender (a between-subjects IV) might be added to type of educational therapy (another between-subjects IV) with both academic achievement and attention span used as DVs. In this case the analysis is a two-way between-subjects factorial MANOVA that provides tests of the main effects of gender and type of educational therapy and their interaction on the centroids of the DVs.

Or duration of therapy (3, 6, and 9 months) might be added to the design as a within-subjects IV with type of educational therapy a between-subjects IV to examine the effects of duration, type of educational therapy, and their interaction on the DVs. In this case, the analysis is a factorial MANOVA with one between- and one within-subjects IV.

Comparisons can be made among margins or cells in the design, and influence on combined or individual DVs can be assessed. For instance, the researcher might plan (or decide post hoc) to look for linear trend in scores associated with duration of therapy for each type of therapy separately (the cells) or across all types of therapy (the margins). The search for linear trend could be conducted among the combined DVs or separately for each DV with appropriate adjustments for Type I error rate.

Virtually any complex ANOVA design (cf. Chapter 3) with multiple DVs can be analyzed through MANOVA, given access to appropriate computer programs. Factorial MANOVA is covered in Chapter 9.

2.1.2.9 Factorial MANCOVA   It is sometimes desirable to incorporate one or more covariates into a factorial MANOVA design to produce factorial MANCOVA. For example, pretest scores on academic achievement and attention span could serve as covariates for the two-way between-subjects design with gender and type of educational therapy serving as IVs and posttest scores on academic achievement and attention span serving as DVs. The two-way between-subjects MANCOVA provides tests of gender, type of educational therapy and their interaction on adjusted, combined centroids for the DVs.

Here again procedures are available for comparisons among groups or cells and for evaluating the influences of IVs and their interactions on the various DVs. Factorial MANCOVA is discussed in Chapter 9.

2.1.2.10 Profile Analysis   A special form of MANOVA is available when all of your DVs are measured on the same scale (or on scales with the same psychometric properties) and you want to know if groups differ on the scales. For example, you might use the subscales of the Profile of Mood States as DVs to assess whether mood profiles differ between a group of belly dancers and a group of ballet dancers. There are two ways to conceptualize this design. The first is as a one-way between-subjects design where the IV is type of dancer and the DVs are the Mood States subscales; one-way MANOVA provides a test of the main effect of type of dancer on the combined DVs. The second way is as a profile study with one grouping variable (type of dancer) and several repeated measures (the subscales); profile analysis provides tests of the main effects of type of dancer and subscales as well as their interaction, frequently the effect of greatest interest to the researcher.

If there is a grouping variable and a repeated measure such as trials where the same DV is measured at several intervals in time, there are three ways to conceptualize the design. The first is as a one-way between-subjects design with several DVs (the score on each trial); MANOVA provides a test of the main effect of the grouping variable. The second is as a two-way between- and within-subjects design; ANOVA provides tests of groups, trials, and their interaction, but with some very restrictive assumptions that are likely to be violated. Third is as a profile study where profile analysis provides tests of the main effects of groups and trials and their interaction, but without the restrictive assumptions. This is sometimes called the "multivariate approach to repeated measures ANOVA."

Finally, you might have a between- and within-subjects design (groups and trials) where several DVs are measured on each trial. For example, you might assess groups of belly and ballet dancers on the Mood State subscales at various points in their training. This application of profile analysis is frequently referred to as a doubly multivariate design. Chapter 10 deals with all these forms of profile analysis.

2.1.3 Prediction of Group Membership

In survey or correlational research where groups are identified, the emphasis is frequently on predicting group membership from a set of variables. Discriminant function analysis is designed to accomplish this.

2.1.3.1 One-Way Discriminant Function   In one-way discriminant function analysis, the goal is to predict membership in groups (the DV) from a set of IVs. For example, the researcher might want to predict a child's membership in one of three groups (two types of educational therapy or control) from posttreatment scores on academic achievement and attention span. The analysis tells us if group membership is predicted reliably. Or, the researcher might try to predict religious affiliation from scores on a number of attitudinal measures such as attitude toward abortion, liberalism vs. conservatism, and attitude toward ecological issues. Or an attempt could be made to discriminate belly dancers from ballet dancers from scores on Mood State subscales.

These are the same questions as those addressed by MANOVA, but turned around. Group membership serves as the IV in MANOVA and the DV in discriminant function analysis. If groups differ significantly on a set of variables in MANOVA, the set of variables reliably predicts group membership in discriminant function analysis. One-way between-subjects designs can be fruitfully analyzed through either procedure and are often best analyzed with a combination of both procedures.

As in MANOVA, there are techniques for assessing the contribution of various IVs to prediction of group membership. For example, the major source of discrimination among religious groups might be abortion attitude, with little predictability contributed by political and ecological attitudes.

In addition, discriminant function analysis offers classification procedures to evaluate how well individual cases are classified into their appropriate groups on the basis of their scores on the IVs. One-way discriminant function analysis is covered in Chapter 11.

2.1.3.2 Hierarchical One-Way Discriminant Function   Sometimes IVs are assigned priorities by the researcher so their effectiveness as predictors of group membership is evaluated sequentially in hierarchical discriminant function analysis. For example, when attitudinal variables are predictors of religious affiliation, variables might be prioritized according to their expected contribution to prediction, with abortion attitude given highest priority, political liberalism vs. conservatism second priority, and ecological attitude lowest priority. Hierarchical discriminant function analysis first assesses the degree to which religious affiliation is reliably predicted from abortion attitude. Gain in prediction is, then, assessed with the addition of political attitude, and then with the addition of ecology attitude.

Hierarchical analysis provides two types of useful information. First, it is helpful in eliminating predictors that do not contribute more than predictors already in the analysis. For example, if political and ecological attitudes do not add appreciably to abortion attitude in predicting religious affiliation, they can be dropped from further analysis. Second, hierarchical discriminant function analysis is a covariance analysis. At each step of the hierarchy, higher priority predictors are covariates for lower priority predictors. Thus, the analysis permits you to assess the contribution of a predictor with the influence of other predictors removed.

Hierarchical discriminant function analysis is also useful for evaluating sets of predictors. For example, if a demographic set of variables is given higher priority than an attitudinal set in prediction of group membership, one can see if attitudes reliably add to prediction after adjustment for demography. Hierarchical discriminant function analysis is covered in Chapter 11.

2.1.3.3 Factorial Discriminant Function   If groups are formed on the basis of more than one attribute, prediction of group membership from a set of IVs is possible through factorial discriminant function analysis. For example, respondents might be classified on the basis of both gender and religious affiliation. One could use attitudes toward abortion, politics, and ecology to predict gender (ignoring religion), or religion (ignoring gender), or both gender and religion. But this is the same problem as addressed by factorial MANOVA. For a number of reasons, programs designed for discriminant function analyis do not readily extend to factorial arrangements of groups. Unless some special conditions are met (cf. Chapter 11), it is usually better to rephrase the research question so that factorial MANOVA can be used.

2.1.3.4 Hierarchical Factorial Discriminant Function Difficulties inherent in factorial discriminant function analysis extend to hierarchical arrangements of predictors. Usually, however, questions of interest can be readily rephrased in terms of factorial MANCOVA.

2.1.4 Structure

A final set of questions is concerned with the latent structure underlying a set of variables. Depending on whether the search for structure is empirical or theoretical, the choice is principal components or factor analysis.

2.1.4.1 Principal Components If scores on numerous variables are available from a group of subjects, the researcher might ask how the variables cluster together. Can the variables be condensed into a smaller number of characteristics on which the subjects differ? Do the large number of observed DVs actually represent a smaller number of latent IVs? For example, suppose people are asked to rate the effectiveness of numerous behaviors for coping with stress (e.g., ''talking to a friend,'' ''going to a movie,'' ''jogging,'' ''making lists of ways to solve the problem''). The numerous behaviors may be condensed into just a few basic coping mechanisms, such as increasing or decreasing social contact, engaging in physical activity, and instrumental manipulation of stress producers.

Principal components analysis develops a small set of uncorrelated components based on the scores on the variables. The components empirically summarize the correlations among the variables. If there are no hypotheses about the components prior to data collection, principal components analysis is often the appropriate strategy, as discussed in Chapter 12.

2.1.4.2 Factor Analysis When there are hypotheses about underlying structure, factor analysis is often used to develop the structure and assess the fit between the data and the hypothetical structure. In this case the underlying IVs are called factors. In the example of mechanisms for coping with stress, one might hypothesize ahead of time that there are two major factors: nature of dealing with the problem (escape vs. direct confrontation) and degree of use of social supports (withdrawing from people vs. seeking them out).

It is often useful to explore differences between groups in terms of latent structure. For example, young college students might use the two coping mechanisms hyothesized above, whereas older adults may have a substantially different factor structure for coping styles.

As implied in this discussion, factor analysis is useful in developing and testing theories. What is the structure of personality? Are there some basic dimensions of personality on which people differ? By collecting scores from many people on numerous variables that may reflect different aspects of personality, we can address questions about underlying structure through factor analysis, as discussed in Chapter 12.

2.2 A DECISION TREE

A decision tree starting with major research questions appears in Table 2.1. For each question, choice among techniques depends on number of IVs and DVs and whether some variables are usefully viewed as covariates. The table also briefly describes analytic goals associated with some techniques.

The paths in Table 2.1 are only recommendations concerning an analytic strategy. Researchers frequently discover that they need two or more of these procedures or, even more frequently, a judicious mix of univariate and multivariate procedures to answer their research questions fully. In short, we recommend a flexible approach to data analysis where both univariate and multivariate procedures are used to clarify the results.

2.3 TECHNIQUE CHAPTERS

Chapters 5 through 12, the basic technique chapters, follow a common format. First, the technique is described and the general purpose briefly discussed. Then the specific kinds of questions that can be answered through application of that technique are listed. Next both the theoretical and practical limitations of the technique are discussed; this section lists assumptions particularly associated with the technique, describes methods for checking the assumptions for your data set, and gives suggestions for dealing with violations. Then a small hypothetical data set is used to illustrate the statistical development of the procedure. Simple analyses by programs from four computer packages are illustrated.

The next section describes the major types of the technique, where appropriate. Then some of the most important issues to be considered when using the technique are covered, including special statistical tests, data snooping, and the like. A direct comparison of the programs available in SPSS, BMDP, SAS, and SYSTAT is then made.

The next section shows, step by step, application of the technique to actual data gathered as described in Appendix B. Assumptions are tested and violations dealt with, where necessary. Major hypotheses are evaluated and follow-up analyses are performed as indicated. Then a Results section is developed, as might be appropriate for submission to a professional journal. When more than one major type of technique is available, there are additional complete examples using real data.

Finally, each technique chapter ends with a description of a near-random selection of articles from the literature that use the technique. The only criteria for selection of examples were that they were recent and covered a variety of fields. They

TABLE 2.1 CHOOSING AMONG STATISTICAL TECHNIQUES

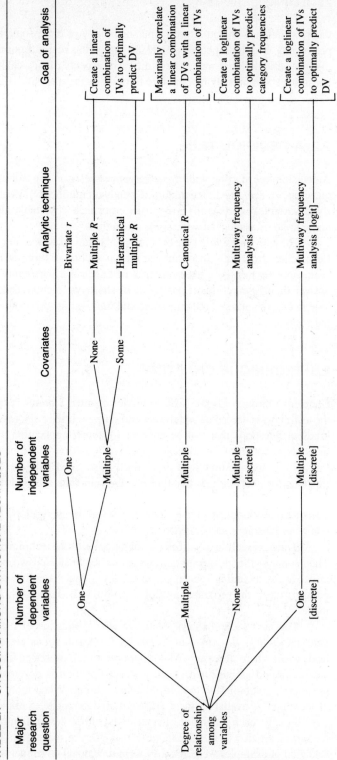

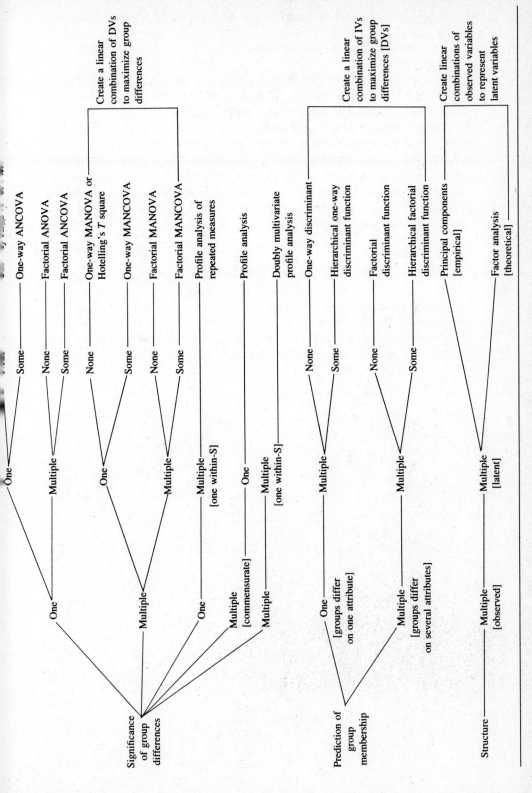

				One-way ANCOVA
			Some	
		One		
			None	Factorial ANOVA
	Multiple			
			Some	Factorial ANCOVA
			None	One-way MANOVA or Hotelling's T square	Create a linear combination of DVs to maximize group differences
		One		
	Multiple		Some	One-way MANCOVA
			None	Factorial MANOVA
Significance of group differences		Multiple		
			Some	Factorial MANCOVA
	One			Profile analysis of repeated measures
	Multiple [commensurate]	One		Profile analysis
	Multiple	Multiple [one within-S]		Doubly multivariate profile analysis

			None	One-way discriminant	Create a linear combination of IVs to maximize group differences [DVs]
	One [groups differ on one attribute]	Multiple		
Prediction of group membership			Some	Hierarchical one-way discriminant function
			None	Factorial discriminant function
	Multiple [groups differ on several attributes]	Multiple		
			Some	Hierarchical factorial discriminant function

			Principal components [empirical]	Create linear combinations of observed variables to represent latent variables
Structure	Multiple [observed]	Multiple [latent]		
			Factor analysis [theoretical]	

31

are not meant to be taken as models, but rather as examples of the way techniques are currently being used.

In working with these technique chapters, it is suggested that the student/ researcher apply the various analyses to some interesting large data set. Many data banks are readily accessible through computer installations.

Further, although we have tried to recommend methods of reporting multivariate results, it may be inappropriate to report them fully in all publications. Certainly one would always want to mention that univariate results were supported and guided by multivariate inference. But the gory details associated with a full disclosure of multivariate results at a colloquium, for instance, might require more attention than one could reasonably expect from an audience. Likewise, a full multivariate analysis may be more than some journals are willing to print.

2.4 PRELIMINARY CHECK OF THE DATA

Before applying any technique, or sometimes even before choosing a technique, you should determine the fit between your data and some very basic assumptions underlying the multivariate statistics. Though each technique has specific assumptions, as well, all require consideration of material in Chapter 4.

Chapter 3

Review of Univariate and Bivariate Statistics

3.1 HYPOTHESIS TESTING

Statistics are used to make rational decisions under conditions of uncertainty. We make inferences (decisions) about populations based on data from samples that contain incomplete information. Samples taken from the same population differ from one another and from the population. Therefore, inferences regarding the population are made with reservation.

The traditional solution to this problem is statistical decision theory. Two hypothetical states of reality are set up, each represented by a probability distribution. Each distribution represents an alternative hypothesis about the true nature of events. Given sample results, we make a best guess as to which distribution the sample was taken from using formalized statistical rules to define "best."

3.1.1 One-Sample z Test as Prototype

Statistical decision theory is most easily illustrated through a one-sample z test, using the standard normal distribution as our model for the two hypothetical states of reality. Suppose we have a sample of 25 IQ scores and need to decide whether this sample of scores is a random sample of a "normal" population with $\mu = 100$ and $\sigma = 15$, or a random sample from a population with $\mu = 108$ and $\sigma = 15$.

First, note that we are testing hypotheses about *means,* not individual scores. Therefore the distributions representing hypothetical states of reality reflect populations of means rather than populations of individual scores. Populations of means produce "sampling distributions of means" that differ systematically from distributions of individual scores; the mean of a population distribution, μ, is equal to the mean of a sampling distribution, μ, but the standard deviation of a population of individual scores, σ, is *not* equal to the standard deviation of a sampling distribution, $\sigma_{\bar{y}}$. Sampling distributions have smaller standard deviations than distributions of scores, and the decrease is related to N, the sample size.

$$\sigma_{\bar{y}} = \frac{\sigma}{\sqrt{N}} \tag{3.1}$$

For our sample, then,

$$\sigma_{\bar{y}} = \frac{15}{\sqrt{25}} = 3$$

The question we are really asking is, Does our mean, taken from a sample of size 25, come from a sampling distribution with $\mu_{\bar{y}} = 100$ and $\sigma_{\bar{y}} = 3$ or does it come from a sampling distribution with $\mu_{\bar{y}} = 108$ and $\sigma_{\bar{y}} = 3$? Figure 3.1(a) shows the first sampling distribution, defined as the null hypothesis, H_0, that is the sampling distribution of means calculated from all possible samples of size 25 taken from a population where $\mu = 100$ and $\sigma = 15$.

The sampling distribution for the null hypothesis has a special, fond place in statistical decision theory because it alone is used to define "best guess." A decision axis for retaining or rejecting H_0 cuts through the distribution so that the probability of rejecting H_0 by mistake is small. "Small" is defined probabilistically as α; an

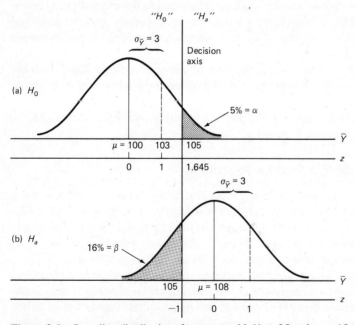

Figure 3.1 Sampling distributions for means with $N = 25$ and $\sigma = 15$ under two hypotheses: (a) H_0:$\mu = 100$ and (b) H_a:$\mu = 108$.

error in rejecting the null hypothesis is referred to as an α, or Type I, error. We have little choice in picking α. Tradition and journal editors decree that it is .05 or smaller, meaning that no more than 5% of the time do we reject the null hypothesis when it is true.

With a table of areas under the standard normal distribution (the table of z scores or standard normal deviates), we decide where to put the decision axis so that the probability of obtaining a sample mean above that point is 5% or less. Looking up 5% in Table C.1, we find that the z corresponding to a 5% cutoff is 1.645. Notice that the z scale is one of two abscissas in Figure 3.1(a). If the decision axis is placed where $z = 1.645$, we translate from the z scale to the $\overline{Y}$ scale to locate the decision axis along $\overline{Y}$. The transformation equation is

$$\overline{Y} = \mu + z\sigma_{\overline{y}} \tag{3.2}$$

Equation 3.2 is a rearrangement of terms from the z test for a single sample:[1]

$$z = \frac{\overline{Y} - \mu}{\sigma_{\overline{y}}} \tag{3.3}$$

Applying Equation 3.2 to our example,

$$\overline{Y} = 100 + (1.645)(3) = 104.935$$

Therefore, we reject the null hypothesis that the mean IQ of the sampling distribution is 100 if the mean IQ of the sample is equal to or greater than 104.935, call it 105.

Frequently this is as far as the model is taken—the null hypothesis is rejected or not. However, if the null hypothesis is rejected, it is rejected in favor of an alternative hypothesis, H_a. The alternative hypothesis is not always stated explicitly,[2] but when it is we can evaluate the probability of retaining the null hypothesis when it *should* be rejected because H_a is true.

This second type of error is called a β, or Type II, error. Because in our example the μ for H_a is 108, we can graph the sampling distribution of means for H_a, as in Figure 3.1(b). The decision axis is positioned by H_0, so we need to find the probability associated with the place it crosses H_a. We first find z corresponding to an IQ score of 105 in a distribution with $\mu_y = 108$ and $\sigma_y = 3$. Applying Equation 3.3, we find that

$$z = \frac{105 - 108}{3} = -1.00$$

[1] The more usual procedure for testing a hypothesis about a single mean is to solve for z on the basis of the sample mean and standard deviation to see if the sample mean is sufficiently far away from the mean of the sampling distribution under the null hypothesis. If z is 1.645 or larger, the null hypothesis is rejected.

[2] Often the alternative hypothesis is simply that the sample is taken from a population that is not equal to the population represented by the null hypothesis. There is no attempt to specify "not equal to."

By looking up $z = -1.00$, we find that about 16% of the time sample means are equal to or less than 105 when the population $\mu = 108$ and the alternative hypothesis is true. Therefore our $\beta = .16$.

H_0 and H_a represent alternative realities, only one of which is true. When the researcher is forced to decide whether to retain or reject H_0, four things can happen. If the null hypothesis is true, a correct decision is made if the researcher retains H_0 and an error is made if the researcher rejects it. If the probability of making the wrong decision is α, the probability of making the right decision is $1 - \alpha$. If, on the other hand, H_a is true, the probability of making the right decision by rejecting H_0 is $1 - \beta$ and the probability of making the wrong decision is β. This information is summarized in a "confusion matrix" (aptly named, according to beginning statistics students) showing the probability of each of these four outcomes:

	Reality	
	H_0	H_a
"H_0"	$1 - \alpha$	β
"H_a"	α	$1 - \beta$
	1.00	1.00

Statistical decision

For our example, the probabilities are

	Reality	
	H_0	H_a
"H_0"	.95	.16
"H_a"	.05	.84
	1.00	1.00

Statistical decision

3.1.2 Power

The lower right-hand cell of the confusion matrix represents the most desirable outcome and the *power* of the research. Usually the researcher believes that H_a is true and hopes that the sample data lead to rejection of H_0. Power is the probability of rejecting H_0 when H_a is true. In Figure 3.1(b), power is the portion of the H_a distribution that falls above the decision axis. Many of the choices in designing research are made with an eye toward increasing power because research with low power usually isn't worth the effort.

Figure 3.1 and Equations 3.1 and 3.2 suggest some ways to enhance power. One obvious way to increase power is to move the decision axis to the left. However, it can't be moved far or Type I error rates reach an unacceptable level. Given the choice between .05 and .01 for α error, though, a decision in favor of .05 increases power. A second strategy is to move the curves farther apart by applying a stronger treatment. Other strategies involve decreasing the standard deviation of the sampling distributions either by decreasing variability in scores (e.g., exerting greater experimental control) or by increasing sample size, N.

This model and these strategies for increasing power generalize to other sampling distributions and to tests of hypotheses other than a single sample mean against a hypothesized population mean.

There is also the danger of *too much power*. The null hypothesis is probably never exactly true and any sample is likely to be slightly different from the population value. With a large enough sample, rejection of H_0 is virtually certain. For that reason, a "minimal meaningful difference" should guide the selection of sample size. The sample size should be large enough to reveal a minimal meaningful difference but not so large as to reveal any difference whatever. Then rejection of H_0 is nontrivial. This issue is considered further in Section 3.4.

3.1.3 Extensions of the Model

The z test for the difference between a sample mean and a population mean readily extends to a z test of the difference between two sample means. A sampling distribution is generated for the difference between means under the null hypothesis that $\mu_1 = \mu_2$ and is used to position the decision axis. The power of an alternative hypothesis is calculated with reference to the decision axis, just as before.

When population variances are unknown, it is desirable to evaluate the probabilities using Student's t rather than z, even for large samples. Numerical examples of use of t to test differences between two means are available in most univariate statistics books and are not presented here. The logic of the process, however, is identical to that described in Section 3.1.1.

3.2 ANALYSIS OF VARIANCE

Analysis of variance is used when two or more means are compared to see if there are any reliable differences among them. Distributions of scores for three hypothetical samples are provided in Figure 3.2. Analysis of variance evaluates the differences among means relative to the overlap in the sampling distributions. The null hypothesis is that $\mu_1 = \mu_2 = \cdots = \mu_k$ as estimated from $\overline{Y}_1, \overline{Y}_2, \ldots, \overline{Y}_k$, with k equal to the number of means being compared.

Analysis of variance (ANOVA) is really a set of analytic procedures based on a comparison of two estimates of variance. One estimate comes from differences among scores within each group; this estimate is considered random or error variance. The second estimate comes from differences in group means and is considered a reflection of group differences or treatment effects plus error. If these two estimates

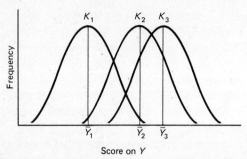

Figure 3.2 Idealized frequency distribution of three samples and their means.

of variance do not differ appreciably, one concludes that all of the group means come from the same sampling distribution of means and that the slight differences among them are due to random error. If, on the other hand, the group means differ more than expected, it is concluded that they were drawn from different sampling distributions of means and the null hypothesis that the means are the same is rejected.

Differences among variances are evaluated as ratios, where the variance associated with difference among sample means is in the numerator, and the variance associated with error is in the denominator. The ratio between these two variances forms an F distribution. F distributions change shape depending on degrees of freedom in both numerator and denominator of the F ratio. Thus tables of critical F, for testing the null hypothesis, depend on two degree-of-freedom parameters (cf. Appendix C, Table C.3).

The many varieties of analysis of variance are conveniently summarized in terms of the partition of *sums of squares,* that is, sums of squared differences between scores and their means. A sum of squares (SS) is simply the numerator of a variance, S^2.

$$S^2 = \frac{\Sigma(Y - \overline{Y})^2}{N - 1}$$ (3.4)

$$SS = \Sigma(Y - \overline{Y})^2$$ (3.5)

The square root of variance is standard deviation, S, the measure of variability that is in the metric of the original scores.

$$S = \sqrt{S^2}$$ (3.6)

3.2.1 One-Way Between-Subjects ANOVA

DV scores appropriate to one-way between-subjects ANOVA with equal n are presented in a table, with k columns representing groups (levels of the IV) and n scores within each group.[3] Table 3.1 shows how subjects are assigned to groups within this design.

Each column has a mean, $\overline{Y}_j$, where $j = 1, 2, . . . , k$. Each score is designated

---

[3] Throughout the book, n is used for sample size within a single group or cell and N is used for total sample size.

TABLE 3.1 ASSIGNMENT OF
SUBJECTS IN
A ONE-WAY
BETWEEN-
SUBJECTS ANOVA

Treatment		
K_1	K_2	K_3
S_1	S_4	S_7
S_2	S_5	S_8
S_3	S_6	S_9

Y_{ij}, where $i = 1, 2, \ldots, n$. The symbol GM represents the grand mean of all scores over all groups.

The difference between each score and the grand mean ($Y_{ij} - $ GM) is considered the sum of two component differences, the difference between the score and its own group mean and the difference between that mean and the overall mean.

$$Y_{ij} - \mathrm{GM} = (Y_{ij} - \overline{Y}_j) + (\overline{Y}_j - \mathrm{GM}) \tag{3.7}$$

This result is achieved by first subtracting and then adding the group mean to the equation. We then square and sum each term separately to produce the sum of squares for error and the sum of squares for treatment, respectively. The basic partition holds because, conveniently, the cross-product terms produced by squaring and summing cancel each other out. Across all scores, the partition is

$$\sum_i \sum_j (Y_{ij} - \mathrm{GM})^2 = \sum_i \sum_j (Y_{ij} - \overline{Y}_j)^2 + n\sum_j (\overline{Y}_j - \mathrm{GM})^2 \tag{3.8}$$

Each of these terms is a sum of squares (SS)—a sum of squared differences between scores (with means sometimes treated as scores) and their associated means. That is, each term is a special case of Equation 3.5.

The term on the left of the equation is the total sum of squared differences between scores and the grand mean, ignoring groups with which scores are associated, designated $\mathrm{SS}_{\text{total}}$. The first term on the right is the sum of squared deviations between each score and its group mean. When summed over all groups, it becomes sum of squares within groups, SS_{wg}. The last term is the sum of squared deviations between each group mean and the grand mean, the sum of squares between groups, SS_{bg}. Equation 3.8 is also symbolized as

$$\mathrm{SS}_{\text{total}} = \mathrm{SS}_{wg} + \mathrm{SS}_{bg} \tag{3.9}$$

Degrees of freedom in ANOVA partition the same way as sums of squares:

$$\mathrm{df}_{\text{total}} = \mathrm{df}_{wg} + \mathrm{df}_{bg} \tag{3.10}$$

Total degrees of freedom are the number of scores minus 1, lost when the grand mean is estimated. Therefore

$$df_{total} = N - 1 \tag{3.11}$$

Within-groups degrees of freedom are the number of scores minus k, lost when the means for each of the k groups are estimated. Therefore

$$df_{wg} = N - k \tag{3.12}$$

Between-groups degrees of freedom are k "scores" (each group mean treated as a score) minus one, lost when the grand mean is estimated, so that

$$df_{bg} = k - 1 \tag{3.13}$$

Verifying the equality proposed in Equation 3.10, we get

$$N - 1 = N - k + k - 1$$

As in the partition of sums of squares, the term associated with group means is subtracted out of the equation and then added back in.

Another common notation for the partition of Equation 3.7 is

$$SS_{total} = SS_K + SS_{S(K)} \tag{3.14}$$

as shown in Table 3.2(a). In this notation, the total sum of squares is partitioned into a sum of squares due to the k groups, SS_K, and a sum of squares due to subjects within the groups, $SS_{S(K)}$. (Notice that the order of terms on the right side of the equation is the reverse of that in Equation 3.9.)

The division of a sum of squares by degrees of freedom produces variance, called *mean square*, MS, in ANOVA. Variance, then, is an average sum of squares. ANOVA produces three variances: one associated with total variability among scores, MS_{total}; one associated with variability within groups, MS_{wg} or $MS_{S(K)}$; and one with variability between groups, MS_{bg} or MS_K.

MS_K and $MS_{S(K)}$ provide the variances for an F ratio to test the null hypothesis that $\mu_1 = \mu_2 = \cdots = \mu_k$.

$$F = \frac{MS_K}{MS_{S(K)}} \qquad df = (k - 1), N - k \tag{3.15}$$

Once F is computed, it is tested against critical F obtained from a table, such as Table C.3, with numerator $df = k - 1$ and denominator $df = N - k$ at desired alpha level. If obtained F exceeds critical F, the null hypothesis is rejected in favor of the hypothesis that there is a difference among the means in the k groups.

Anything that increases obtained F increases power. Power is increased by decreasing the error variability in the denominator ($MS_{S(K)}$) or by increasing differences among means in the numerator (MS_K).

TABLE 3.2 PARTITION OF SUMS OF SQUARES AND DEGREES OF FREEDOM FOR SEVERAL ANOVA DESIGNS

(a) One-way between-subjects ANOVA

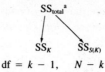

$$df = k - 1, \quad N - k$$

(b) Factorial between-subjects ANOVA

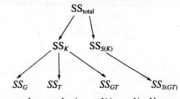

$$df = g - 1, \; t - 1, \; (g - 1)(t - 1), \; N - gt$$

(c) One-way within-subjects ANOVA

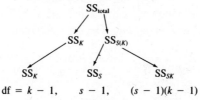

$$df = k - 1, \quad s - 1, \quad (s - 1)(k - 1)$$

(d) One-way matched-randomized ANOVA

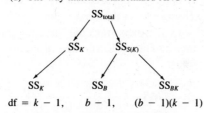

$$df = k - 1, \quad b - 1, \quad (b - 1)(k - 1)$$

(e) Factorial within-subjects ANOVA

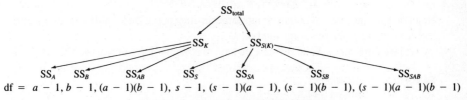

$$df = a - 1, b - 1, (a - 1)(b - 1), \; s - 1, \; (s - 1)(a - 1), \; (s - 1)(b - 1), \; (s - 1)(a - 1)(b - 1)$$

(f) Mixed within-between-subjects ANOVA

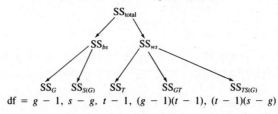

$$df = g - 1, \; s - g, \; t - 1, \; (g - 1)(t - 1), \; (t - 1)(s - g)$$

---

[a] For all SS_{total}, df $= N - 1$.

41

3.2.2 Factorial Between-Subjects ANOVA

If groups are formed along more than one dimension, differences among means are attributable to more than one source. Consider an example with six groups, three of women and three of men, where the DV is scores on a final examination in statistics. One source of variation in means is due to gender, SS_G. If the three groups within each gender are exposed to three different methods of teaching statistics, a second source of differences among means is teaching method, SS_T. The final source of known differences among means is the interaction between gender and teaching methods, SS_{GT}. The interaction tests whether effectiveness of teaching methods varies with gender.

Allocation of subjects in this design is shown in Table 3.3. Sums of squares and degrees of freedom are partitioned as in Table 3.2(b). Error is estimated by variation in scores within each of the six cells, $SS_{S(GT)}$. Three null hypotheses are tested using the F distribution.

The first test asks if means for men and women are likely to have come from the same sampling distribution of means. Scores are averaged across teaching methods to eliminate that source of variability. Gender differences are tested in the F ratio:

$$F = \frac{MS_G}{MS_{S(GT)}} \qquad df = (g - 1), N - gt \qquad (3.16)$$

Rejection of the null hypothesis supports an interpretation of differences between women and men in performance on the final exam.

The second test asks if means from the three teaching methods are likely to have come from the same sampling distribution of means, averaged across women and men. This is tested as:

$$F = \frac{MS_T}{MS_{S(GT)}} \qquad df = (t - 1), N - gt \qquad (3.17)$$

Rejection of the null hypothesis supports an interpretation of differences in effectiveness of the three teaching methods.

The third test asks if the cell means, the means for women and the means for

TABLE 3.3 ASSIGNMENT OF SUBJECTS IN A FACTORIAL BETWEEN-SUBJECTS DESIGN

		Teaching techniques		
		T_1	T_2	T_3
Gender	G_1	S_1 S_2	S_5 S_6	S_9 S_{10}
	G_2	S_3 S_4	S_7 S_8	S_{11} S_{12}

men within each teaching method, are likely to have come from the same sampling distribution of *differences* between means.

$$F = \frac{MS_{GT}}{MS_{S(GT)}} \qquad df = (g - 1)(t - 1), N - gt \qquad (3.18)$$

Rejection of the null hypothesis supports an interpretation that different teaching approaches are effective for men and women.

In each case, the estimate of normally occurring variability in test scores, error, is $MS_{S(GT)}$, or within-cell variance. In each case, critical F is read from a table with appropriate degrees of freedom and desired alpha, and if obtained F (Equations 3.16, 3.17, or 3.18) is greater than critical F, the null hypothesis is rejected. When there is an equal number of scores in each cell, the three tests are independent (except for use of a common error term): the test of each main effect (gender and teaching method) is not related to the test of the other main effect or the test of the interaction.

In a one-way between-subjects design (Section 3.2.1), $k - 1$ degrees of freedom are used to test the null hypothesis of differences among groups. If k is equal to the number of cells in a two-way design, tests of G, T, and GT use up the $k - 1$ degrees of freedom. With proper partitioning, then, of a two-way factorial design, you get three tests for the price of one.

With higher-order between-subjects factorial designs, variation due to differences among groups is partitioned into main effects for each IV, two-way interactions between each pair of IVs, three-way interactions among each trio of IVs, and so on. In any between-subjects factorial design, error sum of squares is the sum of squared differences within each cell of the design.

3.2.3 Within-Subjects ANOVA

In some designs, the means that are tested are derived from the same subjects measured on different occasions, as shown in Table 3.4, rather than from different groups of subjects.[4]

In these designs, computation of sum of squares and mean square for the effect of the IV is the same as for the between-subject designs. However, the error term is

TABLE 3.4 ASSIGNMENT OF SUBJECTS IN A ONE-WAY WITHIN-SUBJECTS DESIGN

		Treatment		
		K_1	K_2	K_3
Subjects	S_1	S_1	S_1	S_1
	S_2	S_2	S_2	S_2
	S_3	S_3	S_3	S_3

[4] This design is also called repeated measures, one-score-per-cell, randomized blocks, matched-randomized changeover, or crossover.

further partitioned into individual differences due to subjects, SS_S, and interaction of individual differences with treatment, SS_{SK}. Because subjects are measured repeatedly, their effect as a source of variability in scores is estimated and subtracted from $SS_{S(K)}$, the error term in a corresponding between-subjects design. The interaction of individual differences with treatment, MS_{SK}, is used as the error term;

$$F = \frac{MS_K}{MS_{SK}} \qquad df = (k - 1),(k - 1)(s - 1) \qquad (3.19)$$

The partition of sums of squares for a one-way within-subjects design with k levels is shown in Table 3.2(c), where s is the number of subjects.

MS_{SK} is used as the error term because, once SS_S is subtracted, no variation is left within cells of the design—there is, in fact, only one score per cell. The interaction of individual differences with treatment is all that remains to serve as an estimate of error variance. If there are individual differences among subjects in scores, and if individuals react similarly to the IV, the interaction is a good estimate of error. Once individual differences are subtracted, the error term is usually smaller than the error term in a corresponding between-subjects design, so the within-subjects design is more sensitive than the between-subjects design.

But if there are no consistent individual differences in scores,[5] or if there is an interaction between subjects and treatment, the error term is larger than that of a between-subjects design. The statistical test is conservative, it is more difficult to reject the null hypothesis of no difference between means, and the power of the test is reduced. Because Type I error is unaffected, the statistical test is not in disrepute, but, in this case, a within-subjects design is a poor choice of research design.

A within-subjects analysis is also used with a *matched-randomized blocks* design, as shown in Table 3.5 and Table 3.2(d). Subjects are first matched on the basis of variables thought to be highly related to the DV. Subjects are then divided into blocks, with as many subjects within each block as there are levels of the IV. Finally, members of each block are randomly assigned to levels of the IV. Although the subjects in each block are actually different people, they are treated statistically

TABLE 3.5 ASSIGNMENT OF SUBJECTS IN A MATCHED-RANDOMIZED DESIGN[a]

		Treatment		
		A_1	A_2	A_3
	B_1	S_1	S_2	S_3
Blocks	B_2	S_4	S_5	S_6
	B_3	S_7	S_8	S_9

[a] Where subjects in the same block have been matched on some relevant variable.

[5] Notice that the degrees of freedom for error, $(k - 1)(s - 1)$, are fewer than in the between-subjects design. Unless the reduction in error variance due to subtraction of SS_S is substantial, the loss of degrees of freedom may overcome the gain due to smaller SS when MS_{SK} is computed.

as if they were the same person. In the analysis, there is a test of the IV, and a test of blocks (the same as the test of subjects in the within-subjects design), with the interaction of blocks and treatment used as the error term. Because matching is used to produce consistency in performance within blocks and the effect of blocks is subtracted from the error term, this design should also be more sensitive than the between-subjects design. It will not be, however, if the matching fails.

Factorial within-subjects designs, as shown in Table 3.6, are an extension of the one-way within-subjects design. The partition of a two-way within-subjects design is in Table 3.2(e). In the analysis, the error sum of squares is partitioned into a number of "subjects by effects" interactions just as the sum of squares for effects is partitioned into numerous sources. It is common (though not universal) to develop a separate error term for each F test; for instance, the test of the main effect of A is

$$F = \frac{MS_A}{MS_{SA}} \qquad df = (a - 1), (s - 1)(a - 1) \qquad (3.20)$$

For the main effect of B, the test is

$$F = \frac{MS_B}{MS_{SB}} \qquad df = (b - 1), (s - 1)(b - 1) \qquad (3.21)$$

and for the interaction,

$$F = \frac{MS_{AB}}{MS_{SAB}} \qquad df = (a - 1)(b - 1), (s - 1)(a - 1)(b - 1) \qquad (3.22)$$

For higher-order factorial designs, the partition into sources of variability grows prodigiously, with an error term developed for each main effect and interaction tested.

There is controversy in within-subject analyses concerning conservatism of the F tests and whether or not separate error terms should be used. In addition, if the repeated measurements are on single subjects, there are often carry-over effects that limit generalizability to situations where subjects are tested repeatedly. Finally, when there are more than two levels of the IV, the analysis has the assumption of homogeneity

TABLE 3.6 ASSIGNMENT OF SUBJECTS IN
A FACTORIAL WITHIN-
SUBJECTS DESIGN

		Treatment A		
		A_1	A_2	A_3
Treatment B	B_1	S_1 S_2	S_1 S_2	S_1 S_2
	B_2	S_1 S_2	S_1 S_2	S_1 S_2

of covariance. Homogeneity of covariance is, roughly speaking, the assumption that subjects have the same rankings in scores for all pairs of levels of the IV. If some pairs of levels are close in time (e.g., trial 2 and trial 3), while other pairs are distant in time (e.g., trial 1 and trial 10), the assumption is often violated. Such violation is serious because Type I error rate is affected. Homogeneity of covariance is discussed in greater detail in Chapters 8, 9, and 10, and in Frane (1980).

For these reasons, within-subjects ANOVA is sometimes replaced by profile analysis, where repetitions of DVs are transformed into separate DVs (Chapter 10).

3.2.4 Mixed Between-Within-Subjects ANOVA[6]

Often in factorial designs, one or more IVs are measured between subjects while other IVs are measured within subjects.[7] The simplest example of the mixed between-within-subjects design involves one between-subjects and one within-subjects IV, as seen in Table 3.7.[8]

To show the partition, the total SS is divided into a source attributable to the between-subjects part of the design (*Groups*), and a source attributable to the within-subjects part (*Trials*), as shown in Table 3.2(f). Each source is then further partitioned into effects and error components: between-subjects into groups and subjects-within-groups error term; and within-subjects into trials, the group-by-trials interaction; and the trial-by-subjects-within-groups error term.

As more between-subjects IVs are added, between-subjects main effects and interactions expand the between-subjects part of the partition. For all the effects, there is a single error term consisting of subjects confined to each combination of the between-subjects IVs. As more within-subjects IVs are added, the within-subjects portion of the design expands. Sources of variability include main effects and interactions of within-subjects IVs and interactions of between- and within-subjects

TABLE 3.7 ASSIGNMENT OF
SUBJECTS IN A
BETWEEN-WITHIN-
SUBJECTS DESIGN

		Trials		
		T_1	T_2	T_3
Groups	G_1	S_1 S_2	S_1 S_2	S_1 S_2
	G_2	S_3 S_4	S_3 S_4	S_3 S_4

[6] This design is also called split plot, repeated measures, and randomized block factorial design.

[7] Mixed designs can also have "blocks" rather than repeated measures on individual subjects as the within-subjects segment of the design.

[8] When the between-subjects variables are based on naturally occurring differences among subjects (e.g., age, sex), the design is said to be "blocked" on the subject variables. This is a different use of the term *blocks* from that of the preceding section. In a mixed design, both kinds of blocking can occur.

IVs. Separate error terms are developed for each source of variability in the within-subjects segment of the design.[9]

Problems associated with within-subjects designs (e.g., homogeneity of co-variance) carry over to mixed designs, and profile analysis is sometimes used to circumvent some of these problems.

3.2.5 Design Complexity

Discussion of analysis of variance has so far been limited to factorial designs where there are equal numbers of scores in each cell and levels of each IV are purposely chosen by the researcher. Several deviations from these straightforward designs are possible. A few of the more common types of design complexity are mentioned below, but the reader actually faced with use of these designs is referred to one of the more complete analysis of variance texts such as Winer (1971), Keppel (1973, 1982), or Myers (1979).

3.2.5.1 Nesting In between-subjects designs, subjects are said to be nested within levels of the IV. That is, each subject is confined to only one level of each IV or combination of IVs. Nesting also occurs with IVs when levels of one IV are confined to only one level of another IV, rather than factorially crossing over the levels of the other IV.

Take the example where the IV is various levels of teaching methods. Children within the same classroom cannot be randomly assigned to different methods but whole classrooms can be so assigned. The design is one-way between-subjects where teaching methods is the IV and classrooms serve as subjects. For each classroom, the mean score for all children on the test is obtained, and the means serve as DVs in one-way ANOVA.

If the effect of classroom is also assessed, the design is nested or hierarchical, as seen in Table 3.8(a). Classrooms are randomly assigned to and nested in teaching methods, and children are nested in classrooms. The error term for the test of classroom is subjects within classrooms and teaching method, while the error term for the test of teaching method is classrooms within teaching technique. The test of teaching method is precisely the one used if classroom effects are ignored in a one-way ANOVA.

TABLE 3.8 SOME COMPLEX ANOVA DESIGNS

(a) Nested designs Teaching techniques				(b) Latin square designs[a] Order		
T_1	T_2	T_3		1	2	3
Classroom 1	Classroom 2	Classroom 3		S_1 A_2	A_1	A_3
Classroom 4	Classroom 5	Classroom 6	Subjects	S_2 A_1	A_3	A_2
Classroom 7	Classroom 8	Classroom 9		S_3 A_3	A_2	A_1

[a] Where the three levels of treatment A are experienced by different subjects in different orders, as indicated.

[9] "Subjects" is no longer available as a source of variance for analysis. Because subjects are confined to levels of the between-subjects variable(s), differences between subjects in each group are used to estimate error for testing variance associated with between-subjects variables.

3.2.5.2 Latin Square Designs   The order of presentation of levels of an IV often produces differences in the DV. In within-subjects designs, subjects become practiced or fatigued or experiment-wise as they experience more levels of the IV. In between-subjects designs, there are often time of day or experimenter effects that change scores on the DV. To get an uncontaminated look at the effects of the IV, it is important to counterbalance the effects of increasing experience, time of day, and the like, so that they are independent of levels of the IV. If the within-subjects IV is something like trials, counterbalancing is not possible because the order of trials cannot be changed. But when the IV is something like background color of slide used to determine if background color affects memory for material on the slide, a Latin square arrangement is often used to control order effects.

A Latin square design is shown in Table 3.8(b). If A_1 is a yellow background, A_2 a blue background, and A_3 a red background, then subjects are presented the slides in the order specified by the Latin square. The first subject is presented the slide with blue background, then yellow, then red. The second subject is presented yellow, then red, then blue, etc. The yellow slide (A_1) appears once in first position, once in second, and once in third, and so on for the other colors, so that order effects are distributed evenly across the levels of the IV.

The simple design of Table 3.8(b) produces a test of the IV (A), a test of subjects (if desired), and a test of order. The effect of order (like the effect of subjects) is subtracted out of the error term, leaving it smaller than it is when order effects are not analyzed. The error term itself is composed of fragments of interactions that are not available for analysis because effects are not fully crossed in the design. Thus, the design is more sensitive when there are order effects and no interactions and less sensitive when there are no order effects but there are interactions. Consult Keppel (1982) or one of the other ANOVA texts for greater detail on this fascinating topic.

3.2.5.3 Unequal *n* and Nonorthogonality   In a simple one-way between-subjects ANOVA, problems created by unequal group sizes are relatively minor. Computation is slightly more difficult, but that is no real disaster, especially if analysis is by computer. However, if sample sizes are very different, the assumption of homogeneity of variance is more difficult to meet. If the group with the smaller *n* has the larger variance, the *F* test is too liberal, leading to increased Type I error rate and an inflated α level.

In factorial designs with more than one between-subjects IV, unequal sample sizes in each cell create difficulty in computation and ambiguity of results. With unequal *n,* a factorial design is nonorthogonal. Hypotheses about main effects and interactions are not independent, and sums of squares are not additive. The various sources of variability contain overlapping variance and the same variance can be attributed to more than one source, as discussed in Chapter 1. If effects are tested without taking the overlap into account, the probability of a Type I error increases because systematic variance contributes to more than one test. A variety of strategies is available to deal with the problem, none of them completely satisfactory.

The simplest strategy is to randomly delete cases from cells with greater *n* until all cells are equal. If unequal *n* is due to random loss of a few subjects in an experimental design originally set up for equal n, deletion is often a good choice. An alternative strategy with random loss of subjects in an experimental design is an unweighted-means analysis, described in Chapter 8 and ANOVA textbooks such as Winer (1971) and Keppel (1982). The unweighted-means approach has greater power than random deletion of cases and is the preferred approach as long as computational aids are available.

But in nonexperimental work, unequal n often results from the nature of the population. Differences in sample sizes reflect true differences in nature. To artificially equalize the n's is to distort the differences and lose generalizability. In these situations, decisions are made as to how tests of effects are to be adjusted for overlapping variance. Standard methods for adjusting tests of effects with unequal n are discussed in Chapter 8.

3.2.5.4 Fixed and Random Effects   In all the ANOVA designs discussed so far, levels of each IV are selected by the researcher on the basis of their interest in testing significance of the IV. This is the usual *fixed* effects model. Sometimes, however, there is a desire to generalize to a population of levels of an IV. In order to generalize to the population of levels of the IV, levels are randomly selected from the population, just as subjects are randomly selected from the population of subjects when the desire is to generalize results to the population. Consider, for example, an experiment to study effects of word familiarity[10] on recall where the desire is to generalize results to all levels of word familiarity. A finite set of familiarity levels is randomly selected from the population of familiarity levels. Word familiarity is considered a *random-*effects IV.

The analysis is set up so that results generalize to levels other than those selected for the experiment—generalize to the population of levels from which the sample was selected. During analysis, alternative error terms for evaluating the statistical significance of random effects IVs are used. Although computer programs are available for analysis of random effects IVs, use of them is rare. The interested reader is referred to one of the more sophisticated ANOVA texts, such as Winer (1971), for a full discussion of the random-effects model.

3.2.6 Specific Comparisons

When an IV has more than one degree of freedom (more than two levels) or when there is an interaction between two or more IVs, the overall test of the effect is ambiguous. The overall test, with $k - 1$ degrees of freedom, is pooled over $k - 1$ single degree of freedom subtests. If the overall test is significant, so are one or more of the subtests, but there is no way to tell which one(s). To find out which single degree of freedom subtests are significant, comparisons are performed.

In analysis, degrees of freedom are best thought of as a nonrenewable resource. They are analyzed once with conventional α levels, but further analyses require very stringent α levels. For this reason, the best strategy is to *plan* your analysis very

---

[10] Word familiarity is usually operationalized by frequency of usage of words in the English language.

carefully so that the most interesting comparisons are tested with conventional α levels. Unexpected findings or less interesting effects are tested later with stringent α levels. This is the strategy used by the researcher who has been working in an area for a while and knows precisely what to look for.

Regrettably, research is often more tentative; so the research "spends" the degrees of freedom on omnibus (routine) ANOVA testing main effects and interactions at conventional α levels, and then snoops the single degree of freedom comparisons of significant effects at stringent α levels. Snooping through data after results of ANOVA are known is called "conducting post hoc comparisons."

We present here the most flexible method of conducting comparisons, with mention of other methods as they are appropriate. The procedure for conducting comparisons is the same for planned and post hoc comparisons up to the point where an obtained F is evaluated against a critical F.

3.2.6.1 Weighting Coefficients for Comparisons Comparison of treatment means begins by assigning a weighting factor (w) to each of the cell or marginal means so the weights reflect your null hypotheses. Suppose you have a one-way design with k means and you want to make comparisons. For each comparison, a weight is assigned to each mean. Weights of zero are assigned to means (groups) that are left out of a comparison, although at least two of the means have to have nonzero weights. Means that are contrasted with each other are assigned weights with opposite signs (positive or negative) with the constraint that the weights sum to zero, that is,

$$\sum_{1}^{k} w_j = 0$$

For example, consider an IV with four levels, producing $\overline{Y}_1$, $\overline{Y}_2$, $\overline{Y}_3$, and $\overline{Y}_4$. If you want to test the hypothesis that $\mu_1 - \mu_3 = 0$, weighting coefficients are 1, 0, -1, 0 producing $1\overline{Y}_1 + 0\overline{Y}_2 + (-1)\overline{Y}_3 + 0\overline{Y}_4$. $\overline{Y}_2$ and $\overline{Y}_4$ are left out while Y_1 is compared with $\overline{Y}_3$. Or, if you want to test the null hypothesis that $(\mu_1 + \mu_2)/2 - \mu_3 = 0$ (to compare the average mean from the first two groups with the mean of the third group leaving out the fourth group), weighting coefficients are $\frac{1}{2}$, $\frac{1}{2}$, -1, 0, (or any multiple of them, such as 1, 1, -2, 0), respectively. Or, if you want to test the null hypothesis that $(\mu_1 + \mu_2)/2 - (\mu_3 + \mu_4)/2 = 0$ (to compare the average mean of the first two groups with the average mean of the last two groups), the weighting coefficients are $\frac{1}{2}$, $\frac{1}{2}$, $-\frac{1}{2}$, $-\frac{1}{2}$, (or 1, 1, -1, -1).

The idea behind the test is that the sum of the weighted means is equal to zero when the null hypothesis is true. The more the sum diverges from zero, the greater the confidence with which the null hypothesis is rejected.

3.2.6.2 Orthogonality of Weighting Coefficients Any pair of comparisons is orthogonal if the sum of the cross products of the weights for the two comparisons is equal to zero. For example, in the three comparisons below:

	w_1	w_2	w_3
Comparison 1	1	-1	0
Comparison 2	1/2	1/2	-1
Comparison 3	1	0	-1

the sum of the cross products of weights for comparison 1 and comparison 2 is

$$(1)(1/2) + (-1)(1/2) + (0)(-1) = 0$$

Therefore the two comparisons are orthogonal.

Comparison 3, however, is orthogonal to neither of the first two comparisons. For instance, checking it against comparison 1,

$$(1)(1) + (-1)(0) + (0)(-1) = 1$$

In general, there are as many orthogonal comparisons as there are degrees of freedom. Because $k = 3$ in the example, df $= 2$. There are only two orthogonal comparisons when there are three levels of an IV, and only three orthogonal comparisons when there are four levels of an IV.

There are advantages to use of orthogonal comparisons, if they suit the needs of the research. First, there are only as many of them as there are degrees of freedom, so the temptation to "overspend" degrees of freedom is avoided. Second, orthogonal comparisons analyze nonoverlapping variance. If one of them is significant, it has no bearing on the significance of another of them. Last, because they are independent, if all $k - 1$ orthogonal comparisons are performed, the sum of the sum of squares for the comparisons is the same as the sum of squares for the IV in omnibus ANOVA. That is, the sum of squares for the effect is completely broken down into the $k - 1$ orthogonal comparisons that comprise it.

3.2.6.3 Obtained F for Comparisons   Once the weighting coefficients are chosen, the following equation is used to obtain F for the comparison if sample sizes are equal in each group.

$$F = \frac{n_c(\Sigma w_j \overline{Y}_j)^2 \, / \, \Sigma w_j^2}{\text{MS}_{\text{error}}} \qquad (3.23)$$

where n_c is the number of scores in each of the means to be compared, $(\Sigma w_j \overline{Y}_j)^2$ is the squared sum of the weighted means Σw_j^2 is the sum of the squared coefficients, and MS_{error} is the mean square for error in the ANOVA.

The numerator of Equation 3.23 is both sum of squares and mean square for the comparison because a comparison has only one degree of freedom.

For factorial designs, comparisons are done on either marginal or cell means, corresponding to comparisons on main effects or interactions, respectively. The number of scores per mean and the error term follow from the ANOVA design used.

However, if comparisons are made on within-subjects effects, it is customary to develop a separate error term for each comparison, just as separate error terms are developed for omnibus tests of within-subjects IVs.

Chapter 10 has a lot more information on comparisons of both main effects and interactions, including control language for performing them through some of the more popular computer programs.

Once you have obtained F for a comparison, whether by hand calculation or computer, the obtained F is compared with critical F to see if it is reliable. If obtained F exceeds critical F, the null hypothesis for the comparison is rejected. But which critical F is used depends on whether the comparison is planned or performed post hoc.

3.2.6.4 Critical F for Planned Comparisons If you are in the enviable position of having planned your comparisons prior to data collection, and if you have planned no more of them than you have degrees of freedom for effect, critical F is obtained from the tables just as in routine ANOVA. Each comparison is tested against critical F at routine α with one degree of freedom in the numerator and degrees of freedom associated with the MS_{error} in the denominator. If obtained F is larger than critical F, the null hypothesis represented by the weighting coefficients is rejected.

With planned comparisons, omnibus ANOVA is not performed;[11] the researcher moves straight to comparisons. Once the degrees of freedom are spent on the planned comparisons, however, it is perfectly acceptable to snoop the data at more stringent α levels (Section 3.2.6.5), including main effects and interactions from omnibus ANOVA if they are appropriate.

Sometimes, however, the researcher can't resist the temptation to plan more comparisons than degrees of freedom for effect. When there are too many tests, even if comparisons are planned, the α level across all tests exceeds the α level for any one test and some adjustment of α for each test is needed. It is common practice to use a Bonferroni-type adjustment where slightly more stringent α levels are used with each test to keep α across all tests at reasonable levels. For instance, when 5 comparisons are planned, if each one of them is tested at $\alpha = .01$, the α across all tests is an acceptable .05 (roughly .01 times 5, the number of tests). However, if 5 comparisons are each tested at $\alpha = .05$, the α across all tests is approximately .25 (roughly .05 times 5), unacceptable by most standards.

If you want to keep overall α at, say, .10, and you have 5 tests to perform, you can assign each of them $\alpha = .02$, or you can assign two of them $\alpha = .04$ with the other three evaluated at $\alpha = .01$, for an overall Type I error rate of roughly .11. The decision about how to apportion α through the tests is also made prior to data collection.

As an aside, it is important to realize that routine ANOVA designs with numerous main effects and interactions suffer from the same problem of inflated Type I error rate across all tests as planned comparisons where there are too many tests. Some adjustment of α for separate tests is needed in big ANOVA problems, as well.

3.2.6.5 Critical F for Post Hoc Comparisons If you are unable to plan your comparisons and choose to start with routine ANOVA instead, you want to follow up

---

[11] You might perform routine ANOVA to compute the error term(s).

significant main effects (with more than two levels) and interactions with post hoc comparisons to find the treatments that are different from one another. Because you have already spent your degrees of freedom in routine ANOVA and are likely to capitalize on chance differences among means that you notice in the data, some form of adjustment of alpha is necessary.

Many procedures for dealing with inflated Type I error rate are available as described in standard ANOVA texts such as Keppel (1982). The tests differ in the number and type of comparisons they permit and the amount of adjustment required of α. The tests that permit more numerous comparisons have correspondingly more stringent adjustments to critical F. For instance, the Dunnett test, which compares the mean from a single control group with each of the means of the other groups, has a less stringent correction than the Tukey test which allows all pairwise comparisons of means. The name of this game is to choose the most liberal test that permits you to perform the comparisons of interest.

The test described here (Scheffé, 1953) is the most conservative and most flexible of the popular methods. Once critical F is computed with the Scheffé adjustment, there is no limit to the number and complexity of comparisons that can be performed. You can perform all pairwise comparisons and all combinations of treatment means pooled and contrasted with other treatment means, pooled or not, as desired. Some possibilities for pooling are illustrated in Section 3.2.6.1. And once you pay the "price" in conservatism for this flexibility, you might as well conduct all the comparisons that make sense given your research design.

The Scheffé method for computing critical F for a comparison on marginal means is

$$F' = (k - 1)F_c \tag{3.24}$$

where F' is adjusted critical F, $(k - 1)$ is degrees of freedom for the effect, and F_c is tabled F with $k - 1$ degrees of freedom in the numerator and degrees of freedom associated with the error term in the denominator.

If obtained F is larger than critical F', the null hypothesis represented by the weighting coefficients for the comparison is rejected.

For tests of cell means in a factorial design, the adjustment is degrees of freedom for the interaction. For example, with two IVs, A (with a levels) and B (with b levels), a comparison of cell means has as adjustment for critical F

$$F' = (a - 1)(b -)F_c$$

where F_c has as degrees of freedom $(a - 1)(b - 1)$ and df error. (See Chapter 10 for a more extended discussion of the appropriate correction)

3.3 PARAMETER ESTIMATION

If a reliable difference among means is found, an estimate of the means in the population is needed to indicate the size and direction of the differences in means. Sample means are unbiased estimators of population means; the best guess about the size of a population mean (μ) is the mean of the sample randomly selected from that population. In most reports of research, therefore, sample mean values are reported along with statistical results.

Sample means are only approximations of population means. They are unbiased because they are systematically neither large nor small, but they are rarely precisely at the population value—and there is no way to know when they are. Thus the error in estimation, the familiar confidence interval of introductory statistics, is often reported along with the estimated means. The size of the confidence interval depends on sample size, the estimation of population variability, and the degree of confidence one wishes to have in estimating μ. Or, sometimes, cell standard deviations or standard errors are presented along with sample means.

3.4 STRENGTH OF ASSOCIATION[12]

Although significance testing, comparisons, and parameter estimation help illuminate the nature of group differences, they do not assess the degree to which the IV(s) and DV are related. It is important to assess the degree of relationship to avoid publicizing trivial results as though they had practical utility. As discussed in Section 3.1.2, overly powerful research sometimes produces results that are statistically significant but realistically meaningless.

Strength of association assesses the proportion of variance in the DV that is associated with levels of an IV. How much of the total variance in the DV is predictable from knowledge of the levels of the IV? If the total variances of the DV and the IV are represented by circles, how much do the circles overlap? Statistical significance testing assesses the *reliability* of the association between the IV and DV. Strength of association measures *how much* association there is.

A rough estimate of strength of association is available for any ANOVA through η^2 (eta squared).

$$\eta^2 = \frac{SS_{effect}}{SS_{total}} \tag{3.25}$$

When there are 2 levels of the IV, η^2 is the (squared) point biserial correlation between the continuous variable (the DV) and the dichotomous variable (the two levels of the IV).[13] After finding a significant main effect or interaction, η^2 shows the proportion of variance in the DV (SS_{total}) attributable to the effect (SS_{effect}). In a balanced, equal-n design, η^2's are additive; the sum of η^2 for all significant effects is the proportion of variation in the DV that is predictable from knowledge of the IVs.

This simple, popular measure of strength of association is flawed for two reasons. The first is that η^2 for a particular IV depends on the number and significance of other IVs in the design. η^2 for an IV tested in a one-way design is likely to be larger than η^2 for the same IV in a two-way design where the other IV and the interaction

---

[12] This is also called effect size or treatment magnitude.

[13] All strength of association values are associated with the particular levels of the IV used in the research and do not generalize to other levels.

add to the total variance, especially if one or both of the additional effects is large. This is because the denominator of η^2 contains systematic variance for other effects in addition to error variance and systematic variance for the effect of interest.

Therefore, an alternative form of η^2 is available where the denominator contains only variance attributable to the effect of interest plus error

$$\eta^2_{alt} = \frac{SS_{effect}}{SS_{effect} + SS_{error}} \qquad (3.26)$$

With this alternative, η^2's for all significant effects in the design *do not* sum to proportion of systematic variance in the DV. Indeed, the sum is sometimes greater than 1.00. It is imperative, therefore, to be clear in your report when this version of η^2 is used.

A second flaw is that η^2 describes proportion of systematic variance in a sample with no attempt to estimate proportion of systematic variance in the population. A statistic developed to estimate strength of association between IV and DV in the population is ω^2 (omega squared).

$$\omega^2 = \frac{SS_{effect} - (df_{effect})(MS_{error})}{MS_{error} + SS_{total}} \qquad (3.27)$$

This is the additive form of ω^2 where the denominator represents total variance, not just variance due to effect plus error, and is *limited to between-subjects analysis of variance designs with equal n*. Forms of ω^2 are available for designs containing repeated measures (or randomized blocks) as described by Vaughn and Corballis (1969).

A separate measure of strength of association is computed and reported for each statistically significant main effect and interaction in a design.

3.5 BIVARIATE STATISTICS: CORRELATION AND REGRESSION

Strength of association as described in Section 3.4 is assessed between a continuous DV and discrete levels of an IV. Frequently, however, a researcher wants to measure the strength of association between two continuous variables where the IV-DV distinction is blurred. For instance, the association between years of education and income is of interest even though neither is manipulated and inferences regarding causality are not possible. Correlation is the measure of the size and direction of relationship between the two variables, and squared correlation is the measure of strength of association between them.

Correlation is used to measure the association between variables; regression is used to predict one variable from the other (or many others). However, the equations for correlation and bivariate regression are very similar, as indicated below.

3.5.1 Correlation

The Pearson product-moment correlation coefficient, r, is easily the most frequently used measure of association and the basis of many multivariate calculations. The most interpretable equation for Pearson r is

$$r = \frac{\Sigma Z_X Z_Y}{N - 1} \tag{3.28}$$

where Pearson r is the average cross product of standardized X and Y variable scores.

$$Z_Y = \frac{Y - \bar{Y}}{S} \quad \text{and} \quad Z_X = \frac{X - \bar{X}}{S}$$

and S is as defined in Equations 3.4 and 3.6.

Pearson r is independent of scale of measurement (because both X and Y scores are converted to standard scores) and independent of sample size (because of division by $N - 1$). The value of r ranges between -1.00 and $+1.00$ where .00 represents no relationship or predictability between the X and Y variables. An r value of -1.00 or $+1.00$ indicates perfect predictability of one score when the other is known. When correlation is perfect, scores in the X distribution have the same relative positions as corresponding scores in the Y distribution.[14]

The raw score form of Equation 3.28 also sheds light on the meaning of r:

$$r = \frac{N\Sigma XY - (\Sigma X)(\Sigma Y)}{\sqrt{[N\Sigma X^2 - (\Sigma X)^2][N\Sigma Y^2 - (\Sigma Y)^2]}} \tag{3.29}$$

Pearson r is the covariance between X and Y relative to (the square root of the product of) the X and Y variances. Only the numerators of variance and covariance equations appear in Equation 3.29 because the denominators cancel each other out.

3.5.2 Regression

Whereas correlation is used to measure the size and direction of the relationship between two variables, regression is used to predict a score on one variable from a score on the other. In bivariate (two-variable) regression where Y is predicted from X, a straight line between the two variables is found. The best-fitting straight line goes through the means of X and Y and minimizes the sum of the squared distances between the data points and the line.

To find the best-fitting straight line, an equation is solved of the form

$$Y' = A + BX \tag{3.30}$$

---

[14] When correlation is perfect, $Z_X = Z_Y$ for each pair and the numerator of Equation 3.28 is, in effect, $\Sigma Z_X Z_X$. Because $\Sigma Z_X^2 = N - 1$, Equation 3.28 reduces to $(N - 1)/(N - 1)$, or 1.00.

where Y' is the predicted score, A is the value of Y when X is 0.00, B is the slope of the line (change in Y divided by change in X), and X is the value from which Y is to be predicted.

The difference between the predicted and the obtained values of Y at each value of X represents error of prediction. The best-fitting straight line is the line that minimizes the squared errors of prediction.

To solve Equation 3.30 both B and A are found.

$$B = \frac{N\Sigma XY - (\Sigma X)(\Sigma Y)}{N\Sigma X^2 - (\Sigma X)^2} \tag{3.31}$$

The bivariate regression coefficient, B, is the ratio of the covariance of the variables and the variance of the one from which predictions are made.

Note the differences and similarities between Equation 3.29 (for correlation), and Equation 3.31 (for the regression coefficient). Both have the covariance between the variables as a numerator but differ in denominator. In correlation, the variances of both are used in the denominator. In regression, the variance of the predictor variable serves as the denominator. If Y is predicted from X, X variance is the denominator. If X is predicted from Y, Y variance is the denominator.

To complete the solution, the value of the intercept, A, is also calculated.

$$A = \overline{Y} - B\overline{X} \tag{3.32}$$

The intercept is the mean of the predicted variable minus the product of the regression coefficient times the mean of the predictor variable.

The techniques discussed in this chapter for making decisions about differences, estimating population means, assessing association between variables, and predicting a score on one variable from a score on another are important to and widely used in the social and behavioral sciences. They form the basis for most undergraduate—and some graduate—statistics courses. It is hoped that this brief review reminds you of material already mastered so that, with common background and language, we begin in earnest the study of multivariate statistics.

Chapter 4

Cleaning Up Your Act: Screening Data Prior to Analysis

This chapter deals with a whole set of issues that are considered before the main analysis is run. Careful consideration of these issues is time-consuming and sometimes tedious; it is very common, for instance, to spend several days in careful examination of data prior to running the main analysis. But consideration and resolution of these issues is fundamental to an honest analysis of the data.

A first issue concerns the accuracy with which data have been entered into the data file along with factors that produce distorted correlations. Missing data, the bane of (almost) every researcher, are assessed and dealt with. Outliers, cases that are extreme, create other headaches because solutions are unduly influenced by them. Next, many multivariate procedures are based on assumptions; the fit between your data set and the assumptions is assessed before the procedure is applied. Transformations of variables to bring them into compliance with requirements of analysis are considered. Finally, perfect or near-perfect correlations among variables can threaten a multivariate analysis.

This chapter deals with issues that are relevant to most analyses. However, they are not all applicable to all analyses all the time; further, some analyses have assumptions not mentioned here. For that reason, assumptions and limitations specific to each analysis are reviewed in the third section of the chapter describing the analysis.

There are differences in data screening for grouped and ungrouped data. If you are performing multiple regression, canonical correlation, or factor analysis, where subjects are not subdivided into groups, there is one procedure for screening data. If you are performing analysis of covariance, multivariate analysis of variance or covariance, profile analysis, or discriminant function analysis, where subjects are in groups, there is another procedure for screening data. Differences in these procedures are illustrated by example in Section 4.2.

You may find the material in this chapter difficult from time to time. It is necessary sometimes to refer ahead to material covered in Chapters 5 to 12 to explain some issue, material that is more understandable after Chapters 5 through 12 are studied. You may want to read this chapter now to get an overview of the tasks to be accomplished prior to the main data analysis and then read it again after mastering the remaining chapters.

4.1 IMPORTANT ISSUES IN DATA SCREENING

4.1.1 Accuracy of Data File

A very important first step is to examine univariate descriptive statistics for accuracy of input through one of the descriptive programs such as SPSS FREQUENCIES, BMDP1D or 2D, SYSTAT STATS, or SAS MEANS or UNIVARIATE. Are all the values on all the variables within range? Are means and standard deviations plausible? If you have discrete variables (such as categories of religious affiliation), are there any out-of-range numbers? Have you accurately programmed your codes for missing values?

4.1.2 Honest Correlations

Most multivariate procedures analyze patterns of correlation (or covariance) among variables. It is important that the correlations, whether between two continuous variables or between a discrete and continuous variable, be as accurate as possible. Under some rather common research conditions, correlations are inflated, deflated, or simply inaccurately computed.

4.1.2.1 Inflated Correlation   When composite variables are constructed from several individual items by pooling responses to individual items, correlations are inflated if some items are reused. Scales on personality inventories, measures of socioeconomic status, health indices, and many other variables in social and behavioral studies are composites of several items. If composite variables are used and they contain, in part, the same items, correlations are inflated. Don't overinterpret a high correlation between two measures composed, in part, of the same items.

4.1.2.2 Deflated Correlation   Correlations are too low under several conditions. Problems with distributions that lead to lowered correlations are discussed in Section 4.1.5. However, restricted range in sampling of cases, very uneven splits in discrete variables, and computational inaccuracy with small coefficients of variation also lead to deflated correlation.

A falsely small correlation between two continuous variables is obtained if the range of responses to one of the variables is restricted in the sample. Correlation is a measure of the way variability in one variable goes with variability in another variable. If the range of variability in a variable is narrow because of an error in sampling, then it is effectively a constant and cannot correlate highly with another variable. In a study of success in graduate school, for instance, quantitative ability could not emerge as highly correlated with other variables if all students had about the same high scores in quantitative skills.

If a correlation is too small because of restricted range in sampling, you can estimate its magnitude in a nonrestricted sample by using Equation 4.1 if you can estimate the standard deviation in the nonrestricted sample. The standard deviation in the nonrestricted sample is estimated from prior data or from knowledge of the population distribution.

$$\tilde{r}_{xy} = \frac{r_{t(xy)} \left[\dfrac{S_x}{S_{t(x)}} \right]}{\sqrt{1 + r_{t(xy)}^2 \left[\dfrac{S_x^2}{S_{t(x)}^2} \right] - r_{t(xy)}^2}} \tag{4.1}$$

where $\tilde{r}_{xy}$ is adjusted correlation, $r_{t(xy)}$ is the correlation between X and Y with the range of X truncated, S_x is the unrestricted standard deviation of X, and $S_{t(x)}$ is the truncated standard deviation of X.

Many programs allow input of a correlation matrix instead of raw data. The estimated correlation is inserted in place of the truncated correlation prior to analysis of the correlation matrix. (However, insertion of estimated correlations may create internal inconsistencies in the correlation matrix, as discussed in Section 4.1.3.3.)

The correlation between a continuous variable and a dichotomous variable, or between two dichotomous variables, is also too low if most (say 80 to 90%) responses to the dichotomous variable fall into one category. Even if the variables are perfectly related, the highest correlation that could be obtained is well below 1. Some recommend dividing the obtained (but deflated) correlation by the maximum it could achieve given the split between the categories and then using the resulting figure in subsequent analyses. This procedure is attractive, but not without hazard, as discussed by Comrey (1973).

When means are very large numbers and standard deviations are very small, the values in the correlation matrix are also sometimes too small. The programs encode the first several digits of a very large number and then round off the rest. If the variability is in the digits that are dropped, then correlations between the variable and others are inaccurately computed. The coefficient of variation (the standard deviation divided by the mean) is an indicator of this problem that is available in many descriptive statistics programs. When the coefficient of variation is very small (0.0001 or less), computational inaccuracy may occur because roundoff errors are too influential to the solution. The solution is to subtract a large constant from every score of a variable with very large numbers before calculating **R**. Subtracting a constant from every score does not affect the size of r.

4.1.3 Missing Data

One of the most pervasive problems in data analysis is that of missing data. The problem occurs when rats die, respondents become recalcitrant, or somebody goofs. Its seriousness depends on how much is missing and why it is missing.

If only a few data points are missing in a random pattern from a large data set, the problems are usually not serious and almost any procedure for handling them yields similar results. If, however, a lot of data are missing from a small- to a moderate-sized data set, the problems can be very serious. Unfortunately, there are as yet no firm guidelines for how much missing data can be tolerated for a sample of a given size.

The pattern of missing data is more important than the amount missing. Missing values scattered randomly through a data matrix rarely pose serious problems. Non-randomly missing values, on the other hand, are serious no matter how few of them there are because they affect the generalizability of results. Suppose that in a questionnaire with both attitudinal and demographic questions several respondents refuse to answer questions about income. It is likely that refusal to answer questions about income is related to attitude. If respondents with missing data on income are deleted, the sample is distorted. Some method of estimating income is needed to retain the cases in the analysis.

Although the temptation to assume that data are missing randomly is nearly overwhelming, the safest thing to do is to test it. Use the information you have to test for patterns in missing data. For instance, construct a variable with two groups, missing and nonmissing values on income, and perform a test of mean differences in attitude between the groups. If there are no differences, decisions about how to handle missing data are not so critical (except, of course, for inferences about income). If there are differences, care is needed to preserve the cases with missing values for other analyses. If this test is done with large N, check eta square to make sure the differences are meaningful (cf. Equation 3.25).

BMDPAM is a program designed specifically to investigate cases with missing values. Table 4.1 shows selected PAM output for a data set with missing values on ATTHOUSE ("1" is missing) and INCOME ("99" is missing). Case label 67 (case number 52), among others, is missing INCOME; Case label 338 (case number 253) is missing ATTHOUSE. Although the program does not reveal the relationships between variables with missing values and variables with complete values, it does reveal the relationships between variables with missing values. For this example, the correlation between a $0 - 1$ ($0 =$ case has missing value, $1 =$ case complete) distribution for ATTHOUSE and a $0 - 1$ distribution for INCOME is $- .011$; there is no association between failure to respond regarding INCOME or ATTHOUSE.

The decision about how to handle missing data is important. At best, the decision is among several bad alternatives, four of which are discussed in the subsections that follow. The alternatives are listed with the most frequently used alternatives first. For greater detail on these alternatives and others, consult Rummel (1970) and Cohen and Cohen (1975).

4.1.3.1 Deleting Cases or Variables

One procedure for handling missing values is simply to drop any cases with them. If only a few cases have missing data and they seem to be a random subsample of the whole sample, deletion is a good alternative. Deletion of cases with missing values is the default option for most programs in the BMDP, SPSS^x, SAS, and SYSTAT packages.[1]

---

[1] Because this is the default option, numerous cases can be deleted without the researcher's knowledge. For this reason, it is important to check the number of cases in your analyses to make sure that all of the cases are used.

TABLE 4.1 BMDPAM SETUP AND SELECTED OUTPUT FOR
MISSING DATA

```
            /PROBLEM    TITLE IS 'BMDPAM RUN FOR MISSING DATA'.
            /INPUT      VARIABLES ARE 8.  FORMAT IS '(A4,7F4.0)'.
                        FILE='SCREEN.DAT'.
            /VARIABLE   NAMES ARE SUBNO,TIMEDRS,ATTDRUG,ATTHOUSE,
                            INCOME,EMPLMNT,MSTATUS,RACE.
                        LABEL = SUBNO.
                        MISSING = (4)1, (5)99.
            /END

NUMBER OF CASES READ. . . . . . . . . . . . . . . . . .      465

   NUMBER OF CASES WITH NO DATA MISSING AND WITH
        POSITIVE CASE WEIGHT. . . . . . . . . . . . . .      438

PERCENTAGES OF MISSING DATA FOR EACH VARIABLE IN EACH GROUP
-----------------------------------------------------------

THESE PERCENTAGES ARE BASED ON SAMPLE SIZES AND GROUP SIZES
REPORTED WITH THE UNIVARIATE SUMMARY STATISTICS BELOW.
VARIABLES WITHOUT MISSING DATA ARE NOT INCLUDED.

                        1

ATTHOUSE    4      0.2
INCOME      5      5.6


SAMPLE SIZES FOR EACH PAIR OF VARIABLES
---------------------------------------
(NUMBER OF TIMES BOTH VARIABLES ARE AVAILABLE)
IN ORDER TO SAVE SPACE, VARIABLES WITH NO MISSING
DATA OR THAT HAVE NO DATA ARE NOT INCLUDED.


              ATTHOUSE INCOME

                   4       5
ATTHOUSE    4     464
INCOME      5     438     439


PAIRWISE PERCENTAGES OF MISSING DATA
------------------------------------
DIAGONAL ELEMENTS ARE THE PERCENTAGES THAT EACH VARIABLE
IS MISSING.  OFF-DIAGONAL ELEMENTS ARE THE PERCENTAGES
EITHER VARIABLE IS MISSING.  THESE PERCENTAGES DO NOT INCLUDE
CASES WITH MISSING GROUP OR WEIGHT VARIABLES, CASES WITH
ZERO WEIGHTS, CASES EXCLUDED BY SETTING USE EQUAL TO A
NON-POSITIVE VALUE BY TRANSFORMATIONS, OR CASES WITH GROUPING
VALUES NOT USED.  VARIABLES WITH NO MISSING DATA OR THAT
HAVE NO DATA ARE NOT INCLUDED HERE.

              ATTHOUSE INCOME

                   4       5
ATTHOUSE    4     0.2
INCOME      5     5.8     5.6


CORRELATIONS OF THE DICHOTOMIZED VARIABLES
------------------------------------------
WHERE FOR EACH VARIABLE ZERO INDICATES THAT THE VALUE WAS
MISSING AND ONE INDICATES THAT THE VALUE WAS PRESENT.
VARIABLES WITH NO MISSING DATA OR THAT ARE COMPLETELY
MISSING ARE NOT INCLUDED.


              ATTHOUSE INCOME

                   4       5
ATTHOUSE    4    1.000
INCOME      5   -0.011   1.000
```

TABLE 4.1 (Continued)

```
PATTERN OF MISSING DATA AND DATA BEYOND LIMITS
------------------------------------------------
COUNT OF MISSING VARIABLES INCLUDES DATA BEYOND LIMITS.
THE COLUMN LABELED WT. IS FOR THE CASE WEIGHT, IF ANY.
M REPRESENTS A MISSING VALUE.  B REPRESENTS A VALUE GREATER
THAN THE MAXIMUM LIMIT.  S REPRESENTS A VALUE LESS THAN THE
MINIMUM LIMIT.
```

CASE LABEL	CASE NO.	NO OF MISS. VARS.	GROUP	WT.	ATTHOUSE	INCOME
67	52	1				M
79	64	1				M
84	69	1				M
95	77	1				M
138	118	1				M
156	135	1				M
228	161	1				M
239	172	1				M
240	173	1				M
241	174	1				M
248	181	1				M
265	196	1				M
272	203	1				M
317	236	1				M
321	240	1				M
338	253	1			M	
343	258	1				M
420	304	1				M
447	321	1				M
453	325	1				M
486	352	1				M
512	378	1				M
513	379	1				M
552	409	1				M
568	419	1				M
570	421	1				M
584	435	1				M

Or, if missing values are concentrated in a few variables and they are not critical to the analysis, or are highly correlated with other, complete variables, the variable(s) with missing values are profitably dropped.

But if missing values are scattered throughout cases and variables, deletion of cases can mean substantial loss of data. This is particularly serious when data are grouped in an experimental design because loss of even one case requires adjustment for unequal n in cells (see Chapter 8). Further, the researcher who has expended considerable time and energy collecting data is not likely to be eager to toss some out. Moreover, if cases with missing values are not randomly distributed through the data, distortions of the sample occur if they are deleted.

4.1.3.2 Estimating Missing Data A second option is to estimate missing values and then use the estimates during data analysis. There are at least three popular schemes for doing so; using prior knowledge, inserting mean values, and using regression.

Prior knowledge is used when a researcher replaces a missing value with a value from a well-educated guess. If the researcher has been working in an area for a while, and if the sample is large and the number of missing values small, this is often a reasonable procedure. The researcher is often confident that the value would have been about at the median, or whatever, for a particular case. Alternatively, the researcher can downgrade a continuous variable to a dichotomous variable (e.g., "high" vs. "low") to predict with confidence which category a case with a missing value falls into. The dichotomous variable replaces the continuous variable in the analysis, but it often has less predictive power than the continuous variable.

Or, *means* are calculated from available data and used to replace missing values prior to analysis. In the absence of all other information, the mean is your best guess about the value of a variable. Part of the attraction of this procedure is that it is conservative; the mean for the distribution as a whole does not change and the researcher is not required to guess at missing values. On the other hand, the variance of the variable is reduced because the mean is closer to itself than to the missing value it replaces, and the correlation the variable has with other variables is reduced because of the reduction in variance. The extent of loss in variance depends on the amount of missing data.

A compromise is to insert the group mean for a missing value. If, for instance, the case with a missing value is a Republican, the mean value for Republicans is computed and inserted in place of the missing value. This procedure has a lot to recommend it; it is not as conservative as inserting overall mean values and not as liberal as using prior knowledge.

The BMDPAM program has provision for inserting the overall mean value or the mean value for a group if the cases are classified into groups through use of METHOD IS MEAN in the ESTIMATE paragraph. The same goal is accomplished through the MEANSUB option in some of the SPSS[x] programs or with variable modification (e.g., RECODE, IF . . . THEN) language in the other packages. SAS STANDARD allows a data set to be created with missing values replaced by the variable mean.

A sophisticated method for estimating missing values uses *regression* (see Chapter 5). Other variables are used as IVs to write a regression equation for the variable with missing data serving as DV. Cases with complete data generate the regression equation; the equation is then used to predict missing values for incomplete cases. Sometimes the predicted values from a first round of regression are inserted for missing values and then all the cases are used in a second regression. The predicted values for the variable with missing data from round two are used for a third equation, and so forth until the predicted values from one step to the next are similar. The predictions from the last round are the ones used to replace missing values.

Advantages to regression are that it is more objective than the researcher's guess but not as blind as simply inserting the grand mean. One disadvantage to use of regression is that the scores fit together better than they should: because the missing value is predicted from other variables, it is likely to be more consistent with them than a real score is. A second problem is reduced variance because the estimate is probably too close to the mean. A third problem is the requirement that good IVs be available in the data set; if none of the other variables is a good predictor of the one with missing data, the estimate from regression is about the same as simply inserting the mean. Finally, estimates from regression are used only if the estimated value falls within the range of values for complete cases; out of range estimates are not acceptable.

Using regression to estimate missing values is convenient in BMDPAM through the METHOD IS sentence of the ESTIMATE paragraph. If SINGLE is specified, the missing value is predicted from bivariate regression with the best complete variable. TWOSTEP uses the best two variables, STEP only uses variables with lots of unique relationship with the DV, and REGR uses all variables available for the case. The data file with estimates inserted is created via the SAVE paragraph.

4.1.3.3 Using a Missing Data Correlation Matrix Another option involves analysis of a missing data correlation matrix, prepared through the TYPE = ALLVALUE option in the ESTIMATE paragraph of BMDPAM, or through the PEARSON/PAIRWISE option of the SYSTAT CORR program, or offered as the PAIRWISE deletion option in some of the SPSS[x] programs. This is the default option for SAS CORR. If this is not an option of the program you want to run (as in the SPSS[x] package), then you generate a missing data correlation matrix through another program for input to the one you are using.

In this option, all available pairs of values are used to calculate each of the correlations in **R**. A variable with 10 missing values has all its correlations with other variables based on 10 fewer pairs of numbers. If some of the other variables also have missing values, but in different cases, the number of complete pairs of variables is further reduced. Thus each correlation in **R** can be based on a different number and a different subset of cases, depending on the pattern of missing values. Because the standard error of the sampling distribution for r is based on N, some correlations are less stable than others in the same correlation matrix.

But that is not the only problem. In a correlation matrix based on complete data, the sizes of some correlations place constraints on the sizes of others. In particular,

$$r_{13}r_{23} - \sqrt{(1 - r_{13}^2)(1 - r_{23}^2)} \leq r_{12} \leq r_{13}r_{23} + \sqrt{(1 - r_{13}^2)(1 - r_{23}^2)} \qquad (4.2)$$

The correlation between variables 1 and 2, r_{12}, cannot be smaller than the value on the left or larger than the value on the right in a three-variable correlation matrix. If $r_{13} = .60$ and $r_{23} = .40$, then r_{12} cannot be smaller than $-.49$ or larger than .97. If, however, r_{12}, r_{23}, and r_{13} are all based on different subsets of cases because of missing data, the value for r_{12} can go out of range.

Most multivariate statistics calculate eigenvalues (and their corresponding eigenvectors) from correlation matrices (see Appendix A). With loosened constraints on size of correlations in a missing data correlation matrix, eigenvalues can become negative. Because eigenvalues represent variance, negative eigenvalues represent something akin to negative variance. Moreover, because the total variance that is partitioned among eigenvalues is constant (usually equal to the number of variables), positive eigenvalues are inflated by the size of negative eigenvalues resulting in inflation of variance. The statistics derived under these conditions are very likely distorted.

However, with a large sample and only a few missing values, eigenvalues are often all positive even if some correlations are based on slightly different pairs of cases. Under these conditions, a missing data correlation matrix provides a reasonable multivariate solution and has the advantage of using all available data. Use of this option for the missing data problem should not be rejected out of hand but should be used cautiously with a wary eye to negative eigenvalues.

4.1.3.4 Treating Missing Data as Data It is possible that failure to respond to a question is itself a very good predictor of the behavior of interest in your research. If a dummy variable is created where cases with complete data are assigned 0 and cases with missing data 1, the liability of missing data could become an asset. The mean is inserted for missing values so that all cases are analyzed, and the dummy variable is used as simply another variable in analysis, as discussed by Cohen and Cohen (1975, pp. 265–290).

4.1.3.5 Repeating Analyses with and without Missing Data If you use some method of estimating missing values or a missing data correlation matrix, consider repeating your analyses using only complete cases. This is particularly important if the data set is small, the proportion of missing values high, or data are missing in a nonrandom pattern. If the results are similar, you can have confidence in them. If they are different, however, you need to investigate the reasons for change, and evaluate which result more nearly approximates ''reality.''

4.1.4 Outliers

Outliers are cases with such extreme values on one variable or a combination of variables that they unduly influence statistics. For example, consider the regression coefficient; outliers, more than other cases, determine which one of a number of possible regression lines is chosen. In the bivariate scatterplot of Figure 4.1, several regression lines, all at slightly different tilts, provide a good fit to the data points

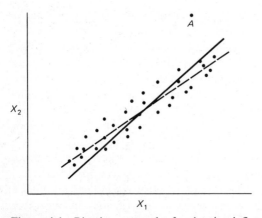

Figure 4.1   Bivariate scatterplot for showing influence of outliers.

inside the oval. But when the data point in the upper right-hand portion of the scatterplot is also considered, the regression coefficient that is computed is the one from among the several good alternatives that provides the best fit to the extreme case. The case is an outlier because it has much more impact on the regression coefficient than any of those inside the oval.

There are four reasons for the presence of an outlier. First is incorrect data entry. Cases that are extreme should be checked carefully to see that data are correctly entered. Second is failure to specify missing value codes in computer control language so that missing value indicators are read as real data. Third is that the outlier is not a member of the population from which you intended to sample. If the case should not have been sampled, it is deleted once it is detected. Fourth is that the case is from the intended population but that the distribution for the variable in the population has more extreme values than a normal distribution. In this event, the researcher retains the case but considers changing the value on the variable(s) so that the case is no longer unduly influential. Although errors in data entry and missing values specification are easily found and remedied, deciding between alternatives three and four, between deletion and retention with alteration, is difficult.

Outliers are found in both univariate and multivariate situations, among both dichotomous and continuous variables, among both IVs and DVs, and in both data and results of analyses. They lead to both Type I and Type II errors, with no clue as to which effect they have in a particular analysis. And they lead to results that do not generalize because the results are overly determined by the outlier(s). Outliers are a pervasive problem in the social, behavioral, biological, and medical sciences.

4.1.4.1 Detecting Univariate and Multivariate Outliers Univariate outliers are cases with an extreme value on one variable; multivariate outliers are cases with an unusual combination of two or more scores. For example, a 15-year-old is perfectly within bounds regarding age, and someone who earns $45,000 a year is in bounds regarding income, but a 15-year-old who earns $45,000 a year is very unusual and is likely to show up as a multivariate outlier.

Among dichotomous variables, those with very uneven splits between two categories are outliers. Rummel (1970) suggests deleting dichotomous variables with 90-10 splits between categories because the correlation coefficients between these variables and others are truncated and the scores in the category with 10% of the cases are more influential than those in the category with 90% of the cases. Dichotomous variables with extreme splits are easily found in the programs for frequency distributions (SPSS^x FREQUENCIES, BMDP2D, SYSTAT TABLES, or SAS UNIVARIATE) used during routine preliminary data screening.

The search for outliers *among continuous variables* depends on whether or not data are grouped. If you are going to perform one of the analyses with ungrouped data (regression, canonical correlation, factor analysis) outliers, both univariate and multivariate, are sought among all cases at once, as illustrated in Section 4.2.1.4. If you are going to perform one of the analyses with grouped data (ANCOVA, MANOVA or MANCOVA, discriminant function analysis, or profile analysis) outliers are sought separately within each group, as illustrated in Section 4.2.2.3.

Among continuous variables, *univariate outliers* are cases with very large standardized scores on one or more variables. Cases with standardized scores in excess of ± 3.00 are potential outliers. However, the extremeness of a standardized score depends on the size of the sample; with a large N, a few standardized scores in excess of 3.00 are expected. Z scores are available through BMDP1D, SPSS^x CONDESCRIPTIVE (plus Option 3), SYSTAT DATA, and SAS STANDARD (with MEAN = 0 and STD = 1). In addition, many of the BMDP programs give the z score associated with the highest and lowest raw scores as part of routine output.

As an alternative or in addition to inspection of z scores, there are graphical methods for finding univariate outliers. Helpful plots are histograms, box plots, normal probability plots, or detrended normal probability plots. Histograms are readily understood and available. Histograms for each variable show scores that are some distance from other scores in the distribution. There is usually a pileup of cases near the mean with cases trailing off away from the mean. An outlier is a case (or a very few cases) that seems to be unattached to the rest of the distribution. Histograms for continuous variables are produced by BMDP2D (ungrouped) and BMDP5D or 7D (grouped data), SPSS^x FREQUENCIES (plus SORT and SPLIT for grouped data), SYSTAT GRAPH (HISTOGRAM and BOX PLOT), and SAS UNIVARIATE or CHART (with BY for grouped data).

Box plots are simpler but less commonly encountered and less readily available. Box plots literally box in observations that are around the median; cases that fall far away from the box are extreme. Normal probability plots and detrended normal probability plots are also very useful for assessing normality of distributions of variables and are discussed in that context in Section 4.1.5.1. However, points that lie a considerable distance from others on these plots are also potential univariate outliers.

Once potential univariate outliers are located, the search for multivariate outliers begins. It is a good idea to see if univariate outliers are also multivariate outliers before deciding what to do with them. Often the same cases show up in both analyses; sometimes they do not. It is better to decide what to do with outliers when the total extent of the problem is known.

Multivariate outliers are cases that have an unusual pattern of scores. Their standardized scores on variables, considered singly, are sometimes within expected ranges, and it is only when you consider two or more variables that the unusual combination of scores becomes apparent. It is almost impossible to screen for multivariate outliers by hand, but both statistical methods and graphical methods are available, particularly among the BMDP programs. Of the two methods, the statistical procedure is preferred.

The statistical procedure is computation of Mahalanobis distance for each case. Mahalanobis distance is the distance of a case from the centroid of the remaining cases where the centroid is the point created by the means of all the variables. Mahalanobis distance is also a discriminant function analysis (see Chapter 11) where an equation is computed that best separates one case from the rest of the cases. If a case has an unusual combination of scores, then those scores are weighted heavily in the equation (the discriminant function) and the Mahalanobis distance of the case from the rest of the cases is significant.

Mahalanobis distance of each case, in turn, from all other cases, is available in several of the BMDP programs (4M, 7M, AM, and 9R) and in SYSTAT MGLH (where the value produced is the square root of the Mahalanobis distance produced by the other programs). Multivariate outliers are also found through SPSS NEW REGRESSION (beginning with Release 9.0) and SPSS^x REGRESSION by requesting Mahalanobis distance within the CASEWISE subcommand. The 10 worst outliers are available through the RESIDUALS subcommand with the OUTLIERS specification. Mahalanobis distances are requested and interpreted in Sections 4.2.1.4 and 4.2.2.3, and numerous other places throughout the book. A very conservative probability estimate for a case being an outlier, say $p < .001$, is appropriate with Mahalanobis distance.

When multivariate outliers are sought in grouped data, they are sought within each group. Mahalanobis distance as a test for within-group multivariate outliers is provided in BMDPAM, BMDP7M, and SYSTAT MGLH. Use of other programs, including SPSS^x REGRESSION, requires separate runs for each group. In these runs you must specify a dummy DV, such as the case number, to find outliers among the set of IVs. Separate runs for each group produce different distance values than a single run for within-group outliers because a different "error term" is used.

Graphical methods for identification of multivariate outliers are available through regression programs that plot results of residuals analyses. In these plots, like univariate plots, outliers are cases that produce points distant from those of other cases in the plots. However, the search for outliers is multivariate in residuals analysis because all the variables are considered when computing residuals. These methods are discussed in Chapter 5.

Frequently some outliers hide behind other outliers. If a few cases identified as outliers are deleted, other cases are then extreme with respect to the central tendency of the group. Sometimes it is a good idea to screen for multivariate outliers several times, each time dealing with cases identified as outliers on the last run, until finally no new outliers are identified.

4.1.4.2 Describing Outliers Once multivariate outliers are identified, you want to discover why the cases are extreme. (You already know why univariate outliers are extreme.) It is important to identify the variables on which the cases are deviant for two reasons. First, it provides an indication of the kinds of cases to which your results do not generalize. Second, if you are going to modify scores instead of delete cases, you have to know which scores to modify.

If there are only a few multivariate outliers, it is reasonable to examine them individually. If there are several, you may choose to examine them as a group to see if there are any variables that separate the group of outliers from the rest of the cases.

Whether you are examining one or a group of outliers, the trick is to create a dummy grouping variable where the outlier(s) have a value of one and the rest of the cases a value of zero. The dummy variable is then used as the grouping variable in discriminant function analysis (Chapter 11) or the dependent variable in regression (Chapter 5). If it is available to you, BMDP9R, the program that performs all possible subsets regression, is the program of choice for investigating outliers. Variables on which the outlier(s) differ from the rest of the cases show up again and again in the various subsets. Once those variables are identified, means for outlying and non-outlying cases are found through any of the routine descriptive programs. Description of outliers is illustrated in Sections 4.2.1.4 and 4.2.2.3.

4.1.4.3 Reducing the Influence of Outliers Once outliers have been identified, there are several strategies for reducing their influence. But before you use one of them, *check the data for the case* to make sure that they are accurately entered into the data file. If the data are accurate, consider the possibility that one variable is responsible for most of the outliers. If so, *elimination of the variable* would reduce the number of outliers. If the variable is highly correlated with others or is not critical to the analysis, deletion of it is a good alternative.

If neither of these simple alternatives is reasonable, you have to decide whether the cases that are outliers are properly part of the population from which you intended to sample. *If the cases are not part of the population, they are deleted* with no loss of generalizability of results to that population. The description of the outliers is a description of the kinds of cases to which the results do not apply.

If you decide that *the outliers are sampled from your target population, they remain in the analysis, but* steps are taken to reduce their influence—*variables are transformed or scores changed.* In this case, outliers are considered part of a nonnormal distribution with tails that are too high so that there are too many cases at extreme values of the distribution. Variable transformation is undertaken to change the shape of the distribution to more nearly normal. Cases that were outliers are still on the tails of the transformed distribution, but their impact is reduced. Transformation of variables has other salutary effects, as described in Section 4.1.6.

A second option is to change the score(s) on the important variable(s) for the outlying case(s) so that they are deviant, but not as deviant as they were. For instance, assign the outlying case(s) a raw score on the offending variable(s) that is one unit larger (or smaller) than the next most extreme score in the distribution. Because measurement of variables is sometimes rather arbitrary anyway, this alternative is often attractive. Any change of scores is reported in the Results section together with the rationale for the change.

4.1.5 Normality, Linearity, and Homoscedasticity

Underlying some multivariate procedures and most statistical tests of their outcomes is the assumption of multivariate normality. Multivariate normality is the assumption that each variable and all linear combinations of the variables are normally distributed. When the assumption is met, the residuals[2] of analysis are also normally distributed and independent. The assumption of multivariate normality is not readily tested because it is impractical to test an infinite number of linear combinations of variables for normality.

Although the assumption of multivariate normality is made in derivation of multivariate significance tests, its importance in analysis of a data set is not, at

---

[2] Residuals are left-overs. They are the segments of scores *not* accounted for by the multivariate analysis. They are also called "errors" between predicted and obtained scores where the analysis provides the predicted scores.

present, known. It is tempting to conclude that the statistics are robust[3] to violations of the assumption, but that conclusion is not, at present, warranted. The univariate F test of mean differences, for example, is frequently said to be robust to violation of assumptions of normality and homogeneity of variance with large and equal samples, but Bradley (1984) questions this generalization. Multivariate statistics probably are sometimes robust to violation of their assumptions but, unfortunately, the literature on robustness is far from conclusive regarding when assumptions can be violated with impunity. The safest tack is to use transformations of variables to get your data to fit the assumption unless there is some compelling reason not to, although even this does not guarantee multivariate normality.

The assumption of multivariate normality applies differently to different multivariate statistics. Multivariate normality almost always applies when statistical inference is used. Fleming and Pinneau (1988) find that both the Type I error rate and the stability of estimation of coefficients are affected by distortion of the distribution of a variable. Bradley (1982) reports that statistical inference becomes less and less robust as distributions depart from normality, rapidly so under many conditions. And even when the procedures are used purely descriptively, normality, linearity, and homoscedasticity of variables enhance the analysis although they are not assumptions of it.

Further, the assumption of multivariate normality applies either to the distributions of the variables themselves (in ungrouped data) or to the sampling distributions[4] of means of variables (in grouped data). For ungrouped data, if there is multivariate normality, each variable is itself normally distributed and the relationships between pairs of variables are linear and homoscedastic (i.e., the variance of one variable is the same at all values of the other variable). The assumption of multivariate normality can be checked partially through normality, linearity, and homoscedasticity of variables or through examination of residuals. The assumption is certainly violated, at least to some extent, if the variables or the residuals are not normally distributed, and do not have pairwise linearity and homoscedasticity.

For grouped data, it is the sampling distributions of the means of variables that are to be normally distributed. The central limit theorem proves that, with large sample sizes, sampling distributions of means are normally distributed regardless of the shapes of the distributions of variables. For example, if there are at least 20 degrees of freedom for error in a univariate ANOVA, the F test is said to be robust to violations of normality of variables (provided that there are no outliers). The degree to which robustness extends to multivariate analyses is not yet clear, but the larger the sample size the less effect nonnormality of variables is likely to have on your conclusions. With grouped data with large samples, transformation of variables is less imperative.

---

[3] Robust means that the researcher is led to correctly reject the null hypothesis at a given α level the right number of times even if the distributions do not meet the assumptions of analysis. Often Monte Carlo procedures are used where a distribution with some known properties is put into a computer, sampled from repeatedly, and repeatedly analyzed; the researcher studies the rates of retention and rejection of the null hypothesis against the known properties of the distribution in the computer.

[4] A sampling distribution is a distribution of statistics (not of raw scores) computed from random samples of a given size taken repeatedly from a population. For example, in univariate ANOVA, hypotheses are tested with respect to the sampling distribution of means (Chapter 3).

These issues are discussed again in the third sections of Chapters 5 through 12 as they apply directly to one or another of the multivariate procedures.

4.1.5.1 Normality   Screening for normality is an important early step in almost every multivariate analysis, particularly when inference is a goal. If there is normality, the residuals are normally and independently distributed. That is, the differences between predicted and obtained scores—the errors—are symmetrically distributed around a mean value of zero and there are no contingencies among the errors. Thus, one way to screen for normality is to examine residuals.

The other way is to examine the distributions of the variables themselves. Although normality of the variables is not always required for analysis, the solution is usually quite a bit better if the variables are all normally distributed. If the variables are not normally distributed, and particularly if they are nonnormal in very different ways (e.g., one positively and one negatively skewed), the solution is degraded. Normality of variables is assessed by either statistical or graphical methods; normality of residuals is usually assessed by graphical methods. Statistically, there are two components to normality, skewness and kurtosis. Skewness has to do with the symmetry of the distribution; a skewed variable is a variable whose mean is not in the center of the distribution. Kurtosis has to do with the peakedness of a distribution; a distribution is either too peaked (with too few cases out in the tails) or too flat (with too many cases out in the tails).[5] Figure 4.2 shows a normal distribution, distributions with skewness, and distributions with nonnormal kurtosis. A variable can have significant skewness, kurtosis, or both.

When a distribution is normal, the values of skewness and kurtosis are zero.[6] If there is positive skewness, there is a pileup of cases to the left and the right tail is too long; with negative skewness, there is a pileup of cases to the right and the left tail is too long. Kurtosis values above zero indicate a distribution that is too peaked (too few cases in the tails), while kurtosis values below zero indicate a distribution that is too flat (too many cases in the tails).

There are significance tests for both skewness and kurtosis that test the obtained value against null hypotheses of zero. For instance, the standard error for skewness is approximately

$$s_s = \sqrt{\frac{6}{N}} \qquad (4.3)$$

where N is the number of cases. The obtained skewness value is then compared with zero using the z distribution, where

$$z = \frac{S - 0}{s_s} \qquad (4.4)$$

---

[5] If you decide that outliers are sampled from the intended population but that there are too many cases in the tails, you are saying that the distribution from which the outliers are sampled has the wrong kurtosis, that it is too flat.

[6] The equation for kurtosis gives a value of 3 when the distribution is normal, but all of the statistical packages subtract 3 before printing kurtosis so that the expected value is zero.

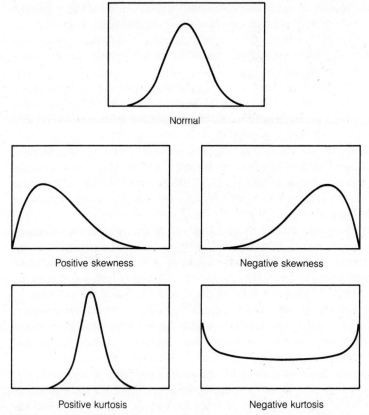

Figure 4.2   Normal distribution, distribution with skewness, and distribution with kurtosis.

and S is the value reported for skewness. The standard error for kurtosis is approximately

$$s_k = \sqrt{\frac{24}{N}} \qquad (4.5)$$

and the obtained kurtosis value is compared with zero using the z distribution, where

$$z = \frac{K - 0}{s_k} \qquad (4.6)$$

and K is the value reported for kurtosis.

Conventional but conservative (.01 or .001) alpha levels are used to evaluate the significance of skewness and kurtosis with small to moderate samples, but if the sample is large, it is a good idea to look at the shape of the distribution. Because the standard errors for both skewness and kurtosis contain N, with large samples the null hypothesis is likely to be rejected when there are only minor deviations from normality. In a large sample, a variable with significant skewness (or kurtosis) often does not deviate enough from normality to make a realistic difference in the analysis. In other words, with large samples the significance levels of skewness and kurtosis are not as important as their actual sizes (worse the farther from 0) and the visual appearance of the distribution.

Values for skewness and kurtosis are available in several programs. SPSS^x FREQUENCIES, for instance, prints as options skewness, kurtosis, and their standard errors and, in addition, superimposes a normal distribution over a frequency histogram for a variable if HISTOGRAM = NORMAL is specified. BMDP2D gives both skewness and kurtosis values and their significance levels, along with a histogram of the variable. SYSTAT STATS produces skewness and kurtosis as options; the distribution is inspected through GRAPH. SAS MEANS and UNIVARIATE provide skewness and kurtosis values. A histogram or stem and leaf plot is also available in SAS UNIVARIATE.

Frequency histograms are an important graphical device for assessing normality, especially with the normal distribution as an overlay, but even more helpful than frequency histograms are expected normal probability plots and detrended expected normal probability plots. In these plots, the scores are ranked and sorted; then an expected normal value is computed and compared with the actual normal value for each case. The expected normal value is the position a case with that rank holds in a normal distribution; the normal value is the position it holds in the actual distribution. If the actual distribution is normal, then the points for the cases fall along the diagonal running from lower left to upper right, with some minor deviations due to random processes. Deviations from normality shift the points away from the diagonal.

Consider the expected normal probability plots for TIMEDRS and ATTDRUG through BMDP5D in Figure 4.3. As reported in Section 4.2.1.1, ATTDRUG is reasonably normally distributed (kurtosis = −.447, skewness = −.123) while TIMEDRS is too peaked and positively skewed (kurtosis = 13.101, skewness = 3.248, both significantly different from 0). The cases for ATTDRUG line up along the diagonal while those for TIMEDRS do not. At low values of TIMEDRS, there are too many cases above the diagonal and at high values there are too many cases below the diagonal, reflecting the patterns of skewness and kurtosis.

Detrended normal probability plots for TIMEDRS and ATTDRUG are also in Figure 4.3. These plots are similar to expected normal probability plots except that deviations from the diagonal are plotted instead of values along the diagonal. In other words, the linear trend from lower left to upper right is removed. If the distribution of a variable is normal, as is ATTDRUG, the cases distribute themselves evenly above and below the horizontal line that intersects the Y axis at 0.0, the line of zero deviation from expected normal values. The skewness and kurtosis of TIMEDRS are again apparent from the cluster of points above the line at low values of TIMEDRS and below the line at high values of TIMEDRS. Normal probability plots for variables are also available in SAS UNIVARIATE, SPSS^x MANOVA, and SYSTAT GRAPH. Many of these programs also produce detrended normal plots.

Univariate outliers show up on these plots as cases that are far away from the other cases. Note, for instance, the two cases in the upper right-hand corner of the expected normal probability plot for TIMEDRS. The cases are a considerable distance

```
/PROBLEM      TITLE IS 'NORMALITY AS ASSESSED THROUGH EXPECTED NORMAL
                         PROBABILITY PLOTS'.
/INPUT        VARIABLES ARE 8.   FORMAT IS '(A4,7F4.0)'.
              FILE='SCREEN.DAT'.
/VARIABLE     NAMES ARE SUBNO,TIMEDRS,ATTDRUG,ATTHOUSE,
                         INCOME,EMPLMNT,MSTATUS,RACE.
              LABEL = SUBNO.
              MISSING = (4)1, (5)99.
              USE = TIMEDRS, ATTDRUG.
/PLOT   TYPE = NORM, DNORM.
/END
```

```
       NORMAL PLOT OF VARIABLE   2 TIMEDRS
           .+........+........+........+........+........+........+........+........+........+.
       3.0 +                              /                            *       +
           -                             /                                     -
           -                            /                                      -
           -                           /                              *        -
           -                          //                     *                 -
       2.4 +                         /                      *                   +
           -                        /                     **                    -
           -                       /                     **                     -
           -                      /              * *                            -
       1.8 +                    /⌐   ** *                                       +
           -                    /* **                                           -
           -                  *** *                                             -
           -                 ***/                                               -
     E     -               **** /                                               -
     X 1.2 +              **  /                                                  +
     P     -             **  /                                                   -
     E     -            *  //                                                    -
     C     -          *** /                                                      -
     T     -          ** /                                                       -
     E .60 +         ** /                                                        +
     D     -        ** /                                                         -
           -        ** /                                                         -
     N     -       ** //                                                         -
     O     -       * /                                                           -
     R 0.0 +       * /                                                           +
     M     -      ** /                                                           -
     A     -      * /                                                            -
     L     -      */                                                             -
           -      **                                                             -
     V -.60 +    /*                                                              +
     A     -   -/ *                                                              -
     L     -   **                                                                -
     U     -   *                                                                 -
     E     -   *                                                                 -
      -1.2 +  *                                                                  +
           -  -**                                                                -
           -  -*                                                                 -
           -  -*                                                                 -
           -  -*                                                                 -
      -1.8 + +*                                                                  +
           -  -*                                                                 -
           -  -*                                                                 -
           -  -*                                                                 -
           -  -*                                                                 -
      -2.4 + +*                                                                  +
           -  -*                                                                 -
           -  -*                                                                 -
           -                                                                     -
      -3.0 + +*                                                                  +
           .+........+........+........+........+........+........+........+........+........+.
              9.        27        45        63        81
```

Figure 4.3 Expected normal probability plot and detrended normal probability plot for TIMEDRS and ATTDRUG through BMDP5D.

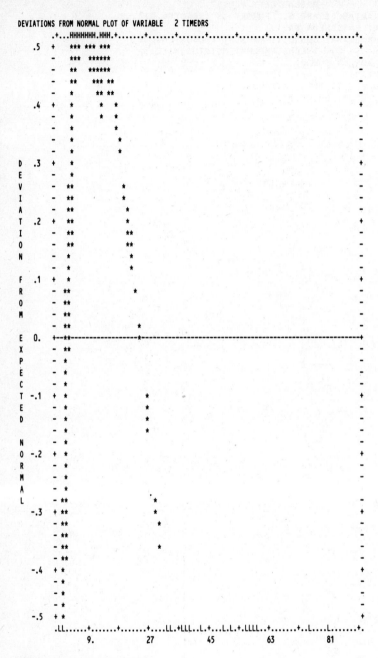

Figure 4.3 *(Continued)*

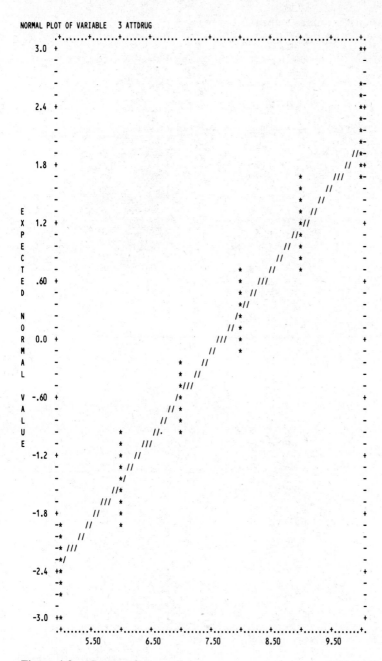

NORMAL PLOT OF VARIABLE   3 ATTDRUG

Figure 4.3   *(Continued)*

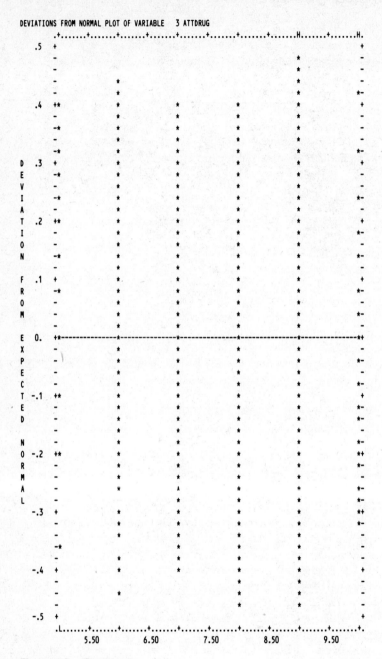

Figure 4.3 *(Continued)*

from the others and do not appear to be connected to them. These cases are most likely to show up as univariate outliers when z scores are considered.

In multiple regression, residuals are also screened for normality through the expected normal probability plot and the detrended normal probability plot.[7] All the BMDP regression programs and SPSS^x REGRESSION provide this diagnostic technique (and others, as discussed in Chapter 5). If the residuals are normally distributed, the expected normal probability plot and the detrended normal probability plot look just the same as they do if a variable is normally distributed. In regression, if the residuals plot looks normal, there is no reason to screen the individual variables for normality.

These plots applied to residuals through the regression programs also provide a screening device for multivariate outliers by revealing cases that are far away from the other cases. Multivariate outliers show up because the analysis uses all the variables and either a real DV (if it is a regression problem) or a dummy DV devised to assist in the screening for outliers, as illustrated in Section 4.2.1.4.

If nonnormality is found among either variables or residuals, transformation of variables is considered. Common transformations are described in Section 4.1.6. Unless there are compelling reasons not to transform, it is safer to do so. However, realize that even if each of the variables is normally distributed, or transformed to normality, there is no guarantee that all linear combinations of the variables are normally distributed. That is, if variables are each univariate normal, they do not necessarily have a multivariate normal distribution. However, it is more likely that the assumption of multivariate normality is met if all the variables are normally distributed.

4.1.5.2 Linearity The assumption of linearity is that there is a straight line relationship between two variables (where one or both of the variables can itself be a combination of several variables). Equations for straight lines do not include cross products of variables or terms with variables raised to powers (squared, cubed, etc.).

Linearity is fundamental to multivariate statistics because the solutions are based on the general linear model. Chapter 13 describes the general linear model and the assumption of linearity in it. Second, the assumption of multivariate normality implies that there is linearity between all pairs of variables; significance tests are based on the assumption. Finally, linearity is important in a very practical sense because only the linear relationships among variables are analyzed and if there are substantial nonlinear relationships among variables, they are ignored unless the variables are transformed so as to capture the nonlinear relationship.

Nonlinearity is diagnosed either from residuals plots or from bivariate scatterplots between pairs of variables. In plots where residuals are plotted against predicted values, nonlinearity is indicated when most of the residuals are above the zero line on the plot at some predicted values and below the zero line at other predicted values (see Chapter 5).

---

[7] For grouped data, residuals have the same shape as within-group distributions because the predicted value is the mean and subtracting a constant does not change the shape of the distribution. Many of the programs for grouped data plot the within-group distribution as an option, as discussed in the next few chapters when relevant.

Linearity between two variables is assessed by inspection of bivariate scatterplots. If both variables are normally distributed and linearly related, the scatterplot is oval-shaped. If one of the variables is nonnormal, then the scatterplot between this variable and the other is not oval. Consider, for instance, the two bivariate scatterplots in Figure 4.4. Figure 4.4 (a) is between ATTDRUG, a variable that is normally distributed, and TIMEDRS, a variable with significant skewness and kurtosis. ATTDRUG is along the Y axis; turn the page so that the Y axis becomes the X axis and you can see the symmetry of the ATTDRUG distribution. TIMEDRS is along the X axis. The asymmetry of the distribution is apparent from the pileup of scores at low values of the variable. The overall shape of the scatterplot is not oval; the variables are not linearly related.

Figure 4.4 (b) is a bivariate scatterplot between ATTDRUG and LTIMEDRS, TIMEDRS transformed to nearly normal by taking the log (see Section 4.1.6). Although still not perfect, the overall shape of the scatterplot is more nearly oval. The nonlinearity that is created by nonnormality of one of the variables is "fixed" by transformation of the variable.

However, sometimes the relationship between variables is simply not linear. Consider, for instance, number of symptoms and dosage of drug, as shown in Figure 4.5 (a). It seems likely that there are lots of symptoms when the dosage is low, only a few symptoms when the dosage is moderate, and lots of symptoms again when the dosage is high. Number of symptoms and drug dosage are curvilinearly related. Probably the best one can do in this case is to do a median split on number of symptoms—code them into high and low on a dummy variable—and then use the dummy variable in place of number of symptoms in analysis.[8] The dichotomous dummy variable can only have a linear relationship with other variables, if, indeed, it has any relationship at all after transformation.

Often two variables have a mix of linear and curvilinear relationships, as shown in Figure 4.5 (b). One variable generally gets smaller (or larger) as the other gets larger (or smaller) but there is also a curve to the relationship. For instance, symptoms might drop off with increasing dosage, but only to a point, increasing dosage beyond the point does not result in further reduction or increase in symptoms. In this case, the linear component is usually strong enough that not much is added by trying to capture the curvilinear component.

Assessing linearity through bivariate scatterplots is a procedure that is reminiscent of reading tea leaves, especially with small samples. And there are many cups of tea if there are several variables and all possible pairs are examined, especially when subjects are grouped and the analysis is done separately within each group. If there are only a few variables, screening all possible pairs is not burdensome; if there are numerous variables, you may want to use statistics on skewness and kurtosis to screen only pairs that are likely to depart from linearity. Think, also, about pairs of variables that might have true nonlinearity and examine them through bivariate scatterplots.

Bivariate scatterplots are produced by BMDP6D, SPSS^x PLOT or SPSS SCATTERGRAM, SYSTAT GRAPH, and SAS PLOT, among other programs.

---

[8] A nonlinear analytic strategy is most appropriate here, such as nonlinear regression through BMDP3R, but such strategies are well beyond the scope of this book.

```
/PROBLEM    TITLE IS 'LINEARITY AS ASSESSED BY SCATTERGRAMS'.
/INPUT      VARIABLES ARE 8.  FORMAT IS '(A4,7F4.0)'.
            FILE='SCREEN.DAT'.
/VARIABLE   NAMES ARE SUBNO,TIMEDRS,ATTDRUG,ATTHOUSE,
                      INCOME,EMPLMNT,MSTATUS,RACE,LTIMEDRS.
            ADD = 1.
            LABEL = SUBNO.
            MISSING = (4)1, (5)99.
/TRANSFORM  LTIMEDRS = LOG(TIMEDRS + 1).
/PLOT   YVAR = ATTDRUG.
        XVAR = TIMEDRS, LTIMEDRS.
        CROSS.
        SIZE IS 40, 25.
/END
```

```
          .+.......+.......+.......+.......+.......+.
      10  +3323124 2 1             1                +
          -                                         -
          -                                         -
          -                                         -
          -                                         -
      9.  +FLK9656131 111 2     1 1      1          +
          -                                         Y
          -                                         -
 A        -                                         -
 T        -                                         -
 T    8.  +T*VCB7212 132  1    1111        1        +
 D        -                                         -
 R        Y                                         -
 U        -                                         -
 G        -                                         -
      7.  +QXR97474121   111       11               +
          -                                         -
          -                                         -
          -                                         -
          -                                         -
      6.  +8IB58332 1        1                      +
          -                                         -
          -                                         -
          -                                         -
          -                                         -
      5.  +523   2     1                            +
          .+.......+.......+.......+.......+.......+.
                  20.             60.             100
          0.0             40.             80.
 N = 465
 R = .104                TIMEDRS
 P = .024
                                       MEAN    S.D.
 --REGRESSION LINE--  -RES.MS-    X  7.9011  10.948
 Y= 7.5990 +.01101*X    1.3249    Y  7.6860  1.1561

            (a)  TIMEDRS  VERSUS ATTDRUG
```

```
          .+.....+.....+.....+.....+.....+.....+.....
      10  +1     2   2 1   2 211  213111        1   +
          -                                         -
          -                                         -
          -                                         -
          -                                         -
      9.  +5     A   A B 7 588121535213 1112 11  1  +
          -                                         -
          -                                         -
 A        -                                         -
 T        -                                         -
 T    8.  +D     G   N J I 94842493 311132 1 22  1  Y
 D        -                                         -
 R        -                                         -
 U        Y                                         -
 G        -                                         -
      7.  +G     A   I F E 4954324253412  12  11    +
          -                                         -
          -                                         -
          -                                         -
          -                                         -
      6.  +4     4   C 6 1 642324323 2 1     1      +
          -                                         -
          -                                         -
          -                                         -
          -                                         -
      5.  +3     2   2   1 11      11    1          +
          .+.....+.....+.....+.....+.....+.....+.....
                .30         .90         1.5
          0.0         .60         1.2         1.8
 N = 465
 R = .110                        LTIMEDRS
 P = .017
                                        MEAN    S.D.
 --REGRESSION LINE--  -RES.MS-     X  .74128  .41525
 Y= 7.4587 +.30660*X    1.3232     Y  7.6860  1.1561

            (b)  LTIMEDRS VERSUS ATTDRUG
```

Figure 4.4 Assessment of linearity through bivariate scatterplots, as produced by BMDP6D; (a) ATTDRUG (normal) versus TIMEDRS (nonnormal) and (b) ATTDRUG (normal) versus LTIMEDRS (transformed).

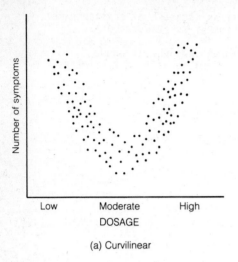

(a) Curvilinear

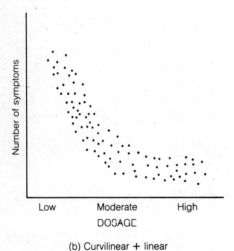

(b) Curvilinear + linear

Figure 4.5 Curvilinear relationship and mixed linear and curvilinear relationship.

4.1.5.3 Homoscedasticity The assumption of homoscedasticity is that the variability in scores for one variable is roughly the same at all values of the other variable. Homoscedasticity is related to the assumption of normality because when the assumption of multivariate normality is met, the relationships between variables are homoscedastic. The bivariate scatterplots between two variables or of residuals are of roughly the same width all over. Homoscedasticity is illustrated in Figure 4.6 (a).

Heteroscedasticity is caused either by nonnormality of one of the variables or by the fact that one variable is not related to the other directly, but rather to some transformation of the other. Consider, for example, the relationship between age (X_1) and income (X_2), as depicted in Figure 4.6 (b). Young people make about the same salaries, but with increasing age, people spread farther apart on income. The rela-

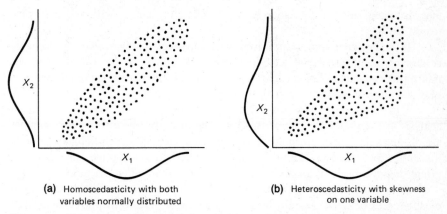

(a) Homoscedasticity with both variables normally distributed

(b) Heteroscedasticity with skewness on one variable

Figure 4.6 Bivariate scatterplots under conditions of homoscedasticity and heteroscedasticity.

tionship is perfectly lawful, but it is not homoscedastic and it is not fully captured by the analysis. In this example, income is likely to be positively skewed and transformation of income is likely to improve the homoscedasticity of the relationship.

It should be noted that heteroscedasticity is not fatal to an analysis. The linear relationship between variables is captured by the analysis but there is even more predictability if the heteroscedasticity is accounted for, as well. If it is not, the analysis is weakened, but not invalidated.

Homoscedasticity is evaluated through bivariate scatterplots and corrected, if possible, by transformation of variables.

4.1.6 Common Data Transformations

Although data transformations are recommended as a remedy for outliers and for failures of normality, linearity, and homoscedasticity, they are not universally recommended. The reason is that an analysis is interpreted from the variables that are in it and transformed variables are sometimes harder to interpret. For instance, although IQ scores are widely understood and meaningfully interpreted, the logarithm of IQ scores may not be understood or easy to interpret.

Whether transformation increases difficulty of interpretation often depends on the scale in which the variable is measured. If the scale is meaningful, transformation often hinders interpretation but if the scale is somewhat arbitrary anyway (as is often the case), transformation does not notably increase the difficulty of interpretation.

With ungrouped data, it is probably best to transform variables to normality unless interpretation is not feasible with the transformed scores. With grouped data, the assumption of normality concerns the sampling distribution of means (not the distribution of scores) and the central limit theorem predicts normality with decently sized samples. However, transformations may improve the analysis, even so, and may have the further advantage of reducing the influence of outliers. Our recommendation, then, is to consider transformation of variables in all situations unless there is some reason not to.

If you decide to transform, it is important to check that the variable is normally or near-normally distributed after transformation. Often you need to try first one transformation and then another until you find the transformation that produces skewness and kurtosis values nearest zero, or the fewest outliers.

With almost every data set where we have used transformations, the results of analysis are substantially improved. This is particularly true when some variables are skewed and others are not, or variables are skewed very differently prior to transformation. However, if all the variables are skewed to about the same moderate extent, improvements of analysis with transformation are often marginal.

With grouped data, the test of mean differences after transformation is a test of differences between medians in the original data.[9] Transformation is undertaken because the distribution is skewed and the mean is not a good indicator of the central tendency of the scores in the distribution. For skewed distributions, the median is often a more appropriate measure of central tendency than the mean, anyway, so interpretation of results after transformation is appropriate.

Variables differ in the extent to which they diverge from normal. Figure 4.7 presents several distributions together with the transformations that are likely to render them normal. If the distribution differs moderately from normal, a square root transformation is tried first. If the distribution differs substantially, a log transformation is tried. If the distribution differs severely, the inverse is tried. According to Bradley (1982) the inverse is the best of several alternatives for J shaped distributions, but even it may not render the distribution normal. Finally, if the departure from normality is severe and no transformation seems to help, you may want to try dichotomizing the recalcitrant variable.

The direction of the deviation is also considered. When distributions have positive skewness, as discussed above, the long tail is to the right. When they have negative skewness, the long tail is to the left. If there is negative skewness, the best strategy is to "reflect" the variable and then apply the appropriate transformation for positive skewness.[10] To reflect a variable, find the largest score in the distribution and add 1 to it to form a constant that is larger than any score in the distribution. Then create a new variable by subtracting each score from the constant. In this way, a variable with negative skewness is converted to one with positive skewness prior to transformation.

Remember to check your transformations after applying them. If a variable is only moderately positively skewed, for instance, a square root transformation may make the variable moderately negatively skewed, and there is no advantage to transformation. Often you have to try several transformations before you find the most helpful one.

---

[9] After a distribution is normalized by transformation, the mean is equal to the median. The transformation affects the mean but not the median because the median depends only on rank order of cases. Therefore, conclusions about means of transformed distributions apply to medians of untransformed distributions.

[10] Remember, however, that the interpretation of a reflected variable is just the opposite of what it was; if big numbers meant good things prior to reflecting the variable, big numbers mean bad things afterwards.

TRANSFORMATION

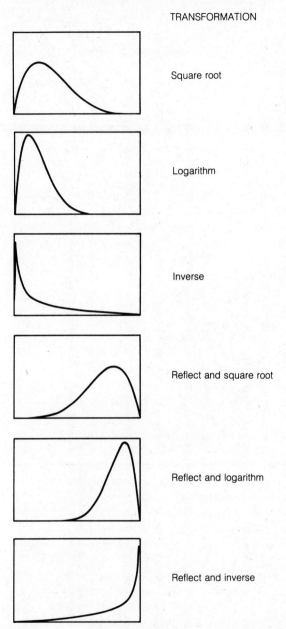

Square root

Logarithm

Inverse

Reflect and square root

Reflect and logarithm

Reflect and inverse

Figure 4.7 Distributions and common transformations to produce normality.

Control language for transforming variables in the four packages we consider is given in Table 4.2. Notice that a constant is also added if the distribution contains a value less than 1. A constant (to bring the smallest value to at least 1) is added to each score to avoid taking the log, square root, or inverse of zero, a fraction, or a negative number.

TABLE 4.2 CONTROL LANGUAGE FOR COMMON DATA TRANSFORMATIONS

	BMDP /TRANSFORM	SPSS[x] COMPUTE	SAS DATA Procedure	SYSTAT Data Module
Moderate positive skewness	NEWX = SQRT(X).	NEWX = SQRT(X)	NEWX = SQRT(X)	LET NEWX = SQR(X)
Substantial positive skewness	NEWX = LOG(X).	NEWX = LG10(X)	NEWX = LOG10(X)	LET NEWX = LOG(X)/LOG(10)
With zero	NEWX = LOG(X + C).	NEWX = LG10(X + C)	NEWX = LOG10(X + C)	LET NEWX = LOG(X + C)/LOG(10)
Severe positive skewness, L-shaped	NEWX = 1/X.	NEWX = 1/X	NEWX = 1/X	LET NEWX = 1/X
With zero	NEWX = 1/(X + C).	NEWX = 1/(X + C)	NEWX = 1/(X + C)	LET NEWX = 1/(X + C)
Moderate negative skewness	NEWX = SQRT(K − X).	NEWX = SQRT(K − X)	NEWX = SQRT(K − X)	LET NEWX = SQR(K − X)
Substantial negative skewness	NEWX = LOG(K − X).	NEWX = LG10(K − X)	NEWX = LOG10(K − X)	LET NEWX = LOG(K − X)/LOG(10)
Severe negative skewness, J-shaped	NEWX = 1/(K − X).	NEWX = 1/(K − X)	NEWX = 1/(K − X)	LET NEWX = 1/(K − X)

C = a constant added to each score so that the smallest score is 1.
K = a constant from which each score is subtracted so that the smallest score is 1; usually equal to the largest score + 1.

It should be clearly understood that this section merely scratches the surface of the topic of transformations, about which a great deal more is known. The interested reader is referred to Box and Cox (1964) or Mosteller and Tukey (1977) for a more flexible and challenging approach to the problem of transformation.

4.1.7 Multicollinearity and Singularity

Multicollinearity and singularity are problems with a correlation matrix that occur when variables are too highly correlated. With multicollinearity, the variables are very highly correlated (.90 and above); with singularity, the variables are perfectly correlated and one of the variables is a combination of one or more of the other variables.

For example, scores on the Wechsler Adult Intelligence Scale (the WAIS) and scores on the Stanford-Binet Intelligence Scale are likely to be *multicollinear* because they are two similar measures of the same thing. But the total WAIS IQ score is *singular* with its subscales because the total score is found by combining subscale scores. When variables are multicollinear or singular, they contain redundant information and they are not all needed in the same analysis. In other words, there are fewer variables than you think and the correlation matrix is not ''of full rank'' because there are not really as many variables as columns.

Either bivariate or multivariate correlations can create multicollinearity or singularity. If a bivariate correlation is too high, it shows up in a correlation matrix as a correlation above .90, and, after deletion of one of the two redundant variables, the problem is solved. If it is a multivariate correlation that is too high, diagnosis is slightly more difficult because you need to examine multivariate statistics to find the offending variable. For example, although the WAIS IQ is a combination of its subscales, the bivariate correlations between total IQ and each of the subscale scores are not all that high. You would not know there was singularity by examination of the correlation matrix.

Multicollinearity and singularity cause both logical and statistical problems. The logical problem is that, unless you are doing factor analysis, it is not a good idea to include redundant variables in the same analysis. They are not needed and, because they reduce degrees of freedom for error, they actually weaken an analysis. Unless you are doing factor analysis, think carefully before including two variables with a bivariate correlation of .70 or more in the same analysis.

The statistical problems created by singularity and multicollinearity occur at much higher correlations. The problem is that singularity prohibits, and multicollinearity renders unstable, matrix inversion. Matrix inversion is the logical equivalent of division; calculations requiring division (and there are many of them—see the fourth sections of Chapter 5 through 12) cannot be performed on singular matrices because they produce determinants equal to zero that cannot be used as divisors (see Appendix A).

With multicollinearity, the determinant is not exactly zero, but it is zero to several decimal places. Division by a near-zero determinant produces very large and unstable numbers in the inverted matrix. The sizes of numbers in the inverted matrix fluctuate wildly with only minor changes (say, in the second or third decimal place) in the sizes of the correlations in **R**. The portions of the multivariate solution that flow from an inverted matrix that is unstable are also unstable.

Most programs protect against multicollinearity and singularity by computing SMCs for the variables. SMC is the squared multiple correlation of a variable where it serves as DV with the rest as IVs in multiple correlation (see Chapter 5). If the SMC is high, the variable is highly related to the others in the set and you have multicollinearity. If the SMC is 1, the variable is perfectly related to others in the set and you have singularity. Many programs convert the SMC values for each variable to tolerance (1 − SMC) and deal with tolerance instead of SMC.

If you use BMDPAM during preliminary data screening for missing values, multicollinearity shows up as very high SMC values given as routine printout. If there is singularity, it is so stated and the offending variable is identified.

If you do not use BMDPAM, screening for singularity takes the form of running your main analysis to see if the computer balks. Singularity aborts most runs except those for principal components analysis (see Chapter 12) where matrix inversion is not required. If the run aborts, you need to identify and delete the offending variable. A first step is to think about the variables. Did you create any of them from others of them; for instance, did you create one of them by adding together two others? If so, deletion of one removes singularity. If not, then use BMDPAM to identify the offending variable.

Screening for multicollinearity that causes statistical instability is also routine with most programs because they have tolerance criteria for inclusion of variables. If the tolerance (1 − SMC) is too low, the variable does not enter the analysis. Default tolerance levels range between .01 and .0001, so SMCs are .99 to .9999 before variables are excluded. You may wish to take control of this process, however, by adjusting the tolerance level (an option with many programs) or deciding yourself which variable(s) to delete instead of letting the program make the decision on purely statistical grounds. For this you need SMCs for each variable. SMCs are given as routine output by many BMDP programs, and are also available through factor analysis programs in all packages.

4.1.8 A Checklist and Some Practical Recommendations

A checklist for screening is in Table 4.3. It is important to consider all the issues prior to the fundamental analysis, lest you be tempted to make some of your decisions on the basis of how they influence the analysis. In some cases, screening involves consideration of residuals and so you can't avoid doing an analysis first; however, in these cases you concentrate on the residuals and not on the other features of the analysis while making your screening decisions.

The order in which screening takes place is important because the decisions that you make at one step influence the outcomes of later steps. In a situation where you have both nonnormal variables and potential outliers, a fundamental decision is whether you would prefer to transform variables or to delete cases or change scores on cases. If you transform variables first, you are likely to find fewer outliers. If you

TABLE 4.3 CHECKLIST FOR SCREENING DATA

1. Inspect univariate descriptive statistics for accuracy of input
   a. Out-of-range values
   b. Plausible means and standard deviations
   c. Coefficient of variation
2. Evaluate amount and distribution of missing data: deal with problem
3. Identify and deal with nonnormal variables
   a. Check skewness and kurtosis, probability plots
   b. Transform variables (if desirable)
   c. Check results of transformation
4. Identify and deal with outliers
   a. Univariate outliers
   b. Multivariate outliers
5. Check pairwise plots for nonlinearity and heteroscedasticity
6. Evaluate variables for multicollinearity and singularity

delete or modify the outliers first, you are likely to find fewer variables with non-normality.

Of the two choices, transformation of variables is usually preferable. It typically reduces the number of outliers. It is likely to produce normality, linearity, and homoscedasticity among the variables. It increases the likelihood of multivariate normality to bring the data into conformity with one of the fundamental assumptions of most inferential tests. And on a very practical level, it usually enhances the analysis even if inference is not a goal. On the other hand, transformation may threaten interpretation, in which case all the statistical niceties are of no avail.

Or, if the impact of outliers is reduced first, you are less likely to find variables that are skewed because significant skewness is sometimes caused by extreme cases on the tails of the distributions. If you have cases that are outliers because they are not part of the population from which you intended to sample, by all means delete them before checking distributions.

Lastly, as will become obvious in the next two sections, although the issues are different, the runs on which they are screened are not necessarily different. That is, the same run often provides you with information regarding two or more issues.

4.2 COMPLETE EXAMPLES OF DATA SCREENING

Evaluation of assumptions is somewhat different for ungrouped and grouped data. That is, if you are going to perform multiple regression, canonical correlation, or factor analysis on ungrouped data, there is one approach to screening. If you are going to perform univariate or multivariate analysis of variance (including profile analysis) or discriminant function analysis on grouped data, there is another approach to screening.

Therefore, two complete examples are presented that use the same set of variables taken from the research described in Appendix B; number of visits to health professionals (TIMEDRS), attitudes toward drug use (ATTDRUG), attitudes toward housework (ATTHOUSE), INCOME, marital status (MSTATUS), and RACE. The grouping variable used in the analysis of grouped data is current employment status (EMPLMNT).

Where possible in these examples, and for illustrative purposes, screening for ungrouped data is performed using SPSS^x, and screening for grouped data is performed using BMDP programs. BMDP programs are generally richer in screening procedures, especially when the data are grouped. (SAS has better screening procedures for ungrouped than grouped data while SYSTAT has better procedures for grouped data. However, both SAS and SYSTAT are generally less useful than BMDP and SPSS^x for data screening because they do not provide significance tests for identifying multivariate outliers through Mahalanobis distance.)

4.2.1 Screening Ungrouped Data

A flow diagram for screening ungrouped data appears as Figure 4.8. The direction of flow assumes that data transformation is undertaken, as necessary. If transformation is not acceptable, then other procedures for handling outliers are used.

4.2.1.1 Accuracy of Input, Missing Data, and Distributions A check on accuracy of data entry, missing data, skewness, and kurtosis for the data set is done through SPSS^x FREQUENCIES, as seen in Table 4.4.

The minimum and maximum values, means, and standard deviations of each of the variables are inspected for plausibility. For instance, the minimum number of visits to health professionals (TIMEDRS) is 0 and the maximum is 81, higher than expected but found to be accurate upon checking the data sheets.[11] The MEAN for the variable is 7.901, higher than the national average but not extremely so, and the standard deviation (STD DEV) is 10.948. These values are all reasonable, as are the values on the other variables. For instance, the ATTDRUG variable is constructed with a range between 5 and 10 so it reassuring to find these values as minima and maxima.

TIMEDRS shows no MISSING CASES but has strong positive skewness (3.248). The significance of SKEWNESS is evaluated by dividing it by S E SKEW, as in Equation 4.4,

$$z = \frac{3.248}{.113} = 28.74$$

to reveal a clear departure from symmetry. The distribution also has significant kurtosis as evaluated by Equation 4.6,

$$z = \frac{13.101}{.226} = 57.97$$

The departures from normality are obvious from inspection of the difference between frequencies expected under the normal distribution (the dots) and obtained frequencies.

---

[11] The woman with this number of visits was terminally ill at the time that she gave us the interview.

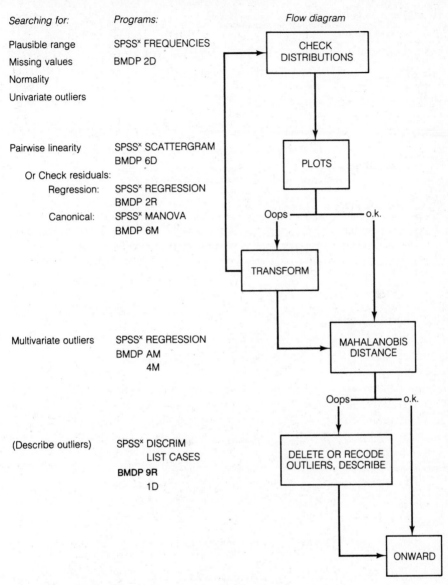

Searching for:	Programs:	Flow diagram
Plausible range	SPSS^x FREQUENCIES	
Missing values	BMDP 2D	
Normality		
Univariate outliers		
Pairwise linearity	SPSS^x SCATTERGRAM	
	BMDP 6D	
Or Check residuals:		
Regression:	SPSS^x REGRESSION	
	BMDP 2R	
Canonical:	SPSS^x MANOVA	
	BMDP 6M	
Multivariate outliers	SPSS^x REGRESSION	
	BMDP AM	
	4M	
(Describe outliers)	SPSS^x DISCRIM	
	LIST CASES	
	BMDP 9R	
	1D	

Figure 4.8 Flow diagram for screening ungrouped data.

ATTDRUG, on the other hand, is well behaved. There are no MISSING CASES and SKEWNESS and KURTOSIS are well within expected values. ATTHOUSE has a single missing value but is otherwise well distributed except for the extremely low scores that may be outliers (investigated later). It is reasonable to replace the single missing value with the mean, 23.541. On INCOME, however, there are 26 MISSING CASES—more than 5% of the sample. Because INCOME is not critical to the present hypothesis, we choose to delete it in subsequent analyses. If INCOME had been

```
        TITLE          DESCRIPTION OF UNGROUPED DATA
        FILE HANDLE    SCREEN
        DATA LIST      FILE=SCREEN    FREE
                       /SUBNO,TIMEDRS,ATTDRUG,ATTHOUSE,
                       INCOME,EMPLMNT,MSTATUS,RACE
        MISSING VALUES ATTHOUSE(1), INCOME(99)
        VAR LABELS     TIMEDRS 'VISITS TO HEALTH PROFESSIONALS'/
                       MSTATUS 'MARITAL STATUS'
        VALUE LABELS   EMPLMNT 0 'PAIDWORK' 1 'HOUSEWIFE'/
                       MSTATUS 1 'SINGLE' 2 'MARRIED'/
                       RACE 1 'WHITE' 2 'NONWHITE'/
        FREQUENCIES    VARIABLES = TIMEDRS TO INCOME, MSTATUS, RACE
                       /HISTOGRAM=NORMAL/FORMAT=NOTABLE/STATISTICS=ALL
```

TIMEDRS   VISITS TO HEALTH PROFESSIONALS

```
    COUNT   MIDPOINT   ONE SYMBOL EQUALS APPROXIMATELY  4.00 OCCURRENCES

     86        0     *************:*********
    187        4     ***************:****************************************
     74        8     ****************:**
     44       12     ***********   .
     27       16     *******    .
     13       20     ***    .
      7       24     **   .
      6       28     **.
      3       32     *.
      4       36     :
      2       40     *
      1       44
      1       48
      2       52     *
      3       56     *
      3       60     *
      0       64
      0       68
      0       72
      1       76
      1       80
                   I....+....I....+....I....+....I....+....I....+....I
                   0        40       80       120      160      200
                            HISTOGRAM FREQUENCY
```

```
MEAN          7.901     STD ERR      .508     MEDIAN        4.000
MODE          2.000     STD DEV    10.948     VARIANCE    119.870
KURTOSIS     13.101     S E KURT     .226     SKEWNESS      3.248
S E SKEW       .113     RANGE      81.000     MINIMUM           0
MAXIMUM      81.000     SUM      3674.000
VALID CASES     465     MISSING CASES     0
```

ATTDRUG

```
    COUNT   VALUE    ONE SYMBOL EQUALS APPROXIMATELY  4.00 OCCURRENCES

     13     5.00    **:
     60     6.00    *************:*
    126     7.00    *******************************.
    149     8.00    ***********************************.
     95     9.00    *********************:***
     22    10.00    *****:
                  I.........I.........I.........I.........I.........I
                  0        40       80       120      160      200
                           HISTOGRAM FREQUENCY
```

```
MEAN          7.686     STD ERR      .054     MEDIAN        8.000
MODE          8.000     STD DEV     1.156     VARIANCE      1.337
KURTOSIS      -.447     S E KURT     .226     SKEWNESS      -.123
S E SKEW       .113     RANGE       5.000     MINIMUM       5.000
MAXIMUM      10.000     SUM      3574.000

VALID CASES     465     MISSING CASES     0
```

TABLE 4.4   (Continued)

ATTHOUSE

```
   COUNT   MIDPOINT     ONE SYMBOL EQUALS APPROXIMATELY  2.00 OCCURRENCES

      0       -2
      0        0
      2        2   *
      0        4
      0        6
      0        8
      0       10
      2       12   *.
      5       14   ***.
     15       16   ******* .
     33       18   **************** .
     53       20   *************************** .
     73       22   ************************************* .
     83       24   *****************************************:*
     80       26   ************************************:*****
     54       28   ************************:**
     41       30   **************:******
     15       32   ******:*
      6       34   **:
      2       36   :
      0       38
               I....+....I....+....I....+....I....+....I....+....I
               0        20        40        60        80       100
                          HISTOGRAM FREQUENCY
```

MEAN	23.541	STD ERR	.208	MEDIAN	24.000
MODE	23.000	STD DEV	4.484	VARIANCE	20.102
KURTOSIS	1.556	S E KURT	.226	SKEWNESS	-.457
S E SKEW	.113	RANGE	33.000	MINIMUM	2.000
MAXIMUM	35.000	SUM	10923.000		

VALID CASES   464   MISSING CASES   1

INCOME

```
   COUNT    VALUE      ONE SYMBOL EQUALS APPROXIMATELY  2.00 OCCURRENCES

     71      1.00   **************:*********************
     39      2.00   ******************** .
     79      3.00   ****************************:********
     84      4.00   ***************************************:*****
     46      5.00   ***********************  .
     36      6.00   ****************** .
     36      7.00   ******************.
     19      8.00   **********.
     14      9.00   ****:**
     15     10.00   *:******
             I.........I.........I.........I.........I.........I
             0        20        40        60        80       100
                        HISTOGRAM FREQUENCY
```

MEAN	4.210	STD ERR	.115	MEDIAN	4.000
MODE	4.000	STD DEV	2.419	VARIANCE	5.851
KURTOSIS	-.359	S E KURT	.226	SKEWNESS	.582
S E SKEW	.117	RANGE	9.000	MINIMUM	1.000
MAXIMUM	10.000	SUM	1848.000		

VALID CASES   439   MISSING CASES   26

MSTATUS   MARITAL STATUS

```
   COUNT    VALUE      ONE SYMBOL EQUALS APPROXIMATELY  8.00 OCCURRENCES

    103      1.00   ************* .
    362      2.00   ****************************************:****
             I.........I.........I.........I.........I.........I
             0        80       160       240       320       400
                        HISTOGRAM FREQUENCY
```

MEAN	1.778	STD ERR	.019	MEDIAN	2.000
MODE	2.000	STD DEV	.416	VARIANCE	.173
KURTOSIS	-.190	S E KURT	.226	SKEWNESS	-1.346
S E SKEW	.113	RANGE	1.000	MINIMUM	1.000
MAXIMUM	2.000	SUM	827.000		

VALID CASES   465   MISSING CASES   0

TABLE 4.4   (Continued)

```
RACE

        COUNT      VALUE    ONE SYMBOL EQUALS APPROXIMATELY 10.00 OCCURRENCES

          424      1.00     ******************************************:
           41      2.00     **:*
                            I.........I.........I.........I.........I.........I
                            0        100       200       300       400       500
                                      HISTOGRAM FREQUENCY

    MEAN         1.088    STD ERR      .013     MEDIAN        1.000
    MODE         1.000    STD DEV      .284     VARIANCE       .081
    KURTOSIS     6.521    S E KURT     .226     SKEWNESS      2.914
    S E SKEW      .113    RANGE       1.000     MINIMUM       1.000
    MAXIMUM      2.000    SUM       506.000

VALID CASES      465    MISSING CASES      0
```

important to the hypothesis, we could have replaced the missing values or deleted the cases that failed to provide scores (if we determined that they were a random subsample).

The two remaining variables are dichotomous and not well split. MSTATUS has a 362 to 103 split, roughly a $3\frac{1}{2}$ to 1 ratio, and is not particularly disturbing. But RACE, with split greater than 10 to 1 is marginal. For this analysis, we choose to retain the variable, realizing that its association with other variables is deflated because of the uneven split.

At this point we have investigated the accuracy of data entry and the distributions of all variables, determined the number of missing values, and found two potential outliers. No cases have been deleted yet, so we still have the full sample of 465 cases.

4.2.1.2 Linearity   Because of nonnormality on at least one variable, SPSS^x RE-GRESSION is run to check the residuals plot, as reproduced in Figure 4.9. The instruction MISSING = MEANS is to replace the missing value on ATTHOUSE with the ATTHOUSE mean.

If the fundamental analysis is multiple regression, then the DV is used as the DV in this analysis also. In this example (for illustrative purposes), case labels (SUBNO) are used as a dummy DV as might be appropriate for factor analysis or canonical correlation; SUBNO has a rectangular distribution and cannot add non-normality, curvilinearity, or heteroscedasticity to the residuals. Any violation of assumptions is attributable to the set of IVs, listed in the VARS statement. Locate the zero-zero point in the center of the residuals plot. Note that the values taper off to the left of the 0 point, while they end abruptly on the right, suggesting nonnormality for at least one variable.

4.2.1.3 Transformation   In this case, the decision is made to transform variables prior to searching for outliers. With strong skewness for TIMEDRS and nonnormality of residuals, a logarithmic transformation is applied to TIMEDRS. Because the smallest value on the variable is zero, one is added to each score before the transformation is performed, as indicated in the COMPUTE statement. Table 4.5 shows

```
TITLE            RESIDUALS FOR UNGROUPED DATA
FILE HANDLE      SCREEN
DATA LIST        FILE=SCREEN      FREE
                 /SUBNO,TIMEDRS,ATTDRUG,ATTHOUSE,
                 INCOME,EMPLMNT,MSTATUS,RACE
MISSING VALUES ATTHOUSE(1), INCOME(99)
VAR LABELS       TIMEDRS 'VISITS TO HEALTH PROFESSIONALS'/
                 MSTATUS 'MARITAL STATUS'
VALUE LABELS     EMPLMNT O 'PAIDWORK' 1 'HOUSEWIFE'/
                 MSTATUS 1 'SINGLE' 2 'MARRIED'/
                 RACE 1 'WHITE' 2 'NONWHITE'/
REGRESSION       MISSING=MEANS/
                 VARIABLES = SUBNO TO ATTHOUSE, MSTATUS, RACE/
                 DEPENDENT=SUBNO/ENTER/
                 SCATTERPLOT (*RESID,*PRED)/
```

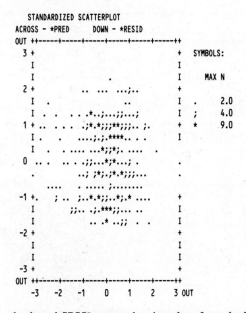

Figure 4.9 Setup and selected SPSS^x output showing plot of standardized residuals.

the distribution of TIMEDRS as transformed to LTIMEDRS. Figure 4.10 shows the residuals plot produced by SPSS^x REGRESSION after substitution of LTIMEDRS for TIMEDRS. SKEWNESS is reduced from 3.248 to .221 and KURTOSIS reduced from 13.101 to − .183 by the transformation. The frequency plot is not exactly pleasing (the frequencies are still too high at small scores) but the statistical evaluation of the distribution is much improved. Note also that in Figure 4.10 the residuals no longer end so abruptly on the right.

Notice in the control language for these runs that limits are placed on the extremity of values for ATTHOUSE; SELECT IF is used to remove the two cases with extreme values of 2 on the variable. This is done in anticipation of their identification as outliers.

TABLE 4.5 INPUT AND SPSS^x OUTPUT SHOWING DESCRIPTIVE
STATISTICS AND HISTOGRAM FOR TIMEDRS
AFTER LOGARITHMIC TRANSFORMATION

```
TITLE           DESCRIPTION OF UNGROUPED DATA SET
FILE HANDLE     SCREEN
DATA LIST       FILE=SCREEN      FREE
                /SUBNO,TIMEDRS,ATTDRUG,ATTHOUSE,
                INCOME,EMPLMNT,MSTATUS,RACE
VAR LABELS      TIMEDRS 'VISITS TO HEALTH PROFESSIONALS'/
                MSTATUS 'MARITAL STATUS'
VALUE LABELS    EMPLMNT 0 'PAIDWORK' 1 'HOUSEWIFE'/
                MSTATUS 1 'SINGLE' 2 'MARRIED'/
                RACE 1 'WHITE' 2 'NONWHITE'/
COMPUTE         LTIMEDRS = LG10(TIMEDRS + 1)
SELECT IF       ATTHOUSE NE 2
RECODE            ATTHOUSE(1=23.541)
FREQUENCIES     VARIABLES = LTIMEDRS
                /HISTOGRAM=NORMAL/FORMAT=NOTABLE/STATISTICS=ALL
```

```
LTIMEDRS
    COUNT   MIDPOINT    ONE SYMBOL EQUALS APPROXIMATELY  1.50 OCCURRENCES

      42      0      *****:**********************
       0      .1             .
       0      .2             .
      44      .3      ****************:************
       0      .4                .
      65      .5      ********************************:********************
      52      .6      *********************************:*******
      41      .7      ***************************** .
      53      .8      *******************************:******
      25      .9      *****************          .
      34     1.0      ************************* .
      33     1.1      *******************:**
      24     1.2      ***************:
      16     1.3      ***********.
      11     1.4      *******.
       6     1.5      **** .
       6     1.6      ***:
       3     1.7      *:
       6     1.8      :***
       2     1.9      :
       0     2.0
               I....+....I....+....I....+....I....+....I....+....I
               0        15       30       45       60       75
                          HISTOGRAM FREQUENCY
```

MEAN	.742	STD ERR	.019	MEDIAN	.699
MODE	.477	STD DEV	.416	VARIANCE	.173
KURTOSIS	-.183	S E KURT	.226	SKEWNESS	.221
S E SKEW	.113	RANGE	1.914	MINIMUM	0
MAXIMUM	1.914	SUM	343.743		
VALID CASES	463	MISSING CASES	0		

4.2.1.4 Outliers   The two extremely low values on the histogram for the ATTHOUSE variable in Table 4.4 are examined further. The score of 2 is 4.8 standard deviations below the mean of ATTHOUSE. Because $z = -4.8$ is well beyond the $p = .001$ criterion of 3.67 (2-tailed), the decision is made to delete from further analysis the data from the two women with extremely favorable attitudes toward housework. Information about these deletions is included in the report of results.

The remaining cases, with transformation applied to LTIMEDHS, are screened for multivariate outliers through SPSS^x REGRESSION (Table 4.6). Outliers among the variables listed in the INDEPENDENT statement are identified by using case labels as the dummy DV. Note that this identification of outliers is available in the same run that produced the plot of residuals in Figure 4.10.

The criterion for multivariate outliers is Mahalanobis distance at $p < .001$. Mahalanobis distance is evaluated as χ^2 with degrees of freedom equal to the number

```
TITLE          OUTLIERS AND RESIDUALS CHECK FOR UNGROUPED DATA
FILE HANDLE    SCREEN
DATA LIST      FILE=SCREEN        FREE
               /SUBNO,TIMEDRS,ATTDRUG,ATTHOUSE,
               INCOME,EMPLMNT,MSTATUS,RACE
VAR LABELS     TIMEDRS 'VISITS TO HEALTH PROFESSIONALS'/
               MSTATUS 'MARITAL STATUS'
VALUE LABELS   EMPLMNT O 'PAIDWORK' 1 'HOUSEWIFE'/
               MSTATUS 1 'SINGLE' 2 'MARRIED'/
               RACE 1 'WHITE' 2 'NONWHITE'/
COMPUTE        LTIMEDRS = LG10(TIMEDRS + 1)
SELECT IF      ATTHOUSE NE 2
RECODE         ATTHOUSE(1=23.541)
REGRESSION     VARS=SUBNO,LTIMEDRS,ATTDRUG,ATTHOUSE,MSTATUS,RACE/
               DEPENDENT=SUBNO/ENTER/
               RESIDUALS=OUTLIERS(MAHAL)/
               SCATTERPLOT (*RESID,*PRED)/
```

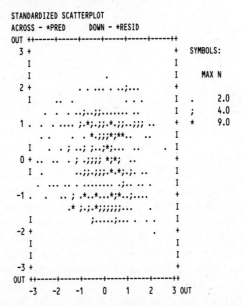

Figure 4.10   Setup and selected SPSS[x] output showing plot of standardized residuals after transformation of TIMEDRS.

of variables, in this case five: LTIMEDRS, ATTDRUG, ATTHOUSE, MSTATUS, and RACE. Any case with a Mahalanobis distance greater than $\chi^2(5) = 20.515$ (cf. Appendix C, Table C.4), then, is a multivariate outlier that has too much influence in the analysis. As seen in Table 4.6, cases 117 and 193 are outliers among these variables in the data set.

There are 463 cases remaining after deletion of the two univariate outliers on ATTHOUSE, and 461 cases remaining if the two multivariate outliers are deleted. Little is lost by deleting all four outliers from the sample. However, it is also necessary to determine why the two cases are multivariate outliers, to know how their deletion limits generalizability, and to include that information in the Results section.

SPSS[x] does not currently offer a program designed to test regressions with all

TABLE 4.6 SELECTED SPSS' OUTPUT FOR MULTIVARIATE OUTLIERS

```
   TITLE           OUTLIERS AND RESIDUALS CHECK FOR UNGROUPED DATA SET
   FILE HANDLE     SCREEN
   DATA LIST       FILE=SCREEN       FREE
                   /SUBNO,TIMEDRS,ATTDRUG,ATTHOUSE,
                    INCOME,EMPLMNT,MSTATUS,RACE
   VAR LABELS      TIMEDRS 'VISITS TO HEALTH PROFESSIONALS'/
                   MSTATUS 'MARITAL STATUS'
   VALUE LABELS    EMPLMNT 0 'PAIDWORK' 1 'HOUSEWIFE'/
                   MSTATUS 1 'SINGLE' 2 'MARRIED'/
                   RACE 1 'WHITE' 2 'NONWHITE'/
   COMPUTE         LTIMEDRS = LG10(TIMEDRS + 1)
   SELECT IF       ATTHOUSE NE 2
   RECODE          ATTHOUSE(1=23.541)
   REGRESSION      VARS=SUBNO,LTIMEDRS,ATTDRUG,ATTHOUSE,MSTATUS,RACE/
                   DEPENDENT=SUBNO/ENTER/
                   RESIDUALS=OUTLIERS(MAHAL)/
                   SCATTERPLOT (*RESID,*PRED)/

OUTLIERS - MAHALANOBIS' DISTANCE

 CASE #       *MAHAL

    117      21.83688
    193      20.65094
    433      19.96552
     99      18.49813
    333      18.46810
    291      17.52021
     58      17.37362
     71      17.17252
    102      16.94146
    196      16.72488
```

subsets of IVs. Therefore BMDP9R, which does all possible subsets regression, is used to identify the combination of variables on which case 117 and case 193 deviate from the remaining 462 cases. Each outlying case is evaluated in a separate BMDP9R run where a dummy variable is created to separate the outlying case from the remaining cases. In Table 4.7, the dummy variable for case 117 is created in the TRANSFORM paragraph with DUMMY = 0 and IF (KASE EQ 117) THEN DUMMY = 1.0.[12] With the dummy variable as the DV and the remaining variables as IVs, you can find the variables that distinguish the outlier from the other cases.

For the 117th case, LTIMEDRS, ATTDRUG, and RACE show up consistently in the best subsets of any size. The biggest R-SQUARED for regressions with one variable (RACE) is .022279. With two variables, the best combination is ATTDRUG and RACE, R-SQUARED = .034562. With three variables, the best combination is LTIMEDRS, ATTDRUG, and RACE, R-SQUARED = .046132. With four variables, the best combination is LTIMEDRS, ATTDRUG, MSTATUS, and RACE, but R-SQUARED is only .047366, not substantially better than the three variable solution. The three variable solution is picked as the 'BEST' SUBSET for distinguishing this outlier. For the 193rd case (Table 4.8), the 'BEST' SUBSET is LTIMEDRS, ATTHOUSE, and RACE.

The final step in evaluating outlying cases is to determine how their scores on the variables that cause them to be outliers differ from the scores of the remaining sample. The SPSS[x] LIST and CONDESCRIPTIVE procedures are used, as seen in Table 4.9. The LIST procedure is run for each outlying case to show its values on all the variables of interest (including a newly created variable called SEQ to verify

---

[12] KASE is a variable that BMDP uses to identify the sequential order of a case in the data file.

```
/PROBLEM     TITLE IS 'DESCRIBE OUTLIER FOR UNGROUPED DATA'.
/INPUT       VARIABLES ARE 8.  FORMAT IS '(A4,7F4.0)'.
             FILE='SCREEN.DAT'.
/VARIABLE    NAMES ARE SUBNO,TIMEDRS,ATTDRUG,ATTHOUSE,
                  INCOME,EMPLMNT,MSTATUS,RACE.
             LABEL = SUBNO.
/TRANSFORM   USE = ATTHOUSE NE 2.
             IF (ATTHOUSE EQ 1) THEN ATTHOUSE = 23.541.
             LTIMEDRS = LOG(TIMEDRS + 1).
             DUMMY = 0.
             IF (KASE EQ 117) THEN DUMMY = 1.0.
/REGRESS     DEPENDENT IS DUMMY.
             INDEPENDENT ARE LTIMEDRS,ATTDRUG,ATTHOUSE,MSTATUS,RACE.
/END
```

SUBSETS WITH   1 VARIABLES

R-SQUARED	ADJUSTED R-SQUARED	CP	
0.022279	0.020158	10.04	RACE
0.011651	0.009507	15.13	ATTDRUG
0.007038	0.004884	17.35	LTIMEDRS
0.000604	-0.001564	20.43	MSTATUS
0.000016	-0.002153	20.72	ATTHOUSE

SUBSETS WITH   2 VARIABLES

R-SQUARED	ADJUSTED R-SQUARED	CP	
0.034562	0.030365	6.14	

VARIABLE	COEFFICIENT	T-STATISTIC
3 ATTDRUG	-0.00444724	-2.42
8 RACE	0.0247388	3.30
INTERCEPT	0.00940593	

R-SQUARED	ADJUSTED R-SQUARED	CP	
0.031190	0.026977	7.76	LTIMEDRS RACE
0.023193	0.018946	11.60	MSTATUS RACE
0.022395	0.018145	11.98	ATTHOUSE RACE
0.020957	0.016700	12.67	LTIMEDRS ATTDRUG
0.000638	-0.003707	22.42	ATTHOUSE MSTATUS

SUBSETS WITH   3 VARIABLES

R-SQUARED	ADJUSTED R-SQUARED	CP	
0.046132	0.039897	2.59	

VARIABLE	COEFFICIENT	T-STATISTIC
10 LTIMEDRS	0.0121284	2.36
3 ATTDRUG	-0.00493636	-2.68
8 RACE	0.0259985	3.48
INTERCEPT	0.00278895	

R-SQUARED	ADJUSTED R-SQUARED	CP	
0.035456	0.029152	7.71	ATTDRUG MSTATUS RACE
0.034763	0.028455	8.05	ATTDRUG ATTHOUSE RACE
0.032404	0.026080	9.18	LTIMEDRS MSTATUS RACE
0.031197	0.024865	9.76	LTIMEDRS ATTHOUSE RACE
0.023366	0.016983	13.51	ATTHOUSE MSTATUS RACE
0.012316	0.005860	18.82	ATTDRUG ATTHOUSE MSTATUS
0.007887	0.001402	20.94	LTIMEDRS ATTHOUSE MSTATUS

TABLE 4.7   (Continued)

```
                                     SUBSETS WITH   4 VARIABLES
                                     ---------------------------
                ADJUSTED
   R-SQUARED    R-SQUARED      CP

   0.047366     0.039046      4.00   VARIABLE        COEFFICIENT   T-STATISTIC
                                     10 LTIMEDRS       0.0123197       2.39
                                      3 ATTDRUG       -0.00493969     -2.68
                                      7 MSTATUS        0.00395668      0.77
                                      8 RACE           0.0262357       3.51
                                        INTERCEPT     -0.00463605

   0.046133     0.037802      4.59   VARIABLE        COEFFICIENT   T-STATISTIC
                                     10 LTIMEDRS       0.0121441       2.34
                                      3 ATTDRUG       -0.00493578     -2.68
                                      4 ATTHOUSE      -0.000010828    -0.02
                                      8 RACE           0.0259927       3.47
                                        INTERCEPT      0.00303508

   0.035729     0.027307      9.58   ATTDRUG  ATTHOUSE MSTATUS   RACE

   0.032404     0.023953     11.18   LTIMEDRS ATTHOUSE MSTATUS   RACE

   0.021796     0.013253     16.27   LTIMEDRS ATTDRUG  ATTHOUSE MSTATUS


                                     SUBSETS WITH   5 VARIABLES
                                     ---------------------------
                ADJUSTED
   R-SQUARED    R-SQUARED      CP

   0.047368     0.036946      6.00   VARIABLE        COEFFICIENT   T-STATISTIC
                                     10 LTIMEDRS       0.0122964       2.36
                                      3 ATTDRUG       -0.00494058     -2.68
                                      4 ATTHOUSE       0.000016491     0.03
                                      7 MSTATUS        0.00396849      0.77
                                      8 RACE           0.0262453       3.50
                                        INTERCEPT     -0.00503308
```

STATISTICS FOR 'BEST' SUBSET

```
MALLOWS' CP                          2.59
SQUARED MULTIPLE CORRELATION       0.04613
MULTIPLE CORRELATION               0.21478
ADJUSTED SQUARED MULT. CORR.       0.03990
RESIDUAL MEAN SQUARE              0.002074
STANDARD ERROR OF EST.            0.045537
F-STATISTIC                          7.40
NUMERATOR DEGREES OF FREEDOM            3
DENOMINATOR DEGREES OF FREEDOM        459
SIGNIFICANCE (TAIL PROB.)          0.0001
```

NOTE THAT THE ABOVE F-STATISTIC AND
ASSOCIATED SIGNIFICANCE TEND TO BE
LIBERAL WHENEVER A SUBSET OF VARIABLES
IS SELECTED BY THE CP OR ADJUSTED
R-SQUARED CRITERIA.

VARIABLE NO. NAME	REGRESSION COEFFICIENT	STANDARD ERROR	STAND. COEF.	T-STAT.	2TAIL SIG.	TOL-ERANCE	CONTRI-BUTION TO R-SQ
INTERCEPT	0.00278895	0.0164722	0.060	0.17	0.866		
10 LTIMEDRS	0.0121284	0.00514020	0.109	2.36	0.019	0.982635	0.01157
3 ATTDRUG	-0.00493636	0.00184093	-0.123	-2.68	0.008	0.986955	0.01494
8 RACE	0.0259985	0.00746970	0.159	3.48	0.001	0.994523	0.02517

THE CONTRIBUTION TO R-SQUARED FOR EACH VARIABLE IS THE AMOUNT
BY WHICH R-SQUARED WOULD BE REDUCED IF THAT VARIABLE WERE
REMOVED FROM THE REGRESSION EQUATION.

```
/PROBLEM      TITLE IS 'DESCRIBE OUTLIER FOR UNGROUPED DATA'.
/INPUT        VARIABLES ARE 8.  FORMAT IS '(A4,7F4.0)'.
              FILE='SCREEN.DAT'.
/VARIABLE     NAMES ARE SUBNO,TIMEDRS,ATTDRUG,ATTHOUSE,
                       INCOME,EMPLMNT,MSTATUS,RACE.
              LABEL - SUBNO.
/TRANSFORM    USE = ATTHOUSE NE 2.
              IF (ATTHOUSE EQ 1) THEN ATTHOUSE = 23.541.
              LTIMEDRS = LOG(TIMEDRS + 1).
              DUMMY = 0.
              IF (KASE EQ 193) THEN DUMMY = 1.0.
/REGRESS      DEPENDENT IS DUMMY.
              INDEPENDENT ARE LTIMEDRS,ATTDRUG,ATTHOUSE,MSTATUS,RACE.
/END
```

```
                              SUBSETS WITH   1 VARIABLES
                              ---------------------------

           ADJUSTED
R-SQUARED  R-SQUARED      CP

0.022279   0.020158    8.77 RACE

0.012096   0.009953   13.64 LTIMEDRS

0.006492   0.004337   16.33 ATTHOUSE

0.002797   0.000634   18.09 ATTDRUG

0.000604  -0.001564   19.14 MSTATUS
```

```
                              SUBSETS WITH   2 VARIABLES
                              ---------------------------

           ADJUSTED
R-SQUARED  R-SQUARED      CP

0.036811   0.032623    3.82  VARIABLE      COEFFICIENT  T-STATISTIC
                             10 LTIMEDRS    0.0135064      2.63
                             8 RACE         0.0257504      3.44
                             INTERCEPT     -0.0358984

0.029925   0.025708    7.11 ATTHOUSE RACE

0.024781   0.020541    9.58 ATTDRUG   RACE

0.023193   0.018946   10.33 MSTATUS   RACE

0.016349   0.012073   13.61 LTIMEDRS ATTHOUSE

0.003411  -0.000922   19.80 ATTDRUG   MSTATUS
```

TABLE 4.8 (Continued)

```
                                    SUBSETS WITH   3 VARIABLES
                                    ---------------------------
             ADJUSTED
R-SQUARED   R-SQUARED       CP

  0.041837    0.035575     3.41   VARIABLE       COEFFICIENT   T-STATISTIC
                                  10 LTIMEDRS     0.0123558       2.39
                                   4 ATTHOUSE     0.000782753     1.55
                                   8 RACE         0.0261670       3.49
                                     INTERCEPT   -0.0539972

  0.038156    0.031870     5.18   VARIABLE       COEFFICIENT   T-STATISTIC
                                  10 LTIMEDRS     0.0130407       2.53
                                   3 ATTDRUG      0.00148125      0.80
                                   8 RACE         0.0255874       3.41
                                     INTERCEPT   -0.0467581

  0.038115    0.031828     5.20  LTIMEDRS MSTATUS  RACE

  0.032170    0.025844     8.04  ATTDRUG  ATTHOUSE RACE

  0.031296    0.024965     8.46  ATTHOUSE MSTATUS  RACE

  0.025705    0.019337    11.13  ATTDRUG  MSTATUS  RACE

  0.014645    0.008205    16.42  LTIMEDRS ATTDRUG  MSTATUS

  0.009990    0.003519    18.65  ATTDRUG  ATTHOUSE MSTATUS

                                    SUBSETS WITH   4 VARIABLES
                                    ---------------------------
             ADJUSTED
R-SQUARED   R-SQUARED       CP

  0.043534    0.035180     4.60   VARIABLE       COEFFICIENT   T-STATISTIC
                                  10 LTIMEDRS     0.0125325       2.42
                                   4 ATTHOUSE     0.000814742     1.61
                                   7 MSTATUS      0.00465001      0.90
                                   8 RACE         0.0264624       3.53
                                     INTERCEPT   -0.0634916

  0.043108    0.034751     4.81   VARIABLE       COEFFICIENT   T-STATISTIC
                                  10 LTIMEDRS     0.0119116       2.29
                                   3 ATTDRUG      0.00143976      0.78
                                   4 ATTHOUSE     0.000777007     1.54
                                   8 RACE         0.0260055       3.47
                                     INTERCEPT    0.0044100

  0.039454    0.031065     6.56  LTIMEDRS ATTDRUG  MSTATUS RACE

  0.033544    0.025103     9.38  ATTDRUG  ATTHOUSE MSTATUS RACE

  0.019115    0.010548    16.29  LTIMEDRS ATTDRUG  ATTHOUSE MSTATUS

                                    SUBSETS WITH   5 VARIABLES
                                    ---------------------------
             ADJUSTED
R-SQUARED   R-SQUARED       CP

  0.044795    0.034344     6.00  LTIMEDRS ATTDRUG  ATTHOUSE MSTATUS  RACE
```

TABLE 4.8 (Continued)

```
STATISTICS FOR 'BEST' SUBSET
------------------------------
MALLOWS' CP                        3.41
SQUARED MULTIPLE CORRELATION    0.04184
MULTIPLE CORRELATION            0.20454
ADJUSTED SQUARED MULT. CORR.    0.03557
RESIDUAL MEAN SQUARE            0.002083
STANDARD ERROR OF EST.          0.045640
F-STATISTIC                        6.68
NUMERATOR DEGREES OF FREEDOM          3
DENOMINATOR DEGREES OF FREEDOM      459
SIGNIFICANCE (TAIL PROB.)       0.0002

NOTE THAT THE ABOVE F-STATISTIC AND
ASSOCIATED SIGNIFICANCE TEND TO BE
LIBERAL WHENEVER A SUBSET OF VARIABLES
IS SELECTED BY THE CP OR ADJUSTED
R-SQUARED CRITERIA.
```

VARIABLE NO.	NAME	REGRESSION COEFFICIENT	STANDARD ERROR	STAND. COEF.	T- STAT.	2TAIL SIG.	TOL- ERANCE	CONTRI- BUTION TO R-SQ
	INTERCEPT	-0.0539972	0.0150204	-1.162	-3.59	0.000		
10	LTIMEDRS	0.0123558	0.00517242	0.111	2.39	0.017	0.974803	0.01191
4	ATTHOUSE	0.000782753	0.000504434	0.072	1.55	0.121	0.977429	0.00503
8	RACE	0.0261670	0.00748856	0.160	3.49	0.001	0.993976	0.02549

```
THE CONTRIBUTION TO R-SQUARED FOR EACH VARIABLE IS THE AMOUNT
BY WHICH R-SQUARED WOULD BE REDUCED IF THAT VARIABLE WERE
REMOVED FROM THE REGRESSION EQUATION.
```

TABLE 4.9 SETUP AND SPSS[x] OUTPUT SHOWING VARIABLE
SCORES FOR MULTIVARIATE OUTLIERS AND
DESCRIPTIVE STATISTICS FOR ALL CASES

```
TITLE           DESCRIPTION OF UNGROUPED DATA SET
FILE HANDLE     SCREEN
DATA LIST       FILE=SCREEN      FREE
                /SUBNO,TIMEDRS,ATTDRUG,ATTHOUSE,
                INCOME,EMPLMNT,MSTATUS,RACE
VAR LABELS      TIMEDRS 'VISITS TO HEALTH PROFESSIONALS'/
                MSTATUS 'MARITAL STATUS'
VALUE LABELS    EMPLMNT 0 'PAIDWORK' 1 'HOUSEWIFE'/
                MSTATUS 1 'SINGLE' 2 'MARRIED'/
                RACE 1 'WHITE' 2 'NONWHITE'/
COMPUTE         LTIMEDRS = LG10(TIMEDRS + 1)
RECODE          ATTHOUSE(1=23.541)
SELECT IF       ATTHOUSE NE 2
COMPUTE         SEQ=$CASENUM
LIST            CASES FROM 117 TO 117
LIST            CASES FROM 193 TO 193
CONDESCRIPTIVE  ATTDRUG,ATTHOUSE,MSTATUS,RACE,LTIMEDRS
```

SUBNO	TIMEDRS	ATTDRUG	ATTHOUSE	INCOME	EMPLMNT	MSTATUS	RACE	LTIMEDRS	SEQ
137.00	30.00	5.00	24.00	10.00	0	2.00	2.00	1.49	117.00

```
NUMBER OF CASES READ =     117   NUMBER OF CASES LISTED =      1
```

TABLE 4.9 (Continued)

```
   SUBNO  TIMEDRS  ATTDRUG ATTHOUSE   INCOME  EMPLMNT  MSTATUS     RACE LTIMEDRS     SEQ

   262.00    52.00     9.00    31.00     4.00     1.00     2.00     2.00     1.72  193.00

  NUMBER OF CASES READ =     193    NUMBER OF CASES LISTED =        1
```

```
  NUMBER OF VALID OBSERVATIONS (LISTWISE) =      463.00

  VARIABLE      MEAN    STD DEV    MINIMUM   MAXIMUM VALID N   LABEL

  ATTDRUG      7.685      1.158      5.000    10.000     463
  ATTHOUSE    23.634      4.258     11.000    35.000     463
  MSTATUS      1.782       .413      1.000     2.000     463   MARITAL STATUS
  RACE         1.089       .284      1.000     2.000     463
  LTIMEDRS      .742       .416          0     1.914     463
```

that the appropriate case is chosen from the data file). Then CONDESCRIPTIVE is used to show the average values for the remaining sample against which the outlying cases are compared.[13]

The 117th case is nonwhite on RACE, has very unfavorable attitudes regarding use of drugs (the lowest possible score on ATTDRUG), and a high score on LTIMEDRS. The 193rd case is also nonwhite on RACE, has very unfavorable attitudes toward housework (a high score on ATTHOUSE), and a very high score on LTIMEDRS. There is some question, then, about the generalizability of subsequent findings to nonwhite women who make numerous visits to physicians, especially in combination with either unfavorable attitude toward use of drugs or housework.

Note that, with outliers deleted, the mean for ATTHOUSE changes to 23.634, the value used to replace the missing ATTHOUSE score in subsequent analyses.

Screening information as it might be described in a Results section of a journal article appears next.

Results

Prior to analysis, number of visits to health professionals, attitude toward drug use, attitude toward housework, income, marital status, and race were examined through various SPSS^x programs for accuracy of data entry, missing values, and fit between their distributions and the assumptions of multivariate analysis. The single missing value on attitude toward housework was replaced by the mean for all cases, while income, with missing values on

---

[13] These values are equal to those shown in the earlier FREQUENCIES runs but for deletion of the two univariate outliers.

```
more than 5% of the cases, was deleted. The
poor split on race (424 to 41) truncates its
correlations with other variables, but it
was retained for analysis. To improve
pairwise linearity and residuals and to
reduce the extreme skewness and kurtosis,
visits to health professionals was
logarithmically transformed.
     Two cases with extremely low z scores on
attitude toward housework were found to be
univariate outliers; two other cases were
identified through Mahalanobis distance as
multivariate outliers with p < .001.¹⁴ All
four outliers were deleted, leaving 461
cases for analysis.
```

4.2.2 Screening Grouped Data

For these analyses, the cases are divided into two groups according to the EMPLMNT (employment) variable; there are 246 cases who have PAIDWORK, and 219 cases who are HOUSEWFEs. For illustrative purposes, variable transformation is considered inappropriate for this example, to be undertaken only reluctantly, if proved necessary. A flow diagram for screening grouped data appears in Figure 4.11.

4.2.2.1 Accuracy of Input, Missing Data, and Distributions   BMDP7D provides histograms and descriptive statistics for each group separately, as seen in Table 4.10. As with ungrouped data, accuracy of input is judged by plausible MEANs and STD.DEV.s and reasonable MAXIMUM and MINIMUM values. The distributions are judged by their overall shapes within each group. TIMEDRS is just as badly skewed when grouped as when ungrouped, but this is of less concern when dealing with sampling distributions based on over 200 cases unless the skewness causes nonlinearity among variables or there are outliers. ATTDRUG remains well distributed within each group.

As seen in Table 4.10, the ATTHOUSE variable is nicely distributed, as well, but the two cases in the PAIDWORK group with very low scores are likely to be outliers. There is also a score missing within the group of women with PAIDWORK. ATTDRUG and most of the other variables have 246 cases in this group, but ATT-

¹⁴ Case 117 was nonwhite with very unfavorable attitudes regarding use of drugs but numerous visits to physicians. Case 193 was also nonwhite with very favorable attitudes toward housework and numerous visits to physicians. Results of analyses may not generalize to nonwhite women with numerous visits to physicians, if they have either very unfavorable attitude toward use of drugs or very favorable attitude toward housework.

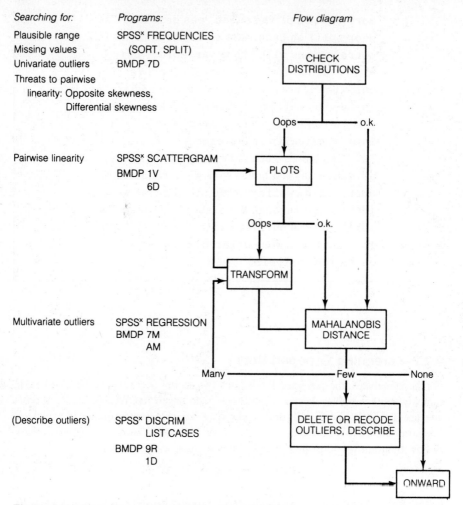

Searching for: *Programs:*   *Flow diagram*

Plausible range   SPSS^x FREQUENCIES
Missing values   (SORT, SPLIT)
Univariate outliers   BMDP 7D
Threats to pairwise
   linearity: Opposite skewness,
                 Differential skewness

Pairwise linearity   SPSS^x SCATTERGRAM
                         BMDP 1V
                              6D

Multivariate outliers   SPSS^x REGRESSION
                         BMDP 7M
                              AM

(Describe outliers)   SPSS^x DISCRIM
                              LIST CASES
                         BMDP 9R
                              1D

Figure 4.11   Flow diagram for screening grouped data.

HOUSE has only 245 cases. Because the case with the missing value is from the larger group, it is decided to delete the case from subsequent analyses.

On INCOME, however, it is the smaller group, HOUSEWFEs, with the greater number of missing values; within that group almost 7% of the cases do not have INCOME scores. INCOME, then, is a good candidate for variable deletion, although other remedies are available should deletion seriously interfere with hypothesis testing.

The splits in the two dichotomous variables, MSTATUS and RACE, are about the same for grouped as for ungrouped data, but the appearance of the graphs is misleading because lengths of the lines do not reflect the actual numbers in the groups. For instance, on the MSTATUS variable among HOUSEWFEs there are 193 cases with value 2 (married) and only 26 cases with value 1 (unmarried) despite the fact that the frequency bars look about the same length for values 1 and 2. Similarly for

RACE; there are only 27 cases with value 2 (nonwhite) among women with PAID-WORK and 14 cases with value 2 among HOUSEWFEs. But despite the misleading appearance of the graphs, the influence of poor dichotomous splits is less serious with analyses based on groups because inferences are made about sampling distributions of means that are likely to be normal with sample sizes this large.

For the remaining analyses, INCOME is deleted as a variable, and the case

TABLE 4.10 SETUP AND SELECTED BMDP7D OUTPUT SHOWING DESCRIPTIVE STATISTICS AND HISTOGRAMS FOR GROUPED DATA

```
            /PROBLEM    TITLE IS 'DISTRIBUTION CHECK FOR GROUPED DATA'.
            /INPUT      VARIABLES ARE 8.  FORMAT IS '(A4,7F4.0)'.
                        FILE='SCREEN.DAT'.
            /VARIABLE   NAMES ARE SUBNO,TIMEDRS,ATTDRUG,ATTHOUSE,
                             INCOME,EMPLMNT,MSTATUS,RACE.
                        LABEL = SUBNO.
                        MISSING = (4)1, (5)99.
            /GROUP      CODES(EMPLMNT) = 0,1.
                        NAMES(EMPLMNT) = PAIDWORK,HOUSEWFE.
            /HISTOGRAM  GROUPING = EMPLMNT.
                        VARIABLE = TIMEDRS TO INCOME, MSTATUS,RACE.
            /END

                ************                    ************
HISTOGRAM OF  * TIMEDRS  * (  2)   GROUPED  BY  * EMPLMNT  * (   6)
                ************                    ************
            PAIDWORK                          HOUSEWFE
MID-
POINTS..................................+..................................+
  87.500)
  84.000)
  80.500)*
  77.000)
  73.500)*
  70.000)
  66.500)
  63.000)
  59.500)**                              *
  56.000)                                ***
  52.500)*                               *
  49.000)*
  45.500)
  42.000)                                *
  38.500)***                             **
  35.000)                                *
  31.500)*                               **
  28.000)**                              ****
  24.500)***                             **
  21.000)****                            ****
  17.500)*******                         ***********
  14.000)********                        ******************
  10.500)*****************               *************************
   7.000)M****************************************40 M**********************
   3.500)************************************108 ****************************79
   0.000)************************************47 *******************************

            GROUP MEANS ARE DENOTED BY M'S IF THEY COINCIDE WITH *'S, N'S OTHERWISE
MEAN            7.293                       8.584
STD.DEV.       11.066                      10.800
S. E. M.        0.706                       0.730
MAXIMUM        81.000                      60.000
MINIMUM         0.000                       0.000
CASES INCL.     246                         219
```

TABLE 4.10 (Continued)

```
            ***********                    ***********
HISTOGRAM OF * ATTDRUG * (   3)   GROUPED  BY   * EMPLMNT * (   6)
            ***********                    ***********

          PAIDWORK                    HOUSEWFE
MID-
POINTS.................................................+................................................+
  10.000)******                    ****************
   9.800)
   9.600)
   9.400)
   9.200)
   9.000)***************************************49 *******************************************46
   8.800)
   8.600)
   8.400)
   8.200)
   8.000)***********************************78 ****************************************71
   7.800)                              N
   7.600)N
   7.400)
   7.200)
   7.000)*********************************72 *************************************54
   6.800)
   6.600)
   6.400)
   6.200)
   6.000)********************************   *************************
   5.800)
   5.600)
   5.400)
   5.200)
   5.000)*******                    ******
        GROUP MEANS ARE DENOTED BY M'S IF THEY COINCIDE WITH *'S, N'S OTHERWISE

MEAN          7.593                     7.790
STD.DEV.      1.113                     1.197
S. E. M.      0.071                     0.081
MAXIMUM      10.000                    10.000
MINIMUM       5.000                     5.000
CASES INCL.   246                       219

            ***********                    ***********
HISTOGRAM OF * ATTHOUSE * (   4)   GROUPED  BY   * EMPLMNT * (   6)
            ***********                    ***********

          PAIDWORK                    HOUSEWFE
MID-
POINTS.................................................+................................................+
  37.500)
  36.000)
  34.500)**                       **
  33.000)****
  31.500)*********                ******
  30.000)**********               ****
  28.500)************************ *************************
  27.000)****************         ************
  25.500)***************************44 ***********************************
  24.000)M***************         M*******************
  22.500)*******************************42 *****************************************43
  21.000)*****************        ****************
  19.500)***************************  *************************
  18.000)************             ********
  16.500)************             ***********
  15.000)**                       **
  13.500)***                      **
  12.000)*
  10.500)                         *
   9.000)
   7.500)
   6.000)
   4.500)
   3.000)
   1.500)**
   0.000)
        GROUP MEANS ARE DENOTED BY M'S IF THEY COINCIDE WITH *'S, N'S OTHERWISE

MEAN         23.641                    23.429
STD.DEV.      4.831                     4.068
S. E. M.      0.309                     0.275
MAXIMUM      34.000                    35.000
MINIMUM       2.000                    11.000
CASES EXCL.  (   1)                    (   0)
CASES INCL.   245                       219
```

TABLE 4.10 (Continued)

```
             ************                         ************
HISTOGRAM OF * INCOME   * (   5)    GROUPED  BY   * EMPLMNT  * (   6)
             ************                         ************

         PAIDWORK                            HOUSEWFE
MID-
POINTS.......................................+..................................+
  10.800)
  10.400)
  10.000)*******                              *******
   9.600)
   9.200)
   8.800)**********                           ****
   8.400)
   8.000)************                         *******
   7.600)
   7.200)
   6.800)****************                     ********************
   6.400)
   6.000)*******************                  *****************
   5.600)
   5.200)
   4.800)***************************          *****************
   4.400)N
   4.000)****************************** M******************************************46
   3.600)
   3.200)
   2.800)******************************46 ***********************************
   2.400)
   2.000)********************                 ******************
   1.600)
   1.200)
   0.800)************************************ ************************************
          GROUP MEANS ARE DENOTED BY M'S IF THEY COINCIDE WITH *'S, N'S OTHERWISE

MEAN           4.238                          4.176
STD.DEV.       2.440                          2.400
S. E. M.       0.159                          0.168
MAXIMUM       10.000                         10.000
MINIMUM        1.000                          1.000
CASES EXCL.  ( 11)                          (  15)
CASES INCL.    235                            204

             ************                         ************
HISTOGRAM OF * MSTATUS  * (   7)    GROUPED  BY   * EMPLMNT  * (   6)
             ************                         ************

         PAIDWORK                            HOUSEWFE
MID-
POINTS.......................................+..................................+
   2.000)***********************************169 ***********************************193
   1.960)
   1.920)
   1.880)                                     N
   1.840)
   1.800)
   1.760)
   1.720)
   1.680)N
   1.640)
   1.600)
   1.560)
   1.520)
   1.480)
   1.440)
   1.400)
   1.360)
   1.320)
   1.280)
   1.240)
   1.200)
   1.160)
   1.120)
   1.080)
   1.040)
   1.000)**********************************77 **************************
          GROUP MEANS ARE DENOTED BY M'S IF THEY COINCIDE WITH *'S, N'S OTHERWISE

MEAN           1.687                          1.881
STD.DEV.       0.465                          0.324
S. E. M.       0.030                          0.022
MAXIMUM        2.000                          2.000
MINIMUM        1.000                          1.000
CASES INCL.    246                            219
```

TABLE 4.10   (Continued)

```
                ************                    ************
HISTOGRAM OF * RACE        * (   8)   GROUPED  BY   * EMPLMNT  * (   6)
                ************                    ************

             PAIDWORK                           HOUSEWFE
MID-
POINTS...........................................+....................................+
    2.000)**************************             **************
    1.960)
    1.920)
    1.880)
    1.840)
    1.800)
    1.760)
    1.720)
    1.680)
    1.640)
    1.600)
    1.560)
    1.520)
    1.480)
    1.440)
    1.400)
    1.360)
    1.320)
    1.280)
    1.240)
    1.200)
    1.160)
    1.120)N
    1.080)                                       N
    1.040)
    1.000)********************************219 *************************************205
           GROUP MEANS ARE DENOTED BY M'S IF THEY COINCIDE WITH *'S, N'S OTHERWISE

MEAN          1.110                              1.064
STD.DEV.      0.313                              0.245
S. E. M.      0.020                              0.017
MAXIMUM       2.000                              2.000
MINIMUM       1.000                              2.000
CASES INCL.    246                                219
```

with the missing value on ATTHOUSE is deleted, leaving a sample size of 464, 245 cases in the PAIDWORK group and 219 cases in the HOUSEWFE group.

4.2.2.2 Linearity   Because of the poor distribution on TIMEDRS, a check of scatterplots is warranted to see if TIMEDRS has a linear relationship with other variables. There is no need to check for linearity with MSTATUS and RACE because variables with two levels have only linear relationships with other variables. Of the two remaining variables, ATTHOUSE and ATTDRUG, the distribution of ATTHOUSE differs most from that of TIMEDRS.

Most appropriately checked first, then, are within-group scatterplots of ATTHOUSE versus TIMEDRS. Within-group scatterplots are similar to residuals plots with ungrouped data. In residuals plots, each case has a predicted score and the residuals plot is a plot of the difference between the predicted score and the actual score for the case. With grouped data, the predicted score for each case is the mean of the group the case is in. But if the mean is subtracted from each case, the cases have the same relative positions that they had before subtraction so nothing is gained. So within-group scatterplots are used, instead.

In the within-group scatterplots of Figure 4.12, there is ample evidence of skewness in the bunching up of scores at low values of TIMEDRS, but no suggestion of nonlinearity for these variables in the group of HOUSEWFEs ("H" on the scatterplots). Among those with PAIDWORK ("P"), a hint of nonlinearity is produced

```
/PROBLEM     TITLE IS 'LINEARITY CHECK FOR GROUPED DATA'.
/INPUT       VARIABLES ARE 8.   FORMAT IS '(A4,7F4.0)'.
             FILE='SCREEN.DAT'.
/VARIABLE    NAMES ARE SUBNO,TIMEDRS,ATTDRUG,ATTHOUSE,
                 INCOME,EMPLMNT,MSTATUS,RACE.
             LABEL = SUBNO.
             MISSING = (4)1, (5)99.
             GROUPING IS EMPLMNT.
/GROUP       CODES(EMPLMNT) = 0,1.
             NAMES(EMPLMNT) = PAIDWORK,HOUSEWFE.
/PLOT        XVAR = ATTHOUSE.   YVAR = TIMEDRS.
             GROUP = EACH.     SIZE=40, 25.
/END
```

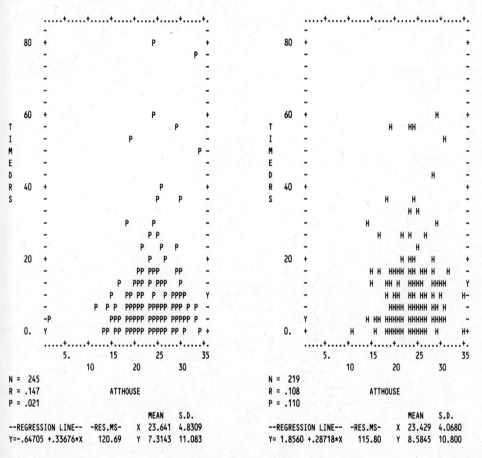

N = 245
R = .147   ATTHOUSE
P = .021

		MEAN	S.D.
--REGRESSION LINE--	-RES.MS-	X 23.641	4.8309
Y=-.64705 +.33676*X	120.69	Y 7.3143	11.083

N = 219
R = .108   ATTHOUSE
P = .110

		MEAN	S.D.
--REGRESSION LINE--	-RES.MS-	X 23.429	4.0680
Y= 1.8560 +.28718*X	115.80	Y 8.5845	10.800

ATTHOUSE VERSUS TIMEDRS   (4 VS. 2) GROUP=PAIDWORK, SYMBOL=P   ATTHOUSE VERSUS TIMEDRS   (4 VS. 2) GROUP=HOUSEWFE, SYMBOL=H

Figure 4.12 Setup and partial BMDP6D output showing within-group scatterplot of ATTHOUSE vs. TIMEDRS.

TABLE 4.11 SETUP AND SELECTED BMDPAM OUTPUT FOR IDENTIFICATION OF MULTIVARIATE OUTLIERS

```
/PROBLEM     TITLE IS 'MULTIVARIATE OUTLIERS FOR GROUPED DATA'.
/INPUT       VARIABLES ARE 8.  FORMAT IS '(A4,7F4.0)'.
             FILE='SCREEN.DAT'.
/VARIABLE    NAMES ARE SUBNO,TIMEDRS,ATTDRUG,ATTHOUSE,
                       INCOME,EMPLMNT,MSTATUS,RACE.
             LABEL = SUBNO.
             USE=2,3,4,6,7,8.
             GROUPING IS EMPLMNT.
/GROUP       CODES(EMPLMNT) = 0,1.
             NAMES(EMPLMNT) = PAIDWORK,HOUSEWFE.
/TRANSFORM   USE = ATTHOUSE GT 2.
/EST         METHOD=REGR.
/PRINT       MATR=DIS.
/END
```

UNIVARIATE SUMMARY STATISTICS

GROUP IS PAIDWORK   SIZE IS   243

VARIABLE	SAMPLE SIZE	MEAN	STANDARD DEVIATION	COEFFICIENT OF VARIATION	SMALLEST VALUE	LARGEST VALUE	SMALLEST STANDARD SCORE	LARGEST STANDARD SCORE	SKEWNESS	KURTOSIS
2 TIMEDRS	243	7.05800	11.11842	1.511060	0.000000	81.000000	-0.66	6.62	3.80	17.29
3 ATTDRUG	243	7.59871	1.11429	0.146681	5.000000	10.000000	-2.33	2.16	-0.14	-0.51
4 ATTHOUSE	243	23.81893	4.43038	0.186002	12.000000	34.000000	-2.67	2.30	-0.04	-0.49
7 MSTATUS	243	1.69547	0.46116	0.271993	1.000000	2.000000	-1.51	0.66	-0.84	-1.29
8 RACE	243	1.10700	0.30975	0.279808	1.000000	2.000000	-0.35	2.88	2.53	4.40

VALUES FOR KURTOSIS GREATER THAN ZERO INDICATE DISTRIBUTIONS WITH HEAVIER TAILS THAN THE NORMAL DISTRIBUTION.

UNIVARIATE SUMMARY STATISTICS

GROUP IS HOUSEWFE   SIZE IS   219

112

VARIABLE	SAMPLE SIZE	MEAN	STANDARD DEVIATION	COEFFICIENT OF VARIATION	SMALLEST VALUE	LARGEST VALUE	SMALLEST STANDARD SCORE	LARGEST STANDARD SCORE	SKEWNESS	KURTOSIS
2 TIMEDRS	219	8.58447	10.79950	1.258027	0.00000	60.00000	-0.79	4.76	2.53	7.56
3 ATTDRUG	219	7.78995	1.19687	0.153656	5.00000	10.00000	-2.33	1.85	-0.14	-0.48
4 ATTHOUSE	219	23.42922	4.06803	0.173631	11.00000	35.00000	-3.06	2.84	-0.06	-0.04
7 MSTATUS	219	1.88128	0.32420	0.172330	1.00000	2.00000	-2.72	0.37	-2.34	3.50
8 RACE	219	1.06393	0.24518	0.230451	1.00000	2.00000	-0.26	3.82	3.54	10.59

VALUES FOR KURTOSIS GREATER THAN ZERO INDICATE DISTRIBUTIONS
WITH HEAVIER TAILS THAN THE NORMAL DISTRIBUTION.
MAHALANOBIS DISTANCES ARE COMPUTED FROM EACH CASE TO THE
CENTROID OF ITS GROUP. ONLY THOSE VARIABLES WHICH WERE
ORIGINALLY AVAILABLE ARE USED--ESTIMATED VALUES ARE NOT USED.
FOR LARGE MULTIVARIATE NORMAL SAMPLES, THE MAHALANOBIS
DISTANCES HAVE AN APPROXIMATELY CHI-SQUARE DISTRIBUTION WITH
THE NUMBER OF DEGREES OF FREEDOM EQUAL TO THE NUMBER OF
NONMISSING VARIABLES. SIGNIFICANCE LEVELS REPORTED BELOW
THAT ARE LESS THAN .001 ARE FLAGGED WITH AN ASTERISK.

ESTIMATES OF MISSING DATA. MAHALANOBIS D-SQUARED (CHI-SQUARED)
AND SQUARED MULTIPLE CORRELATIONS WITH AVAILABLE VARIABLES

CASE LABEL	CASE NUMBER	MISSING VARIABLE	R-SQUARED ESTIMATE	GROUP	CHI-SQ	CHISQ/DF	D.F.	SIGNIFICANCE
259	190			PAIDWORK	5.155	1.031	5	0.3973
260	191			PAIDWORK	3.396	0.679	5	0.6392
261	192			PAIDWORK	5.325	1.065	5	0.3775
262	193			HOUSEWFE	30.453	6.091	5	0.0000 *
263	194			PAIDWORK	1.269	0.254	5	0.9381
264	195			HOUSEWFE	1.525	0.305	5	0.9102
265	196			HOUSEWFE	18.503	3.701	5	0.0024
266	197			HOUSEWFE	1.131	0.226	5	0.9514
267	198			PAIDWORK	13.920	2.784	5	0.0161
268	199			HOUSEWFE	5.202	1.040	5	0.3918
269	200			HOUSEWFE	3.917	0.783	5	0.5615
270	201			PAIDWORK	1.295	0.259	5	0.9355
271	202			HOUSEWFE	0.974	0.195	5	0.9646
272	203			HOUSEWFE	5.460	1.092	5	0.3623
273	204			PAIDWORK	6.912	1.382	5	0.2273
274	205			HOUSEWFE	3.796	0.759	5	0.5792
276	206			HOUSEWFE	20.530	4.106	5	0.0010 *
277	207			PAIDWORK	3.764	0.753	5	0.5839
278	208			HOUSEWFE	1.196	0.239	5	0.9453
279	209			PAIDWORK	3.948	0.790	5	0.5569
280	210			HOUSEWFE	14.519	2.904	5	0.0126

TABLE 4.12 SETUP AND SELECTED BMDPAM OUTPUT AFTER TRANSFORMATION OF TIMEDRS

```
/PROBLEM      TITLE IS 'MULTIVARIATE OUTLIERS FOR GROUPED DATA'.
/INPUT        VARIABLES ARE 8.  FORMAT IS '(A4,7F4.0)'.
              FILE='SCREEN.DAT'.
/VARIABLE     NAMES ARE SUBNO,TIMEDRS,ATTDRUG,ATTHOUSE,
              INCOME,EMPLMNT,MSTATUS,RACE.
              LABEL = SUBNO.
              USE=3,4,6,7,8,9.
              GROUPING IS EMPLMNT.
/GROUP        CODES(EMPLMNT) = 0,1.
              NAMES(EMPLMNT) = PAIDWORK,HOUSEWFE.
/TRANSFORM    USE = ATTHOLSE GT 2.
              LTIMEDRS = LOG(TIMEDRS + 1 ).
/EST          METHOD=REGR.
/PRINT        MATR=DIS.
/END
```

UNIVARIATE SUMMARY STATISTICS

GROUP IS PAIDWORK   SIZE IS   243

VARIABLE	SAMPLE SIZE	MEAN	STANDARD DEVIATION	COEFFICIENT OF VARIATION	SMALLEST VALUE	LARGEST VALUE	SMALLEST STANDARD SCORE	LARGEST STANDARD SCORE	SKEWNESS	KURTOSIS
3 ATTDRUG	243	7.59671	1.11429	0.146681	5.00000	10.00000	-2.33	2.16	-0.14	-0.51
4 ATTHOUSE	243	23.81893	4.43038	0.186002	12.00000	34.00000	-2.67	2.30	-0.04	-0.49
7 MSTATUS	243	1.69547	0.46116	0.271993	1.00000	2.00000	-1.51	0.66	-0.84	-1.29
8 RACE	243	1.10700	0.30975	0.279808	1.00000	2.00000	-0.35	2.88	2.53	4.40
9 LTIMEDRS	243	0.72247	0.39158	0.542001	0.00000	1.91381	-1.85	3.04	0.41	0.31

VALUES FOR KURTOSIS GREATER THAN ZERO INDICATE DISTRIBUTIONS
WITH HEAVIER TAILS THAN THE NORMAL DISTRIBUTION.

UNIVARIATE SUMMARY STATISTICS

GROUP IS HOUSEWFE   SIZE IS   219

VARIABLE	SAMPLE SIZE	MEAN	STANDARD DEVIATION	COEFFICIENT OF VARIATION	SMALLEST VALUE	LARGEST VALUE	SMALLEST STANDARD SCORE	LARGEST STANDARD SCORE	SKEWNESS	KURTOSIS
3 ATTDRUG	219	7.78995	1.19697	0.153656	5.00000	10.00000	-2.33	1.85	-0.14	-0.48
4 ATTHOUSE	219	23.42922	4.06803	0.173631	11.00000	35.00000	-3.06	2.84	-0.06	-0.04
7 MSTATUS	219	1.88128	0.32420	0.172330	1.00000	2.00000	-2.72	0.37	-2.34	3.50
8 RACE	219	1.06393	0.24518	0.230451	1.00000	2.00000	-0.26	3.82	3.54	10.59
9 LTIMDRS	219	0.76578	0.44141	0.576423	0.00000	1.78533	-1.73	2.31	0.03	-0.60

VALUES FOR KURTOSIS GREATER THAN ZERO INDICATE DISTRIBUTIONS WITH HEAVIER TAILS THAN THE NORMAL DISTRIBUTION.

MAHALANOBIS DISTANCES ARE COMPUTED FROM EACH CASE TO THE CENTROID OF ITS GROUP. ONLY THOSE VARIABLES WHICH WERE ORIGINALLY AVAILABLE ARE USED--ESTIMATED VALUES ARE NOT USED. FOR LARGE MULTIVARIATE NORMAL SAMPLES, THE MAHALANOBIS DISTANCES HAVE AN APPROXIMATELY CHI-SQUARE DISTRIBUTION WITH THE NUMBER OF DEGREES OF FREEDOM EQUAL TO THE NUMBER OF NONMISSING VARIABLES. SIGNIFICANCE LEVELS REPORTED BELOW THAT ARE LESS THAN .001 ARE FLAGGED WITH AN ASTERISK.

ESTIMATES OF MISSING DATA, MAHALANOBIS D-SQUARED (CHI-SQUARED) AND SQUARED MULTIPLE CORRELATIONS WITH AVAILABLE VARIABLES

CASE LABEL	CASE NUMBER	MISSING VARIABLE	R-SQUARED ESTIMATE	GROUP	CHI-SQ	CHISQ/DF	D.F.	SIGNIFICANCE
259	190			PAIDWORK	5.058	1.012	5	0.4089
260	191			PAIDWORK	3.464	0.693	5	0.6289
261	192			PAIDWORK	5.886	1.177	5	0.3175
262	193			HOUSEWFE	20.695	4.139	5	0.0009 *
263	194			PAIDWORK	1.349	0.270	5	0.9298
264	195			HOUSEWFE	2.396	0.479	5	0.7921
265	196			HOUSEWFE	19.353	3.871	5	0.0017
266	197			HOUSEWFE	3.766	0.753	5	0.5835
267	198			PAIDWORK	14.166	2.833	5	0.0146
268	199			HOUSEWFE	5.967	1.193	5	0.3094
269	200			HOUSEWFE	4.013	0.803	5	0.5475
270	201			PAIDWORK	1.897	0.379	5	0.8632
271	202			HOUSEWFE	0.891	0.178	5	0.9709
272	203			HOUSEWFE	5.667	1.133	5	0.3399
273	204			PAIDWORK	9.668	1.934	5	0.0852
274	205			HOUSEWFE	5.966	1.193	5	0.3095
276	206			HOUSEWFE	6.667	1.133	5	0.2466
277	207			PAIDWORK	4.098	0.820	5	0.5353
278	208			HOUSEWFE	1.914	0.383	5	0.8609
279	209			PAIDWORK	4.860	0.972	5	0.4333
280	210			HOUSEWFE	14.912	2.982	5	0.0107

```
          /PROBLEM      TITLE IS 'DESCRIBE OUTLIER FOR GROUPED DATA'.
          /INPUT        VARIABLES ARE 8.  FORMAT IS '(A4,7F4.0)'.
                        FILE='SCREEN.DAT'.
          /VARIABLE     NAMES ARE SUBNO,TIMEDRS,ATTDRUG,ATTHOUSE,
                               INCOME,EMPLMNT,MSTATUS,RACE.
                        LABEL = SUBNO.
                        MISSING = (4)1, (5)99.
          /TRANSFORM    USE = EMPLMNT EQ 1.0.
                        LTIMEDRS = LOG(TIMEDRS + 1).
                        DUMMY = 0.
                        IF (KASE EQ 193) THEN DUMMY = 1.0.
          /REGRESS      DEPENDENT IS DUMMY.
                        INDEPENDENT ARE LTIMEDRS,ATTDRUG,ATTHOUSE,MSTATUS,RACE.
          /END
```

```
NUMBER OF CASES READ. . . . . . . . . . . . . .     465
   CASES WITH USE SET TO ZERO . . . . . . . . .     246
       REMAINING NUMBER OF CASES . . . . . . . .    219
```

```
                                    SUBSETS WITH   1 VARIABLES
                                    ---------------------------
            ADJUSTED
 R-SQUARED  R-SQUQRED    CP

 0.067169   0.062870    8.88 RACE

 0.021728   0.017220   19.79 LTIMEDRS

 0.015960   0.011426   21.17 ATTHOUSE

 0.004709   0.000123   23.87 ATTDRUG

 0.000618  -0.003987   24.85 MSTATUS


                                    SUBSETS WITH    2 VARIABLES
                                    ---------------------------
            ADJUSTED
 R-SQUARED  R-SQUQRED    CP

 0.097723   0.089369    3.55  VARIABLE       COEFFICIENT  T-STATISTIC
                              9 LTIMEDRS      0.0268997      2.70
                              8 RACE          0.0763766      4.27
                              INTERCEPT      -0.0972922

 0.083742   0.075258    6.90 ATTHOUSE RACE

 0.070865   0.062262    9.99 ATTDRUG  RACE

 0.068064   0.059435   10.66 MSTATUS  RACE

 0.031735   0.022769   19.38 LTIMEDRS ATTHOUSE

 0.005346  -0.003864   25.72 ATTDRUG  MSTATUS
```

TABLE 4.13 (Continued)

```
                                    SUBSETS WITH   3 VARIABLES
                                    ---------------------------

              ADJUSTED
 R-SQUARED    R-SQUQRED      CP

 0.107160     0.094702      3.28   VARIABLE       COEFFICIENT   T-STATISTIC
                                   9 LTIMEDRS       0.0239913      2.37
                                   4 ATTHOUSE       0.00164393     1.51
                                   8 RACE           0.0760936      4.26
                                     INTERCEPT     -0.133280

 0.100689     0.088141      4.83   VARIABLE       COEFFICIENT   T-STATISTIC
                                   9 LTIMEDRS       0.0280625      2.79
                                   7 MSTATUS        0.0114618      0.84
                                   8 RACE           0.0768860      4.29
                                     INTERCEPT     -0.120288

 0.098504     0.085925      5.36   LTIMDRS ATTDRUG   RACE

 0.087407     0.074670      8.02   ATTDRUG  ATTHOUSE  RACE

 0.086116     0.073364      8.33   ATTHOUSE MSTATUS   RACE

 0.071778     0.058826     11.77   ATTDRUG  MSTATUS   RACE

 0.025439     0.011841     22.89   LTIMEDRS ATTDRUG   MSTATUS

 0.022543    -0.008904     23.59   ATTDRUG  ATTHOUSE MSTATUS


                                    SUBSETS WITH   4 VARIABLES
                                    ---------------------------

              ADJUSTED
 R-SQUARED    R-SQUUQRED     CP

 0.111586     0.094980      4.22   VARIABLE       COEFFICIENT   T-STATISTIC
                                   9 LTIMEDRS       0.0251827      2.48
                                   4 ATTHOUSE       0.00177923     1.62
                                   7 MSTATUS        0.0141022      1.03
                                   8 RACE           0.0766971      4.29
                                     INTERCEPT     -0.164535

 0.108148     0.091478      5.04   VARIABLE       COEFFICIENT   T-STATISTIC
                                   9 LTIMEDRS       0.0230161      2.23
                                   3 ATTDRUG        0.00180983     0.49
                                   4 ATTHOUSE       0.00166301     1.52
                                   8 RACE           0.0756492      4.22
                                     INTERCEPT     -0.146606

 0.101404     0.084608      6.66   LTIMDRS ATTDRUG   MSTATUS  RACE

 0.089809     0.072796      9.45   ATTDRUG  ATTHOUSE MSTATUS   RACE

 0.037019     0.019019     22.12   LTIMDRS ATTDRUG   ATTHOUSE MSTATUS

                                    SUBSETS WITH   5 VARIABLES
                                    ---------------------------
              ADJUSTED
 R-SQUARED    R-SQUQRED      CP

 0.112501     0.091667      6.00   LTIMDRS ATTDRUG   ATTHOUSE MSTATUS   RACE


STATISTICS FOR 'BEST' SUBSET
----------------------------

                                                                       CONTRI-
   VARIABLE      REGRESSION     STANDARD   STAND.    T-    2TAIL    TOL-  BUTION
 NO.   NAME     COEFFICIENT      ERROR    COEF.   STAT.   SIG.   ERANCE TO R-SQ

    INTERCEPT    -0.133280     0.0321994  -1.972  -4.14  0.000
  9 LTIMEDRS      0.0239913    0.0101029   0.157   2.37  0.018 0.953473 0.02342
  4 ATTHOUSE      0.00164393   0.00109055  0.099   1.51  0.133 0.963449 0.00944
  8 RACE          0.0760936    0.0178549   0.276   4.26  0.000 0.989452 0.07543


THE CONTRIBUTION TO R-SQUARED FOR EACH VARIABLE IS THE AMOUNT
BY WHICH R-SQUARED WOULD BE REDUCED IF THAT VARIABLE WERE
REMOVED FROM THE REGRESSION EQUATION.
```

TABLE 4.14 SETUP AND BMDP1D OUTPUT SHOWING VARIABLE
SCORES FOR MULTIVARIATE OUTLIERS AND
DESCRIPTIVE STATISTICS FOR ALL CASES

```
/PROBLEM      TITLE IS 'DESCRIPTION OF GROUPED DATA SET'.
/INPUT        VARIABLES ARE 8.  FORMAT IS '(A4,7F4.0)'.
              FILE='SCREEN.DAT.
/VARIABLE     NAMES ARE SUBNO,TIMEDRS,ATTDRUG,ATTHOUSE,
              INCOME EMPLMNT,MSTATUS,RACE.
              LABEL = SUBNO.
              GROUPING IS EMPLMNT.
              USE = 1,3,4,6 TO 10.
/GROUP        CODES(EMPLMNT) = 0,1.
              NAMES(EMPLMNT) = PAIDWORK,HOUSEWFE.
/TRANSFORM    USE = ATTHOUSE GT 2.
              LTIMEDRS = LOG(TIMEDRS + 1).
              DUMMY = 0, IF (KASE EQ 117 OR KASE EQ 193) THEN DUMMY=XMIS.
/PRINT        MISS.
/END
```

CASE		3	4	6	7	8	9	10
NO.	LABEL	ATTDRUG	ATTHOUSE	EMPLMNT	MSTATUS	RACE	LTIMEDRS	DUMMY
117	137	5	24	PAIDWORK	2	2	1.491	MISSING
193	262	9	31	HOUSEWFE	2	2	1.724	MISSING

```
NUMBER OF CASES READ. . . . . . . . .       465
CASES WITH USE SET TO ZERO. . . . . .         3
   REMAINING NUMBER OF CASES                 462
```

VARIABLE NO. NAME	GROUPING VARIABLE	LEVEL	TOTAL FREQUENCY	MEAN	STANDARD DEVIATION	ST.ERR OF MEAN	COEFF. OF VARIATION	SMALLEST VALUE	SMALLEST Z-SCORE	LARGEST VALUE	LARGEST Z-SCORE	RANGE
3 ATTDRUG	EMPLMNT		462	7.688	1.157	0.0538	0.15049	5.000	-2.32	10.000	2.00	5.000
		PAIDWORK	243	7.597	1.114	0.0715	0.14668	5.000	-2.33	10.000	2.16	5.000
		HOUSEWFE	219	7.790	1.197	0.0809	0.15366	5.000	-2.33	10.000	1.85	5.000
4 ATTHOUSE	EMPLMNT		462	23.634	4.262	0.1983	0.18035	11.000	-2.96	35.000	2.67	24.000
		PAIDWORK	243	23.819	4.430	0.2842	0.18600	12.000	-2.67	34.000	2.30	22.000
		HOUSEWFE	219	23.429	4.068	0.2749	0.17363	11.000	-3.06	35.000	2.84	24.000
7 MSTATUS	EMPLMNT		462	1.784	0.412	0.0192	0.23115	1.000	-1.90	2.000	0.53	1.000
		PAIDWORK	243	1.695	0.461	0.0296	0.27199	1.000	-1.51	2.000	0.66	1.000
		HOUSEWFE	219	1.881	0.324	0.0219	0.17233	1.000	-2.72	2.000	0.37	1.000
8 RACE	EMPLMNT		462	1.087	0.282	0.0131	0.25909	1.000	-0.31	2.000	3.24	1.000
		PAIDWORK	243	1.107	0.310	0.0199	0.27981	1.000	-0.35	2.000	2.88	1.000
		HOUSEWFE	219	1.064	0.245	0.0166	0.23045	1.000	-0.26	2.000	3.82	1.000
9 LTIMEDRS	EMPLMNT		462	0.743	0.416	0.0194	0.55996	0.000	-1.79	1.914	2.81	1.914
		PAIDWORK	243	0.722	0.392	0.0251	0.54200	0.000	-1.85	1.914	3.04	1.914
		HOUSEWFE	219	0.766	0.441	0.0298	0.57642	0.000	-1.73	1.785	2.31	1.785
10 DUMMY	EMPLMNT		460	0.000	0.000	0.0000	0.00000	0.000	0.00	0.000	0.00	0.000
		PAIDWORK	242	0.000	0.000	0.0000	0.00000	0.000	0.00	0.000	0.00	0.000
		HOUSEWFE	218	0.000	0.000	0.0000	0.00000	0.000	0.00	0.000	0.00	0.000

by the data point in the lower left corner of the plot. However, according to the ATTHOUSE scale, the point is produced by the two cases with very low scores that are probably univariate outliers. Because the plot is otherwise acceptable, there is no evidence that the extreme skewness of TIMEDRS produces a harmful departure from linearity. Nor is there any reason to expect nonlinearity with the symmetrically distributed ATTDRUG.

4.2.2.3 Outliers   Two univariate outliers are seen in Table 4.10, on ATTHOUSE in the PAIDWORK group. With scores of 2, each case is 4.48 standard deviations below the mean for her group—beyond the $\alpha = .001$ criterion of 3.67 for a two-tailed test. These cases also cause departure from linearity (Figure 4.12). Because there are more cases in the PAIDWORK group, it is decided to delete these two women with extremely favorable attitudes toward housework from further analysis, and to report the deletion in the Results section.

Multivariate outliers with the groups are sought using BMDPAM, one of two convenient BMDP programs (the other is BMDP7M) for this purpose. Table 4.11 shows univariate summary statistics for both groups and a selected portion of the section which provides Mahalanobis distance for each case from its group centroid. In the TRANSFORM paragraph, only cases with ATTHOUSE GT 2 are selected to eliminate the case with the missing value (coded 1) and the two univariate outliers (value 2). In BMDPAM, if missing values are declared, cases with missing data are included during the search for multivariate outliers. Because we have decided to eliminate those cases in any event, TRANSFORM is used to omit them.

Because $\alpha = .001$ is the criterion chosen for identifying multivariate outliers, they are readily found through BMDPAM because the program places an asterisk after each case that is an outlier at that α level. Table 4.11 shows that cases 193 and 206 are outliers in the HOUSEWFE group (SUBNO 262 and 276, respectively).

Altogether, nine cases (about 2%) about evenly distributed in the two groups are identified as multivariate outliers. While this is not an exceptionally large number of cases to delete, it is worth investigating alternative strategies for dealing with outliers. The univariate summary statistics in Table 4.11 show a LARGEST STANDARD SCORE of 6.62 among those with PAIDWORK and 4.76 among HOUSEWFEs on TIMEDRS; the poorly distributed variable produces univariate outliers in both groups. The skewed histograms of Table 4.10 suggest a logarithmic transformation of TIMEDRS.

Table 4.12 shows output from a second run of BMDPAM identical to the run in Table 4.11 except that TIMEDRS is replaced by LTIMEDRS, its logarithmic transform. The 193rd case remains extreme, but case 206 is no longer an outlier. With the transformed variable, the entire data set contains only two multivariate outliers (the same two identified in ungrouped data).

Identification of the variables causing outliers to be extreme proceeds in the same manner as for ungrouped data except that the values for the case are compared with the means for the group the case comes from. For case 193, a housewife, the BMDP9R all-subsets-regression run is limited to housewives, as seen in the TRANSFORM paragraph USE = EMPLMNT EQ 1.0 sentence of Table 4.13. As seen in the table, the same variables cause this woman to be an outlier from her group as from the entire sample: she differs on the combination of RACE, ATTHOUSE, and LTIMEDRS. Similarly, case 117 differs from her group on the same variables that make her extreme with respect to the entire sample (output not shown).

As with ungrouped data, identification of variables on which cases are outliers is followed by an analysis of the scores on the variables for those cases. This analysis is done through BMDP1D where outlying cases are declared to have a missing value on a dummy variable (Table 4.14) and /PRINT MISS is used to provide scores on all variables for the outliers. The two outliers are nonwhite with frequent visits to health professionals. Case 117, in addition, has very unfavorable attitudes toward drug use, while case 193 has favorable attitudes toward housework.

Screening information as it might be described in a Results section of a journal article appears next.

Results

Prior to analysis, number of visits to health professionals, attitude toward drug use, attitude toward housework, income, marital status, and race were examined through various BMDP programs for accuracy of data entry, missing values, and fit between their distributions and the assumptions of multivariate analysis. The variables were examined separately for the 246 employed women and the 219 housewives.

A case with a single missing value on attitude toward housework was deleted from the group of employed women, leaving 245 cases in that group. Income, with missing values on more than 5% of the cases, was deleted. Pairwise linearity was checked using within-group scatterplots and found to be satisfactory.

Two cases in the employed group were univariate outliers because of their extremely low z scores on attitude toward housework; these cases were deleted. Using Mahalanobis distance with $p < .001$, 9 cases (about 2%) were identified as multivariate outliers in their own groups. Because several of these cases had extreme z scores on visits to health professionals and because that variable was severely skewed, a logarithmic transformation was applied. With the transformed variable in the variable set, only two cases were identified as multivariate outliers. One multivariate outlier was from the employed group and the other from the housewife group.[15] With all four outliers and the case with missing values deleted, 242 cases remained in the employed group and 218 in the group of housewives.

---

[15] Case 117, an employed woman, was nonwhite with very unfavorable attitudes regarding use of drugs but numerous visits to physicians. Case 193, a housewife, was also nonwhite with very favorable attitudes toward housework and numerous visits to physicians. Results of analyses may not generalize to nonwhite women with numerous visits to physicians, if they have either very unfavorable attitudes toward use of drugs or very favorable attitudes toward housework.

Chapter 5

Multiple Regression

5.1 GENERAL PURPOSE AND DESCRIPTION

Regression analyses are a set of statistical techniques that allow one to assess the relationship between one DV and several IVs. For example, can reading ability in primary grades (the DV) be predicted from several IVs such as gender and preschool measures of perceptual and motor development? The terms regression and correlation are used more or less interchangeably to label these procedures although the term regression is often used when the intent of the analysis is prediction, and the term correlation is used when the intent is simply to measure the degree of association between the DV and the IVs.

Regression techniques can be applied to a data set in which the IVs are correlated with one another and with the DV to varying degrees. One can, for instance, assess the relationship between IVs such as education, income, and socioeconomic status and a DV such as occupational prestige. Because regression techniques can be used when the IVs are correlated, they are helpful both in experimental research (when, for instance, correlation among IVs is created by unequal numbers of cases in cells) and in observational or survey research where nature has "manipulated" correlated variables. The flexibility of regression techniques is, then, of special importance to the researcher who is interested in real-world or very complicated problems that cannot be meaningfully reduced to orthogonal designs in a laboratory setting.

Multiple regression is an extension of bivariate regression (see Chapter 3) in which several IVs instead of just one are combined to predict a value on a DV for each subject. The result of regression is an equation that represents the best prediction of a DV from several continuous or dichotomous IVs. The regression equation takes the following form:

$$Y' = A + B_1X_1 + B_2X_2 + \cdots + B_kX_k$$

where Y' is the predicted value on the DV, A is the Y intercept (the value of Y when all the X values are zero), the X's represent the various IVs (of which there are k), and the Bs are the coefficients assigned to each of the IVs during regression. Although the intercept and the coefficients are the same for a whole sample, a different Y' value is predicted for each subject as a result of inserting the subject's own X values into the equation.

The goal of regression is to arrive at the set of B values, called regression coefficients, for the IVs that bring the Y values predicted from the equation as close as possible to the Y values obtained by measurement. The regression coefficients that are computed accomplish two intuitively appealing and highly desirable goals: they minimize (the sum of the squared) deviations between predicted and obtained Y values and they optimize the correlation between the predicted and obtained Y values for the data set. In fact, one of the important statistics derived from a regression analysis is the multiple correlation coefficient, the Pearson product moment correlation co-efficient between the obtained and predicted Y values (see Section 5.4.1).

Regression techniques consist of standard multiple regression, hierarchical regression, and statistical (stepwise) or setwise regression. Differences between these techniques involve the way variables enter the equation. What happens to variance shared by variables and who determines the order in which variables enter the equation?

5.2 KINDS OF RESEARCH QUESTIONS

The primary goal of regression analysis is usually to investigate the relationship between a DV and several IVs. As a preliminary step, one determines how strong the relationship is between DV and IVs; then, with some ambiguity, one assesses the importance of each of the IVs to the relationship.

A more complicated goal might be to investigate the relationship between a DV and some IVs with the effect of other IVs statistically eliminated. Researchers often use regression to perform what is essentially a covariates analysis in which they ask if some critical variable (or variables) adds anything to a prediction equation for a DV after other IVs—the covariates—have already entered the equation. For example, does gender add to prediction of mathematical performance after extent and type of mathematical training have been accounted for?

Another strategy is to compare the ability of several competing sets of IVs to predict a DV. Can use of Valium be predicted better by a set of health variables or by a set of attitudinal variables?

All too often, regression is used to find the best prediction equation for some phenomenon regardless of the meaning of the equation, a goal met by statistical (stepwise) regression. In statistical regression, of which there are several varieties, statistical criteria alone, computed from a single sample, determine which IVs enter the equation and the order in which they enter.

Regression analyses can be used with either continuous or dichotomous IVs. But a variable that is initially discrete can also be used if it is converted into a set of dichotomous variables by dummy variable coding. For example, consider an initially discrete religious affiliation variable on which 1 stands for Protestant, 2 for Catholic, 3 for Jewish, and 4 for none or other. The variable may be converted into a set of three new IVs (Protestant vs. non-Protestant, Catholic vs. non-Catholic, Jewish vs. non-Jewish), one IV for each degree of freedom. When the new set is entered as a group, the variance due to the original discrete variable is analyzed, and, in addition, one can examine effects of the newly created dichotomous components. Dummy variable coding is covered in glorious detail by Cohen and Cohen (1975, pp. 173–188).

ANOVA (Chapter 3) is a special case of regression where main effects and interactions have been dummy variable coded. ANOVA problems can be handled through multiple regression, but multiple regression problems often cannot readily be converted into ANOVA because of correlations among IVs and the presence of continuous IVs. If analyzed through ANOVA, continuous IVs have to be rendered discrete (e.g., high, medium, and low), a process that often results in loss of information. In regression, the full range of continuous IVs is maintained.

As a statistical tool, regression is very helpful in answering a number of practical questions, as discussed in Sections 5.2.1 through 5.2.8.

5.2.1 Degree of Relationship

How good is the regression equation? Does the regression equation really provide better-than-chance prediction? Is the multiple correlation really any different from zero when allowances for naturally occurring fluctuations in such correlations are made? For example, can one reliably predict reading ability given knowledge of perceptual development, motor development, and gender? The statistical procedures described in Section 5.6.2.1 allow you to determine if your multiple correlation is reliably different from zero.

5.2.2 Importance of IVs

If the multiple correlation is different from zero, you may want to ask which of the IVs is important in the equation and which of the IVs can be deleted. For example, is knowledge of motor development helpful in predicting reading ability, or can we do just as well with knowledge of only gender and perceptual development? The methods in Section 5.6.1 help you to evaluate the relative importance of various IVs to a regression solution.

5.2.3 Adding IVs

Suppose that you have just computed a regression equation and you want to know whether or not you can improve your prediction of the DV by addition of one or more IVs to the equation. For example, is prediction of a child's reading ability enhanced by adding a variable reflecting parental interest in reading to the three IVs already included in the equation? A test for improvement of the multiple correlation after addition of one new variable is given in Section 5.6.1.2 and for improvement after addition of several new variables in Section 5.6.2.3.

5.2.4 Changing IVs

Although the regression equation is a linear equation (that is, it does not contain squared values, cubed values, cross products of variables, and the like), the researcher may include nonlinear relationships in the analysis by redefining IVs. Curvilinear relationships can be made available for analysis by squaring or raising to higher powers original IVs. Interaction can be made available for analysis by creating a new IV that is a cross product of two or more original IVs.

For an example of a curvilinear relationship, suppose a child's reading ability increases with increasing parental interest up to a point, and then levels off. Greater parental interest does not result in greater reading ability. If the square of parental interest were added as an IV, better prediction of a child's reading ability could be achieved.

Inspection of a scatterplot between predicted and obtained Y values (known as residuals analysis—see Section 5.3.2.4) may reveal that the relationship between the DV and the IVs is not exclusively linear but has more complicated components such as curvilinearity. To improve prediction, one may have to include new IVs as described above.

There is danger, however, in too liberal use of powers or cross products of IVs; the sample data may be overfit to the extent that results no longer generalize to a population. Procedures for using regression for nonlinear curve fitting are discussed in Cohen and Cohen (1975) and McNeil, Kelly, and McNeil (1975).

5.2.5 Contingencies among IVs

You may be interested in the way that one IV behaves in the context of one, or a set, of the other IVs. Hierarchical regression can be used to hold the effects of several IVs statistically "constant" while examining the relationship between an especially interesting IV and the DV. For example, after adjustment for differences in perceptual and motor development, does gender predict reading ability? This procedure is described in Section 5.5.2.

5.2.6 Comparing Sets of IVs

Is prediction of a DV from one set of IVs better than prediction from another set of IVs? For example, is prediction of reading ability based on perceptual and motor development and gender as good as prediction from family income and parental educational attainments? Section 5.6.2.5 demonstrates a method for comparing the solutions given by two sets of predictors.

5.2.7 Predicting DV Scores for Members of a New Sample

One of the more important applications of regression involves predicting scores on a DV for subjects for whom only data on IVs are available. This application is fairly frequent in personnel selection for employment or graduate training and the like. Over a fairly long period, a researcher collects data on a DV, say, success in graduate school, and on several IVs, say, undergraduate GPA, GRE verbal scores, and GRE math scores. If the IVs are strongly related to the DV, then for a new sample of applicants to graduate school, regression coefficients are applied to IV scores to predict success in graduate school ahead of time. Admission to the graduate school may, in fact, be based on prediction of success from regression.

The generalizability of a regression solution to a new sample may also be checked within a single large sample by a procedure called cross validation. A regression equation is developed from a portion of a sample and then applied to the other portion of the sample. If the solution generalizes, the regression equation predicts DV scores better than chance for the new cases, as well.

5.2.8 Causal Modeling

Path analysis, or causal analysis, is a special application of regression analysis in which questions about the minimum number of relationships (causal influences) and their directions are asked by observing the sizes of regression coefficients with and without certain variables entered into equations. For example, one can investigate the direct and indirect influence of intelligence on reading ability in the context of perceptual and motor development and gender. We have not included a description of path analysis in this book but refer the interested reader to Asher (1976) or Heise (1975).

An increasingly popular set of new techniques, called structural analyses (Bentler, 1980), involves path analysis of a set of IVs and DVs as latent variables (factors) all considered simultaneously. Analysis is facilitated by computer programs such as LISREL (Joreskog and Sorbom, 1978) and EQS (Bentler, 1985). Although the method is not included in this book, you should be aware that many researchers are finding this combination of canonical (Chapter 6) and factor analysis (Chapter 12) useful in attempting causal inference from nonexperimental or quasi-experimental research.

5.3 LIMITATIONS TO REGRESSION ANALYSES

5.3.1 Theoretical Issues

Regression analyses reveal relationships among variables but do not imply that the relationships are causal. Demonstration of causality is a logical and experimental, rather than statistical, problem. An apparently strong relationship between variables could stem from many sources, including the influence of other, currently unmeasured variables. One can make an airtight case for causal relationship among variables only by showing that manipulation of some of them is followed inexorably by change in others when all other variables are controlled.

Another problem for logic rather than statistics is inclusion of variables. Which DV should be used and how it is to be measured? Which IVs should be examined and how are they to be measured? If one already has some IVs in an equation, which IVs should be added to the equation for the most improvement in prediction? The answers to these questions can be provided by theory, astute observation, or good hunches, but they will not be provided by statistics.

There are, however, some general considerations for choosing IVs. Regression will be best when each IV is strongly correlated with the DV but uncorrelated with other IVs. A general goal of regression, then, might be to select the fewest IVs necessary to provide good prediction of a DV where each IV predicts a substantial and independent segment of the variability in the DV.

There are other considerations to selection of variables. If the goal of research is manipulation of some DV (say body weight), it is strategic to use as IVs those variables that can be manipulated (e.g., caloric intake, physical activity) rather than those that cannot (e.g., genetic predisposition). Or, if one is interested in predicting a variable such as annoyance caused by noise for a neighborhood, it is strategic to use cheaply obtained sets of IVs (e.g., neighborhood characteristics published by the Census Bureau) rather than expensively obtained ones (e.g., attitudes from in-depth interviews) if both sets of variables predict equally well.

It should be clearly understood that a regression solution is extremely sensitive to the combination of variables that is included in it. Whether or not an IV appears particularly important in a solution depends on the nature of other IVs in the set. If the IV of interest is the only one that assesses some important facet of the DV, the IV will appear important; if the IV of interest is only one of several that assess the same important facet of the DV, it will appear less important. If some important facet of the DV is assessed by none of the IVs, the solution is weakened. An optimal set of IVs is the smallest, uncorrelated set that "covers the waterfront" with respect to the DV.

5.3.2 Practical Issues

In addition to theoretical considerations, use of multiple regression requires that several practical matters be attended, as described in Sections 5.3.2.1 through 5.3.2.4.

5.3.2.1 Ratio of Cases to IVs The cases-to-IVs ratio has to be substantial or the solution will be perfect—and meaningless. With more IVs than cases, one can find a regression solution that completely predicts the DV for each case, but only as an artifact of the cases-to-IV ratio. If either standard multiple or hierarchical regression is used, one would like to have 20 times more cases than IVs. That is, if you plan to include 5 IVs, it would be lovely to measure 100 cases. In fact, because of the width of the errors of estimating correlation with small samples, power may be unacceptably low no matter what the cases-to-IVs ratio if you have fewer than 100 cases. However, *a bare minimum requirement is to have at least 5 times more cases than IVs*—at least 25 cases if 5 IVs are used.

Be sure to verify that you have as many cases as you think you have. By default, regression programs delete cases for which there are missing values on any of the variables. Consult Chapter 4 if you have missing values and wish to rescore rather than delete them.

A higher cases-to-IV ratio is needed when the DV is skewed, effect size is anticipated to be small, or substantial measurement error is expected from unreliable variables. If the DV is not normally distributed and transformations are not undertaken, more cases are required. The size of anticipated effect is also relevant because more cases are needed to demonstrate a small effect than a large one. Finally, if substantial measurement error is expected from somewhat unreliable variables, more cases are needed.

It is also possible to have too many cases, however. As the number of cases becomes quite large, almost any multiple correlation will depart significantly from zero, even one that predicts negligible variance in the DV. For both statistical and practical reasons, then, one wants to measure the smallest number of cases that has a decent chance of revealing a significant relationship if, indeed, one is there.

If stepwise regression is to be used, more cases are needed. A cases-to-IV ratio of 40 to 1 is reasonable because stepwise regression often produces a solution that does not generalize beyond the sample unless the sample is large. An even larger sample is needed in stepwise regression if cross validation (deriving the solution with some of the cases and testing it on the remaining cases) is to be used to test the generalizability of the solution.

If you cannot measure as many as five cases for each IV, there are some strategies that may help. You can delete some IVs or create one (or more than one) IV that is a composite of several others. The new, composite IV is used in the analysis in place of the original IVs. Lastly, you can employ a setwise regression strategy, as available in BMDP9R and SAS RSQUARE and described in Section 5.5.3. In this frankly exploratory strategy, regressions are computed for all possible subsets of IVs and the researcher decides which regression is best.

5.3.2.2 Outliers Extreme cases have too much impact on the regression solution and should be deleted or rescored to reduce their influence. Consult Chapter 4 for a summary of procedures for detecting and dealing with univariate and multivariate outliers.

In regression, cases are evaluated for univariate extremeness with respect to the DV and each IV. Univariate outliers show up in initial screening runs (e.g., with BMDP2D or SPSS^x FREQUENCIES) as cases far from the mean on either plots or z scores. Multivariate outliers are sought using either statistical methods such as Mahalanobis distance which looks at the combination of IVs (through BMDPAM, SPSS^x REGRESSION, or SYSTAT MGLH as described in Chapter 4) or graphical methods that look at the combination of IVs in the context of the DV.

Graphical methods for identifying multivariate outliers use residuals to identify cases for which there is a poor fit between the obtained and predicted DV scores. A residuals plot (described in Section 5.3.2.4) is requested in an initial regression run;

extreme cases produce very large residuals which are some distance from the rest of the cases. That is, multivariate outliers produce points well outside the swarm of points for the remaining cases. Offending cases are then identified by requesting case by case output of the measures plotted.

Another graphical method based on residuals uses leverage on the X axis and residuals on the Y axis; the combination of leverage and residuals is called influence. *Residuals* identify outliers in the solution; *leverage* identifies outliers among the IVs; *influence* identifies cases that are too influential. As in routine residuals plots, outlying cases fall outside the swarm of points produced by the remainder of the cases.

Several varieties of residuals, leverage, and influence are available for plotting in the four statistical packages. For example, *residuals* are available in raw or standardized form—with or without the outlying case deleted. *Leverage* measures are Mahalanobis distance or variations of the diagonal elements of a "hat" matrix, called HATDIAG, RHAT, and h_i. The diagonal elements of the hat matrix are similar to Mahalanobis distance, but on a different scale so that significance tests based on a χ^2 distribution do not apply. Most *influence* measures are variations on Cook's distance, including modified Cook's distance, DFFITS, and DBETAS. Cook's distance is a measure of the change in regression coefficients produced by leaving out a case; cases with scores larger than 1.00 are suspected of being outliers. Tables 5.11 through 5.13 show which programs in the four computer packages produce these measures; consult the program manuals for a more comprehensive list.

5.3.2.3 Multicollinearity and Singularity

Calculation of regression coefficients requires inversion of the matrix of correlations among the IVs (Equation 5.6), an inversion that is impossible if IVs are singular and unstable if they are multicollinear, as discussed in Chapter 4. Singularity and multicollinearity can be identified through perfect or very high squared multiple correlations (SMC) among IVs, where each IV in turn serves as DV while the others are IVs, or very low tolerances $(1 - \text{SMC})$. In regression, these conditions are also signaled by a very large (relative to the scale of the variable) standard error for a regression coefficient.

Most multiple regression programs have default values for tolerance that protect the user against inclusion of multicollinear variables. If the default values for the programs are in place, variables that are very highly correlated with variables already in the equation are not entered. This makes sense both statistically and logically because the variables threaten the analysis due to instability of regression coefficients and also are not needed because of their correlations with other variables.

However, you may want to make your own choice about which variable is deleted on logical rather than statistical grounds by considering such issues as the reliability of the variables. You may want to delete the least reliable variable rather than the variable identified by the program with very low tolerance. With a less reliable IV deleted from the set of IVs, the tolerance for the IV in question may be sufficient for entry.

If multicollinearity is detected but you want to maintain your set of IVs anyway, ridge regression might be considered. Ridge regression is a controversial procedure that attempts to stabilize estimates of regression coefficients by inflating the variance that is analyzed. For a more thorough description of ridge regression, see Chapter

7 of Dillon and Goldstein (1984). Although originally greeted with enthusiasm (cf. Price, 1977), serious questions about the procedure have been raised by Rozeboom (1979) and others. If, after consulting this literature, you still want to employ ridge regression, it is available through BMDP2R (see Appendix C–7 of the BMDP manual, Dixon, 1985).

5.3.2.4 Normality, Linearity, Homoscedasticity, and Independence of Residuals
Examination of residuals scatterplots provides a test of assumptions of normality, linearity, and homoscedasticity between predicted DV scores and errors of prediction. Assumptions of analysis are that the residuals (differences between obtained and predicted DV scores) are normally distributed about the predicted DV scores, that residuals have a straight line relationship with predicted DV scores, and that the variance of the residuals about predicted DV scores is the same for all predicted scores. When these assumptions are met, the residuals will appear as plot (a) in Figure 5.1.

Residuals scatterplots may be examined in lieu of initial screening runs or after them. If residuals scatterplots are examined in lieu of initial screening, and the assumptions of analysis are deemed met, further screening of variables is unnecessary. That is, if the residuals show normality, linearity, and homoscedasticity, if no outliers are in evidence, if the number of cases is sufficient, and if there is no evidence of multicollinearity or singularity, then regression requires only one run. (Parenthetically, we might note that we have never, in many years of multivariate analyses with many data sets, found this to be the case.) If, on the other hand, the residuals scatterplot from an initial run looks yucky, further screening via the procedures in Chapter 4 is warranted.

Residuals scatterplots are provided by all the statistical programs discussed in this chapter. All provide a scatterplot in which one axis is predicted scores and the other axis is errors of prediction. Which axis is which, however, and whether or not the predicted scores and residuals are standardized differ from program to program.

SPSS and BMDP provide the plots directly in their regression programs. In SPSS, both predicted scores and errors of prediction are standardized; in BMDP they are not. For SYSTAT and SAS, you have to save the predicted values and residuals, and then plot them using SYSTAT GRAPH or SAS PLOT, in either standardized or unstandardized form. In any event, it is the overall shape of the scatterplot that is of interest. If all assumptions are met, the residuals will be nearly rectangularly distributed with a concentration of scores along the center. As mentioned above, Figure 5.1(a) illustrates a distribution in which all assumptions are met.

The assumption of normality is that errors of prediction are normally distributed around each and every predicted DV score. The residuals scatterplot should reveal a pileup of residuals in the center of the plot at each value of predicted score, and a normal distribution of residuals trailing off symmetrically from the center. Figure 5.1(b) illustrates a failure of normality, with a skewed distribution of residuals.

A second test for the normality of the distribution of residuals is available through SPSS and BMDP programs for regression. The test is a normal probability plot of residuals in which their expected normal values are plotted against their actual normal values. Expected normal values are estimates of the z score a score should

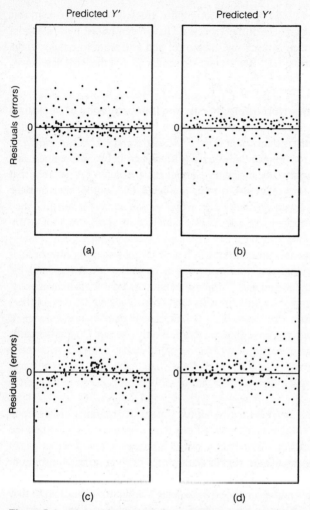

Figure 5.1 Plots of predicted values of the DV (Y') against residuals, showing (a) assumptions met, (b) failure of normality, (c) nonlinearity, and (d) heteroscedasticity.

have, given its rank in the original distribution if the original distribution is normal. If the expected normal values of residuals correspond to actual normal values (i.e., if the distribution of residuals is normal), the points will fall along a straight line running from the bottom left to the upper right corners of the graph. Distributions that are not normal will deviate from the straight line by curving above or below it in specific ways, depending on how the residuals are skewed. Bock (1975, pp. 156–160) illustrates the effects of several deviations from normality in the normal probability plot of residuals, as well as other details for the interested reader. The "fix" for the failure of normality of residuals is transformation of variables, after viewing distributions of individual variables.

Linearity of relationship between predicted DV scores and errors of prediction is also assumed. If nonlinearity is present, the overall shape of the scatterplot will be curved instead of rectangular, as seen in Figure 5.1(c). In this illustration, errors of prediction are generally in a negative direction for low and high predicted scores and in a positive direction for medium predicted scores. Typically, nonlinearity of residuals can be made linear by transforming IVs (or the DV) so that there is a linear relationship between each IV and the DV. If, however, there is a curvilinear relationship between an IV and the DV, it may be necessary to include the square of the IV in the set of IVs.

Failure of linearity of residuals in regression does not invalidate an analysis so much as weaken it. A curvilinear relationship between the DV and an IV is a perfectly good relationship that is not completely captured by a linear correlation coefficient. The power of the analysis is reduced to the extent that the analysis does not have available the full extent of the relationships among the IVs and the DV.

The assumption of homoscedasticity is the assumption that the standard deviations of errors of prediction are approximately equal for all predicted DV scores. Heteroscedasticity also does not invalidate the analysis so much as weaken it. Homoscedasticity means that the band enclosing the residuals is approximately equal in width at all values of the predicted DV. Typical heteroscedasticity is a case in which the band becomes wider at larger predicted values, as illustrated in Figure 5.1(d). In this illustration, the errors of prediction increase as the size of the prediction increases. Heteroscedasticity may occur when some of the variables are skewed and others are not. Transformation of the variables may reduce or eliminate heteroscedasticity.

Another assumption of regression is that errors of prediction are independent of one another. In some instances, this assumption is violated as a function of something associated with the order of cases. For example, subjects who are interviewed early in a survey might be expected to exhibit more variability of response because of interviewer inexperience with the questionnaire.

Nonindependence of errors associated with order of cases can be checked in SPSS[x] REGRESSION, BMDP9R, or SYSTAT MGLH by entering cases in order and requesting a plot of residuals against sequence of cases. The Durbin-Watson statistic is a measure of autocorrelation of errors over the sequence of cases, and, if significant, indicates nonindependence of errors. Details on the use of this statistic and a test for its significance are given by Wesolowsky (1976). If nonindependence is found, consult Dillon and Goldstein (1984) for the options available to you.

5.4 FUNDAMENTAL EQUATIONS FOR MULTIPLE REGRESSION

A data set appropriate for multiple regression consists of a sample of research units (e.g., graduate students) for whom scores are available on a number of IVs and on one DV. A small sample of hypothetical data with three IVs and one DV is illustrated in Table 5.1.

Table 5.1 contains six students, with scores on three IVs: a measure of professional motivation (MOTIV), a composite rating of qualifications for admissions

TABLE 5.1 SMALL SAMPLE OF HYPOTHETICAL DATA FOR ILLUSTRATION OF
MULTIPLE CORRELATION

Case No.	IVs			DV
	MOTIV (X_1)	QUAL (X_2)	GRADE (X_3)	COMPR (Y)
1	14	19	19	18
2	11	11	8	9
3	8	10	14	8
4	13	5	10	8
5	10	9	8	5
6	10	7	9	12
Mean	11.00	10.17	11.33	10.00
Standard deviation	2.191	4.834	4.367	4.517

to graduate training (QUAL), and a composite rating of performance in graduate courses (GRADE). The DV is a rating of performance on graduate comprehensive exams (COMPR). We ask how well we can predict COMPR from scores on MOTIV, QUAL, and GRADE.

In actually addressing that question, a sample of six cases is highly inadequate, but the sample is adequate to illustrate calculation of multiple correlation and to demonstrate some analyses by canned computer programs. The reader is encouraged to work problems involving these data by hand as well as by available computer programs. Setup and selected output for this example through SPSS^x REGRESSION, BMDP1R, SAS REG, and SYSTAT MGLH appear in Section 5.4.3.

A variety of ways is available to develop the "basic" equation for multiple correlation.

5.4.1 General Linear Equation

One way of developing multiple correlation is to obtain the prediction equation for Y' in order to compare the predicted value of the DV with obtained Y.

$$Y' = A + B_1X_1 + B_2X_2 + \cdots + B_kX_k \qquad (5.1)$$

where Y' is the predicted value of Y, A is the value of Y' when all X's are zero, B_1 to B_k represent regression coefficients, and X_1 to X_k represent the IVs.

The best-fitting regression coefficients produce a prediction equation for which squared differences between Y and Y' are at a minimum. Because squared errors of prediction—$(Y - Y')^2$—are minimized, this solution is called a least-squares solution.

In the sample problem, $k = 3$. That is, there are three IVs available to predict the DV, COMPR.

$$(COMPR)' = A + B_M(MOTIV) + B_Q(QUAL) + B_G(GRADE)$$

To predict a student's COMPR score, the available IV scores (MOTIV, QUAL, and GRADE) are multipled by their respective regression coefficients. The coefficient-by-score products are summed and added to the intercept, or baseline, value (A).

Differences among the observed values of the DV (Y), the mean of Y ($\overline{Y}$), and the predicted values of Y (Y') are summed and squared, yielding estimates of variation attributable to different sources. Total sum of squares for Y is partitioned into a sum of squares due to regression and a sum of squares left over or residual.

$$SS_y = SS_{reg} + SS_{res} \qquad (5.2)$$

Total sum of squares of Y

$$SS_y = \Sigma\,(Y - \overline{Y})^2$$

is, as usual, the sum of squared differences between each individual's observed Y score and the mean of Y over all N cases. The sum of squares for regression

$$SS_{reg} = \Sigma\,(Y' - \overline{Y})^2$$

is the portion of the variation in Y that is explained by use of the IVs as predictors. That is, it is the sum of squared differences between predicted Y' and the mean of Y because the mean of Y is the best prediction for the value of Y in the absence of any useful IVs. Sum of squares residual

$$SS_{res} = \Sigma\,(Y - Y')^2$$

is the sum of squared differences between observed Y and the predicted scores, Y', and represents errors in prediction.

The squared multiple correlation is

$$R^2 = \frac{SS_{reg}}{SS_y} \qquad (5.3)$$

The squared multiple correlation, R^2, is the proportion of sum of squares for regression in the total sum of squares for Y.

The squared multiple correlation is, then, the proportion of variation in the DV that is predictable from the best linear combination of the IVs. The multiple correlation itself is the correlation between the obtained and predicted Y values; that is, $R = r_{yy'}$.

Total sum of squares (SS_y) is calculated directly from the observed values of the DV. For example, in the sample problem, where the mean on the comprehensive examination is 10,

$$
\begin{aligned}
SS_C &= (18 - 10)^2 + (9 - 10)^2 + (8 - 10)^2 \\
&\quad + (8 - 10)^2 + (5 - 10)^2 + (12 - 10)^2 \\
&= 102
\end{aligned}
$$

To find the remaining sources of variation, it is necessary to solve the prediction equation (Equation 5.1) for Y', which means finding the best-fitting A and B_i. The most direct method of deriving the equation involves thinking of multiple correlation in terms of individual correlations.

5.4.2 Matrix Equations

Another way of looking at R^2 is in terms of the correlations between each of the IVs and the DV. The squared multiple correlation is the sum across all IVs of the product of the correlation between the DV and IV and the (standardized) regression coefficient for the IV.

$$R^2 = \sum_{i=1}^{k} r_{yi} \beta_i \qquad (5.4)$$

where each r_{yi} is the correlation between the DV and the ith IV, and β_i is the standardized regression coefficient, or beta weight. The standardized regression coefficient is the regression coefficient that would be applied to the standardized X_i value—the z score of the X_i value—to predict standardized Y'.

Because r_{yi} are calculated directly from the data, computation of R^2 involves finding the standardized regression coefficients (β_i) for the k IVs.

Derivation of the k equations in k unknowns is beyond the scope of this book. However, solution of these equations is easily illustrated using matrix algebra. For those who are not familiar with matrix algebra, the rudiments of it are available in Appendix A. Sections A.4 (matrix multiplication) and A.5 (matrix inversion or division) are the only portions of matrix algebra necessary to follow the next few steps.

In matrix form:

$$R^2 = \mathbf{R}_{yi}\mathbf{B}_i \qquad (5.5)$$

where $\mathbf{R}_{yi}$ is the row matrix of correlations between the DV and the k IVs, and $\mathbf{B}_i$ is a column matrix of standardized regression coefficients for the same k IVs.

The standardized regression coefficients can be found by inverting the matrix of correlations among IVs and multiplying that inverse by the matrix of correlations between the DV and the IVs.

$$\mathbf{B}_i = \mathbf{R}_{ii}^{-1}\mathbf{R}_{iy} \qquad (5.6)$$

$\mathbf{B}_i$ is the column matrix of standardized regression coefficients, $\mathbf{R}_{ii}^{-1}$ is the inverse of the matrix of correlations among the IVs, and $\mathbf{R}_{iy}$ is the column matrix of correlations between the DV and the IVs. Because multiplication by an inverse is the same as division, the column matrix of correlations between the IVs and the DV is divided by the correlation matrix of IVs.

These equations,[1] then, are used to calculate R^2 for the sample COMPR data from Table 5.1. All the required correlations are in Table 5.2.

---

[1] The equations can be solved in terms of $\mathbf{\Sigma}$ (variance-covariance) matrices or $\mathbf{S}$ (sum-of-squares and cross-product) matrices as well as correlation matrices. If $\mathbf{\Sigma}$ or $\mathbf{S}$ matrices are substituted for corresponding $\mathbf{R}$ matrices, the regression coefficients are nonstandardized coefficients, as in Equation 5.1.

TABLE 5.2   CORRELATIONS AMONG IVs AND THE DV FOR
SAMPLE DATA IN TABLE 5.1

		R_{ii}			R_{iy}
		MOTIV	QUAL	GRADE	COMPR
	MOTIV	1.00000	.39658	.37631	.58613
	QUAL	.39658	1.00000	.78329	.73284
	GRADE	.37631	.78329	1.00000	.75043
R_{yi}	COMPR	.58613	.73284	.75043	1.00000

TABLE 5.3   INVERSE OF MATRIX OF INTERCOR-
RELATIONS AMONG IVs FOR SAM-
PLE DATA IN TABLE 5.1

	MOTIV	QUAL	GRADE
MOTIV	1.20255	− 0.31684	− 0.20435
QUAL	− 0.31684	2.67113	− 1.97305
GRADE	− 0.20435	− 1.97305	2.62238

Procedures for inverting a matrix are amply demonstrated elsewhere (e.g., Cooley and Lohnes, 1971; Harris, 1975) and are typically available in computer installations. Because the procedure is extremely tedious by hand, and becomes increasingly so as the matrix becomes larger, the inverted matrix for the sample data is presented without calculation in Table 5.3.

Using Equation 5.6, the B_i matrix is found by postmultiplying the R_{ii}^{-1} matrix by the R_{iy} matrix.

$$\mathbf{B}_i = \begin{bmatrix} 1.20255 & -0.31684 & -0.20435 \\ -0.31684 & 2.67113 & -1.97305 \\ -0.20435 & -1.97305 & 2.62238 \end{bmatrix} \begin{bmatrix} .58613 \\ .73284 \\ .75043 \end{bmatrix} = \begin{bmatrix} 0.31931 \\ 0.29117 \\ 0.40221 \end{bmatrix}$$

so that $\beta_M = 0.319$, $\beta_Q = 0.291$, and $\beta_G = 0.402$. Then, using Equation 5.5, we obtain

$$R^2 = [.58613 \quad .73284 \quad .75043] \begin{bmatrix} 0.31931 \\ 0.29117 \\ 0.40221 \end{bmatrix} = .70237$$

In this example, 70% of the variance in graduate comprehensive exam scores is predictable from knowledge of motivation, admission qualifications, and graduate course performance.

Once the standardized regression coefficients are available, they are used to write the equation for the predicted values of COMPR (Y'). If z scores are used throughout, the beta weights (β_i) are used to set up the prediction equation. The equation is similar to Equation 5.1 except that there is no A (intercept).

If the equation is needed in raw score form, the coefficients must first be transformed to unstandardized B_i coefficients.

$$B_i = \beta_i \left(\frac{S_y}{S_i} \right) \tag{5.7}$$

Unstandardized coefficients (B_i) are found by multiplying standardized coefficients (beta weights $-\beta_i$) by the ratio of standard deviations of the DV and IV, where S_i is the standard deviation of the ith IV and S_y is the standard deviation of the DV.
and

$$A = \overline{Y} - \sum_{i=1}^{k} (B_i \overline{X}_i) \tag{5.8}$$

The intercept is the mean of the DV less the sum of the means of the IVs multiplied by their respective unstandardized coefficients.

For the sample problem of Table 5.1:

$$B_M = 0.319 \left(\frac{4.517}{2.191} \right) = 0.658$$

$$B_Q = 0.291 \left(\frac{4.517}{4.834} \right) = 0.272$$

$$B_G = 0.402 \left(\frac{4.517}{4.366} \right) = 0.416$$

$$A = 10 - [(0.658)(11.00) + (0.272)(10.17) + (0.416)(11.33)] = -4.72$$

The prediction equation for raw COMPR scores, once scores on MOTIV, QUAL, and GRADE are known, is

$$(COMPR)' = -4.72 + 0.658 \, (MOTIV) + 0.272 \, (QUAL) + 0.416 \, (GRADE)$$

If a graduate student has ratings of 12, 14, and 15, respectively, on MOTIV, QUAL, and GRADE, the predicted rating on COMPR is

$$(COMPR)' = -4.72 + 0.658(12) + 0.272(14) + 0.416(15) = 13.22$$

5.4.3 Computer Analyses of Small Sample Example

Tables 5.4 through 5.7 demonstrate setup and selected output for computer analyses of the data in Table 5.1, using default values. Table 5.4 illustrates SPSS^x REGRESSION. The simplest BMDP program, BMDP1R, is illustrated in Table 5.5. Tables 5.6 and 5.7 show runs through SAS REG and SYSTAT MGLH, respectively.

In SPSS^x REGRESSION all of the variables, IVs and DV, are listed in the

TABLE 5.4   SETUP AND SELECTED SPSS' REGRESSION OUTPUT
FOR STANDARD MULTIPLE REGRESSION ON SAMPLE
DATA IN TABLE 5.1

```
            TITLE          SMALL SAMPLE MULTIPLE REGRESSION
            FILE HANDLE    TAPE50
            DATA LIST      FILE=TAPE50    FREE
                           /SUBJNO,MOTIV,QUAL,GRADE,COMPR
            REGRESSION     VARS=MOTIV TO COMPR/
                           DEP=COMPR/ENTER/



            * * * *   M U L T I P L E   R E G R E S S I O N   * * * *

VARIABLE LIST NUMBER 1    LISTWISE DELETION OF MISSING DATA

EQUATION NUMBER 1    DEPENDENT VARIABLE..  COMPR

BEGINNING BLOCK NUMBER  1.  METHOD:  ENTER


VARIABLE(S) ENTERED ON STEP NUMBER   1..    GRADE
                                     2..    MOTIV
                                     3..    QUAL


MULTIPLE R         .83807    ANALYSIS OF VARIANCE
R SQUARE           .70235                      DF    SUM OF SQUARES    MEAN SQUARE
ADJUSTED R SQUARE  .25588    REGRESSION         3          71.64007       23.88002
STANDARD ERROR    3.89615    RESIDUAL           2          30.35993       15.17997

                             F =      1.57313      SIGNIF F =   .4114



----------------- VARIABLES IN THE EQUATION -----------------

VARIABLE            B        S B      BETA         T   SIG T

GRADE            .41603    .64619    .40221      .644  .5857
MOTIV            .65827    .87213    .31931      .755  .5292
QUAL             .27205    .58911    .29116      .462  .6896
(CONSTANT)     -4.72180   9.06565               -.521  .6544


FOR BLOCK NUMBER  1    ALL REQUESTED VARIABLES ENTERED.
```

VARS $=$ sentence of the REGRESSION instruction. Then the DV is specified. ENTER is the instruction that specifies standard multiple regression. Because this is standard multiple regression, results are given for only one step. At the left are R, R^2, adjusted R^2 (see Section 5.6.3) and STANDARD ERROR, the standard error of the predicted score, Y'. To the right is the ANOVA table, showing details of the F test of the hypothesis that multiple regression is zero (see Section 5.6.2.1). Below are the regression coefficients and their significance tests, including B weights, the standard error of B (S B), β weights (BETA), t-tests for the coefficients, and their significance levels (SIG T). The term (CONSTANT) refers to the intercept (A).

BMDP1R is limited to standard multiple regression so no special instruction is necessary to specify that procedure. The REGRESS instruction identifies the DV and the IVs. BMDP1R provides the same information as SPSS[x] but in slightly different format and with some different labels. For example, β is referred to as STD. REG COEFF and B weights are in the column labeled COEFFICIENT. Other differences are seen by comparing values in Tables 5.4 and 5.5. An additional statistic is TOLERANCE (cf. Chapter 4), but adjusted R^2 is not reported. BMDP1R also provides

TABLE 5.5 SETUP AND SELECTED BMDP1R OUTPUT FOR
STANDARD MULTIPLE REGRESSION ON SAMPLE DATA IN
TABLE 5.1

```
          /PROBLEM      TITLE IS 'SMALL SAMPLE MULTIPLE REGRESSION'.
          /INPUT        VARIABLES=5.  FORMAT IS FREE.
                        FILE = 'TAPE50'.
          /VARIABLE     NAMES ARE SUBJNO,MOTIV,QUAL,GRADE,COMPR.
          /REGRESS      DEPENDENT=COMPR.
                        INDEPENDENT=MOTIV TO GRADE.
          /END
```

VARIABLE	MEAN	STANDARD DEVIATION	COEFFICIENT OF VARIATION	MINIMUM	MAXIMUM
1 SUBJNO	3.50000	1.87083	.53452	1.00000	6.00000
2 MOTIV	11.00000	2.19089	.19917	8.00000	14.00000
3 QUAL	10.16667	4.83391	.47547	5.00000	19.00000
4 GRADE	11.33333	4.36654	.38528	8.00000	19.00000
5 COMPR	10.00000	4.51664	.45166	5.00000	18.00000

```
MULTIPLE R           .8381         STD. ERROR OF EST.         3.8961
MULTIPLE R-SQUARE    .7024
```

ANALYSIS OF VARIANCE

	SUM OF SQUARES	DF	MEAN SQUARE	F RATIO	P(TAIL)
REGRESSION	71.6401	3	23.8800	1.573	.4114
RESIDUAL	30.3599	2	15.1800		

VARIABLE		COEFFICIENT	STD. ERROR	STD. REG COEFF	T	P(2 TAIL)	TOLERANCE
INTERCEPT		-4.72180					
MOTIV	2	.65827	.87213	.319	.755	.5292	.83157
QUAL	3	.27205	.58911	.291	.462	.6896	.37437
GRADE	4	.41603	.64619	.402	.644	.5857	.38133

TABLE 5.6 SETUP AND SAS REG OUTPUT FOR STANDARD
MULTIPLE REGRESSION ON SAMPLE DATA OF
TABLE 5.1

```
          DATA SSAMPLE;
          INFILE TAPE50;
          INPUT SUBJNO MOTIV QUAL GRADE COMPR;
          PROC REG;
          MODEL COMPR = MOTIV QUAL GRADE;
```

DEP VARIABLE: COMPR

ANALYSIS OF VARIANCE

SOURCE	DF	SUM OF SQUARES	MEAN SQUARE	F VALUE	PROB>F
MODEL	3	71.64007	23.88002	1.573	0.4114
ERROR	2	30.35993	15.17997		
C TOTAL	5	102			

ROOT MSE	3.896148	R-SQUARE	0.7024	
DEP MEAN	10	ADJ R-SQ	0.2559	
C.V.	38.96148			

PARAMETER ESTIMATES

VARIABLE	DF	PARAMETER ESTIMATE	STANDARD ERROR	T FOR H0: PARAMETER=0	PROB > |T|
INTERCEP	1	-4.7218	9.065646	-0.521	0.6544
MOTIV	1	0.6582658	0.8721309	0.755	0.5292
QUAL	1	0.2720479	0.5891141	0.462	0.6896
GRADE	1	0.4160342	0.6461905	0.644	0.5857

univariate statistics for each of the variables (see Chapter 4 for definition of the coefficient of variation).

SAS REG is also a program designed specifically for standard multiple regression. The variables for the regression equation are specified in the MODEL

TABLE 5.7 SETUP AND SYSTAT MGLH OUTPUT FOR STANDARD MULTIPLE REGRESSION ON SAMPLE DATA IN TABLE 5.1

```
USE TAPE50
MODEL COMPR=CONSTANT+MOTIV+QUAL+GRADE
ESTIMATE
```

```
DEP VAR:   COMPR     N:    6    MULTIPLE R:  .838    SQUARED MULTIPLE R:  .702
ADJUSTED SQUARED MULTIPLE R:  .256    STANDARD ERROR OF ESTIMATE:      3.896

  VARIABLE.   COEFFICIENT    STD ERROR     STD COEF TOLERANCE     T     P(2 TAIL)

  CONSTANT       -4.722        9.066       0.000 1.0000000    -0.521    0.654
    MOTIV         0.658        0.872       0.319  .8315651     0.755    0.529
    QUAL          0.272        0.589       0.291  .3743735     0.462    0.690
    GRADE         0.416        0.646       0.402  .3813334     0.644    0.586

                          ANALYSIS OF VARIANCE

   SOURCE    SUM-OF-SQUARES   DF   MEAN-SQUARE    F-RATIO       P

  REGRESSION      71.640       3     23.880        1.573      0.411
   RESIDUAL       30.360       2     15.180
```

statement, with the DV on the left side of the equation and the IVs on the right. In the ANOVA table the sum of squares for regression is called MODEL and residual is called ERROR. Total SS and df are in the row labeled C TOTAL. Below the ANOVA is the standard error of the estimate, shown as the square root of the error term, MS_{res} (ROOT MSE). Also printed are the mean of the DV (DEP MEAN), R^2, adjusted R^2, and the coefficient of variation (C.V.—defined as 100 times the standard error of the estimate divided by the mean of the DV). The section labeled PARAMETER ESTIMATES includes the usual B coefficients in the PARAMETER ESTIMATE column, their standard errors, t-tests for those coefficients (T FOR HO: PARAMETER = 0) and significance levels (PROB > |T|). Standardized regression coefficients are not printed unless requested.

Setup for SYSTAT MGLH is similar to SAS REG, except that the IV side of the equation includes CONSTANT (specifying the intercept) and terms are connected by + signs. The listing of information about R is fully labeled, followed by information about the regression coefficients—B (COEFFICIENT), standard error of B (STD ERROR), β (STD COEF), TOLERANCE, t-test and significance level of t. The ANOVA table is printed last.

Additional features of these programs are discussed in Section 5.7.

5.5 MAJOR TYPES OF MULTIPLE REGRESSION

There are three major analytic strategies in multiple regression: standard multiple regression, hierarchical regression, and statistical (stepwise and setwise) regression. Differences among the strategies involve what happens to overlapping variability due to correlated IVs and who determines the order of entry of IVs into the equation.

Consider the Venn diagram in Figure 5.2(a) in which there are three IVs and one DV. IV_1 and IV_2 both correlate substantially with the DV and with each other. IV_3 correlates to a lesser extent with the DV and to a negligible extent with IV_2. R^2

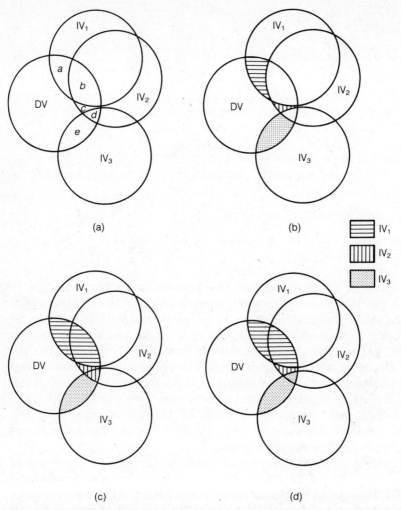

Figure 5.2   Venn diagrams illustrating (a) overlapping variance sections; and allocation of overlapping variance in (b) standard multiple regression, (c) hierarchical regression, and (d) stepwise regression.

for this situation is the area $a + b + c + d + e$. Area a comes unequivocally from IV_1; area c unequivocally from IV_2; area e from IV_3. However, there is ambiguity regarding areas b and d. Both areas could be predicted from either of two IVs; area b from either IV_1 or IV_2, area d from either IV_2 or IV_3. To which IV should the contested area be assigned? The interpretation of analysis can be drastically affected by choice of strategy because the apparent importance of the various IVs to the solution changes.

5.5.1 Standard Multiple Regression

The standard model is the one used in the solution for the small sample graduate student data in Table 5.1. In the standard, or simultaneous, model all IVs enter into the regression equation at once; each one is assessed as if it had entered the regression after all other IVs had entered. Each IV is evaluated in terms of what it adds to prediction of the DV that is different from the predictability afforded by all the other IVs.

Consider Figure 5.2(b). The darkened areas of the figure indicate the variability accorded each of the IVs when the procedure of Section 5.6.1.1 is used. IV_1 "gets credit" for area a, IV_2 for area c, and IV_3 for area e. That is, each IV is assigned only the area it contributes uniquely. The overlapping areas, b and d, contribute to R^2, but are not assigned to any of the individual IVs.

In standard multiple regression, it is possible for a variable like IV_2 to appear unimportant in the solution when it actually is highly correlated with the DV. If the area of that correlation is whittled away by other IVs, the unique contribution of the IV is often very small despite a substantial correlation with the DV. For this reason, both the full correlation and the unique contribution of the IV need to be considered in interpretation.

Standard multiple regression is handled in the SPSS[x] package by the RE-GRESSION program (NEW REGRESSION in the update of the original SPSS program), as are all other types of multiple regression. A selected part of output is given in Table 5.4 for the sample problem of Table 5.1. (Full interpretation of program output is available in substantive examples presented later in this chapter.)

Within the BMD series, standard multiple regression is routinely handled by BMDP1R. A selected sample of output from that program is given in Table 5.5. Standard multiple regression is also available through the more elaborate stepwise or setwise programs—BMDP2R or BMDP9R. SAS REG and SYSTAT MGLH are used for standard analyses, as illustrated in Tables 5.6 and 5.7.

5.5.2 Hierarchical Multiple Regression

In hierarchical regression, IVs enter the equation in an order specified by the researcher. Each IV is assessed in terms of what it adds to the equation at its own point of entry. Consider the example in Figure 5.2(c). Assume that the researcher assigns IV_1 first entry, IV_2 second entry, and IV_3 third entry. In assessing importance of variables by the procedure of Section 5.6.1.2, IV_1 "gets credit" for areas a and b, IV_2 for areas c and d, and IV_3 for area e. Each IV is assigned the variability, unique and overlapping, left to it at its own point of entry. Notice that the apparent importance of IV_2 would increase dramatically if it were assigned first entry and, therefore, "got credit" for b, c, and d.

The researcher normally assigns order of entry of variables according to logical or theoretical considerations. For example, IVs that are presumed (or manipulated) to be causally prior are given higher priority of entry. For instance, height might be considered prior to amount of training in assessing success as a basketball player and accorded a higher priority of entry. Variables with greater theoretical importance could also be given early entry.

Or the opposite tack could be taken. The research could enter manipulated or other variables of major importance on later steps, with "nuisance" variables given higher priority for entry. The lesser, or nuisance, set is entered first; then the major set is evaluated for what it adds to the prediction over and above the lesser set. For example, we might want to see how well we can predict reading speed (the DV) from intensity and length of a speed reading course (the major IVs) while holding constant initial differences in reading speed (the nuisance IV). This is the basic analysis of covariance problem in regression format.

IVs can be entered one at a time or in blocks. The analysis proceeds in steps, with information about variables both in and out of the equation given in computer output at each step. Finally, after all variables are entered, summary statistics are provided along with the information available at the last step.

In the SPSS^x package, hierarchical regression is performed by the REGRESSION program (NEW REGRESSION in earlier SPSS versions). In Table 5.8, selected output is shown for the sample problem of Table 5.1, with higher priority given to admission qualifications and course performance, and lower priority given to motivation.

In the BMDP package, hierarchical regression is run with BMDP2R. Output from BMDP2R is illustrated in the next section. Neither SAS or SYSTAT provide programs for developing hierarchical models in a single run. IVs can be forced into the regression, but their order cannot be specified. However, each step in the hierarchy can be evaluated by a separate run using either SAS REG or SYSTAT MGLH.

5.5.3 Statistical (Stepwise) and Setwise Regression

Statistical regression (sometimes generically called stepwise regression) is a rather controversial procedure, in which order of entry of variables is based solely on statistical criteria. The meaning or interpretation of the variables is not relevant. Decisions about which variables are included and which omitted from the equation are based solely on statistics computed from the particular sample drawn; minor differences in these statistics can have profound effect on the apparent importance of an IV.

Consider the example in Figure 5.2(d). IV_1 and IV_2 both correlate substantially with the DV; IV_3 correlates less strongly. The choice between IV_1 and IV_2 for first entry is based on which of the two IVs has the higher full correlation with the DV, even if the higher correlation shows up in the second or third decimal place. Let's say that IV_1 has the higher correlation with the DV and enters first. It "gets credit" for areas a and b. At the second step, IV_2 and IV_3 are compared, where IV_2 has available to add to prediction areas c and d, and IV_3 has areas e and d. At this point IV_3 contributes more strongly to R^2 and enters the equation. IV_2 is now assessed for whether or not its remaining area, c, contributes significantly to R^2. If it does, IV_2 enters the equation; otherwise it does not despite the fact that it is almost as highly correlated with the DV as the variable that entered first. For this reason, interpretation of a statistical regression equation is hazardous unless the researcher takes special care to remember the message of the initial DV–IV correlations.

There are actually three versions of statistical regression, forward selection, backward deletion, and stepwise regression. In forward selection, the equation starts out empty and IVs are added one at a time provided they meet the statistical criteria for entry. Once in the equation, an IV stays in. In backward deletion, the equation starts out with all IVs entered and they are deleted one at a time if they do not contribute significantly to regression. Stepwise regression is a compromise between

TABLE 5.8 SETUP AND SELECTED SPSS' REGRESSION OUTPUT
FOR HIERARCHICAL MULTIPLE REGRESSION ON
SAMPLE DATA IN TABLE 5.1

```
TITLE          SMALL SAMPLE HIERARCHICAL MULTIPLE REGRESSION
FILE HANDLE    TAPE50
DATA LIST      FILE=TAPE50    FREE
               /SUBJNO MOTIV QUAL GRADE COMPR
REGRESSION     VARIABLES=MOTIV TO COMPR/
               STATISTICS=DEFAULTS,HISTORY/
               DEPENDENT=COMPR/
               ENTER QUAL GRADE/ENTER MOTIV/
```

** * * *   M U L T I P L E   R E G R E S S I O N   * * * ***

VARIABLE LIST NUMBER 1   LISTWISE DELETION OF MISSING DATA

EQUATION NUMBER 1   DEPENDENT VARIABLE.. COMPR

BEGINNING BLOCK NUMBER 1. METHOD: ENTER   QUAL   GRADE

VARIABLE(S) ENTERED ON STEP NUMBER   1..   GRADE
                                      2..   QUAL

MULTIPLE R .78586	ANALYSIS OF VARIANCE	

	DF	SUM OF SQUARES	MEAN SQUARE
R SQUARE .61757			
ADJUSTED R SQUARE .36262	REGRESSION	2	62.99218	31.49609
STANDARD ERROR 3.60591	RESIDUAL	3	39.00782	13.00261

F =   2.42229   SIGNIF F =   .2365

------------ VARIABLES IN THE EQUATION ------------   ------------ VARIABLES NOT IN THE EQUATION ------------

VARIABLE	B	SE B	BETA	T	SIG T		VARIABLE	BETA IN	PARTIAL	MIN TOLER	T	SIG T
GRADE	.59408	.47216	.45647	.37529	.795		MOTIV	.4848	.31931	.47085	.37437	.755
.5292												
QUAL	.35065	.53664		.653	.5601							
(CONSTANT)	4.40060	1.08387		.244	.8229							

FOR BLOCK NUMBER 1   ALL REQUESTED VARIABLES ENTERED.

145

TABLE 5.8 (Continued)

```
                        * * * *   M U L T I P L E   R E G R E S S I O N   * * * *

EQUATION NUMBER 1     DEPENDENT VARIABLE..  COMPR

BEGINNING BLOCK NUMBER  2.  METHOD:  ENTER    MOTIV

VARIABLE(S) ENTERED ON STEP NUMBER   3..    MOTIV

MULTIPLE R          .83807         ANALYSIS OF VARIANCE
R SQUARE            .70235                          DF     SUM OF SQUARES     MEAN SQUARE
ADJUSTED R SQUARE   .25588         REGRESSION       3        71.64007         23.88002
STANDARD ERROR     3.89615         RESIDUAL         2        30.35993         15.17997

                                        F =    1.57313      SIGNIF F =  .4114


------------ VARIABLES IN THE EQUATION ------------------

VARIABLE        B          SE B      BETA        T     SIG T
GRADE         .41603      .64619    .40221      .644   .5857
QUAL          .27205      .58911    .29116      .462   .6896
MOTIV         .65827      .87213    .31931      .755   .5292
(CONSTANT)  -4.72180     9.06535              -.521   .6544


FOR BLOCK NUMBER  2   ALL REQUESTED VARIABLES ENTERED.

            * * * * * * * * * * * * * * * * * * * * * * * * * * * * * * * *

                                     SUMMARY TABLE
                                     -------------
```

STEP	MULTR	RSQ	ADJRSQ	F(EQU)	SIGF	RSQCH	FCH	SIGCH		VARIABLE	BETAIN	CORREL	LABEL
1	.7859	.6176	.3626	2.422	.236	.6176	2.422	.236	IN:	GRADE	.7504	.7504	
2	.8381	.7024	.2559	1.573	.411	.0848	.570	.529	IN:	QUAL	.3753	.7328	
3									IN:	MOTIV	.3193	.5861	

the two other procedures in which the equation starts out empty and IVs are added one at a time if they meet statistical criteria, but they may also be deleted at any step where they no longer contribute significantly to regression. Of the three procedures, stepwise regression is considered the surest path to the best prediction equation.

Statistical regression is typically used to develop a subset of IVs that is useful in predicting the DV, and to eliminate those IVs that do not provide additional prediction to the IVs already in the equation. For this reason, statistical regression may have some utility if the only aim of the researcher is a prediction equation. Even so, the sample from which the equation is drawn should be large and representative because of the following considerations.

The procedure's controversy lies primarily in capitalization on chance and overfitting of data. Decisions about which variables are included and which omitted from the equation are dependent on potentially minor differences in statistics computed from a single sample, where some variability in the statistics from sample to sample is expected. In this way, the technique capitalizes on chance differences within a single sample. For similar reasons, the equation derived from a sample is too close to the sample and may not generalize well to the population. In this way, statistical regression may overfit the data.

For statistical regression, cross validation with a second sample is highly recommended. At the very least, separate analyses of two halves of an available sample should be conducted, with conclusions limited to results that are consistent for both analyses. (This caution applies to setwise regression as well.)

Further, a statistical analysis may not lead to the optimum solution in terms of R^2. Several IVs considered together may increase R^2, while any one of them considered alone does not. In simple statistical regression, none of the IVs enters.

By specifying that IVs enter in blocks, one can set up combinations of hierarchical and statistical regression. A block of high-priority IVs is set up to compete among themselves stepwise for order of entry; then a second block of IVs compete among themselves for order of entry. The regression is hierarchical over blocks, but statistical within blocks.

An alternative to statistical regression in choosing an optimal subset of variables is setwise regression. In setwise regression, separate regressions are computed for all IVs singly, all possible pairs of IVs, all possible trios of IVs and so forth until the best subset of IVs is identified according to some criterion (such as maximum R^2) from among all possible subsets. One computer program in the BMD series, BMDP9R, is designed to do this, as is one program in the SAS system, RSQUARE. This alternative enables the researcher to find IVs that significantly improve R^2 only when considered as a group.

The SPSS[x] REGRESSION program handles stepwise regression in a manner similar to that of hierarchical regression. Both SPSS[x] REGRESSION and BMDP2R provide a variety of statistical criteria and stepping options for statistical regression. For the data in Table 5.1, output from a BMDP2R run with default values is shown in Table 5.9.

SAS STEPWISE is the procedure provided by the SAS system for statistical regression. SYSTAT MGLH allows statistical regression as well, but minimal output is provided. For full information at each step, it is necessary to do separate runs.

TABLE 5.9 SETUP AND SELECTED BMDP2R OUTPUT FOR STEPWISE REGRESSION ON SAMPLE DATA N TABLE 5.1

```
/PROBLEM      TITLE IS 'SMALL SAMPLE STEPWISE REGRESSION'.
/INPUT        VARIABLES ARE 5.  FORMAT IS FREE.
              FILE IS 'TAPE50'.
/VARIABLE     NAMES ARE SUBJNO,MOTIV,QUAL,GRADE,COMPR.
              USE = MOTIV TO COMPR.
/REGRESS      DEPENDENT=COMPR.
              INDEPENDENT=MOTIV TO GRADE.
/END
```

STEP NO. 0

STD. ERROR OF EST.   4.5166

ANALYSIS OF VARIANCE

	SUM OF SQUARES	DF	MEAN SQUARE
RESIDUAL	102.00000	5	20.40000

VARIABLES IN EQUATION FOR COMPR

VARIABLE	COEFFICIENT	STD. ERROR OF COEFF	STD REG COEFF	TOLERANCE
(Y-INTERCEPT	10.00000)			

VARIABLES NOT IN EQUATION

	F		PARTIAL		F	
TO REMOVE LEVEL.	VARIABLE	CORR.	TOLERANCE	TO ENTER LEVEL
.	MOTIV	2	0.58613	1.00000	2.09	1

STEP NO. 1

VARIABLE ENTERED   4 GRADE

MULTIPLE R   0.7504
MULTIPLE R-SQUARE   0.5631
ADJUSTED R-SQUARE   0.4539

STD. ERROR OF EST.   3.3376

ANALYSIS OF VARIANCE

	SUM OF SQUARES	DF	MEAN SQUARE	F RATIO
REGRESSION	57.440560	1	57.44056	5.16
RESIDUAL	44.559420	4	11.13986	

VARIABLES IN EQUATION FOR COMPR   VARIABLES NOT IN EQUATION

148

STEP NO.   1

VARIABLE ENTERED   4 GRADE

MULTIPLE R   0.7504
MULTIPLE R-SQUARE   0.5631
ADJUSTED R-SQUARE   0.4539

STD. ERROR OF EST.   3.3376

ANALYSIS OF VARIANCE

	SUM OF SQUARES	DF	MEAN SQUARE	F RATIO
REGRESSION	57.440560	1	57.44056	5.16
RESIDUAL	44.559420	4	11.13986	

VARIABLES IN EQUATION FOR COMPR

VARIABLE	COEFFICIENT	STD. ERROR OF COEFF	STD REG COEFF	TOLERANCE	F TO REMOVE LEVEL.
(Y-INTERCPT	1.20280)				
GRADE 4	0.77622	0.3418	0.750	1.00000	5.16 1 .

***** F LEVEL(4.000, 3.900) OR TOLERANCE INSUFFICIENT FOR FURTHER STEPPING

VARIABLES NOT IN EQUATION

VARIABLE	PARTIAL CORR.	TOLERANCE	F TO ENTER LEVEL
MOTIV	2 0.49600	0.85839	0.98 1

STEPWISE REGRESSION COEFFICIENTS

STEP	VARIABLES 0 Y-INTCPT	2 MOTIV	3 QUAL	4 GRADE
0	10.0000**	1.2083	0.6847	0.7762
1	1.2028*	0.7295	0.3507	0.7762*

NOTE - 1) REGRESSION COEFFICIENTS FOR VARIABLES IN THE
          EQUATION ARE INDICATED BY AN ASTERISK
       2) THE REMAINING COEFFICIENTS ARE THOSE WHICH WOULD
          BE OBTAINED IF THAT VARIABLE WERE TO ENTER IN THE
          NEXT STEP

SUMMARY TABLE

STEP NO.	VARIABLE ENTERED	VARIABLE REMOVED	MULTIPLE R	RSQ	CHANGE IN RSQ	F TO ENTER	F TO REMOVE	NO.OF VAR. INCLUDED
1	4 GRADE		0.7504	0.5631	0.5631	5.16		1

149

5.5.4 Choosing among Regression Strategies

To simply assess relationships among variables, and answer the basic question of multiple correlation, the method of choice is standard multiple regression. As a matter of fact, unless there is good reason to use some other technique, standard multiple regression is recommended. Reasons for using hierarchical regression are theoretical or for testing hypotheses.

Hierarchical regression allows the researcher to control the advancement of the regression process. Importance of IVs in the prediction equation is determined by the researcher according to logic or theory. Explicit hypotheses are tested about proportion of variance attributable to some IVs after variance due to IVs already in the equation is accounted for.

Although there are similarities in programs used and output produced for hierarchical and statistical regression, there are fundamental differences in the way that IVs enter the prediction equation and in the interpretations that can be made from the results. In hierarchical regression the researcher controls entry of variables, while in statistical and setwise regression statistics computed from sample data control order of entry. Statistical and setwise regression are, therefore, model-building rather than model-testing procedures. As exploratory techniques, they may be useful for such purposes as eliminating variables that are clearly superfluous in order to tighten up future research. However, clearly superfluous IVs will show up in any of the procedures.

When multicollinearity or singularity are present, setwise regression can be indispensable in identifying multicollinear variables, as indicated in Chapter 4.

For the example of Section 5.4, in which performance on graduate comprehensive exam (COMPR) is predicted from professional motivation (MOTIV), qualifications for graduate training (QUAL), and performance in graduate courses (GRADE), the differences among regression strategies might be phrased as follows. If standard multiple regression is used, two fundamental questions are asked: (1) What is the size of the overall relationship between COMPR and MOTIV, QUAL, and GRADE? and (2) How much of the relationship is contributed uniquely by each IV? If hierarchical regression is used, with QUAL and GRADE entered before MOTIV, the question is, Does MOTIV significantly add to prediction of COMPR after differences among students in QUAL and GRADE have been statistically eliminated? If statistical regression is used, one asks, What is the best linear combination of IVs to predict the DV in this sample? And use of setwise regression leads to the query, What is the size of R^2 from each one of the IVs, from all possible combinations of two IVs, and from all three IVs for this sample?

5.6 SOME IMPORTANT ISSUES

5.6.1 Importance of IVs

If the IVs are uncorrelated with each other, assessment of the contribution of each of them to multiple regression is straightforward. IVs with bigger correlations or higher standardized regression coefficients are more important to the solution than those with lower (absolute) values. (Because unstandardized regression coefficients are in a metric that depends on the metric of the original variables, their sizes are harder to interpret.)

If the IVs are correlated with each other, assessment of the importance of each of them to regression is more ambiguous. When IVs are correlated with each other, correlations and regression coefficients are somewhat redundant and misleading. The correlation between an IV and the DV reflects variance shared with the DV, but some of that variance may be predictable from other IVs. Regression coefficients can be misleading when an IV has a large regression coefficient not because it directly predicts the DV, but because it predicts the DV well after another IV suppresses irrelevant variance (see Section 5.6.4).

To get the most straightforward answer regarding the importance of an IV to regression, one needs to consider the type of regression it is, and both the full and unique relationship between the IV and the DV. This section reviews several of the issues to be considered when assessing the importance of an IV to standard multiple, hierarchical, or statistical regression. In all cases, one needs to compare the total relationship of the IV with the DV, the unique relationship of the IV with the DV, and the correlations of the IVs with each other in order to get a complete picture of the function of an IV in regression. The total relationship of the IV with the DV (correlation) and the correlations of the IVs with each other are given in the correlation matrix. The unique contribution of an IV to predicting a DV is generally assessed by either partial or semipartial correlation.

For standard multiple and hierarchical regression, the relationships between correlation, partial correlation, and semipartial correlation are given in Figure 5.3 for a simple case of one DV and two IVs. In the figure, (squared) correlation, partial correlation, and semipartial correlation coefficients are defined as areas created by overlapping circles. Area $a + b + c + d$ is the total area of the DV and reduces to a value of 1 in many equations. Area b is the segment of the variability of the DV that can be explained by either IV_1 or IV_2, and is the segment that creates the ambiguity.

In partial correlation, the contribution of the other IVs is taken out of both the IV and the DV. In semipartial correlation, the contribution of other IVs is taken out of only the IV. Thus (squared) semipartial correlation expresses the unique contribution of the IV to the total variance of the DV. Squared semipartial correlation (sr_i^2) is a very useful measure of importance of an IV. The interpretation of sr_i^2 differs, however, depending on the type of multiple regression employed.

5.6.1.1 Standard Multiple and Setwise Regression

In standard multiple regression and setwise regression sr_i^2 for an IV is the amount by which R^2 is reduced if that IV is deleted from the regression equation. That is, sr_i^2 represents the unique contribution of the IV to R^2 in that set of IVs.

When the IVs are correlated, squared semipartial correlations do not necessarily sum to multiple R^2. The sum of sr_i^2 is usually smaller than R^2 (although under some rather extreme circumstances, the sum can be larger than R^2). When the sum is smaller, the difference between R^2 and the sum of sr_i^2 for all IVs represents shared

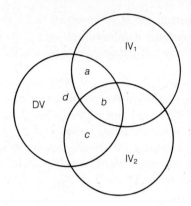

	Standard Multiple	Heirarchical
r_i^2	IV$_1$ $(a + b)/(a + b + c + d)$	$(a + b)/(a + b + c + d)$
	IV$_2$ $(c + b)/(a + b + c + d)$	$(c + b)/(a + b + c + d)$
sr_i^2	IV$_1$ $a/(a + b + c + d)$	$(a + b)/(a + b + c + d)$
	IV$_2$ $c/(a + b + c + d)$	$c/(a + b + c + d)$
pr_i^2	IV$_1$ $a/(a + d)$	$(a + b)/(a + b + d)$
	IV$_2$ $c/(c + d)$	$c/(c + d)$

Figure 5.3 Areas representing squared correlation, squared semipartial correlation, and squared partial correlation in standard multiple and hierarchical regression (where IV$_1$ is given priority over IV$_2$).

variance, variance that is contributed to R^2 by two or more IVs. It is rather common to find substantial R^2 with sr_i^2 for all IVs quite small.

Table 5.10 summarizes procedures for finding sr_i^2 (and pr_i^2) for both standard multiple and hierarchical regression through SPSS^x, BMDP, SAS, and SYSTAT.

SPSS^x and SAS provide sr_i^2 as part of their output. If you use SPSS^x RE-GRESSION or SPSS NEW REGRESSION, sr_i is optionally available as PART COR by requesting STATISTICS = ZPP. SAS REG provides two semipartial correlations with different error terms. For standard multiple regression, the appropriate form is SCORR2, which uses the Type II error term.

For BMDP1R, BMDP9R, or SYSTAT MGLH, sr_i^2 is most easily calculated using Equation 5.9.

$$sr_i^2 = \frac{T_i^2}{df_{res}} (1 - R^2) \tag{5.9}$$

The squared semipartial correlation (sr_i^2) for the ith IV is calculated from the T_i ratio for that IV, the residual degrees of freedom (df_{res}), and R^2.

In BMDP1R, R^2 is labeled MULTIPLE R-SQUARE and df_{res} is found in the analysis of variance table (cf. Table 5.5). With BMDP9R, information for calculating sr_i^2 is available for the best subset, with df_{res} shown as DENOMINATOR DEGREES OF FREEDOM. R^2 is shown as SQUARED MULTIPLE CORRELATION. SYSTAT

TABLE 5.10   PROCEDURES FOR FINDING sr_i^2 AND pr_i^2 THROUGH SPSS^x, SAS, BMDP, AND SYSTAT FOR STANDARD MULTIPLE AND HIERARCHICAL REGRESSION

	sr_i^2	pr_i^2
Standard Multiple Regression		
SPSS^x REGRESSION	STATISTICS = ZPP PART COR	STATISTICS = ZPP PARTIAL
SAS REG	SCORR2	PCORR2
BMDP1R, 9R	Use Eq. 5.9	Use Eq. 5.10
SYSTAT MGLH	Use Eq. 5.9	Use Eq. 5.10
Hierarchical Regression		
SPSS^x REGRESSION	RSQCH in SUMMARY TABLE	Use Eq. 5.10
BMDP2R	CHANGE IN RSQ	Use Eq. 5.10
SAS STEPWISE REG	PARTIAL R**2 SCORR1	PCORR1
SYSTAT MGLH	Get by subtraction of R^2 in sequential runs	Use Eq. 5.10

MGLH offers R^2 as SQUARED MULTIPLE R (cf. Table 5.7) and shows df_{res} in the analysis of variance table. (SAS RSQUARE offers neither significance tests for IVs nor residual degrees of freedom for any equation.)

From the output shown in Table 5.5 the squared semipartial correlation for GRADE is

$$sr_i^2 = \frac{(0.644)^2}{2}(1 - .7024) = .06$$

In this set of IVs, then, R^2 would be reduced by .06 if GRADE were deleted from the equation.

If partial correlations squared (pr_i^2) are desired, they are available (in unsquared form) in output in SPSS and SAS. The statistic PARTIAL is reported in SPSS^x REGRESSION or SPSS NEW REGRESSION. SAS REG provides two types of partial correlation coefficients, differing by the error term chosen. As for sr_i^2, the Type II error term is appropriate for standard multiple regression, and pr_i^2 is requested as PCORR2.

For BMDP and SYSTAT, squared partial correlations are calculated from T_i and df_{res} as shown in Equation 5.10.

$$pr_i^2 = \frac{T_i^2}{T_i^2 + df_{res}} \tag{5.10}$$

In all standard multiple regression programs, T_i is the significance test for sr_i^2, B_i, and β_i, as discussed in Section 5.6.2.

5.6.1.2 Hierarchical or Statistical Regression In these two forms of regression, sr_i^2 is interpreted as the amount of variance added to R^2 by each IV at the point that it enters the equation. The research question is, How much does this IV add to multiple R^2 after IVs with higher priority have contributed their share to prediction of the DV? Thus, the apparent importance of an IV is very likely to depend on its point of entry into the equation.

The sr_i^2 do sum to R^2 in hierarchical and statistical regression. Consult Figure 5.2 if you want to review this point.

As reviewed in Table 5.10, SPSS, BMDP, and SAS programs provide squared semipartial correlations as part of routine output for hierarchical and statistical regression. For SPSS, sr_i^2 is RSQCH for each IV in the SUMMARY TABLE (see Table 5.8). For BMDP2R, sr_i^2 is CHANGE IN RSQ for each IV (see Table 5.9). SAS STEPWISE provides sr_i^2 as PARTIAL R**2. If you use SAS REG for hierarchical regression by ordering IVs in successive MODEL statements, you can find the appropriate type of sr_i^2 by requesting SCORR1.

For SYSTAT MGLH, R^2 is provided for each sequential run. You can calculate sr_i^2 by subtraction between subsequent runs.

If desired, partial correlations can be found from T_i, as in Equation 5.10. These partial correlations are the same as for standard multiple regression. SAS REG offers PCORR1 as the form of partial correlation appropriate for hierarchical multiple regression.

5.6.2 Statistical Inference

This section covers significance tests for multiple regression and for regression coefficients for individual IVs. A test, F_{inc}, is also described for evaluating the statistical significance of adding two or more IVs to a prediction equation in hierarchical or statistical analysis. Calculations of confidence limits for unstandardized regression coefficients and procedures for comparing the predictive capacity of two different sets of IVs conclude the section.

It should be noted that when the researcher is using either statistical or setwise regression as an exploratory tool, inferential procedures of any kind may be inappropriate. Inferential procedures require that the researcher have a hypothesis to test. When statistical or setwise regression is used to snoop data, there may be no hypothesis, even though the statistics themselves are available.

5.6.2.1 Test for Multiple R The overall inferential test in multiple regression is whether the sample of scores is drawn from a population in which multiple R is zero. This is equivalent to the null hypothesis that all correlations between DV and IVs and all regression coefficients are zero. With large N, the test of this hypothesis becomes trivial because it is almost certain to be rejected. Testing multiple R is more interesting with fewer cases.

For standard multiple and hierarchical regression, the test of this hypothesis is presented in all computer outputs as analysis of variance. For hierarchical regression (and for standard multiple regression performed through stepwise programs), the analysis of variance table at the last step gives the relevant information. The F ratio for mean square regression over mean square residual tests the significance of multiple R. Mean square regression is the sum of squares for regression in Equation 5.2 divided by k degrees of freedom; mean square residual is the sum of squares for residual in the same equation divided by $(N - k - 1)$ degrees of freedom.

If you insist on inference in statistical or setwise regression, adjustments are necessary because all potential IVs do not enter the equation and sample R^2 is not distributed as F. Therefore the analysis of variance table at the last step (or for the "best" equation) is misleading; the reported F is biased so that the F ratio actually reflects a Type I error rate in excess of alpha.

Wilkinson and Dallal (1981) have developed tables for critical R^2 when forward selection procedures are used for statistical addition of variables and the selection stops when the F-to-enter for the next variable falls below some preset value. Appendix C contains Table C.5 that shows how large multiple R^2 must be to be statistically significant at .05 or .01 levels, given N, k, and F, where N is sample size, k is the number of potential IVs, and F is the minimum F-to-enter that is specified. F-to-enter values that can be chosen are 2, 3, or 4.

For example, for a statistical regression in which there are 100 subjects, 20 potential IVs, and an F-to-enter value of 3 chosen for the solution, a multiple R^2 of approximately .19 is required to be considered significantly different from zero at $\alpha = .05$ (and approximately .26 at $\alpha = .01$). Wilkinson and Dallal report that linear interpolation on N and k works well. However, they caution against extensive extrapolation for values of N and k beyond those given in the table. In hierarchical regression this table is also used to find critical R^2 values if a post hoc decision is made to terminate regression as soon as R^2 reaches statistical significance, if the appropriate F-to-enter is substituted for R^2 probability value as the stopping rule.

Wilkinson and Dallal recommend forward selection procedures in favor of other selection methods (e.g., stepwise selection, subset regression). They argue that, in practice, results using different procedures are not likely to be substantially different. Further, forward selection is computationally simple and efficient, and allows straightforward specification of stopping rules. If you are able to specify in advance the number of variables you wish to select, an alternative set of tables is provided by Wilkinson (1979) to evaluate significance of multiple R^2 with forward selection.

After looking at the data, you may wish to test the significance of some subsets of IVs in predicting the DV where a subset may even consist of a single IV. If several a posteriori tests like this are desired, Type I errors become increasingly likely. Larzelere and Mulaik (1977) recommend the following conservative F test to keep Type I error rate below α for all combinations of IVs:

$$F = \frac{R_s^2/k}{(1 - R_s^2)/(N - k - 1)} \tag{5.11}$$

where R_s^2 is the squared multiple (or bivariate) correlation to be tested for significance, and k is the total number of IVs. Obtained F is compared with tabled F, with k and $(N - k - 1)$ degrees of freedom. That is, the critical value of F for each subset is the same as the critical value for the overall multiple R.

In the sample problem of Section 5.4, the bivariate correlation between MOTIV and COMPR is tested a posteriori as follows:

$$F = \frac{.58613^2/3}{(1 - .58613^2)/2} = 0.349, \quad \text{with df} = 3, 2$$

5.6.2.2 Test of Regression Components In standard multiple regression, for each IV the same significance test evaluates B_i, β_i, sr_i and pr_i. The test is straightforward, and results are given in computer output. In SPSS NEW REGRESSION or SPSS^x, F_i tests the unique contribution of each IV and appears in the output section labeled VARIABLES IN THE EQUATION (see Table 5.4). Degrees of freedom are 1 and df_{res}, which appears in the accompanying analysis of variance table. In BMDP1R, SYSTAT MGLH, and SAS REG, T_i values are given for each IV, tested with df_{res} from the analysis of variance table (see Tables 5.5 to 5.7). If BMDP2R is used for standard multiple regression, F_i values are available at the last step of the analysis as F-TO-REMOVE; they are interpreted as for SPSS.

It is important in interpretation to recall the limitations of these significance tests. The significance tests are sensitive only to the unique variance an IV adds to R^2. A very important IV that shares variance with another IV in the analysis may be nonsignificant although the two IVs in combination are responsible in large part for the size of R^2. An IV that is highly correlated with the DV but has a nonsignificant regression coefficient may have suffered just such a fate. For this reason, it is important to report and interpret r_{iy} in addition to sr_i^2 and F_i for each IV, as shown in Table 5.18.

For setwise regression through BMDP9R the individual IVs are best interpreted as if for standard multiple regression where each variable enters last. The T_i values given for each variable in the "best" subset are interpreted the same as those for BMDP1R noted earlier.

For statistical and hierarchical regression, assessment of contribution of variables is more complex, and appropriate significance tests may not appear in the computer output. First, there is inherent ambiguity in the testing of each variable. In statistical and hierarchical regression, tests of sr_i^2 and pr_i^2 are not the same as tests of the regression coefficients (B_i and β_i). Regression coefficients are independent of order of entry of the IVs, whereas sr_i^2 and pr_i^2 depend directly on order of entry. Because sr_i^2 reflects "importance" as typically of interest in hierarchical or statistical regression, tests based on sr_i^2 are discussed here.[2]

SPSS, BMDP, and SAS provide significance tests for sr_i^2 in summary tables. For SPSS, the test is FCH—F ratio for change in R^2—that is accompanied by a significance value, SIGCH. In BMDP2R, the F ratio for sr_i^2 is reported in the summary table as F TO ENTER. No significance value is given, but you can evaluate this F with numerator df $= 1$ and denominator df_{res} for the last step in the equation. For SAS, the relevant values are F in the summary table and, for significance, PROB>F.

---

[2] For combined standard-hierarchical regression, it might be desirable to use the "standard" method for all IVs simply to maintain consistency. If so, be sure to report that the F test is for regression coefficients.

If you use SYSTAT MGLH, you need to calculate F for your sr_i^2 as found by subtraction (cf. Section 5.6.1.2) using the following equation:

$$F_i = \frac{sr_i^2}{(1 - R^2)/\text{df}_\text{res}} \qquad (5.12)$$

The F_i for each IV is based on sr_i^2 (the squared semipartial correlation), multiple R^2 at the final step; and residual degrees of freedom from the analysis of variance table for the final step.

5.6.2.3 Test of Added Subset of IVs For hierarchical and statistical regression, one can test whether a block of two or more variables significantly increases R^2 above the R^2 predicted by a set of variables already in the equation.

$$F_\text{inc} = \frac{(R_{wi}^2 - R_{wo}^2)/M}{(1 - R_{wi}^2)/\text{df}_\text{res}} \qquad (5.13)$$

where F_inc is the incremental F ratio; R_{wi}^2 is the multiple R^2 achieved with the new block of IVs in the equation; R_{wo}^2 is the multiple R^2 without the new block of IVs in the equation; M is the number of IVs in the block; and $\text{df}_\text{res} = (N - k - 1)$ is residual degrees of freedom in the final analysis of variance table. Both R_{wi}^2 and R_{wo}^2 are found in the SUMMARY TABLE of any computer output. If you are using SPSS^x REGRESSION or SPSS NEW REGRESSION, F_inc is available through the TEST command.

The null hypothesis of no increase in R^2 is tested as F with M and df_res degrees of freedom. If the null hypothesis is rejected, the new block of IVs does significantly increase the explained variance.

Although this is a poor example because only one variable is in the new block, we can use the hierarchical example in Table 5.8 to test whether MOTIV adds significantly to the variance contributed by the first two variables to enter the equation, QUAL and GRADE.

$$F_\text{inc} = \frac{(.70235 - .61757)/1}{(1 - .70235)/2} = 0.570 \qquad \text{with df} = 1, 2$$

(Because only one variable was entered, this test is the same as the test provided for MOTIV in Table 5.8 under the heading VARIABLES NOT IN THE EQUATION. When the T value of .755 is squared, it is the same as F_inc above. Thus, F_inc can be used when there is only one variable in the block, but the information is already provided in the output.)

5.6.2.4 Confidence Limits around B To estimate population values, confidence limits for unstandardized regression coefficients (B_i) are calculated. Standard errors of unstandardized regression coefficients, unstandardized regression coefficients, and the critical value of t for the desired confidence level (based on $N - 2$ degrees of freedom, where N is the sample size) are used in Equation 5.14.

$$CL_{B_i} = B_i \pm SE_{B_i} (t_{\alpha/2}) \tag{5.14}$$

The $(1 - \alpha)$ confidence limits for the unstandardized regression coefficient for the ith IV (CL_{B_i}) is the regression coefficient (B_i) plus or minus the standard error of the regression coefficient (SE_{B_i}) times the critical value of t, with ($N - 2$) degrees of freedom at the desired level of α.

If 95% confidence limits are desired, they are given in SPSS[x] REGRESSION output as 95 CONFDNCE INTRVL B under the segment of output titled VARIABLES IN THE EQUATION. With other output or when 99% confidence limits are desired, Equation 5.14 is used. Unstandardized regression coefficients and the standard errors of those coefficients appear in the sections labeled VARIABLES IN THE EQUATION. Standard errors are called STD ERROR OF COEFF in BMDP2R and are called STD ERROR in SYSTAT MGLH, SAS, BMDP1R, and BMDP9R.

For the example in Table 5.4, the 95% confidence limits for GRADE, with df = 4, are

$$CL_{B_G} = 0.416 \pm 0.646(2.78) = 0.416 \pm 1.796 = -1.380 \leftrightarrow 2.212$$

If the confidence interval contains zero, one retains the null hypothesis that the population regression coefficient is zero.

5.6.2.5 Comparing Two Sets of Predictors
It is sometimes of interest to know whether one set of IVs predicts a DV better than another set of IVs. For example, can ratings of current belly dancing ability be better predicted by personality tests or by past dance and musical training?

The procedure for finding out is fairly convoluted, but if you have a large sample and you have access to BMDP or are willing to enter a pair of predicted scores for each subject in your sample, a test for the significance of the difference between two "correlated correlations" (both correlations are based on the same sample and share a variable) is available (Steiger, 1980). (If sample size is small, nonindependence among predicted scores for cases can result in serious violation of the assumptions underlying the test.)

As suggested in Section 5.4.1, a multiple correlation can be thought of as a simple correlation between obtained DVs and predicted DVs; that is, $R = r_{yy'}$. If there are two sets of predictors, $Y_{a'}$ and $Y_{b'}$ (where, for example $Y_{a'}$ is prediction from personality scores and $Y_{b'}$ is prediction from past training), a comparison of their relative effectiveness in predicting Y is made by testing for the significance of the difference between $r_{yy'_a}$ and $r_{yy'_b}$. For simplicity, let's call these r_{ya} and r_{yb}.

The trick is that to test the difference we need to know the correlation between the predicted scores from set A (personality) and those from set B (training), that is, $r_{ya'yb'}$ or, simplified, r_{ab}. This is where the BMDP file manipulation procedures or hand entering become necessary.

The z test for the difference between r_{ya} and r_{yb} is

$$\bar{Z}^* = (z_{ya} - z_{yb}) \sqrt{\frac{N-3}{2 - 2\bar{s}_{ya,yb}}} \tag{5.15}$$

where N is, as usual, the sample size,

$$z_{ya} = \tfrac{1}{2} \ln \left(\frac{1 + r_{ya}}{1 - r_{ya}} \right) \qquad \text{and} \qquad z_{yb} = \tfrac{1}{2} \ln \left(\frac{1 + r_{yb}}{1 - r_{yb}} \right)$$

and

$$\bar{s}_{ya,yb} = \frac{(r_{ab})(1 - \bar{r}^2 - \bar{r}^2) - \tfrac{1}{2}(\bar{r}^2)(1 - \bar{r}^2 - \bar{r}^2 - r_{ab}^2)}{(1 - \bar{r}^2)^2}$$

where $\bar{r} = \tfrac{1}{2}(r_{ya} + r_{yb})$.

So, for the example, if the correlation between currently measured ability and ability as predicted from personality scores is .40 ($R_a = r_{ya} = .40$), the correlation between currently measured ability and ability as predicted from past training is .50 ($R_b = r_{yb} = .50$), and the correlation between ability as predicted from personality and ability as predicted from training is .10 ($r_{ab} = .10$), and $N = 103$,

$$\bar{r} = \tfrac{1}{2}(.40 + .50) = .45$$

$$\bar{s}_{ya,yb} = \frac{(.10)(1 - .45^2 - .45^2) - 1/2(.45)^2(1 - .45^2 - .45^2 - .10^2)}{(1 - .45^2)^2}$$

$$= .0004226$$

$$z_{ya} = \tfrac{1}{2} \ln \left(\frac{1 + .40}{1 - .40} \right) = .42365$$

$$z_{yb} = \tfrac{1}{2} \ln \left(\frac{1 + .50}{1 - .50} \right) = .54931$$

and, finally,

$$\bar{z}^* = (.42365 - .54931) \sqrt{\frac{103 - 3}{2 - .000845}} = -0.88874$$

Because $\bar{z}^*$ is within the critical values of ± 1.96 for a two-tailed test, there is no statistically significant difference between multiple R when predicting Y from Y'_a or

Y'_b. That is, there is no statistically significant difference in predicting current belly dancing ability from past training vs. personality tests.

Steiger (1980) presents additional significance tests for situations where both the DV and the IVs are different, but from the same sample, and for comparing the difference between any two correlations within a correlation matrix.

5.6.3 Adjustment of R^2

Just as simple r_{xy} from a sample is expected to fluctuate around the value of the correlation in the population, sample R is expected to fluctuate around the population value. But multiple R never takes on negative value, so all chance fluctuations are in the positive direction and add to the magnitude of R. As in any sampling distribution, the magnitude of chance fluctuations increases as the sample size decreases. Therefore, R tends to be overestimated, and the smaller the sample the greater the overestimation. For this reason, in estimating the population value of R, adjustment is made for expected inflation in sample R.

All the programs discussed in Section 5.7 routinely provide adjusted R^2. Wherry (1931) provides a simple equation for this adjustment, which he calls $\tilde{R}^2$:

$$\tilde{R}^2 = 1 - (1 - R^2)\left(\frac{N - 1}{N - k - 1}\right) \tag{5.16}$$

where N is the sample size, k is the number of IVs, and R^2 is the squared multiple correlation. For the same sample problem,

$$\tilde{R}^2 = 1 - (1 - .70235)(5/2) = .25588$$

as printed out for SPSS^x, Table 5.4.

For statistical regression, Cohen and Cohen (1975) recommend k based on the number of IVs considered for inclusion, rather than on the number of IVs selected by the program. They also suggest the convention of reporting $\tilde{R}^2 = 0$ when the value spuriously becomes negative.

When the number of subjects is 60 or fewer and there are numerous IVs (say, more than 20), Equation 5.16 may provide inadequate adjustment for R^2. The adjusted value may be off by as much as .10 (Cattin, 1980). In these situations of small N and numerous IVs, Equation 5.17 (Browne, 1975) provides further adjustment.

$$R_s^2 = \frac{(N - k - 3)\,\tilde{R}^4 + \tilde{R}^2}{(N - 2k - 2)\,\tilde{R}^2 + k} \tag{5.17}$$

The adjusted R^2 for small samples is a function of the number of cases, N, the number of IVs, k, and the $\tilde{R}^2$ value as found from Equation 5.16.

When N is less than 50, Cattin (1980) provides an equation that produces even less bias but requires far more computation.

5.6.4 Suppressor Variables

Occasionally you may find an IV that is useful in predicting the DV and in increasing the multiple R^2 by virtue of its correlations with other IVs. This IV is called a suppressor variable because it "suppresses" variance that is irrelevant to prediction of the DV. In a full discussion of suppressor variables, Cohen and Cohen (1975) describe and provide examples of several varieties of suppression.

In psychology, for instance, one might administer as IVs two paper-and-pencil tests, a test of ability to list dance patterns and a test of test-taking ability. By itself the first test poorly predicts the DV (say, belly dancing ability). The second IV is a suppressor variable because it removes variance due to ability in taking tests, and thus enhances prediction of the DV by the first IV.

In output, the presence of a suppressor variable is identified by the pattern of regression coefficients and correlations of each IV with the DV. Compare the simple correlation between each IV and the DV in the correlation matrix available from output with the standardized regression coefficient (beta weight) for the IV. If the beta weight is significantly different from zero, either one of the following two conditions signals the presence of a suppressor variable: (1) the absolute value of the simple correlation between IV and DV is substantially smaller than the beta weight for the IV, or (2) the simple correlation and beta weight have opposite signs.

If you find that a suppressor variable is present, you need to search for it among IVs for which regression coefficients and correlations are congruent. You do this by systematically leaving each congruent IV out of the equation and examining the changes in regression coefficients for the IV(s) that had discrepancy between regression coefficient and correlation in the original equation.

Once suppressor variables are identified, they are properly interpreted as variables that enhance importance of other IVs by virtue of suppression of irrelevant variance in other IVs or in the DV.

5.7 COMPARISON OF PROGRAMS

The popularity of multiple regression is reflected in the abundance of applicable programs. BMDP and SAS each have three separate programs for standard, statistical, and setwise regression. SPSS^x has a single, highly flexible program for multiple regression. In SYSTAT, multiple regression is handled through a multivariate general linear model program.

Direct comparisons of multiple and statistical/hierarchical programs are summarized in Tables 5.11 and 5.12, respectively. Table 5.13 compares the two programs designed for setwise (subsets) regression. Some of these features are elaborated in Sections 5.7.1 through 5.7.4.

5.7.1 SPSS Package

The original SPSS program for regression was SPSS REGRESSION. Beginning with Release 9.0, the more elaborate NEW REGRESSION program became available (Hull

TABLE 5.11 COMPARISON OF PROGRAMS FOR STANDARD MULTIPLE REGRESSION

Feature	SPSS REGRESSION[a]	BMDP1R	SYSTAT MGLH	SAS REG
Input				
Correlation matrix input	Yes	Yes	Yes	Yes
Covariance matrix input	Yes	Yes	Yes	Yes
Missing data options	Yes	Yes[b]	No	No
Regression through the origin	Yes	No[b]	Yes	Yes
Tolerance option	Yes	Yes	No	Yes
Test of equality of subsamples	No	Yes	No	No
Post hoc hypotheses[c]	Yes	No	Yes	Yes
Optional error terms	No	No	Yes	No
Weighted least squares	No	Yes	Yes	Yes
Regression output				
Expanded variable labels	Yes	No	No	No
Analysis of variance for regression	Yes	Yes	Yes	Yes
Multiple R	Yes	Yes	Yes	No
R^2	Yes	Yes	Yes	Yes
Standard error for R	STANDARD ERROR	STD. ERROR OF ESTIMATE	STANDARD ERROR OF ESTIMATE	ROOT MSE
Adjusted R^2	Yes	Yes	Yes	Yes
Coefficient of variation	No	No	No[d]	Yes
Correlation matrix	Yes	Yes	No[d]	Yes
Significance levels of correlation matrix	Yes	No	No	Yes
Sum-of-squares and cross-products (SSCP) matrix	Yes	No	No[d]	Yes
Inverse of SSCP matrix	No	No	No	Yes
Covariance matrix	Yes	Yes	No[d]	No
Means and standard deviations	Yes	Yes	No[e]	Yes
Matrix of correlation coefficients if some not computed	Yes	No[b]	N.A.	N.A.
N for each correlation coefficient	Yes	No[b]	N.A.	N.A.

Minimum and maximum of variables	No	Yes	No[e]	Yes
Sums for each variable	No	No	No	Yes
Coefficient of variation for each variable	No	Yes	No	No
Unstandardized regression coefficients	B	COEFFICIENT	COEFFICIENT	PARAMETER ESTIMATE
Standard error of regression coefficient	SE B	STD. ERROR	STD. ERROR	STANDARD ERROR
F or t test of regression coefficient	T(F optional)	T	T	T
Intercept (constant)	Yes	Yes	Yes	Yes
Standardized regression coefficient	BETA	STD. REG. COEFF.	STD. COEFF.	STANDARDIZED ESTIMATE
Approx. standard error of β	Yes	No	No	No
Partial correlation	Yes	No	Yes	Yes[f]
Semipartial correlation	PART COR	No	No	Yes[f]
Tolerance	Yes	Yes	Yes	Yes
Sums of squares for B	No	No	No	Yes[g]
Variance inflation factors	No	No	No	Yes
Variance-covariance matrix for unstandardized B coefficients	Yes	No	No	Yes
Correlation matrix for unstandardized B coefficients	No	Yes	Yes	Yes
95% confidence interval for B	Yes	No	No	No
Sweep matrix	Yes	No	No	Yes
Condition number bounds	Yes	No	No	No
Eigenvalues	No	No	Yes	Yes
Condition indices	No	No	Yes	Yes
Variance proportions	No	No	Yes	Yes
Hypothesis matrices	No	No	Yes	Yes
Residuals				
Predicted scores, residuals and standardized residuals	Yes	Yes	Yes[h]	Yes
95% confidence interval for predicted value	No	No	No	Yes[g]

TABLE 5.11 (Continued)

Feature	SPSS[x] REGRESSION[a]	BMDP1R	SYSTAT MGLH	SAS REG
Plot of standardized residuals against predicted scores	Yes	Yes	No[i]	No[j]
Normal plot of residuals	Yes	Yes	No[i]	No[j]
Durbin-Watson statistic	Yes	Yes	Yes	Yes
Standard error of predicted values	Yes	No	Yes[h]	Yes
Mahalanobis and/or Cook's distance	Yes	No[k]	Yes[h]	Yes
Influence diagnostics	No	No	No	Yes
Heteroscedasticity test	No	No	No[i]	Yes
Histograms	Yes	No[l]	No	No
Casewise plots	Yes	No	No	Yes
Partial plots	Yes[m]	Yes	No	Yes
Other plots available	Yes	Yes	No[i]	No[j]
Leverage	Yes	No	Yes[h]	No
Summary statistics for residuals	Yes	No	No	Yes
Save predicted values/residuals	Yes	Yes	Yes	Yes

Note: BMDP2R (cf. Table 5.12) and SAS GLM (cf. Table 9.11) can also be used for standard multiple regression.
[a] These features are available in SPSS NEW REGRESSION except as noted.
[b] Available through BMDPAM or BMDP8D.
[c] Does *not* use Larzelere and Mulaik (1977) correction.
[d] Available through SYSTAT CORR.
[e] Available through SYSTAT STATS.
[f] Use Type II for standard MR. Values in output are *squared* correlations.
[g] Optional types.
[h] Saved into a file.
[i] Available through SYSTAT GRAPH.
[j] Available through SAS PLOT.
[k] Available through BMDPAM, BMDP4M, or BMDP2R.
[l] Available through BMDP5D.
[m] Not available in SPSS NEW REGRESSION.

TABLE 5.12 COMPARISON OF PROGRAMS FOR STATISTICAL AND/OR HIERARCHICAL REGRESSION

Feature	SAS STEPWISE	SPSS[x] REGRESSION[a]	BMDP2R	SYSTAT MGLH
Input				
Correlation matrix input	No	Yes	Yes	Yes
Covariance matrix input	No	Yes	Yes	Yes
Missing data options	No	Yes	No[b]	No
Specify stepping algorithm	Yes	Yes	Yes	No
Specify tolerance	No	Yes	Yes	No
Specify F to enter	No	Yes	Yes	No
Specify F to remove	No	Yes	Yes	No
Specify probability of F to enter and/or remove	Yes	Yes	Yes	Yes
Specify maximum number of steps	No	Yes	Yes	No
Regression through the origin	No	Yes	Yes	Yes
Force variables into equation	INCLUDE	ENTER	FORCE	FORCE
Weighted least squares	No[c]	No	Yes	Yes
Specify order of entry (hierarchy)	No	Yes	Yes	No
IV sets for entry in single step	No	Yes	Yes	No
Regression output				
Expanded variable labels	No	Yes	No	No
Analysis of variance for regression, each step	Yes	Yes	Yes	No[d]
Multiple R, each step	No	Yes	Yes	No[d]
R^2, each step	R SQUARE	R SQUARE	MULTIPLE R-SQUARE	No[d]
Standard error for R, each step	No[c]	STANDARD ERROR	STD. ERROR OF EST.	No[d]
Adjusted R^2, each step	No	Yes	Yes	No[d]
Mallow's C_p, each step	Yes	No	No	No
Sum-of-squares and cross-products matrix	No[c]	Yes	No	No[e]
Correlation matrix	No[c]	Yes	Yes	No[e]

TABLE 5.12 (Continued)

Feature	SAS STEPWISE	SPSS^x REGRESSION[a]	BMDP2R	SYSTAT MGLH
Correlation significance levels	No[c]	Yes	No	No
Covariance matrix	No	Yes	Yes	No
Correlation matrix of regression coefficients	No[c]	No	Yes	Yes
Covariance matrix of regression coefficients	No[c]	Yes	No	No
Means and standard deviations	No[c]	Yes	Yes	No
Matrix of correlation coefficients if some not computed	N.A.	Yes	No[b]	N.A.
N for each correlation coefficient	N.A.	Yes	No[b]	N.A.
Coefficients of variation	No[c]	No	Yes	Yes
Minimums and maximums	No	No	Yes	No
Smallest and largest z	No	No	Yes	No
Skewness and kurtosis	No	No	Yes	No
95% confidence interval for B	No	Yes	No	No
Sweep matrix	No[c]	Yes	No	No
Condition number bounds	Yes	Yes	No	No
Eigenvalues	No	No	No	Yes
Condition indices	No	No	No	Yes
Variance proportions	No	No	No	Yes
Variables in equation (each step)				
Unstandardized regression coefficients	B VALUE	B	COEFFICIENT	No[d]
Standard error of regression coefficient	STD ERROR B	SE B	STD. ERROR OF COEFF	No[d]
95% confidence interval for B	No	Yes	No	No
Standard regression coefficient	No[c]	BETA	STD REG COEFF	No[d]

F (or T) to remove	F	T (F optional)	F TO REMOVE	No
Intercept	INTERCEPT	(CONSTANT)	(Y-INTERCEPT)	No[d]
Variables not in equation (each step)				
Standardized regression coefficient for entering	No	BETA IN	No	No
Partial correlation coefficient for entering	No	PARTIAL	PARTIAL CORR.	No
Squared partial correlation for entering	MODEL R**2	No	No	No[d]
Tolerance	Yes	Yes	Yes	No
F (or T) to enter	F	T	F TO ENTER	No[d]
Table of stepwise regression coefficients, F to enter and remove, and partial correlations	No	No	Yes	Yes
Summary table				
Multiple R	No	MULTR	Yes	
R^2	MODEL R**2	RSQ	MULTIPLE RSQ	RSQUARE
Change in R^2 (squared semipartial correlation)	PARTIAL R**2	RSQCH	CHANGE IN RSQ	No
Adjusted R^2	No	ADJRSQ	No	No
F to enter equation	No	F(EQU)	F TO ENTER	No
F to remove from equation	F	FCH	F TO REMOVE	No
Significance of T to enter or remove	Yes	Yes	No	No
Standardized regression coefficient	No	BETA IN	No	No
Mallow's C_p	Yes	No	No	
Residuals				
Predicted scores, residuals, and standardized residuals	No[c]	Yes	Yes	Yes[f]
Plot of standardized residuals against predicted scores	No	Yes	Yes	No[g]
Normal plot of residuals	No	Yes	Yes	No[g]

TABLE 5.12 (Continued)

Feature	SAS STEPWISE	SPSS^x REGRESSION^a	BMDP2R	SYSTAT MGLH
Durbin-Watson statistic	No^c	Yes	No	Yes
Standardized error of predicted values	No^c	Yes	No	Yes^f
Mahalanobis and/or Cook's distance	No^c	Yes	Yes	Yes^f
Summary statistics for residuals	No^c	Yes	No	No
Leverage	No	Yes	Yes	Yes
Casewise plots	No^c	Yes	Yes	No
Partial plots	No^c	Yes	Yes	No
Save predicted values/residuals	No^c	Yes	Yes	Yes
Misc. additional diagnostics	No	Yes	Yes	No
Other plots available	No	Yes	Yes	No^g

[a] These features are available in SPSS NEW REGRESSION except as noted.
[b] Available through BMDPAM or BMDP8D.
[c] Available through SAS REG.
[d] Available by running separate standard multiple regressions for each step.
[e] Available through SYSTAT CORR.
[f] Saved to file.
[g] Available through SYSTAT GRAPH.

TABLE 5.13 COMPARISON OF PROGRAMS FOR SETWISE REGRESSION

Feature	BMDP9R	SAS RSQUARE
Input		
Correlation matrix input	Yes	Yes
Covariance matrix input	Yes	Yes
SSCP matrix input	No	Yes
Specify tolerance	Yes	No
Specify model criterion	Yes	No
Regression through the origin	Yes	Yes
Include variables in all models	No	Yes
Weighted least squares	Yes	Yes
Specify penalty for additional variables	Yes	No
Specify number of subsets	Yes	Yes
Specify minimum subset size	No	Yes
Specify maximum subset size	Yes	Yes
Regression output		
Means and standard deviations	Yes	Yes
Largest and smallest values	Yes	No
Skewness and kurtosis	Yes	No
Correlation matrix	Yes	Yes
Covariance matrix	Yes	No
Correlation of regression coefficients	Yes	No
Covariance of regression coefficients	Yes	No
Shaded, sorted correlation matrix	Yes	No
For each subset		
Variables in subset	Yes	Yes
R^2	Yes	Yes
Adjusted R^2	Yes	Yes
Model criteria	Yes[a]	Yes
Regression coefficients (B)	COEFFICIENT	PARAMETER ESTIMATES
T statistics for each variable	Yes	No
Residual mean square	Yes[b]	Yes[c]
Error sum of squares	No	Yes
For best subset (additional to above)		
Standard errors of B	Yes	No
Standardized coefficients (β)	Yes	No
Tolerances	Yes	No
Contributions to R^2 (sr^2)	Yes	No
Multiple R	Yes	No
F statistics with df and significance	Yes	No
Residuals		
Histograms of standardized residuals	Yes	No
Plots of residuals against predicted scores	Yes	No
Normal plot of residuals	Yes	No
Plots of variables against residuals	Yes	No

[a] Chosen criterion only.

[b] For "best" subset only.

[c] Several varieties.

and Nie, 1981). With few modifications, the latter program appears as REGRESSION in the SPSS^x series (SPSS, Inc., 1986). The distinctive feature of these programs is flexibility. The major improvement in SPSS^x is that the manual now approaches comprehensibility for this program. SPSS^x REGRESSION is the program summarized in Tables 5.11 and 5.12. With noted exceptions, the summary applies to SPSS NEW REGRESSION as well.

The SPSS programs offer four options for treatment of missing data (described in the manuals). Provision for expanded variable labels enhances recall of output that has not been looked at in a while. Data can be input raw or as correlation matrices. And as in all SPSS procedures, analysis can be limited to subsets of cases.

A special option is available so that correlation matrices are printed only when one or more of the correlations cannot be calculated. Also convenient is the optional printing of semipartial correlations for standard multiple regression and 95% confidence intervals for regression coefficients.

The statistical procedure offers forward, backward, and stepwise selection of variables. User-modifiable statistical criteria for statistical selection are (1) the maximum number of steps for model building, (2) the minimum F ratio for adding or maximum F for deleting a variable in the equation, (3) the minimum probability of F for adding or maximum probability of F for removing a variable in the equation, and (4) tolerance. Tolerance is the proportion of variance of a potential IV that is not explained by IVs already in the equation. The smaller the tolerance value, the less the restriction placed on entering variables. One avoids multicollinearity by maintaining reasonable tolerance levels for entry, and thereby disallowing entry of variables that add virtually nothing to predictability.

A series of ENTER subcommands can be used for hierarchical regression. Each ENTER subcommand is evaluated in order; the IV or IVs listed after each ENTER subcommand are evaluated in that order. Within a single subcommand, variables are entered in order of decreasing tolerance. If there is more than one IV in the subcommand, they are treated as a block in the summary table where changes in the equation are evaluated.

Extensive analysis of residuals is available. For example, a table of predicted scores and residuals can be requested and accompanied by a plot of standardized residuals against standardized predicted values of the DV (z scores of the Y' values). Plots of standardized residuals against sequenced cases are also available. For a sequenced file, one can request a Durbin-Watson statistic, which is used for a test of autocorrelation between adjacent cases. In addition, you can request Mahalanobis distance for the ten most extreme cases as a convenient way of evaluating outliers. This is the only program within the SPSS packages that offers Mahalanobis distance. Partial residual plots (partialing out all but one of the IVs) are available in SPSS^x REGRESSION but not in SPSS NEW REGRESSION.

The flexibility of input for the SPSS^x REGRESSION program does not carry over into output. The only difference between standard and statistical or hierarchical regression is in the printing of a progression of steps. Otherwise the statistics and parameter estimates are identical. These values, however, have different meanings, depending on the type of analysis. For example, you can request semipartial correlations (called PART COR) through the ZPP statistics. But these apply only to standard multiple regression. As pointed out in Section 5.6.1, semipartial correlations for statistical or hierarchical analysis appear in the summary table (obtained through the HISTORY statistic) as RSQ CHANGE. It is best not to request a summary table for standard multiple regression because it contains no useful information that does not appear elsewhere.

5.7.2 BMD Series

The BMDP package (Dixon, 1985) offers separate programs for the various types of regression. All the programs do standard multiple regression, but the simplest applicable program is BMDP1R. For statistical and hierarchical regression, the program is BMDP2R. BMDP9R is available as a setwise program. All these programs exclude all cases with missing data, unless the values are estimated by, perhaps, some of the procedures available through BMDPAM or BMDP8D.

For standard multiple regression, BMDP1R offers analysis of "case combinations" or subsamples. Reduction of residuals due to grouping serves as a test of the equality of regression among the groups. Covariance and correlation matrices are available as output or can be used as input. In addition to tables of predicted values and residuals, the BMDP1R program allows a number of plots of residuals. Although not available in BMDP1R, Mahalanobis distance as well as other measures of extreme cases can be found using any one of several other BMDP programs, including BMDP2R.

The statistical/hierarchical regression program in the BMDP series, BMDP2R, combines several features of other statistical programs and adds some that are unavailable elsewhere. The user can specify the stepping algorithm (how the program adds or deletes variables in the equation as it progresses) in addition to specifying statistical criteria, as described in the BMDP manual (Dixon, 1985). Specifiable statistical criteria are tolerance, F to enter, F to remove, and maximum number of steps, as for SPSS. At the end of the output for all steps, a table of stepwise regression coefficients is available with the full regression equation for each step. The intercept (A) is given, along with unstandardized regression coefficients (B_i) for each of the variables in the equation at the next step. For each regression coefficient, F to enter and remove, and partial correlations are given.

In BMDP2R, the same input and output matrices are available as in BMDP1R. Residuals tables and plots, however, are far more extensive than BMDP1R. There are 21 diagnostic variables (including Mahalanobis distance) that can be printed or plotted, and 17 of these can be saved. For each plot, you can specify through options the number of extreme cases for which values are printed.

BMDP9R is a setwise program that searches among all possible subsets of variables for those subsets that are best. "Best" is defined by the user from among criteria such as maximum R^2. The researcher can specify the number of best subsets to be identified. Because of extensive output about the contribution of IVs to various subsets, BMDP9R is especially useful in identifying a combination of variables causing outlying cases to be unusual, as illustrated in Section 4.2.1.4. The extent of residuals analysis in BMDP9R is somewhere between that for BMDP1R and BMDP2R. A more extensive list of residuals and other diagnostics (including Mahalanobis distance) can be saved to a file than is printed through BMDP9R.

Cross validation is simplified in BMDP2R and BMDP9R by a case-weighting procedure, where cases to be omitted from computation of the regression equation are given a weight of zero.

BMDP offers a number of additional programs for regression analysis. These cover procedures such as nonlinear regression, analysis of variance and covariance, periodic regression, and asymptotic regression.

5.7.3 SAS System

Like BMDP, SAS has separate programs for different types of regression analysis (SAS Institute, Inc., 1985). Standard multiple regression is handled through REG, statistical (though not hierarchical) through STEPWISE, and setwise through RSQUARE. In addition, GLM can be used for regression analysis; it is more flexible and powerful than any of the other SAS programs, but also more difficult to use.

SAS REG handles correlation or covariance matrix input but has no options for dealing with missing data. A case is deleted if it contains any missing values.

In SAS REG, two types of partial and semipartial correlations are available. But beware; the value reported as SEMI-PARTIAL CORR is actually squared semipartial correlation (sr^2). The sr_i^2 and pr_i^2 appropriate for standard multiple regression are those that use TYPE II (partial) sums of squares (which can also be printed). TYPE I sums of squares are used for evaluating hierarchical models where order of entry of IVs follows order in the MODEL statement. Since STEPWISE does not handle hierarchical regression, SAS REG is used for hierarchical analysis within SAS.

Tables of residuals and other diagnostics (including Cook's value but not Mahalanobis distance) are extensive but are saved to file rather than printed. After being saved to file, the diagnostics and residuals can be plotted through SAS PLOT. The only plots printed within REG are partial residual plots where all IVs except one are partialed out.

SAS STEPWISE is a limited program for statistical multiple regression. The program does not accept matrix input, nor can order of entry of IVs be specified. The usual forward, backward, and stepwise criteria for selection are available, in addition to two enhancements on forward selection. The only statistical criteria available are the probability of F to enter and F to remove an IV from the equation. Summary table information is adequate but misleading; squared semipartial correlations are labeled PARTIAL R**2.

Few of the optional features of SAS REG are available in STEPWISE. For example, descriptive statistics and matrices cannot be printed out. One useful feature is Mallow's C_p which, as described in the manual, can be used as a criterion for selecting a model produced by statistical regression. No diagnostics or residuals are available.

SAS RSQUARE does setwise regression but differs in several ways from BMDP9R. Instead of specifying a criterion for selecting subsets, RSQUARE evaluates all subsets, and then, for each subset, prints out values for the various optimizing criteria. No "best" subset is identified. You make the choice of the best subset by comparing values on your chosen criterion. You can reduce output by specifying the number of subsets to be evaluated as well as minimum and maximum subset sizes.

In SAS RSQUARE, descriptive statistics and matrix output are limited. The only information given about each IV in a subset is the regression coefficient (B_i). As a result, the program is not useful for identifying the combination of IVs on which outlying cases differ. No residuals/diagnostics analysis is available.

5.7.4 SYSTAT System

In SYSTAT (Wilkinson, 1986), multiple regression is done through MGLH, a multivariate general linear program that handles ANOVA, MANOVA, canonical correlation, discriminant function analysis, and several types of regression. While SYSTAT MGLH has a variety of features for standard multiple regression, handling of statistical regression is limited, and there is no direct way to specify hierarchical or setwise regression.

Matrix input is accepted in SYSTAT MGLH, but processing of output of matrices or descriptive statistics requires the use of other programs in the SYSTAT package. The program is especially convenient for testing post hoc hypotheses, allowing you to specify your own error terms should you be unhappy with the ones normally provided. Residuals and other diagnostics are handled by saving values to a file, which can then be printed out through SYSTAT DATA or plotted through SYSTAT GRAPH. In this way, you can take a look at Cook's value, but not Mahalanobis distance, for each case. Or, you can find summary values for residuals and diagnostics through SYSTAT STATS, the program for descriptive statistics.

For statistical regression, SYSTAT MGLH provides a summary table and instructions to use the chosen predictors in a model to estimate coefficients. If you follow the instructions, you have a standard multiple regression at the last step. If you want to see results at any other step, you request a standard multiple regression for that step. The only stepping criterion available is the one that other programs label STEPWISE.

Although you can force inclusion of some IVs in the equation, you cannot specify their order of entry. To do hierarchical analysis, then, you must run a separate standard multiple regression for each step in the hierarchy and then compute summary statistics (cf. Sections 5.6.1 and 5.6.2).

5.8 COMPLETE EXAMPLES OF REGRESSION ANALYSIS

To illustrate applications of regression analysis, variables are chosen from among those measured in the research described in Appendix B, Section B.1. Two analyses are reported here, both with number of visits to health professionals (TIMEDRS) as the DV and both using the SPSS^x REGRESSION program.

The first example is a standard multiple regression between the DV and number of physical health symptoms (PHYHEAL), number of mental health symptoms (MENHEAL), and stress from acute life changes (STRESS). From this analysis, one can assess the degree of relationship between the DV and IVs, the proportion of variance in the DV predicted by regression, and the relative importance of the various IVs to the solution.

The second example demonstrates hierarchical regression with the same DV and IVs. The first step of the analysis is entry of PHYHEAL to determine how much variance in number of visits to health professionals can be accounted for by differences in physical health. The second step is entry of STRESS to determine if there is a significant increase in R^2 when differences in stress are added to the equation. The final step is entry of MENHEAL to determine if differences in mental health are related to number of visits to health professionals after differences in physical health and stress are statistically accounted for.

5.8.1 Evaluation of Assumptions

Because both analyses use the same variables, this evaluation is appropriate for both.

5.8.1.1 Ratio of Cases to IVs With 465 respondents and 3 IVs, the cases-to-IV ratio is 155:1, well above the minimum requirements for regression. There are no missing data.

5.8.1.2 Normality, Linearity, Homoscedasticity, and Independence of Residuals In these examples, we choose to examine the residuals to see if screening is necessary, rather than screening prior to examination of residuals. The initial run through SPSS^x REGRESSION uses untransformed variables in a standard multiple regression to produce the scatterplot of residuals against predicted DV scores that appears in Figure 5.4.

Notice the execrable overall shape of the scatterplot that violates all the assumptions regarding distribution of residuals. Comparison of Figure 5.4 with Figure 5.1(a) (in Section 5.3.2.4) suggests further analysis of the distributions of the variables. (It was noticed in passing, although we tried not to look, that R^2 for this analysis was significant, but only .22).

SPSS^x FREQUENCIES is used to examine the distributions of the variables, as seen in Table 5.14. All the variables have significant positive skewness (see Chapter 4), which explains, at least in part, the problems in the residuals scatterplot. Logarithmic and square root transformations are applied as appropriate, and the transformed distributions checked once again for skewness. Thus TIMEDRS and PHYHEAL, with logarithmic transformations, become LTIMEDRS and LPHYHEAL, and STRESS, with square root transformation, becomes SSTRESS. In the case of MENHEAL, application of the milder square root transformation makes the variable significantly negatively skewed, so no transformation is undertaken.

Table 5.15 shows output from FREQUENCIES for one of the transformed variables, LPHYHEAL, the worst prior to transformation. Transformations similarly reduce skewness in the other two transformed variables.

The residuals scatterplot from SPSS^x REGRESSION following regression with

```
TITLE          LARGE SAMPLE REGRESSION-OUTLIERS & RESIDUALS
FILE HANDLE    REGRESS
DATA LIST      FILE=REGRESS       FREE
               /SUBJNO,TIMEDRS,PHYHEAL,MENHEAL,STRESS
VAR LABELS     TIMEDRS 'VISITS TO HEALTH PROFESSIONALS'
REGRESSION     DESCRIPTIVES/
               VARS=TIMEDRS TO STRESS/
               DEP=TIMEDRS/ENTER/
               RESIDUALS=OUTLIERS(MAHAL)/
               SCATTERPLOT (*RESID,*PRED)/
```

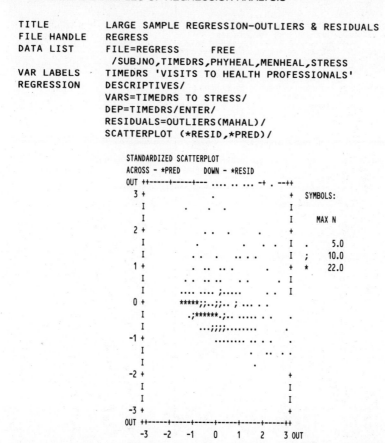

Figure 5.4 SPSS^x REGRESSION setup and residuals scatterplot for original variables.

the transformed variables appears as Figure 5.5. Notice that, although the scatterplot is still not perfectly rectangular, its shape is considerably improved over that in Figure 5.4.

5.8.1.3 Outliers Multivariate outliers are sought using the transformed IVs as part of an SPSS^x REGRESSION run in which the Mahalanobis distance of each case to the centroid of all cases is computed. The ten cases with the largest distance are printed (see Table 5.16). Mahalanobis distance is distributed as a chi square (χ^2) variable, with degrees of freedom equal to the number of IVs. To determine which cases are multivariate outliers, one looks up critical χ^2 at the desired alpha level (Table C.4). In this case, critical χ^2 at $\alpha = .001$ for 3 df is 16.266. Any case with a value larger than 16.266 in the *MAHAL column of the output is a multivariate outlier among the IVs. None of the cases has a value in excess of 16.266. (If outliers are found, the procedures detailed in Chapter 4 are followed to reduce their influence.)

5.8.1.4 Multicollinearity and Singularity The REGRESSION run of Table 5.17 resolves doubts about possible multicollinearity and singularity among the transformed

TABLE 5.14 SETUP AND OUTPUT FOR EXAMINING DISTRIBUTIONS
OF VARIABLES THROUGH SPSS* FREQUENCIES

```
        TITLE        LARGE SAMPLE REGRESSION-FREQUENCIES
        FILE HANDLE  REGRESS
        DATA LIST    FILE=REGRESS    FREE
                        /SUBJNO,TIMEDRS,PHYHEAL,MENHEAL,STRESS
        FREQUENCIES  VARIABLES=TIMEDRS TO STRESS/
                     HISTOGRAM/
                     FORMAT=NOTABLE/
                     STATISTICS=ALL/
```

TIMEDRS   VISITS TO HEALTH PROFESSIONALS

```
     COUNT    MIDPOINT    ONE SYMBOL EQUALS APPROXIMATELY  4.00 OCCURRENCES

        86         0    **********************
       187         4    **************************************************
        74         8    ******************
        44        12    ***********
        27        16    *******
        13        20    ***
         7        24    **
         6        28    **
         3        32    *
         4        36    *
         2        40    *
         1        44
         1        48
         2        52    *
         3        56    *
         3        60    *
         0        64
         0        68
         0        72
         1        76
         1        80
                         I....+....I....+....I....+....I....+....I....+....I....+....I
                         0        40       80       120      160      200
                                    HISTOGRAM FREQUENCY
```

MEAN	7.901	STD ERR	.508	MEDIAN	4.000
MODE	2.000	STD DEV	10.948	VARIANCE	119.870
KURTOSIS	13.101	S E KURT	1.996	SKEWNESS	3.248
S E SKEW	.113	RANGE	81.000	MINIMUM	0
MAXIMUM	81.000	SUM	3674.000		

VALID CASES   465   MISSING CASES   0

PHYHEAL

```
     COUNT    VALUE    ONE SYMBOL EQUALS APPROXIMATELY  2.00 OCCURRENCES

        59      2.00    *******************************
        93      3.00    ***********************************************
        78      4.00    ***************************************
        69      5.00    **********************************
        64      6.00    ********************************
        33      7.00    *****************
        29      8.00    ***************
        18      9.00    *********
         8     10.00    ****
         6     11.00    ***
         4     12.00    **
         2     13.00    *
         1     14.00    *
         1     15.00    *
                         I.........I.........I.........I.........I.........I
                         0        20       40       60       80       100
                                    HISTOGRAM FREQUENCY
```

MEAN	4.972	STD ERR	.111	MEDIAN	5.000
MODE	3.000	STD DEV	2.388	VARIANCE	5.704
KURTOSIS	1.124	S E KURT	1.996	SKEWNESS	1.031
S E SKEW	.113	RANGE	13.000	MINIMUM	2.000
MAXIMUM	15.000	SUM	2312.000		

VALID CASES   465   MISSING CASES   0

TABLE 5.14 (Continued)

MENHEAL

```
     COUNT      VALUE    ONE SYMBOL EQUALS APPROXIMATELY  1.00 OCCURRENCE

        27          0    ****************************
        41       1.00    *****************************************
        35       2.00    ***********************************
        37       3.00    *************************************
        46       4.00    **********************************************
        45       5.00    *********************************************
        48       6.00    ************************************************
        33       7.00    *********************************
        31       8.00    *******************************
        20       9.00    ********************
        24      10.00    ************************
        15      11.00    ***************
        23      12.00    ***********************
        13      13.00    *************
         6      14.00    ******
         9      15.00    *********
         7      16.00    *******
         2      17.00    **
         3      18.00    ***
                         I.........I.........I.........I.........I.........I
                         0        10        20        30        40        50
                                       HISTOGRAM FREQUENCY

     MEAN         6.123    STD ERR        .194    MEDIAN         6.000
     MODE         6.000    STD DEV       4.194    VARIANCE      17.586
     KURTOSIS     -.292    S E KURT      1.996    SKEWNESS        .602
     S E SKEW      .113    RANGE       18.000    MINIMUM            0
     MAXIMUM    18.000    SUM       2847.000

     VALID CASES    465    MISSING CASES      0
```

STRESS

```
     COUNT   MIDPOINT    ONE SYMBOL EQUALS APPROXIMATELY  1.50 OCCURRENCES

        36         20    ************************
        63         64    ******************************************
        59        108    ***************************************
        65        152    *******************************************
        53        196    ***********************************
        48        240    ********************************
        41        284    ***************************
        38        328    *************************
        21        372    **************
        16        416    **********
         6        460    ****
         5        504    ***
         8        548    *****
         3        592    **
         1        636    *
         0        680
         1        724    *
         0        768
         0        812
         0        856
         1        900    *
                         I....+....I....+....I....+....I....+....I....+....I
                         0        15        30        45        60        75
                                       HISTOGRAM FREQUENCY

     MEAN       204.217    STD ERR       6.297    MEDIAN       178.000
     MODE             0    STD DEV     135.793    VARIANCE   18439.662
     KURTOSIS     1.801    S E KURT      1.996    SKEWNESS       1.043
     S E SKEW      .113    RANGE       920.000    MINIMUM            0
     MAXIMUM    920.000    SUM       94961.000

     VALID CASES    465    MISSING CASES      0
```

TABLE 5.15 SETUP AND OUTPUT FOR EXAMINING DISTRIBUTION
OF TRANSFORMED VARIABLE THROUGH
SPSS' FREQUENCIES

```
        TITLE          LARGE SAMPLE REGRESSION-FREQUENCIES
        FILE HANDLE    REGRESS
        DATA LIST      FILE=REGRESS      FREE
                         /SUBJNO,TIMEDRS,PHYHEAL,MENHEAL,STRESS
        VAR LABELS     TIMEDRS 'VISITS TO HEALTH PROFESSIONALS'
        COMPUTE        LTIMEDRS = LG10(TIMEDRS + 1)
        COMPUTE        LPHYHEAL = LG10(PHYHEAL)
        COMPUTE        SSTRESS = SQRT(STRESS)
        FREQUENCIES    VARIABLES=LTIMEDRS,LPHYHEAL,SSTRESS/
                       HISTOGRAM/
                       FORMAT=NOTABLE/
                       STATISTICS=ALL/
```

```
LTIMEDRS

     COUNT   MIDPOINT   ONE SYMBOL EQUALS APPROXIMATELY  1.50 OCCURRENCES

        42       0      ******************************
         0      .1
         0      .2
        44      .3      ******************************
         0      .4
        67      .5      ***************************************************
        52      .6      ***********************************
        41      .7      ****************************
        53      .8      ************************************
        25      .9      *****************
        34     1.0      ***********************
        33     1.1      **********************
        24     1.2      ****************
        16     1.3      ***********
        11     1.4      *******
         6     1.5      ****
         6     1.6      ****
         3     1.7      **
         6     1.8      ****
         2     1.9      *
         0     2.0
                        I....+....I....+....I....+....I....+....I....+....I
                        0        15       30       45       60       75
                                 HISTOGRAM FREQUENCY
```

```
MEAN        .741     STD ERR      .019     MEDIAN       .699
MODE        .477     STD DEV      .415     VARIANCE     .172
KURTOSIS   -.177     S E KURT    1.996     SKEWNESS     .228
S E SKEW    .113     RANGE       1.914     MINIMUM        0
MAXIMUM    1.914     SUM       344.698
```

IVs. All variables enter the equation without violating the default value for tolerance (cf. Chapter 4). Further, the highest correlation, between MENHEAL and LPHY-HEAL, is .511. (If multicollinearity is indicated, SMCs between each IV serving as DV with the other two as IVs are found through REGRESSION or any of the other appropriate programs and redundant IVs dealt with as discussed in Chapter 4.)

5.8.2 Standard Multiple Regression

SPSS^x REGRESSION is used to compute a standard multiple regression between LTIMEDRS (the transformed DV), and MENHEAL, LPHYHEAL, and SSTRESS (the IVs).

The output of this REGRESSION run appears as Table 5.17. Included are descriptive statistics, the values of R, R^2, and adjusted R^2, and a summary of the analysis of variance for regression. The significance level for R is found in the source table with $F(3, 461) = 92.90$, $p < .001$. In the portion labeled VARIABLES IN

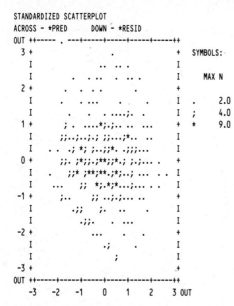

```
STANDARDIZED SCATTERPLOT
ACROSS - *PRED      DOWN - *RESID
OUT ++----- . ---+-----+-----+-----+-----++
  3 +             .              +   SYMBOLS:
  I                 .. ... .     I
  I            .  . ...  . ...   I     MAX N
  2 +         .  .  . .        . +
  I          .  . ...    .   .   I   .   2.0
  I           .   . .....;. .    I   ;   4.0
  1 +        ; . ....*;.;. .. ...  +   *   9.0
  I          ;;..;..;.; ;;....;*.. ..  I
  I      .. .; *; ;..;;*. .;;;...   I
  0 +       ;;. ;*;;.;**;;*.; ;.;.... .   +
  I      .   ;;* ;**;**.;*;..; ... . .  I
  I      ...   ;; *;.*;*...;... . .  I
 -1 +      ;..    ;; ..;.;.... ..   +
  I        .;;   ;. ..    .     I
  I        .;;.   . ...          I
 -2 +           ...   .  .      +
  I             .;    .         I
  I              ;             I
 -3 +                          .+
OUT ++-----+-----+-----+-----+-----++
    -3   -2   -1    0    1    2   3 OUT
```

Figure 5.5 Residuals scatterplot following regression with transformed variables. Output from SPSS^X REGRESSION. See Table 5.16 for setup.

THE EQUATION are printed unstandardized and standardized regression coefficients with their significance levels and 95% confidence intervals, and three correlations: total (CORREL), semipartial (PART COR), and partial.

The significance levels for the regression coefficients (see Table 5.17, VARIABLES IN THE EQUATION) are evaluated against 1 and 461 df. Only two of the IVs, SSTRESS and LPHYHEAL, contribute significantly to regression, with F values

TABLE 5.16 SETUP AND OUTPUT FROM SPSS^x REGRESSION SHOWING MULTIVARIATE OUTLIERS

```
TITLE          LARGE SAMPLE STANDARD MULTIPLE REGRESSION
FILE HANDLE    REGRESS
DATA LIST      FILE=REGRESS     FREE
                 /SUBJNO ,TIMEDRS ,PHYHEAL ,MENHEAL ,STRESS
VAR LABELS     TIMEDRS 'VISITS TO HEALTH PROFESSIONALS'
COMPUTE        LTIMEDRS = LG10(TIMEDRS + 1)
COMPUTE        LPHYHEAL = LG10(PHYHEAL)
COMPUTE        SSTRESS = SQRT(STRESS)
REGRESSION     DESCRIPTIVES/
               VARS=MENHEAL ,LTIMEDRS TO SSTRESS/
               STATISTICS=DEFAULTS,ZPP,CI,F/
               DEP=LTIMEDRS/ENTER/
               RESIDUALS=OUTLIERS(MAHAL)/
               SCATTERPLOT (*RESID,*PRED)/


OUTLIERS - MAHALANOBIS' DISTANCE

  CASE #        *MAHAL

     403      14.13530
     125      11.64940
     198      10.56870
      52      10.54759
     446      10.22483
     159       9.35066
      33       8.62835
     280       8.58686
     405       8.43122
     113       8.35261
```

TABLE 5.17 STANDARD MULTIPLE REGRESSION ANALYSIS OF
LTIMEDRS (THE DV) WITH MENHEAL, SSTRESS, AND
LPHYHEAL (THE IVs). SELECTED OUTPUT. SETUP SHOWN
IN TABLE 5.16

```
               MEAN   STD DEV   LABEL

MENHEAL        6.123    4.194
LTIMEDRS        .741     .415
LPHYHEAL        .648     .206
SSTRESS       13.400    4.972

N OF CASES =    465

CORRELATION:

             MENHEAL   LTIMEDRS   LPHYHEAL   SSTRESS

MENHEAL       1.000      .355       .511       .383
LTIMEDRS       .355     1.000       .586       .359
LPHYHEAL       .511      .586      1.000       .317
SSTRESS        .383      .359       .317      1.000


            * * * *   M U L T I P L E   R E G R E S S I O N   * * * *

EQUATION NUMBER 1    DEPENDENT VARIABLE..  LTIMEDRS

BEGINNING BLOCK NUMBER  1.  METHOD:  ENTER

VARIABLE(S) ENTERED ON STEP NUMBER  1..    SSTRESS
                                    2..    LPHYHEAL
                                    3..    MENHEAL


MULTIPLE R           .61382       ANALYSIS OF VARIANCE
R SQUARE             .37678                      DF     SUM OF SQUARES    MEAN SQUARE
ADJUSTED R SQUARE    .37272       REGRESSION      3          30.14607       10.04869
STANDARD ERROR       .32888       RESIDUAL      461          49.86410         .10817

                     F =     92.90143      SIGNIF F = -.0000


--------------------------------------- VARIABLES IN THE EQUATION ---------------------------------------

                                        95 CONFDNCE                          PART
VARIABLE         B          SE B        INTRVL BB    BETA    CORREL    COR    PARTIAL        F    SIG F

SSTRESS        .01571      .00336    .00910   .02232  .18811  .35905  .17173  .21256     21.815   .0000
LPHYHEAL      1.03997      .08718    .86864  1.21130  .51641  .58575  .43859  .48565    142.289   .0000
MENHEAL        .00188      .00440   -.00677   .01053  .01902  .35513  .01573  .01993       .183   .6689
(CONSTANT)    -.15504      .05826   -.26952  -.04056                                     7.083   .0081


FOR BLOCK NUMBER  1    ALL REQUESTED VARIABLES ENTERED.
```

of 21.82 and 142.29, respectively. The 95% confidence intervals for these IVs also do not contain zero as similar indications of their significance.

Semipartial correlations (cf. Equation 5.9) are labeled PART COR in the VARIABLES IN THE EQUATION section. These values, when squared, indicate the amount by which R^2 would be reduced if an IV were omitted from the equation. The sum for the two significant IVs ($.17173^2 + .43859^2 = .2219$) is the amount of R^2 attributable to unique sources. The difference between R^2 and unique variance ($.37678 - .2219 = .1548$) represents variance that SSTRESS, LPHYHEAL, and MENHEAL jointly contribute to R^2.

Information from this analysis is summarized in Table 5.18, in a form that might be appropriate for publication in a professional journal.

It is noted from the correlation matrix in Table 5.17 that MENHEAL correlates with LTIMDRS ($r = .355$) but does not contribute significantly to regression. If

TABLE 5.18 STANDARD MULTIPLE REGRESSION OF HEALTH AND STRESS VARIABLES ON NUMBER OF VISITS TO HEALTH PROFESSIONALS

Variables	LTIMEDRS (DV)	LPHYHEAL	SSTRESS	MENHEAL	B	β	sr^2 (unique)
LPHYHEAL	.59				1.040**	0.52	.19
SSTRESS	.36	.32			0.016**	0.19	.03
MENHEAL	.36	.51	.38		0.002	0.02	
				Intercept = − 0.155			
Means	0.74	0.65	13.40	6.12			
Standard deviations	0.42	0.21	4.97	4.19			
						R^2 = .38ª	
					Adjusted R^2 = .37		
						R = .61**	

** p < .01.

ª Unique variability = .22; shared variability = .16.

TABLE 5.19   CHECKLIST FOR STANDARD MULTIPLE
                 REGRESSION

1. Issues
   a. Ratio of cases to IVs
   b. Normality, linearity, and homoscedasticity of residuals
   c. Outliers
   d. Multicollinearity and singularity
2. Major analyses
   a. Multiple R^2, F ratio
   b. Adjusted multiple R^2, overall proportion of variance accounted for
   c. Significance of regression coefficients
   d. Squared semipartial correlations
3. Additional analyses
   a. Post hoc significance of correlations
   b. Unstandardized (B) weights, confidence limits
   c. Standardized (β) weights
   d. Unique versus shared variability
   e. Suppressor variables
   f. Prediction equation

Equation 5.11 is used a posteriori to evaluate the significance of the correlation coefficient,

$$ F = \frac{(.355)^2/3}{(1 - .355^2)/(465 - 3 - 1)} = 22.16 $$

the correlation between MENHEAL and LTIMDRS differs reliably from zero; $F(3, 461) = 22.16$, $p < .01$.

Thus, although the bivariate correlation between MENHEAL and LTIMEDRS is reliably different from zero, the relationship seems to be mediated by, or redundant to, the relationship between LTIMDRS and other IVs in the set. Had the researcher measured only MENHEAL and LTIMEDRS, however, the significant correlation might have led to stronger conclusions than are warranted about the relationship between mental health and number of visits to health professionals.

Table 5.19 contains a checklist of analyses performed in standard multiple regression. An example of a Results section in journal format appears next.

```
              Results
    A standard multiple regression was
performed between number of visits to health
professionals as the dependent variable and
physical health, mental health, and stress
as independent variables. Analysis was
performed using SPSSˣ REGRESSION with an
```

assist from SPSS[x] FREQUENCIES for evaluation
of assumptions.

Results of evaluation of assumptions
led to transformation of the variables to
reduce skewness in their distributions,
reduce the number of outliers, and improve
the normality, linearity, and
homoscedasticity of residuals. A square
root transformation was used on the measure
of stress. Logarithmic transformations were
used on number of visits to health
professionals and on physical health. One
IV, mental health, was positively skewed
without transformation and negatively
skewed with it; it was not transformed. With
the use of a $p < .001$ criterion for
Mahalanobis distance no outliers among the
cases were found. No cases had missing data
and no suppressor variables were found, N =
465.

Table 5.18 displays the correlations
between the variables, the unstandardized
regression coefficients (B) and intercept,
the standardized regression coefficients
(β), the semipartial correlations (sr^2) and
R, R^2, and adjusted R^2. R for regression was
significantly different from zero, $F(3,
461) = 92.90$, $p < .001$. For the two
regression coefficients that differed
significantly from zero, 95% confidence
limits were calculated. The confidence
limits for square root of stress were 0.0091
to 0.0223, and those for log of physical
health were 0.8686 to 1.2113.

Insert Table 5.18 about here

Only two of the IVs contributed
significantly to prediction of number of
visits to health professionals as
logarithmically transformed, log of
physical health scores ($sr^2 = .19$) and
square root of acute stress scores ($sr^2 =
.03$). The three IVs in combination

> contributed another .15 in shared
> variability. Altogether, 38% (37% adjusted)
> of the variability in visits to health
> professionals was predicted by knowing
> scores on these three IVs.
>
>   Although the correlation between log of
> visits to health professionals and mental
> health was .36, mental health did not
> contribute significantly to regression.
> Post hoc evaluation of the correlation
> revealed that it was significantly
> different from zero, $\underline{F}$(3, 461) = 22.16, $\underline{p}$ <
> .01. Apparently the relationship between
> the number of visits to health professionals
> and mental health is an indirect result of
> the relationships between physical health,
> stress, and visits to health professionals.

5.8.3 Hierarchical Regression

The second example involves the same three IVs entered one at a time in an order determined by the researcher. LPHYHEAL is the first IV to enter, followed by SSTRESS and then MENHEAL. The main research question is whether or not information regarding differences in mental health can be used to predict visits to health professionals after differences in physical health and in acute stress are statistically eliminated. In other words, do people go to health professionals for more numerous mental health symptoms if they have physical health and stress similar to other people?

Table 5.20 shows setup and selected portions of the output for hierarchical analysis using the SPSS^x REGRESSION program. Notice that a complete regression solution is provided at the end of each of steps 1 through 3. The significance of the bivariate relationship between LTIMEDRS and LPHYHEAL is assessed at the end of step 1, $F(1, 463) = 241.83, p < .001$. The bivariate correlation is .59, accounting for 34% of the variance. After step 2, with both LPHYHEAL and SSTRESS in the equation, $F(2, 462) = 139.51, p < .01, R = .61$, and $R^2 = .38$. With the addition of MENHEAL, $F(3, 461) = 92.90, R = .61$, and $R^2 = .38$. Increments in R^2 at each step are read directly from the RSQCH column of the SUMMARY TABLE. Thus, $sr^2_{LPHYHEAL} = .34$ $sr^2_{SRAHE} = .03$, and $sr^2_{MENHEAL} = .00$

Significance of the addition of SSTRESS to the equation is indicated in the output at STEP NUMBER 2 (the step where SSTRESS entered) as F for SSTRESS in the segment labeled VARIABLES IN THE EQUATION. Because the F value of 24.772 exceeds critical F with 1 and 461 df, SSTRESS is making a significant contribution to the equation at this step.

Similarly, significance of the addition of MENHEAL to the equation is indicated at STEP NUMBER 3, where the F for MENHEAL is .183. Because this F value does not exceed critical F with 1 and 461 df, MENHEAL is not significantly improving R^2 at its point of entry.

The significance levels of the squared semipartial correlations are also available in the summary table as FCH, with probability value SIGCH for evaluating the significance of the added IV.

Thus there is no reliable increase in prediction of LTIMEDRS by addition of MENHEAL to the equation if differences in LPHYHEAL and SSTRESS are already accounted for. Apparently the answer is no to the question: Do people go to health professionals for more numerous mental health symptoms if they have physical health and stress similar to others? A summary of information from this output appears in Table 5.21.

Table 5.22 is a checklist of items to consider in hierarchical regression. An example of a Results section in journal format appears next.

TABLE 5.20 SETUP AND SELECTED OUTPUT FOR SPSS*
HIERARCHICAL REGRESSION

```
         TITLE          LARGE SAMPLE HIERARCHICAL MULTIPLE REGRESSION
         FILE HANDLE    REGRESS
         DATA LIST      FILE=REGRESS      FREE
                          /SUBJNO,TIMEDRS,PHYHEAL,MENHEAL,STRESS
         VAR LABELS     TIMEDRS 'VISITS TO HEALTH PROFESSIONALS'
         COMPUTE        LTIMEDRS = LG10(TIMEDRS + 1)
         COMPUTE        LPHYHEAL = LG10(PHYHEAL)
         COMPUTE        SSTRESS = SQRT(STRESS)
         REGRESSION     VARS=MENHEAL,LTIMEDRS TO SSTRESS/
                        STATISTICS=DEFAULTS,HISTORY,CI,F/
                        DEP=LTIMEDRS/ENTER LPHYHEAL/
                                     ENTER SSTRESS/
                                     ENTER MENHEAL/
```

VARIABLE LIST NUMBER 1   LISTWISE DELETION OF MISSING DATA

EQUATION NUMBER 1   DEPENDENT VARIABLE.. LTIMEDRS

BEGINNING BLOCK NUMBER 1. METHOD: ENTER   LPHYHEAL

VARIABLE(S) ENTERED ON STEP NUMBER 1..   LPHYHEAL

		ANALYSIS OF VARIANCE			
MULTIPLE R	.58575		DF	SUM OF SQUARES	MEAN SQUARE
R SQUARE	.34310	REGRESSION	1	27.45151	27.45151
ADJUSTED R SQUARE	.34168	RESIDUAL	463	52.55866	.11352
STANDARD ERROR	.33692				

F =   241.82595   SIGNIF F = .0000

---------------------------- VARIABLES IN THE EQUATION ----------------------------

VARIABLE	B	SE B	95 CONFDNCE INTRVL BB		BETA	F	SIG F
LPHYHEAL	1.17961	.07586	1.03055	1.32868	.58575	241.826	.0000
(CONSTANT)	-.02354	.05160	-.12495	.07786		.208	.6484

------------- VARIABLES NOT IN THE EQUATION -------------

VARIABLE	BETA IN	PARTIAL	MIN TOLER	F	SIG F
MENHEAL	.07528	.07982	.73850	2.963	.0859
SSTRESS	.19278	.22559	.89956	24.772	.0000

FOR BLOCK NUMBER 1   ALL REQUESTED VARIABLES ENTERED.

TABLE 5.20 (Continued)

```
EQUATION NUMBER 1    DEPENDENT VARIABLE..  LTIMEDRS

BEGINNING BLOCK NUMBER  2.  METHOD:  ENTER       SSTRESS


VARIABLE(S) ENTERED ON STEP NUMBER  2..    SSTRESS

MULTIPLE R            .61362         ANALYSIS OF VARIANCE
R SQUARE             .37653                        DF    SUM OF SQUARES    MEAN SQUARE
ADJUSTED R SQUARE    .37383         REGRESSION       2          30.12626       15.06313
STANDARD ERROR       .32859         RESIDUAL       462          49.88391         .10797

                                    F =    139.50724     SIGNIF F = -.0000
```

```
----------------------------- VARIABLES IN THE EQUATION -----------------------------

VARIABLE            B          SE B     95 CONFDNCE INTRVL BB       BETA       F  SIG F

LPHYHEAL       1.05658       .07800      .90330     1.20986      .52465   183.486  .0000
SSTRESS         .01610       .00323      .00974      .02246      .19278    24.772  .0000
(CONSTANT)     -.15950       .05726     -.27203     -.04697                7.758  .0056
```

```
------------ VARIABLES NOT IN THE EQUATION -------------

VARIABLE     BETA IN  PARTIAL  MIN TOLER     F  SIG F

MENHEAL       .01902   .01993     .68426    .183  .6689
```

```
FOR BLOCK NUMBER  2   ALL REQUESTED VARIABLES ENTERED.
              * * * *   M U L T I P L E   R E G R E S S I O N   * * * *

EQUATION NUMBER 1    DEPENDENT VARIABLE..  LTIMEDRS

BEGINNING BLOCK NUMBER  3.  METHOD:  ENTER       MENHEAL


VARIABLE(S) ENTERED ON STEP NUMBER  3..    MENHEAL


MULTIPLE R            .61382         ANALYSIS OF VARIANCE
R SQUARE             .37678                        DF    SUM OF SQUARES    MEAN SQUARE
ADJUSTED R SQUARE    .37272         REGRESSION       3          30.14607       10.04869
STANDARD ERROR       .32888         RESIDUAL       461          49.86410         .10817

                                    F =     92.90142     SIGNIF F = -.0000
```

```
----------------------------- VARIABLES IN THE EQUATION -----------------------------

VARIABLE            B          SE B     95 CONFDNCE INTRVL BB       BETA       F  SIG F

LPHYHEAL       1.03997       .08718      .86864     1.21130      .51641   142.289  .0000
SSTRESS         .01571       .00336      .00910      .02232      .18811    21.815  .0000
MENHEAL         .00188       .00440     -.00677      .01053      .01902      .183  .6689
(CONSTANT)     -.15504       .05826     -.26952     -.04056                7.083  .0081
```

```
FOR BLOCK NUMBER  3   ALL REQUESTED VARIABLES ENTERED.



          * * * * * * * * * * * * * * * * * * * * * * * * * * * * *


                                    SUMMARY TABLE
                                    -------------


STEP  MULTR   RSQ   ADJRSQ    F(EQU)  SIGF  RSQCH     FCH  SIGCH       VARIABLE  BETAIN  CORREL
  1  .5857  .3431   .3417   241.826  .000  .3431  241.826  .000  IN:  LPHYHEAL   .5857   .5857
  2  .6136  .3765   .3738   139.507 -.000  .0334   24.772  .000  IN:  SSTRESS    .1928   .3590
  3  .6138  .3768   .3727    92.901 -.000  .0002     .183  .669  IN:  MENHEAL    .0190   .3551
```

TABLE 5.21 HIERARCHICAL REGRESSION OF HEALTH AND STRESS VARIABLES ON NUMBER OF VISITS TO HEALTH PROFESSIONALS

Variables	LTIMEDRS (DV)	LPHYHEAL	SSTRESS	MENHEAL	B	β	sr^2 (incremental)
LPHYHEAL	.59				1.040	0.52	.34**
SSTRESS	.36	.32			0.016	0.19	.03**
MENHEAL	.36	.51	.38		0.002	0.02	.00
				Intercept =	−0.155		
Means	0.74	0.65	13.40	6.12			
Standard deviation	0.42	0.21	4.97	4.19			

$$R^2 = .38$$
$$\text{Adjusted } R^2 = .37$$
$$R = .61**$$

** $p < .01$.

TABLE 5.22   CHECKLIST OF HIERARCHICAL
                 REGRESSION ANALYSIS

1. Issues
    a. Ratio of cases to IVs
    b. Normality, linearity, and homoscedasticity of residuals
    c. Outliers
    d. Multicollinearity and singularity
2. Major analyses
    a. Multiple R^2, F ratio
    b. Adjusted R^2, proportion of variance accounted for
    c. Squared semipartial correlations
    d. Significance of regression coefficients
    e. Incremental F
3. Additional analyses
    a. Unstandardized (B) weights, confidence limits
    b. Standardized (β) weights
    c. Prediction equation
    d. Post hoc significance of correlations
    e. Suppressor variables

Results

Hierarchical regression was employed to determine if addition of information regarding stress and then mental health symptoms improved prediction of visits to health professionals beyond that afforded by differences in physical health. Analysis was performed using SPSS^x REGRESSION with an assist from SPSS^x FREQUENCIES in evaluation of assumptions.

Results of evaluation of assumptions led to transformation of the variables to reduce skewness in their distributions, reduce the number of outliers, and improve the normality, linearity, and homoscedasticity of residuals. A square root transformation was used on the measure of stress. Logarithmic transformations were used on the number of visits to health professionals and physical health. One IV, mental health, was positively skewed without transformation and negatively skewed with it; it was not transformed. With

the use of a $p < .001$ criterion for
Mahalanobis distance no outliers among the
cases were identified. No cases had missing
data and no suppressor variables were found,
$N = 465$.

Table 5.21 displays the correlations
between the variables, the unstandardized
regression coefficients (B) and intercept,
the standardized regression coefficients
(β), the semipartial correlations (sr^2), and
R, R^2, and adjusted R^2 after entry of all
three IVs. R was significantly different
from zero at the end of each step. After step
3, with all IVs in the equation, $R = .61$,
$F(3, 461) = 92.90$, $p < .01$.

Insert Table 5.21 about here

After step 1, with log of physical
health in the equation, $R^2 = .34$, F_{inc} (1,
461) $= 241.83$, $p < .001$. After step 2, with
square root of stress added to prediction of
log of visits to health professionals by log
of physical health, $R^2 = .38$, F_{inc} (1, 461) $=$
24.77, $p < .01$. Addition of square root of
stress to the equation results in a
significant increment in R^2. After step 3,
with mental health added to prediction of
visits by log of physical health and square
root of stress, $R^2 = .38$ (adjusted $R^2 = .37$),
F_{inc} (1, 461) $= 0.18$. Addition of mental
health to the equation did not reliably
improve R^2.

5.9 SOME EXAMPLES FROM THE LITERATURE

The following examples have been included to give you a taste for the ways in which
regression can be, and has been, used by researchers.

Multiple regression is the statistical tool used in judgment analysis, a procedure
developed to identify important variables underlying policy decisions. In a demonstration involving diagnosis of learning disabilities, Beatty (1977) gave diagnosticians
(judges) test profiles for a sample of children. Judges ranked the children in order

of severity of learning disability, and these rankings served as the DV.[3] Beatty used as IVs the age of the child and the set of five test scores on which judges had based their rankings.

Judges were analyzed individually, and then in combination, so that judges using similar policies could be grouped. For each group of judges with a distinct policy, it was possible to evaluate how successfully the rankings could be predicted and the relative importance of IVs in influencing the diagnostic rankings. In Beatty's demonstration, the three scores on the Wechsler Intelligence Scale for Children (Verbal, Performance, and Full Scale) contributed the most to multiple correlation for all judges. Some judges seemed to have more predictable policies than others; for one judge, 96% of the variance in rankings was explained by knowledge of the IVs.

In evaluating a negative income tax project in New Jersey, Nicholson and Wright (1977) argue for inclusion of "participant's understanding of the treatment" as one of the IVs in policy experimentation. At each stage of their analysis, separate multiple regression equations were set up for each of six DVs (all related to a family's earnings and hours worked). (These six DVs could have been combined into a single canonical correlation evaluated against the IVs; see Chapter 6.) Separate equations were also run on the basis of ethnicity. IVs for all equations included (1) pretreatment value of the DV; (2) variables related to family age, education, and size; (3) variables related to site and whether or not families received welfare; (4) a dummy variable indicating whether a family was in the treatment or the control group; and (5) variables related to specification of which experimental plan a family was in, with eight combinations of guaranteed income and tax levels.

As the main focus of the evaluation, the regression coefficient for the difference between experimental and control groups was evaluated for statistical significance in each of the regression equations. Then Nicholson and Wright reran the regression equations with 11 binary variables related to participant's knowledge about the tax plan. Finally, a third set of equations was run that included terms related to interactions between knowledge levels and estimated effects of the guarantee and tax levels. It was found that the multiple correlation increased significantly with the addition of variables representing participant knowledge, and increased again when interaction variables were included.

In general, estimates of disincentive effects for families with high understanding were larger than for those who lacked understanding of the system. That is, among families who understood the system, those in the experimental group worked fewer hours and earned less than those in the control group. The difference between experimental and control was less, and in some cases in the opposite direction, for families who showed little understanding of the policy.

S. Fidell (1978) applied stepwise regression procedures to predict annoyance among groups of people. Respondents were interviewed at 24 sites across the United States, stratified by noise exposure and population density. The proportion of respondents at each site who had been highly annoyed by neighborhood noise exposure was predicted from both situational (demographic and physical) variables and atti-

---

[3] Opinions differ regarding application of regression to rank order data. However, since rank order data produce rectangular distributions with neither skewness nor outliers, the application may be considered justified.

tudinal variables. Since the prevalence of shared attitudes in the community was of greater interest than the intensity of individual beliefs, the unit of analysis was the site rather than the individual respondent. Primary interest in prediction also dictated stepwise entry of variables, many of which were available from census data. Excellent prediction was available from several combinations of variables, with one equation of 3 IVs accounting for 88% of the variance in community annoyance at exposure to noise.

Species density as a function of environmental measures in the United States was studied by Schall and Pianka (1978) in statistical regression analysis. Separate analyses were run with birds and lizards as DVs. The 11 IVs were the same for both analyses, consisting of such environmental variables as highest and lowest elevation, sunfall, average annual precipitation and several temperature measures. Units of measurement (cases) were 895 quadrants, each $1°$ of latitude by $1°$ of longitude. Environmental measures were found to be much better predictors of lizard density (multiple $R^2 = .82$) than of bird density (multiple $R^2 = .34$). There is also suggestion that different environmental factors predict density for lizards than for birds. Sunfall, entering the regression first for lizards, by itself accounted for .67 of the variance in density. For birds, sunfall appeared to be acted on by a suppressor variable. Simple correlation with bird density was virtually zero, but sunfall entered fifth in the regression and added 9% of predictable variance. Although no cross validation was reported, discrepancies in the two analyses are large enough to be considered important. (Hierarchical regression would have produced results that could be interpreted with more confidence.)

Maki, Hoffman, and Berk (1978) used standard multiple regression to study the effectiveness of a water conservation campaign. Two parallel analyses were done on sales and production of water (DVs) over 126 months, with months acting as the unit of study. IVs were moratorium (whether or not it and/or conservation measures were in effect), rainfall, season, population, and dollars spent on public relations (as a measure of intensity of campaign).

Results of the two analyses were similar, with both showing significant effects of the moratorium-conservation efforts. Money spent on the campaign was not statistically significant in either campaign. Rainfall was a significant predictor of production, but not of sales. Population and season predicted both sales and production.

Chapter 6

Canonical Correlation

6.1 GENERAL PURPOSE AND DESCRIPTION

The goal of canonical correlation is to analyze the relationships between two sets of variables. It may be useful to think of one set of variables as IVs and the other set as DVs, or it may not. In any event, canonical correlation provides a statistical analysis for research where each subject is measured on two sets of variables and the researcher wants to know how the two sets relate to each other.

Suppose, for instance, a researcher is interested in the relationship between a set of variables measuring medical compliance (willingness to buy drugs, to make return office visits, to use drugs, to restrict activity) and a set of demographic characteristics (educational level, religious affiliation, income, medical insurance). Canonical analysis might reveal that there are two reliable ways that the two sets of variables are related. The first way is between income and insurance on the demographic side and purchase of drugs and willingness to make return office visits on the medical compliance side. Together, these results indicate a relationship between compliance and demography based on ability to pay for medical services. The second way is between willingness to use drugs and restrict activity on the compliance side and religious affiliation and educational level on the demographic side, interpreted, perhaps, as a tendency to accede to authority (or not).

The easiest way to understand canonical correlation is to think in terms of multiple regression. In regression, there are several variables on one side of the equation and a single variable on the other side. The several variables are combined into a predicted value to produce, across all subjects, the highest correlation between the predicted value and the single variable. The combination of variables can be thought of as a dimension among the many variables that predicts the single variable.

In canonical correlation, the same thing happens except that there are several variables on both sides of the equation. Sets of variables on each side are combined to produce, for each side, a predicted value that has the highest correlation with the predicted value on the other side. The combination of variables on each side can be thought of as a dimension that relates the variables on one side to the variables on the other.

There is a complication, however. In multiple regression, there is only one combination of variables because there is only a single variable to predict on the other side of the equation. In canonical correlation, however, there are several variables on both sides and there may be several ways to recombine the variables on both sides to relate them to each other. In the example, the first way of combining the variables had to do with money and the second way had to do with authority. There are as many ways to recombine the variables as there are variables in the smaller set. Usually, however, only the first two or three combinations are reliable and need to be interpreted.

A good deal of the difficulty with canonical correlation is due to jargon. First there are variables, then there are canonical variates, and finally there are pairs of canonical variates. Variables refers to the variables measured in research (e.g., income). Canonical variates are linear combinations of variables, one combination on the IV side (e.g., income and medical insurance) and a second combination on the DV side (e.g., purchase of drugs and willingness to make return office visits). These two combinations form a pair of canonical variates. However, there may be more than one reliable pair of canonical variates (e.g., a pair associated with money and a pair associated with authority).

Canonical analysis is the most general of the multivariate techniques. In fact, the other procedures—multiple regression, discriminant function analysis, MANOVA—are all special cases of it. But it is also the least used and most impoverished of the techniques, for reasons that are discussed below.

6.2 KINDS OF RESEARCH QUESTIONS

Although a large number of research questions are answered by canonical analysis in one of its specialized forms (such as discriminant analysis), relatively few intricate research questions are readily answered through direct application of computer programs currently available for canonical correlation. In part this has to do with the programs themselves and in part it has to do with the kinds of questions researchers consider appropriate in canonical correlation.

In its present stage of development, canonical correlation is best considered a descriptive technique or a screening procedure rather than a hypothesis-testing procedure. The following sections, however, contain questions that can be addressed with the aid of SPSS, BMDP, SAS, and SYSTAT programs.

6.2.1 Number of Canonical Variate Pairs

How many reliable canonical variate pairs are there in the data set? Or, Along how many dimensions are the variables in one set related to the variables in the other? In the example, Is the pair associated with income reliable? And if it is, Is the pair associated with authority also reliable? Because canonical variate pairs are computed in descending order of magnitude, the first one or two pairs are often reliable while remaining ones are not. Significance tests for canonical variate pairs are described in Sections 6.4 and 6.5.1.

6.2.2 Nature of Canonical Variates

How are the dimensions that relate two sets of variables to be interpreted? What is the meaning in the pattern of variables that compose one variate in conjunction with the pattern composing the other in the same pair? In the example, the variables that are important to the first pair of canonical variates all have to do with money, so the pair is interpreted as an economic dimension. Interpretation of pairs of canonical variates usually proceeds from matrices of correlations between variables and canonical variates, as described in Sections 6.4 and 6.5.2.

6.2.3 Importance of Canonical Variates

There are several ways to assess the importance of canonical variates. The first is to ask how strongly the variate on one side of the equation relates to the variate on the other side of the equation; that is, how strong is the correlation between variates in a pair? The second is to ask how strongly the variate on one side of the equation relates to the variables on its own side of the equation. The third is to ask how strongly the variate on one side of the equation relates to the variables on the other side of the equation. Finally, if variates are not interpretable, they are not retained, no matter how important they seem to be from a variance perspective.

For the example, What is the correlation between the income variate on the compliance side and the income variate on the demographic side? Then, How much variance does the income variate on the compliance side extract from the compliance variables? Finally, How much variance does the income variate on the compliance side extract from the demographic variables? These questions are answered by the procedures described in Sections 6.4 and 6.5.1.

6.2.4 Canonical Variate Scores

Had the canonical variates from both sets of variables been measured directly, what scores would subjects have received on them? For instance, What scores would the first subject have received had it been possible to measure directly the income variate from the compliance side and the income variate from the demographic side? Examination of canonical variate scores reveals deviant cases, the shape of the relationship between two canonical variates, and the shape of the relationships between canonical variates and the original variables, as discussed briefly in Sections 6.3 and 6.4.

If canonical variates are interpretable, scores on them might be useful as IVs or DVs in other analyses. For instance, the researcher might use scores on the income variate from the compliance side to examine the effects of publicly supported medical facilities.

6.3 LIMITATIONS

6.3.1 Theoretical Limitations[1]

Canonical correlation has several important theoretical limitations that help explain its rarity in the literature. Perhaps the most critical limitation is interpretability; procedures that maximize correlation do not necessarily maximize interpretation of pairs of canonical variates. Therefore a canonical solution may be mathematically elegant but uninterpretable. And, although it is common in some multivariate procedures (e.g., factor analysis, Chapter 12) to rotate the solution to improve interpretation, rotation of canonical variates is not common practice or even available in some computer programs.

The algorithm used for canonical correlation maximizes the linear relationship between two sets of variables. If the relationship is curvilinear the analysis misses some or most of the relationship. If you suspect a curvilinear relationship between dimensions in a pair, you may want to reconsider use of canonical correlation or transform the variables to capture the nonlinear relationship.

The algorithm also computes pairs of canonical variates that are independent of all other pairs. In factor analysis (Chapter 12) one has a choice between an orthogonal (uncorrelated) and an oblique (correlated) solution, but in canonical analysis only the orthogonal solution is routinely available.

An important concern is the sensitivity of the solution in one set of variables to the variables included *in the other set*. In canonical analysis, the solution depends both on correlations among variables in each set and on correlations among variables between sets. Changing the variables in one set may markedly alter the composition of canonical variates in the other set. To some extent, this is expected given the goals of analysis, yet the sensitivity of the procedure to apparently minor changes is a cause for concern.

6.3.2 Practical Issues

6.3.2.1 Ratio of Cases to IVs   The number of cases needed for analysis depends on the reliability of the variables. For variables in the social sciences where reliability is often around .80, 10 cases are needed for every variable. However, if reliability is very high, as, for instance, in political science where the variables are the economic performance of countries and the like, then a much lower ratio of cases to variables is acceptable.

6.3.2.2 Normality, Linearity, and Homoscedasticity   Although there is no requirement that the variables be normally distributed when canonical correlation is used descriptively, the analysis is enhanced if they are. However, inference regarding number of significant canonical variate pairs proceeds on the assumption of multivariate normality. Multivate normality is the assumption that all variables and all

---

[1] The authors are indebted to James Fleming for many of the insights of this section.

linear combinations of variables are normally distributed. It is not itself a testable hypothesis, but the likelihood of multivariate normality is increased if the variables are all normally distributed.

Linearity is important to canonical analysis in at least two ways. The first is that the analysis is performed on correlation or variance-covariance matrices that are sensitive to linear, but not higher order, relationship. If the relationship between two variables is curvilinear, it is not "captured" by these statistics and the canonical result misses the nonlinear part of the relationship unless the variables are transformed. The second is that canonical correlation maximizes the linear relationship between a variate from one set of variables and a variate from the other set. If the relationship between variates is not linear, canonical analysis misses it.

Finally, canonical analysis is best when relationships among pairs of variables are homoscedastic, that is, when the variance of one variable is the same at all levels of the other variable. Heteroscedastic relationships are also not fully captured by Pearson r, so the analysis is degraded when they are present but not represented.

There are two ways to assess the normality, linearity, and homoscedasticity of variables. The first way is to perform a canonical correlation and look at the distributions of canonical variate scores. If the scores show normality, linearity, and homoscedasticity, there is no reason to examine the original variables. Scatterplots where pairs of canonical variates are plotted against each other or variables are plotted against variates are available through BMDP6M, and through SAS CANCORR and SYSTAT MGLH if canonical variates scores are written to a file for processing through a scatterplot program. This analysis is illustrated in Section 6.7.1.2.

If, in the scatterplots, there is evidence of failure of normality, linearity, and/ or homoscedasticity, or if the second way of screening is used, investigation of the variables is undertaken. Variables are examined individually for skewness through one of the descriptive programs such as BMDP2D or SPSS^x FREQUENCIES. Variables are examined pairwise, both within sets and across sets, for nonlinearity or heteroscedasticity through programs such as BMDP6D or SPSS^x PLOT or SPSS SCATTERGRAM. If one or more of the variables is in violation of the assumptions, transformation is considered, as discussed in Chapter 4 and illustrated in Section 6.7.1.2.

6.3.2.3 Missing Data   Levine (1977) gives an example of a dramatic change in a canonical solution with a change in procedures for handling missing data. Because canonical correlation is quite sensitive to minor changes in a data set, consider carefully the methods of Chapter 4 for estimating or eliminating cases with missing data.

6.3.2.4 Outliers   Cases that are unusual often have undue influence on canonical analysis just as they have on other statistical analyses. The search for univariate and multivariate outliers is conducted separately within each set of variables. Consult Chapter 4 and Section 6.7.1.3 for methods of detecting and reducing the effects of both univariate and multivariate outliers.

6.3.2.5 Multicollinearity and Singularity   For both logical and computational reasons, it is important that the variables in each set and across sets are not too highly correlated with each other. This restriction applies to values in $\mathbf{R}_{xx}$, $\mathbf{R}_{yy}$, and $\mathbf{R}_{xy}$ (see

Equation 6.1). BMDP6M provides a direct test of multicollinearity and singularity, as illustrated in Section 6.7.1.4. If you are using one of the other programs, consult Chapter 4 for methods of identifying and eliminating multicollinearity and singularity in correlation matrices.

6.4 FUNDAMENTAL EQUATIONS FOR CANONICAL CORRELATION

A data set that is appropriately analyzed through canonical correlation has several subjects, each measured on four or more variables. The variables form two sets with at least two variables in the smaller set. A hypothetical data set, appropriate for canonical correlation, is presented in Table 6.1. Five intermediate- and advanced-level belly dancers are rated on the quality of their "top" shimmies (TS), "top" circles (TC), "bottom" shimmies (BS), and "bottom" circles (BC). Each characteristic of the dance is rated by two judges on a 7-point scale (with larger numbers indicating higher quality) and the ratings averaged. The goal of analysis is to discover patterns, if any, between the quality of the movements on top and the quality of movements on bottom.

TABLE 6.1 SMALL HYPOTHETICAL DATA SET FOR ILLUSTRATION OF CANONICAL CORRELATION

	Set 1 (X)		Set 2 (Y)	
Belly dancers	TS	TC	BS	BC
S_1	6.3	6.6	6.0	5.9
S_2	3.0	5.8	3.2	5.5
S_3	4.1	3.9	4.0	3.9
S_4	5.5	3.5	5.7	3.3
S_5	3.1	2.9	2.8	3.0
Mean	4.4	4.5	4.3	4.3
Standard deviation	1.46	1.58	1.45	1.31

You are cordially invited to follow this example by hand and by computer at your own computer center. Examples of setup and output for this analysis using several popular computer programs appear at the end of this section.

The first step in a canonical analysis is generation of a correlation matrix. In this case, however, the correlation matrix is subdivided into four parts: the correlations between the DVs ($\mathbf{R}_{yy}$), the correlations between the IVs ($\mathbf{R}_{xx}$), and the two matrices of correlations between DVs and IVs ($\mathbf{R}_{xy}$ and $\mathbf{R}_{yx}$).[2] Table 6.2 contains the correlation matrices for the data in the example.

There are several ways to write the fundamental equation for canonical cor-

---

[2] Although in this example the sets of variables are neither IVs nor DVs, it is useful to use the terms when explaining the procedure.

TABLE 6.2 CORRELATION MATRICES FOR THE DATA SET IN TABLE 6.1

| | | R_{xx} | R_{xy} | |
| | | R_{yx} | R_{yy} | |
	TS	TC	BS	BC
TS	1.0000	.3597	.9852	.2704
TC	.3597	1.0000	.3613	.9947
BS	.9852	.3613	1.0000	.2725
BC	.2704	.9947	.2725	1.0000

relation, some more intuitively appealing than others. The equations are all variants on the following equation.

$$\mathbf{R} = \mathbf{R}_{yy}^{-1}\mathbf{R}_{yx}\mathbf{R}_{xx}^{-1}\mathbf{R}_{xy} \tag{6.1}$$

The canonical correlation matrix is a product of four correlation matrices, between DVs (inverted), between IVs (inverted), and between DVs and IVs.

It is helpful conceptually to compare Equation 6.1 with Equation 5.6 for regression. Equation 5.6 indicates that regression coefficients for predicting Y from a set of X's are a product of (the inverse of) the matrix of correlations among the X's and the matrix of correlations between the X's and Y. Equation 6.1 can be thought of as a product of regression coefficients for predicting X's from Y's ($\mathbf{R}_{yy}^{-1}\mathbf{R}_{yx}$) and regression coefficients for predicting Y's from X's ($\mathbf{R}_{xx}^{-1}\mathbf{R}_{xy}$).

Canonical analysis proceeds by solving for the eigenvalues and eigenvectors of the matrix $\mathbf{R}$ of Equation 6.1. As discussed in Chapter 12 and in Appendix A, solving for the eigenvalues of a matrix is a process that redistributes the variance in the matrix, consolidating it into a few composite variates rather than many individual variables. The eigenvector that corresponds to each eigenvalue is transformed into the coefficients (e.g., regression coefficients, canonical coefficients) used to combine the original variables into the composite variate.

Calculation of eigenvalues and corresponding eigenvectors is demonstrated in Appendix A but is difficult and not particularly enlightening. For this example, the task is accomplished with an assist from SPSS^x MANOVA (see Table 6.5 and Section 6.6.1). The goal is to redistribute the variance in the original variables into a very few pairs of canonical variates, each pair capturing a large share of variance and defined by linear combinations of IVs on one side and DVs on the other. Linear combinations are chosen to maximize the canonical correlation for each pair of canonical variates.

Although computing eigenvalues and eigenvectors is best left to the computer, the relationship between a canonical correlation and an eigenvalue[3] is simple, namely,

$$\lambda_i = r_{ci}^2 \tag{6.2}$$

---

[3] SPSS^x and SAS use the terms SQUARED CORR and SQUARED CANONICAL CORRELATION, respectively, in place of eigenvalue and use the term "eigenvalue" in a different way.

Each eigenvalue, λ_i, is equal to the squared canonical correlation, r_{ci}^2, for the pair of canonical variates.

Once the eigenvalue is calculated for each pair of canonical variates, canonical correlation is found by taking the square root of the eigenvalue. Canonical correlation, r_{ci}, is interpreted as an ordinary Pearson product-moment correlation coefficient. When r_{ci} is squared, it represents, as usual, overlapping variance between two variables, or, in this case, variates. Because $r_{ci}^2 = \lambda_i$, the eigenvalues themselves represent overlapping variance between pairs of canonical variates.

For the data set of Table 6.1, two eigenvalues are calculated, one for each variable in the smaller set (both sets in this case). The first eigenvalue is .99820, which corresponds to a canonical correlation of .99910. The second eigenvalue is .96747, so canonical correlation is .98360. That is, the first pair of canonical variates correlate .99910 and overlap 99.82% in variance, while the second pair correlate .98360 and overlap 96.75% in variance.[4]

Significance tests (Bartlett, 1941) are available to test whether one or a set of r_c's differs from zero.

$$\chi^2 = -\left[N - 1 - \left(\frac{k_x + k_y + 1}{2}\right)\right] \ln \Lambda_m \qquad (6.3)$$

The significance of one or more canonical correlations is evaluated as a chi square variable where N is the number of cases, k_x is the number of variables in the IV set, k_y is the number in the DV set, and the natural logarithm of lambda, Λ, is defined in Equation 6.4. This chi square has $(k_x)(k_y)$ df.

$$\Lambda_m = \prod_{i=1}^{m} (1 - \lambda_i) \qquad (6.4)$$

Lambda, Λ, is the product of differences between eigenvalues and unity, generated across m canonical correlations.

For the example, to test if the canonical correlations as a set differ from zero:

$$\Lambda_2 = (1 - \lambda_1)(1 - \lambda_2) = (1 - .99820)(1 - .96747) = .000058554$$

$$\chi^2 = -\left[5 - 1 - \left(\frac{2 + 2 + 1}{2}\right)\right] \ln .000058554$$

$$= -(1.5)(-9.74556)$$

$$= 14.618$$

This χ^2 is evaluated with $(k_x)(k_y) = 4$ df. The two canonical correlations differ from zero: $\chi^2(4) = 14.62$, $p < .006$. The results of this test are interpreted to mean that

---

[4] Aren't made-up data wonderful!

there is significant overlap in variability between the variables in the IV set and the variables in the DV set, that is, that there is some reliable relationship between quality of top movements and of bottom movements. The result is taken as evidence that at least the first canonical correlation is significant.

With the first canonical correlate removed, is there still a reliable relationship between the two sets of variables?

$$\Lambda_1 = (1 - .96747) = .03253$$

$$\chi^2 = -\left[5 - 1 - \frac{2 + 2 + 1}{2}\right] \ln .03253$$

$$= -(1.5)(-3.42559)$$

$$= 5.138$$

This chi square has $(k_x - 1)(k_y - 1) = 1$ df and also differs significantly from zero: $\chi^2(1) = 5.14, p < .02$. This result indicates that there is still significant overlap between the two sets of variables after the first pair of canonical variates is removed. It is taken as evidence that the second canonical correlation is also significant.

Significance of canonical correlations is also evaluated using the F distribution as, for example, in SPSS^x MANOVA.

Two sets of canonical coefficients (analogous to regression coefficients) are required for each canonical correlation, one set to combine the DVs and the other to combine the IVs. The canonical coefficients for the DVs are found as follows:

$$\mathbf{B}_y = (\mathbf{R}_{yy}^{-1/2})'\hat{\mathbf{B}} \tag{6.5}$$

Canonical coefficients for the DVs are a product of (the transpose of the inverse of the square root of) the matrix of correlations between DVs and the matrix of eigenvectors, $\hat{\mathbf{B}}$, for the DVs.

For the example:[5]

$$\mathbf{B}_y = \begin{bmatrix} 1.02 & -0.14 \\ -0.14 & 1.02 \end{bmatrix} \begin{bmatrix} 0.18 & 0.99 \\ 0.99 & -0.19 \end{bmatrix} = \begin{bmatrix} 0.05 & 1.04 \\ 0.98 & -0.33 \end{bmatrix}$$

Once the canonical coefficients for the DVs are computed, coefficients for the IVs are found using the following equation:

$$\mathbf{B}_x = \mathbf{L}\mathbf{R}_{xx}^{-1}\mathbf{R}_{xy}\mathbf{B}_y \tag{6.6}$$

Coefficients for the IVs are a product of four matrices: $\mathbf{L}$, a diagonal matrix of reciprocals of eigenvalues; $\mathbf{R}_{xx}^{-1}$, the inverse of the matrix of correlation between IVs; $\mathbf{R}_{xy}$, the matrix of correlations between IVs and DVs; and $\mathbf{B}_y$, the coefficients for the DVs.

---

[5] These calculations, like others in this section, were carried to several decimal places and then rounded back. The results agree with computer analyses of the same data but the rounded-off figures presented here do not always check out to both decimals.

For the example:

$$\mathbf{B}_x = \begin{bmatrix} \dfrac{1}{.99} & 0 \\ 0 & \dfrac{1}{.98} \end{bmatrix} \begin{bmatrix} 1.15 & -0.41 \\ -0.41 & 1.15 \end{bmatrix} \begin{bmatrix} .99 & .27 \\ .36 & .99 \end{bmatrix} \begin{bmatrix} 0.05 & 1.04 \\ 0.98 & -0.33 \end{bmatrix}$$

$$= \begin{bmatrix} -0.05 & 1.07 \\ 1.02 & -0.34 \end{bmatrix}$$

The two matrices of canonical coefficients are used to estimate scores on canonical variates:

$$\mathbf{X} = \mathbf{Z}_x \mathbf{B}_x \tag{6.7}$$

and

$$\mathbf{Y} = \mathbf{Z}_y \mathbf{B}_y \tag{6.8}$$

Scores on canonical variates are estimated as the product of the standardized scores on the original variates, $\mathbf{Z}_x$ and $\mathbf{Z}_y$, and the canonical coefficients used to weight them, $\mathbf{B}_x$ and $\mathbf{B}_y$.

For the example:

$$\mathbf{X} = \begin{bmatrix} 1.30 & 1.33 \\ -0.96 & 0.80 \\ -0.21 & -0.40 \\ 0.75 & -0.66 \\ -0.89 & -1.04 \end{bmatrix} \begin{bmatrix} -0.05 & 1.07 \\ 1.02 & -0.34 \end{bmatrix} = \begin{bmatrix} 1.26 & 0.95 \\ 0.86 & -1.30 \\ -0.04 & -0.09 \\ -0.71 & 1.03 \\ -1.02 & -0.60 \end{bmatrix}$$

$$\mathbf{Y} = \begin{bmatrix} 1.15 & 1.21 \\ -0.79 & 0.90 \\ -0.23 & -0.32 \\ 0.94 & -0.78 \\ -1.06 & -1.01 \end{bmatrix} \begin{bmatrix} 0.05 & 1.04 \\ 0.98 & -0.33 \end{bmatrix} = \begin{bmatrix} 1.24 & 0.80 \\ 0.84 & -1.12 \\ -0.33 & -0.13 \\ -0.72 & 1.23 \\ -1.04 & -0.77 \end{bmatrix}$$

The first belly dancer, in standardized scores (and appropriate costume), has a z score of 1.30 on TS, 1.33 on TC, 1.15 on BS, and 1.21 on BC. When these z scores are weighted by canonical coefficients, she is estimated to have a score of 1.26 on the first canonical variate and a score of 0.95 on the second canonical variate for the IVs (the X's), and scores of 1.24 and 0.80 on the first and second canonical variates, respectively, for the DVs (the Y's).

The sum for all belly dancers on each canonical variate is zero, within rounding error. These scores, like factor scores (Chapter 12) represent estimates of scores the dancers would receive if they were judged directly on the canonical variates.

Matrices of correlations between the variables and the canonical coefficients, called loading matrices, are used to interpret the canonical variates.

$$\mathbf{A}_x = \mathbf{R}_{xx} \mathbf{B}_x \tag{6.9}$$

and

$$A_y = R_{yy}B_y \qquad (6.10)$$

Correlations between variables and canonical variates are found by multiplying the matrix of correlations between variables by the matrix of canonical coefficients.

For the example:

$$A_x = \begin{bmatrix} 1.00 & .36 \\ .36 & 1.00 \end{bmatrix} \begin{bmatrix} -0.05 & 1.07 \\ 1.02 & -0.34 \end{bmatrix} = \begin{bmatrix} .32 & .95 \\ 1.00 & .05 \end{bmatrix}$$

$$A_y = \begin{bmatrix} 1.00 & .27 \\ .27 & 1.00 \end{bmatrix} \begin{bmatrix} 0.05 & 1.04 \\ 0.98 & -0.33 \end{bmatrix} = \begin{bmatrix} .31 & .95 \\ .99 & -.05 \end{bmatrix}$$

The loading matrices for these data are summarized in Table 6.3. Results are intepreted down columns across sets of variables. For the first canonical variate pair (the first column), TS correlates .32, TC 1.00, BS .31, and BC .99. The first canonical variates are primarily TC for the first set and BC for the second set, indicating that quality of circles on top is highly related to the quality of circles on bottom.

For the second canonical variate pair (the second column), TS correlates .95, TC .05, BS .95, and BC −.05, indicating that the quality of shimmies on top is related to the quality of shimmies on bottom. Thus the first pair of variates is primarily a "circles" (top and bottom) pair that correlates .99910. The second pair is primarily a "shimmies" pair that correlates .98360. Ability to do circles well is apparently one dimension of belly dancing, while ability to do shimmies well is another, independent dimension.

But how much variance does each of the canonical variates extract from the variables on its own side of the equation? The proportion of variance extracted from the IVs by the canonical variates of the IVs is

$$pv_{xc} = \sum_{i=1}^{k_x} \frac{a_{ixc}^2}{k_x} \qquad (6.11)$$

TABLE 6.3 LOADING MATRIX FOR THE DATA SET OF TABLE 6.1

	Variable sets	Canonical variate pairs	
		First	Second
First	TS	.32	.95
	TC	1.00	.05
Second	BS	.31	.95
	BC	.99	−.05

and

$$pv_{yc} = \sum_{i=1}^{k_y} \frac{a_{iyc}^2}{k_y}$$ (6.12)

The proportion of variance extracted from a set of variables by a canonical variate of the set (called *averaged squared loading* by BMDP6M) is the sum of the squared correlations divided by the number of variables in the set.

Thus for the first canonical variate in the set of IVs,

$$pv_{x1} = \frac{.32^2 + 1.00^2}{2} = .5512$$

and for the second canonical variate of the IVs,

$$pv_{x2} = \frac{.95^2 + .05^2}{2} = .4525$$

The first canonical variate, representing quality of top circles, extracts 55.12% of the variance in judgments of top movements while the second canonical variate, representing quality of top shimmies, extracts 45.25% of the variance in judgments of top movements. In summing for the two variates, 100% of the variance in the IVs is extracted by the two canonical variates.

For the DVs and the first canonical variate,

$$pv_{y1} = \frac{.31^2 + .99^2}{2} = .5381$$

and for the second canonical variate,

$$pv_{y2} = \frac{.95^2 + (-.05)^2}{2} = .4525$$

That is, the first canonical variate, representing quality of bottom circles, extracts 53.81% of the variance in judgments of bottom movements, while the second canonical variate, representing quality of bottom shimmies, extracts 45.25% of variance in judgments of bottom movements. Together the two canonical variates extract 99.06% of the variance in the DVs.

Often, however, one is interested in knowing how much variance the canonical variates from the IVs extract from the DVs, and vice versa. In canonical analysis, this variance is called redundancy.

$$rd = (pv)(r_c^2)$$ (6.13)

The redundancy in a canonical variate is the percent of variance it extracts from its own set of variables times the canonical correlation squared for the pair.

Thus for the example:

$$rd_{x1} = \left(\frac{.32^2 + 1.00^2}{2}\right)(.99820) = .5502$$

$$rd_{x2} = \left(\frac{.95^2 + .05^2}{2}\right)(.96747) = .4378$$

$$rd_{y1} = \left(\frac{.31^2 + .99^2}{2}\right)(.99820) = .5371$$

and

$$rd_{y2} = \left(\frac{.95^2 + (-.05)^2}{2}\right)(.96747) = .4378$$

So, the first canonical variate from the IVs, representing quality of top circles, extracts 55.02% of the variance in judgments of quality of bottom movements. The second canonical variate of the IVs, representing quality of top shimmies, extracts 43.78% of the variance in judgments of quality of bottom movements. Together the two variates extract 98.80% of the variance in the DVs, variance in judgments of quality of bottom movements.

The first and second canonical variates for the DVs, representing quality of bottom circles and bottom shimmies, extract 53.71 and 43.78% of the variance in judgments of quality of top movements, respectively. Together they extract 97.49% of the variance in judgments of quality of top movements.

Tables 6.4 through 6.7 show analyses of this data set by BMDP6M, SPSS[x] MANOVA, SYSTAT MGLH, and SAS CANCORR, respectively.

BMDP6M allows specification of the first and second sets of variables in a CANON paragraph (Table 6.4). The output includes numerous univariate descriptive statistics for each variable (cf. Chapter 4), a matrix of correlations among the variables, and SMCs where each variable, in turn, serves as DV while other variables in the same set are IVs.

The table in which the first column is labeled EIGENVALUE allows you to determine the number of significant canonical variate pairs. The first row tests the significance of all pairs considered together through CHI-SQUARE ($=$ 14.62 in the example). The second row shows the EIGENVALUE (canonical correlation in BMDP6M) and the squared CANONICAL CORRELATION for the first canonical variate pair. The ''1'' refers to the fact that the first canonical variate pair is ''peeled off'' for the significance test (CHI SQUARE $=$ 5.14) of the remaining canonical variate(s) (in this case only one). (With more than two variables in both sets, the second test includes all the remaining canonical correlations after the first canonical variate pair is peeled off.) The third row shows the canonical correlation and squared canonical correlation for the second canonical variate pair.

Next are two tables for the first set of variables, and in identical format, two tables for the second set of variables. CANONICAL VARIABLE LOADINGS are the correlations of variables in the sets with the canonical variates. In the tables labeled

TABLE 6.4 SETUP AND SELECTED BMDP6M OUTPUT FOR CANONICAL CORRELATION ANALYSIS ON SAMPLE DATA IN TABLE 6.1

```
/PROBLEM      TITLE IS 'SMALL SAMPLE CANONICAL CORRELATION'.
/INPUT        VARIABLES ARE 5.  FILE=SSCANON.
              FORMAT IS FREE.
/VARIABLES    NAMES ARE ID, TS, TC, BS, BC.
              LABEL IS ID.
/CANON        FIRST=TS,TC.  SECOND=BS,BC.
/END
```

UNIVARIATE SUMMARY STATISTICS

VARIABLE	MEAN	STANDARD DEVIATION	COEFFICIENT OF VARIATION	SMALLEST VALUE	LARGEST VALUE	SMALLEST STANDARD SCORE	LARGEST STANDARD SCORE	SKEWNESS	KURTOSIS
2 TS	4.40000	1.46287	0.332471	3.00000	6.30000	-0.96	1.30	0.21	-2.07
3 TC	4.54000	1.58209	0.348477	2.90000	6.60000	-1.04	1.30	0.25	-2.07
4 BS	4.34000	1.44845	0.333744	2.80000	6.00000	-1.06	1.15	0.13	-2.17
5 BC	4.32000	1.30843	0.302878	3.00000	5.90000	-1.01	1.21	0.19	-2.16

VALUES FOR KURTOSIS GREATER THAN ZERO INDICATE DISTRIBUTIONS WITH HEAVIER TAILS THAN THE NORMAL DISTRIBUTION.

CORRELATIONS

	TS	TC	BS	BC
	2	3	4	5
TS	1.000			
TC	0.360	1.000		
BS	0.985	0.361	1.000	
BC	0.270	0.995	0.273	1.000

SQUARED MULTIPLE CORRELATIONS OF EACH VARIABLE IN SECOND SET WITH ALL OTHER VARIABLES IN SECOND SET

VARIABLE NUMBER	NAME	R-SQUARED
4	BS	0.07427
5	BC	0.07427

205

TABLE 6.4 (Continued)

SQUARED MULTIPLE CORRELATIONS OF EACH VARIABLE IN
FIRST SET WITH ALL OTHER VARIABLES IN FIRST SET

```
        VARIABLE
    NUMBER   NAME      R-SQUARED

       2  TS           0.12939
       3  TC           0.12939
```

```
                   NUMBER OF                          CHI-            TAIL
  CANONICAL       EIGENVALUES     BARTLETT'S TEST FOR  SQUARE  D.F.   PROB.
EIGENVALUE  CORRELATION  REMAINING EIGENVALUES

0.99820   0.99910              1              14.62    4    0.0056
0.96747   0.98360                              5.14    1    0.0234
```

BARTLETT'S TEST ABOVE INDICATES THE NUMBER OF CANONICAL
VARIABLES NECESSARY TO EXPRESS THE DEPENDENCY BETWEEN THE
TWO SETS OF VARIABLES. THE NECESSARY NUMBER OF CANONICAL
VARIABLES IS THE SMALLEST NUMBER OF EIGENVALUES SUCH THAT
THE TEST OF THE REMAINING EIGENVALUES IS NON-SIGNIFICANT.
FOR EXAMPLE, IF A TEST AT THE .01 LEVEL WERE DESIRED,
THEN 1 VARIABLES WOULD BE CONSIDERED NECESSARY.
HOWEVER, THE NUMBER OF CANONICAL VARIABLES OF PRACTICAL
VALUE IS LIKELY TO BE SMALLER.

CANONICAL VARIABLE LOADINGS

(CORRELATIONS OF CANONICAL VARIABLES WITH ORIGINAL VARIABLES)
FOR FIRST SET OF VARIABLES

```
        CNVRF1   CNVRF2
          1        2

TS    2   0.319    0.948
TC    3   0.999    0.044
```

SQUARED MULTIPLE CORRELATIONS OF EACH VARIABLE IN THE FIRST SET WITH ALL VARIABLES IN THE SECOND SET.

VARIABLE	R-SQUARED	ADJUSTED R-SQUARED	F STATISTIC	DEGREES OF FREEDOM		P-VALUE
2 TS	0.970589	0.941178	33.00	2	2	0.0294
3 TC	0.998144	0.996288	537.83	2	2	0.0019

CANONICAL VARIABLE LOADINGS

(CORRELATIONS OF CANONICAL VARIABLES WITH ORIGINAL VARIABLES)
FOR SECOND SET OF VARIABLES

		CNVRS1 1	CNVRS2 2
BS	4	0.321	0.947
BC	5	0.999	-0.051

SQUARED MULTIPLE CORRELATIONS OF EACH VARIABLE IN THE SECOND SET WITH ALL VARIABLES IN THE FIRST SET.

VARIABLE	R-SQUARED	ADJUSTED R-SQUARED	F STATISTIC	DEGREES OF FREEDOM		P-VALUE
4 BS	0.970641	0.941282	33.06	2	2	0.0294
5 BC	0.998123	0.996247	531.89	2	2	0.0019

CANON. VAR.	AVERAGE SQUARED LOADING FOR EACH CANONICAL VARIABLE (1ST SET)	AV. SQ. LOADING TIMES SQUARED CANON. CORREL. (1ST SET)	AVERAGE SQUARED LOADING FOR EACH CANONICAL VARIABLE (2ND SET)	AV. SQ. LOADING TIMES SQUARED CANON. CORREL. (2ND SET)	SQUARED CANON. CORREL.
1	0.54977	0.54878	0.55027	0.54928	0.99820
2	0.45023	0.43559	0.44973	0.43510	0.98747

THE AVERAGE SQUARED LOADING TIMES THE SQUARED CANONICAL CORRELATION IS THE AVERAGE SQUARED CORRELATION OF A VARIABLE IN ONE SET WITH THE CANONICAL VARIABLE FROM THE OTHER SET. IT IS SOMETIMES CALLLED A REDUNDANCY INDEX.

```
          TITLE     SMALL SAMPLE CANONICAL CORRELATION
          FILE HANDLE   SSCANON
          DATA LIST    FILE=SSCANON    FREE
                        /ID,TS,TC,BS,BC
          MANOVA     TS, TC WITH BS, BC/
                      NOPRINT=PARAMETERS(ESTIM)/
                      PRINT=DISCRIM(RAW,STAN,ESTIM,COR)/
                      DESIGN
```

* * * * * * * * * * * * * * * * * A N A L Y S I S   O F   V A R I A N C E * * * * * * * * * * * * * * * * *

```
     5 CASES ACCEPTED.
     0 CASES REJECTED BECAUSE OF OUT-OF-RANGE FACTOR VALUES.
     0 CASES REJECTED BECAUSE OF MISSING DATA.
     1 NON-EMPTY CELLS.
     1 DESIGN WILL BE PROCESSED.
```

- -

CORRESPONDENCE BETWEEN EFFECTS AND COLUMNS OF BETWEEN-SUBJECTS DESIGN 1

```
 STARTING  ENDING
  COLUMN   COLUMN   EFFECT NAME

    1        1      CONSTANT
```

* * * * * * * * * * * * * * * * * A N A L Y S I S   O F   V A R I A N C E * * * * * * * * * * * * * * * * *

EFFECT .. WITHIN CELLS   REGRESSION

MULTIVARIATE TESTS OF SIGNIFICANCE (S = 2, M = -1/2, N = -1/2)

TEST NAME	VALUE	APPROX. F	HYPOTH. DF	ERROR DF	SIG. OF F
PILLAIS	1.96567	57.26549	4.00	4.00	.001
HOTELLINGS	585.22958	0	4.00	0	1.000
WILKS	.00006	64.89803	4.00	2.00	.015
ROYS	.99820				

- -

EIGENVALUES AND CANONICAL CORRELATIONS

ROOT NO.	EIGENVALUE	PCT.	CUM. PCT.	CANON. COR.	SQUARED COR.
1	555.48745	94.91787	94.91787	.99910	.99820
2	29.74213	5.08213	100.00000	.98360	.96747

DIMENSION REDUCTION ANALYSIS

ROOTS	WILKS LAMBDA	F	HYPOTH. DF	ERROR DF	SIG. OF F
1 TO 2	.00006	64.89803	4.00	2.00	.015
2 TO 2	.03253	59.48426	1.00	2.00	.016

UNIVARIATE F-TESTS WITH (2,2) D. F.

VARIABLE	SQ. MUL. R	MUL. R	ADJ. R-SQ.	HYPOTH. MS	ERROR MS	F	SIG. OF F
TS	.97059	.98518	.94118	4.15412	.12588	33.00099	.029
TC	.99814	.99907	.99629	4.99671	.00929	537.82711	.002

RAW CANONICAL COEFFICIENTS FOR DEPENDENT VARIABLES

          FUNCTION NO.

VARIABLE	1	2
TS	-.03207	.73192
TC	.64214	-.21577

- -

TABLE 6.5 (Continued)

STANDARDIZED CANONICAL COEFFICIENTS FOR DEPENDENT VARIABLES

FUNCTION NO.

VARIABLE	1	2
TS	-.04692	1.07071
TC	1.01592	-.34136

- -

CORRELATIONS BETWEEN DEPENDENT AND CANONICAL VARIABLES

FUNCTION NO.

VARIABLE	1	2
TS	.31851	.94792
TC	.99904	.04378

- -

VARIANCE EXPLAINED BY CANONICAL VARIABLES OF DEPENDENT VARIABLES

CAN. VAR.	PCT VAR DEP	CUM PCT DEP	PCT VAR COV	CUM PCT COV
1	54.97670	54.97670	54.87791	54.87791
2	45.02330	100.00000	43.55875	98.43666

RAW CANONICAL COEFFICIENTS FOR COVARIATES

FUNCTION NO.

COVARIATE	1	2
BS	.03652	.71663
BC	.75227	-.25510

- -

STANDARDIZED CANONICAL COEFFICIENTS FOR COVARIATES

CAN. VAR.

COVARIATE	1	2
BS	.05289	1.03800
BC	.98429	-.33378

- -

CORRELATIONS BETWEEN COVARIATES AND CANONICAL VARIABLES

CAN. VAR.

COVARIATE	1	2
BS	.32114	.94703
BC	.99870	-.05089

- -

VARIANCE EXPLAINED BY CANONICAL VARIABLES OF THE COVARIATES

CAN. VAR.	PCT VAR DEP	CUM PCT DEP	PCT VAR COV	CUM PCT COV
1	54.92822	54.92822	55.02710	55.02710
2	43.50999	98.43821	44.97290	100.00000

REGRESSION ANALYSIS FOR WITHIN CELLS   ERROR TERM

DEPENDENT VARIABLE .. TS

COVARIATE	B	BETA	STD. ERR.	T-VALUE	SIG. OF T	LOWER 95 CL	UPPER 95 CL
BS	.9944383521	.9846314887	.12729	7.81225	.016	.44674	1.54214
BC	.0022620275	.0020232189	.14091	.01605	.989	-.60405	.60857

DEPENDENT VARIABLE .. TC

COVARIATE	B	BETA	STD. ERR.	T-VALUE	SIG. OF T	LOWER 95 CL	UPPER 95 CL
BS	.1064848423	.0974900087	.03458	3.07922	.091	-.04231	.25528
BC	1.1705610858	.9680895623	.03828	30.57713	.001	1.00584	1.33528

TABLE 6.6 SETUP AND SELECTED SYSTAT MGLH OUTPUT FOR
CANONICAL CORRELATION ANALYSIS OF SAMPLE DATA
OF TABLE 6.1

```
USE SSCANON
MODEL TS TC = CONSTANT + BS + BC
ESTIMATE
PRINT = LONG
HYPOTHESIS
STANDARDIZE = TOTAL
EFFECT = BS & BC
TEST
MODEL BS BC = CONSTANT + TS + TC
ESTIMATE
PRINT = LONG
HYPOTHESIS
STANDARDIZE = TOTAL
EFFECT = TS & TC
TEST
```

NUMBER OF CASES PROCESSED: 5

DEPENDENT VARIABLE MEANS

	TS	TC
	4.400	4.540

REGRESSION COEFFICIENTS $B = (X'X)^{-1} X'Y$

	TS	TC
CONSTANT	0.074	-0.979
BS	0.994	0.106
BC	0.002	1.171

MULTIPLE CORRELATIONS

	TS	TC
	0.985	0.999

SQUARED MULTIPLE CORRELATIONS

	TS	TC
	0.971	0.998

ADJUSTED $R^2 = 1-(1-R^2)*(N-1)/DF$, WHERE N=   5, AND DF= 2.

	TS	TC
	0.941	0.996

- -

TEST FOR EFFECT CALLED:
                    BS
            AND
                    BC

UNIVARIATE F TESTS

VARIABLE	SS	DF	MS	F	P
TS	8.308	2	4.154	33.001	0.029
ERROR	0.252	2	0.126		
TC	9.993	2	4.997	537.827	0.002
ERROR	0.019	2	0.009		

TABLE 6.6 (Continued)

```
MULTIVARIATE TEST STATISTICS

            WILKS' LAMBDA =      0.000
              F-STATISTIC =     64.898      DF =   4,   2      PROB =      0.015

             PILLAI TRACE =      1.966
              F-STATISTIC =     57.265      DF =   4,   4      PROB =      0.001

   HOTELLING-LAWLEY TRACE =    585.230
              F-STATISTIC =      0.000      DF =   4,   0      PROB =      1.000

                    THETA =  0.998  S =  2, M = -.5, N =  -.5  PROB =      0.004

TEST OF RESIDUAL ROOTS

   ROOTS  1 THROUGH  2
     CHI-SQUARE STATISTIC =     14.621      DF =        4      PROB =      0.006

   ROOTS  2 THROUGH  2
     CHI-SQUARE STATISTIC =      5.138      DF =        1      PROB =      0.023


CANONICAL CORRELATIONS

                   1           2

                 0.999       0.984

DEPENDENT VARIABLE CANONICAL COEFFICIENTS
STANDARDIZED BY SAMPLE STANDARD DEVIATIONS

                   1           2

       TS        -0.047       1.071
       TC         1.016      -0.341


CANONICAL LOADINGS (CORRELATIONS BETWEEN
DEPENDENT VARIABLES AND DEPENDENT CANONICAL FACTORS)

                   1           2

       TS         0.319       0.948
       TC         0.999       0.044
--------------------------------------------------------------------------------
NUMBER OF CASES PROCESSED:      5

DEPENDENT VARIABLE MEANS

                  BS          BC

                 4.340       4.320

                              -1
REGRESSION COEFFICIENTS  B = (X'X)   X'Y

                  BS          BC

   CONSTANT       0.027       0.845

       TS         0.973      -0.090

       TC         0.007       0.852



STANDARDIZED REGRESSION COEFFICIENTS

                  BS          BC

   CONSTANT       0.000       0.000

       TS         0.982      -0.100

       TC         0.008       1.031


MULTIPLE CORRELATIONS

                  BS          BC

                 0.985       0.999
```

TABLE 6.6 (Continued)

```
SQUARED MULTIPLE CORRELATIONS

                        BS          BC

                      0.971       0.998


          2        2
ADJUSTED R = 1-(1-R )*(N-1)/DF, WHERE N=   5, AND DF=  2.

                        BS          BC

                      0.941       0.996
```

```
TEST FOR EFFECT CALLED:
                        TS
            AND
                        TC


UNIVARIATE F TESTS

     VARIABLE      SS        DF       MS            F            P

        BS       8.146       2      4.073        33.061       0.029
      ERROR      0.246       2      0.123
        BC       6.835       2      3.418       531.886       0.002
      ERROR      0.013       2      0.006

CANONICAL CORRELATIONS

                        1           2

                      0.999       0.984


DEPENDENT VARIABLE CANONICAL COEFFICIENTS
STANDARDIZED BY SAMPLE STANDARD DEVIATIONS

                        1           2

        BS            0.053       1.038
        BC            0.984      -0.334


CANONICAL LOADINGS (CORRELATIONS BETWEEN
DEPENDENT VARIABLES AND DEPENDENT CANONICAL FACTORS)

                        1           2

        BS            0.321       0.947
        BC            0.999      -0.051
```

TABLE 6.7 SETUP AND SAS CANCORR OUTPUT FOR CANONICAL
CORRELATION ANALYSIS OF SAMPLE DATA OF TABLE 6.1

```
            DATA SSAMPLE;
            INFILE SSCANON;
            INPUT ID TS TC BS BC;
            PROC CANCORR;
            VAR TS TC;
            WITH BS BC;

                    CANCORR PROCEDURE

                CANONICAL CORRELATION ANALYSIS

                        ADJUSTED      APPROX       SQUARED
            CANONICAL   CANONICAL    STANDARD     CANONICAL
            CORRELATION CORRELATION    ERROR      CORRELATION

        1   0.999101    0.998850     0.000898     0.998203
        2   0.983601       .         0.016264     0.967471
```

TABLE 6.7 (Continued)

```
                      EIGENVALUES OF INV(E)*H
                       = CANRSQ/(1-CANRSQ)
```

	EIGENVALUE	DIFFERENCE	PROPORTION	CUMULATIVE
1	555.4875	525.7453	0.9492	0.9492
2	29.7421	.	0.0508	1.0000

```
       TEST OF HO: THE CANONICAL CORRELATIONS IN THE CURRENT ROW
                  AND ALL THAT FOLLOW ARE ZERO
```

	LIKELIHOOD RATIO	APPROX F	NUM DF	DEN DF	PR > F
1	0.00005845	64.8980	4	2	0.0152
2	0.03252865	59.4843	1	2	0.0164

```
              S=2    M=-0.5    N=0
```

STATISTIC	VALUE	F	NUM DF	DEN DF	PR > F
WILKS' LAMBDA	0.00005845	64.8980	4	2	0.0152
PILLAI'S TRACE	1.96567437	57.2655	4	4	0.0009
HOTELLING-LAWLEY TRACE	585.22958400	.	4	0	0.0
ROY'S GREATEST ROOT	555.48745270	555.4875	2	2	0.0018

```
     NOTE: F STATISTIC FOR ROY'S GREATEST ROOT IS AN UPPER BOUND.
     NOTE: F STATISTIC FOR WILKS' LAMBDA IS EXACT.
```

```
        RAW CANONICAL COEFFICIENTS FOR THE 'VAR' VARIABLES
```

	V1	V2
TS	-.0320726572	.73192101537
TC	.64213781688	-.2157666441

```
        RAW CANONICAL COEFFICIENTS FOR THE 'WITH' VARIABLES
```

	W1	W2
BS	.03651648835	.71662625086
BC	.75226519977	-.2550956707

```
     STANDARDIZED CANONICAL COEFFICIENTS FOR THE 'VAR' VARIABLES
```

	V1	V2
TS	-0.0469	1.0707
TC	1.0159	-0.3414

```
     STANDARDIZED CANONICAL COEFFICIENTS FOR THE 'WITH' VARIABLES
```

	W1	W2
BS	0.0529	1.0380
BC	0.9843	-0.3338

```
   CORRELATIONS BETWEEN THE 'VAR' VARIABLES AND THEIR CANONICAL VARIABLES
```

	V1	V2
TS	0.3185	0.9479
TC	0.9990	0.0438

```
   CORRELATIONS BETWEEN THE 'WITH' VARIABLES AND THEIR CANONICAL VARIABLES
```

	W1	W2
BS	0.3211	0.9470
BC	0.9987	-0.0509

SQUARED MULTIPLE CORRELATIONS OF EACH VARIABLE . . . are R-SQUARED, ADJUSTED R-SQUARED, and an F STATISTIC where each variable in one set serves as DV with the variables in the other set as IVs.

The final table (AVERAGE SQUARED LOADING . . .) shows, for each canonical variate, the proportion of variance extracted from each set of variables (AVERAGED SQUARED LOADING FOR EACH CANONICAL VARIABLE) and redundancy (AV. SQ. LOADINGS . . .), as described earlier in this section and in the output. The squared canonical correlation is repeated for each canonical variate in the last column.

In SPSS[x] MANOVA (Table 6.5), one set of variables is called DEPENDENT VARIABLES while the other is called COVARIATES. In the MANOVA instruction, the set listed before WITH is the DV set. The NOPRINT instruction suppresses unwanted output, and the PRINT statements provide information useful for canonical analysis.

After various tests of the within cells correlation matrix are several significance tests in the section labeled EFFECT . . . WITHIN CELLS REGRESSION. The first three (PILLAIS, HOTTELLINGS, and WILKS) test all canonical variates together; ROYS tests the first canonical variate by itself (see Section 9.5.1). Information from this table is repeated in subsequent sections. EIGENVALUES AND CANONICAL CORRELATIONS are printed in the next section. In SPSS[x] MANOVA, the terms eigenvalue and canonical correlation are not used interchangeably—here the eigenvalue is $R^2/(1 - R^2)$. The CANON. COR. and SQUARED canonical COR. are given for each canonical variate pair in rows labeled ROOT. The percent of variance in the solution contributed by the canonical variate pair is listed in the column labeled PCT. CUM. PCT. cumulates the percent of variance over the successive roots. In this example, the first canonical variate pair accounts for 94.92% of the variance in the solution. Both canonical variate pairs considered together account for 100% of the variance in the solution (because there are only two pairs).

The DIMENSION REDUCTION ANALYSIS follows the same "peel-off" procedure noted in BMDP6M above but tested as F and labeled a bit better. The first row tests both variate pairs (ROOTS 1 TO 2) and the second row (2 TO 2) tests variate pairs remaining after the first is peeled off. (The section on UNIVARIATE F-TESTS is of limited utility because it evaluates only the set of variables labeled DEPENDENT; remember this is a MANOVA program at heart).

Four tables are then given for each set of variables, labeled DEPENDENT VARIABLES and COVARIATES. The first two tables show RAW and STANDARDIZED CANONICAL COEFFICIENTS (see Equations 6.5 and 6.6). The loading matrix follows (CORRELATIONS BETWEEN . . . AND CANONICAL VARIABLES). The last table is VARIANCE EXPLAINED BY . . . and contains percent of variance (PCT VAR same set) and redundancy measures (PCT VAR other set). Considering the DEPENDENT VARIABLE set for the example, the first canonical variate accounts for 54.97670% of the variance in that set (PCT VAR DEP) while redundancy is 54.87791% (PCT VAR COV). Cumulative percents of variance and redundancy over successive canonical variates are also given.

The final MANOVA table shows bivariate regression analyses (cf. Chapter 3) for each of the DEPENDENT variables with each of the COVARIATE variables.

Setup for SYSTAT MGLH is fairly complex for canonical analysis, but the procedures are clearly spelled out in the manual. Basically you set up two multiple-multiple regression equations where each set of variables takes a turn as DVs (left side of the equation). As you might guess, this produces voluminous output with confusing labels. Comparison of results with the hand-worked example and other output helps clarify them, however.

The first segment of output shows means for the variables labeled DEPENDENT. These are followed by multiple regression information for predicting each DV from the set IVs—unstandardized coefficients (*B* weights), multiple correlations, SMCs, adjusted SMCs, and univariate *F* tests for the significance of the prediction.

This is followed by MULTIVARIATE TEST STATISTICS for the canonical correlation considering all variates together (compare with SPSS[x] MANOVA section labeled EFFECT . . . WITHIN CELLS REGRESSION). The TEST OF RESIDUAL ROOTS provides the "peel off" chi square tests of the significance of pairs of canonical variates. Canonical correlations for each pair of canonical variates are then shown.

The next two segments of output give standardized canonical coefficients and the loading matrix for the DVs. The remaining output is based on the second MODEL statement in the input where the second set of variables serves as DVs. The output parallels that of the first MODEL although the MULTIVARIATE TEST STATISTICS and TEST OF RESIDUAL ROOTS are deleted because they are exactly the same for both MODELs.

In SAS CANCORR (Table 6.7), the first set of variables is listed in the input statement that begins VAR, the second set in the statement that begins WITH. The first segment of output contains the canonical correlations for each of the canonical variates (labeled 1 and 2), including adjusted and squared correlations as well as standard errors for the correlations. The next table shows the eigenvalue (calculated as for SPSS[x] MANOVA, above), the difference between eigenvalues, the proportion and the cumulative proportion of variance in the solution accounted for by each canonical variate pair. The TEST OF HO: . . . table shows "peel off" significance tests for canonical variate pairs evaluated through approximate *F* followed in the next table by several multivariate significance tests. Matrices of raw and standardized canonical coefficients for each canonical variate labeled 'VAR' and 'WITH' in the setup follow; loading matrices are labeled CORRELATIONS BETWEEN THE . . . VARIABLES AND THEIR CANONICAL VARIABLES.

6.5 SOME IMPORTANT ISSUES

6.5.1 Importance of Canonical Variates

As in most statistical procedures, establishing significance is usually the first step in evaluating a solution. Relationships that are not statistically significant are not reliable, and should not be interpreted. Conventional statistical procedures apply to significance tests for number of reliable canonical variate pairs. The results of Equations 6.3 and 6.4, or a corresponding *F* test, are available in all programs reviewed in Section 6.6. But the number of statistically significant pairs of canonical variates is often larger than the number of interpretable pairs if N is at all sizable.

The only potential source of confusion is the meaning of the chain of significance tests. The first test is for all pairs taken together and is a test of independence between the two sets of variables. The second test is for all pairs of variates with the first and most important pair of canonical variates removed; the third is done with the first two pairs removed, and so forth. If the first test, but not the second, reaches significance, then only the first pair of canonical variates is interpreted. If the first and second tests are significant but the third is not, then the first two pairs of variates are interpreted, and so on. Because canonical correlations are reported out in descending order of importance, usually only the first few pairs of variates are significant.

Once significance is established, amount of variance accounted for is of critical importance. Because there are two sets of variables, several assessments of variance are relevant. First, there is variance overlap between variates in a pair. Second is variance overlap between a variate and its own set of variables. Third is variance overlap between a variate and the other set of variables.

The first, and easiest because it is available in all programs reviewed, is the canonical correlation between each pair of significant variates. As indicated in Equation 6.2, the squared canonical correlation (or eigenvalue in BMDP) is the overlapping variance between a pair of canonical variates. Because r_c values of .30 or lower represent, squared, less than a 10% overlap in variance, most researchers do not interpret pairs with a canonical correlation lower than .30 even if significant (because significance depends, to a large extent, on N).

The next consideration is the variance a canonical variate extracts from its own set of variables. A pair of canonical variates may extract very different amounts of variance from their respective sets of variables. Equations 6.11 and 6.12 indicate that the variance extracted, pv, is the sum of squared loadings on a variate divided by the number of variables in the sets.[6] Because canonical variates are independent of one another (orthogonal), pv's are summed across all reliable variates to arrive at the total variance extracted from the variables by all the variates of the set.

The last consideration is the variance a variate from one set extracts from the variables in the other set, called redundancy (Stewart and Love, 1968; Miller and Farr, 1971). Equation 6.13 shows that redundancy is the percent of variance extracted by a canonical variate times the canonical correlation for the pair. It is possible for a canonical variate from the IVs to be an important dimension in its own set of variables, but correlated with an unimportant dimension among the DVs (and vice versa). Therefore the redundancies for a pair of canonical variates are usually not equal. Because canonical variates are orthogonal, redundancies for a set of variables are also added across canonical variates to get a total for the variables and, thus, total redundancy for the DVs relative to the IVs, and vice versa.

6.5.2 Interpretation of Canonical Variates

As with all multivariate procedures, canonical correlation creates linear combinations of variables, called canonical variates in this case, which represent mathematically

---

[6] This calculation is identical to the one used in factor analysis for the same purpose, as shown in Table 12.4.

viable dimensions of the variables. But, although the dimensions are mathematically viable, they are not necessarily understandable. A major task for the researcher is to discern, if possible, the meaning of pairs of canonical variates.

Interpretation of reliable pairs of canonical variates is based on the loading matrices, A_x and A_y (Equations 6.9 and 6.10, respectively). Each pair of canonical variates is interpreted as a pair, with a variate from one set of variables interpreted vis-à-vis the variate from the other set. A variate is interpreted by considering the pattern of variables highly correlated (loaded) with it. Because the loading matrices contain correlations, and because squared correlations measure overlapping variance, variables with correlations of .30 (10% of variance) and above are usually interpreted as part of the variate, while variables with loadings below .30 are not. Deciding on a cutoff for interpreting correlations is, however, somewhat a matter of taste, although guidelines are presented in Section 12.6.5.

6.6 COMPARISON OF PROGRAMS

One program each in the SAS, SYSTAT, and BMDP series and two in the SPSS package are available for canonical analyses. Table 6.8 provides a comparison of important features of the programs. If available, the program of choice is BMDP6M. Second choices are SAS CANCORR and, with limitations, SPSS^x MANOVA.

6.6.1 SPSS Package

The two programs in the SPSS series for canonical analysis are CANCORR (Nie et al., 1975) and MANOVA (Hull and Nie, 1981; SPSS, Inc., 1986). Although the CANCORR program in SPSS was designed specifically for canonical correlation, it provides less information than MANOVA and is not available in the SPSS^x package.

A major strength of SPSS CANCORR is its flexibility in handling missing data and multicollinearity. Weaknesses include unavailability of loading matrices, percents of variance, and redundancies. Although these statistics can be hand-calculated by methods shown in Section 6.4, a minimal amount of matrix multiplication (see Appendix A) is required.

A more complete canonical analysis is available through SPSS^x MANOVA (or its predecessor, SPSS MANOVA), which provides loadings, percents of variance, redundancy and much more. But problems arise with reading the results, because MANOVA is not designed specifically for canonical analysis and some of the labels are misleading.

Canonical analysis is requested through MANOVA by calling one set of variables DVs and the other set covariates; no IVs are listed. An example of the MANOVA setup for canonical analysis appears in Table 6.5. Interpretation of the output is facilitated if parameter estimates are suppressed and DISCRIM statistics are requested, as in Table 6.5, simply because there is less output and much of what there is pertains directly to canonical analysis.

Remembering that one set of variables is identified as DVs and the other set as covariates assists in reading the output. Immediately following the first section, labeled MULTIVARIATE TESTS OF SIGNIFICANCE, is output labeled EIGEN-

TABLE 6.8 COMPARISON OF SPSS, BMDP, SAS, AND SYSTAT PROGRAMS FOR CANONICAL CORRELATION

Feature	SPSS CANCORR[a]	SPSS[x] MANOVA[b]	BMDP6M	SAS CANCORR	SYSTAT MGLH[b]
Input					
Correlation matrix	Yes	Yes[c]	Yes	Yes	Yes
Covariance matrix	No	No	Yes	Yes	Yes
SSCP matrix	No	No	No	No	Yes
Number of canonical variates	Yes	No	Yes	Yes	No
Tolerance	Yes	No	Yes.	Yes	No
Minimum canonical correlation	Specify significance	Specify alpha	CONSTANT	No	No
Labels for canonical variates	No	No	No	Yes	No
Error df if residuals input	No	No	No	Yes	No
Output					
Univariate:					
Means	Yes	Yes	Yes	Yes	Yes
Standard deviations	Yes	Yes	Yes	Yes	No
Number of cases	Yes	Yes	Yes	Yes	Yes
Coefficients of variation	No	No	Yes	No	No
Smallest value	No	No	Yes	No	No
Largest value	No	No	Yes	No	No
Smallest standard score	No	No	Yes	No	No
Largest standard score	No	No	Yes	No	No
Skewness	No	No	Yes	No	No
Kurtosis	No	No	Yes	No	No
Normal plots	No	Yes	No	No	No

Multivariate:

Canonical correlations	Yes	Yes	Yes	Yes	Yes
Eigenvalues (r_c^2)	Yes	Yes	Yes	Yes	No
Significance test	χ^2	F	χ^2	F	χ^2
Lambda	Yes	Yes	No	Yes	Yes
Additional test criteria	Yes	Yes	No	Yes	Yes
Correlation matrix	Yes	No	Yes	No	No
Covariance matrix	Yes	No	Yes	No	No
Loading matrix	No	Yes	Yes	Yes	Yes
Raw canonical coefficients	No	Yes	Yes	Yes	Yes
Standardized canonical coefficients	Yes	Yes	Yes	Yes	Yes
Canonical variate scores	Yes	No	Yes	Yes[d]	Yes[d]
Percent of variance	No	Yes	Yes	No	No
Redundancy	No	Yes	Yes	Yes	No
Variable-variable plots	No	No	Yes	No	No
Variable-variate plots	No	No	Yes	No	No
Variate-variate plots	No	No	Yes	No	No
Numerical consistency	No	No	Yes	No	No
Multicollinearity (within-sets SMCs)	Yes	No	Yes	No	No
Multiple analyses	Yes	Yes	No	No	No
Residuals plots	No	Yes	No	No	No

[a] This program is not available in the newer SPSS^x package.
[b] Additional features are listed in Sections 8.6 and 9.6.
[c] Matrix input is not allowed in some earlier versions of SPSS MANOVA.
[d] Written in file.

VALUES AND CANONICAL CORRELATIONS. Both r_c (CANON. CORR.) and r_c^2 (SQUARED CORR.) are given for each pair of canonical variates (ROOT NO.). The DIMENSION REDUCTION ANALYSIS that follows contains significance tests for the canonical correlations, evaluated as F. RAW CANONICAL COEFFICIENTS FOR DEPENDENT VARIABLES, STANDARDIZED CANONICAL COEFFICIENTS FOR DEPENDENT VARIABLES, and CORRELATIONS BETWEEN DEPENDENT AND CANONICAL VARIABLES (loadings) follow unambiguously. Under VARIANCE EXPLAINED BY CANONICAL VARIATES OF THE DEPENDENT VARIABLES one finds percent of variance (PCT. VAR. DEPENDENTS) and redundancy (PCT. VAR. COVARIATES). These statistics are all repeated immediately for the second set of variables, labeled COVARIATES.

Although SPSS^x MANOVA provides a rather complete canonical analysis, it does not calculate canonical variate scores, nor does it test for multivariate outliers.

6.6.2 BMD Series

BMDP6M (Dixon, 1985) provides a complete canonical analysis and is the program of choice, if available. In addition to checks on accuracy of input are checks for univariate outliers (LARGEST and SMALLEST VALUES), multicollinearity (WITHIN SET SMCs), linearity (variable-variable, variable-variate, and variate-variate PLOTS), and normality (SKEWNESS and KURTOSIS). Flexibility of input, control over the progression of the analysis, and flexibility of output are other desirable features of the program. Percent of variance and redundancy are given. And, as for many of the BMDP programs, the output is replete with helpful interpretive comments.

6.6.3 SYSTAT System

Canonical analysis is done through the multivariate general linear hypothesis MGLH program in SYSTAT (Wilkinson, 1986). But to get all the output, the analysis must be done twice, once with the first set of variables defined as the DVs, and a second time with the other set of variables defined as DVs. Although tests of significance and canonical correlations are the same for both analyses, coefficients and loading matrices differ. And, of course, you must keep track of which set was called "DVs" in which analysis.

The program provides several test criteria for overall significance of the canonical analysis and this is the only program that accepts a sum-of-squares and cross-products matrix as input. However, few other amenities are available in this program. As for all SYSTAT procedures, additional output is produced by writing case statistics to file and analyzing them through modules such as STAT for univariate statistics or GRAPH for plots.

6.6.4 SAS System

SAS CANCORR (SAS Institute, 1985) is a flexible program second only to BMDP6M in features and ease of interpretation. Along with the basics, you can specify easily interpretable labels for canonical variates; multivariate output is more detailed than in BMDP6M, with several test criteria and voluminous redundancy analyses. Univariate output is minimal, however, and if plots are desired, case statistics such as canonical scores are written to a file to be analyzed by the SAS PLOT procedure.

6.7 COMPLETE EXAMPLE OF CANONICAL CORRELATION

For an example of canonical correlation, variables are selected from among those made available by research described in Appendix B, Section B.1. The goal of analysis is to discover the dimensions, if any, along which certain attitudinal variables are related to certain health characteristics.

Selected attitudinal variables (Set 1) include attitudes toward the role of women (ATTROLE), toward locus of control (CONTROL), toward current marital status (ATTMAR), and toward self (ESTEEM). Larger numbers indicate increasingly conservative attitudes about the proper role of women, increasing feelings of powerlessness to control one's fate (external as opposed to internal locus of control), increasing dissatisfaction with current marital status, and increasingly poor self-esteem.

Selected health variables (Set 2) include mental health (MENHEAL), physical health (PHYHEAL), number of visits to health professionals (TIMEDRS), attitude toward use of medication (ATTDRUG), and a frequency-duration measure of use of psychotropic drugs (DRUGUSE). Larger numbers reflect poorer mental and physical health, more visits, greater willingness to use drugs, and more use of them.

Canonical correlation provides a means for studying the relationships among these two sets of variables and for studying the number and nature of dimensions of correspondence.

6.7.1 Evaluation of Assumptions

6.7.1.1 Missing Data A screening run through BMDP2D, partially illustrated in Table 6.9, finds missing data for 6 of the 465 cases. One woman lacks a score on CONTROL, and five lack scores on ATTMAR. With deletion of these cases, remaining $N = 459$.

6.7.1.2 Normality, Linearity, and Homoscedasticity BMDP6M provides a particularly flexible scheme for assessing normality, linearity, and homoscedasticity between pairs of canonical variates, between pairs of variables, and between pairs of variables and variates. The PLOT paragraph allows one to request any number of scatterplots of pairs of variables and variates as well as to control the size of the plot.

Figure 6.1 shows two scatterplots produced by BMDP6M for the example using default size values for the plots. The scatterplots are between the first and second pairs of canonical variates, respectively. CNVRF1 is canonical variate scores, first set, first variate; CNVRS1 is canonical variate scores, second set, first variate. CNVRF2 is canonical variate scores, first set, second variate; CNVRS2 is canonical variate scores, second set, second variate.

TABLE 6.9 INPUT AND SELECTED BMDP2D OUTPUT FOR INITIAL SCREENING OF CANONICAL CORRELATION DATA SET

```
/PROBLEM    TITLE IS 'INITIAL SCREENING WITH BMDP2D'.
/INPUT      VARIABLES=10.  FILE=CANON.   FORMAT='(A4,9F6.0)'.
/VARIABLE   NAMES ARE SUBNO,TIMEDRS,ATTDRUG,PHYHEAL,MENHEAL,ESTEEM,
            CONTROL,ATTMAR,DRUGUSE,ATTROLE.

            MISSING ARE (7) 0, 0.
            LABEL = SUBNO.
/PRINT      NO COUNT.
/END
```

```
***********
* ATTDRUG *
***********
```

MAXIMUM	10.0000000
MINIMUM	5.0000000
RANGE	5.0000000
VARIANCE	1.3365410
ST.DEV.	1.1560920
(Q3-Q1)/2	1.00000000
MX.ST.SC.	2.00
MN.ST.SC.	-2.32

VARIABLE NUMBER	3
NUMBER OF DISTINCT VALUES . .	6
NUMBER OF VALUES COUNTED. .	465
NUMBER OF VALUES NOT COUNTED	0

	ESTIMATE	ST.ERROR	95% CONFIDENCE LOWER	UPPER
MEAN	7.6860210	0.0536125	7.5806680	7.7913750
MEDIAN	8.0000000	0.0000000		
MODE	8.0000000			

```
                                         H   H
                                         H   H
                             H   H   H   H   H
                             H   H   H   H   H
                             H   H   H   H   H
                         H   H   H   H   H   H
                         H   H   H   H   H   H
                   H-------------------------------------U
EACH '-' ABOVE =   0.2000
               L=  5.0000
               U= 10.2000

CASE NO. OF MIN. VAL. =  15
CASE NO. OF MAX. VAL. =  20
```

EACH 'H' REPRESENTS 15 COUNT(S)

	VALUE	VALUE/S.E.		
SKEWNESS	-0.12	-1.07	Q1=	7.0000000
KURTOSIS	-0.47	-2.05	Q3=	9.0000000
			S-=	6.5299290
			S+=	8.8421140

```
**********
* ATTMAR *
**********

VARIABLE NUMBER . . . . .          8          MAXIMUM        58.0000000
NUMBER OF DISTINCT VALUES .        37         MINIMUM        11.0000000
NUMBER OF VALUES COUNTED. .       460         RANGE          47.0000000
NUMBER OF VALUES NOT COUNTED       5          VARIANCE       73.1608900
                                              ST.DEV.        8.5534100
                                              (Q3-Q1)/2      6.0000000
                                              MX.ST.SC.      4.09
                                              MN.ST.SC.      -1.40

            ESTIMATE      ST.ERROR
MEAN        22.9804300    0.3988048
MEDIAN      21.0000000    0.5773506              95^ CONFIDENCE
MODE        16.0000000                         LOWER         UPPER
                                               22.1967200    23.7641400


                                       H
                                       H H
                                       HHHH
                                       HHHH H                                 EACH 'H'
                                       HHHHHHH                                 REPRESENTS
                                       HHHHHHHH H   H                              6
                                       HHHHHHHHH    H                          COUNT(S)
                                       HHHHHHHHH    H
                                       HHHHHHHHHH HH H H
                                       HHHHHHHHHHHHHHHHH   H H H
                                       L----------------------------------U

                                       EACH '-' ABOVE =      2.0000
                                                       L=    8.0000
                                                       U=    60.0000

                                       CASE NO. OF MIN. VAL. =   11
                                       CASE NO. OF MAX. VAL. =  279

                           VALUE   VALUE/S.E.          Q1= 16.0000000
                 SKEWNESS   1.00     8.73              Q3= 28.0000000
                 KURTOSIS   0.77     3.39              S-= 14.4270200
                                                       S+= 31.5338400


***********
* DRUGUSE *
***********

VARIABLE NUMBER . . . . .          9          MAXIMUM        66.0000000
NUMBER OF DISTINCT VALUES .        42         MINIMUM        0.0000000
NUMBER OF VALUES COUNTED. .       465         RANGE          66.0000000
NUMBER OF VALUES NOT COUNTED       0          VARIANCE       102.2737000
                                              ST.DEV.        10.1130500
                                              (Q3-Q1)/2      6.5000000
                                              MX.ST.SC.      5.64
                                              MN.ST.SC.      -0.89

            ESTIMATE      ST.ERROR
MEAN        9.0021500     0.4689813
MEDIAN      5.0000000     0.5773506              95^ CONFIDENCE
MODE        0.0000000                          LOWER         UPPER
                                               8.0805610     9.9237400


                                       H
                                       H
                                       H                                      EACH 'H'
                                       H                                      REPRESENTS
                                       H                                         19
                                       HH                                     COUNT(S)
                                       HHH
                                       HHHHH H
                                       HHHHHHHHHHH   HH H
                                       L-----------------------------U

                                       EACH '-' ABOVE =      3.0000
                                                       L=    0.0000
                                                       U=    78.0000

                                       CASE NO. OF MIN. VAL. =   2
                                       CASE NO. OF MAX. VAL. =  35

                           VALUE   VALUE/S.E.          Q1= 1.0000000
                 SKEWNESS   1.75     15.40             Q3= 14.0000000
                 KURTOSIS   4.17     18.36             S-= -1.1108950
                                                       S+= 19.1152000
```

223

```
/PROBLEM      TITLE IS 'BMDP6M SCREENING RUN'.
/INPUT        VARIABLES=10.  FILE=CANON.    FORMAT='(A4,9F6.0)'.
/VARIABLE     NAMES ARE SUBNO,TIMEDRS,ATTDRUG,PHYHEAL,MENHEAL,ESTEEM,
                    CONTROL,ATTMAR,DRUGUSE,ATTROLE.
              MISSING ARE (7) 0, 0.
              LABEL = SUBNO.
/CANONICAL    FIRST ARE ESTEEM, CONTROL, ATTMAR, ATTROLE.
              SECOND ARE TIMEDRS,ATTDRUG,PHYHEAL,MENHEAL,DRUGUSE.
/PLOT         XVARS ARE CNVRF1,CNVRF2.
              YVARS ARE CNVRS1,CNVRS2.
/END
```

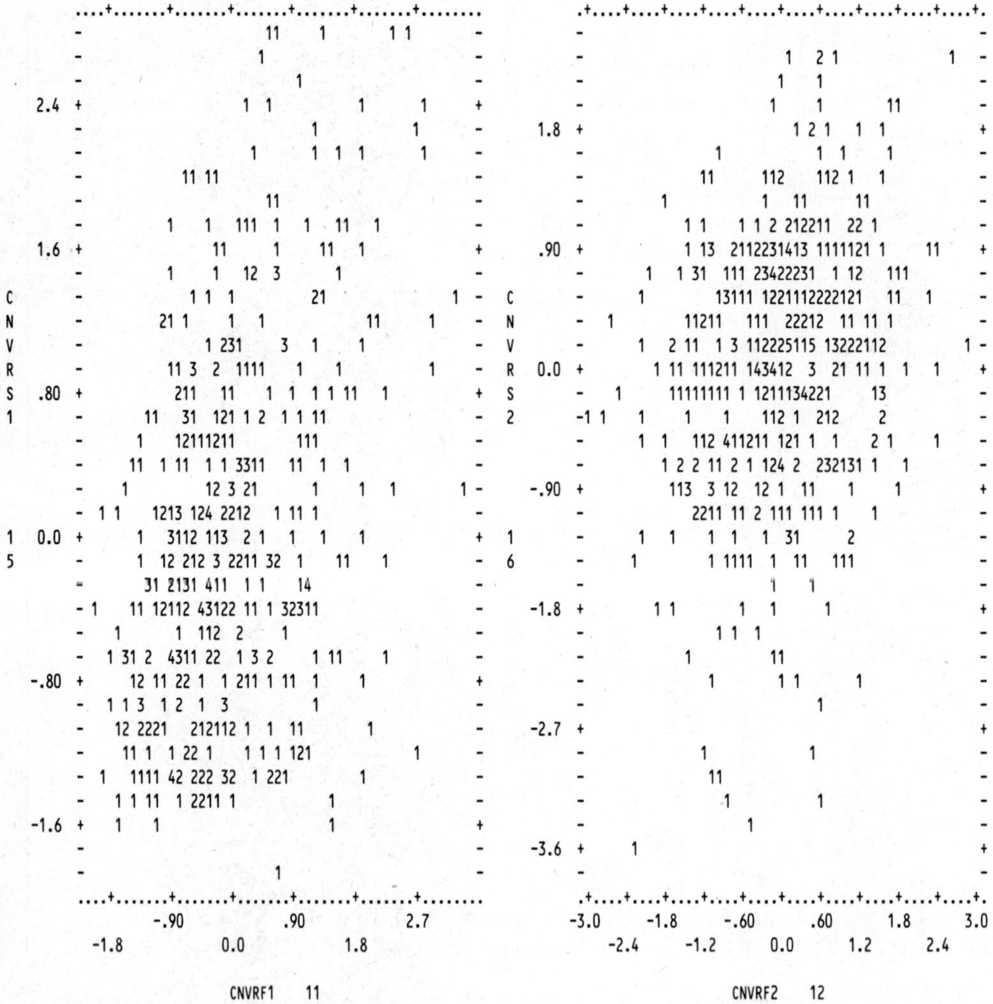

Figure 6.1 Setup and selected BMDP6M output showing scatterplots between first and second pairs of canonical variates.

The shapes of the scatterplots reflect the low canonical correlations for the solution (see the next section), particularly for the second pair of variates where the overall shape is nearly circular except for a few extreme values in the lower third of the plot. There are no obvious departures from linearity or homoscedasticity because the overall shapes do not curve and they are of about the same width throughout.

Deviation from normality is evident, however, for both pairs of canonical variates. Note how on both plots the 0–0 point departs from the center of the vertical and horizontal axes. If the points are projected as a frequency distribution to the vertical or horizontal axes of the plots, there is further evidence of skewness. For the first plot there is a pileup of cases at low scores and a smattering of cases at high scores on both axes, indicating positive skewness. In plot 2, there are widely scattered cases with extremely low scores on CNVRS2, with no corresponding high scores, indicating negative skewness.

Departure from normality is confirmed by the output of BMDP2D, where Table 6.9 shows positive skewness for ATTMAR and, especially, DRUGUSE. Logarithmic transformation of these variables, plus TIMEDRS and PHYHEAL (output containing their distributions is not shown) results in variables that are far less skewed. The transformed variables are named LATTMAR, LDRUGUSE, LTIMEDRS, and LPHYHEAL. A second BMDP6M run provides univariate statistics for both transformed and untransformed variables. Compare the skewness and kurtosis of ATTMAR and DRUGUSE in Table 6.9 with that of LATTMAR and LDRUGUSE in Table 6.10. BMDP6M plots based on transformed variables (not shown) confirm improvement in normality with transformed variables, particularly for the second pair of canonical variates.

Note in the BMDP6M descriptive output of Table 6.10 that coefficients of variation, and kurtosis are also reasonable for a data set of this size.

6.7.1.3 Outliers UNIVARIATE SUMMARY STATISTICS in Table 6.10 provide information on univariate outliers. Note that the smallest and largest standard scores are within a range anticipated in a sample of over 400 cases. The extremely large standard scores (MX.ST.SC. and MN.ST.SC) on ATTMAR and DRUGUSE (before transformation, Table 6.9) are not seen on LATTMAR and LDRUGUSE.

Analysis of multivariate outliers is not available through BMDP6M. Therefore BMDPAM is used to screen for multivariate outliers separately among the two sets of variables.[7] Outliers appear in the BMDPAM program as cases with large Mahalanobis distances, shown as χ^2, from the centroid of the cases for original data. BMDPAM highlights with an asterisk any SIGNIFICANCE value for a case where $p < .001$. A segment of the output for the first set of variables appears in Table 6.11. There are no outliers in this segment or any other in either set of variables.

Cases with missing values on one variable show three rather than four degrees of freedom for χ^2. The full run confirms that there are indeed six cases with missing data, rather than fewer as there would be if a case were missing data on two variables.

6.7.1.4 Multicollinearity and Singularity BMDP6M provides a direct test of multicollinearity/singularity, as shown in Table 6.12. SMCs are reported where each

[7] BMDP4M could also be used.

TABLE 6.10 INPUT AND SELECTED BMDP6M OUTPUT SHOWING DESCRIPTIVE STATISTICS FOR TRANSFORMED AND UNTRANSFORMED VARIABLES

```
/PROBLEM    TITLE IS 'CANONICAL CORRELATION THROUGH BMDP6M'.
/INPUT      VARIABLES=10.  FILE=CANON.  FORMAT='(A4,9F6.0)'.
/VARIABLE   NAMES ARE SUBNO,TIMEDRS,ATTDRUG,PHYHEAL,MENHEAL,ESTEEM,
            CONTROL,ATTMAR,DRUGUSE,ATTROLE,
            LTIMEDRS,LPHYHEAL,LATTMAR,LDRUGUSE.  ADD=4.

            MISSING ARE (7) 0, 0.
            LABEL = SUBNO.
/TRANSFORM  LTIMEDRS = LOG(TIMEDRS + 1).
            LPHYHEAL = LOG(PHYHEAL).
            LATTMAR = LOG(ATTMAR).
            LDRUGUSE = LOG(DRUGUSE + 1).
/CANONICAL  FIRST ARE ESTEEM,CONTROL,LATTMAR,ATTROLE.
            SECOND ARE LTIMEDRS,ATTDRUG,LPHYHEAL,MENHEAL,LDRUGUSE.
/PRINT      MATRICES ARE CORR,COEF,LOAD.
/PLOT       XVARS ARE CNVRF1,CNVRF2.
            YVARS ARE CNVRS1,CNVRS2.
/END
```

UNIVARIATE SUMMARY STATISTICS

VARIABLE	MEAN	STANDARD DEVIATION	COEFFICIENT OF VARIATION	SMALLEST VALUE	LARGEST VALUE	SMALLEST STANDARD SCORE	LARGEST STANDARD SCORE	SKEWNESS	KURTOSIS
6 ESTEEM	15.81699	3.94798	0.249604	8.00000	29.00000	-1.98	3.34	0.48	0.27
7 CONTROL	6.74946	1.27127	0.188352	5.00000	10.00000	-1.38	2.56	0.48	-0.44
13 LATTMAR	1.33374	0.15408	0.115522	1.04139	1.76343	-1.90	2.79	0.23	-0.61
10 ATTROLE	35.19608	6.75141	0.191823	18.00000	55.00000	-2.55	2.93	0.05	-0.42
11 LTIMEDRS	0.74139	0.41677	0.562148	0.00000	1.91381	-1.78	2.81	0.23	-0.21
3 ATTDRUG	7.67756	1.15433	0.150352	5.00000	10.00000	-2.32	2.01	-0.11	-0.45
12 LPHYHEAL	0.64766	0.20675	0.319231	0.30103	1.17609	-1.68	2.56	-0.01	-0.71
5 MENHEAL	6.11329	4.14532	0.678083	0.00000	18.00000	-1.47	2.87	-0.57	-0.35
14 LDRUGUSE	0.76372	0.48867	0.639851	0.00000	1.82607	-1.56	2.17	-0.15	-1.10

TABLE 6.11 INPUT AND SELECTED BMDPAM OUTPUT FOR
IDENTIFICATION OF MULTIVARIATE OUTLIERS FOR FIRST
SET OF VARIABLES

```
/PROBLEM      TITLE IS 'BMDPAM ON FIRST SET OF VARIABLES'.
/INPUT        VARIABLES=10, FILE=CANON,  FORMAT='(A4,9F6.0)'.
/VARIABLE     NAMES ARE SUBNO,TIMEDRS,ATTDRUG,PHYHEAL,MENHEAL,ESTEEM,
                  CONTROL,ATTMAR,DRUGUSE,ATTROLE,
                  LTIMEDRS,LPHYHEAL,LATTMAR,LDRUGUSE,  ADD=4.
              MISSING ARE (7) 0, 0.
              LABEL = SUBNO.
              USE = 6, 7, 10, 13.
/TRANSFORM    LTIMEDRS = LOG(TIMEDRS + 1).
              LPHYHEAL = LOG(PHYHEAL).
              LATTMAR = LOG(ATTMAR).
              LDRUGUSE = LOG(DRUGUSE + 1).
/EST          METHOD=REGR.
/PRINT        MATR=DIS.
/END
```

MAHALANOBIS DISTANCES ARE COMPUTED FROM EACH CASE TO THE
CENTROID OF ITS GROUP. ONLY THOSE VARIABLES WHICH WERE
ORIGINALLY AVAILABLE ARE USED--ESTIMATED VALUES ARE NOT USED.
FOR LARGE MULTIVARIATE NORMAL SAMPLES, THE MAHALANOBIS
DISTANCES HAVE AN APPROXIMATELY CHI-SQUARE DISTRIBUTION WITH
THE NUMBER OF DEGREES OF FREEDOM EQUAL TO THE NUMBER OF
NONMISSING VARIABLES. SIGNIFICANCE LEVELS REPORTED BELOW
THAT ARE LESS THAN .001 ARE FLAGGED WITH AN ASTERISK.

ESTIMATES OF MISSING DATA, MAHALANOBIS D-SQUARED (CHI-SQUARED)
AND SQUARED MULTIPLE CORRELATIONS WITH AVAILABLE VARIABLES

CASE LABEL	CASE NUMBER	MISSING VARIABLE	ESTIMATE	R-SQUARED	GROUP	CHI-SQ	CHISQ/DF	D.F.	SIGNIFICANCE
121	101					0.572	0.143	4	0.9661
122	102					1.178	0.294	4	0.8818
123	103					1.105	0.276	4	0.8934
124	104					3.109	0.777	4	0.5397
125	105					2.130	0.533	4	0.7118
126	106					9.019	2.255	4	0.0606
127	107					2.459	0.615	4	0.6520
128	108					2.282	0.570	4	0.6841
129	109					4.520	1.130	4	0.3402
130	110					3.101	0.775	4	0.5410
131	111					1.026	0.256	4	0.9059
132	112					3.404	0.851	4	0.4926
133	113					4.624	1.541	3	0.2015
134	114					3.312	0.828	4	0.5070
135	115					3.073	0.768	4	0.5456

TABLE 6.12 SELECTED BMDP6M OUTPUT FOR ASSESSMENT OF
MULTICOLLINEARITY. SETUP SHOWN IN TABLE 6.10

SQUARED MULTIPLE CORRELATIONS OF EACH VARIABLE IN
SECOND SET WITH ALL OTHER VARIABLES IN SECOND SET

VARIABLE NUMBER	NAME	R-SQUARED
11	LTIMEDRS	0.36864
3	ATTDRUG	0.08002
12	LPHYHEAL	0.47204
5	MENHEAL	0.29564
14	LDRUGUSE	0.30872

SQUARED MULTIPLE CORRELATIONS OF EACH VARIABLE IN
FIRST SET WITH ALL OTHER VARIABLES IN FIRST SET

VARIABLE NUMBER	NAME	R-SQUARED
6	ESTEEM	0.22705
7	CONTROL	0.13316
13	LATTMAR	0.13184
10	ATTROLE	0.06751

variable in a set serves, in turn, as DV with the others as IVs. If R-SQUARED values become very large (say, .99 or above), then one variable in the set is nearly a linear combination of others. In this case, low R-SQUARED values indicate absence of singularity or multicollinearity and, indeed, considerable heterogeneity in the sets of variables, particularly the attitudinal set.

6.7.2 Canonical Correlation

The number and importance of canonical variates are determined using procedures from Section 6.5.1. Significance of the relationships between the sets of variables is reported directly by BMDP6M, as shown in Table 6.13. With all four canonical correlations included, $\chi^2(20) = 108.19, p < .001$. With the first and second canonical correlations removed, χ^2 values are not significant; $\chi^2(6) = 4.11, p = .66$. Therefore only the first two pairs of canonical variates are reliable and to be interpreted.

Canonical correlations (r_c) and eigenvalues (r_c^2) are also in Table 6.13. The first canonical correlation is .38, representing 14% overlapping variance for the first pair of canonical variates (see Equation 6.2). The second canonical correlation is .27, representing 7% overlapping variance for the second pair of canonical variates. Although highly significant, neither of these two canonical correlations represents a substantial relationship between pairs of canonical variates. Interpretation of the second canonical correlation and its corresponding pair of canonical variates is especially marginal.

Loading matrices between canonical variates and original variables are in Table 6.14. Interpretation of the two significant pairs of canonical variates from loadings follows procedures mentioned in Section 6.5.2. Correlations between variables and variates (loadings) in excess of .3 are interpreted. Both the direction of correlations in the loading matrices and the direction of scales of measurement have to be considered for a correct interpretation of the pattern of the canonical variates.

The first pair of canonical variates has high loadings on ESTEEM, CONTROL, and LATTMAR (.596, .784, and .730, respectively) on the attitudinal set and on

TABLE 6.13 SELECTED PORTION OF BMDP6M OUTPUT SHOWING
CANONICAL CORRELATIONS AND SIGNIFICANCE LEVELS
FOR SETS OF CANONICAL CORRELATIONS

EIGENVALUE	CANONICAL CORRELATION	NUMBER OF EIGENVALUES	BARTLETT'S TEST FOR REMAINING EIGENVALUES		
			CHI-SQUARE	D.F.	TAIL PROB.
			108.19	20	0.0000
0.14358	0.37892	1	37.98	12	0.0002
0.07203	0.26839	2	4.11	6	0.6613
0.00787	0.08873	3	0.53	2	0.7661
0.00118	0.03429				

BARTLETT'S TEST ABOVE INDICATES THE NUMBER OF CANONICAL
VARIABLES NECESSARY TO EXPRESS THE DEPENDENCY BETWEEN THE
TWO SETS OF VARIABLES. THE NECESSARY NUMBER OF CANONICAL
VARIABLES IS THE SMALLEST NUMBER OF EIGENVALUES SUCH THAT
THE TEST OF THE REMAINING EIGENVALUES IS NON-SIGNIFICANT.
FOR EXAMPLE, IF A TEST AT THE .01 LEVEL WERE DESIRED,
THEN   2 VARIABLES WOULD BE CONSIDERED NECESSARY.
HOWEVER, THE NUMBER OF CANONICAL VARIABLES OF PRACTICAL
VALUE IS LIKELY TO BE SMALLER

TABLE 6.14 SELECTED BMDP6M OUTPUT OF LOADING MATRICES FOR THE TWO SETS OF VARIABLES IN THE EXAMPLE

```
CANONICAL VARIABLE LOADINGS
---------------------------
(CORRELATIONS OF CANONICAL VARIABLES WITH ORIGINAL VARIABLES)
FOR FIRST SET OF VARIABLES

            CNVRF1    CNVRF2    CNVRF3    CNVRF4
              1         2         3         4

ESTEEM    6   0.596     0.601    -0.286    -0.450
CONTROL   7   0.784     0.148    -0.177     0.577
LATTMAR  13   0.730    -0.317     0.434    -0.422
ATTROLE  10  -0.094     0.783     0.605     0.113
```

```
CANONICAL VARIABLE LOADINGS
---------------------------
(CORRELATIONS OF CANONICAL VARIABLES WITH ORIGINAL VARIABLES)
FOR SECOND SET OF VARIABLES

            CNVRS1    CNVRS2    CNVRS3    CNVRS4
              1         2         3         4

LTIMEDRS 11   0.123    -0.359    -0.860     0.249
ATTDRUG   3   0.077     0.559    -0.033     0.405
LPHYHEAL 12   0.408    -0.048    -0.640    -0.505
MENHEAL   5   0.968    -0.143    -0.189     0.066
LDRUGUSE 14   0.276    -0.548     0.016    -0.005
```

LPHYHEAL and MENHEAL (.408 and .968) on the health side. Thus, low self esteem, external locus of control, and dissatisfaction with marital state are related to poor physical and mental health.

The second pair of canonical variates has high loadings on ESTEEM, LATTMAR and ATTROLE (.601, − .317, and .783) on the attitudinal side and LTIMEDRS, ATTDRUG, and LDRUGUSE (− .359, .559, and − .548) on the health side. Big numbers on ESTEEM, little numbers on LATTMAR, and big numbers on ATTROLE go with little numbers on LTIMEDRS, big numbers on ATTDRUG, and little numbers on LDRUGUSE. That is, low self-esteem, satisfaction with marital state, and conservative attitudes toward the proper role of women in society go with few visits to physicians, favorable attitudes toward use of drugs, and little actual use of them. (Figure that one out!)

Loadings are converted to *pv* values by application of Equations 6.11 and 6.12. These values are shown in the BMDP6M output in columns labeled AVERAGE SQUARED LOADING FOR EACH CANONICAL VARIATE (Table 6.15). The values for the first pair of canonical variates are .38 for the first set of variables and .24 for the second set of variables. That is, the first canonical variate pair extracts 38% of variance from the attitudinal variables and 24% of variance from the health variables. The values for the second pair of canonical variates are .27 for the first set of variables and .15 for the second set; the second canonical variate pair extracts 27% of variance from the attitudinal variables and 15% of variance from the health variables. Together, the two canonical variates account for 65% of variance (38% plus 27%) in the attitudinal set, and 39% of variance (24% and 15%) in the health set.

Redundancies for the canonical variates are found in BMDP6M under columns labeled AV. SQ. LOADING TIMES SQUARED CANON. CORREL. (Table 6.15). That is, the first attitudinal variate accounts for 5% of variance in the health variables,

TABLE 6.15 SELECTED BMDP6M OUTPUT SHOWING PERCENTS OF VARIANCE AND REDUNDANCIES FOR FIRST AND SECOND SET OF CANONICAL VARIATES

CANON. VAR.		AVERAGE SQUARED LOADING FOR EACH CANONICAL VARIABLE (1ST SET)	AV. SQ. LOADING TIMES SQUARED CANON. CORREL. (1ST SET)	AVERAGE SQUARED LOADING FOR EACH CANONICAL VARIABLE (2ND SET)	AV. SQ. LOADING TIMES SQUARED CANON. CORREL. (2ND SET)	SQUARED CANON. CORREL.
1		0.37773	0.05424	0.24008	0.03447	0.14358
2		0.27392	0.01973	0.15294	0.01102	0.07203
3		0.16678	0.00131	0.23720	0.00187	0.00787
4		0.18157	0.00021	0.09702	0.00011	0.00118

THE AVERAGE SQUARED LOADING TIMES THE SQUARED CANONICAL CORRELATION IS THE AVERAGE SQUARED CORRELATION OF A VARIABLE IN ONE SET WITH THE CANONICAL VARIABLE FROM THE OTHER SET. IT IS SOMETIMES CALLLED A REDUNDANCY INDEX.

TABLE 6.16 SELECTED BMDP6M OUTPUT OF UNSTANDARDIZED AND STANDARDIZED CANONICAL VARIATE COEFFICIENTS

COEFFICIENTS FOR CANONICAL VARIABLES FOR FIRST SET OF VARIABLES

		CNVRF1 1	CNVRF2 2	CNVRF3 3	CNVRF4 4
ESTEEM	6	0.061949	0.155148	-0.158971	-0.172697
CONTROL	7	0.465185	0.021395	-0.086712	0.699597
LATTMAR	13	3.401715	-2.916037	4.756204	-2.413351
ATTROLE	10	-0.012933	0.091924	0.118283	0.030312

STANDARDIZED COEFFICIENTS FOR CANONICAL VARIABLES FOR FIRST SET OF VARIABLES

(THESE ARE THE COEFFICIENTS FOR THE STANDARDIZED VARIABLES - MEAN ZERO, STANDARD DEVIATION ONE.)

		CNVRF1 1	CNVRF2 2	CNVRF3 3	CNVRF4 4
ESTEEM	6	0.245	0.613	-0.628	-0.682
CONTROL	7	0.591	0.027	-0.110	0.889
LATTMAR	13	0.524	-0.449	0.733	-0.372
ATTROLE	10	-0.087	0.621	0.788	0.205

COEFFICIENTS FOR CANONICAL VARIABLES FOR SECOND SET OF VARIABLES

		CNVRS1 1	CNVRS2 2	CNVRS3 3	CNVRS4 4
LTIMEDRS	11	-0.643182	-0.925366	-2.051052	1.873687
ATTDRUG	3	0.039664	0.673311	-0.039252	0.388091
LPHYHEAL	12	0.208193	2.159183	-2.144725	-5.739981
MENHEAL	5	0.256358	0.008586	0.036875	0.091726
LDRUGUSE	14	-0.122005	-1.693256	1.048614	-0.102427

STANDARDIZED COEFFICIENTS FOR CANONICAL VARIABLES FOR SECOND SET OF VARIABLES

(THESE ARE THE COEFFICIENTS FOR THE STANDARDIZED VARIABLES - MEAN ZERO, STANDARD DEVIATION ONE.)

		CNVRS1 1	CNVRS2 2	CNVRS3 3	CNVRS4 4
LTIMEDRS	11	-0.268	-0.386	-0.855	0.781
ATTDRUG	3	0.046	0.777	-0.045	0.448
LPHYHEAL	12	0.043	0.446	-0.443	-1.187
MENHEAL	5	1.063	0.036	0.153	0.380
LDRUGUSE	14	-0.060	-0.827	0.512	-0.050

TABLE 6.17   CORRELATIONS, STANDARDIZED CANONICAL COEFFICIENTS, CANONICAL CORRELATIONS, PERCENTS OF VARIANCE, AND REDUNDANCIES BETWEEN ATTITUDINAL AND HEALTH VARIABLES AND THEIR CORRESPONDING CANONICAL VARIATES

	First canonical variate		Second canonical variate		
	Correlation	Coefficient	Correlation	Coefficient	
Attitudinal set					
CONTROL	.78	.59	.15	.03	
LATTMAR	.73	.52	−.32	−.45	
ESTEEM	.60	.25	.60	.61	
ATTROLE	−.09	−.09	.78	.62	
Percent of variance	.38		.27		Total = .65
Redundancy	.05		.02		Total = .07
Health set					
MENHEAL	.97	1.06	−.14	.04	
LPHYHEAL	.41	.04	−.53	.45	
LTIMEDRS	.12	−.27	−.36	−.39	
ATTDRUG	.08	.05	.56	.78	
LDRUGUSE	.28	−.06	−.55	−.83	
Percent of variance	.24		.15		Total = .39
Redundancy	.03		.01		Total = .04
Canonical correlation	.38		.27		

while the second attitudinal variate accounts for 2% of the variance. Together, two attitudinal variates "explain" 7% of the variance in health variables. The first health variate accounts for 3% and the second 1% of the variance in the attitudinal set. Together the two health variates overlap the variance in the attitudinal set 4%.

If a goal of analysis is production of scores on canonical variates, coefficients for them are readily available. Table 6.16 shows both standardized and unstandardized coefficients for production of canonical variates. Scores on the variates themselves

TABLE 6.18   CHECKLIST FOR CANONICAL CORRELATION

1. Issues
   a. Missing data
   b. Normality, linearity, homoscedasticity
   c. Outliers
   d. Multicollinearity and singularity
2. Major analyses
   a. Significance of canonical correlations
   b. Correlations of variables and variates
   c. Variance accounted for
      (1) By canonical correlations
      (2) By same-set canonical variates
      (3) By other-set canonical variates (redundancy)
3. Additional analyses
   a. Canonical coefficients
   b. Canonical variates scores

for each case are also produced by BMDP6M if CANV is requested in the /PRINT paragraph. A summary table of information appropriate for inclusion in a journal article appears in Table 6.17.

A checklist for canonical correlation appears in Table 6.18. An example of a Results section in journal format follows for the complete analysis described in Section 6.7.

Results

Canonical correlation was performed between a set of attitudinal variables and a set of health variables with the use of BMDP6M (Dixon, 1985). The attitudinal set included attitudes toward the role of women, toward locus of control, toward current marital status, and toward self. The health set measured mental health, physical health, visits to health professionals, attitude toward use of medication, and use of psychotropic drugs. Increasingly large numbers reflected more conservative attitudes toward women's role, external locus of control, dissatisfaction with marital status, low self-esteem, poor mental health, poor physical health, more numerous health visits, favorable attitudes toward drug use, and more drug use.

To improve linearity of relationship between variables and normality of their distributions, logarithmic transformations were applied to attitude toward marital status, visits to health professionals, physical health, and drug use. No within-set multivariate outliers were identified at $p < .001$, although six cases were found to be missing data on locus of control or attitude toward marital status and were deleted, leaving $N = 459$. Assumptions regarding within-set multicollinearity were met.

The first canonical correlation was .38 (14% of variance); the second was .27 (7% of

variance). The remaining two canonical
correlations were effectively zero. With
all four canonical correlations included,
$\chi^2(20) = 108.19$, $\underline{p} < .001$, and with the first
canonical correlation removed, $\chi^2(12) =$
37.98, $\underline{p} < .001$. Subsequent χ^2 tests were not
statistically significant. The first two
pairs of canonical variates, therefore,
accounted for the significant relationships
between the two sets of variables.

Data on the first two pairs of canonical
variates appear in Table 6.17. Shown in the
table are correlations between the
variables and the canonical variates,
standardized canonical variate
coefficients, within-set variance
accounted for by the canonical variates
(percent of variance), redundancies, and
canonical correlations. Total percent of
variance and total redundancy indicate that
the first pair of canonical variates was
moderately related, but the second pair was
only minimally related; interpretation of
the second pair is marginal.

Insert Table 6.17 about here

With a cutoff correlation of .3, the
variables in the attitudinal set that were
correlated with the first canonical variate
were locus of control, log of attitude
toward marital status, and self-esteem.
Among the health variables, mental health
and log of physical health correlated with
the first canonical variate. The first pair
of canonical variates indicate that those
with external locus of control (.78),
feelings of dissatisfaction toward marital
status (.73), and lower self-esteem (.60)
also tended to have more numerous mental
health symptoms (.97) and more numerous
physical health symptoms (.41).

The second canonical variate in the

attitudinal set was composed of attitude
toward role of women, self-esteem, and
negative of log of attitude toward marital
status, while the corresponding canonical
variate from the health set was composed of
negative of log of drug use, attitude toward
drugs, and negative of log of visits to
health professionals. Taken as a pair, these
variates suggest that a combination of more
conservative attitudes toward the role of
women (.78), lower self-esteem (.60), but
relative satisfaction with marital status
(−.32) corresponds with a combination of
more favorable attitudes toward use of drugs
(.56), but lower psychotropic drug use
(−.55), and fewer visits to health
professionals (−.36).

6.8 SOME EXAMPLES FROM THE LITERATURE

Wingard, Huba, and Bentler (1979) report a canonical analysis of relationships
between personality variables and use of various licit and illicit drugs among junior
high school students in a metropolitan area. The sample was sufficiently large
($N = 1634$) to be randomly divided into groups for assessment of the stability of the
canonical solution. (When a large enough sample is available, cross validation between
randomly selected halves of the sample is highly desirable.) At least two reliable
pairs of canonical variates were discovered for both samples; the first pair, but not
the second, was similar for the samples. The first canonical variate in the drug use
set seemed to reflect early patterns of experimentation with the relatively less dangerous, more readily available legal and illegal drugs. The corresponding variate
from the personality set reflected "non-abidance with the law, liberalism, leadership,
extraversion, lack of diligence, and lack of deliberateness" (p. 139). An attempt to
rotate the two dimensions to facilitate interpretation of the second pair of variates
failed and also degraded the correspondence on the first dimension for the subsamples.
Problems with variate skewness were noted as potentially responsible for difficulties
in comparing the second pair of canonical variates for the two subsamples. Although
canonical correlations and significance levels were high, redundancies were not.

        Cohen, Gaughran, and Cohen (1979) examined the relationships between pat-
terns of fertility across six different age groups (as DVs) and five different sets of
demographic characteristics as IVs (education and occupation; income-labor force;
ethnicity; marriage life-cycle; and housing and occupancy) in five separate canonical
analyses. The units of analysis were 338 New York City health areas. At least three

of a possible six pairs of canonical variates were interprteted for each set of IVs, with the first pair showing substantial r_c values in all cases.

On the DV side, the first canonical variates represented substantial teenage and early twenties childbearing. On the IV side, the first canonical variates represented poorly educated, poverty-level, minority, unmarried persons living in overcrowded housing. Redundancy levels for predicting birthrates from each of the five sets of IVs were about 30%.

On the DV side, the second canonical variates were associated with low rates of childbearing in middle years (20 to 39). Canonical correlations were around .65 and redundancies around 14% for second pairs of variates. On the IV side, low middle-age childbearing was associated with constellations of demographic characteristics associated with an unmarried, affluent-singles life-style and working, educated women. The third pairs of canonical variates had modest correlations and redundancies. Childbearing among the oldest group of women (over 40) was associated with certain ethnic and religious characteristics. Cohen and colleagues direct those who seek to understand and control population expansion to the relationships found in the first pairs of canonical variates, where the absolute level of childbearing was highest.

In a paper comparing canonical analysis with a method known as external single-set components analysis (ESSCA), Fornell (1979) describes relationships between a set of variables measuring characteristics of 128 consumer affairs departments and a set measuring the ability of the departments to influence management decision making. Two significant pairs of canonical variates were discovered, but the second pair had distinctly marginal redundancy (7%) in the direction of interest (predicting influence on decision making from characteristics of departments). Neither pair was deemed interpretable without rotation.

The rotated solution provided by ESSCA did, however, permit interpretation. The first component predicted impact on consumer service and information, and the second predicted impact on marketing decisions. Fornell recommends ESSCA, which maximizes the sum of squared loadings between IV variates and DV variables, over canonical analysis, which maximizes correlation between pairs of canonical variates, when the distinction between IV and DV is clear, so that only a set of variates from the IVs is required. ESSCA may also prove more interpretable when values in $\mathbf{R}_{xy}$ differ greatly in magnitude.

Chapter 7

Multiway Frequency Analysis

7.1 GENERAL PURPOSE AND DESCRIPTION

Relationships among three or more discrete (categorical, qualitative) variables are studied through multiway frequency analysis or an extension of it called loglinear analysis. Most of us are familiar with the two-way χ^2 test of association between two discrete variables where, for example, a researcher studies the simple two-way association between area of psychology (clinical, general experimental, developmental) and average number of publications a year (0, 1, 2, 3, and 4 or more). Is there a relationship between area of psychology and number of publications? If a third variable is added, such as number of statistics courses taken (two or fewer vs. more than two), associations are sought through multiway frequency analysis. The questions, however, remain the same. Is number of publications related to area of psychology and to number of statistics courses taken? Is number of statistics courses taken related to area of psychology?

To do a multiway frequency analysis, tables are formed that contain the one-way, two-way, three-way, and higher-order associations. A linear model of (the logarithm of) expected cell frequencies is developed. The loglinear model starts with all of the one-, two-, three-, and higher-way associations and then eliminates as many of them as possible while still maintaining an adequate fit between expected and observed cell frequencies. In the example above, the three-way association between number of publications, area of psychology, and number of statistics courses is tested first and then eliminated if not statistically significant. Then the two-way associations (number of publications and area of psychology, number of publications and number of statistics courses, area of psychology and number of statistics courses) are tested and eliminated, if not significant. Finally, there is a one-way test for each of the variables against the hypothesis that frequencies are equal in each cell (e.g., that there are equal numbers of psychologists in each area—a test analogous to equal frequency goodness-of-fit tests in χ^2 analysis).

In the model, one of the variables can be considered a DV while the others are considered IVs. For example, a psychologist's success as a professional (successful vs. unsuccessful) is studied as a function of number of publications, area of psychology, number of statistics courses taken, and their interactions. Used this way, multiway frequency analysis is like a multiple regression or a nonparametric analysis of variance with a discrete DV as well as discrete IVs.

7.2 KINDS OF RESEARCH QUESTIONS

The purpose of multiway frequency analysis is to discover associations among discrete variables. Once a preliminary search for associations is complete, a model is fit that includes only the associations that are necessary to reproduce the observed frequencies. Separately for each cell, parameter estimates are derived for the associations retained in the model. The parameter estimates are used to predict cell frequency, and they also reflect the importance of each effect to the frequency in that cell. If one of the variables is a DV, the odds that a case falls into one of its categories can be predicted from the estimates. The following questions, then, are addressed by multiway frequency analysis.

7.2.1 Associations among Variables

Which variables are associated with one another? By knowing which category a case falls into on one variable, can you predict the category it falls into on another? The procedures of Section 7.4 show, for a simple data set, how to determine statistically which variables are associated and how to decide on the level of complexity of associations necessary to describe the relationships.

As the number of variables increases so do the number of potential associations and their complexity. With three variables there are seven potential associations: one three-way association, three two-way associations, and three one-way associations. With four variables there is a potential four-way association, four three-way associations, and so on. With more variables, then, the highest-level associations are tested and eliminated until a preliminary model is found with the fewest required associations.

In the example above, the three-way association between number of publications, number of statistics courses, and area in psychology might be ruled out in preliminary analysis. The set of two-way associations is then tested to see which of these might be ruled out. Number of statistics courses and number of publications might be associated, as well as area of psychology and number of statistics courses, but not area of psychology and number of publications. One-way associations may or may not automatically be included in the model depending on whether or not a hierarchical approach is taken (cf. Section 7.5.1).

7.2.2 Effect on a Dependent Variable

In the usual multiway frequency table, cell frequency is the DV that is influenced by one or more categorical variables and their associations. Sometimes, however, one of the variables is considered a DV. In this case, questions about association are translated into tests of main effects (associations between the DV and each IV) and interactions (association between the DV and the joint effects of two or more IVs).

For example, suppose we investigate a dichotomous measure of a psychologist's success as a function of number of statistics courses, number of publications, and area of psychology. The only associations of research interest are those that include the success variable. The analysis reveals the odds of being in various DV categories as a function of the levels of the independent variables: what, for example, are the odds of success if a psychologist takes two or fewer vs. more than two statistics courses? This multiple regression–like application of multiway frequency analysis is called logit analysis and is discussed in Section 7.5.3.

7.2.3 Parameter Estimates

What is the expected frequency for a particular combination of categories of variables? Or, given the level(s) of one or more variables, what are the odds of being at a particular level on another variable? First, reliable effects are identified, and then coefficients, called parameter estimates, are found *for each cell* for all the reliable effects. Section 7.4.3.2 shows how to calculate parameter estimates and use them to find expected frequencies. Interpretation of parameter estimates as odds in logit analysis (where one variable is considered a DV) is described in Section 7.5.3. For example, given information about number of statistics courses and number of publications, what are the odds that a psychologist is successful?

7.2.4 Importance of Effects

Since parameter estimates are developed for each effect in each cell of the frequency table, the relative importance of each effect to the cell can be evaluated. Effects with larger standardized parameter estimates are more important in predicting that cell's frequency than effects with smaller standardized parameter estimates. If, for instance, number of statistics courses has a higher standardized parameter estimate than number of publications for successful psychologists, it is the more important effect.

7.2.5 Specific Comparisons and Trend Analysis

If a significant association is found, it may be of interest to decompose the association to find its significant components. For example, if area of psychology and number of publications, both with several levels, are associated, which areas differ in number of publications? These questions are analogous to those of analysis of variance where a many-celled interaction is investigated in terms of simpler interactions or in terms of simple effects (cf. Section 10.5.1). Similarly, if the categories of one of the variables differ in quantity (e.g, number of publications), a trend analysis often helps one understand the nature of its relationship with other variables. Planned and post hoc comparisons are discussed in Section 7.5.5.

7.2.6 Causal Modeling

If time is one of the discrete variables, models involving various hypotheses about causal or changing relationships among other discrete variables can be tested. These models are similar to those of analysis of covariance structures applied to continuous variables (e.g., through LISREL or EQS). Examples of path and panel studies with discrete variables are presented in detail by Knoke and Burke (1980).

7.3 LIMITATIONS TO MULTIWAY FREQUENCY ANALYSIS

7.3.1 Theoretical Issues

As a nonparametric statistical technique with no assumptions about population distributions, multiway frequency analysis is remarkably free of limitations. The technique can be applied almost universally, even to continuous variables which fail to meet distributional assumptions of parametric statistics, if the variables are cut into discrete categories.

With the enormous flexibility of current programs for loglinear analysis, many of the questions posed by highly complex data sets can be answered. However, the greatest danger in the use of this analysis is inclusion of so many variables that interpretation boggles the mind—a danger frequently noted in multifactorial analysis of variance, as well.

In logit analyses where one of the variables is considered a DV, the usual cautions about causal inference apply. Validity of causal inference is determined by manipulation of IVs, random assignment of subjects to conditions, and experimental control, not by statistical technique.

7.3.2 Practical Issues

The only limitation to using multiway frequency analysis is the size of expected frequency in each cell. During interpretation, however, certain cells may turn out to be poorly predicted by the solution.

7.3.2.1 Adequacy of Expected Frequencies The fit between observed and expected frequencies is an empirical question in tests of association among discrete variables. Sample cell sizes are observed frequencies; statistical tests compare them with expected frequencies derived from some hypothesis, such as independence between variables. The requirement in multiway frequency analysis is that expected frequencies are large enough. Two conditions produce expected frequencies that are too small: a small sample in conjunction with too many variables with too many levels, and rare events.

The greater the number of variables and their levels, the greater the overall sample size needed to assure adequate expected frequencies in every cell. If marginal frequencies are roughly evenly distributed among various levels of the variables, consider the two-way association that has the largest number of cells—the two-way association between the two variables with the most categories. You should have about five times the number of cases as cells in that two-way association to end up with adequate expected frequencies. In the example, area of psychology has 3 levels and number of publications 5 levels, so $3 \times 5 \times 5$ or 75 cases are needed for adequate expected frequencies.

Rarity of events, however, also needs to be considered. When events are rare, the marginal frequencies are not evenly distributed among the various levels of the variables. For example, there are likely to be few psychologists who average four or more publications a year. A cell from a low-probability row and/or a low-probability column will have a very low expected frequency. The best way to avoid low expected frequencies is to attempt to determine in advance of data collection which cells will be rare, and then sample until those cells are adequately filled.

In any event, *examine expected cell frequencies for all two-way associations to assure that all are greater than one, and that no more than 20% are less than five*. Inadequate expected frequencies generally do not lead to increased Type I error (except in some cases with use of the Pearson χ^2 statistic, cf. Section 7.5.2). But power can be so drastically reduced with inadequate expected frequencies that the analysis is worthless. Reduction of power becomes notable as expected frequencies for two-way associations drop below five in some cells (Milligan, 1980).

If low expected frequencies are encountered despite care in obtaining your sample, several choices are available. First, you can simply choose to accept reduced power for testing effects associated with low expected frequencies. Second, you can collapse categories for variables with more than two levels. For example, you could collapse the three and four or more categories for number of publications into one category with three or more. The categories you collapse depend on theoretical considerations as well as practical ones because it is quite possible that associations will disappear as a result. Because this is equivalent to a complete reduction in power for testing those associations, nothing has been gained.

Finally,[1] you can delete variables to reduce the number of cells. Care is taken to delete only variables that are not associated with the remaining variables. For example, in a three-way table you might consider deleting a variable if there is no three-way association and if at least one of the two-way associations with the variable is nonsignificant (Milligan, 1980).

Section 7.7.1 demonstrates procedures for screening a multidimensional frequency table for expected cell frequencies.

---

[1] The common practice of adding a constant to each cell is not recommended because it has the effect of further reducing power. Its purpose is to stabilize Type I error rate, but as noted above, that is generally not the problem and when it is, other remedies are available (Section 7.5.2).

7.3.2.2 Outliers in the Solution Sometimes there are substantial differences between observed and expected frequencies derived from the best fitting model for some cells. If the differences are large enough, there may be no model that adequately fits the data. Levels of variables may have to be deleted or collapsed before a model is fit. But whether or not a model is fit, examination of residuals in search of discrepant cells leads to better interpretation of the data set. Analysis of residuals is discussed in Sections 7.4.3.1, 7.7.2.3, and 7.7.3.3.

7.4 FUNDAMENTAL EQUATIONS FOR MULTIWAY FREQUENCY ANALYSIS

Analysis of multiway frequency tables typically requires three steps: (1) screening, (2) choosing and testing appropriate models, and (3) evaluating and interpreting the selected model. A small sample example of hypothetical data with three categorical variables is illustrated in Table 7.1. The first variable is type of reading material preferred, READTYP, with two levels: science fiction (SCIFI) and spy novels (SPY). The second variable is SEX; the third variable is represented by three levels of profession, PROFESS: politicians (POLITIC), administrators (ADMIN), and belly dancers (BELLY).

In this section, the simpler calculations are illustrated in detail, while the more complex arithmetic is covered only enough to provide some idea of the methods used to model multidimensional data sets. The computer packages used in this section are also the most straightforward. With real data sets, the various computer packages allow choice of strategy on the basis of utility rather than simplicity. Computer analyses of this data set through BMDP4F, SAS CATMOD, and SPSS^x LOGLINEAR and HILOGLINEAR are in Section 7.4.4.

TABLE 7.1 SMALL SAMPLE OF HYPOTHETICAL DATA FOR ILLUSTRATION OF MULTIWAY FREQUENCY ANALYSIS

			Reading type		
PROFESSION	SEX		SCIFI	SPY	Total
Politicians	Male		15	15	30
	Female		10	15	25
		Total	25	30	55
Administrators	Male		10	30	40
	Female		5	10	15
		Total	15	40	55
Belly dancers	Male		5	5	10
	Female		10	25	35
		Total	15	30	45

If only a single association is of interest, as is usually the case in the analysis of a two-way table, the familiar χ^2 statistic is used:

$$\chi^2 = \sum_{ij} (fo - Fe)^2/Fe \qquad (7.1)$$

where fo represents observed frequencies in each cell of the table and Fe represents the expected frequencies in each cell under the null hypothesis of independence (no association) between the two variables. Summation is over all cells in the two-way table.

If the goodness of fit tests for the two marginal effects are computed, the usual χ^2 tests for the three effects do not sum to total χ^2. This situation is similar to that of unequal-n ANOVA where F tests of main effects and interactions are not independent (cf. Chapter 8). Because overlapping variance cannot be unambiguously assigned to effects, and because overlapping variance is repeatedly analyzed, interpretation of results is not clear-cut. In multiway frequency tables, as in ANOVA, nonadditivity of χ^2 becomes more serious as additional variables produce higher-order (e.g., three-way and four-way) associations.

An alternative strategy is to use the likelihood ratio statistic, G^2. The likelihood ratio statistic is distributed as χ^2, so the χ^2 tables can be used to evaluate significance. However, under conditions to be described later, G^2 has the property of additivity of effects. For example, in a two-way analysis,

$$G_T^2 = G_A^2 + G_B^2 + G_{AB}^2 \qquad (7.2)$$

The test of overall association within a two-way table, G_T^2, is the sum of the first-order goodness-of-fit tests, G_A^2 and G_B^2, and the test of association, G_{AB}^2.

G^2, like χ^2, has a single equation for its various manifestations which differ among themselves only in how the expected frequencies are found.

$$G^2 = 2 \Sigma(fo) \ln(fo/Fe) \qquad (7.3)$$

For each cell, the natural logarithm of the ratio of obtained to expected frequency is multipled by the obtained frequency. These values are summed over cells, and the sum is doubled to produce the likelihood ratio statistics.

7.4.1 Screening

The first step in screening is to determine if there are any effects to be investigated. If there are, then screening progresses to a computation of Fe for each effect, a test of the reliability (significance) of each effect (finding G^2 for the first-order effects, the second-order or two-way associations, the third-order or three-way associations, and so on), and an estimation of the size of the reliable effects. Because Equation 7.3 is used for all tests of the observed frequencies (fo), the trick is to find the Fe necessary to test the various hypotheses, as illustrated below using the data of Table 7.1.

7.4.1.1 Total Effect If done by hand, the process starts by calculation of overall G_T^2, which is used to test the hypothesis of no effects in the table (the hypothesis that all cells have equal frequencies). If this hypothesis cannot be rejected, there is no point to proceeding further. (Note that when all effects are tested simultaneously, as in computer programs, one can test either G_T^2 or G^2 for each of the effects, but not both, because degrees of freedom limit the number of hypotheses to be tested.)

For the test of total effect,

$$Fe = N/rsp \tag{7.4}$$

Expected frequencies, Fe, for testing the hypothesis of no effects, are the same for each cell in the table and are found by dividing the total frequency (N) by the number of cells in the table, i.e., the number of levels of READTYP (represented by r) times the number of levels of SEX (s) times the number of levels of PROFESS (p).

For these data, then,

$$Fe = 155/(2)(2)(3) = 12.9167$$

Applying Equation 7.3 for the test of overall effect,

$$G_T^2 = 2 \sum_{ijk} (fo) \ln(fo/Fe) \qquad df = rsp - 1$$

where $i = 1, 2, \ldots, r;$ $j = 1, 2, \ldots, s;$ and $k = 1, 2, \ldots, p$. Filling in frequencies for each of the cells in Table 7.1, then,

$$
\begin{aligned}
G^2 &= 2[15 \ln(15/12.9167) + 15 \ln(15/12.9167) + 10 \ln(10/12.9167) \\
&\quad + 15 \ln(15/12.9167) + 10 \ln(10/12.9167) + 30 \ln(30/12.9167) \\
&\quad + 5 \ln(5/12.9167) + \cdots + 25 \ln(25/12.9167)] \\
&= 2[2.243 + 2.243 + (-2.559) + 2.243 + (-2.559) + 25.280 \\
&\quad + (-4.745) + (-2.559) + (-4.745) + (-4.745) \\
&\quad + (-2.559) + 16.509] \\
&= 48.09
\end{aligned}
$$

With $df = 12 - 1 = 11$ and critical χ^2 at $\alpha = .05$ equal to 19.6751, there is a statistically reliable departure from equal frequencies among the 12 cells.[2] Further analysis is now required to screen the table for sources of this departure. In the normal course of data analysis the highest-order association is tested first, and so on. Because, however, the complexity for finding expected frequencies is greater with higher-order associations, the presentation here is in the reverse direction, from the first-order to highest-order associations.

---

[2] Throughout this section calculations may differ slightly from those produced by computer programs due to rounding error.

7.4.1.2 First-Order Effects There are three first-order effects to test, one for each of the categorical variables. Starting with READTYP, a goodness-of-fit test evaluates the equality of preference for science fiction and spy novels. Only the marginal sums for the two types of reading material are relevant, producing the following observed frequencies:

$$fo$$

SCIFI	SPY
55	100

Expected frequencies are found by dividing the total frequency by the number of relevant "cells," i.e., $r = 2$, yielding $Fe = 155/2 = 77.5$. The expected frequencies, then, are

$$Fe$$

SCIFI	SPY
77.5	77.5

and the test for goodness of fit is

$$G_R^2 = 2 \sum_i (fo) \ln(fo/Fe) \qquad df = r - 1$$

$$= 2 [55 \ln(55/77.5) + 100 \ln(100/77.5)]$$

$$= 13.25 \qquad df = 1$$

Because critical χ^2 with df = 1 at $\alpha = .05$ is 3.84, a significant preference for spy novels is suggested. As in ANOVA, significant lower-order (main) effects cannot be interpreted unambiguously if there are higher-order (interaction) effects.

Similar tests for main effects of SEX and PROFESS produce $G_S^2 = 0.16$ with 1 df and $G_P^2 = 1.32$ with 2 df, suggesting no statistically significant difference in the number of men (80) and women (75), nor a significant difference in the numbers of politicians (55), administrators (55), and belly dancers (45).

7.4.1.3 Second-order Effects Complications arise in how to compute intermediate level associations. In a three-way design, such as the present example, there are two methods of computing the two-way associations. The simpler method is the marginal (or unconditional) test, in which each two-way association is analyzed ignoring the other two-way and higher-order associations. A two-way table is formed by summing over (deleting) the third factor and ignoring the presence or absence of other reliable effects.

In the more complicated partial (or conditional) test, each two-way association is adjusted for all the other associations. This second method of analysis requires computation of higher-order associations before lower-order ones, to provide for that adjustment.[3]

The two procedures test different hypotheses about the associations, akin to the various strategies for dealing with unequal sample sizes in ANOVA (see Section 8.5.2.2). There is no consensus as to which procedure is more valid, or even consensus regarding the circumstances under which one procedure is more appropriate than the other. One currently popular strategy (Brown, 1976) is to compute intermediate-order associations both ways and make decisions about them with a conservative statistical criterion to avoid inflated α levels due to multiple tests. Because these computations typically are only part of the initial screening process, the results of both sets of calculations are often useful in the second step of choosing and evaluating a model.

For the simpler *marginal* tests of association in the small sample data of Table 7.1, the three-way table is collapsed into three two-way tables, one for each two-way interaction. For the $R \times S$ association, for instance, the cells for each combination of reading type and sex are summed over the three levels of profession (P), forming as the observed frequencies:

fo

	SCIFI	SPY	
MEN	30	50	80
WOMEN	25	50	75
	55	100	155

The expected frequencies are found as in the usual way for a two-way χ^2 test of association:

$$\text{Cell } Fe = (\text{row sum})(\text{column sum})/N \qquad (7.5)$$

for the appropriate row and column for each cell; that is, for the first cell, men preferring science fiction,

$$Fe = (80)(55)/155 = 28.3871$$

After the computations are completed for the remaining cells, the following table of expected frequencies is found:

Fe

	SCIFI	SPY	
MEN	28.3871	51.6129	80
WOMEN	26.6129	48.3871	75
	55	100	155

The only difference between the usual two-way test of association and this test is use of the likelihood ratio statistic instead of χ^2.

---

[3] Recall hierarchical multiple regression (Chapter 5). A marginal effect is the test of a term entering the equation before other terms with lower priority; therefore the other terms are not partialed out of the effect. A partial effect is one that enters the equation after other terms with higher priority; therefore, effects of those other terms are partialed out.

$$G^2_{RS(marg)} = 2 \sum_{ij} (fo) \ln(fo/Fe) \qquad df = (r - 1)(s - 1)$$

$$= 2[30 \ln(30/28.3871) + 50 \ln(50/51.6129)$$
$$+ 25 \ln(25/26.6129) + 50 \ln(50/48.3871)]$$

$$= 0.29 \qquad df = 1$$

The result is obviously not statistically significant.[4]

The same procedure for the remaining two-way associations leads to a non-significant result for the READTYP $\times$ PROFESS interaction with $G^2_{RP(marg)} = 4.09$ and df $= (r - 1)(p - 1) = 2$. For the SEX $\times$ PROFESS interaction, $G^2_{SP(marg)}$ $= 26.77$ with df $= (s - 1)(p - 1) = 2$, which exceeds critical $\chi^2 = 5.99$ at α $= .05$, suggesting a significant association. A look back at the original data in Table 7.1 reveals a dearth of male belly dancers and female administrators.

Tests of the more complicated *partial* associations use an iterative procedure to develop a full set of expected frequencies in which all marginal sums (except the one to be tested) match the observed marginal frequencies.[5] The first iteration uses Equation 7.5 to compute expected frequencies as in the marginal test. Once found, the expected frequencies are duplicated at each level of the other variable. The results of this iteration for the partial test of the $R \times S$ association appear in Table 7.2. Notice that the same expected frequencies as computed for the marginal test are repeated for politicians, administrators, and belly dancers.

TABLE 7.2 FIRST ITERATION ESTIMATES OF EXPECTED
FREQUENCIES FOR THE PARTIAL TEST OF THE
READTYP X SEX ASSOCIATION

			Reading type		
PROFESSION	SEX		SCIFI	SPY	Total
Politicians	Male		28.3871	51.6129	80
	Female		26.6129	48.3871	75
		Total	55	100	155
Administrators	Male		28.3871	51.6129	80
	Female		26.6129	48.3871	75
		Total	55	100	155
Belly dancers	Male		28.3871	51.6129	80
	Female		26.6129	48.3871	75
		Total	55	100	155

[4] The test just illustrated is the simplest marginal test of association using the loglinear model (Marascuilo & Levin, 1983) and, unlike the algorithms in some computer packages, does not use an iterative process.

[5] Other methods for finding partial association are based on differences in G^2 between hierarchical models.

All the entries are too large because the two-way table has simply been duplicated three times. That is, $N = 465$ instead of 155, there are 80 male politicians instead of 30, and so on. A second iteration is performed to adjust the values in Table 7.2 for another two-way association, in this case the $R \times P$ association. This iteration begins with the $R \times P$ table of observed frequencies and relevant marginal sums:

	fo	
	SCIFI	SPY
POLITIC	25	30
ADMIN	15	40
BELLY	15	30
	55	100

Note that the actual number of politicians preferring science fiction is 25, while after the first iteration (Table 7.2) the number is $(28.3871 + 26.6129) = 55$. The goal is to compute a proportion that, when applied to the relevant numbers in Table 7.2 (in this case, both male and female politicians who prefer science fiction), eliminates the effects of any $R \times P$ interaction:

$$fo/Fe^{\#1} = 25/55 = 0.454545455$$

producing

$$Fe^{\#2} = Fe^{\#1}(0.45455) = (28.3871)(0.45455) = 12.9032$$

and

$$Fe^{\#2} = Fe^{\#1}(0.45455) = (26.6129)(0.45455) = 12.0968$$

for male and female politicians preferring science fiction, respectively.

To find second iteration expected frequency for female belly dancers preferring spy stories, the last cell in the table,

$$fo/Fe^{\#1} = 30/100 = 0.3$$
$$Fe^{\#2} = (48.3871)(0.3) = 14.5161$$

Table 7.3 shows the results of applying this procedure to all cells of the data matrix.

Notice that correct totals have been produced for overall N, for R, P, and S, and for $R \times P$, but that the $S \times P$ values are incorrect. The third and final iteration, then, adjusts the $S \times P$ expected values from the second iteration for the $S \times P$ matrix of observed values. These $S \times P$ matrices are:

	fo				$Fe^{\#2}$	
	Men	Women			Men	Women
POLITIC	30	25		POLITIC	28.3871	26.6129
ADMIN	40	15		ADMIN	28.3871	26.6129
BELLY	10	35		BELLY	23.2258	21.7742

TABLE 7.3 SECOND ITERATION ESTIMATES OF EXPECTED
FREQUENCIES FOR THE PARTIAL TEST OF THE
READTYP BY SEX ASSOCIATION

			Reading type		
PROFESSION	SEX		SCIFI	SPY	Total
Politicians	Male		12.9032	15.4839	28.3871
	Female		12.0968	14.5161	26.6129
		Total	25	30	55
Administrators	Male		7.7419	20.6452	28.3871
	Female		7.2581	19.3548	26.6129
		Total	15	40	55
Belly dancers	Male		7.7419	15.4839	23.2258
	Female		7.2581	14.5161	21.7742
		Total	15	30	45

For the first cell, male politicians preferring to read science fiction, the proportional adjustment (rounded off) is

$$fo/Fe^{\#2} = 30/28.3871 = 1.0568$$

to produce

$$Fe^{\#3} = Fe^{\#2}(1.0568) = (12.9032)(1.0568) = 13.6363$$

And for the last cell, female belly dancers who prefer spy stories,

$$fo/Fe^{\#2} = 35/21.7742 = 1.6074$$
$$Fe^{\#3} = (14.5161)(1.6074) = 23.3333$$

Following this procedure for the remaining 10 cells of the matrix produces the third iteration estimates as shown in Table 7.4. These values fulfill the requirement that all expected marginal frequencies are equal to observed marginal frequencies except for $R \times S$, the association to be tested.

At this point we have the Fe necessary to calculate G^2_{RS}.

$$G^2_{RS} = 2 \sum_{ij} (fo) \ln(fo/Fe)$$
$$= 2[15 \ln(15/13.6363) + \cdots + 25 \ln(25/23.3333)]$$
$$= 2.47$$

However, a final adjustment is made for the three-way association, G^2_{RSP} (as computed below). The partial likelihood ratio statistic for the association between READTYP and SEX, then, is

TABLE 7.4   THIRD ITERATION ESTIMATES OF EXPECTED
FREQUENCIES FOR THE PARTIAL TEST OF THE
READTYP BY SEX ASSOCIATION

		Reading type		
PROFESSION	SEX	SCIFI	SPY	Total
Politicians	Male	13.6363	16.3637	30
	Female	11.3637	13.6363	25
	Total	25	30	55
Administrators	Male	10.9090	29.0910	40
	Female	4.0909	10.9091	15
	Total	15	40	55
Belly dancers	Male	3.3333	6.6666	10
	Female	11.6667	23.3333	35
	Total	15	30	45

$$G^2_{RS(\text{part})} = G^2_{RS} - G^2_{RSP} \qquad df = (r-1)(s-1)$$
$$= 2.47 - 1.85 = 0.62 \qquad df = 1$$

This partial test shows a lack of association, as does the corresponding marginal test of the $R \times S$ association.

The same process is followed for the partial tests of the $R \times P$ and the $S \times P$ associations. The resultant partial likelihood ratio statistic for the $R \times P$ association is

$$G^2_{RP(\text{part})} = 4.42 \qquad df = 2$$

showing lack of association between reading preferences and profession, a result consistent with that of the marginal test. For the $S \times P$ association, the partial likelihood ratio result is

$$G^2_{SP(\text{part})} = 27.12 \qquad df = 2$$

a statistically significant association that is also consistent with the corresponding marginal test.

In this example, corresponding marginal and partial tests of intermediate associations produce the same conclusions and interpretation is clear-cut: there is a reliable association between sex and profession and no evidence of association between sex and reading preferences or between reading preferences and profession. In some situations, however, interpretation is more problematic because the results of marginal and partial tests differ. Procedures for dealing with such situations are discussed in Section 7.5.4.

7.4.1.4 Third-order Effect   The test for the three-way $R \times S \times P$ association requires a much longer iterative process because *all* marginal expected frequencies must match

TABLE 7.5 SUMMARY OF SCREENING TESTS FOR SMALL SAMPLE EXAMPLE OF
MULTIWAY FREQUENCY ANALYSIS

EFFECT	df	G^2	PROB	G^2	PROB
ALL (total)	11	48.09	<.05		
READTYP	1	13.25	<.05		
SEX	1	0.16	>.05		
PROFESS	2	1.32	>.05		
		(partial)		(marginal)	
$R \times S$	1	0.62	>.05	0.29	>.05
$R \times P$	2	4.42	>.05	4.09	>.05
$S \times P$	2	27.12	<.05	26.79	<.05
$R \times S \times P$	2	1.85	>.05		
Sums	11	48.74		45.90	

observed frequencies (R, S, P, $R \times S$, $R \times P$, and $S \times P$). Ten iterations are required
to compute the appropriate Fe for the 12 cells (not shown in the interests of brevity
and avoidance of terminal boredom), producing

$$G^2_{RSP} = 2 \sum_{ijk} (fo) \ln (fo/Fe) \qquad df = (r-1)(s-1)(p-1)$$

$$= 1.85 \qquad df = 2$$

The three-way association, then, shows no statistical significance.

A summary of the results of the calculations for all effects appears in Table
7.5. At the bottom of the table is the sum of all one-, two-, and three-way effects
using both the marginal and partial methods for calculating G^2. As can be seen,
neither of these matches G^2_T. If the simple marginal method is applied to the two-
way associations, the sum is too large. If the more complicated method is applied,
the sum is too small. Further, depending on the data, either over- or underadjustment
of each effect may occur with either method. Therefore, Brown (1976) recommends
attention to both sets of results in subsequent modeling decisions (see Section 7.5.4).

7.4.2 Modeling

In some applications of multiway frequency analysis, results of screening provide
sufficient information for the researcher. In the current example, for instance, the
results are clear-cut. One first-order effect, preference for reading type, is statistically
significant, as is the sex by profession association. Often, however, the results are
not so evident and consistent, and/or the goal is to find the best model for predicting
frequencies in each cell of the design.

A loglinear model is developed in which an additive regression-type equation
is written for (the log of) expected frequency as a function of the effects in the design.
The procedure is similar to multiple regression where a predicted DV is obtained by
combining the effects of several IVs.

If all possible effects are included in multiway frequency analysis, the model
is said to be saturated. The saturated model for the 3-way design of the example is

$$\ln Fe_{ijk} = \theta + \lambda_{A_i} + \lambda_{B_j} + \lambda_{C_k} + \lambda_{AB_{ij}} + \lambda_{AC_{ik}} + \lambda_{BC_{jk}} + \lambda_{ABC_{ijk}} \quad (7.6)$$

For each cell (the natural logarithm of) the expected frequency, $\ln Fe$, is an additive sum of the effect parameters, λ's, and a constant, θ.

For each effect in the design, there are as many values of λ—called effect parameters—as there are levels in the effect, and these values sum to zero. In the example there are two levels of READTYP, so there is a value of λ_R for SCIFI and for SPY, and the sum of these two values is zero. For each cell, then, the expected frequency is derived from a different combination of effect parameters.

A saturated model always provides a perfect fit to data so that expected frequencies exactly equal observed frequencies. The purpose of modeling is to find the *unsaturated* model with the fewest effects that still closely mimics the observed frequencies. Screening is done to avoid the necessity of exploring all possible unsaturated models, an inhumane effort with large designs, even with computers. Effects that are found to be unreliable during the screening process are often omitted during modeling.

Model fitting is accomplished by finding G^2 for a particular unsaturated model and evaluating its significance. Because G^2 is a test of fit between observed and expected frequencies, a good model is one with a *nonsignificant* G^2. Because there are often many "good" models, however, there is a problem in choosing among them. The task is to compare nonsignificant models with one another, a process that is easier with one type of model than with another.

Models come in two flavors, hierarchical and nonhierarchical. Hierarchical models include the highest-order reliable association and all its component parts; nonhierarchical models do not necessarily include all the components (see Section 7.5.1). For *hierarchical* models, the optimal model is one that is *not significantly worse* than the next most complex one. Therefore the choice among hierarchical models is made with reference to statistical criteria. There are no statistical criteria for choosing among nonhierarchical models.

Several methods for comparing models are available, as discussed in Section 7.5.4. In the simplest method, illustrated here, a few hierarchical models are selected on the basis of screening results and compared using the significance of the difference in G^2 between them. When the models are hierarchical (or nested), the difference between the two G^2's is itself a G^2. That is,

$$G_1^2 - G_2^2 = G^2 \quad (7.7)$$

if Model 1 is a subset of Model 2 in which all the effects in Model 1 are included in Model 2. For the example, a Model 1 with $R \times P$, R, and P effects is nested with a Model 2 with $R \times S$, $R \times P$, R, S, and P effects.

To simplify description of models, Model 1 above is designated (RP) and Model 2 (RS,RP). This is a fairly standard notation for hierarchical models. Each association term (e.g., RS), implies that all lower-order effects (R and S) are included in the model. In the example, the most obvious model to choose is (SP,R), which includes the $S \times P$ association and all three first-order effects.

In practice, the first step is to evaluate the highest-order effect before sequentially testing lower-order effects. During screening on the example, the three-way association is ruled out but at least one of the two-way associations is reliable. Because there are only three effects in the design, it would not be difficult by computer to try out a model with all three two-way associations (RS,RP,SP), and compare that with models with all pairwise combinations of two-way associations. If there are discrepancies between marginal and partial tests of effects, models with and without the ambiguous effects are compared.

In the example, lack of significance for either the marginal or partial tests of the RP and RS effects would ordinarily preclude their consideration in the set of models to be tested. The RP effect is included in a model to be tested here for illustrative purposes only.

For each of the models to be tested, expected frequencies and G^2 are found. To obtain G^2 for a model, the G^2 for each of the effects is subtracted from total G^2 to yield a test of residual frequency that is *not* accounted for by effects in the model. If the residual frequencies are not significant, there is a good fit between frequencies obtained and frequencies expected from the reduced model.

For the example, G^2 values for the (SP,R) model are available from the screening run shown in Table 7.5. For the two-way effects, the G^2 values from the marginal tests are used because in this case they are smaller and more conservative than the partial values. G^2 for the (SP,R) model is, then

$$G^2_{(SP,R)} = G^2_T - G^2_{SP} - G^2_S - G^2_P - G^2_R$$
$$= 48.09 - 26.79 - 0.16 - 1.32 - 13.25$$
$$= 6.57$$

Degrees of freedom are those associated with each of the effects as in Section 7.4.1, so that df $= 11 - 2 - 1 - 2 - 1 = 5$. Because residuals from this model are not statistically significant, the model is adequate.

For the example, a more complex model includes the $R \times P$ association. Following the above procedures, the (SP,RP) model produces $G^2 = 2.48$ with 3 df. The test of the difference between (SP,R) and (SP,RP) is simply the difference between G^2's (Equation 7.4) for the two models, using the difference between degrees of freedom to test for significance:

$$G^2_{(diff)} = G^2_{(SP,R)} - G^2_{(SP,RP)}$$
$$= 6.57 - 2.48 = 4.09 \qquad \text{with df} = 5 - 3 = 2$$

a nonsignificant result. (In this simple case, G^2 for the difference is the same as G^2 for the partial test of the $S \times P$ association.) Because the difference between models is not statistically significant, the more parsimonious (SP,R) model is preferred over the more complex (SP,RP) model. The model of choice, then, is

$$\ln Fe = \theta + \lambda_R + \lambda_S + \lambda_P + \lambda_{SP}$$

7.4.3 Evaluation and Interpretation

Once the optimal model is chosen, it is evaluated in terms of both the degree of fit to the overall data matrix (as discussed in the previous section) and the amount of deviation from fit in each cell.

7.4.3.1 Residuals Once a model is chosen, expected frequencies for each cell are computed and then the deviation between the expected and observed frequencies in each cell (the residual) is used to assess the adequacy of the model for fitting the observed frequency in that cell. In some cases, a model predicts the frequencies in some cells well, and in others very poorly, to give an indication of the combination of levels of variables for which the model is and is not adequate.

For the example, the observed frequencies are in Table 7.1. Expected frequencies under the (SP,R) model, derived through an iterative procedure as demonstrated in Section 7.4.1.3, are shown in Table 7.6. Residuals are computed as the cell-by-cell differences between the values in the two tables.

Rather than trying to interpret raw differences, residuals are usually standardized by dividing the difference between observed and expected frequencies by the square root of the expected frequency to produce a z value. Both raw differences and standardized residuals for the example are in Table 7.7. The most deviant cell is for male politicians preferring science fiction, with 4.4 fewer cases expected than observed and a standardized residual of $z = 1.3$. Although the discrepancies for men are larger than those for women, none of the cells is terribly discrepant; so this seems to be an acceptable model.

7.4.3.2 Parameter Estimates There is a different linear combination of parameters for each cell and the sizes of the parameters in a cell reflect the contribution of each of the effects in a model to the frequency found in that cell. One can evaluate, for example, how important READTYP is to the number of cases found in the cell for female politicians who read science fiction.

TABLE 7.6 EXPECTED FREQUENCIES UNDER THE MODEL (SP,R)

| PROFESSION | SEX | | Reading type | | |
			SCIFI	SPY	Total
Politicians	Male		10.6	19.4	30.0
	Female		8.9	16.1	25.0
		Total	19.5	35.5	55.0
Administrators	Male		14.2	25.8	40.0
	Female		5.3	9.7	15.0
		Total	19.5	35.5	55.0
Belly dancers	Male		3.5	6.5	10.0
	Female		12.4	22.6	35.0
		Total	16.0	29.0	45.0

TABLE 7.7 RAW AND STANDARDIZED RE-
SIDUALS FOR HYPOTHETICAL
DATA SET UNDER MODEL (SP,R)

		Reading type	
PROFESSION	SEX	SCIFI	SPY
Raw residuals ($fo - Fe$):			
Politicians	Male	4.4	−4.4
	Female	1.1	−1.1
Administrators	Male	−4.2	4.2
	Female	−0.3	0.3
Belly dancers	Male	1.5	−1.5
	Female	−2.4	2.4
Standardized residuals ($fo - Fe$)/$Fe^{1/2}$:			
Politicians	Male	1.3	−1.0
	Female	0.4	−0.3
Administrators	Male	−1.1	0.8
	Female	−0.1	0.1
Belly dancers	Male	0.8	−0.6
	Female	−0.7	0.5

Parameters are estimated for the model from the Fe in Table 7.6 in a manner that closely follows ANOVA. In ANOVA, the size of an effect for a cell is expressed as a deviation from the grand mean. Each cell has a different combination of deviations that correspond to the particular combinations of levels of the reliable effects for that cell.

In MFA deviations are derived from natural logarithms of proportions: $\ln(P_{ijk})$. Expected frequencies for the model (Table 7.6) are converted to proportions by dividing Fe for each cell by $N = 155$, and then the proportions are changed to natural logarithms. For example, for the first cell, male politicians who prefer science fiction:

$$\ln(P_{ijk}) = \ln(Fe_{ijk}/155)$$
$$= \ln(10.6/155)$$
$$= -2.6825711$$

Table 7.8 gives all the resulting values.

The values in Table 7.8 are then used in a three-step process that culminates in parameter estimates, expressed in standard deviation units, for each effect for each cell. The first step is to find both the overall mean and the mean (in natural logarithm units) for each level of each of the effects in the model. The second step is to express each level of each effect as a deviation from the overall mean. The third step is to convert the deviations to standard scores to compare the relative contributions of various parameters to the frequency in a cell.

In the first step, various means are found by summing $\ln(P_{ijk})$ across appropriate

TABLE 7.8 EXPECTED ln(P_{ijk}) FOR MODEL (SP,R)

PROFESSION	SEX	Reading type	
		SCIFI	SPY
Politicians	Male	−2.6825711	−2.0781521
	Female	−2.8573738	−2.2646058
Administrators	Male	−2.3901832	−1.7930506
	Female	−3.3757183	−2.7712992
Belly dancers	Male	−3.7906621	−3.1716229
	Female	−2.5257286	−1.9254752

Note: $\ln(P_{ijk}) = \ln(Fe_{ijk}/155) = \ln(Fe_{ijk}) - \ln(155)$.

cells and dividing each sum by the number of cells involved. For example, to find the overall mean,

$$\bar{x}_{...} = (1/rsp) \sum_{ijk} \ln(P_{ijk})$$

$$= (1/12) \left[-2.6825711 + (-2.0781521) + (-2.8573738) \right.$$
$$\left. + \cdots + (-1.9254752) \right]$$

$$= -2.6355346$$

To find the mean for SCIFI, the first level of READTYP:

$$\bar{x}_{1..} = (1/sp) \sum_{jk} \ln(P_{ijk})$$

$$= (1/6) \left[-2.6825711 + (-2.8573738) + (-2.3901832) \right.$$
$$\left. + (-3.3757183) + (-3.7906621) + (-2.5257286) \right]$$

$$= -2.9370395$$

The mean for belly dancers is

$$\bar{x}_{..3} = (1/rs) \sum_{ij} \ln(P_{ijk})$$

$$= (1/4) \left[-3.7906621 + (-3.1716229) + (-2.5257286) \right.$$
$$\left. + (-1.9254752) \right]$$

$$= -2.8533722$$

and so on for the first-order effects.

The means for second-order effects are found in a similar manner. For instance, for the $S \times P$ association, the mean for male politicians is

$$\bar{x}_{.11} = (1/r) \sum_{i} \ln(P_{ijk})$$

$$= (1/2) \left[-2.6825711 + (-2.0781521) \right]$$

$$= -2.3803616$$

In the second step, parameter estimates are found by subtraction. For first-order effects, the overall mean is subtracted from the mean for each level. For example, λ_{R_i}, the parameter for SCIFI, the first level of READTYP is

$$\begin{aligned} \lambda_{R_1} &= \bar{x}_1 .. - \bar{x}... \\ &= -2.9370395 - (-2.6355346) \\ &= -.302 \end{aligned}$$

For belly dancers, the third level of PROFESS

$$\begin{aligned} \lambda_{P_3} &= \bar{x}._{.3} - \bar{x}... \\ &= -2.8533722 - (-2.63555346) \\ &= -.218 \end{aligned}$$

and so on.

To find λ for a cell in two-way effect, the two appropriate main effect means are subtracted from the two-way mean, and the overall mean is added (in a pattern that is also familiar from ANOVA). For example, $\lambda_{SP_{23}}$, the parameter for female belly dancers (second level of sex, third level of profession), is found by subtracting from the female belly dancer mean (averaged over the two types of reading material) the mean for women and the mean for belly dancers, and then adding the overall mean.

$$\begin{aligned} \lambda_{SP_{23}} &= \bar{x}._{23} - \bar{x}._{2.} - \bar{x}._{.3} + \bar{x}... \\ &= -2.2256019 - (-2.6200335) - (-2.8533722) + (-2.6355346) \\ &= .612 \end{aligned}$$

All the λ values, as shown in Table 7.9, are found in a similar, if tedious, fashion. In the table, θ is the conversion of the overall mean from proportion to frequency units by addition of $\ln(N)$:

$$\begin{aligned} \theta &= \bar{x}... + \ln(155) \\ &= 2.4079 \end{aligned}$$

The expected frequency generated by the model for each cell is then expressed as a function of the appropriate parameters. For example, the expected frequency (19.40) for male politicians who read spy novels is

$$\begin{aligned} \ln Fe &= \theta + \lambda_{R_2} + \lambda_{S_1} + \lambda_{P_1} + \lambda_{SP_{11}} \\ \ln Fe &= 2.4079 + .302 + (-.015) + .165 + .106 \\ &= 2.9659 \approx \ln(19.40) \end{aligned}$$

within rounding error.

These parameters are used to find expected frequencies for each cell but are not interpreted in terms of magnitude until step 3 is taken. During step 3 parameters

TABLE 7.9 PARAMETER ESTIMATES FOR MODEL (SP,R)
           θ (MEAN) = 2.4079

EFFECT	LEVEL	λ	λ/SE
READTYP	SCIFI	$-.302$	-3.598
	SPY	$.302$	3.598
SEX	MALE	$-.015$	-0.186
	FEMALE	$.015$	0.186
PROFESSION	POLITICIAN	$.165$	2.045
	ADMINISTRATOR	$.053$	0.657
	BELLY DANCER	$-.218$	-2.702
SEX BY PROFESS	MALE POLITICIAN	$.106$	1.154
	FEMALE POLITICIAN	$-.106$	-1.154
	MALE ADMINISTRATOR	$.506$	5.510
	FEMALE ADMINISTRATOR	$-.506$	-5.510
	MALE BELLY DANCER	$.612$	7.200
	FEMALE BELLY DANCER	$-.612$	-7.200

are divided by their respective standard errors to form standard normal deviates that are interpreted according to their relative magnitudes. Therefore the parameter values in Table 7.9 are given both in their λ form and after division by their standard errors.

Standard errors of parameters, SE, are found by squaring the reciprocal of the number of levels for the set of parameters, dividing by the observed frequencies, and summing over the levels. For example, for READTYP:

$$SE^2 = \Sigma \, (1/r_i)^2/fo$$
$$= (1/2)^2/55 + (1/2)^2/100$$
$$= (.25)/55 + (.25)/100$$
$$= .0070455$$

and

$$SE = .0839372$$

Note that this is the simplest method for finding SE (Goodman, 1978) and does not weight the number of levels by unequal marginal frequencies, as do methods such as the one used in BMDP.

To find the standard normal deviate for SCIFI (the first level of READTYP), λ for SCIFI is divided by its standard error

$$\lambda_{R_1}/SE = -.302/.0839372$$
$$= -3.598$$

This ratio is interpreted as a standard normal deviate (z) and compared with critical z to assess the contribution of an effect to a cell. The relative importance of the various effects to a cell is also derived from these values. For female belly dancers preferring spy novels, for example, the standard normal deviates for the parameters are 3.598 (SPY), 0.186 (FEMALE), -2.702 (BELLY), -7.200 (FEMALE BELLY). The most important influences on cell frequency are, in order, the sex by profession association, preferred type of reading material, and profession—all statistically significant at $p < .01$ because they exceed 2.58. Sex contributes little to the expected frequency in this cell and is not statistically significant.

Because of the large number of effects produced in typical loglinear models, a conservative criterion should be used if statistical significance is evaluated. A criterion z of 4.00 is often considered reasonable.

Further insights into interpretation are provided in Sections 7.7.2.4 and 7.7.3.3. Conversion of parameters to odds when one variable is a DV is discussed in Section 7.5.3.2.

7.4.4 Computer Analyses of Small Sample Example

Setup and selected output for computer analyses of the data in Table 7.1 appear in Tables 7.10 through 7.14. SPSS^x HILOGLINEAR and LOGLINEAR are in Tables 7.10 and 7.11, respectively. BMDP4F is illustrated in Table 7.12, SAS CATMOD in Table 7.13, and SYSTAT TABLES in Table 7.14.

This setup of SPSS^x HILOGLINEAR (Table 7.10) produces output appropriate for screening a hierarchical multiway frequency analysis. Additional code is necessary to test models. The instruction PRINT = FREQ produces the OBSERVED, EXPECTED FREQUENCIES AND RESIDUALS. Because no model is specified in the setup, a saturated model (all effects included in the model) is produced in which expected and observed frequencies are identical. The GOODNESS-OF-FIT STATISTICS also reflect a perfectly fitting model.

The next three tables are produced by the ASSOCIATION instruction and consist

TABLE 7.10 MULTIWAY FREQUENCY ANALYSIS OF SMALL SAMPLE
EXAMPLE THROUGH SPSS^x HILOGLINEAR (SETUP AND
SELECTED OUTPUT)

```
          TITLE   SMALL SAMPLE MULTIWAY ANALYSIS
          FILE HANDLE  SSMFA
          DATA LIST   FILE=SSMFA
                  / PROFESS SEX READTYP FREQ
          WEIGHT BY FREQ
          VALUE LABELS  PROFESS 1 'POLITIC' 2 'ADMIN' 3 'BELLY'/
                        SEX 1 'MALE' 2 'FEMALE'/
                        READTYP 1 'SCIFI' 2 'SPY'/
          HILOGLINEAR   PROFESS(1,3) SEX READTYP (1,2)/
                        PRINT=FREQ, ASSOCIATION

>NOTE    12717
>THE LAST COMMAND IS NOT A DESIGN/MODEL SPECIFICATION.  A SATURATED MODEL IS
>GENERATED FOR THIS PROBLEM.

DATA    INFORMATION

          12 UNWEIGHTED CASES ACCEPTED.
           0 CASES REJECTED BECAUSE OF OUT-OF-RANGE FACTOR VALUES.
           0 CASES REJECTED BECAUSE OF MISSING DATA.
         155 WEIGHTED CASES WILL BE USED IN THE ANALYSIS.

FACTOR INFORMATION

       FACTOR  LEVEL  LABEL
       PROFESS    3
       SEX        2
       READTYP    2
```

TABLE 7.10 (Continued)

```
* * * * * * * *  H I E R A R C H I C A L   L O G   L I N E A R  * * * * * * * *

DESIGN 1 HAS GENERATING CLASS

     PROFESS*SEX*READTYP

THE ITERATIVE PROPORTIONAL FITTING CONVERGED AT ITERATION 1.

- - - - - - - - - - - - - - - - - - - - - - - - - - - - - - - - - - - - -
   OBSERVED, EXPECTED FREQUENCIES AND RESIDUALS.

         FACTOR          CODE        OBS. COUNT  & PCT.    EXP. COUNT  & PCT.


   PROFESS       POLITIC
     SEX           MALE
       READTYP       SCIFI          15.00 ( 9.68)        15.00 ( 9.68)
       READTYP       SPY            15.00 ( 9.68)        15.00 ( 9.68)
     SEX           FEMALE
       READTYP       SCIFI          10.00 ( 6.45)        10.00 ( 6.45)
       READTYP       SPY            15.00 ( 9.68)        15.00 ( 9.68)
   PROFESS       ADMIN
     SEX           MALE
       READTYP       SCIFI          10.00 ( 6.45)        10.00 ( 6.45)
       READTYP       SPY            30.00 (19.35)        30.00 (19.35)
     SEX           FEMALE
       READTYP       SCIFI           5.00 ( 3.23)         5.00 ( 3.23)
       READTYP       SPY            10.00 ( 6.45)        10.00 ( 6.45)

   PROFESS       BELLY
     SEX           MALE
       READTYP       SCIFI           5.00 ( 3.23)         5.00 ( 3.23)
       READTYP       SPY             5.00 ( 3.23)         5.00 ( 3.23)
     SEX           FEMALE
       READTYP       SCIFI          10.00 ( 6.45)        10.00 ( 6.45)
       READTYP       SPY            25.00 (16.13)        25.00 (16.13)

- - - - - - - - - - - - - - - - - - - - - - - - - - - - - - - - - - - - -
   GOODNESS-OF-FIT TEST STATISTICS

       LIKELIHOOD RATIO CHI SQUARE =        0    DF = 0  P = 1.000
                  PEARSON CHI SQUARE =       0    DF = 0  P = 1.000

* * * * * * * *  H I E R A R C H I C A L   L O G   L I N E A R  * * * * * * * *

   TESTS THAT K-WAY AND HIGHER ORDER EFFECTS ARE ZERO.

       K    DF   L.R. CHISQ    PROB  PEARSON CHISQ    PROB   ITERATION

       3    2        1.848    .3969        1.920    .3828        3
       2    7       33.353    .0000       32.994    .0000        2
       1   11       48.089    .0000       52.097    .0000        0


- - - - - - - - - - - - - - - - - - - - - - - - - - - - - - - - - - - - -

   TESTS THAT K-WAY EFFECTS ARE ZERO.

       K    DF   L.R. CHISQ    PROB  PEARSON CHISQ    PROB   ITERATION

       1    4       14.737    .0053       19.103    .0008        0
       2    5       31.505    .0000       31.073    .0000        0
       3    2        1.848    .3969        1.920    .3828        0

* * * * * * * *  H I E R A R C H I C A L   L O G   L I N E A R  * * * * * * * *

   TESTS OF PARTIAL ASSOCIATIONS.

   EFFECT NAME                      DF   PARTIAL CHISQ      PROB   ITER

   PROFESS*SEX                       2        27.122      .0000      2
   PROFESS*READTYP                   2         4.416      .1099      2
   SEX*READTYP                       1          .621      .4308      2
   PROFESS                           2         1.321      .5166      2
   SEX                               1          .161      .6879      2
   READTYP                           1        13.255      .0003      2
```

of tests of all effects individually, effects combined at each order, and effects combined at each order and higher orders. The TESTS OF PARTIAL ASSOCIATION table shows tests of each two-way and one-way effect. Note the correspondence between these values and those of Table 7.5, produced by hand calculation. In the TESTS THAT K-WAY EFFECTS ARE ZERO table, the combined associations at each order are tested. That is, the row labeled 2 evaluates the three two-way associations combined and, in this case, shows statistical significance using both the likelihood ratio (L.R.) and Pearson chi-square criteria. This output suggests that at least one of the two-way associations is significant by both criteria. The test of the single three-way association is also provided in this table when $k = 3$; it is not significant. In the TESTS THAT K-WAY AND HIGHER ASSOCIATIONS ARE ZERO table, the row labeled 1 tests that the combination of all one-way, two-way, and three-way associations is significant against both likelihood ratio and Pearson chi-square criteria. The row labeled 2 evaluates the combination of all two- and three-way associations, and so on.

Table 7.11 shows the results of an unspecified, and therefore saturated, model run through SPSS[x] LOGLINEAR, a nonhierarchical program. After some descriptive information about how the program internally processes the model, a table of OB-SERVED, EXPECTED FREQUENCIES AND RESIDUALS is printed out, similar to that of the HILOGLINEAR program but with the RESIDUALS columns included along with GOODNESS-OF-FIT STATISTICS. The remaining output tests effects in

TABLE 7.11 MULTIWAY FREQUENCY ANALYSIS OF SMALL SAMPLE EXAMPLE THROUGH SPSS[x] LOGLINEAR (SETUP AND SELECTED OUTPUT)

```
        TITLE   SMALL SAMPLE MULTIWAY ANALYSIS
        FILE HANDLE   SSMFA
        DATA LIST   FILE=SSMFA
                / PROFESS SEX READTYP FREQ
        WEIGHT BY FREQ
        VALUE LABELS   PROFESS 1 'POLITIC' 2 'ADMIN' 3 'BELLY'/
                       SEX 1 'MALE' 2 'FEMALE'/
                       READTYP 1 'SCIFI' 2 'SPY'/
        LOGLINEAR      PROFESS(1,3) SEX READTYP (1,2)/


>NOTE    12717
>THE LAST COMMAND IS NOT A DESIGN/MODEL SPECIFICATION.  A SATURATED MODEL IS
>GENERATED FOR THIS PROBLEM.

* * * * * * * L O G   L I N E A R   A N A L Y S I S * * * * * * *

DATA    INFORMATION

        12 UNWEIGHTED CASES ACCEPTED.
         0 CASES REJECTED BECAUSE OF OUT-OF-RANGE FACTOR VALUES.
         0 CASES REJECTED BECAUSE OF MISSING DATA.
       155 WEIGHTED CASES WILL BE USED IN THE ANALYSIS.


FACTOR INFORMATION

      FACTOR  LEVEL  LABEL
      PROFESS    3
      SEX        2
      READTYP    2

DESIGN INFORMATION

    1 DESIGN/MODEL WILL BE PROCESSED.
- - - - - - - - - - - - - - - - - - - - - - - - - - - - - - - -
```

TABLE 7.11 (Continued)

```
* * * * * * * L O G   L I N E A R   A N A L Y S I S * * * * * * * *
```

CORRESPONDENCE BETWEEN EFFECTS AND COLUMNS OF DESIGN/MODEL 1

STARTING COLUMN	ENDING COLUMN	EFFECT NAME
1	2	PROFESS
3	3	SEX
4	4	READTYP
5	6	PROFESS BY SEX
7	8	PROFESS BY READTYP
9	9	SEX BY READTYP
10	11	PROFESS BY SEX BY READTYP

- -

*** ML CONVERGED AT ITERATION 3. THE CONVERGE CRITERION =   .00000

- -

OBSERVED, EXPECTED FREQUENCIES AND RESIDUALS

FACTOR	CODE	OBS. COUNT & PCT.	EXP. COUNT & PCT.	RESIDUAL	STD. RESID.	ADJ. RESID.
PROFESS	POLITIC					
SEX	MALE					
READTYP	SCIFI	15.00 (9.68)	15.00 (9.68)	0	0	0
READTYP	SPY	15.00 (9.68)	15.00 (9.68)	0	0	0
SEX	FEMALE					
READTYP	SCIFI	10.00 (6.45)	10.00 (6.45)	0	0	0
READTYP	SPY	15.00 (9.68)	15.00 (9.68)	0	0	0
PROFESS	ADMIN					
SEX	MALE					
READTYP	SCIFI	10.00 (6.45)	10.00 (6.45)	0	0	0
READTYP	SPY	30.00 (19.35)	30.00 (19.35)	0	0	0
SEX	FEMALE					
READTYP	SCIFI	5.00 (3.23)	5.00 (3.23)	0	0	0
READTYP	SPY	10.00 (6.45)	10.00 (6.45)	0	0	0
PROFESS	BELLY					
SEX	MALE					
READTYP	SCIFI	5.00 (3.23)	5.00 (3.23)	0	0	0
READTYP	SPY	5.00 (3.23)	5.00 (3.23)	0	0	0
SEX	FEMALE					
READTYP	SCIFI	10.00 (6.45)	10.00 (6.45)	0	0	0
READTYP	SPY	25.00 (16.13)	25.00 (16.13)	0	0	0

- -

GOODNESS-OF-FIT TEST STATISTICS

```
      LIKELIHOOD RATIO CHI SQUARE =        0   DF = 0   P = 1.000
                 PEARSON CHI SQUARE =        0   DF = 0   P = 1.000
```

- -

ESTIMATES FOR PARAMETERS

PROFESS

PARAMETER	COEFF.	STD. ERR.	Z-VALUE	LOWER 95 CI	UPPER 95 CI
1	.2081107641	.12285	1.69400	-.03268	.44890
2	.0053782101	.13368	.04023	-.25663	.26739

SEX

PARAMETER	COEFF.	STD. ERR.	Z-VALUE	LOWER 95 CI	UPPER 95 CI
3	-.0087800430	.09404	-.09337	-.19309	.17553

READTYP

PARAMETER	COEFF.	STD. ERR.	Z-VALUE	LOWER 95 CI	UPPER 95 CI
4	-.2594596091	.09404	-2.75918	-.44377	-.07515

TABLE 7.11 (Continued)

PROFESS BY SEX

PARAMETER	COEFF.	STD. ERR.	Z-VALUE	LOWER 95 CI	UPPER 95 CI
5	.1101463200	.12285	.89658	-.13064	.35094
6	.4567199103	.13368	3.41651	.19471	.71873

PROFESS BY READTYP

PARAMETER	COEFF.	STD. ERR.	Z-VALUE	LOWER 95 CI	UPPER 95 CI
7	.1580933321	.12285	1.28686	-.08270	.39888
8	-.1884802582	.13368	-1.40993	-.45049	.07353

SEX BY READTYP

PARAMETER	COEFF.	STD. ERR.	Z-VALUE	LOWER 95 CI	UPPER 95 CI
9	.0763575610	.09404	.81201	-.10795	.26067

PROFESS BY SEX BY READTYP

PARAMETER	COEFF.	STD. ERR.	Z-VALUE	LOWER 95 CI	UPPER 95 CI
10	.0250087160	.12285	.20357	-.21578	.26580
11	-.1777238380	.13368	-1.32947	-.43974	.08429

a different way. Instead of a partial (or marginal) test for each effect, parameter estimates for the effect are tested by dividing each λ by its standard error (STD. ERR.) to produce a Z-VALUE and a 95% confidence interval (LOWER 95 CI and UPPER 95 CI).[6] Note that if an effect has more than 1 df, a single test for the effect is not provided because the parameter estimate for each df is tested separately.

BMDP4F output in Table 7.12 is similar to that of SPSS[x] HILOGLINEAR. The program begins by printing out univariate descriptive statistics for each variable, useful for ensuring against miscoded values. The observed frequency table shows cell frequencies and marginal sums. The FIT paragraph with ASSOCIATION IS 3 provides the remaining output: tests of all terms, individually, combined at each order, and combined at each order and higher orders, as in SPSS[x] HILOGLINEAR. The table labeled ***** ASSOCIATION OPTION SELECTED FOR ALL TERMS OF ORDER LESS THAN OR EQUAL TO 3 shows both marginal and partial significance tests of the individual effects. In the *****SIMULTANEOUS TEST ..... table, the combined associations at each order are tested (i.e., the row labeled 2 shows statistical significance when all two-way associations are tested together). In the ***** THE RESULTS OF FITTING ALL K-FACTOR MARGINALS table, 0-MEAN contains output for testing whether there are any effects considering all associations together. The row labeled 1 contains output for testing whether any of the two- or higher-way associations are statistically significant, and so on. Tests of specific models require additional instructions.

SAS CATMOD setup and output for nonhierarchical MFA appear in Table 7.13. The saturated model is specified by listing the three-way association, PRO-FESS*SEX*READTYP equal to –RESPONSE–, a keyword that induces a loglinear

---

[6] These parameter estimates differ somewhat from those produced by hand calculation—Table 7.9—because of the different algorithm used by this program.

```
                /PROBLEM    TITLE IS 'SMALL SAMPLE MULTIWAY FREQUENCY ANALYSIS'.
                /INPUT      VARIABLES ARE 4.
                            CASES ARE 12.
                            FORMAT IS FREE.
                /VARIABLE   NAMES ARE PROFESS, SEX, READTYP, FREQ.
                /CATEGORY   CODES(SEX,READTYP) ARE 1, 2.
                            NAMES(READTYP) ARE SCIFI, SPY.
                            NAMES(SEX) ARE MALE, FEMALE.
                            CODES(PROFESS) ARE 1, 2, 3.
                            NAMES(PROFESS) ARE POLITIC, ADMIN, BELLY.
                /TABLE      INDICES ARE READTYP, SEX, PROFESS.
                            COUNT = FREQ.
                /FIT        ASSOCIATION IS 3.
                /END
```

						NO. OF VALUES		
VARIABLE	MEAN	STD.	MIN.	MAX.	TOTAL		LT GT NE	
NO. NAME		DEV.	VALUE	VALUE	FREQ. MISSING	MIN MAX CODES	
1 PROFESS	2.00	0.85	1.00	3.00	12	0	0 0 0	
2 SEX	1.50	0.52	1.00	2.00	12	0	0 0 0	
3 READTYP	1.50	0.52	1.00	2.00	12	0	0 0 0	

```
        ***********************
        *  TABLE PARAGRAPH   1  *
        ***********************

*****  OBSERVED FREQUENCY TABLE  1

VARIABLE    4   FREQ       USED AS COUNT VARIABLE.
                ********

PROFESS  SEX        READTYP
------   ------     ------
                    SCIFI    SPY    TOTAL
---------------------------------------------

POLITIC  MALE        15       15 :    30
         FEMALE      10       15 :    25
         ----------------------------:---------
         TOTAL       25       30 :    55

ADMIN    MALE        10       30 :    40
         FEMALE       5       10 :    15
         ----------------------------:---------
         TOTAL       15       40 :    55

BELLY    MALE         5        5 :    10
         FEMALE      10       25 :    35
         ----------------------------:---------
         TOTAL       15       30 :    45

        TOTAL OF THE OBSERVED FREQUENCY TABLE IS      155

        ALL CASES HAD COMPLETE DATA FOR THIS TABLE.
```

```
*****  THE RESULTS OF FITTING ALL K-FACTOR MARGINALS.
       SIMULTANEOUS TEST THAT ALL K+1 AND HIGHER FACTOR INTERACTIONS ARE ZERO.
```

K-FACTOR	D.F.	LR CHISQ	PROB.	PEARSON CHISQ	PROB.	ITERATION
0-MEAN	11	48.09	0.00000	52.10	0.00000	
1	7	33.35	0.00002	32.99	0.00003	2
2	2	1.85	0.39694	1.92	0.38257	4
3	0	0.	1.	0.	1.	

TABLE 7.12 (Continued)

*****SIMULTANEOUS TEST THAT ALL K-FACTOR INTERACTIONS ARE SIMULTANOUSLY ZERO.
THE CHI-SQUARES ARE DIFFERENCES IN THE ABOVE TABLE.

K-FACTOR	D.F.	LR CHISQ	PROB.	PEARSON CHISQ	PROB.
1	4	14.74	0.00528	19.10	0.00075
2	5	31.50	0.00001	31.07	0.00001
3	2	1.85	0.39694	1.92	0.38257

***** ASSOCIATION OPTION SELECTED FOR ALL TERMS OF ORDER LESS THAN OR EQUAL TO 3

		PARTIAL ASSOCIATION			MARGINAL ASSOCIATION			
EFFECT	D.F.	CHISQUARE	PROB	ITER	D.F.	CHISQUARE	PROB	ITER
R.	1	13.25	0.0003					
S.	1	0.16	0.6880					
P.	2	1.32	0.5166					
RS.	1	0.62	0.4308	2	1	0.29	0.5878	2
RP.	2	4.42	0.1099	2	2	4.09	0.1294	2
SP.	2	27.12	0.0000	2	2	26.79	0.0000	2
RSP.	2	1.85	0.3969					

model. ML requests maximum likelihood estimates, and NOGLS suppresses unneeded output. The REPEATED instruction is used unless a logit model is to be tested. The instructions in this setup specify that all variables—PROFESS, SEX, READTYP— are to be treated the same, that none is the DV.

The RESPONSE label in the output refers to cells; levels 1 to 12 refer to the 12 ($2 \times 2 \times 3$) cells in the frequency table. The WEIGHT VARIABLE refers to the variable in the data set in which cell frequencies are encoded (FREQ). After information on description of the design, CATMOD provides details about the iterative process in fitting the model and estimating the parameters. The ANALYSIS OF VARIANCE TABLE contains likelihood ratio CHI-SQUARE tests of each effect individually. Note that due to differences in the algorithms used, these estimates do not match those produced by hand calculation, SPSS^x HILOGLINEAR, or BMDP4F, all of which are based on a different testing strategy.

There are also tests of individual parameter estimates in the section labeled ANALYSIS OF INDIVIDUAL PARAMETERS, although some of these differ from both the ones shown for hand calculation (Table 7.9) and those produced by SPSS^x LOGLINEAR (Table 7.11). The next page of output provides a clue to the order in which parameter estimates are given. The first effect in _RESPONSE_ = is PROFESS. Because PROFESS has 2 df, parameters 1 and 2 refer to parameter estimates for PROFESS. The second effect is SEX, with 1 df, so parameter 3 refers to parameter estimates for SEX, and so on. CHI-SQUARE tests (rather than z) are given for each of the parameter estimates. The last table shows observed and predicted frequencies and residuals for each of the 12 cells. The relevant information appears at the bottom portion of the table. For the first cell (level 1 of each of the variables), both the observed and predicted frequencies are 15 and the residual is effectively zero—again because this is a saturated, perfectly fitting model.

SYSTAT TABLES, in Table 7.14, produces much less output than the other programs. To illustrate this program, a model is specified in the setup because no

TABLE 7.13 MULTIWAY FREQUENCY ANALYSIS OF SMALL SAMPLE EXAMPLE THROUGH SAS CATMOD (SETUP AND SELECTED OUTPUT)

```
DATA SSAMPLE;
INFILE SSMFA;
INPUT PROFESS SEX READTYP FREQ;
PROC CATMOD;
  WEIGHT FREQ;
  MODEL PROFESS*SEX*READTYP=_RESPONSE_/
    PRED=FREQ ML NOGLS;
  REPEATED/_RESPONSE_=PROFESS!SEX!READTYP;
```

CATMOD PROCEDURE

RESPONSE: PROFESS*SEX*READTYP	RESPONSE LEVELS (R== 12
WEIGHT VARIABLE: FREQ	POPULATIONS (S)= 1
DATA SET: SSAMPLE	TOTAL FREQUENCY (N)= 155
	OBSERVATIONS (OBS)= 12

	SAMPLE SIZE
SAMPLE	
1	155

RESPONSE PROFILES

RESPONSE	PROFESS	SEX	READTYP
1	1	1	1
2	1	1	2
3	1	2	1
4	1	2	2
5	2	1	1
6	2	1	2
7	2	2	1
8	2	2	2
9	3	1	1
10	3	1	2
11	3	2	1
12	3	2	2

MAXIMUM LIKELIHOOD ANALYSIS

ITERATION	SUB ITERATION	-2 LOG LIKELIHOOD	CONVERGENCE CRITERION
0	0	770.321	1
1	0	727.844	0.055142
2	0	722.249	.00768748
3	0	722.232	2.36E-05
4	0	722.232	9.65E-10

TABLE 7.13 (Continued)

PARAMETER ESTIMATES

ITERATION	1	2	3	4	5	6	7	8	9	10	11
0	0	0	0	0	0	0	0	0	0	0	0
1	0.0645161	0.0645161	0.0107527	0.0860215	0.408602	-0.268817	0.172043	-0.150538	0.0322581	0.0645161	-0.322581
2	0.21072	0.00898561	-.0155149	0.117777	0.442347	-0.190744	0.0884821	-.0532267	0.0744033	0.0278585	-0.179455
3	0.208078	.00539716	.0171113	0.118477	0.440068	-0.19012	0.0887541	-.0497742	0.0763498	0.0250158	-0.177766
4	0.208111	.00537821	-.0171163	0.118483	0.440047	-0.190089	0.0887231	-.0497231	0.0763576	0.0250087	-0.177724

ANALYSIS OF VARIANCE TABLE

SOURCE	DF	CHI-SQUARE	PROB
PROFESS	2	3.46	0.1777
SEX	1	0.03	0.8719
PROFESS*SEX	2	14.49	0.0007
READTYP	1	3.70	0.0546
PROFESS*READTYP	2	0.52	0.7696
SEX*READTYP	1	0.66	0.4168
PROFESS*SEX*READTYP	2	1.89	0.3894
LIKELIHOOD RATIO	0	0.00	1.0000

ANALYSIS OF INDIVIDUAL PARAMETERS

EFFECT	PARAMETER	ESTIMATE	STANDARD ERROR	CHI-SQUARE	PROB
RESPONSE	1	0.208111	0.122854	2.87	0.0903
	2	.0053782	0.133682	0.00	0.9679
	3	-.017116	0.106116	0.03	0.8719
	4	0.118483	0.135783	0.76	0.3829
	5	0.440047	0.166515	6.98	0.0082
	6	-.190089	.0988892	3.70	0.0546
	7	.0887231	0.122896	0.52	0.4703
	8	-0.04974	0.146392	0.12	0.7340
	9	.0763576	.0940364	0.66	0.4168
	10	.0250087	0.122854	0.04	0.8387
	11	-.177724	0.133682	1.77	0.1837

NOTE: _RESPONSE_ = PROFES SEX PROFES*SEX READTYP
PROFES*READTYP SEX*READTYP
PROFES*SEX*READTYP

PREDICTED VALUES FOR RESPONSE FUNCTIONS AND FREQUENCIES

SAMPLE	PROFESS	SEX	READTYP	FUNCTION NUMBER	OBSERVED FUNCTION	OBSERVED STANDARD ERROR	PREDICTED FUNCTIN	PREDICTED STANDARD ERROR	RESIDUAL
1				1	-0.510826	0.326599	-0.510826	0.326601	0
				2	-0.510826	0.326599	-0.510826	0.326601	0
				3	-0.916291	0.374166	-0.916291	0.37417	2.433E-10
				4	-0.510826	0.326599	-0.510826	0.326601	0
				5	-0.916291	0.374166	-0.916291	0.37417	2.433E-10
				6	0.182322	0.270801	0.182322	0.27079	-4.685E-09
				7	-1.6094	0.489898	-1.6094	0.489899	1.427E-09
				8	-0.916291	0.374166	-0.916291	0.37417	2.433E-10
				9	-1.6094	0.489898	-1.6094	0.889899	1.427E-09
				10	-1.6094	0.489898	-1.6094	0.489899	1.427E-09
				11	-0.916291	0.374166	-0.916291	00.37417	2.433E-10
1	1	1	1	F1	15	3.68081	15	3.6809	1.163E-08
1	1	1	2	F2	15	3.68081	15	3.6809	1.163E-08
1	1	2	1	F3	10	3.05857	10	3.05864	9.203E-09
2	1	2	2	F4	15	3.68081	15	3.6809	1.163E-08
2	2	1	1	F5	10	3.05857	10	3.05864	9.203E-09
2	2	1	2	F6	30	4.91869	30	4.91848	-1.202E-07
2	2	2	1	F7	5	2.19971	5	2.19972	1.052E-08
3	2	2	2	F8	10	3.05857	10	3.05864	9.203E-09
3	3	1	1	F9	5	2.19971	5	2.19972	1.052E-08
3	3	1	2	F10	5	2.19971	5	2.19972	1.052E-08
	3	2	1	F11	10	3.05857	10	3.05864	9.203E-09
	3	2	2	F12	25	4.57905	25	4.57895	1.694E-08

TABLE 7.14   MULTIWAY FREQUENCY ANALYSIS OF SMALL SAMPLE EXAMPLE THROUGH SYSTAT TABLES (SETUP AND OUTPUT)

```
        WEIGHT = FREQ
        TABULATE PROFESS*SEX*READTYP
        MODEL SEX*PROFESS + READTYP


TABLE OF     SEX    (ROWS) BY  READTYP      (COLUMNS)
    FOR THE FOLLOWING VALUES:
        PROFESS      =        1

FREQUENCIES

               1          2      TOTAL

     1        15         15       30

     2        10         15       25

TOTAL         25         30       55


TABLE OF     SEX    (ROWS) BY  READTYP      (COLUMNS)
    FOR THE FOLLOWING VALUES:
        PROFESS      =        2

FREQUENCIES

               1          2      TOTAL

     1        10         30       40

     2         5         10       15

TOTAL         15         40       55


TABLE OF     SEX    (ROWS) BY  READTYP      (COLUMNS)
    FOR THE FOLLOWING VALUES:
        PROFESS      =        3

FREQUENCIES

               1          2      TOTAL

     1         5          5       10

     2        10         25       35

TOTAL         15         30       45



MODEL WAS FIT AFTER  2 ITERATIONS.

TEST OF FIT OF MODEL

    DEGREES OF FREEDOM =     5
    PEARSON CHI-SQUARE =           6.59    PROBABILITY =  .253
    LIKELIHOOD RATIO CHI-SQUARE =      6.56    PROBABILITY = .256
```

information is gained from fitting a saturated model with this program. The model specified in Table 7.14 is the one chosen as optimal in Section 7.4.2. The MODEL instruction in the setup, SEX*PROFESS + READTYP, specifies a hierarchical model based on the two-way sex by profession association (which includes the sex and profession first-order effects) plus the first-order effect of reading type. The observed frequencies are first printed out, along with marginal sums. This is followed by a TEST OF FIT OF MODEL specified in the MODEL instruction.

7.5 SOME IMPORTANT ISSUES

7.5.1 Hierarchical and Nonhierarchical Models

A model is hierarchical, or nested, if it includes all the lower effects contained in the highest-order association that is retained in the model. A four-way design, *ABCD*, with a significant three-way association, *ABC*, has as a hierarchical model $A \times B \times C, A \times B, A \times C, B \times C$, and *A, B*, and *C*. The hierarchical model might or might not also include some of the other two-way associations and the *D* first-order effect. A nonhierarchical model derived from the same four-way design includes only the significant two-way associations and first-order effects along with the significant three-way association; that is, a nonsignificant $B \times C$ association is included in a hierarchical model that retains the *ABC* effect but probably is not included in a nonhierarchical model.

In loglinear analysis of multiway frequency tables, hierarchical models are the norm (e.g., Goodman, 1978; Knoke and Burke, 1980). At present the analysis of nonhierarchical models requires use of the Newton-Raphson algorithm which makes greater demands on the resources of both computers and humans.

The major advantage of hierarchical models is the availability of a significance test for the difference between nested models, so that the most parsimonious adequately fitting model can be identified using inferential procedures. With nonhierarchical models, a statistical test for the difference between models is not available.

Nonhierarchical models are more advantageous when the highest-order association is significant. If the highest-order interaction is significant, a hierarchical strategy is inadequate because the model is saturated and reproduces the observed frequencies perfectly because it includes all the effects in the design. Adequacy of fit cannot be assessed; fit is perfect and residuals are nonexistent. With a nonhierarchical approach, one or more of the lower-order effects can be excluded. And, although nonhierarchical models cannot be statistically compared, the adequacy of fit of any one of them can be assessed.

In summary, if your purpose is to identify the best model, the hierarchical approach is a good choice. On the other hand, if the highest-order interaction is significant or if you are simply interested in finding and interpreting the associations that are statistically significant, nonhierarchical models are helpful.

Only SPSS^x LOGLINEAR and SAS CATMOD use the Newton-Raphson algorithm; hence only they can be used to develop both hierarchical and nonhierarchical models. BMDP4F, SPSS^x HILOGLINEAR, and SYSTAT TABLES are restricted to hierarchical models.

7.5.2 Statistical Criteria

Confusion is rampant in MFA because tests of models look for statistical nonsignificance while tests of effects look for statistical significance. Worse, both kinds of tests commonly use the same statistics—forms of χ^2.

7.5.2.1 Models For screening for the complexity of model necessary to fit data and for testing overall fit of models, both Pearson χ^2 and the likelihood ratio statistic G^2 are routinely available. Of the two, consistency favors use of G^2 because it is available for testing overall fit, screening, and testing for differences among hierarchical models. Also, under some conditions, inadequate expected frequencies can inflate Type I error rate when Pearson χ^2 is used (Milligan, 1980).

In assessing goodness-of-fit for a model, you look for a *nonsignificant* G^2 where the frequencies estimated from the model are similar to the observed frequencies. Thus, retention of the null hypothesis is the desired outcome—an unhappy state of affairs for choosing an appropriate α level. In order to avoid finding too many ''good'' models, you need a higher criterion for α, say .10 or .25.

Further, with very large samples, small discrepancies often result in statistical significance. A significant model, even at $\alpha = .05$, may actually have adequate fit. With very small samples, on the other hand, large discrepancies often fail to reach statistical significance so that a nonsignificant model, even at $\alpha = .25$, actually has a poor fit. Choice of a significance level, then, is a matter of considering both sample size and the nature of the test. With larger samples, smaller tail probability values are chosen.

For logit models, in which one of the variables is viewed as a DV, SPSS[x] LOGLINEAR offers two additional tests based on Haberman (1982). Both are ''analyses of dispersion,'' an analog to analysis of variance for linear models. As in ANOVA, dispersion and degrees of freedom are partitioned into sources due to fit (systematic ''variance''), conditional dispersion (error ''variance''), and total dispersion. An F ratio is formed in the usual fashion for ANOVA. Both tests have corresponding measures of association (effect size) equal to the proportion of total dispersion that is due to fit—a direct analogy to eta square.

The two measures of dispersion are entropy and concentration (the latter called concordance by Haberman). Shannon (1948) defined entropy as the negative of the sum of a set of probabilities times their logarithm. Gini (1912) defined concentration as one minus the sum of a set of squared probabilities. Both measures are used to construct an analysis of dispersion for cases in which one or more IVs are used to predict a categorical DV.

For these measures, a statistically *significant* result indicates that the model fits the observed frequencies better than expected by chance; i.e., at $\alpha = .05$, you expect this good a fit by chance only 5% of the time. Significant dispersion tests, then, correspond to nonsignificant chi-square tests in evaluating a model. If dispersion due to fit is significant, measures of association for entropy and concentration indicate how good the fit is.

7.5.2.2 Effects Two types of tests are available for testing individual effects in multiway frequency tables, chi-square tests of marginal and partial first-order and higher-order effects, and z tests for single df parameter estimates.

During screening with a saturated model, BMDP4F provides partial G^2 tests of all first- and higher-order effects and both partial and marginal tests of all intermediate effects. Therefore statistically significant effects can be identified (though not always unambiguously, cf. Section 7.4.1) without resort to modeling. SPSS^x HILOGLINEAR provides only partial G^2 tests of all effects in a saturated model. SAS CATMOD provides Wald's chi-square test of association for all components whether or not the model is saturated. In addition, all three programs print parameter estimates and their standard errors, which are easily converted to z tests of parameters.

SPSS^x LOGLINEAR provides parameter estimates and their associated z tests, but no omnibus test for any effect that has more than one degree of freedom. If an effect has more than two levels, then, you do not get a single inferential test of that effect. Although one can attribute statistical significance to an effect if any of its single df tests is significant, no overall tail probability level is available. Also, an effect may be statistically significant even though none of its single df parameters reaches significance. With only the single df z tests of parameters, such an effect is not identified.

7.5.3 One Variable as DV (Logit Analysis)

In many applications of multiway frequency analysis one of the categorical variables is considered a DV and the others IVs. When loglinear analysis is used with one variable considered DV, it is called logit analysis. The questions are the same as in multiple regression or ANOVA; the significance of the relationship with the DV is assessed overall and for all IVs and their interactions.

If the DV has only two categories and the responses are split relatively evenly between the two categories (say no more extreme than 25/75%), either multiple regression/ANOVA or logit analysis is an appropriate analysis and the results of logit analysis are usually quite close to those of multiple regression/ANOVA with a dichotomous DV (Goodman, 1978). If, on the other hand, the splits are more extreme or there are more than two categories for the DV, logit analysis is the better choice because it has no parametric assumptions that are likely to be violated.

When devising a model for a logit problem, the rationale behind picking terms is fairly straightforward if one thinks in terms of regression/ANOVA. The main effect of any IV in ANOVA is the IV-DV association in logit analysis. Take, for example, prediction of vegetarianism (yes or no) by gender, age category (older, younger), and geographic region (several levels representing west coast to east coast). The main effect of gender in ANOVA is tested by the vegetarianism by gender association in logit analysis. The gender by age interaction in ANOVA is the three-way association between gender, age, and vegetarianism, and so on. The test of equality of frequency for vegetarianism appears in the ANOVA model as a constant or "correction for the mean." In logit analysis, only effects with vegetarianism as a component are of research interest.

If modeling is a goal, however, reliable first- and higher-order effects of IVs are included, so that adjustments are made for them. Modeling follows the usual hierarchical or nonhierarchical strategy. Can the highest-order interaction (vegetarianism by gender by age by geographic region) be eliminated without significantly degrading the fit of the model to the data? Can any or all of the three-way associations be eliminated? And so on.

7.5.3.1 Programs for Logit Analysis   SPSS^x LOGLINEAR and SAS CATMOD are well suited to logit analysis with their simple instructions for requesting such models and provision for running contrasts on an effect with more than two levels (cf. Section 7.5.5). For example, if geographic region is associated with vegetarianism, one might want to look at trends (linear, quadratic, etc.) to identify the source of geographic differences. If vegetarianism is more prevalent along the coasts than in the interior, for instance, a quadratic trend is present.

Additionally, SPSS^x LOGLINEAR allows inclusion of continuous covariates, should they be of interest. If age is viewed as a nuisance variable, for instance, it could serve as a continuous covariate rather than as one of the IVs. SPSS^x LOGLINEAR also provides measures of strength of association, in the form of entropy (Theil, 1970) and concentration (Haberman, 1982), as discussed in Section 7.5.2.1.

Logit analysis with a dichotomous DV can also be performed through logistic regression, and logistic regression is the more general procedure because it allows continuous as well as categorical IVs. BMDPLR, SPSS^x LOGLINEAR, SAS CATMOD, and SYSTAT LOGIT (a supplementary program) perform logistic regression.

The hierarchical loglinear programs—BMDP4F and SPSS^x HILOGLINEAR, and SYSTAT TABLES—are less convenient for logit analysis.

7.5.3.2 Odds Ratios   Once a model is identified and the parameters for each cell computed, the parameters can be converted to odds. The term "logit," in fact, refers to interpretation of the parameters as the log of odds ratios. The general equation for converting a parameter, λ, to an odds ratio is

$$\text{Odds ratio} = e^{2\lambda} \qquad (7.8)$$

In the research described above, for example, suppose that two effects are required for an optimal model, vegetarianism (the DV, required in all models) and the association between gender and vegetarianism (the main effect of gender on vegetarianism). Suppose the λ_V value for nonvegetarianism in that optimal model is 1.15. Applying Equation 7.8:

$$\text{Odds ratio} = e^{2(1.15)}$$
$$= 9.97$$

The a priori odds for being a nonvegetarian, then, are approximately 10 to 1 (an odds ratio of 10).

Parameters for significant effects change the odds. If, for example, the parameter for gender by vegetarianism (the effect of gender on vegetarianism), λ_{SV}, is -0.2 for women, the odds ratio for that parameter is

$$\text{Odds ratio (women)} = e^{2(-0.2)} = 0.67$$

To determine the odds associated with being a female nonvegetarian, one multiplies the odds for the level of the DV (10 for nonvegetarians) by the odds for the significant effect (.67 for the association between female and nonvegetarianism). (Odds for parameters are multiplicative since log odds are additive.) Thus, the odds against vegetarianism are about 6.7 to 1 for women. For men, the associated λ_{SV} of 0.2 leads to odds of about 15 to 1 against vegetarianism, since

$$\text{Odds ratio (men)} = e^{2(0.2)} = 1.50$$
$$\text{Odds ratio (nonvegetarians)} = e^{2(1.15)} = 9.97$$

and

$$\text{Odds ratio (male nonvegetarians)} = (1.50)(9.97) \approx 15$$

7.5.4 Strategies for Choosing a Model

Strategies for choosing a model differ depending on whether hierarchical or nonhierarchical models are to be considered, and whether you are using SPSS[x], SYSTAT, SAS, or BMDP. Options and features differ among programs for both hierarchical and nonhierarchical models. For example, of the six programs discussed in Section 7.6, only SPSS[x] LOGLINEAR and SAS CATMOD can be used for nonhierachical models. And, while BMDP4F provides the information necessary to closely follow model screening strategies discussed by Benedietti and Brown (1978), other strategies are available for the remaining programs. You may find it handy to mix programs—use one program to screen and another to evaluate models. This is the tack taken in Section 7.7 for one of the large sample examples.

7.5.4.1 BMDP4F (Hierarchical) For the saturated model, this program provides a test of each individual effect (with both marginal and partial χ^2 reported where appropriate), simultaneous tests of all k-way effects (all one-way effects taken together, all two-way effects taken together, and so on), and simultaneous tests of all k- and higher-way effects (with a four-way model, all three- and four-way effects, all two-, three-, and four-way effects, and so on). Both Pearson and likelihood ratio χ^2 (G^2) are reported. A strategy that follows the recommendations of Benedetti and Brown (1978) proceeds as follows.

Consider the *ABC* effect in a four-way design with *ABCD*. First, look at the tests of all three-way effects combined and three-way and four-way effects combined because combined results take precedence over tests of individual effects. If both combined tests are nonsignificant, the *ABC* association is deleted regardless of its partial and marginal tests unless this specific three-way interaction has been hypothesized beforehand. If both combined tests are significant, and the *ABC* effect is significant, the *ABC* effect is retained in the final model. If some of the tests are significant while others are not, further screening is recommended. This process is demonstrated in Section 7.7.2.1. (Recall that the cutoff p values for assessing significance depend on sample size. Larger samples are tested with smaller p values to avoid including statistically significant but trivial effects.)

Further screening of effects with mixed results proceeds stepwise. BMDP4F allows two ways of adding or deleting terms, simple and multiple. In simple stepping, models are changed (whether by addition or deletion) one term at a time. For example, model (AB,C) with effects $A \times B$, A, B, and C is compared with model (AB) with effects $A \times B$, A, and B. In multiple stepping, models that differ by more than one effect are compared; model (AB,C) with effects $A \times B$, A, B, and C is compared with model (C) with effect C only. Fine points of multiple and simple effects assessment are covered in Dixon (1985).

BMDP4F also provides two stepping options, forward and backward. Forward stepping starts with the simplest model that includes no ambiguous effects and adds terms. The "best" term, whether simple or multiple, is added first where "best" is the term that produces the largest difference in G^2 between the simpler and the more complex model. Then other terms of the same order are added, with G^2 for difference reported as each one is added. Terms that add significantly to the model are retained in the final model.

In backward stepping, one starts with all the unambiguously significant effects plus all ambiguous effects from the initial screening of the saturated model. The term (either simple or multiple) that is least helpful to the model is deleted first, followed by assessment of the remaining terms of the same order. Again χ^2 for the difference between simpler and more complex models is reported. Terms that do not significantly degrade the model are excluded.

The safest procedure is a combination of *backward stepping and simple deletion* (Dixon, 1985). However, if forward stepping is used, multiple rather than simple addition is recommended. Additional choices for stepping include specification of the criterion probability for significance of fit (default is .05) and of the maximum number of steps for adding or deleting effects.

Note that this stepwise procedure, like others, violates rules of hypothesis testing. Therefore, don't take the χ^2 and probability values produced by the stepping procedure too seriously. View this as a search for the most reasonable model, with χ^2 providing guidelines for choosing among models, as opposed to a stricter view that some models are truly significantly better or worse than others.

7.5.4.2 SPSS^x HILOGLINEAR (Hierarchical) There is less information and fewer choices for screening with SPSS^x HILOGLINEAR, but the strategy is basically the same as for BMDP4F. The simultaneous tests for k-way and k- and higher-way associations for the saturated model are identical to those of BMP4F, but only partial tests of individual effects are reported. If all tests of an effect are unambiguously significant, the effect is retained. If all tests are unambiguously nonsignificant, the effect is deleted. Ambiguity for any effect leads to further stepwise screening.

Only one stepping method is available—backward deletion of simple effects. Therefore the starting model includes all ambiguous effects as well as those to be definitely retained. Rules for interpreting results of stepping are the same as when BMDP4F is used, including the caveat with respect to overinterpreting significance.

7.5.4.3 SYSTAT TABLES (Hierarchical) This program provides only χ^2 and G^2 tests of specified models; neither parameter estimates nor tests of effects, alone or in combination, are available. Nor is stepwise selection available. As a result, SYSTAT TABLES is an inappropriate program for model screening and is suitable only to test the difference between two hypothesized models; you run each of them and test for the significance of the difference between them through the $G^2_{(\text{diff})}$ procedure in Section 7.4.2.

7.5.4.4 SPSS^x LOGLINEAR (Nonhierarchical)

This program provides neither simultaneous tests for associations nor a stepping algorithm. Therefore the procedure for choosing an appropriate model is simpler but less flexible.

A preliminary run with a saturated model is used to identify effects whose parameters differ significantly from zero. Recall that each cell of a design has a parameter for each effect and that, if the effect has more than two levels, the size of the parameter for the same effect may be different in the different cells. If an effect has a parameter that is highly significant for any cell, the effect is retained. If all the parameters for an effect are clearly nonsignificant, the effect is deleted.

Ambiguous cases occur when some parameters are marginally significant. Subsequent runs are made with and without ambiguous effects. In these runs, the significance of parameters is assessed along with the fit of the overall model. The strategy of backward elimination of simple effects, as described above, is followed for the safest trip to the most reasonable model. Since the final model need not be hierarchical, decisions as to which terms to delete on successive runs are fairly straightforward. Judgment is required to decide on the "best" model, since no statistical test is available to compare nonhierarchical models. Look for the model with fewest components that is not statistically significant.

If you have access to both SPSS^x LOGLINEAR and HILOGLINEAR, consider a preliminary screening run with HILOGLINEAR before proceeding to tests of nonhierarchical models with LOGLINEAR.

7.5.4.5 SAS CATMOD (Nonhierarchical)

Although this program has no provision for stepwise model building and no simultaneous tests of association for each order, it does provide tests of association for all effects in a model. A preliminary run with a saturated model, then, is used to identify candidates for model testing through Wald's chi-square test of association. Evaluation of models follows the spirit of backward elimination of simple effects as described in Section 7.5.4.1. Recall that statistical tests are available only to compare hierarchical models. Choice among nonhierarchical models is based on judgment; you search for the model with the fewest effects that fits "decently"—say, $p > .10$.

7.5.5 Contrasts

Just as in ANOVA, you may be interested in trends across levels of an ordered categorical variable, or in associations among categorical variables when only some levels of variables are considered, or when some levels are combined. These are often particularly interesting when one of your variables is considered a DV.

The distinction between planned and post hoc analysis applies to multiway frequency analysis as it does to ANOVA. If contrasts are not planned, adjustment is made for the possibility of capitalizing on chance effects. Unlike ANOVA, multiway frequency analysis has no cornucopia of adjustment strategies. None of the contrast procedures in the programs covered here provides for post hoc testing. About the best you can do is apply a Bonferroni-type adjustment (cf. Section 3.2.6). At the very least, set a strict criterion level for each contrast, say $\alpha = .01$.

SPSS^x LOGLINEAR offers extensive provision for contrasts and trend analysis. The logic and setup closely follow that of SPSS^x MANOVA. You can specify orthogonal polynomial contrasts (trend analysis) or a matrix of special contrasts, applying contrast coefficients as illustrated in Sections 3.2.6.1, 8.5.2.3, and 10.5.1.

TABLE 7.15 CONTRASTS ON SMALL SAMPLE DATA THROUGH SPSS'
LOGLINEAR (SETUP AND PARTIAL OUTPUT)

```
         TITLE        SMALL SAMPLE MULTIWAY ANALYSIS, CONTRASTS
         FILE HANDLE  SSMFA
         DATA LIST    FILE=SSMFA
                      / PROFESS SEX READTYP FREQ
         WEIGHT BY FREQ
         VALUE LABELS PROFESS 1 'POLITIC' 2 'ADMIN' 3 'BELLY'/
                      SEX 1 'MALE' 2 'FEMALE'/
                      READTYP 1 'SCIFI' 2 'SPY'/
         LOGLINEAR    PROFESS(1,3) SEX READTYP (1,2)/
                      PRINT=ESTIM/
                      CONTRAST(PROFESS) = SPECIAL(1  1  1,
                                                   1  1 -2,
                                                   1 -1  0)/
                      DESIGN=READTYP, SEX, PROFESS,
                             SEX BY PROFESS(1), SEX BY PROFESS(2)/
```

```
CORRESPONDENCE BETWEEN EFFECTS AND COLUMNS OF DESIGN/MODEL 1

     STARTING  ENDING
     COLUMN    COLUMN    EFFECT NAME

        1         1      READTYP
        2         2      SEX
        3         4      PROFESS
        5         5      SEX BY PROFESS(1)
        6         6      SEX BY PROFESS(2)
```

- -

```
ESTIMATES FOR PARAMETERS

READTYP
```

PARAMETER	COEFF.	STD. RRR.	Z-VALUE	LOWER 95 CI	UPPER 95 CI
1	-.2989185004	.08394	-3.56122	-.46344	-.13440

```
SEX
```

PARAMETER	COEFF.	STD. ERR.	Z-VALUE	LOWER 95 CI	UPPER 95 CI
2	-.0149353598	.09030	-.16539	-.19193	.16206

```
PROFESS
```

PARAMETER	COEFF.	STD. ERR.	Z-VALUE	LOWER 95 CI	UPPER 95 CI
3	.1084280461	.06868	1.57869	-.02619	.24305
4	.0557858878	.10155	.54934	-.14325	.25482

```
SEX BY PROFESS(1)
```

PARAMETER	COEFF.	STD. ERR.	Z-VALUE	LOWER 95 CI	UPPER 95 CI
5	.3057230622	.06868	4.45126	.17111	.44034

```
SEX BY PROFESS(2)
```

PARAMETER	COEFF.	STD. ERR.	Z-VALUE	LOWER 95 CI	UPPER 95 CI
6	-.1996269241	.10155	-1.96579	-.39867	-.00059

For example, consider the gender by profession association found in the small sample data set of Section 7.4. We can decompose this association into two interesting single df comparisons: one where we combine politicians and administrators and contrast them with belly dancers, and a second where we contrast politicians with administrators. An SPSS^x LOGLINEAR run in which this is done is in Table 7.15.

The first line of the contrast matrix adjusts for the constant. The second line (contrast 1) pits the first two levels of profession, politician and administrator, against belly dancers. The third line (contrast 2) contrasts politicians and administrators. The design statement requests separate parameter estimates for the two contrasts.

The output shows a highly significant association between profession and gender when politicians and administrators are combined and contrasted with belly dancers, $z = 4.45$, $p < .01$. The second contrast barely exceeds the criterion level for $\alpha = .05$. This contrast is statistically significant if it is a planned comparison, but not if it is unplanned.

SAS CATMOD also requires that you specify matrices of coefficients. The manual (SAS, 1985), however, warns against using the same contrast coefficients you would use in an analogous general linear model, because of differences in the way parameters are estimated.

BMDP4F has extensive statistics outside the loglinear model for two-way tables, including, for example, linear trend analysis. These statistics might be applied to portions of a multiway frequency table if contrasts are planned and α levels are stringent. Otherwise, BMDP4F as well as SYSTAT TABLES and SPSS^x HILOGLINEAR have no procedures to test contrasts within the context of a loglinear model.

7.6 COMPARISON OF PROGRAMS

Six programs are available in BMDP, SAS, SPSS, and SYSTAT for analysis of multiway frequency tables. There are two types of programs for loglinear analysis, those that deal exclusively with hierarchical models and those that can handle nonhierarchical models as well (cf. Section 7.5.1). SPSS^x LOGLINEAR and SAS CATMOD are general programs for nonhierarchical as well as hierarchical models (cf. Section 7.5.1). SPSS^x HILOGLINEAR and BMDP4F deal only with hierarchical models, but include features for stepwise model building (cf. Section 7.5.4). All four programs provide observed and expected cell frequencies, tests of fit of unsaturated models, and parameter estimates accompanied by their standard errors. Beyond that, the programs differ widely.

SYSTAT programs are quite different from all the rest. SYSTAT TABLES is a hierarchical program offering limited features and output. SYSTAT LOGIT is a logistic regression program (nonhierarchical) to be used when one of the variables is considered a DV. Features of the six programs appear in Table 7.16.

7.6.1 SPSS Package

Until SPSS^x (SPSS, Inc., 1985), there were no facilities for handling multidimensional frequency tables in the SPSS package. Currently there are two programs:

TABLE 7.16 COMPARISON OF PROGRAMS FOR MULTIWAY FREQUENCY ANALYSIS

Feature	SPSS^x LOGLINEAR	SPSS^x HILOGLINEAR	BMDP4F	SAS CATMOD	SYSTAT TABLES	SYSTAT LOGIT
Input						
Individual case data	Yes	Yes	Yes	Yes	Yes	Yes
Multiway frequency table	No	No	Yes	No	No	No
Cell frequencies and indices	Yes	Yes	Yes	Yes	Yes	Yes
Cell weights (structural zeros)	Yes	Yes	Yes	Yes	Yes	No
Stepping options	No	Yes	Yes	No	No	No
Convergence criteria	Yes	Yes	Yes	Yes	Yes	Yes
Specify maximum no. of iterations	Yes	Yes	Yes	Yes	Yes	Yes
Specify maximum no. of steps	N.A.	Yes	Yes	N.A.	N.A.	N.A.
Specify significance level for adequate fit	N.A.	Yes	Yes	N.A.	N.A.	N.A.
Specify maximum order of terms	N.A.	Yes	Yes	N.A.	N.A.	N.A.
Force terms into stepping model	N.A.	No	Yes	N.A.	N.A.	N.A.
Covariates (continuous)	Yes	No	No	No	No	Yes
Logit model specification	Yes	No	No	Yes	No	Yes
Single df partitions and contrasts	Yes	No	No	Yes	Yes	No
Specify delta for each cell	Yes	No	Yes	Yes	Yes	No
Request automatic collapse of adjacent categories	No	No	Yes	Yes	No	No
Maximum no. of factors/cells	Unspecified	10 factors	10K cells	Unspecified	50 levels per factor	Unspecified
Output						
Nonhierarchical as well as hierarchical models	Yes	No	No	Yes	No	Yes
Tests of partial association	No	Yes[a]	Yes	No	No	No
Tests of marginal association	No	No	Yes	No	No	No
Wald's test of association	No	No	No	Yes	No	No
Tests of k-way effects	No	Yes	Yes	No	No	No
Tests of k-way and higher effects	No	Yes	Yes	No	No	No

Statistic	Col 1	Col 2	Col 3	Col 4	Col 5
Pearson model tests	Yes	Yes	No	Yes	No
Likelihood ratio model tests	Yes	Yes	Yes	Yes	Yes
Observed and expected (predicted) frequencies	Yes	Yes	Yes	Yes	No
Observed and expected probabilities or percentages	Yes	Yes	Yes[b]	No	Yes[d]
Cell percentages of rows and columns	No	Yes	No	No	No[c]
Raw residuals	Yes	Yes	Yes	Yes	No
Standardized residuals	Yes	Yes	No	Yes	No
Adjusted residuals	No	No	No	No	No
Freeman-Tukey residuals	No	Yes	Yes	No	No
Loglinear parameter estimates	Yes[a]	Yes	Yes	No	Yes
Standard error of parameter estimate	Yes[a]	Yes	Yes	No	Yes
Ratio of parameter estimate to standard error (z or t)	Yes[a]	Yes	Yes	No	Yes
Chi-square tests for parameter estimates	No	No	Yes	No	No
Multiplicative parameter estimates	No	Yes	No	No	No
Likelihood ratio tests for model components	No	Yes	Yes	No	No
Correlation matrix for parameter estimate	No	Yes	Yes	No	No
Covariance matrix for parameter estimate	No	Yes	Yes	No	Yes
Design matrix	No	No	Yes	No	No
Marginal subtables	No	Yes	No	No	No
Table of reasons for excluded cases	No	Yes	No	No	No
Plots of standardized residuals	Yes	No	No	No	No
Normal plots of adjusted residuals vs. observed and expected frequency	No	No	No	No	No
Descriptive statistics	No	Yes	No	No	Yes[e]
IV derivatives	No	No	No	No	Yes

[a] Saturated model only.
[b] Predicted only.
[c] Observed frequencies for DV only.
[d] For DV only.
[e] Means for each effect at each level of DV.

279

LOGLINEAR, a general purpose program for both hierarchical and nonhierarchical models, and HILOGLINEAR, which deals with only hierarchical models.

SPSS[x] HILOGLINEAR is well suited to comparisons among hierarchical models, with several options for controlling stepwise selection of effects. Tests of all k-way effects are available, as well as simultaneous tests of all k-way and higher effects, allowing for a quick screening of the complexity of model from which to start stepwise selection. Parameter eliminates and tests of partial association (model components) are available, but only for saturated models. If your purpose is to discover which categorical variables are associated rather than to model, SPSS[x] HILOGLINEAR provides direct and easily interpreted output (although there is some doubt as to the appropriateness of providing only partial tests of association, cf. Section 7.4.1).

SPSS[x] LOGLINEAR does not provide stepwise selection of hierarchical models, although it can be used to compare user-specified models of any sort. Of the five programs, only this one allows adjustment for continuous covariates. Also available are a simple specification of a logit model (in which one factor is a DV), and single-degree-of-freedom partitions for contrasts among specified cells.

No inferential tests of model components are provided. Parameter estimates and their z tests are available for any specified model, along with their 95% confidence intervals. However, the parameter estimates are reported by single degrees of freedom, so that a factor with more than two categories has no omnibus significance test reported for either its main effect or its association with other effects (cf. Section 7.5.2). No quick screening for k-way tests is available. Screening information can be gleaned from a saturated model run, but identifying an appropriate model may be tedious with a large number of factors. Both SPSS[x] programs offer residuals plots.

7.6.2 BMDP Series

BMDP4F (Dixon, 1985) is a highly flexible program with useful features for evaluating model components, and for selecting and comparing hierarchical models (cf. Section 7.5.1).

Several stepping options are available. Logit models are dealt with by forcing specified terms into the stepping model. Screening is facilitated by use of tests of k-way and k-way-and-higher effects. Extensive output of parameter estimation is available for any selected model.

BMDP4F offers the most comprehensive evaluation of effects of any k-way model of association. Likelihood ratio statistics are given for each effect. For intermediate level effects (other than first- or highest-order) both partial and marginal tests of association are given (cf. Section 7.4.1.3). If you are simply interested in associations among effects rather than modeling, BMDP4F provides the most direct output. Thus BMDP4F is also useful in screening for nonhierarchical models.

BMDP4F offers the most extensive descriptive output, with cell percentages of rows and columns, marginal subtables, and reasons for excluded cases available on request. Several forms of residuals analysis but no residuals plots are available.

7.6.3 SAS System

SAS CATMOD (SAS Institute Inc., 1985) is a highly general program for modeling of categorical data of which loglinear modeling is only one type. The program is primarily set up for logit analyses where some variables are IVs and another the DV but provision is made for loglinear models where no such distinction is made. The program offers both hierarchical and nonhierarchical modeling, simple designation of logit models, contrasts, and single df tests of parameters as well as Wald's chi-square tests of more complex components. The program lacks provision for continuous covariates and stepwise model building procedures.

SAS CATMOD uses different algorithms from the other three programs for both parameter estimation and model testing. Although parameter estimation uses the Newton-Raphson algorithm, as does SPSS[x] LOGLINEAR, and iterative proportional fitting, as does BMDP4F, resulting parameters differ somewhat from those in SPSS[x] and BMDP4F, although parameters generated by SPSS[x] and BMDP4F agree. The test for model components as well as overall model fit similarly do not always agree with SPSS[x] and BMDP4F, although the latter programs agree with each other. The output in Table 7.13 compared with that of Tables 7.10 to 7.12 demonstrates some of the inconsistencies.

7.6.4 SYSTAT System

SYSTAT TABLES (Wilkinson, 1986) is a minimal hierarchical program for loglinear analysis of categorical data. Designed for model testing but not screening, the program provides only overall tests of fit for a specified model in terms of both χ^2 and G^2. There are no tests of components nor are there parameter estimates.

Fit of a specified model is evaluated by requesting tables of fitted values (expected frequencies), differences between observed and expected values, and standardized residuals. The program also offers the option of adding a constant to each cell, and you can declare some cells to be structural zeros.

Although SYSTAT TABLES does not provide explicitly for specification of one variable as DV, a supplementary program is available for such analyses. SYSTAT LOGIT is a logistic regression program for evaluating the effect on a categorical DV of continuous IVs. The program can also be used with categorical IVs if they are dichotomous with fairly even splits, or if they are discrete but have been recoded into dichotomous dummy variables, or if they have ordered categories that you are willing to treat as continuous.

SYSTAT LOGIT is set up like SYSTAT MGLH (multivariate general linear hypothesis). Output includes the observed frequencies for each category of the DV and, for each of the IVs and interaction effects, a "mean" for each category of the DV. Parameter estimates, their standard errors, and t statistics are given for each effect, along with G^2 as a test of fit of the model. In common with other nonhierarchical programs, screening aids are not provided.

7.7 COMPLETE EXAMPLES OF MULTIWAY FREQUENCY ANALYSIS

Data to illustrate multiway frequency analyses were taken from the survey of clinical psychologists described in Appendix B, Section B.2. The first example is a hierarchical analysis of five dichotomous variables: whether or not the therapists thought (1) that their clients were aware of the therapist's attraction to them (AWARE), (2) the attraction was beneficial to the therapy (BENEFIT), and (3) the attraction was harmful to the therapy (HARM), as well as whether or not the therapists had (4) sought consultation when attracted to a client (CONSULT), or (5) felt uncomfortable as a result of the attraction (DISCOMF).

The second example is a nonhierarchical logit analysis with one variable considered a DV and three other IVs. The DV is whether or not the therapist ever engaged in sexual intimacies with a client (INTIMACY). The IVs are SEX, AGE (45 or younger, 46 or older), and amount of education about attraction to clients in the therapist's graduate training (none, very little, or some—GRADTRNG).

7.7.1 Evaluation of Assumptions: Adequacy of Expected Frequencies

There are 585 psychologists in the sample. Of these, 151 are excluded from the first (hierarchical) analysis because of missing data and because only therapists who had felt attraction to at least one client answered the questions used for the analysis. Sixteen respondents are excluded from the second (logit) analysis because they failed to answer one or more of the questions used as variables in the example. The usable samples, then, consist of 434 psychologists for the hierarchical analysis and 569 psychologists for the logit analyses.

Sample sizes are adequate for both analyses. The $2 \times 2 \times 2 \times 2 \times 2$ hierarchical analysis contains 32 cells, for which a sample of 434 should be sufficient; more than five cases are expected per cell if the dichotomous splits are not too bad. Similarly, the 24 cells in the $2 \times 2 \times 2 \times 3$ logit analysis should each have expected frequencies of five or greater with 569 respondents unless some of the marginal distributions are very badly split.

To determine the adequacy of expected frequencies, all two-way contingency tables for both analyses are examined. Table 7.17 shows the setup and the two-way tables from the preliminary run through BMDP4F for the hierarchical analysis. Expected frequencies (Fe) are calculated in the usual way for χ^2 analysis:

$$\text{Cell } Fe = (\text{row sum})(\text{column sum})/N$$

For this analysis, the smallest $Fe = (128)(140)/434 = 41.29$ for a cell of the AWARE by BENEFIT association. For both analyses, all $Fe > 5$.

Both BMDP4F and SPSS^x CROSSTABS provide printed output of expected frequencies for two-way tables. However, obtaining the expected frequencies by computer requires an additional run in either case because neither of the SPSS^x multiway

```
/PROBLEM     TITLE IS 'LOGLINEAR ANALYSIS OF PKS QUESTIONNAIRE'.
/INPUT       VARIABLES ARE 10.  FORMAT IS FREE.
             FILE=MFA.
/VARIABLE    NAMES ARE SEX,AGE,AWARE,BENEFIT,HARM,CONSULT,DISCOMF,
                INTIMACY,GRADTRNG,IDNO.
             MISSING = 9*9.
             LABEL=IDNO.
/CATEGORY    CODES(1) ARE 1,2.  NAMES(1) ARE MALE,FEMALE.
             CODES(2) ARE 1,2.  NAMES(2) ARE YOUNGER, OLDER.
             CODES(3) ARE 1,2.  NAMES(3) ARE 'PROB NOT', YES.
             CODES(4 TO 7) ARE 1,2. NAMES(4 TO 7) ARE NEVER, YES.
             CODES(8) ARE 1 TO 4. NAMES(8) ARE NEVER,YES,YES,YES.
             CODES(9) ARE 1 TO 4. NAMES(9) ARE NONE,SOME,SOME,SOME.
/TABLE    INDICES ARE BENEFIT,HARM,AWARE,DISCOMF,CONSULT.
/FIT      ASSOCIATION IS 5.
          ITERATION = 30.
/PRINT    MARGINALS ARE 4.
/END
```

```
*****  MARGINAL SUBTABLE -- TABLE  1

HARM         BENEFIT
------       ------
             NEVER   YES    TOTAL
-------------------------------------
NEVER          90    119 I    209
YES            50    175 I    225
-------------------------I---------
TOTAL         140    294 I    434
```

```
*****  MARGINAL SUBTABLE -- TABLE  1

AWARE        BENEFIT
------       ------
             NEVER   YES    TOTAL
-------------------------------------
PROB NOT      129    177 I    306
YES            11    117 I    128
-------------------------I---------
TOTAL         140    294 I    434
```

```
*****  MARGINAL SUBTABLE -- TABLE  1

DISCOMF      BENEFIT
------       ------
             NEVER   YES    TOTAL
-------------------------------------
NEVER          63     90 I    153
YES            77    204 I    281
-------------------------I---------
TOTAL         140    294 I    434
```

```
*****  MARGINAL SUBTABLE -- TABLE  1

CONSULT      BENEFIT
------       ------
             NEVER   YES    TOTAL
-------------------------------------
NEVER          85     95 I    180
YES            55    199 I    254
-------------------------I---------
TOTAL         140    294 I    434
```

```
*****  MARGINAL SUBTABLE -- TABLE  1

AWARE        HARM
------       ------
             NEVER   YES    TOTAL
-------------------------------------
PROB NOT      173    133 I    306
YES            36     92 I    128
-------------------------I---------
TOTAL         209    225 I    434
```

TABLE 7.17 (Continued)

```
*****   MARGINAL SUBTABLE -- TABLE   1

DISCOMF      HARM
------       ------
            NEVER     YES    TOTAL
-----------------------------------
NEVER        107       46 I   153
YES          102      179 I   281
------------------------I---------
TOTAL        209      225 I   434

*****   MARGINAL SUBTABLE -- TABLE   1

CONSULT      HARM
------       ------
            NEVER     YES    TOTAL
-----------------------------------
NEVER        114       66 I   180
YES           95      159 I   254
------------------------I---------
TOTAL        209      225 I   434

*****   MARGINAL SUBTABLE -- TABLE   1

DISCOMF      AWARE
------       ------
           PROB NOT    YES   TOTAL
-----------------------------------
NEVER        119       34 I   153
YES          187       94 I   281
------------------------I---------
TOTAL        306      128 I   434

*****   MARGINAL SUBTABLE -- TABLE   1

CONSULT      AWARE
------       ------
           PROB NOT    YES   TOTAL
-----------------------------------
NEVER        155       25 I   180
YES          151      103 I   254
------------------------I---------
TOTAL        306      128 I   434

*****   MARGINAL SUBTABLE -- TABLE   1

CONSULT      DISCOMF
------       ------
            NEVER     YES    TOTAL
-----------------------------------
NEVER         94       86 I   180
YES           59      195 I   254
------------------------I---------
TOTAL        153      281 I   434
```

frequency programs offers marginal subtables, and because special control language is needed by BMDP4F. In light of the simplicity of the calculations, it might be as easy to perform them by hand.

Discussion of outliers in the solution appears in the section on adequacy of fit of the selected model that follows the section on selection of a model for both analyses.

7.7.2 Hierarchical Loglinear Analysis

7.7.2.1 Preliminary Model Screening
Table 7.18 contains the information needed to start the model-building procedure; the simultaneous tests for effects of each order, each order and higher, and the tests of individual association.

TABLE 7.18 PARTIAL OUTPUT FOR BMDP4F PRELIMINARY RUN OF SIMULTANEOUS AND COMPONENT ASSOCIATIONS. (SEE TABLE 7.16 FOR SETUP)

```
***** THE RESULTS OF FITTING ALL K-FACTOR MARGINALS.
      THIS IS A SIMULTANEOUS TEST THAT ALL K+1 AND HIGHER FACTOR INTERACTIONS ARE ZERO.
```

K-FACTOR	D.F.	LR CHISQ	PROB.	PEARSON CHISQ	PROB.	ITERATION
0-MEAN	31	436.15	0.00000	491.35	0.00000	
1	26	253.51	0.00000	364.21	0.00000	2
2	16	24.09	.08764	21.82	.14907	7
3	6	10.16	.11793	11.06	.08662	4
4	0	.03	1.00000	.02	1.00000	28
5	0	0.	1.	0.	1.	

```
***** A SIMULTANEOUS TEST THAT ALL K-FACTOR  INTERACTIONS ARE SIMULTANOUSLY ZERO.
      THE CHI-SQUARES ARE DIFFERENCES IN THE ABOVE TABLE.
```

K-FACTOR	D.F.	LR CHISQ	PROB.	PEARSON CHISQ	PROB.
1	5	182.64	0.00000	127.13	0.00000
2	10	229.42	0.00000	342.39	0.00000
3	10	13.92	.17655	10.76	.37624
4	6	10.13	.11921	11.04	.08711
5	0	.03	1.00000	.02	1.00000

```
:
***** ASSOCIATION OPTION SELECTED FOR ALL TERMS OF ORDER LESS THAN OR EQUAL TO  5
```

EFFECT	D.F.	PARTIAL ASSOCIATION CHISQUARE	PROB	ITER	MARGINAL ASSOCIATION CHISQUARE	PROB	ITER
B.	1	55.85	0.0000				
H.	1	.59	.4424				
A.	1	75.20	0.0000				
D.	1	38.32	0.0000				
C.	1	12.68	.0004				
BH.	1	4.69	.0304	7	21.73	.0000	2
BA.	1	31.95	0.0000	7	54.14	0.0000	2
BD.	1	.31	.5759	6	8.47	.0036	2
BC.	1	9.77	.0018	6	31.40	0.0000	2
HA.	1	11.71	.0006	7	30.00	0.0000	2
HD.	1	28.99	0.0000	7	45.79	0.0000	2
HC.	1	4.28	.0385	6	28.67	0.0000	2
AD.	1	.26	.6081	6	6.18	.0129	2
AC.	1	15.95	.0001	6	38.40	0.0000	2
DC.	1	21.47	.0000	6	38.81	0.0000	2
BHA.	1	6.09	.0136	5	7.32	.0068	5
BHD.	1	1.06	.3026	5	2.85	.0916	4
BHC.	1	.41	.5218	4	.41	.5217	5
BAD.	1	.15	.6946	5	2.12	.1453	5
BAC.	1	.66	.4173	5	.02	.8817	4
BDC.	1	.42	.5156	4	.01	.9218	4
HAD.	1	.74	.3885	5	2.16	.1413	5
HAC.	1	.61	.4345	4	.15	.6949	4
HDC.	1	.05	.8166	5	.30	.5852	4
ADC.	1	2.20	.1376	3	2.64	.1040	4
BHAD.	1	3.86	.0495	6	2.12	.3461	5
BHAC.	1	3.33	.0682	6	2.31	.3148	5
BHDC.	1	.49	.4849	30	.38	.5384	4
BADC.	1	1.46	.2264	5	.04	.8423	4
HADC.	1	2.08	.1490	30	3.31	.0690	4
BHADC.	0	.03	1.0000				

Both likelihood ratio and Pearson criteria are used to evaluate the k level and $k + 1$ and higher-level associations. Note that the probability levels for more than two-way associations are greater than 0.05 for the simultaneous tests of both k level and $k + 1$ and higher level. The two sets of simultaneous tests agree that variables are independent in three-way and higher-order effects. Thus the model need contain no associations greater than two-way.[7]

The final portion of the table provides the basis of a search for the best model of one- and two-way effects. Among the two-way effects, several associations are clearly significant ($p < .01$) by both the partial and marginal tests: BENEFIT by AWARE, BENEFIT by CONSULT, HARM by AWARE, HARM by DISCOMF, AWARE by CONSULT, DISCOMF by CONSULT. None of the two-way effects is clearly nonsignificant ($p > .05$) by both partial and marginal tests. The four two-way effects that are ambiguous, BENEFIT by HARM, BENEFIT by DISCOMF, HARM by CONSULT, and AWARE by DISCOMF, are tested through a stepwise analysis.

All first-order effects need to be included in the final hierarchical model, most because they are highly significant, and HARM because it is part of a significant two-way association. Recall that in a hierarchical model a term is included if it is a part of a higher-order association.

7.7.2.2 Stepwise Model Selection

Stepwise selection by simple deletion from the model with all two-way terms is illustrated in the BMDP4F run of Table 7.19. Although 10 steps are permitted by the instruction STEP = 10, the selection process stops after the fourth step because the criterion probability is reached.

Recall that each potential model generates a set of expected frequencies. The goal of model selection is to find the model with the smallest number of effects that still provides a fit between expected frequencies and observed frequencies. First, the optimal model must have a nonsignificant LIKELIHOOD-RATIO CHI-SQUARE value (cf. Section 7.5.2.1 for choice between Pearson and likelihood ratio values). Second, the selected model should not be significantly worse than the next more complicated model. That is, if an effect is deleted from a model, that model should not be significantly different from the model with the term still in it.

Notice first in Table 7.19 that Model 1 includes all the effects, certain and ambiguous, that might be included. Model 1 is not significant, meaning that it provides an acceptable fit between expected and observed frequencies. At step 1, effects are deleted one at a time. The best model at step 1 is the model with the smallest CHI-SQUARE value because that is the model with the best fit between expected and observed frequencies. For these data, the best model at step 1 is the one with *AD* deleted.

The best model at step 1 becomes the basis of testing at step 2 in which, once again, effects are deleted one at a time. The best model at step 2 becomes the basis for tests at step 3 when effects are again deleted one at a time, and so forth.

In Table 7.19, all the models at step 4 are significant. That is, all the models with only six effects provide expected frequencies that differ significantly from observed frequencies. None of them is a good candidate for model of choice. The best model at step 3 has seven effects (*BH, BA, BC, HA, HD, AC, DC*) and, with a

---

[7] Although one three-way effect, BENEFIT by HARM by AWARE, meets the $p < .01$ criterion by the marginal test and $p < .05$ by the partial test, none of the three-way associations is considered for inclusion because the simultaneous tests take precedence over the component associations.

nonsignificant CHI-SQUARE value of 28.92, produces a good fit between observed and expected frequencies.

However, the second criterion is that the model should not be significantly different from the next more complicated model. The next more complicated model is the step 2 model that contains *HC*. Deletion of *HC* at step 3 results in a significant

TABLE 7.19 SETUP AND PARTIAL OUTPUT FROM BMDP4F MODEL SELECTION RUN FOR HIERARCHICAL LOGLINEAR ANALYSIS

```
/PROBLEM     TITLE IS 'LOGLINEAR ANALYSIS OF PKS QUESTIONNAIRE,
                MODEL SELECTION'.
/INPUT       VARIABLES ARE 10.  FORMAT IS FREE.
             FILE=MFA.
/VARIABLE    NAMES ARE SEX,AGE,AWARE,BENEFIT,HARM,CONSULT,DISCOMF,
                INTIMACY,GRADTRNG,IDNO.
             MISSING = 9*9.
             LABEL=IDNO.
/CATEGORY    CODES(1) ARE 1,2.   NAMES(1) ARE MALE,FEMALE.
             CODES(2) ARE 1,2.   NAMES(2) ARE YOUNGER, OLDER.
             CODES(3) ARE 1,2.   NAMES(3) ARE 'PROB NOT', YES.
             CODES(4 TO 7) ARE 1,2. NAMES(4 TO 7) ARE NEVER, YES.
             CODES(8) ARE 1 TO 4. NAMES(8) ARE NEVER,YES,YES,YES.
             CODES(9) ARE 1 TO 4. NAMES(9) ARE NONE,VLITTLE,SOME,SOME.
/TABLE       INDICES ARE BENEFIT,HARM,AWARE,DISCOMF,CONSULT.
/FIT         MODEL IS BH, BA, BD, BC, HA, HD, HC, AD, AC, DC.
             DELETE=SIMPLE.
             STEP=10.
/END
```

```
*****************
*   MODEL  1   *
*****************
```

MODEL	D.F.	LIKELIHOOD-RATIO CHI-SQUARE	PROB	PEARSON CHI-SQUARE	PROB	ITER.
-----	----	----------	----	----------	----	-----
BH,BA,BD,BC,HA,HD,HC,AD,AC,DC.	16	24.09	.0876	21.82	.1491	7

MODELS FORMED BY DELETING TERMS FROM MODEL -- BH,BA,BD,BC,HA,HD,HC,AD,AC,DC.

MODEL	SIMPLE EFFECT	D.F.	LIKELIHOOD-RATIO CHI-SQUARE	PROB	PEARSON CHI-SQUARE	PROB	ITER.
-----	------	----	----------	----	----------	----	-----
BA,BD,BC,HA,HD,HC,AD,AC,DC.		17	28.77	.0367	25.76	.0790	7
DIFF. DUE TO DELETING BH.		1	4.69	.0304			
BH,BD,BC,HA,HD,HC,AD,AC,DC.		17	56.04	.0000	50.44	.0000	7
DIFF. DUE TO DELETING BA.		1	31.95	.0000			
BH,BA,BC,HA,HD,HC,AD,AC,DC.		17	24.40	.1090	22.22	.1764	6
DIFF. DUE TO DELETING BD.		1	.31	.5759			
BH,BA,BD,HA,HD,HC,AD,AC,DC.		17	33.86	.0088	34.93	.0064	6
DIFF. DUE TO DELETING BC.		1	9.77	.0018			
BH,BA,BD,BC,HD,HC,AD,AC,DC.		17	35.79	.0049	34.35	.0076	7
DIFF. DUE TO DELETING HA.		1	11.71	.0006			
BH,BA,BD,BC,HA,HC,AD,AC,DC.		17	53.07	.0000	49.79	.0000	7
DIFF. DUE TO DELETING HD.		1	28.99	.0000			
BH,BA,BD,BC,HA,HD,AD,AC,DC.		17	28.37	.0408	28.39	.0405	6
DIFF. DUE TO DELETING HC.		1	4.28	.0385			
BH,BA,BD,BC,HA,HD,HC,AC,DC.		17	24.35	.1103	21.64	.1990	6
DIFF. DUE TO DELETING AD.		1	.26	.6081			
BH,BA,BD,BC,HA,HD,HC,AD,DC.		17	40.03	.0013	35.06	.0061	6
DIFF. DUE TO DELETING AC.		1	15.95	.0001			
BH,BA,BD,BC,HA,HD,HC,AD,AC.		17	45.56	.0002	42.89	.0005	6
DIFF. DUE TO DELETING DC.		1	21.47	.0000			

STEP 1.   BEST MODEL FOUND IS -- BH,BA,BD,BC,HA,HD,HC,AC,DC.

TABLE 7.19 (Continued)

BA,BD,BC,HA,HD,HC,AC,DC,	18	29.22	.0457	25.84	.1035	6
DIFF, DUE TO DELETING BH,	1	4.87	.0273			
BH,BD,BC,HA,HD,HC,AC,DC,	18	56.19	.0000	49.84	.0001	5
DIFF, DUE TO DELETING BA,	1	31.84	.0000			
BH,BA,BC,HA,HD,HC,AC,DC,	18	24.55	.1379	21.99	.2325	6
DIFF, DUE TO DELETING BD,	1	.20	.6551			
BH,BA,BD,HA,HD,HC,AC,DC,	18	34.39	.0113	34.83	.0099	5
DIFF, DUE TO DELETING BC,	1	10.04	.0015			
BH,BA,BD,BC,HD,HC,AC,DC,	18	35.95	.0072	34.74	.0102	6
DIFF, DUE TO DELETING HA,	1	11.60	.0007			
BH,BA,BD,BC,HA,HC,AC,DC,	18	53.23	.0000	50.19	.0001	6
DIFF, DUE TO DELETING HD,	1	28.88	.0000			
BH,BA,BD,BC,HA,HD,AC,DC,	18	28.86	.0501	27.82	.0648	6
DIFF, DUE TO DELETING HC,	1	4.51	.0336			
BH,BA,BD,BC,HA,HD,HC,DC,	18	40.18	.0020	35.15	.0091	5
DIFF, DUE TO DELETING AC,	1	15.83	.0001			
BH,BA,BD,BC,HA,HD,HC,AC,	18	45.71	.0003	43.23	.0007	5
DIFF, DUE TO DELETING DC,	1	21.36	.0000			

STEP 2.   BEST MODEL FOUND IS -- BH,BA,BC,HA,HD,HC,AC,DC,

BA,BC,HA,HD,HC,AC,DC,	19	30.31	.0480	27.32	.0974	6
DIFF, DUE TO DELETING BH,	1	5.76	.0164			
BH,BC,HA,HD,HC,AC,DC,	19	56.38	.0000	50.66	.0001	6
DIFF, DUE TO DELETING BA,	1	31.83	.0000			
BH,BA,HA,HD,HC,AC,DC,	19	35.82	.0111	36.73	.0086	6
DIFF, DUE TO DELETING BC,	1	11.27	.0008			
BH,BA,BC,HD,HC,AC,DC,	19	36.14	.0102	34.80	.0148	5
DIFF, DUE TO DELETING HA,	1	11.59	.0007			
BH,BA,BC,HA,HC,AC,DC,	19	55.37	.0000	52.33	.0001	2
DIFF, DUE TO DELETING HD,	1	30.82	.0000			
BH,BA,BC,HA,HD,AC,DC,	19	28.92	.0673	28.00	.0835	3
DIFF, DUE TO DELETING HC,	1	4.37	.0366			
BH,BA,BC,HA,HD,HC,DC,	19	40.37	.0029	35.28	.0129	4
DIFF, DUE TO DELETING AC,	1	15.82	.0001			
BH,BA,BC,HA,HD,HC,AC,	19	40.00	.0003	45.55	.0006	5
DIFF, DUE TO DELETING DC,	1	23.48	.0000			

STEP 3.   BEST MODEL FOUND IS -- BH,BA,BC,HA,HD,AC,DC,

BA,BC,HA,HD,AC,DC,	20	36.70	.0127	35.13	.0194	3
DIFF, DUE TO DELETING BH,	1	7.78	.0053			
BH,BC,HA,HD,AC,DC,	20	59.39	.0000	59.33	.0000	4
DIFF, DUE TO DELETING BA,	1	30.47	.0000			
BH,BA,HA,HD,AC,DC,	20	42.21	.0026	43.91	.0015	4
DIFF, DUE TO DELETING BC,	1	13.29	.0003			
BH,BA,BC,HD,AC,DC,	20	43.91	.0015	42.55	.0023	3
DIFF, DUE TO DELETING HA,	1	14.99	.0001			
BH,BA,BC,HA,AC,DC,	20	67.48	.0000	63.93	.0000	2
DIFF, DUE TO DELETING HD,	1	38.57	.0000			
BH,BA,BC,HA,HD,DC,	20	48.14	.0004	42.92	.0021	4
DIFF, DUE TO DELETING AC,	1	19.22	.0000			
BH,BA,BC,HA,HD,AC,	20	60.46	.0000	58.09	.0000	4
DIFF, DUE TO DELETING DC,	1	31.54	.0000			

STEP 4.   BEST MODEL FOUND IS -- BA,BC,HA,HD,AC,DC,

STEPPING STOPS DUE TO CRITERION PROBABILITY (.050).

difference between the models, $\chi^2(1) = 4.37$, $p > .05$. Therefore, the model at step 3 is unsatisfactory because it is significantly different from the next more complicated model. (Use of a more conservative α, for example $p < .01$, would lead to a decision in favor of the best model at step 3 with seven effects.)

The best model at step 2 (*BH, BA, BC, HA, HD, HC, AC, DC*) is satisfactory by all criteria. Observed and expected frequencies based on this model do not differ significantly, nor is there a significant difference between this model and the next more complex one because deletion of *BD* produces a nonsignificant CHI-SQUARE value of .20. Remember that this model includes all one-way effects because all variables are represented in one or more associations.

The model of choice for explaining the observed frequencies, then, includes all first-order effects and the two-way associations between benefit and harm, benefit and awareness, benefit and consultation, harm and awareness, harm and discomfort, harm and consultation, awareness and consultation, and discomfort and consultation. Not required in the model are the two-way associations between benefit and discomfort or awareness and harm.

7.7.2.3 Adequacy of Fit Overall evaluation of the model is made on the basis of the likelihood ratio or Pearson χ^2, both of which, as seen in Table 7.19, indicate a good fit between observed and expected frequencies. For the model of choice (*BH, BA, BC, HA, HD, HC, AC, DC*) the likelihood ratio value is 24.55 with $p = .1379$; χ^2 is 21.99 with $p = .2325$.

Assessment of fit of the model in individual cells proceeds through inspection of the standardized residuals for each cell (cf. Section 7.4.3.1). These residuals, as produced by BMDP4F, are shown in Table 7.20. The table displays the observed frequencies for each cell, the expected frequencies for each cell, the standardized deviates (the standardized residual values from which discrepancies are evaluated), and the differences between observed and expected frequencies.

Most of the standardized residual values are quite small; only one cell has a value that exceeds 1.96. Since the classification table has 32 cells, a residual value of 2.2 (the largest of the standardized residuals) for one of them is not unexpected; this cell is not deviant enough to be considered an outlier. However, the fit of the model is least effective for this cell, which contains therapists who felt their attraction to clients was beneficial to the therapy, who thought their clients aware of the attraction, and who felt uncomfortable about it, but who never felt it harmful to the therapy or sought consultation about it. As seen from the observed frequency table, seven of the 434 therapists responded in this way. The expected frequency table shows that, according to the model, only about three were predicted to provide this response.

7.7.2.4 Interpretation of Selected Model Two types of information are useful in interpreting the selected model: parameter estimates for the model and marginal observed frequency tables for all included effects.

The loglinear parameter estimate, λ, and the ratio λ/SE (cf. Section 7.4.3.2) are shown in Table 7.21 for each effect included in the model. Because there are only two levels of each variable, each effect is summarized by a single parameter value where one level of the effect has the positive value of the parameter and the other the negative value of the parameter.

TABLE 7.20 SETUP AND PARTIAL OUTPUT OF BMDP4F RUN TO
            EVALUATE RESIDUALS

```
/PROBLEM     TITLE IS 'LOGLINEAR ANALYSIS OF PKS QUESTIONNAIRE,
                 RESIDUALS AND PARAMETER ESTIMATES'.
/INPUT       VARIABLES ARE 10.  FORMAT IS FREE.
             FILE=MFA.
/VARIABLE    NAMES ARE SEX,AGE,AWARE,BENEFIT,HARM,CONSULT,DISCOMF,
                 INTIMACY,GRADTRNG,IDNO.
             MISSING = 9*9.
             LABEL=IDNO.
/CATEGORY    CODES(1) ARE 1,2.  NAMES(1) ARE MALE,FEMALE.
             CODES(2) ARE 1,2.  NAMES(2) ARE YOUNGER, OLDER.
             CODES(3) ARE 1,2.  NAMES(3) ARE 'PROB NOT', YES.
             CODES(4 TO 7) ARE 1,2. NAMES(4 TO 7) ARE NEVER, YES.
             CODES(8) ARE 1 TO 4. NAMES(8) ARE NEVER,YES,YES,YES.
             CODES(9) ARE 1 TO 4. NAMES(9) ARE NONE,VLITTLE,SOME,SOME.
/TABLE       INDICES ARE BENEFIT,HARM,AWARE,DISCOMF,CONSULT.
/FIT     MODEL IS BH, BA, BC, HA, HD, HC, AC, DC.
/PRINT   EXPECTED. DIFF. STANDARDIZED. LAMBDA.
/END
```

***** OBSERVED FREQUENCY TABLE 1

CONSULT	DISCOMF	AWARE	HARM	BENEFIT NEVER	YES	TOTAL
NEVER	NEVER	PROB NOT	NEVER	43	27 !	70
			YES	4	11 !	15
			TOTAL	47	38 !	85
		YES	NEVER	0	3 !	3
			YES	1	5 !	6
			TOTAL	1	8 !	9
	YES	PROB NOT	NEVER	20	14 !	34
			YES	14	22 !	36
			TOTAL	34	36 !	70
		YES	NEVER	0	7 !	7
			YES	3	6 !	9
			TOTAL	3	13 !	16
YES	NEVER	PROB NOT	NEVER	10	13 !	23
			YES	4	7 !	11
			TOTAL	14	20 !	34
		YES	NEVER	1	10 !	11
			YES	0	14 !	14
			TOTAL	1	24 !	25
	YES	PROB NOT	NEVER	16	30 !	46
			YES	18	53 !	71
			TOTAL	34	83 !	117
		YES	NEVER	0	15 !	15
			YES	6	57 !	63
			TOTAL	6	72 !	78

TOTAL OF THE OBSERVED FREQUENCY TABLE IS   434

TABLE 7.20 (Continued)

MODEL	D.F.	LIKELIHOOD-RATIO CHI-SQUARE	PROB	PEARSON CHI-SQUARE	PROB	ITER.
BH,BA,BC,HA,HD,HC,AC,DC,	18	24.55	.1379	21.99	.2325	6

***** EXPECTED VALUES USING ABOVE MODEL

CONSULT	DISCOMF	AWARE	HARM	BENEFIT NEVER	YES	TOTAL
NEVER	NEVER	PROB NOT	NEVER	37.1	28.0 !	65.1
			YES	7.5	9.9 !	17.4
			TOTAL	44.6	37.9 !	82.5
		YES	NEVER	1.3	5.4 !	6.6
			YES	.6	4.3 !	4.9
			TOTAL	1.9	9.6 !	11.5
	YES	PROB NOT	NEVER	21.8	16.5 !	38.3
			YES	14.8	19.4 !	34.2
			TOTAL	36.6	35.9 !	72.5
		YES	NEVER	.8	3.1 !	3.9
			YES	1.2	8.4 !	9.6
			TOTAL	1.9	11.6 !	13.5
YES	NEVER	PROB NOT	NEVER	9.8	16.0 !	25.8
			YES	3.2	9.0 !	12.2
			TOTAL	13.0	25.0 !	38.0
		YES	NEVER	.9	8.5 !	9.4
			YES	.7	10.9 !	11.6
			TOTAL	1.6	19.4 !	21.0
	YES	PROB NOT	NEVER	16.7	27.1 !	43.8
			YES	18.1	51.2 !	69.3
			TOTAL	34.8	78.3 !	113.0
		YES	NEVER	1.6	14.4 !	16.0
			YES	4.0	62.0 !	65.9
			TOTAL	5.6	76.4 !	82.0

***** STANDARDIZED DEVIATES = (OBS - EXP)/SQRT(EXP) FOR ABOVE MODEL

CONSULT	DISCOMF	AWARE	HARM	BENEFIT NEVER	YES
NEVER	NEVER	PROB NOT	NEVER	1.0	-.2
			YES	-1.3	.4
		YES	NEVER	-1.1	-1.0
			YES	.5	.3
	YES	PROB NOT	NEVER	-.4	-.6
			YES	-.2	.6
		YES	NEVER	-.9	2.2
			YES	1.7	-.8

TABLE 7.20 (Continued)

YES	NEVER	PROB NOT	NEVER	.1	-.7
			YES	.5	-.7
		YES	NEVER	.1	.5
			YES	-.8	.9

	YES	PROB NOT	NEVER	-.2	.6
			YES	.0	.3
		YES	NEVER	-1.3	.2
			YES	1.0	-.6

***** DIFFERENCES BETWEEN OBSERVED AND EXPECTED USING ABOVE MODEL

CONSULT	DISCOMF	AWARE	HARM	BENEFIT	
				NEVER	YES

NEVER	NEVER	PROB NOT	NEVER	5.9	-1.0
			YES	-3.5	1.1
		YES	NEVER	-1.3	-2.4
			YES	.4	.7

	YES	PROB NOT	NEVER	-1.8	-2.5
			YES	-.8	2.6
		YES	NEVER	-.8	3.9
			YES	1.8	-2.4

YES	NEVER	PROB NOT	NEVER	.2	-3.0
			YES	.8	-2.0
		YES	NEVER	.1	1.5
			YES	-.7	3.1

	YES	PROB NOT	NEVER	-.7	2.9
			YES	-.1	1.8
		YES	NEVER	-1.6	.6
			YES	2.0	-5.0

TABLE 7.21 PARTIAL OUTPUT FOR BMDP4F RUN ON PARAMETER ESTIMATES. (SETUP APPEARS IN TABLE 7.20)

ASYMPTOTIC STANDARD ERRORS OF THE PARAMETER ESTIMATES ARE COMPUTED BY INVERTING THE INFORMATION MATRIX.

ESTIMATES OF THE LOG-LINEAR PARAMETERS (LAMBDA) IN THE MODEL ABOVE
THETA(MEAN)   1.9657

***** ESTIMATES OF THE LOG-LINEAR PARAMETERS (LAMBDA) IN THE MODEL ABOVE

BENEFIT	
NEVER	YES
-.617	.617

***** RATIO OF THE LOG-LINEAR PARAMETER ESTIMATES TO ITS STANDARD ERROR

BENEFIT	
NEVER	YES
-7.228	7.228

TABLE 7.21 (Continued)

```
***** ESTIMATES OF THE LOG-LINEAR PARAMETERS (LAMBDA) IN THE MODEL ABOVE

            HARM
            ------
       NEVER      YES
       -------------------
        .035     -.035


*****              RATIO OF THE LOG-LINEAR PARAMETER ESTIMATES TO ITS STANDARD ERROR

                   HARM
                   ------
                  NEVER      YES
                  -------------------
                   .508     -.508

***** ESTIMATES OF THE LOG-LINEAR PARAMETERS (LAMBDA) IN THE MODEL ABOVE

HARM         BENEFIT
------       ------
             NEVER      YES
--------------------------
NEVER         .138     -.138
YES          -.138      .138

*****              RATIO OF THE LOG-LINEAR PARAMETER ESTIMATES TO ITS STANDARD ERROR

     HARM         BENEFIT
     ------       ------
                  NEVER      YES
     --------------------------
     NEVER        2.397     -2.397
     YES         -2.397      2.397

***** ESTIMATES OF THE LOG-LINEAR PARAMETERS (LAMBDA) IN THE MODEL ABOVE

            AWARE
            ------
       PROB NOT    YES
       -------------------
        .794     -.794


*****              RATIO OF THE LOG-LINEAR PARAMETER ESTIMATES TO ITS STANDARD ERROR

                   AWARE
                   ------
                  PROB NOT    YES
                  -------------------
                   8.815    -8.815

***** ESTIMATES OF THE LOG-LINEAR PARAMETERS (LAMBDA) IN THE MODEL ABOVE

AWARE        BENEFIT
------       ------
             NEVER      YES
--------------------------
PROB NOT      .428     -.428
YES          -.428      .428


*****              RATIO OF THE LOG-LINEAR PARAMETER ESTIMATES TO ITS STANDARD ERROR

     AWARE        BENEFIT
     ------       ------
                  NEVER      YES
     --------------------------
     PROB NOT     4.947     -4.947
     YES         -4.947      4.947
```

TABLE 7.21 (Continued)

```
*****  ESTIMATES OF THE LOG-LINEAR PARAMETERS (LAMBDA) IN THE MODEL ABOVE

AWARE       HARM
------      ------
            NEVER    YES
-----------------------------
PROB NOT    .206    -.206
YES        -.206     .206

*****           RATIO OF THE LOG-LINEAR PARAMETER ESTIMATES TO ITS STANDARD ERROR

        AWARE       HARM
        ------      ------
                    NEVER    YES
        -----------------------------
        PROB NOT   3.355   -3.355
        YES       -3.355    3.355

*****  ESTIMATES OF THE LOG-LINEAR PARAMETERS (LAMBDA) IN THE MODEL ABOVE

            DISCOMF
            ------
            NEVER    YES
        --------------------
            -.302    .302

*****           RATIO OF THE LOG-LINEAR PARAMETER ESTIMATES TO ITS STANDARD ERROR

                DISCOMF
                ------
                NEVER    YES
            --------------------
                -5.445   5.445

*****  ESTIMATES OF THE LOG-LINEAR PARAMETERS (LAMBDA) IN THE MODEL ABOVE

DISCOMF     HARM
------      ------
            NEVER    YES
-----------------------------
NEVER       .302    -.302
YES        -.302     .302


*****           RATIO OF THE LOG-LINEAR PARAMETER ESTIMATES TO ITS STANDARD ERROR

        DISCOMF     HARM
        ------      ------
                    NEVER    YES
        -----------------------------
        NEVER      5.414   -5.414
        YES       -5.414    5.414

*****  ESTIMATES OF THE LOG-LINEAR PARAMETERS (LAMBDA) IN THE MODEL ABOVE

            CONSULT
            ------
            NEVER    YES
        --------------------
            -.166    .166

*****           RATIO OF THE LOG-LINEAR PARAMETER ESTIMATES TO ITS STANDARD ERROR

                CONSULT
                ------
                NEVER    YES
            --------------------
                -2.319   2.319
```

TABLE 7.21 (Continued)

```
*****   ESTIMATES OF THE LOG-LINEAR PARAMETERS (LAMBDA) IN THE MODEL ABOVE

CONSULT    BENEFIT
------     ------
          NEVER     YES
---------------------------
NEVER      .191    -.191
YES       -.191     .191

*****           RATIO OF THE LOG-LINEAR PARAMETER ESTIMATES TO ITS STANDARD ERROR

        CONSULT    BENEFIT
        ------     ------
                  NEVER     YES
        ---------------------------
        NEVER     3.345    -3.345
        YES      -3.345     3.345

*****   ESTIMATES OF THE LOG-LINEAR PARAMETERS (LAMBDA) IN THE MODEL ABOVE

CONSULT    HARM
------     ------
          NEVER     YES
---------------------------
NEVER      .118    -.118
YES       -.118     .118


*****           RATIO OF THE LOG-LINEAR PARAMETER ESTIMATES TO ITS STANDARD ERROR

        CONSULT    HARM
        ------     ------
                  NEVER     YES
        ---------------------------
        NEVER     2.095    -2.095
        YES      -2.095     2.095

*****   ESTIMATES OF THE LOG-LINEAR PARAMETERS (LAMBDA) IN THE MODEL ABOVE

CONSULT    AWARE
------     ------
          PROB NOT   YES
---------------------------
NEVER      .256    -.256
YES       -.256     .256

*****           RATIO OF THE LOG-LINEAR PARAMETER ESTIMATES TO ITS STANDARD ERROR

        CONSULT    AWARE
        ------     ------
                  PROB NOT   YES
        ---------------------------
        NEVER     3.844    -3.844
        YES      -3.844     3.844

*****   ESTIMATES OF THE LOG-LINEAR PARAMETERS (LAMBDA) IN THE MODEL ABOVE

CONSULT    DISCOMF
------     ------
          NEVER     YES
---------------------------
NEVER      .265    -.265
YES       -.265     .265

*****           RATIO OF THE LOG-LINEAR PARAMETER ESTIMATES TO ITS STANDARD ERROR

        CONSULT    DISCOMF
        ------     ------
                  NEVER     YES
        ---------------------------
        NEVER     4.812    -4.812
        YES      -4.812     4.812
```

295

Especially useful for interpretation are the standardized parameter estimates. Effects with the largest standardized parameter estimates are the most important in influencing the frequency in a cell. If the effects are rank ordered by the sizes of their standardized parameter estimates, the relative importance of the various effects becomes apparent. With a standardized parameter estimate of 8.815, the strongest predictor of cell size is whether or not the therapist thought the client was aware of the therapist's attraction. The least predictive of all the effects in the model, with a standardized parameter estimate of .508, is whether or not the therapist's attraction to the client was believed to be harmful to the therapy. (Recall from Table 7.18 that this one-way effect is included in the hierarchical model only because it is a component of at least one two-way association; it was not statistically significant by itself.)

Parameter estimates are useful in determining the relative strength of effects and in creating a prediction equation, but they do not provide a simple view of the

TABLE 7.22 PARTIAL OUTPUT FROM BMDP4F RUN SHOWING
OBSERVED FREQUENCIES FOR SIGNIFICANT EFFECTS.
(SETUP APPEARS IN TABLE 7.17)

```
*****  MARGINAL SUBTABLE -- TABLE  1

           BENEFIT
           ------
           NEVER     YES    TOTAL
       ---------------------------
            140      294 !   434

*****  MARGINAL SUBTABLE -- TABLE  1

           HARM
           ------
           NEVER     YES    TOTAL
       ---------------------------
            209      225 !   434

*****  MARGINAL SUBTABLE -- TABLE  1

           AWARE
           ------
          PROB NOT    YES    TOTAL
       ---------------------------
            306      128 !   434

*****  MARGINAL SUBTABLE -- TABLE  1

           DISCOMF
           ------
           NEVER     YES    TOTAL
       ---------------------------
            153      281 !   434

*****  MARGINAL SUBTABLE -- TABLE  1

           CONSULT
           ------
           NEVER     YES    TOTAL
       ---------------------------
            180      254 !   434


*****  MARGINAL SUBTABLE -- TABLE  1

HARM        BENEFIT
------      ------
           NEVER     YES    TOTAL
       ---------------------------
NEVER       90       119 !   209
YES         50       175 !   225
       ---------------------!---------
TOTAL      140       294 !   434
```

TABLE 7.22 (Continued)

```
*****  MARGINAL SUBTABLE -- TABLE  1

AWARE          BENEFIT
------         ------
               NEVER     YES    TOTAL
---------------------------------------
PROB NOT       129      177 !    306
YES             11      117 !    128
--------------------------!---------
TOTAL          140      294 !    434


*****  MARGINAL SUBTABLE -- TABLE  1

CONSULT        BENEFIT
------         ------
               NEVER     YES    TOTAL
---------------------------------------
NEVER           85       95 !    180
YES             55      199 !    254
--------------------------!---------
TOTAL          140      294 !    434


*****  MARGINAL SUBTABLE -- TABLE  1

AWARE          HARM
------         ------
               NEVER     YES    TOTAL
---------------------------------------
PROB NOT       173      133 !    306
YES             36       92 !    128
--------------------------!---------
TOTAL          209      225 !    434

*****  MARGINAL SUBTABLE -- TABLE  1

DISCOMF        HARM
------         ------
               NEVER     YES    TOTAL
---------------------------------------
NEVER          107       46 !    153
YES            102      179 !    281
--------------------------!---------
TOTAL          209      225 !    434

*****  MARGINAL SUBTABLE -- TABLE  1

CONSULT        HARM
------         ------
               NEVER     YES    TOTAL
---------------------------------------
NEVER          114       66 !    180
YES             95      159 !    254
--------------------------!---------
TOTAL          209      225 !    434


*****  MARGINAL SUBTABLE -- TABLE  1

CONSULT        AWARE
------         ------
               PROB NOT  YES    TOTAL
---------------------------------------
NEVER          155       25 !    180
YES            151      103 !    254
--------------------------!---------
TOTAL          306      128 !    434

*****  MARGINAL SUBTABLE -- TABLE  1

CONSULT        DISCOMF
------         ------
               NEVER     YES    TOTAL
---------------------------------------
NEVER           94       86 !    180
YES             59      195 !    254
--------------------------!---------
TOTAL          153      281 !    434
```

TABLE 7.23 SUMMARY OF HIERARCHICAL MODEL OF THERAPIST'S ATTRACTION TO CLIENTS,

N = 434
Constant = 1.996

Effect	Partial association chi-square		Loglinear parameter estimate (lambda)		Lambda/SE	
			Prob. not	Yes	Prob. not	Yes
First-order effects:						
Aware	75.20**		0.794	-0.794	8.815	-8.815
Benefit	55.85**		-0.617	0.617	-7.228	7.228
Discomfort	38.32**		-0.302	0.302	-5.445	5.445
Consult	12.68**		-0.166	0.166	-2.319	2.319
Harm	0.59		0.035	-0.035	0.508	-0.508
Second-order effects:			Prob. not	Yes	Prob. not	Yes
Benefit by aware	31.95**	Never	0.428	-0.428	4.947	-4.947
		Yes	-0.428	0.428	-4.947	4.947
Harm by discomfort	28.39**	Never	0.302	-0.302	5.414	-5.414
		Yes	-0.302	0.302	-5.414	5.414

Effect	χ^2		Never	Yes		Never	Yes
Discomfort by consult	21.47**	Never	0.265	−0.265	Never	4.812	−4.812
		Yes	−0.265	0.265	Yes	−4.812	4.812
Aware by consult	15.95**	Prob. not	0.256	−0.256	Prob. not	3.844	−3.844
		Yes	−0.256	0.256	Yes	−3.844	3.844
Harm by aware	11.71**	Prob. not	0.206	−0.206	Prob. not	3.355	−3.355
		Yes	−0.206	0.206	Yes	−3.355	3.355
Benefit by consult	9.77**	Never	0.191	−0.191	Never	3.345	−3.345
		Yes	−0.191	0.191	Yes	−3.345	3.345
Benefit by harm	4.69*	Never	0.138	−0.138	Never	2.397	−2.397
		Yes	−0.138	0.138	Yes	−2.397	2.397
Harm by consult	4.28*	Never	0.118	−0.118	Never	2.095	−2.095
		Yes	−0.118	0.118	Yes	−2.095	2.095

* $p < .05$.
** $p < .01$.

TABLE 7.24   CHECKLIST FOR HIERARCHI-
                 CAL MULTIWAY FREQUENCY
                 ANALYSIS

1. Issues
   a. Adequacy of expected frequencies
   b. Outliers in the solution
2. Major analysis
   a. Model screening
   b. Model selection
   c. Evaluation of overall fit. If adequate:
      (1) Significance tests for each model effect
      (2) Parameter estimates
3. Additional analyses
   a. Interpretation via proportions
   b. Identifying extreme cells (if fit inadequate)

direction of effects. For interpretation of direction, the marginal tables of observed frequencies for each effect in the model are useful, as illustrated in Table 7.22.

The results as displayed in Table 7.22 are best interpreted as proportions of therapists responding in a particular way. For example, the BENEFIT marginal sub-table shows that 32% (140/434) of the therapists believe that there was never any benefit to be gained from the therapist being attracted to a client. The BENEFIT by HARM marginal subtable shows that, among those who believe that there was no benefit, 64% (90/140) also believe there was no harm. Of those who believe there was at least some benefit, 59% (175/294) also believe there was at least some harm.

Table 7.23 summarizes the model in terms of significance tests and parameter estimates.

A checklist for hierarchical multiway frequency analysis appears in Table 7.24. A Result section, in journal format, follows for the analysis described.

Results

A five-way frequency analysis was performed to develop a hierarchical loglinear model of attraction of therapists to clients. Dichotomous variables analyzed were whether or not the therapist (1) believed the attraction to be beneficial to the client, (2) believed the attraction to be harmful to the client, (3) thought the client was aware of the attraction, (4) felt discomfort, and (5) sought consultation as a result of the attraction.

Four hundred thirty-four therapists

provided usable data for this analysis. All
two-way contingency tables provided
expected frequencies in excess of five.
After the model was selected, none of the 32
cells was an outlier.

Stepwise selection by simple deletion
of effects using BMDP4F produced a model
that included all first-order effects and
eight of the ten possible two-way
associations. The model had a likelihood
ratio $\chi^2(18) = 24.55$, $\underline{p} = .14$, indicating a
good fit between observed frequencies and
expected frequencies generated by the
model. A summary of the model with results of
tests of significance (partial likelihood
ratio χ^2) and loglinear parameter estimates
in raw and standardized form appears in
Table 7.23.

Insert Table 7.23 about here

Most of the therapists (68%) reported
that the attraction they felt for clients
was at least occasionally beneficial to
therapy, while a slight majority (52%) also
reported that it was at least occasionally
harmful. Seventy-one percent of the
therapists thought that clients were
probably aware of the attraction. Most
therapists (65%) felt at least some
discomfort about the attraction, and more
than half (58%) sought consultation as a
result of the attraction.

Of those therapists who thought the
attraction beneficial to the therapy, 60%
also thought it harmful. Of those who
thought the attraction never beneficial,
38% thought it harmful. Perception of
benefit was also related to client's
awareness. Of those who thought their
clients were aware of the attraction, 91%
thought it beneficial. Among those who
thought clients unaware, only 58% thought it
beneficial.

Those who sought consultation were also more likely to see the attraction as beneficial. Of those seeking consultation, 78% judged the attraction beneficial. Of those not seeking consultation, 53% judged it beneficial.

Lack of harm was associated with lack of awareness. Fifty-seven percent of therapists who thought their clients unaware felt the attraction was never harmful. Only 28% of those who thought their clients aware considered it never harmful. Discomfort was more likely to be felt by those therapists who considered the attraction harmful to therapy (80%) than by those therapists who thought it was not harmful to therapy (49%). Similarly, consultation was more likely to be sought by those who felt the attraction harmful (71%) than by those who did not feel it harmful (45%).

Seeking consultation was also related to client awareness and therapist discomfort. Therapists who thought clients were aware of the attraction were more likely to seek consultation (80%) than those who thought the client unaware (43%). Those who felt discomfort were more likely to seek consultation (69%) than those who felt no such discomfort (39%).

No statistically significant two-way associations were found between benefit and discomfort or between awareness and discomfort. None of the higher-order associations reached statistical significance.

7.7.3 Nonhierarchical Logit Analysis

This example illustrates a nonhierarchical logit analysis. The DV is whether or not therapists report having been intimate with clients (INTIMACY). Predictors are SEX of therapist, AGE of therapist with two levels (younger and older), and how much training in graduate school therapists had in handling feelings of attraction to clients (GRADTRNG) with three levels (none, very little, and some).

7.7.3.1 Model Screening and Selection Nonhierarchical modeling, as you recall, does not require that all lower-order components of a higher-order association be included in the model. For the associations that include the DV as a component, only significant effects need be included. However, all associations among IVs and their lower-order components must also be included.

An attempt was made to run the full, saturated model through SPSS^x LOG-LINEAR but the program failed to converge. (A warning message was printed out before the table of frequencies: ML DID NOT CONVERGE. THE CONVERGE CRITERION = .50001.) This failure did not prevent the program from printing out worthless parameter estimates, however.

In a second attempt to screen for significant effects, the saturated model is run through BMDP4F, partially illustrated in Table 7.25. With the iterative proportional fitting algorithm used by BMDP4F, convergence is accomplished.

As seen in Table 7.25, three effects containing the DV, INTIMACY, INTIMACY BY SEX, and INTIMACY BY SEX BY AGE BY GRADTRNG are statistically significant by both partial and marginal tests. A fourth effect, INTIMACY BY GRADTRNG, is statistically significant by the partial test of association but non-significant by the marginal test. This leads to the decision to try two alternative models, one with and one without the questionable effect.

Because the highest-order association is significant for these data, a nonhier-archical approach is required. In a hierarchical approach, if the highest-order as-sociation is significant, the model contains all effects, and is saturated.

Both of the potential models achieve convergence when run through SPSS^x LOGLINEAR (not shown). Comparison of the goodness-of-fit tests for the models does not clarify the choice between models because both models provide adequate fit to the data, $p > .05$ for both goodness-of-fit tests. Unhappily, application of the second criterion again produces ambiguous results. Using the likelihood ratio test, the fit is significantly worse with the deletion of the INTIMACY BY GRADTRNG interaction:

$$G^2(\text{without } IG) = 10.39425, \text{df} = 6, p > .05,$$
$$G^2(\text{with } IG) = 3.55128, \text{df} = 8, p > .05,$$
$$G^2(\text{difference}) = 6.84, \text{df} = 2, p < .05$$

According to the Pearson χ^2 goodness-of-fit test, however, the difference between the models is not statistically reliable, $\chi^2(2) = 4.05, p > .05$.

In the interests of parsimony and to avoid overfitting, the decision is made to delete the INTIMACY BY GRADTRNG interaction from the final model. SPSS^x LOGLINEAR setup and partial output of the chosen model are illustrated in Table 7.26.

7.7.3.2 Adequacy of Fit The model specified in Table 7.26 provides a reasonable fit to the observed frequencies according to both the Pearson chi square test (χ^2 =

TABLE 7.25 SETUP AND PARTIAL OUTPUT OF BMDP4F RUN TO
SCREEN NONHIERARCHICAL MODEL FOR LOGIT
ANALYSIS

```
/PROBLEM      TITLE IS 'LOGLINEAR ANALYSIS OF PKS QUESTIONNAIRE'.
/INPUT        VARIABLES ARE 10.  FORMAT IS FREE.
              FILE=MFA.
/VARIABLE     NAMES ARE SEX,AGE,AWARE,BENEFIT,HARM,CONSULT,DISCOMF,
                  INTIMACY,GRADTRNG,IDNO.
              MISSING = 9*9.
              LABEL=IDNO.
/CATEGORY     CODES(1) ARE 1,2.  NAMES(1) ARE MALE,FEMALE.
              CODES(2) ARE 1,2.  NAMES(2) ARE YOUNGER, OLDER.
              CODES(3) ARE 1,2.  NAMES(3) ARE 'PROB NOT', YES.
              CODES(4 TO 7) ARE 1,2. NAMES(4 TO 7) ARE NEVER, YES.
              CODES(8) ARE 1 TO 4. NAMES(8) ARE NEVER,YES,YES,YES.
              CODES(9) ARE 1 TO 4. NAMES(9) ARE NONE,VLITTLE,SOME,SOME.
/TABLE        INDICES ARE INTIMACY,SEX,AGE,GRADTRNG.
/FIT          ASSOCIATION IS 4.
/PRINT        MARGINAL=3.
/END
```

```
***** THE RESULTS OF FITTING ALL K-FACTOR MARGINALS.
      THIS IS A SIMULTANEOUS TEST THAT ALL K+1 AND HIGHER FACTOR INTERACTIONS ARE ZERO.
```

K-FACTOR	D.F.	LR CHISQ	PROB.	PEARSON CHISQ	PROB.	ITERATION
0-MEAN	23	677.51	.00000	657.19	.00000	
1	18	43.23	.00074	40.28	.00191	2
2	9	13.51	.14087	12.59	.18194	4
3	2	7.43	.02432	8.94	.01146	4
4	0	0.	1.	0.	1.	

```
***** A SIMULTANEOUS TEST THAT ALL K-FACTOR  INTERACTIONS ARE SIMULTANOUSLY ZERO.
      THE CHI-SQUARES ARE DIFFERENCES IN THE ABOVE TABLE.
```

K-FACTOR	D.F.	LR CHISQ	PROB.	PEARSON CHISQ	PROB.
1	5	634.28	.00000	616.91	.00000
2	9	29.72	.00049	27.69	.00107
3	7	6.08	.53085	3.66	.81853
4	2	7.43	.02432	8.94	.01146

```
***** ASSOCIATION OPTION SELECTED FOR ALL TERMS OF ORDER LESS THAN OR EQUAL TO  4
```

		PARTIAL ASSOCIATION				MARGINAL ASSOCIATION		
EFFECT	D.F.	CHISQUARE	PROB	ITER	D.F.	CHISQUARE	PROB	ITER
I.	1	515.02	.0000					
S.	1	15.27	.0001					
A.	1	.04	.8340					
G.	2	103.95	.0000					
IS.	1	12.72	.0004	3	1	11.94	.0005	2
IA.	1	.22	.6362	3	1	.21	.6491	2
IG.	2	6.08	.0477	3	2	5.45	.0656	2
SA.	1	.26	.6123	4	1	.45	.5014	2
SG.	2	6.09	.0476	3	2	5.67	.0588	2
AG.	2	5.19	.0745	4	2	5.53	.0630	2
ISA.	1	1.90	.1678	3	1	1.64	.1999	3
ISG.	2	.62	.7346	4	2	.85	.6532	3
IAG.	2	1.24	.5370	3	2	.84	.6570	3
SAG.	2	2.73	.2559	4	2	2.06	.3565	3
ISAG.	2	7.43	.0243					

7.13884, $p = .522$) and the more conservative likelihood ratio chi square ($G^2 = 10.39425$, $p = .238$). The sizes of the standardized residuals in the earlier part of the output in Table 7.26 also indicate that the fit between observed and expected frequencies derived from the model is good; only two of the 24 cells show $|z| > 1.00$, and both are below 1.96. The data, then, are adequately modeled, with no outlying cells.

However, the two measures of strength of association—entropy and concentration (cf. Section 7.5.2.1)—indicate that the DV, sexual intimacy with clients, is not well predicted by sex, age, and graduate training involving sexual attraction to clients. Less than 10% of the variance in the DV is predictable by the more generous of these indicators.

7.7.3.3 Interpretation of Selected Model

Marginal subtables for the significant first- and second-order effects appear in Table 7.27, along with a table of observed

TABLE 7.26 SETUP AND PARTIAL OUTPUT FROM SPSS[x] LOGLINEAR RUN OF SELECTED LOGIT MODEL

```
            TITLE        'LOGIT ANALYSIS OF PKS QUESTIONNAIRE'.
            FILE HANDLE  MFA
            DATA LIST    FILE=MFA   FREE
                         /SEX AGE AWARE BENEFIT HARM CONSULT DISCOMF INTIMACY
                         GRADTRNG IDNO
            MISSING VALUES  SEX TO GRADTRNG(9)
            VALUE LABELS    SEX 1 'MALE' 2 'FEMALE'/
                            AGE 1 'YOUNGER' 2 'OLDER'/
                            AWARE 1 'PROB NOT' 2 'YES'/
                            BENEFIT TO INTIMACY 1 'NEVER' 2 'YES'/
                            GRADTRNG 1 'NONE' 2 'VLITTLE' 3 'SOME'/
            RECODE          INTIMACY (3, 4=2)
            RECODE          GRADTRNG (4=3)
            LOGLINEAR       INTIMACY (1,2) BY SEX AGE (1,2) GRADTRNG (1,3)/
                            PRINT=DEFAULT,ESTIM/
                            DESIGN=INTIMACY, INTIMACY BY SEX,
                                    INTIMACY BY SEX BY AGE BY GRADTRNG/

* * * * * * * * * * * * * * * *  L O G  L I N E A R  A N A L Y S I S  * * * * * * * * * * * * * * * * *

DATA   INFORMATION

       569 UNWEIGHTED CASES ACCEPTED.
         2 CASES REJECTED BECAUSE OF OUT-OF-RANGE FACTOR VALUES.
        14 CASES REJECTED BECAUSE OF MISSING DATA.
       569 WEIGHTED CASES WILL BE USED IN THE ANALYSIS.


FACTOR INFORMATION

    FACTOR  LEVEL  LABEL
    INTIMACY   2
    SEX        2
    AGE        2
    GRADTRNG   3


DESIGN INFORMATION

    1 DESIGN/MODEL WILL BE PROCESSED.


- - - - - - - - - - - - - - - - - - - - - - - - - - - - - - - - - - - - - - - - - - - - - - - - - - -


* * * * * * * * * * * * * * * *  L O G  L I N E A R  A N A L Y S I S  * * * * * * * * * * * * * * * * *

CORRESPONDENCE BETWEEN EFFECTS AND COLUMNS OF DESIGN/MODEL 1

   STARTING  ENDING
    COLUMN   COLUMN    EFFECT NAME

       1        1      INTIMACY
       2        2      INTIMACY BY SEX
       3        4      INTIMACY BY SEX BY AGE BY GRADTRNG


- - - - - - - - - - - - - - - - - - - - - - - - - - - - - - - - - - - - - - - - - - - - - - - - - - -

  *** ML CONVERGED AT ITERATION  6. THE CONVERGE CRITERION =   .00000

- - - - - - - - - - - - - - - - - - - - - - - - - - - - - - - - - - - - - - - - - - - - - - - - - - -
```

TABLE 7.26 (Continued)

OBSERVED, EXPECTED FREQUENCIES AND RESIDUALS

FACTOR	CODE	OBS. COUNT & PCT.	EXP. COUNT & PCT.	RESIDUAL	STD RESID.	ADJ. RESID.
INTIMACY	NEVER					
SEX	MALE					
AGE	YOUNGER					
GRADTRNG	NONE	75.00 (92.59)	76.89 (94.93)	-1.8923	-.2158	-1.3448
GRADTRNG	VLITTLE	45.00 (90.00)	45.69 (91.38)	-.6893	-.1020	-.4614
GRADTRNG	SOME	35.00 (94.59)	31.78 (85.89)	3.2202	.5712	2.0195
AGE	OLDER					
GRADTRNG	NONE	71.00 (83.53)	72.96 (85.83)	-1.9585	-.2293	-1.0238
GRADTRNG	VLITTLE	35.00 (89.74)	35.67 (91.45)	-.6668	-.1117	-.4821
GRADTRNG	SOME	39.00 (99.99)	37.01 (94.91)	1.9868	.3266	1.7043
SEX	FEMALE					
AGE	YOUNGER					
GRADTRNG	NONE	56.00 (94.92)	56.55 (95.86)	-.5548	-.0738	-.5012
GRADTRNG	VLITTLE	27.00 (96.43)	27.33 (97.61)	-.3309	-.0633	-.4470
GRADTRNG	SOME	27.00 (99.99)	26.63 (98.61)	.3744	.0726	.6517
AGE	OLDER					
GRADTRNG	NONE	82.00 (98.80)	81.85 (98.62)	.1455	.0161	.1622
GRADTRNG	VLITTLE	19.00 (99.99)	18.54 (97.59)	.4583	.1064	.7281
GRADTRNG	SOME	21.00 (95.45)	21.09 (95.87)	-.0924	-.0201	-.1106
INTIMACY	YES					
SEX	MALE					
AGE	YOUNGER					
GRADTRNG	NONE	6.00 (7.41)	4.11 (5.07)	1.8923	.9336	1.3448
GRADTRNG	VLITTLE	5.00 (10.00)	4.31 (8.62)	.6893	.3320	.4614
GRADTRNG	SOME	2.00 (5.41)	5.22 (14.11)	-3.2202	-1.4094	-2.0195
AGE	OLDER					
GRADTRNG	NONE	14.00 (16.47)	12.04 (14.17)	1.9585	.5644	1.0238
GRADTRNG	VLITTLE	4.00 (10.26)	3.33 (8.55)	.6668	.3652	.4821
GRADTRNG	SOME	0 (0)	1.99 (5.09)	-1.9868	-1.4095	-1.7043
SEX	FEMALE					
AGE	YOUNGER					
GRADTRNG	NONE	3.00 (5.08)	2.45 (4.14)	.5548	.3548	.5012
GRADTRNG	VLITTLE	1.00 (3.57)	.67 (2.39)	.3309	.4045	.4470
GRADTRNG	SOME	0 (0)	.37 (1.39)	-.3744	-.6119	-.6517
AGE	OLDER					
GRADTRNG	NONE	1.00 (1.20)	1.15 (1.38)	-.1455	-.1360	-.1622
GRADTRNG	VLITTLE	0 (0)	.46 (2.41)	-.4583	-.6770	-.7281
GRADTRNG	SOME	1.00 (4.55)	.91 (4.13)	.0924	.0970	.1106

- -

GOODNESS-OF-FIT TEST STATISTICS

```
    LIKELIHOOD RATIO CHI SQUARE =      10.39425    DF = 8   P =  .238
                PEARSON CHI SQUARE =       7.13884    DF = 8   P =  .522
```

- -

ANALYSIS OF DISPERSION

	DISPERSION		
SOURCE OF VARIATION	ENTROPY	CONCENTRATION	DF
DUE TO MODEL	9.691	2.378	
DUE TO RESIDUAL	127.199	66.810	
TOTAL	136.890	69.188	568

TABLE 7.26   (Continued)

- -

```
MEASURES OF ASSOCIATION

        ENTROPY =    .070793
   CONCENTRATION =   .034377
```

* * * * * * * * * * * * * * L O G   L I N E A R   A N A L Y S I S * * * * * * * * * * * * * *

```
ESTIMATES FOR PARAMETERS

INTIMACY
```

PARAMETER	COEFF.	STD. ERR.	Z-VALUE	LOWER 95 CI	UPPER 95 CI
1	1.5176581490	.11885	12.76951	1.28471	1.75060

```
INTIMACY BY SEX
```

PARAMETER	COEFF.	STD. ERR.	Z-VALUE	LOWER 95 CI	UPPER 95 CI
2	-.3348915106	.11420	-2.93241	-.55873	-.11105

```
INTIMACY BY SEX BY AGE BY GRADTRNG
```

PARAMETER	COEFF.	STD. ERR.	Z-VALUE	LOWER 95 CI	UPPER 95 CI
3	.2819983626	.11521	2.44768	.05619	.50781
4	-.0023797221	.13678	-.01740	-.27046	.26571

- -

frequencies. Although this information can be derived from the SPSS[x] output of Table 7.26, it is more conveniently obtained through the earlier BMDP4F run.

Interpretation is simple from the marginal tables. Ninety-three percent (532/569) of the therapists report that they have never been sexually intimate with clients, but, as the significant two-way effect of SEX by INTIMACY indicates, the percentages vary with sex. For male therapists the incidence of reported sexual intimacy is about 9% (31/331) while for female therapists it is about 3% (6/238).

The significant four-way association is best interpreted with the assistance of a figure similar to the figures used to illustrate three-way interactions in ANOVA. In Figure 7.1, percentage of therapists reporting sexual intimacy, the DV, is shown as a function of the three IVs.

Note that special care needs to be taken in interpreting percentages based on cells representing rare events. For example, in Table 7.27 one can see that there are very few female therapists with any graduate training related to sexual attraction for clients. One occurrence of intimacy among these therapists translates to as much as 5%.

Interpretation of the sizes and directions of parameter estimates is often helpful. Parameter estimates for the model, as produced by SPSS[x] LOGLINEAR, appear near the end of the output in Table 7.26. The parameter estimates are called COEFF (while PARAMETER in the output refers to which degree of freedom is being analyzed—most confusing).

The INTIMACY parameter simply reflects the preponderance of "never" answers to the question regarding sexual intimacy. That is, answers coded 1 (never)

TABLE 7.27 PARTIAL OUTPUT FROM BMDP4F RUN SHOWING OBSERVED FREQUENCY TABLES FOR SIGNIFICANT EFFECTS IN LOGIT MODEL. SETUP APPEARS IN TABLE 7.25

```
*****  OBSERVED FREQUENCY TABLE   1

GRADTRNG AGE      SEX          INTIMACY
------   ------   ------       ------
                               NEVER    YES   TOTAL
-------------------------------------------------------

NONE     YOUNGER  MALE           75      6 !    81
                  FEMALE         56      3 !    59
                               -----------------!---------
                  TOTAL         131      9 !   140

         OLDER    MALE           71     14 !    85
                  FEMALE         82      1 !    83
                               -----------------!---------
                  TOTAL         153     15 !   168


VLITTLE  YOUNGER  MALE           45      5 !    50
                  FEMALE         27      1 !    28
                               -----------------!---------
                  TOTAL          72      6 !    78

         OLDER    MALE           35      4 !    39
                  FEMALE         19      0 !    19
                               -----------------!---------
                  TOTAL          54      4 !    58


SOME     YOUNGER  MALE           35      2 !    37
                  FEMALE         27      0 !    27
                               -----------------!---------
                  TOTAL          62      2 !    64

         OLDER    MALE           39      0 !    39
                  FEMALE         21      1 !    22
                               -----------------!---------
                  TOTAL          60      1 !    61

         TOTAL OF THE OBSERVED FREQUENCY TABLE IS     569

*****  MARGINAL SUBTABLE -- TABLE  1

             INTIMACY
             ------
             NEVER    YES   TOTAL
          ------------------------------
             532      37 !   569


*****  MARGINAL SUBTABLE -- TABLE  1

SEX          INTIMACY
------       ------
             NEVER    YES   TOTAL
          ------------------------------
MALE         300      31 !   331
FEMALE       232       6 !   238
          ------------------!---------
TOTAL        532      37 !   569
```

are far more likely than those coded 2 (yes). The coefficient for "yes" is the negative of the coefficient for answers coded "never." Therefore, if we are estimating the frequency of a cell that includes "never" for intimacy, 1.52 is added to the ln Fe (natural logarithm of the expected frequency). If we are estimating the frequency of a cell that includes "yes" for intimacy, 1.52 is subtracted from the ln Fe.

The INTIMACY BY SEX parameter shows the impact of sex on the likelihood

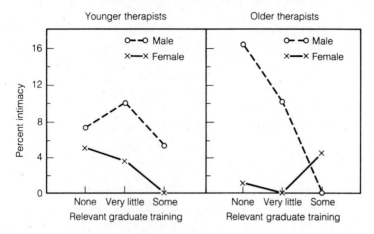

Figure 7.1 Percent of therapists reporting at least one occasion of sexual intimacy with clients.

of intimacy. Notice that the parameter has a negative sign ($-.33$). The direction of the parameter—negative—indicates that small numbers on one dimension go with large numbers on the other dimension (just as a negative correlation indicates that large numbers on one variable go with small numbers on the other). To interpret the parameter, consider the way categories were coded for each of the effects. For this example, the combinations of male and "never" and female and "yes" were coded the same way while the combinations of male and "yes" and female and "never" were coded in opposite directions. The negative sign indicates that the combinations coded in opposite directions are more likely than the combinations coded in the same directions.

The four-way interaction is much more difficult to interpret. First, because GRADTRNG has three levels, two parameters are generated, making it difficult to interpret direction of effect. One obvious finding, however, is the lack of impact of the second level of GRADTRNG, reflected by the size of parameter 4. (Parameter 3 refers to the first level of GRADTRNG and parameter 4 to the second level.) The parameter value of the third level of GRADTRNG is the negative of the coefficients coming before: zero minus parameter 3 minus parameter 4; in this case $0 - .282 - (-.002) = -.28$. This suggests that the category "very little" adds nothing to interpretation and can safely be ignored or, in further analyses, combined with either of the other two categories.

Because of the complexity of the model, in particular the inclusion of the four-way effect, it is decided to forgo interpretation of the model in terms of the impact of effects on odds ratios (cf. Section 7.5.3.2). A summary of significant effects and their parameter estimates appears in Table 7.28.

Table 7.29 contains a checklist for nonhierarchical logit analysis of multiway frequency tables. A Results section follows, in journal format.

TABLE 7.28 SUMMARY OF LOGIT MODEL OF SEXUAL INTIMACY BETWEEN CLIENTS AND THERAPISTS, $N = 569$

Effect	Partial association chi-square	Loglinear parameter estimate (lambda)		Lambda/SE	

Intimacy

	Never	Yes
Loglinear parameter estimate (lambda)	1.518	−1.518
Lambda/SE	12.770	−12.770

Intimacy by sex — Partial association chi-square $12.72**$, $df = 1$

	Loglinear: Never	Loglinear: Yes	Lambda/SE: Never	Lambda/SE: Yes
Male	−0.335	0.335	−2.932	2.932
Female	0.335	−0.335	2.932	−2.932

Intimacy by age by sex by graduate training — Partial association chi-square $7.43*$, $df = 2$

Intimacy	Age	Sex	Graduate training (Loglinear): None	Very little	Some	Graduate training (Lambda/SE): None	Very little
Never	Old	M	−0.282	0.002	0.280	−2.448	0.017
Never	Old	F	0.282	−0.002	−0.280	2.448	−0.017
Never	Young	M	0.282	−0.002	−0.280	2.448	−0.017
Never	Young	F	−0.282	0.002	0.280	−2.448	0.017
Yes	Old	M	0.282	−0.002	−0.280	2.448	−0.017
Yes	Old	F	−0.282	0.002	0.280	−2.448	0.017
Yes	Young	M	−0.282	0.002	0.280	−2.448	0.017
Yes	Young	F	0.282	−0.002	−0.280	2.448	−0.017

$* p < .05$
$** p < .05$

310

TABLE 7.29 CHECKLIST FOR LOGIT
ANALYSIS

1. Issues
   a. Adequacy of expected frequencies
   b. Outliers in the solution
2. Major analysis
   a. Model screening and selection
   b. Evaluation of overall fit. If adequate:
      (1) Significance test for each model effect
      (2) Parameter estimates
      (3) Effect sizes
3. Additional analyses
   a. Interpretation via proportions
   b. Interpretation via odds ratios
   c. Identifying extreme cells (if fit inadequate)

Results

A four-way frequency analysis was performed to develop a logit model of sexual intimacy between therapists and clients. Predictors were therapist sex, age (younger vs. older), and amount of graduate training the therapist had on the subject of attraction to clients (none, very little, some).

Five hundred sixty-nine therapists provided usable data for this analysis. All component two-way contingency tables showed expected frequencies in excess of five. All cells were adequately predicted by the model after analysis.

A nonhierarchical logit analysis produced a model with one two-way association (INTIMACY by SEX), the four-way association (INTIMACY by SEX by AGE BY GRADTRNG), and the first-order effect of the DV, INTIMACY. The model had adequate fit betwen observed and expected frequencies, likelihood ratio $\underline{G}^2(8) = 10.39$, $\underline{p} = .24$. Although the model produced reasonable expected frequencies, reduction in uncertainty in prediction of intimacy by the model was low, with concentration = .03.

A summary of the model with results of tests of significance (partial likelihood ratio $\underline{G}^2$) and loglinear parameter estimates in raw and standardized form appears in Table 7.28.

Insert Table 7.28 about here

Most therapists (93%) reported that they had never been sexually intimate with a client. The percentages vary significantly with sex, however. Male therapists reported about 9% incidence of sexual intimacy with clients while female therapists reported about 3%.

The complex four-way association relating sexual intimacy to a joint function of therapist sex, age, and amount of graduate training is illustrated in Figure 7.1. Younger male and female therapists show similar patterns of sexual intimacy as a function of graduate training. The difference between male and female therapists is much more pronounced for older therapists. Older male therapists show a striking decrease in intimacy when there has been some graduate training in handling attraction to clients. It should be noted that the paucity of female therapists with any relevant training at all makes the corresponding cell percentages highly unstable.

Insert Figure 7.1 about here

None of the remaining two-way associations that included intimacy was statistically significant, nor were any of the three-way associations with intimacy.

7.8 SOME EXAMPLES FROM THE LITERATURE

Stability in types of errors made on math tests by boys and girls was studied by Marshall (1983). Data were derived from the Survey of Basic Skills, Grade 6, a test given annually to all sixth grade children in California. Chosen for study were the 160 math items on the survey that had four alternatives. Only incorrect responses were analyzed. A three-way contingency table was formed for each item: a $2 \times 4 \times 4$ table of sex by year (1976, 1977, 1978, 1979) by distractor (three incorrect alternatives plus no answer).

It was hypothesized that the best-fit model would include the distractor by sex association (indicating that boys and girls make different errors) but not the three-way association (which would indicate that sex differences were not stable but varied from year to year). First-order effects were not of interest. The analysis was also to determine which of the other two-way interactions (sex by year, and distractor by year) were needed if the three-way effect and the distractor by sex effects were as predicted.

The clear winner among the hierarchical models was indeed limited to second-order effects; the distractor by sex interaction was significant as hypothesized as was the distractor by year interaction. This was the best-fit model for 128 of the 160 items. The next best model included the three-way interaction, but only emerged as best for 10 items. The author concluded that sex differences in choice of distractor were indeed stable—boys and girls systematically select different incorrect alternatives.

Porter and Albert (1977) studied attitudes toward women's role among South African and American women. Attitude toward women's role was categorized as traditional or liberal on two scales, familial roles and working mothers. Two $3 \times 2 \times 2 \times 2 \times 2$ frequency tables were formed, one for attitudes on the familial roles dimension and the other for attitudes on the working mother measure. Predictors of attitudes were religion (liberal Christian, orthodox Christian, Jew), country (South Africa, United States), whether or not the mother worked outside the home, and educational level (low, high).

Separate logit analysis was done for each of the two attitude scales. In addition, hierarchical modeling was used to find the best model to predict attitudes. This, then, was a combination of hierarchical modeling with logit analysis.

Attitude toward women's role on both scales was predicted by country and work status; the model for attitude toward familial roles also included the effect of education. The questions of interest were whether religion added predictability and whether religion interacted with the other predictors in enhancing prediction of attitudes.

For both attitude scales, religion significantly enhanced predictability over that of country, work status, and education while religion in association with the other three predictors did not significantly enhance prediction of attitude. For both scales, then, the best fit model was one that included the interaction of each predictor, including religion, with attitude but none of the higher-order interactions.

Relative magnitude of effects was evaluated through logits (log odds) of each effect as computed from effect parameters and then translated into probabilities. For example, South Africans were found to be 52% more likely to be traditional on attitudes toward familial roles than were Americans; this was the single strongest effect in both models. For attitudes toward working mothers, the strongest effects were associated with whether or not the respondent was a working woman. Also for this scale, a particularly good fit was obtained by including the work by religion interaction, although this was not significantly better than the model that excluded this interaction.

Probability of keeping appointments in a community mental health center was analyzed by Rock (1982) in a six-way frequency table. With outcome (show, no show) as the dependent variable, the five predictors were method of contact (in person, telephone), contact person (client, significant other), sex of client, time of day of intake appointment (A.M., P.M.), and number of days between first contact and intake appointment (0 to 7 days). As a result of fitting the saturated model, it was found that four of the six main effects were significant but that none of the interactions was. Although one of the main effects was the outcome, none of the interactions with the outcome reached statistical significance. There was, then, no logit model to test.

This analysis was followed by several bivariate chi-square analyses searching for significant relationships among those variables which produced significant main effects in the loglinear analysis. Since this is akin to performing post hoc interaction contrasts after finding nonsignificant omnibus interaction effects in analysis of variance, the statistical justification for these analyses is not clear. Nevertheless, three "significant" associations were found among the four tested; method of contact by outcome, latency of outcome, and method of contact by contact person. The author interpreted the results as evidence that individuals were more likely to show up for their appointments if contact was made by telephone and the appointment was scheduled relatively soon after initial contact. Also, contact was most likely to be made by telephone if made by a significant other.

Hierarchical modeling was used to investigate change in household composition of the elderly by Fillenbaum and Wallman (1984). Two separate analyses were done, one on a full sample of elderly respondents and another on a subsample of those who had no change in marital status in the 30-month period before first and second interviews. For both analyses, the dependent variable was whether or not there was change in household composition during the 30-month period. As predictors, both analyses used indices of whether or not there was change in economic status and self-care capacity. A third predictor was whether extensive or limited help was available. For the full sample, an additional predictor was whether or not there was change in marital status.

Second-order models, including associations of change in household composition with the remaining variables, were tested individually against a model that included only the main effect of household composition change. Among these, the only model with a single two-way association that provided improvement in fit was the one that included change in marital status. However, the fit was improved by considering a second two-way association—change in household composition by extent of help available. That is, the best fitting model included the associations of change in marital status with change in household composition, and extent of help with change in household composition. No mention was made of testing third-order models.

Change in household composition, then, is most likely by those elderly who experience change in marital status. A small gain in predictability is made by considering extent of help available. Those who perceive extensive availability of help are less likely to change their household composition. This interpretation is supported by separate analysis of the subsample which did not experience change in marital status. The only significant two-way association for this sample was between availability of help and change in household composition.

Chapter 8

Analysis of Covariance

8.1 GENERAL PURPOSE AND DESCRIPTION

Analysis of covariance is an extension of analysis of variance in which main effects and interactions of IVs are assessed after DV scores are adjusted for differences associated with one or more covariates.[1] The major question for ANCOVA is the same as for ANOVA: Are mean differences among groups on the adjusted DV likely to have occurred by chance? For example, is there a mean difference between a treated group and a control group on a posttest (the DV) after posttest scores are adjusted for differences in pretest scores (the covariate)?

Analysis of covariance is used for three major purposes. The first purpose is to increase the sensitivity of the test of main effects and interactions by reducing the error term; the error term is adjusted for, and hopefully reduced by, the relationship between the DV and the covariate(s). The second purpose is to adjust the means on the DV themselves to what they would be if all subjects scored equally on the covariate(s). The third use of ANCOVA occurs in MANOVA where the researcher assesses one DV after adjustment for other DVs that are treated as covariates.

The first use of ANCOVA is the most common. In an experimental setting, ANCOVA increases the power of an F test for main effect or interaction by removing predictable variance associated with covariate(s) from the error term. That is, covariates are used to assess the "noise" in the DV where "noise" is undesirable variance in the DV (e.g., individual differences) that is estimated by scores on covariates (e.g., pretests). Although in most experiments subjects are randomly

---

[1] Strictly speaking, ANCOVA, like multiple regression, is not a multivariate technique because it involves a single DV. For the purposes of this book, however, it is convenient to consider it along with multivariate analyses.

assigned to groups, random assignment in no way assures equality among groups—
it only guarantees, within probability limits, that there are no *systematic* differences
between groups to begin with. Random individual differences, however, can both
spread out scores among subjects within groups (and increase the error term) and
create differences among groups that are not associated with treatment.

Either of these effects makes it hard to show that differences among groups
are due to different experimental treatments. One way to diminish the effect of
individual differences is to adjust for them statistically. Individual differences are
measured as covariates and ANCOVA is used to provide the adjustment. The effect
of the adjustment is to provide a (usually) more powerful test of differences among
groups uncontaminated with differences on covariates.

In this sense, ANCOVA is similar to a within-subjects (repeated measures)
ANOVA (cf. Chapter 3), in which individual differences among subjects are estimated
from consistencies within subjects over treatment and then the variance due to individual differences is removed from the error term. In ANCOVA, variance due to
individual differences is assessed through one or more covariates, usually continuous
variables, that are linearly related to the DV. Then the relationship between the DV
and the covariate(s) is removed from the error term.

In the classical experimental use of ANCOVA, subjects are randomly assigned
to levels of one or more IVs and are measured on one or more covariates before
administration of treatment. A common covariate is a pretest score, measured the
same as the DV but before manipulation of treatment. However, a covariate might
also be some demographic characteristic (educational level, socioeconomic level, or
whatever) or a personal characteristic (anxiety level, IQ, etc.) that is completely
different from the DV. Manipulation of the IV(s) occurs and then the DV is measured.

For example, suppose an experiment is designed to investigate methods for
reducing test anxiety in statistics courses. Volunteers enrolled in statistics courses
are randomly assigned to one of three treatment groups: desensitization, relaxation
training, or a control group (for whom anxiety reduction is offered after the experiment). The three treatment groups are the three levels of the IV. Before treatment,
however, students in all three groups are given a standardized measure of test anxiety
that serves as the covariate. Then treatment, or "waiting-list" control, is carried out
for some specified time. After the treatment period, students in all three groups are
tested again on the measure of text anxiety (preferably an alternate form to avoid
memory of the pretest); this measure serves as the DV. The goal of statistical analysis
is to test the null hypothesis that all three groups have the same mean test anxiety
after treatment after adjusting for preexisting differences in test anxiety.

The second use of ANCOVA commonly occurs in nonexperimental situations
where subjects cannot be randomly assigned to treatments. ANCOVA is used as a
statistical matching procedure, although interpretation is fraught with difficulty, as
discussed in Section 8.3.1. ANCOVA is used primarily to adjust group means to what
they would be if all subjects scored identically on the covariate(s). Differences between
subjects on covariates are removed so that, presumably, the only differences that
remain are related to the effects of the grouping IV(s). (Differences could also, of
course, be due to attributes that have *not* been used as covariates.)

This second application of ANCOVA is primarily for descriptive model building:

the covariate enhances prediction of the DV, but there is no implication of causality. If the research question to be answered involves causality, ANCOVA is no substitute for running an experiment.

As an example, suppose we are looking at regional differences in political attitudes where the DV is some measure of liberalism-conservatism. Regions of the United States form the IV, say, Northeast, South, Midwest, and West. Two variables that are expected to vary with political attitude and with geographical region are socioeconomic status and age. These two variables serve as covariates. The statistical analysis tests the null hypothesis that political attitudes do not differ with geographical region after adjusting for socioeconomic status and age. However, if age and socioeconomic differences are inextricably tied to geography, adjustment for them is not realistic. And, of course, there is no implication that political attitudes are caused in any way by geographic region. Further, unreliability in measurement of the covariate and the DV-covariate relationship may lead to over- or underadjustment of scores and means and, therefore, to misleading results. These issues are discussed in greater detail throughout the chapter.

In the third major application of ANCOVA, discussed more fully in Chapter 9, ANCOVA is used to interpret IV differences when several DVs are used in MANOVA. After a multivariate analysis of variance, it is frequently desirable to assess the contribution of the various DVs to significant differences among IVs. One way to do this is to test DVs, in turn, with the effects of other DVs removed. Removal of the effects of other DVs is accomplished by treating them as covariates in a procedure called a stepdown analysis.

The statistical operations are identical in all three major applications of ANCOVA. As in ANOVA, variance in scores is partitioned into variance due to differences between groups and variance due to differences within groups. Squared differences between scores and various means are summed (see Chapter 3) and these sums of squares, when divided by appropriate degrees of freedom, provide estimates of variance attributable to different sources (main effects of IVs, interactions between IVs, and error). Ratios of variances then provide tests of hypotheses about the effects of IVs on the DV.

However, in ANCOVA, the regression of the DV on one or more covariates is estimated first. Then DV scores and means are adjusted to remove the linear effects of the covariate(s) before analysis of variance is performed on these adjusted values.

Lee (1975) presents an intuitively appealing illustration of the manner in which ANCOVA reduces error variance in a one-way between-subjects design with three levels of the IV (Figure 8.1). Note that the vertical axis on the right-hand side of the figure illustrates scores and group means in ANOVA. The error term is computed from the sum of squared deviations of DV scores around their associated group means. In this case, the error term is substantial because there is considerable spread in scores within each group.

When the same scores are analyzed in ANCOVA, a regression line is found first that relates the DV to the covariate. The error term is based on the (sum of squared) deviations of the DV scores from the regression line running through each group mean instead of from the means themselves. Consider the score in the lower left-hand corner of Figure 8.1. The score is near the regression line (a small deviation

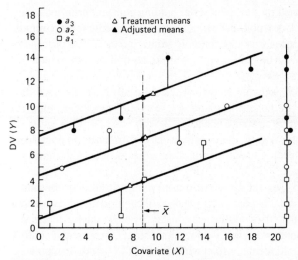

Figure 8.1 Plot of hypothetical data. The straight lines with common slope are those that best fit the data for the three treatments. The data points are also plotted along the single vertical line on the right. (From *Experimental Design and Analysis* by Wayne Lee. W. H. Freeman and Company. Copyright © 1975. Reprinted with permission.)

for error in ANCOVA) but far from the mean for its own group (a large deviation for error in ANOVA). As long as the slope of the regression lines is not zero, ANCOVA produces a smaller sum of squares for error than ANOVA. If the slope is zero, error sum of squares is the same as in ANOVA but error mean square is larger because covariates use up degrees of freedom.

Covariates can be used in all ANOVA designs—factorial between-subjects, within-subjects,[2] mixed within-between, nonorthogonal, and so on. In only a few programs, however, are analyses of these more complex designs readily available. Similarly, specific comparisons and trend analysis are possible in ANCOVA but not always readily available through the programs.

8.2 KINDS OF RESEARCH QUESTIONS

As with ANOVA, the question in ANCOVA is whether mean differences in the DV between groups are larger than expected by chance. In ANCOVA, however, one gets a more precise look at the IV-DV relationship when the effect of covariates is removed.

8.2.1 Main Effects of IVs

Holding all else constant, are changes in behavior associated with different levels of an IV larger than expected through random fluctuations occurring by chance? For

---

[2] If a covariate is measured only once, it does not provide adjustment to the levels of a within-subjects effect because each subject has the same value on the covariate for all levels of the effect. However, if a covariate is measured repeatedly along with the DV, the covariate provides adjustment to the within-subjects effect.

example, is test anxiety affected by treatment, after holding constant prior individual differences in test anxiety? Does political attitude vary with geographical region, after holding constant differences in socioeconomic status and age? The procedures described in Section 8.4 answer this question by testing the null hypothesis that the IV has no systematic effect on the DV.

With more than one IV, separate statistical tests are available for each one. Suppose there is a second IV in the political attitude example, for example, religious affiliation, with four groups: Protestant, Catholic, Jewish, and None-or-other. In addition to the test of geographic region, there is also a test of differences in liberalism-conservatism associated with religious affiliation after adjustment for differences in socioeconomic status and age.

In experimental design, several devices are used to "hold all else constant." One of them, the topic for this chapter, is to measure the influence of extraneous variables (covariates) and then hold their influence constant by adjusting for differences on them in ANCOVA. A second procedure is to institute strict experimental controls over extraneous variables to hold them constant while levels of IV(s) are manipulated. A third procedure is to turn a potentially effective extraneous variable into another IV and hold it constant by crossing the levels of the more interesting IV with it in factorial design. The last procedure is to randomize the assignment of levels of the extraneous variable (such as time of day) throughout the levels of the primary IV in the hopes of distributing the effects of the extraneous variable equally among treatments. A major consideration in experimental design is which of these procedures is more effective or more feasible for a given research program.

8.2.2 Interactions among IVs

Holding all else constant, does change in behavior over levels of one IV depend on levels of another IV? That is, do IVs interact in their effect on behavior? (See Chapter 3 for a discussion of interaction.) For the political attitude example where religious affiliation is added as a second IV, are differences in liberalism-conservatism over geographic region the same for all religions, after adjusting for socioeconomic status and age?

Tests of interactions, while interpreted differently from main effects, are statistically similar, as demonstrated in Section 8.7. With more than two IVs, multiple interactions are generated. Except for common error terms, each interaction is tested separately from other interactions and from main effects. The tests are independent when sample sizes in all groups are equal and the design is balanced.

8.2.3 Specific Comparisons and Trend Analysis

When statistically significant effects are found in a design with more than two levels of a single IV, it is often desirable to evaluate the nature of the differences. Which groups differ significantly from each other? Or, is there a simple trend over sequential levels of an IV? For the test anxiety example, we ask whether (1) the two treatment groups are more effective in reducing test anxiety than the waiting-list control, after adjusting for individual differences in test anxiety; and if (2) among the two treatment groups, desensitization is more effective than relaxation training in reducing test anxiety, again after adjusting for preexisting differences in test anxiety?

These two questions could be asked in planned comparisons *instead* of answering, through routine ANCOVA, the omnibus question of whether means are the same for all three levels of the IV. Or, with some loss in sensitivity, these two questions could be asked post hoc *after* finding a main effect of the IV in ANCOVA. Planned and post hoc comparisons are discussed in Section 8.5.2.3.

8.2.4 Effects of Covariates

Analysis of covariance is based on a linear regression between covariate(s) and the DV, but there is no guarantee that the regression is reliable. The regression can be evaluated statistically by testing the covariate(s) as a source of variance in DV scores, as discussed in Section 8.5.3. For instance, consider the test anxiety example where the covariate is a pretest and the DV a posttest. To what extent is it possible to predict posttest anxiety from pretest anxiety, ignoring effects of differential treatment?

8.2.5 Strength of Association

If a main effect or interaction of IVs is reliably associated with changes in behavior, the next logical question is, How much? How much of the variance in the adjusted DV scores—adjusted for the covariate(s)—is associated with the IV(s)? In the test anxiety example, if a main effect is found between the means for desensitization, relaxation training, and control group, one then asks, What proportion of variance in the adjusted test anxiety scores is attributed to the IV? Simple descriptions of strength of association are demonstrated in Sections 8.4 and 8.5.2.4.

8.2.6 Adjusted Marginal and Cell Means

If any main effects or interactions are statistically significant, what is the estimated population parameter (adjusted mean) for each level of the IV? How do group scores differ, on the average, on the DV, after adjustment for covariates? For the test anxiety example, if there is a main effect of treatment, what is the average adjusted posttest anxiety score in each of the three groups? The reporting of parameter estimates is demonstrated in Section 8.7.

8.3 LIMITATIONS TO ANALYSIS OF COVARIANCE

8.3.1 Theoretical Issues

As with ANOVA, the statistical test in no way assures that changes in the DV were caused by the IV. The inference of causality is a logical rather than a statistical problem that depends on the manner in which subjects are assigned to levels of the IV(s) and the controls used in the research. The statistical test is available to test hypotheses from both nonexperimental and experimental research, but only in the latter case can attribution of causality be justified.

Choice of covariates is a logical exercise as well. As a general rule, one wants a very small number of covariates, all correlated with the DV and none correlated with each other. The goal is to obtain maximum adjustment of the DV with minimum loss of degrees of freedom for error. Calculation of the regression of the DV on the covariates(s) results in the loss of one degree of freedom for error for each covariate. Thus *the gain in power from decreased sum of squares for error may be offset by the loss in degrees of freedom.* When there is a substantial correlation between the DV and a covariate, the gain in reduced error variance offsets the loss of a degree of freedom. With multiple covariates, however, a point of diminishing returns is quickly reached, especially if the covariates correlate with one another (see Section 8.5.4).

In experimental work, a frequent caution is that the covariates must be independent of treatment. It is suggested that data on covariates be gathered before treatment is administered. Violation of this precept results in removal of some portion of the effect of the IV on the DV—that portion of the effect that is associated with the covariate. In this situation, adjusted group means may be closer together than unadjusted means. Further, the adjusted means may be difficult to interpret.

In nonexperimental work, adjustment for prior differences in means associated with covariates is appropriate. If the adjustment reduces mean differences on the DV, so be it—unadjusted differences reflect unwanted influences (other than the IV) on the DV. In other words, mean differences on a covariate associated with an IV are quite legitimately corrected for as long as the covariate differences are not caused by the IV (Overall and Woodward, 1977).

When ANCOVA is used to evaluate a series of DVs after MANOVA, independence of the "covariates" and the IV is not required. Because covariates are actually DVs, it is expected that they be dependent on the IV.

In all uses of ANCOVA, however, adjusted means must be interpreted with some caution. The mean DV score after adjustment for covariates may not correspond to any situation in the real world. Adjusted means are the means that would have occurred *if* all subjects had the same scores on the covariates. Especially in nonexperimental work, such a situation may be so unrealistic as to make the adjusted values meaningless.

Sources of bias in ANCOVA are many and subtle, and can produce either under- or overadjustment of the DV. At best, the nonexperimental use of ANCOVA allows you to look at IV-DV relationships (noncausal) adjusted for the effects of covariates, as measured. If causal inference regarding effects is desired, there is no substitute for random assignment of subjects. Don't expect ANCOVA to permit causal inference of treatment effects with nonequivalent groups. If random assignment is absolutely impossible, or if it breaks down because of nonrandom loss of subjects, be sure to thoroughly ground yourself in the literature regarding use of ANCOVA in such cases, starting with Cooke and Campbell (1979).

Limitations to generalizability apply to ANCOVA as they do to ANOVA or any other statistical test. One can generalize only to those populations from which a random sample is taken. ANCOVA may, in some limited sense, sometimes adjust for a failure to randomly assign the sample to groups, but it does not affect the relationship between the sample and the population to which one can generalize.

8.3.2 Practical Issues

The ANCOVA model assumes reliability of covariates, linearity between covariates and between covariates and the DV, and homogeneity of regression in addition to the usual ANOVA assumptions of normality and homogeneity of variance.

8.3.2.1 Unequal Sample Sizes and Missing Data If scores on the DV are missing in a between-subjects ANCOVA, this is reflected as the problem of unequal n. That is, combinations of IV levels do not contain equal numbers of cases. Consult Section 8.5.2.2 for strategies for dealing with unequal sample sizes.

If some subjects are missing scores on covariate(s), or if, in repeated measures ANOVA, some DV scores are missing for some subjects, this is more clearly a missing-data problem. Consult Chapter 4 for methods of dealing with missing data.

8.3.2.2 Outliers Within each group, univariate outliers can occur in the DV or any one of the covariates. Multivariate outliers can occur in the space of the DV and covariate(s). Multivariate outliers among DV and covariate(s) can produce heterogeneity of regression (Section 8.3.2.7), leading to rejection of ANCOVA or at least unreasonable adjustment of the DV. If the covariates are serving as a convenience in most analyses, rejection of ANCOVA because there are multivariate outliers is hardly convenient.

Consult Chapter 4 for methods dealing with univariate outliers in the DV or covariate(s) and multivariate outliers among the DV and covariate(s). Tests for univariate outliers within each group through BMDP7D and multivariate outliers within each group through BMDPAM are demonstrated in Section 8.7.1.

8.3.2.3 Multicollinearity and Singularity If there are multiple covariates, they should not be highly correlated with each other. Highly correlated covariates should be eliminated, both because they add no adjustment to the DV over that of other covariates and because of potential computational difficulties if they are singular or multicollinear. Programs for ANCOVA in the BMDP series automatically guard against multicollinearity or singularity of covariates, as do the general linear hypothesis programs of SAS and SYSTAT; but the SPSS programs do not. Consult Chapter 4 for methods of testing for multicollinearity and singularity among multiple covariates. A test for multicollinearity and singularity is demonstrated in Section 8.7.1.5.

8.3.2.4 Normality As in all ANOVA, it is assumed that the sampling distributions of means, as described in Chapter 3, are normal within each group. Note that it is the sampling distributions of means and not the raw scores within each cell that need to be normally distributed. Without knowledge of population values, or production of actual sampling distributions of means, there is no way to test this assumption.

However, the central limit theorem suggests that, with large samples, sampling distributions are normal even if raw scores are not. *With relatively equal sample sizes in groups, no outliers, and two-tailed tests, robustness is expected with 20 degrees of freedom for error.* (See Chapter 3 for calculation of error degrees of freedom.)

Larger samples are necessary for one-tailed tests. With small, unequal samples or with outliers present, it may be necessary to consider data transformation (cf. Chapter 4).

8.3.2.5 Homogeneity of Variance

Just as in ANOVA, it is assumed in ANCOVA that the variance of DV scores within each cell of the design is a separate estimate of the same population variance. Because of the robustness of the analysis to violation of this assumption as long as there are no outliers, it is typically unnecessary to test for homogeneity of variance if samples are large and the following checks are met.

Harris (1975) suggests the following checks to ensure robustness. *For two-tailed tests, sample sizes should preferably be equal, but in no event should the ratio of largest to smallest sample size for groups be greater than 4 : 1.* Then examine the variances (standard deviations squared) within each cell *to assure that the ratio between largest and smallest variance is no greater than approximately 20 : 1.* If one-tailed tests are used, use more conservative criteria because the test is less robust.

If any of these conditions is violated, a formal test for homogeneity of variance is called for as described in any standard ANOVA text (e.g., Winer, 1971; Keppel, 1982) and as available in some canned programs (see Table 8.14). Most of these tests are sensitive to nonnormality as well as heterogeneity of variance and can lead to overly conservative rejection of the use of ANCOVA. BMDP7D offers a test of homogeneity of variance that is typically not sensitive to departures from normality. Major heterogeneity should, of course, always be reported by providing within-cell standard deviations.

In any event, gross violations of homogeneity can be corrected by transformation of the DV scores (cf. Chapter 4). Interpretation, however, is then limited to the transformed scores. Add to this the difficulty of interpreting adjusted means, and interpretation becomes increasingly speculative.

8.3.2.6 Linearity

The ANCOVA model is based on the assumption that the relationship between each covariate and the DV and the relationships among pairs of covariates are linear. As with multiple regression (Chapter 5), violation of this assumption reduces the power of the statistical test; errors in statistical decision making are in a conservative direction. Error terms are not reduced as fully as they might be, optimum matching of groups is not achieved, and group means are incompletely adjusted.

To begin screening for linearity, produce residuals plots through BMDP1V, SAS GLM, or SPSS^x MANOVA, and interpret them as described in Section 5.3.2.4. *If there is indication of serious curvilinearity, examine within-cell scatterplots of the DV with each covariate and all covariates with one another.* Any of the programs that produce scatterplots (e.g., BMDP6D and SPSS^x SCATTERGRAM or PLOT) can be used to evaluate linearity. Relationships between the DV and each covariate can be produced as one of the sets of plots in BMDP1V.

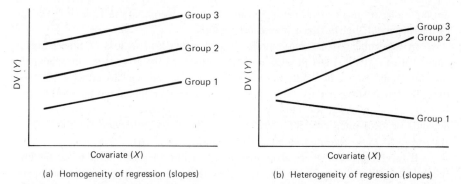

(a) Homogeneity of regression (slopes)   (b) Heterogeneity of regression (slopes)

Figure 8.2   DV-covariate regression lines for three groups plotted on the same coordinates for conditions of (a) homogeneity and (b) heterogeneity of regression.

Where curvilinearity is indicated,[3] it may be corrected by transforming some of the variables. Or, because of the difficulties in interpreting transformed variables, you may consider eliminating a covariate that produces nonlinearity. Or, a higher-order power of the covariate can be used to produce an alternative covariate that incorporates nonlinear influences.

8.3.2.7 Homogeneity of Regression   Adjustment of scores in ANCOVA is made on the basis of an average within-cell regression coefficient. The assumption is that the slope of the regression of the DV on the covariate(s) within each cell is an estimate of the same population regression coefficient, that is, that the slopes are equal for all cells.

Heterogeneity of regression implies that there is an interaction between IV(s) and covariate(s). The interaction means that the relationship between the covariate(s) and the DV is different at different levels of the IV(s), or that the covariate adjustment that is needed for various cells is different. Figure 8.2 illustrates, for three groups, perfect homogeneity of regression (equality of slopes) and extreme heterogeneity of regression (inequality of slopes).

If a between-subjects design is used, test the assumption of homogeneity of regression according to procedures described in Section 8.5.1. If any other design is used, and interaction between IVs and covariates is suspected, ANCOVA is inappropriate. If there is no reason to suspect an IV-covariate interaction with complex designs, it is probably safe to proceed with ANCOVA on the basis of the robustness of the model. Alternatives to ANCOVA are discussed in Section 8.5.5.

8.3.2.8 Reliability of Covariates   It is assumed in ANCOVA that covariates are measured without error; that they are perfectly reliable. In the case of such variables as sex and age, the assumption can usually be justified. With self-report of demographic variables, and with variables measured psychometrically, such assumptions

---

[3] Tests for deviation from linearity are available (cf. Winer, 1971), but there seems to be little agreement as to the appropriate test or the seriousness of significant deviation in the case of ANCOVA. Therefore formal tests are not recommended.

are not so easily made. And variables such as body weight are reliable at the point of measurement, but fluctuate over short periods.

In experimental research, unreliable covariates lead to loss of power and a conservative statistical test through underadjustment of the error term. In nonexperimental applications, however, unreliable covariates can lead to either under- or overadjustment of the means. Group means may be either spread too far apart (Type I error) or compressed too closely together (Type II error). The degree of error depends on how unreliable the covariates are. *In nonexperimental research, limit covariates to those that can be measured reliably ($r_{xx} > .8$).*

If fallible covariates are absolutely unavoidable, they can be adjusted for unreliability. However, there is no one procedure that produces appropriate adjustment under all conditions nor is there even agreement about which procedure is most appropriate for which application. Because of this disagreement, and because procedures for correction require use of sophisticated programs, they are not covered in this book. The interested reader is referred to Cohen and Cohen (1975) (who recommend a strategy in which the analysis is done both with and without correction for unreliability and interpretation is based on the outcome of both analyses) or to other procedures as discussed by St. Pierre (1978).

8.4 FUNDAMENTAL EQUATIONS FOR ANALYSIS OF COVARIANCE

In the simplest application of analysis of covariance there is a DV score, a grouping variable (IV), and a covariate score for each subject. An example of such a small hypothetical data set is in Table 8.1. The IV is type of treatment given to a sample of nine learning-disabled children. Three children are randomly assigned to one of two treatment groups or to a control group, so that sample size of each group is three. For each of the nine children, two scores are measured, covariate and DV. The covariate is a pretest score on the reading subtest of the Wide Range Achievement Test (WRAT-T), measured before the experiment begins. The DV is a posttest score on the same test measured at the end of the experiment.

The research question is, Does differential treatment of learning-disabled children affect reading scores, after adjusting for differences in the children's prior reading

TABLE 8.1 SMALL SAMPLE DATA FOR ILLUSTRATION OF ANALYSIS OF COVARIANCE

| | Groups | | | | | |
| | Treatment 1 | | Treatment 2 | | Control | |
	Pre	Post	Pre	Post	Pre	Post
	85	100	86	92	90	95
	80	98	82	99	87	80
	92	105	95	108	78	82
Sums	257	303	263	299	255	257

ability? The sample size is, of course, inadequate for a realistic tests of the research question but is convenient for illustration of the techniques in ANCOVA. The reader is encouraged to follow this example with hand calculations. Computer analyses using four popular programs follow this section.

Equations for ANCOVA are an extension of those for ANOVA. As discussed in Chapter 3, averaged squared deviations from means—variances—are partitioned into variance associated with different levels of the IV (between-groups variance) and variance associated with differences in scores within groups (unaccounted for or error variance). Variance is partitioned by summing and squaring differences between scores and various means.

$$\sum_i \sum_j (Y_{ij} - GM)^2 = n \sum_j (\bar{Y}_j - GM)^2 + \sum_i \sum_j (Y_{ij} - \bar{Y}_j)^2 \qquad (8.1)$$

or

$$SS_{total} = SS_{bg} + SS_{wg}$$

The total sum of squared differences between scores on Y (the DV) and the grand mean (GM) is partitioned into two components: sum of squared differences between group means ($\bar{Y}_j$) and the grand mean (i.e., systematic or between-groups variability); and sum of squared differences between individual scores (Y_{ij}) and their respective group means.

In ANCOVA, there are two additional partitions. First, the differences in co-variate scores are partitioned into between- and within-groups sums of squares:

$$SS_{total(x)} = SS_{bg(x)} + SS_{wg(x)} \qquad (8.2)$$

The total sum of squared differences on the covariate (X) is partitioned into differences between groups and differences within groups.

Similarly, the covariance (the linear relationship between the DV and the covariate) is partitioned into sums of products associated with covariance between groups and sums of products associated with covariance within groups.

$$SP_{total} = SP_{bg} + SP_{wg} \qquad (8.3)$$

A *sum of squares* involves taking deviations of scores from means (e.g., $X_{ij} - \bar{X}_j$ or $Y_{ij} - \bar{Y}_j$), squaring them, and then summing the squares over all subjects; a *sum of products* involves taking two deviations from the same subject (e.g., both $X_{ij} - \bar{X}_j$ and $Y_{ij} - \bar{Y}_j$), multiplying them together (instead of squaring), and then summing the products over all subjects. As discussed in Chapter 3, the means that are used to produce the deviations are different for the different sources of variance in the research design.

The partitions for the covariate (Equation 8.2) and the partitions for the association between the covariate and DV (Equation 8.3) are used to adjust the sums of squares for the DV according to the following equations:

$$SS'_{bg} = SS_{bg} - \left[\frac{(SP_{bg} + SP_{wg})^2}{SS_{bg(x)} + SS_{wg(x)}} - \frac{(SP_{wg})^2}{SS_{wg(x)}} \right] \qquad (8.4)$$

The adjusted between-groups sum of squares (SS'_{bg}) is found by subtracting from the unadjusted between-groups sum of squares a term based on sums of squares associated with the covariate, X, and sums of products for the linear relationship between the DV and the covariate.

$$SS'_{wg} = SS_{wg} - \frac{(SP_{wg})^2}{SS_{wg(x)}} \qquad (8.5)$$

The adjusted within-groups sum of squares (SS'_{wg}) is found by subtracting from the unadjusted within-groups sum of squares a term based on within-groups sums of squares and products associated with the covariate and with the linear relationship between the DV and the covariate.

This can be expressed in an alternate form. The adjustment for each score consists of subtracting from the deviation of that score from the grand mean a value that is based on the deviation of the corresponding covariate from the grand mean on the covariate, weighted by the regression coefficient for predicting the DV from the covariate. Symbolically, for an individual score:

$$(Y - Y') = (Y - GM_y) - B_{y.x}(X - GM_x) \qquad (8.6)$$

The adjustment for any subject's score $(Y - Y')$ is obtained by subtracting from the unadjusted deviation score $(Y - GM_y)$ the individual's deviation on the covariate $(X - GM_x)$ weighted by the regression coefficient, $B_{y.x}$.

Once adjusted sums of squares are found, mean squares are found as usual by dividing by appropriate degrees of freedom. The only difference in degrees of freedom between ANOVA and ANCOVA is that in ANCOVA the error degrees of freedom is reduced by one for each covariate because a degree of freedom is used up in estimating each regression coefficient.

For computational convenience, raw score equations rather than deviation equations are provided in Table 8.2 for Equations 8.4 and 8.5. Note that these equations apply only to equal-n designs.

When applied to the data in Table 8.1, the six sums of squares and products are as follows:

$$SS_{bg} = \frac{(303)^2 + (299)^2 + (257)^2}{3} - \frac{(859)^2}{(3)(3)} = 432.889$$

$$SS_{wg} = (100)^2 + (98)^2 + (105)^2 + (92)^2 + (99)^2 + (108)^2$$

$$+ (95)^2 + (80)^2 + (82)^2 - \frac{(303)^2 + (299)^2 + (257)^2}{3} = 287.333$$

$$SS_{bg(x)} = \frac{(257)^2 + (263)^2 + (255)^2}{3} - \frac{(775)^2}{(3)\,(3)} = 11.556$$

$$SS_{wg(x)} = (85)^2 + (80)^2 + (92)^2 + (86)^2 + (82)^2 + (95)^2$$

$$+ (90)^2 + (87)^2 + (78)^2 - \frac{(257)^2 + (263)^2 + (255)^2}{3} = 239.333$$

$$SP_{bg} = \frac{(257)(303) + (263)(299) + (255)(257)}{3} - \frac{(775)(859)}{(3)(3)} = 44.889$$

$$SP_{wg} = (85)(100) + (80)(98) + (92)(105) + (86)(92) + (82)(99)$$

$$+ (95)(108) + (90)(95) + (87)(80) + (78)(82)$$

$$- \frac{(257)(303) + (263)(299) + (255)(257)}{3} = 181.667$$

These values are conveniently summarized in a sum-of-squares and cross-products matrix (cf. Chapter 1). For the between-groups sums of squares and cross products,

$$\mathbf{S}_{bg} = \begin{bmatrix} 11.556 & 44.889 \\ 44.889 & 432.889 \end{bmatrix}$$

The first entry (first row and first column) is the sum of squares for the covariate and the last (second row, second column) is the sum of squares for the DV; the sum of products is shown off-diagonal. For the within-groups sums of squares and cross products, arranged similarly,

$$\mathbf{S}_{wg} = \begin{bmatrix} 239.333 & 181.667 \\ 181.667 & 287.333 \end{bmatrix}$$

From these values, the adjusted sums of squares are found as per Equations 8.4 and 8.5.

$$SS'_{bg} = 432.889 - \left[\frac{(44.889 + 181.667)^2}{11.556 + 239.333} - \frac{(181.667)^2}{239.333} \right] = 366.202$$

$$SS'_{wg} = 287.333 - \frac{(181.667)^2}{239.333} = 149.438$$

These values are entered into a source table such as Table 8.3. Degrees of freedom for between-groups variance are $k - 1$, and for within-groups variance $N - k - c$. (N = total sample size, k = number of levels of the IV, and c = number of covariates.)

TABLE 8.2 COMPUTATION EQUATIONS FOR SUMS OF SQUARES AND CROSS PRODUCTS IN ONE-WAY BETWEEN-SUBJECTS ANALYSIS OF COVARIANCE

Source	Sum of squares for Y (DV)	Sum of squares for X (covariate)	Sum of products
Between groups	$SS_{bg} = \dfrac{\sum^k\left(\sum^n Y\right)^2}{n} - \dfrac{\left(\sum^k\sum^n Y\right)^2}{kn}$	$SS_{bg(x)} = \dfrac{\sum^k\left(\sum^n X\right)^2}{n} - \dfrac{\left(\sum^k\sum^n X\right)^2}{kn}$	$SP_{bg} = \dfrac{\sum^k\left(\sum^n Y\right)\left(\sum^n X\right)}{n} - \dfrac{\left(\sum^k\sum^n Y\right)\left(\sum^k\sum^n X\right)}{kn}$
Within groups	$SS_{wg} = \sum^k\sum^n Y^2 - \dfrac{\sum^k\left(\sum^n Y\right)^2}{n}$	$SS_{wg(x)} = \sum^k\sum^n X^2 - \dfrac{\sum^k\left(\sum^n X\right)^2}{n}$	$SP_{wg} = \sum^k\sum^n (XY) - \dfrac{\sum^k\left(\sum^n Y\right)\left(\sum^n X\right)}{n}$

Note: k = number of groups; n = number of subjects per group.

TABLE 8.3 ANALYSIS OF COVARIANCE FOR DATA OF
TABLE 8.1

Source of variance	Adjusted SS	df	MS	F
Between groups	366.202	2	183.101	6.13*
Within groups	149.439	5	29.888	

* $p < .05$.

As usual, mean squares are found by dividing sums of squares by appropriate degrees of freedom. The hypothesis that there are no differences among groups is tested by the F ratio formed by dividing the adjusted mean square between groups by the adjusted mean square within groups.

$$F = \frac{183.101}{29.888} = 6.13$$

From a standard F table, we find that the obtained F of 6.13 exceeds the critical F of 5.79 at $\alpha = .05$ with 2 and 5 df. We therefore reject the null hypothesis of no change in WRAT reading scores associated with the three treatment levels, after adjustment for pretest reading scores.

If a statistically significant effect of the IV on the adjusted DV scores is found, the strength of association for the effect is assessed using η^2.

$$\eta^2 = \frac{SS'_{bg}}{SS'_{bg} + SS'_{wg}} \tag{8.7}$$

For the sample data,

$$\eta^2 = \frac{366.202}{366.202 + 149.438} = .71$$

We conclude that 71% of the variance in the adjusted DV scores (WRAT-R) is associated with treatment.

ANOVA for the same data appears in Table 8.4. Compare the results with those of ANCOVA in Table 8.3. ANOVA produces larger sums of squares, especially for the error term. There is also one more degree of freedom for error because there is no covariate. However, in ANOVA the null hypothesis is retained while in ANCOVA

TABLE 8.4 ANALYSIS OF VARIANCE FOR DATA OF
TABLE 8.1

Source of variance	SS	df	MS	F
Between groups	432.889	2	216.444	4.52
Within groups	287.333	6	47.889	

it is rejected. Thus, use of the covariate has reduced the "noise" in the error term for this example.

ANCOVA extends to factorial and repeated-measures designs (Section 8.5.2.1), unequal n (Section 8.5.2.2), and multiple covariates (Section 8.5.3). In all cases, the analysis is done on adjusted, rather than raw, DV scores.

Tables 8.5 through 8.8 demonstrate analyses of covariance of this small data set using BMDP2V, SPSS^x ANOVA, SYSTAT MGLH, and SAS GLM. Minimal

TABLE 8.5 SETUP AND SELECTED BMDP2V OUTPUT FOR ANALYSIS OF COVARIANCE ON SAMPLE DATA IN TABLE 8.1

```
            /PROBLEM    TITLE IS 'SMALL SAMPLE ANCOVA THROUGH BMDP2V'.
            /INPUT      VARIABLES ARE 4.  FORMAT IS FREE.
                        FILE = TAPE70.
            /VARIABLE   NAMES ARE SUBJNO,TREATMNT,PRE,POST.
            /GROUP      CODES(TREATMNT) ARE 1,2,3.
                        NAMES(TREATMNT) ARE TREAT1,TREAT2,CONTROL.
            /DESIGN     DEPENDENT IS POST.
                        COVARIATE IS PRE.
                        GROUPING IS TREATMNT.
            /END

GROUP STRUCTURE

TREATMNT    COUNT
TREAT1        3
TREAT2        3
CONTROL       3

        CELL MEANS FOR  1-ST COVARIATE

                                           MARGINAL
TREATMNT=  TREAT1      TREAT2      CONTROL

PRE        85.66667    87.66667    85.00000    86.11111

COUNT         3           3           3           9

STANDARD DEVIATIONS  FOR  1-ST COVARIATE

TREATMNT=  TREAT1      TREAT2      CONTROL

PRE         6.02771     6.65833     6.24500

        CELL MEANS FOR  1-ST DEPENDENT VARIABLE

                                           MARGINAL
TREATMNT=  TREAT1      TREAT2      CONTROL

POST       101.00000   99.66667    85.66667    95.44444

COUNT         3           3           3           9

STANDARD DEVIATIONS  FOR  1-ST DEPENDENT VARIABLE

TREATMNT=  TREAT1      TREAT2      CONTROL

POST        3.60555     8.02081     8.14453

ANALYSIS OF VARIANCE FOR  1-ST
DEPENDENT VARIABLE - POST

        SOURCE           SUM OF    DEGREES OF    MEAN       F     TAIL      REGRESSION
                         SQUARES    FREEDOM     SQUARE            PROB.    COEFFICIENTS

        TREATMNT         366.20123      2      183.10061   6.13  0.0452
        PRE              137.89461      1      137.89461   4.61  0.0845     0.75905
   1    ERROR            149.43872      5       29.88774

ADJUSTED CELL MEANS  FOR  1-ST DEPENDENT VARIABLE

TREATMNT=  TREAT1      TREAT2      CONTROL

POST       101.33736   98.48592    86.51006
```

TABLE 8.6 SETUP AND SPSS' ANOVA OUTPUT FOR ANALYSIS OF
COVARIANCE ON SAMPLE DATA IN TABLE 8.1

```
        TITLE          SMALL SAMPLE ANALYSIS OF COVARIANCE
        FILE HANDLE    TAPE70
        DATA LIST      FILE=TAPE70    FREE
                       /SUBJNO,TREATMNT,PRE,POST
        VALUE LABELS   TREATMNT 1 'TREAT1' 2 'TREAT2' 3 'CONTROL'
        ANOVA          POST BY TREATMNT(1,3) WITH PRE


        * * *   A N A L Y S I S   O F   V A R I A N C E   * * *

              POST
        BY    TREATMNT
        WITH  PRE
```

SOURCE OF VARIATION	SUM OF SQUARES	DF	MEAN SQUARE	F	SIGNIF OF F
COVARIATES	204.582	1	204.582	6.845	.047
PRE	204.582	1	204.582	6.845	.047
MAIN EFFECTS	366.201	2	183.101	6.126	.045
TREATMNT	366.201	2	183.101	6.126	.045
EXPLAINED	570.784	3	190.261	6.366	.037
RESIDUAL	149.439	5	29.888		
TOTAL	720.222	8	90.028		

output is requested for each of the programs, although much more is available upon request.

In BMDP2V, a DESIGN paragraph is used to specify the DV (DEPENDENT), the COVARIATE(s), and the IV (GROUPING), as seen in Table 8.5. The GROUP paragraph defines levels of IV(s). For all covariates and DVs, cell and marginal means, standard deviations, and sample sizes are given. The variable is identified at the left of each of these tables—PRE for the covariate and POST for the DV in this example.

These descriptive statistics are followed by a standard ANOVA source table where the first effect is the IV, TREATMNT, adjusted for the covariate(s). The second source listed (PRE) is the covariate. The test is whether the covariate significantly adjusts the DV after effects of the IV(s) are removed. Along with the significance test, the unstandardized regression coefficient (B weight, see Equation 8.6) is provided. The ERROR term ends the table. The one to the left of that line indicates that this is the first (and in this case only) error term in the analysis.

TABLE 8.7 SETUP AND SYSTAT MGLH OUTPUT FOR ANALYSIS OF
COVARIANCE ON SAMPLE DATA IN TABLE 8.1

```
        USE TAPE70
        OUTPUT=OUTFILE
        CATEGORY TREATMNT=3
        MODEL POST = CONSTANT + TREATMNT + PRE
        ESTIMATE
```

DEP VAR: POST   N: 9   MULTIPLE R: .890   SQUARED MULTIPLE R: .793

ANALYSIS OF VARIANCE

SOURCE	SUM-OF-SQUARES	DF	MEAN-SQUARE	F-RATIO	P
TREATMNT	366.201	2	183.101	6.126	0.045
PRE	137.895	1	137.895	4.614	0.084
ERROR	149.439	5	29.888		

TABLE 8.8 INPUT AND SAS GLM OUTPUT FOF ANALYSIS OF COVARIANCE ON SAMPLE DATA IN TABLE 8.1

```
DATA SSAMPLE;
INFILE TAPE70;
INPUT CASENO TREATMNT PRE POST;
PROC GLM;
   CLASS TREATMNT;
   MODEL POST = PRE TREATMNT
```

GENERAL LINEAR MODELS PROCEDURE

CLASS LEVEL INFORMATION

CLASS	LEVELS	VALUES
TREATMNT	3	1 2 3

NUMBER OF OBSERVATIONS IN DATA SET = 9

DEPENDENT VARIABLE: POST

SOURCE	DF	SUM OF SQUARES	MEAN SQUARE	F VALUE	PR > F	R-SQUARE	C.V.
MODEL	3	570.78350356	190.26116785	6.37	0.0369	0.792510	5.7279
ERROR	5	149.43871866	29.88774373		ROOT MSE		POST MEAN
CORRECTED TOTAL	8	720.22222222			5.46696842		95.44444444

SOURCE	DF	TYPE I SS	F VALUE	PR > F	DF	TYPE III SS	F VALUE	PR > F
PRE	1	204.58227537	6.85	0.0473	1	137.89461467	4.61	0.0845
TREATMNT	2	366.20122819	6.13	0.0452	2	366.20122819	6.13	0.0452

334

Finally, adjusted cell means are provided. If required, adjusted marginal means (with appropriate adjustment for unequal sample sizes) can be calculated from these adjusted cell means for factorial designs, as illustrated in Section 8.7.2.

In SPSS[x] ANOVA (Table 8.6) the DV (POST) is specified in the ANOVA paragraph followed BY the IV (TREATMNT, with the range of levels in parentheses) WITH the covariate (PRE). The output is an ANOVA source table with some re-dundancies. First, COVARIATES are tested as a group if there is more than one covariate. (In this example the overall test is also the test of the single covariate, PRE.) Unlike BMDP2V, the test of the covariate(s) is not adjusted for TREATMNT if default values are used and, in this case, produces a significant result for PRE which BMDP2V does not. Because there is only one IV in this example the test of the single main effect for TREATMNT (adjusted for the covariate) is identical to the test for all MAIN EFFECTS considered together. The source of variance labeled EXPLAINED provides a test of all known sources of variability: covariates, main effects and interactions. The source labeled RESIDUAL is the error term. The final source of variance is TOTAL but beware: total is *not* adjusted for the covariate and is not appropriate for the denominator of η^2 (Equation 8.7). Subtract the sum of squares for the covariates to calculate the adjusted total sum of squares.

In SYSTAT MGLH (Table 8.7) the IV and the number of levels of the IV are specified in the CATEGORY instruction. The DV is specified on the left side of the MODEL equation with CONSTANT, plus the IV(s), plus their interactions, and the covariate(s) on the right. Output begins with multiple R and R^2 for predicting the DV from the combination of the IV and covariate. The standard ANOVA table shows tests of the IV and the covariate, each adjusted for the other.

SAS GLM is similar in setup to SYSTAT MGLH, as seen in Table 8.8. Two source tables are provided, one for the problem as a regression and the other for the problem as a more standard ANOVA, both with the same ERROR term.

The first test in the regression table asks if there is significant prediction of the DV by the combination of the IV and the covariate. The output resembles that of standard multiple regression (Chapter 5) and includes R-SQUARE, the unadjusted MEAN on POST (the DV), the ROOT MSE (square root of the error mean square), and C.V., the coefficient of variation (100 times the ROOT MSE divided by the mean of the DV).

In the ANOVA-like table, two forms of tests are given, labeled TYPE I SS and TYPE III SS. The sums of squares for TREATMNT are the same in both forms because both adjust treatment for the covariate. In TYPE I SS the sum of squares for the covariate (PRE) is not adjusted for the effect of treatment while in TYPE III SS the sum of squares for the covariate is adjusted for treatment.

8.5 SOME IMPORTANT ISSUES

8.5.1 Test for Homogeneity of Regression

The assumption of homogeneity of regression is that the slopes of the regression of the DV on the covariate(s) (the regression coefficients or B weights as described in

Chapter 5) are the same for all cells of a design. Both homogeneity and heterogeneity of regression are illustrated in Figure 8.2. Because the average of the slopes for all cells is used to adjust the DV, it is assumed that the slopes do not differ significantly either from one another or from a single estimate of the population value. If the null hypothesis of equality among slopes is rejected, the analysis of covariance is inappropriate and an alternative strategy as described in Sections 8.3.2.7 and 8.5.5 should be used.

Hand calculation of the test of homogeneity of regression is extremely tedious (see, for instance, Keppel, 1982). However, for designs consisting solely of between-subjects IVs, computer programs (cf. Table 8.14) are available for testing the assumption, and their use is recommended. The most straightforward programs for testing homogeneity of regression are BMDP1V, SYSTAT MGLH, and SPSS^x MANOVA.

BMDP1V automatically provides a test for homogeneity of regression, as shown in Table 8.9 for the data of Table 8.1. The obtained F value of .0337 (with 2 and 3 df) for EQUALITY OF SLOPE falls far short of critical $F = 9.55$ at the .95 level of confidence. Therefore the slopes are not reliably different and the assumption of homogeneity of regression is tenable. The slopes themselves are shown for the three groups in the segment of output labeled ESTIMATES OF SLOPES WITHIN EACH GROUP in Table 8.9.[4] The test of EQUALITY OF ADJ. CELL MEANS is a test of the major ANCOVA hypothesis and the results are identical to those in Table 8.3. The test of the efficacy of the covariate, adjusted for treatment, in providing adjustment of the DV is given under ZERO SLOPES. Because the test does not reach significance, PRE is not providing a reliable adjustment to POST in this example, owing, no doubt, to the very small sample size.

But BMDP1V only analyzes one-way between subjects designs. For factorial designs it is necessary to create a variable that corresponds to the cells of the design and to code each case according to the cell in which it falls. For example, in a 3 × 3 factorial design, cells are coded 1 through 9, and analyzed as a one-way design with 9 levels of the IV. This type of recoding is demonstrated in Section 8.7.1.7, Table 8.19. The output contains the assessment of equality of slope in each cell but tests of main effects and interactions require additional runs on programs for factorial ANCOVA such as BMDP2V, BMDP4V, SPSS^x ANOVA and MANOVA, SYSTAT MGLH, and SAS GLM.

Programs based on the general linear model test for homogeneity of regression by evaluating the covariate(s) by IV(s) interaction as the last effect entering a regression model (Chapter 5). This procedure is demonstrated in a straightforward manner in the ANCOVA section of the SYSTAT manual for the MGLH program (Wilkinson, 1986). The same procedure works for SAS GLM but is not shown in the manual.

Hierarchical regression programs such as BMDP2R and SPSS^x REGRESSION can be used to test homogeneity of regression through a slightly more complicated procedure. IVs and their interactions and the IV by covariate interaction are dummy

---

[4] If contrasts of a certain form are desired, tests for homogeneity of regression for each group contrasted with all others are provided in the segment of output labeled EQUALITY OF SLOPES BETWEEN EACH GROUP AND ALL OTHERS.

TABLE 8.9 INPUT AND SELECTED OUTPUT FROM BMDP1V FOR
SAMPLE DATA OF TABLE 8.1

```
/PROBLEM      TITLE IS 'SMALL SAMPLE ANCOVA THROUGH BMDP1V'.
/INPUT        VARIABLES ARE 4.  FORMAT IS FREE.
              FILE = TAPE70.
/VARIABLE     NAMES ARE SUBJNO,TREATMNT,PRE,POST.
              GROUPING IS TREATMNT.
/GROUP        CODES(TREATMNT) ARE 1,2,3.
              NAMES(TREATMNT) ARE TREAT1,TREAT2,CONTROL.
/DESIGN       DEPENDENT IS POST.
              INDEPENDENT IS PRE.
              CONTRAST = 1, 0, -1.
              CONTRAST = 1, -2, 1.
              CONTRAST = 1, 1, -2.
/END
```

ANALYSIS OF COVARIANCE TABLE
============================

SOURCE OF VARIATION	D.F.	SUM OF SQ.	MEAN SQ.	F-VALUE	PROB(TAIL)
EQUALITY OF ADJ. MEANS	2	366.2015	183.1008	6.1263	0.045

ZERO SLOPES

ALL COVARIATES	1	137.8946	137.8946	4.6138	0.084
ERROR	5	149.4387	29.8877		

EQUALITY OF SLOPES

ALL COVARIATES, ALL GROUPS	2	3.2868	1.6434	0.0337	0.967
ERROR	3	146.1519	48.7173		

  BETWEEN EACH GROUP AND ALL OTHERS

GROUP 1 TREAT1 , ALL COVARIATES	1	2.9210	2.9210	0.0797	0.792
ERROR	4	146.5177	36.6294		
GROUP 2 TREAT2 , ALL COVARIATES	1	1.9243	1.9243	0.0522	0.831
ERROR	4	147.5144	36.8786		
GROUP 3 CONTROL , ALL COVARIATES	1	0.0612	0.0612	0.0016	0.970
ERROR	4	149.3775	37.3444		

THE TESTS FOR EQUALITY OF ADJUSTED MEANS AND ZERO SLOPES
ARE BASED ON THE ASSUMPTION THAT THE SLOPES OF THE DEPENDENT
VARIABLE ON THE COVARIATES ARE THE SAME IN EVERY GROUP.
THUS, THE RESULTS OF THESE TESTS MAY BE MISLEADING,
IF ANY OF THE ABOVE TESTS FOR EQUALITY OF SLOPES IS
SIGNIFICANT AT THE 5% LEVEL.

THE ABOVE TESTS FOR PARALLELISM DO NOT INDICATE A PROBLEM
WITH EQUALITY OF SLOPES.

SINCE NONE OF THE TESTS FOR ZERO SLOPE IS SIGNIFICANT
THE USE OF THESE (THIS) COVARIATE(S) MAY NOT BE WARRANTED
IN FUTURE STUDIES.

ESTIMATES OF SLOPES WITHIN EACH GROUP

		TREAT1	TREAT2	CONTROL
		1	2	3
PRE	3	0.5917	0.8759	0.7821

variable coded with the IV by covariate interaction entered as the last step of analysis. Homogeneity of regression is tested as the difference between the last step and the preceding step as discussed in Section 5.6.2.3 and illustrated in Tabachnick and Fidell (1983).

A simpler procedure is available in the SPSS^x MANOVA program. Special language is provided for the test, as shown in Table 8.10. The ANALYSIS statement specifies the DV. Effects are listed, in order, in the DESIGN statement: covariate, IV, IV by covariate interaction. The significance test for homogenity of regression is the last one listed (PRE by TREATMENT) and is the same as the test of equality of slopes in Table 8.9. Section 8.7 illustrates this procedure with multiple IVs and covariates.

8.5.2 Design Complexity

Extension of ANCOVA to factorial between-subjects designs is straightforward as long as sample sizes within cells are equal. Partitioning of sources of variance follows ANOVA (cf. Chapter 3) with "subjects nested within cells" as the simple error term. Sums of squares are adjusted for differences due to the association between the covariate and the DV, just as they are for the one-way design demonstrated in Section 8.4.

There are, however, two major design complexities that arise: within-subjects IVs, and unequal sample sizes in the cells of a factorial design. And, just as in ANOVA, contrasts are appropriate for a significant IV with more than two levels and assessment of strength of association is appropriate for any significant effect.

8.5.2.1 Within-Subjects and Mixed Within-Between Designs
Just as a DV can be measured once or repeatedly, so also a covariate can be measured once or repeatedly. In fact, the same design may contain one or more covariates measured once and other covariates measured repeatedly.

A covariate that is measured only once does not provide adjustment to a within-subjects effect because it provides the same adjustment (equivalent to no adjustment)

TABLE 8.10 SETUP AND SELECTED OUTPUT FOR TEST OF
HOMOGENEITY OF REGRESION THROUGH SPSS' MANOVA

```
          TITLE        SMALL SAMPLE ANALYSIS OF COVARIANCE
          FILE HANDLE  TAPE70
          DATA LIST    FILE=TAPE70      FREE
                       /SUBJNO,TREATMNT,PRE,POST
          VALUE LABELS TREATMNT 1 'TREAT1' 2 'TREAT2' 3 'CONTROL'
          MANOVA       PRE POST BY TREATMNT(1,3)/
                       PRINT=SIGNIF(BRIEF)/
                       NOPRINT=PARAMETERS(ESTIM)/
                       ANALYSIS=POST/
                       DESIGN=PRE, TREATMNT, PRE BY TREATMNT/
```

TESTS OF SIGNIFICANCE FOR POST USING SEQUENTIAL SUMS OF SQUARES

SOURCE OF VARIATION	SUM OF SQUARES	DF	MEAN SQUARE	F	SIG. OF F
WITHIN+RESIDUAL	146.15192	3	48.71731		
CONSTANT	81986.77778	1	81986.77778	1682.90864	.000
PRE	204.58228	1	204.58228	4.19938	.133
TREATMNT	366.20123	2	183.10061	3.75843	.152
PRE BY TREATMNT	3.28680	2	1.64340	.03373	.967

for each level of the effect. The covariate does, however, adjust any between-subjects effects in the same design. Thus ANCOVA with one or more covariates measured once is useful in a design with both between and within-subjects effects for increasing the power of the test of between-subjects IVs. With covariates measured repeatedly, both between- and within-subjects effects are adjusted.

ANOVA and ANCOVA with repeated measures (within-subjects effects) are more complicated than designs with only between-subjects effects. One complication is the assumption of homogeneity of covariance, as discussed in Chapter 3. A second complication (more computational than conceptual) is development of separate error terms for various segments of within-subjects effects.

BMDP2V handles designs with between-subjects effects, within-subjects effects, and both types of effects. Both types of covariates can be entered—those that are measured once and those that are measured repeatedly. However, BMDP2V provides only one method of adjustment in the event of unequal cell sample sizes, as described in the following section. Alternatively, BMDP4V is useful for all varieties of complex ANCOVA and has several different methods of unequal-n adjustment but does not provide adjusted cell or marginal means. Both BMDP2V and BMDP4V provide the Huynh-Feldt correction for failure of the assumption of homogeneity of covariance, a correction that varies with the degree to which the assumption is violated.

Of the SPSS^x programs, only MANOVA can be used for complex designs. MANOVA provides adjusted means and a test for homogeneity of covariance but does not provide the Huynh-Feldt correction for failure of the assumption. SYSTAT MGLH and SAS GLM also handle complex designs. However, SYSTAT MGLH provides neither adjusted means nor adjustment for failure of homogeneity of covariance.[5] Therefore, BMDP2V is the program of choice for applications of ANCOVA to complex experimental data (where equal n was planned), and BMDP4V is the program of choice for applications of ANCOVA to complex nonexperimental data.

By the way, if the assumption of homogeneity of covariance is violated, the researcher has several alternatives. First is to assess the significance of within-subjects effects using the Huynh-Feldt correction. Second is to perform single degree of freedom analyses (such as trend analyses) of the within-subjects IV (Section 8.5.2.3). Third is to substitute profile analysis (Chapter 10) or multivariate analysis of variance (Chapter 9) for repeated-measures ANCOVA.

8.5.2.2 Unequal Sample Sizes

In a factorial design, if cells have unequal numbers of scores, the total sums of squares for all effects is greater than SS_{total} and there is ambiguity regarding assignment of overlapping sums of squares to sources. The hypotheses tested for main effects and interactions are no longer independent; the design has become nonorthogonal (cf. Section 3.2.5.3). The problem generalizes directly to ANCOVA.

---

[5] If there is reason to doubt homogeneity of covariance—if strong or differential carry-over effects are suspected from one level of the within-subjects IV to the other—a conservative adjustment of critical F for the within-subjects effect can be made where critical F (at a given α level) is assessed with 1 instead of $k-1$ degrees of freedom. The correction is applied to the main effect of the within-subjects IV, all interactions that contain the within-subjects IV, and all error terms associated with it. However, this extremely conservative adjustment is not as desirable as the Huynh-Feldt correction because of loss of power.

If artificially equalizing cell sizes by random deletion of cases is inappropriate, there are a number of strategies for dealing with unequal n. The choice among strategies depends on the type of research. Of the three major methods described by Overall and Spiegel (1969), Method 1 is usually appropriate for experimental research, Method 2 for survey or nonexperimental research, and Method 3 for research in which the researcher has clear priorities for effects.

Table 8.11 summarizes research situations calling for different methods and notes some of the jargon used by various sources. As Table 8.11 reveals, there are a number of ways of viewing these methods; the terminology associated with these viewpoints by different authors is quite different and, sometimes, seemingly contradictory.

Differences in these methods are easiest to understand from the perspective of multiple regression (Chapter 5). Method 1 is like standard multiple regression with each main effect and interaction assessed after adjustment is made for all other main effects and interactions, as well as for covariates. In SPSS this is called the *regression* approach.[6] The same hypotheses are tested as in the unweighted-means approach where each cell mean is given equal weight regardless of its sample size. This is the recommended approach for experimental research unless there is a reason for doing otherwise.

Reasons include a desire to give heavier weighting to some cells than to others because of unequal importance or unequal population sizes for treatments when occurring naturally. (If cells initially designed to be equal-n end up grossly unequal, the problem is not one of adjustment but, more seriously, of differential dropout.)

Method 2 imposes a hierarchy of testing of effects where main effects are adjusted only for each other and for covariates, while interactions are adjusted for same and lower-level interactions, for covariates, and for main effects. The order of priority for adjustment emphasizes main effects over interactions and lower-order interactions over higher-order interactions. It is called the *classic experimental* approach by SPSS although it is normally used in nonexperimental work when there is a desire to weight marginal means by the sizes of samples in cells from which they are computed. The adjustment assigns heavier weighting to cells with larger sample sizes when computing marginal means and lower-order interactions.

Method 3 allows the researcher to set up the hierarchy for adjustment of covariates, main effects, and interactions. In addition, BMDP4V gives the researcher the opportunity to weight cells by importance or by population size rather than by sample size.

All programs in the reviewed packages perform ANCOVA with unequal sample sizes. BMDP2V and SYSTAT provide for complex designs and unequal n, but only with Method 1 adjustment. SPSS[x] ANOVA allows Methods 1 and 2, and many varieties of Method 3, but requires a between-subjects design. SPSS[x] MANOVA, SAS GLM, and BMDP4V (for two-way designs) provide for design complexity and flexibility in adjustment for unequal n. If BMDP4V is used for higher-order designs, one main effect can be unadjusted (given highest priority) but all other effects must be fully adjusted.

[6] Note that the terms *regression* and *classic experimental* as used by SPSS do *not* imply that these approaches are most appropriately used for nonexperimental and experimental designs, respectively. If anything, the opposite is true.

TABLE 8.11 TERMINOLOGY FOR STRATEGIES FOR ADJUSTMENT FOR UNEQUAL CELL SIZE

Research type	Overall and Spiegel	SPSS	BMDP	SYSTAT	SAS
1. Experiments designed to be equal-n, with random dropout. All cells equally important	Method 1	*Regression* approach. Option 9 in ANOVA program. In MANOVA, METHOD = SSTYPE(UNIQUE)	*Equal* cell weights Defaults for BMDP2V	Default	TYPE III and TYPE IV[a]
2. Nonexperimental research in which sample sizes reflect importance of cells. Main effects have equal priority[b]	Method 2	*Classic experimental* approach. Default in ANOVA	N.A.	N.A.	TYPE II
3. Like number 2 above, except all effects have unequal priority	Method 3	*Sequential* approach. In MANOVA, METHOD = SSTYPE(SEQUENTIAL)	N.A.	N.A.	TYPE I
4. Like number 2 above, except main effects have unequal priority, interactions have equal priority[c]	Method 3	*Hierarchical* approach. Option 10 in ANOVA	*Cell size* weights[d]	N.A.	N.A.
5. Research in which cells are given unequal weight on basis of prior knowledge	N.A.	N.A.	User-defined cell weights	N.A.	N.A.

[a] Type III and IV differ only if there are missing cells.
[b] The programs take different approaches to adjustment for interaction effects.
[c] The programs take different approaches to adjustment for main effects. For a two-way design, hierarchical = sequential.
[d] For a two-way design, output table in BMDP4V gives information for interpretation as either research type 2 or research type 3.

Unless you choose Method 1, output from various programs may disagree even though ostensibly the same method is chosen. This is because programs differ in algorithms used to generate sums of squares, and in the subtleties of adjustment for various effects. Because of these discrepancies, and of disagreements as to the *best* adjustment method in a research situation, some researchers advocate use of Method 1 always. Because Method 1 is the most conservative, you are unlikely to draw criticism by using it. On the other hand, with a nonexperimental design you risk loss of power and perhaps interpretability by treating all cells as if they had equal sample sizes.

Hierarchical regression programs can be used for unequal *n,* but they require far more sophistication on the part of the user, and there is no significant advantage over the simpler programs for most applications.

8.5.2.3 Specific Comparisons and Trend Analysis   If there are more than two levels of an IV, the finding of a significant main effect or interaction in ANOVA or ANCOVA is often insufficient for full interpretation of the effects of IV(s) on the DV. The omnibus F test of a main effect or interaction gives no information as to which means are significantly different from which other means. Comparisons are formed to single out some adjusted means and contrast them with other adjusted means. With a quantitative IV (whose levels differ in amount rather than in kind), interpretation is frequently enhanced by the discovery of a simple trend of change in adjusted means of the DV over sequential levels of the IV.

As with ANOVA (Chapter 3), specific comparisons or trends can be either planned as part of the research design, or tested post hoc as part of a data-snooping procedure after omnibus analyses are completed. For planned comparisons, protection against inflated Type I error is achieved by running a small number of comparisons instead of omnibus F (where the number does not exceed the available degrees of freedom) and by working with an orthogonal set of coefficients. For post hoc comparisons, the probability of Type I error increases with the number of possible comparisons so adjustment is made for inflated α error.

Comparisons are achieved by specifying coefficients and running analyses based on these coefficients. The comparisons can be simple (between two marginal or cell means with the other means left out) or complicated (where means for some cells or margins are pooled and contrasted with means for other cells or margins, or where coefficients for trend—linear, quadratic, etc.—are used). The difficulty of conducting comparisons depends on the complexity of the design and the effect to be analyzed. Comparisons are more difficult if the design has within-subjects IVs, either alone or in combination with between-subjects IVs, where problems arise from the need to develop a separate error term for each comparison. Comparisons are more difficult for interactions than for main effects because there are several approaches to comparisons for interactions. Some of these issues are reviewed in Section 10.5.1.

If all you want to do is pairwise comparisons of adjusted means (and if all IVs are between subjects and cell sample sizes are equal[7]), they are easily obtained

---

[7] If cell sizes are unequal, BMDP4V can be used for a test of appropriately adjusted (weighted) means (see Section 8.6) but at the present time the program does not print out the adjusted means. Specific comparisons with unequal cell sizes and covariates are also available through SPSS^x MANOVA, as discussed in Chapter 9.

through BMDP1V. The design must be one-way or converted to one-way. The program automatically produces a T-TEST MATRIX FOR ADJUSTED GROUP MEANS where differences among all pairs of adjusted means are evaluated. For the data of Table 8.1, the matrix of t tests is shown in Table 8.12. Evaluation of the significance of any of the t values depends on whether the pairwise comparisons were planned or post hoc.

If only a few pairwise comparisons are planned (as many as degrees of freedom for the IV) and the rest are ignored or evaluated post hoc, then the significance of each planned comparison is tested as t with df $= N - k - c$, where $N =$ total sample size (9 in the example), $k =$ number of cells (3 in the example), and $c =$ number of covariates (1 in the example). BMDP1V prints out both the degrees of freedom and the matrix of probability levels. A planned comparison between the adjusted means for TREAT1 and CONTROL, for instance, with an obtained t value of -3.3171, is statistically significant for a two-tailed test at $\alpha = .05$ because the p value of .0211 is less than .05.

If this same comparison is evaluated post hoc, a Scheffé adjustment (or some other adjustment) is made to counteract inflation of the Type I error rate caused by too many tests. First, obtained t is squared to produce obtained F. For the example, F obtained for TREAT1 versus CONTROL is $(-3.3171)^2 = 11.00$. F obtained is then compared with an adjusted critical F produced by multiplying the tabled F value (in this case 5.79 for 2 and 5 df, $\alpha = .05$) by the degrees of freedom associated with the number of cells, or $k - 1$. For this example, the adjusted critical F value is $2(5.79) = 11.58$, and the difference between TREAT1 and CONTROL now fails to reach statistical significance.

Pairwise tests of adjusted means are also available through options in SPSS[x] MANOVA and SAS GLM. In addition, SAS GLM provides several tests that incorporate post hoc adjustments, such as Scheffé or Tukey.

For more complicated comparisons, one set of coefficients is entered in the BMDP1V setup for each comparison. The program prints out the t-test value, the probability level P(T), and the coefficients for each comparison. As previously, the

---

TABLE 8.12 COMPARISONS OF ALL PAIRS OF CELL MEANS AS PRODUCED BY BMDP1V FOR SAMPLE DATA. (SETUP APPEARS IN TABLE 8.9)

```
T-TEST MATRIX FOR ADJUSTED GROUP MEANS ON     5 DEGREES OF FREEDOM
------------------------------------------------------------------


                TREAT1      TREAT2      CONTROL

                  1           2           3
TREAT1      1   0.0000
TREAT2      2  -0.6309     0.0000
CONTROL     3  -3.3171    -2.6250      0.0000



PROBABILITIES FOR THE T-VALUES ABOVE
------------------------------------


                TREAT1      TREAT2      CONTROL

                  1           2           3
TREAT1      1   1.0000
TREAT2      2   0.5558      1.0000
CONTROL     3   0.0211      0.0468      1.0000
```

TABLE 8.13   CONTRASTS THROUGH BMDP1V FOR SAMPLE DATA
             AS SPECIFIED IN SETUP IN TABLE 8.9

T-VALUES FOR CONTRASTS IN ADJUSTED GROUP MEANS

CONTRAST NUMBER	T	P(T)	GROUP TREAT1	GROUP TREAT2	GROUP CONTROL
1	3.3171	0.0211	1.0000	0.0000	-1.0000
2	-1.1542	0.3006	1.0000	-2.0000	1.0000
3	3.4272	0.0187	1.0000	1.0000	-2.0000

probability level is appropriate for a planned comparison but must be adjusted for a post hoc comparison.

Table 8.13 illustrates more complicated comparisons with the sample data. The first two sets of coefficients are for linear and quadratic trend, respectively.[8] (Because the IV in this sample is not quantitative, trend analysis is inappropriate; trend coefficients are used here for illustration only.) The third comparison contrasts the pooled adjusted means of the two treatment groups against the adjusted mean of the control group to test the null hypothesis of no difference between treatment and control. Consider the first two comparisons planned and the third post hoc. The first two are tested against a critical t with 5 df of 2.57; the linear trend is reliable, but the quadratic trend is not. The third comparison, tested post hoc, is reliable because obtained F of 11.75 exceeds adjusted critical F of 11.58. Thus, even by the conservative post hoc criterion, the adjusted mean for the control group differs from the pooled adjusted means of the two treated groups.

In designs with more than one IV, the size of the Scheffé adjustment to critical F for post hoc comparisons depends on the degrees of freedom for the effect being analyzed. For a two-way design, for example, with IVs A and B, adjusted critical F for A is tabled critical F for A multiplied by $(a - 1)$ (where a is the number of levels of A); adjusted critical F for B is tabled critical F for B multiplied by $(b - 1)$; adjusted critical F for the $A \times B$ interaction is tabled critical F for the interaction multiplied by $(a - 1)(b - 1)$.

Procedures similar to these for BMDP1V are applied when using ANCOVA programs such as BMDP4V, SPSS[x] MANOVA, SYSTAT MGLH and SAS GLM. If appropriate programs are unavailable, hand calculations for specific comparisons are not particularly difficult, as long as sample sizes are equal for each cell. The equation for hand calculation is available as Equation 3.23; to apply the equation, one obtains the adjusted cell or marginal means and the error mean square from an omnibus ANCOVA program.

8.5.2.4 Strength of Association   Once an effect is found reliable, the next logical question is, How important is the effect? Importance is usually assessed by computing the percentage of variance in the DV that is associated with the IV. For one-way designs, the strength of association between effect and DV for adjusted sums of squares is found using Equation 8.7. For factorial designs, one uses an extension of Equation 8.7.

-----

[8] Coefficients for orthogonal polynomials are available in most standard ANOVA texts such as Keppel (1982) or Winer (1971).

The numerator for η^2 is the adjusted sum of squares for the main effect or interaction being evaluated; the denominator is the total adjusted sum of squares. The total adjusted sum of squares includes adjusted sums of squares for all main effects, interactions, and error terms but does *not* include components for covariates or the mean, which are typically printed out by computer programs.[9] To find the strength of association between an effect and the adjusted DV scores, then,

$$\eta^2 = \frac{SS'_{effect}}{SS'_{total}} \tag{8.8}$$

In multifactorial designs, the size of η^2 for a particular effect is, in part, dependent on the strength of other effects in the design. In a design where several main effects and interactions are significant, η^2 for a particular effect is diminished because other significant effects increase the size of the denominator. An alternative method of computing η^2 uses in the denominator only the adjusted sum of squares for the effect being tested and the adjusted sum of squares for the appropriate error term for that effect (see Chapter 3 for appropriate error terms).

$$\eta^2_{alt} = \frac{SS'_{effect}}{SS'_{effect} + SS'_{error}} \tag{8.9}$$

Because this alternative form is not standard, an explanatory footnote is appropriate whenever the technique is used in published results.

8.5.3 Evaluation of Covariates

Covariates in ANCOVA can themselves be interpreted as predictors of the DV. From a hierarchical regression perspective (Chapter 5), each covariate is a high-priority (continuous) IV with remaining IVs (main effects and interactions) evaluated after the relationship between the covariate and the DV is removed.

Significance tests for covariates test their utility in adjusting the DV. If a covariate is significant, it provides adjustment of the DV scores. For the example in Table 8.5, the covariate, PRE, does not provide significant adjustment to the DV, POST, with $F(1, 5) = 4.61, p > .05$. PRE is interpreted in the same way as any IV in multiple regression (Chapter 5).

With multiple covariates, all covariates enter the multiple regression equation at once and, as a set, are treated as a standard multiple regression (Section 5.5.1). Within the set of covariates, the significance of each covariate is assessed as if it entered the equation last; only the unique relationship between the covariate and the DV is tested for significance after overlapping variability with other covariates, in their relationship with the DV, is removed. Therefore, although a covariate may be significantly correlated with the DV when considered individually, it may add no

---

[9] If you are using SPSS[x] ANOVA, such summary terms as *main effects, two-way interactions, explained* variance, and the like, are also omitted. Similarly, the total sum of squares is inappropriate because it is the sum of squares for original DV scores rather than adjusted DV scores.

significant adjustment to the DV when considered last. When interpreting the utility of a covariate, it is necessary to consider both correlations among covariates and the DV and significance levels for each covariate as reported in ANCOVA source tables. Evaluation of covariates is demonstrated in Section 8.7.

Unstandardized regression coefficients, provided by most canned computer programs, have the same meaning as regression coefficients described in Chapter 5. However, with unequal n, interpretation of the coefficients depends on the method used for adjustment. When Method 1 (standard multiple regression—see Table 8.11 and Section 8.5.2.2) is used, the significance of the regression coefficients for covariates is assessed as if the covariate entered the regression equation after all main effects and interactions. With other methods, however, covariates enter the equation first, or after main effects but before interactions. The coefficients are evaluated at whatever point the covariates enter the equation.

8.5.4 Choosing Covariates

If several covariates are available, one wants to use an optimum set. When too many covariates are used and they are correlated with each other, a point of diminishing returns in adjustment of the DV is quickly reached. Power is reduced because numerous correlated covariates subtract degrees of freedom from the error term while not removing commensurate sums of squares for error. Preliminary analysis of the covariates, however, improves chances of picking an optimum set.

Statistically, the goal is to identify a small set of covariates that are uncorrelated with each other but correlated with the DV. Conceptually, one wants to select covariates that adjust the DV for predictable but unwanted sources of variability. It may be possible to pick the covariates on theoretical grounds or on the basis of knowledge of the literature regarding important sources of variability that should be controlled.

If theory is unavailable or the literature is insufficiently developed to provide a guide to important sources of variability in the DV, statistical considerations assist the selection of covariates. Look at correlations among covariates and select one among those that are substantially correlated with each other, perhaps by choosing the one with the highest correlation with the DV. Or, if even fewer covariates are desired, a look at squared multiple correlations may be useful, where each covariate, in turn, serves as DV with all other covariates serving as IVs.[10] BMDPAM and programs for principal components analysis (see Chapter 12) provide squared multiple correlations automatically, if all covariates are entered as a set of variables.

If N is large and power is not a problem, it may still be worthwhile to find a small set of covariates for the sake of parsimony. Useless covariates are identified in the first ANCOVA run. Then further runs are made, each time eliminating covariates, until a set of covariates is found that is expected to be optimal in future research. The analysis with the smallest set of covariates is reported, but mention is made in the Results section of the discarded covariate(s) and the fact that the pattern of results did not change when they were eliminated.

---

[10] This information is also useful for checking multicollinearity among covariates. However, the criterion for choosing a small set of covariates is a much lower R^2 than that which indicates multicollinearity.

8.5.5 Alternatives to ANCOVA

Because of the stringent limitations to ANCOVA and potential ambiguity in interpreting results of ANCOVA, alternative analytical strategies are often sought. The availability of alternatives depends on such issues as the scale of measurement of the covariate(s) and the DV, the time that elapses between measurement of the covariate and assignment to treatment, and the difficulty of interpreting results.

When the covariate(s) and the DV are measured on the same scale, two alternatives are available: use of difference scores and conversion of the pretest and posttest scores into a within-subjects IV. In the first alternative, the difference between a pretest score (the previous covariate) and a posttest score (the previous DV) is computed for each subject and used as the DV in ANOVA. If the research question is phrased in terms of "change," then difference scores provide the answer. For example, suppose self-esteem is measured both before and after a year of either belly dance or aerobic dance classes. If difference scores are used, the research question is, Does one year of belly dance training change self-esteem scores more than participation in aerobic dance classes? If ANCOVA is used, the research question is, Do belly dance classes produce greater self-esteem than aerobic dance classes, if pretreatment self-esteem scores are used as a baseline?

A problem with difference scores is ceiling and floor effects (or, in general, a problem of regression toward the mean). A difference may be small because the pretest score is very near the end of the scale and no treatment effect can change it very much, or it may be small because the effect of treatment is small—the DV is the same in either case and the researcher is hard pressed to decide between them. A second problem is that use of difference scores assumes that covariate and DV are measured with perfect reliability. If either ANCOVA or ANOVA with difference scores could be used, ANCOVA is usually the better approach.

The second alternative, if measurement scales are the same, is to convert pre- and posttest scores into two levels of a within-subjects IV. The problem with this strategy is that the effect of the IV of interest is no longer assessed as a main effect, but rather as an interaction between the IV and the pre- vs. posttest IV.

When covariates are measured on any continuous scale, other alternatives are available: randomized blocks and blocking. In the randomized block design, subjects are matched into blocks—equated—on the basis of scores on what would have been the covariate(s). Each block has as many subjects as the number of levels of the IV in a one-way design or number of cells in a larger design (cf. Section 3.2.3). Subjects within each block are randomly assigned to levels or cells of the IV(s) for treatment. In the analytic phase, subjects in the same block are treated as if they were the same person, in a repeated-measures analysis.

Disadvantages to this approach are the strong assumption of homogeneity of covariance of a within-subjects analysis and the loss of degrees of freedom for error without commensurate loss of sums of squares for error if the variable(s) used to block is not related to the DV. In addition, implementation of the randomized block design requires the added step of equating subjects before randomly assigning them to treatment, a step that may be inconvenient, if not impossible, in some applications.

The last alternative is use of blocking. Subjects are measured on potential covariate(s) and then grouped according to their scores (e.g., into groups of high, medium, and low self-esteem on the basis of pretest scores). The groups of subjects become the levels of another full-scale IV that are crossed with the levels of the IV(s) of interest in factorial design. Interpretation of the main effect of the IV of interest is straightforward and variation due to the potential covariate(s) is removed from the estimate of error variance and assessed as a separate main effect. Furthermore, if the assumption of homogeneity of regression would have been violated in ANCOVA, it shows up as an interaction between the potential covariate and the IV of interest.

Blocking has several advantages over ANCOVA and the other alternatives listed here. First, it has none of the assumptions of ANCOVA or within-subjects ANOVA. Second, the relationship between the potential covariate(s) and the DV need not be linear; curvilinear relationships can be captured in ANOVA when three (or more) levels of an IV are analyzed. Blocking, then, is preferable to ANCOVA in many situations, and particularly for experimental, rather than correlational, research.

Blocking can also be expanded to multiple covariates. That is, several new IVs, one per covariate, can be developed through blocking and, with some difficulty, crossed in factorial design. However, the design rapidly becomes very large and cumbersome to implement as the number of IVs increases.

For some applications, however, ANCOVA is preferable to blocking. When the relationship between the DV and the covariate is linear, ANCOVA is more powerful than blocking. And, if the assumptions of ANCOVA are met, conversion of a continuous covariate to a discrete IV can result in loss of information. Finally, practical limitations may prevent measurement of potential covariate(s) sufficiently in advance of treatment to accomplish random assignment of equal numbers of subjects to the cells of the design. When blocking is attempted after treatment, sample sizes within cells are likely to be highly discrepant, leading to the problems of unequal n.

In some applications, a combination of blocking and ANCOVA may turn out to be best. Some potential covariates are used to create new IVs, while others are analyzed as covariates.

8.6 COMPARISON OF PROGRAMS

For the novice, there is a bewildering array of canned computer programs in the BMDP (BMDP1V, 2R, 2V, and 4V) and SPSS^x (REGRESSION, ANOVA, and MANOVA) packages for ANCOVA. For our purposes, the programs based on regression (BMD2R and SPSS^x REGRESSION) are not discussed because they offer little advantage over the other, more easily used programs.

SYSTAT and SAS each have a single general linear model program designed for use with both discrete and continuous variables. Both of these programs deal well with ANCOVA. Features of seven programs are described in Table 8.14.

8.6.1 BMDP Series

BMDP1V is designed for one-way, between-subjects ANOVA and ANCOVA but offers features unavailable in programs for factorial designs. Most notably, it tests both homogeneity of regression (equality of slopes) and user-designated comparisons of adjusted means. As discussed in Sections 8.5.1 and 8.5.2.3, data from factorial between-subjects designs can be recoded into one-way designs to take advantage of these features and then analyzed through factorial programs.

For factorial ANCOVA, the best all-around program in the BMDP series is BMDP2V. It allows for complex designs (both between- and within-subjects IVs and unbalanced layouts) and unequal n (but only with Method 1 adjustment). The test for homogeneity of covariance in within-subjects designs, labeled *compound symmetry,* is available within the program; only the portion of the compound-symmetry test relevant to homogeneity of covariance is evaluated. If the assumption of homogeneity of covariance is violated, two forms of adjustment (Huynh-Feldt and Greenhouse-Geisser) are available. There is full orthogonal decomposition (trend analysis, but not other forms of comparisons) of within-subjects variables for equal or unequal spacing between levels of the within-subjects variable(s).

BMDP4V, described more fully in Section 8.7, is more comprehensive but is also more difficult to use. It provides a variety of methods for dealing with unequal n. In addition, specific comparisons of both between- and within-subjects IVs are available. A disadvantage of the program is the failure to provide adjusted cell or marginal means, or regression coefficients for covariates.

The combination of P1V (to test for equality of slope and user-specified comparisons in between-subjects designs), P2V (to analyze complex designs and to test for homogeneity of covariance and trend in within-subjects IVs), P4V (to analyze complex designs, provide a variety of adjustments for unequal n, and give complete flexibility in comparisons), and PAM or P7M (to search for outliers in each group) gives the user ability to handle almost any ANCOVA problem.

8.6.2 SPSS Package

Two SPSS^x programs perform ANCOVA, MANOVA, and ANOVA. Of the two, MANOVA is far richer. Because the MANOVA program is so flexible, it takes more time to master than SPSS^x ANOVA but is well worth the effort. SPSS^x MANOVA is useful for nonorthogonal and complex designs, and shines in its ability to test assumptions such as homogeneity of regression (see Section 8.5.1) and of variance and covariance. It provides adjusted cell and marginal means; specific comparisons and trend analysis are readily available. However, SPSS^x does not facilitate the search for multivariate outliers among the DV and covariate(s) in each group.

SPSS^x ANOVA has nowhere near the flexibility of BMDP2V or of SPSS^x MANOVA but with between-subjects designs and an assist from SPSS^x MANOVA, a fairly large subset of ANCOVA problems can be analyzed. The program only works for between-subjects IVs but does offer a wide variety of adjustments for unequal n (see Section 8.5.2.2). Tests of assumptions and specific comparisons, including trend analysis, cannot be accomplished through SPSS^x ANOVA but are available through SPSS^x MANOVA. Another major limitation is lack of adjusted cell means, although adjusted marginal means can be computed from marginal deviations.

8.6.3 SYSTAT System

The SYSTAT package has a single multivariate general linear program for regression, univariate and multivariate analysis of variance and covariance, and discriminant

TABLE 8.14 COMPARISON OF SELECTED PROGRAMS FOR ANALYSIS OF COVARIANCE

Feature	BMDP4V[a]	BMDP1V	BMDP2V	SPSS[x] ANOVA	SPSS[x] MANOVA[a]	SYSTAT MGLH[a]	SAS GML[a]
Input							
Maximum number of IVs	8 B–S + 8 W–S	One	9 B–S + 9 W–S	5	10	No limit	No limit
Choice of unequal-n adjustment	Yes	N.A.	No	Yes	Yes	No	Yes
Within-subjects IVs	Yes	No	Yes	No	Yes	Yes	Yes
Specify tolerance	No	No	Yes	No	No	No	Yes
Output							
Source table	Yes	Yes	Yes	Yes	Yes	Yes	Yes
Unadjusted cell means	Yes	Yes	Yes	Yes	Yes	No	Yes
Unadjusted marginal means	Yes	Yes	No	Yes	Yes	No	Yes
Cell standard deviations	No	No	Yes	No	Yes	No	Yes
Adjusted cell means	No	Yes	Yes	No	Yes	Data file	LSMEANS
Adjusted marginal means	No	Yes	No	No	Yes	No	LSMEANS
Grand means and adjusted marginal deviation	No	No	No	Yes	No	No	No
Test for equality of slope (homogeneity of regression)	No	Yes	No	No[b]	Yes	Yes	No
Within-groups regression coefficients	No	Yes	No	No	Yes	No	No
t test for all pairs of cell means	No	Yes	No	No[c]	Yes	No	Yes
Post hoc tests with adjustment	No	No	No	No[c]	No	No	Yes
Trend analysis (including unequal spacing)	Yes	No	Within-subjects IVs only	No[c]	Yes	Yes	Yes
User-specified contrasts (including equal-space trend analysis)	Yes	Yes	No	No[b]	Yes	Yes	Yes
Hypothesis SSCP matrices	Yes	No	No	No	Yes	Yes	Yes
Pooled within-cell error SSCP matrices	Yes	No	No	No	Yes	Yes	Yes
Hypothesis covariance matrices	No	Yes	No	No	Yes	No	No
Pooled within-cell error covariance	No	Yes	No	No	Yes	Yes	Yes
Group covariance matrices	No	Yes	No	No	Yes	No	No

Pooled within-cell correlation matrix	No	No	No	Yes	Yes	Yes
Group correlation matrices	Yes	No	No	Yes	No	No
Correlation matrix for adjusted group means	Yes	No	No	No	No	No
Correlation matrix for regression coefficients	No	No	No	No	No	No
Regression coefficient for each covariate	Yes	Yes	Yes	Yes	No	No
Regression coefficient for each cell	No	No	No	No	No	Yes
Regression coefficient for each main effect	No	No	Yes	No	No	No
Multiple R and/or R^2	No	No	Yes	Yes	No	Yes
Observed maximums and minimums	Yes	No	No	No	No	No
Test for homogeneity of variance	No	No	No[c]	Yes	No	No
Test for homogeneity of covariance	Sphericity	Yes	N.A.	Yes	No	No
Adjustment for heterogeneity of covariance	Yes	Yes	N.A.	No	No	No
Predicted values and residuals	No	Yes	No	Yes	Data file	Yes
Mahalanobis distance	No	No	No	Yes	Yes	No

[a] Additional features described in Chapter 8 (MANOVA).
[b] Can be done through SPSS REGRESSION.
[c] Through subprogram ONEWAY only.

function analysis. Yet it is surprisingly straightforward and easy to use. The manual (Wilkinson, 1986) is replete with helpful examples. On the other hand, the program has fewer features than some of the other comprehensive programs and full analysis may require use of other SYSTAT programs.

SYSTAT MGLH can handle complex designs and provides a test for homogeneity of regression, as clearly demonstrated in the manual. And this is the only ANCOVA program that provides Mahalanobis distance (actually the square root of the value usually reported) as a test for multivariate outliers. But the program allows only Method 1 adjustment for unequal *n*, and descriptive statistics are harder to come by. Adjusted cell means can be saved into a data file, but unadjusted cell and marginal means and other descriptive statistics are available only through the STATS program. Nor are there regression coefficients for covariates. Nevertheless, one could do a respectable ANCOVA using SYSTAT given equal *n* or a data set in which the Method 1 adjustment for unequal *n* is appropriate (e.g., an experiment).

8.6.4 SAS System

SAS GLM is a program for univariate and multivariate analysis of variance and covariance. The setup is similar to that of SYSTAT MGLH, but there are many more options in the SAS program that make it both more complicated to use and ultimately more useful.

SAS GLM offers analysis of complex designs, several adjustments for unequal-*n*, a test for homogeneity of covariance for within-subjects IVs, a fully array of descriptive statistics (upon request), and a wide variety of post hoc tests in addition to user-specified comparisons and trend analysis. Although there is no example of a test for homogeneity of regression in the SAS manual, the procedures highlighted in the SYSTAT manual can be followed.

8.7 COMPLETE EXAMPLE OF ANALYSIS OF COVARIANCE

The research described in Appendix B, Section B.1, provides the data for this illustration of ANCOVA. The research question is whether attitudes toward drugs are affected by current employment status and/or religious affiliation.

Attitude toward drugs (ATTDRUG) serves as the DV, with increasingly high scores reflecting more favorable attitudes. The two IVs, factorially combined, are current employment status (EMPLMNT) with two levels, (1) employed and (2) unemployed; and religious affiliation (RELIGION) with four levels, (1) None-or-other, (2) Catholic, (3) Protestant, and (4) Jewish.

In examining other data for this sample of women, three variables stand out that could be expected to relate to attitudes toward drugs and might obscure effects of employment status and religion. These variables are general state of physical health, mental health, and the use of psychotropic drugs. In order to control for the effects of these three variables on attitudes toward drugs, they are treated as covariates. Covariates, then, are physical health (PHYHEAL), mental health (MENHEAL), and sum of all psychotropic drug uses, prescription and over-the-counter (PSYDRUG). For all three covariates, larger scores reflect increasingly poor health or more use of drugs.

The 2 × 4 analysis of covariance, then, provides a test of the effects of employment status, religion, and their interaction on attitudes toward drugs after adjustment for differences in physical health, mental health, and use of psychotropic drugs. Note that this is a form of ANCOVA in which no causal inference can be made.

8.7.1 Evaluation of Assumptions

These variables are examined with respect to practical limitations of ANCOVA as described in Section 8.3.2.

8.7.1.1 Unequal *n* and Missing Data   BMDP7D provides an initial screening run to look at histograms and descriptive statistics for DV and covariates for the eight groups. Output for PSYDRUG is shown in Table 8.15. Three women out of 465 failed to provide information on religious affiliation. Because RELIGION is one of the IVs, for which cell sizes are unequal in any event, the three cases are dropped from analysis.

The cell-size approach (Method 2 of Section 8.5.2.2) for dealing with unequal *n* is chosen for this study. This method weights cells by their sample sizes, which, in this study, are meaningful because they represent population sizes for the groups.

8.7.1.2 Normality   The histograms in Table 8.15 reveal obvious (positive) skewness for some variables. Because the assumption of normality applies to the sampling distribution of means, and not to the raw scores, skewness by itself poses no problem. With the large sample size and use of two-tailed tests, there is every reason to expect normality of sampling distributions of means.

8.7.1.3 Linearity   There is no reason to expect curvilinearity considering the variables used and the fact that the variables, when skewed, are all skewed in the same direction. Had there been reason to suspect curvilinearity, plots of residuals would have been examined through the BMDP1V run for homogeneity of regression (see below).

8.7.1.4 Outliers   The histograms as well as maximum values in the BMDP7D run of Table 8.15 show that, although no outliers are evident for the DV, several cases are univariate outliers for two of the covariates, PHYHEAL and PSYDRUG (note the histogram for the UNEMPLYD PROTEST group in Table 8.15). Positive skewness is also visible for these variables.

To facilitate the decision between transformation of variables and deletion of outliers, the groups are recoded into a one-way design and analyzed by BMDPAM. A portion of the output is shown in Table 8.16. Significance levels for skewness are given for each variable in each group. The two covariates in question have significant skewness in the group of unemployed, Protestant women. Skewness is evident for these variables in the other groups as well.

TABLE 8.15 SETUP AND PARTIAL OUTPUT OF SCREENING RUNS FOR UNIVARIATE OUTLIERS USING BMDP7D

```
/PROBLEM    TITLE IS 'LARGE SAMPLE ANCOVA, UNIVARIATE OUTLIERS'.
/INPUT      VARIABLES ARE 7. FORMAT IS '(A4,6F4.0)'.
            FILE = ANCOVA.
/VARIABLE   NAMES ARE SUBJNO, ATTDRUG, PHYHEAL, MENHEAL,
            PSYDRUG, EMPLMNT, RELIGION.
            MISSING = (7)9.
            LABEL = SUBJNO.
/GROUP      CODES(EMPLMNT) ARE 1,2.
            NAMES(EMPLMNT) ARE EMPLYD, UNEMPLYD.
            CODES(RELIGION) ARE 1 TO 4.
            NAMES(RELIGION) ARE NONOTHER-CATHOLIC,PROTEST,JEWISH.
/HISTOGRAM  GROUPING IS EMPLMNT, RELIGION.
/END
```

```
**************                        *************
HISTOGRAM OF * PSYDRUG * (  5)  GROUPED  BY   * EMPLMNT  * (  6)
**************                   AND          * RELIGION * (  7)
                                              *************
```

```
            EMPLYD    EMPLYD    EMPLYD    EMPLYD    UNEMPLYD  UNEMPLYD  UNEMPLYD  UNEMPLYD   CASES WITH
            NONOTHER  CATHOLIC  PROTEST   JEWISH    NONOTHER  CATHOLIC  PROTEST   JEWISH      UNUSED
                                                                                           VALUES FOR
                                                                                            EMPLMNT
                                                                                            RELIGION
MIDPOINTS........+.........+.........+.........+.........+.........+.........+.........+.........+
 45.000)
 43.200)
 41.400)
 39.600)
 37.800)
 36.000)
 34.200)*
 32.400)*
 30.600)
 28.800)
 27.000)                              **
 25.200)                                                                            *..
 23.400)**                  *                              *                   *     ...
 21.600)
 19.800)
 18.000)           *        ***       ***                            *          **     ***
 16.200)**         **       ****      *.        *          *         ***        *.      ***
 14.400)***        ****     *.        **        *.         ***       *.          ***
 12.600)*          ********  *.******  ***      **         ***       ****         ***
 10.800)*          ****     *.*******  ***                 ***       ****        *.
  9.000)*          *.       *.******   *.                  ***      ****          *
  7.200)****       *.       ****       M****     ***       **.      M****        M*
  5.400)M****      ********  M******    ***       ****      M**       M****        *.
  3.600)*          M******   M******    *.        M         *.         *.
  1.800)***        M*         **          *.
  0.000)**********23 *********36 *********50 ********24 ********20 ********37 *********47 ********21 M**
```

GROUP MEANS ARE DENOTED BY M'S IF THEY COINCIDE WITH *'S, N'S OTHERWISE

MEAN	5.348	2.349	4.185	4.932	2.067	3.232	5.096	6.083	0.000
STD.DEV.	7.657	3.413	5.736	7.212	3.629	5.579	8.641	7.582	0.000
S.E.M.	1.129	0.430	0.598	1.087	0.663	0.746	0.949	1.094	0.000
MAXIMUM	32.000	14.000	23.000	27.000	13.000	25.000	43.000	25.000	0.000
MINIMUM	0.000	0.000	0.000	0.000	0.000	0.000	0.000	0.000	0.000
CASES EXCL.	(0)	(0)	(0)	(0)	(0)	(0)	(0)	(0)	(0)
CASES INCL.	46	63	92	44	30	56	83	48	3

TABLE 8.16 TEST FOR MULTIVARIATE OUTLIERS. BMDPAM SETUP AND SELECTED OUTPUT

```
/PROBLEM     TITLE IS 'LARGE SAMPLE ANCOVA, MULTIVARIATE OUTLIERS'.
/INPUT       VARIABLES ARE 7   FORMAT IS '(A4,6F4.0)'.
             FILE = ANCOVA.
/VARIABLE    NAMES ARE SUBJNO,ATTDRUG,PHYHEAL,MENHEAL,PSYDRUG,EMPLMNT,RELIGION.
             MISSING = (7)9.
             LABEL = SUBJNO.
             GROUPING = CELL.
             USE=2 TO 5, CELL.
/TRANSFORM   IF (EMPLMNT EQ . AND RELIGION EQ 1) THEN CELL = 1.
             IF (EMPLMNT EQ . AND RELIGION EQ 2) THEN CELL = 2.
             IF (EMPLMNT EQ . AND RELIGION EQ 3) THEN CELL = 3.
             IF (EMPLMNT EQ . AND RELIGION EQ 4) THEN CELL = 4.
             IF (EMPLMNT EQ 2 AND RELIGION EQ 1) THEN CELL = 5.
             IF (EMPLMNT EQ 2 AND RELIGION EQ 2) THEN CELL = 6.
             IF (EMPLMNT EQ 2 AND RELIGION EQ 3) THEN CELL = 7.
             IF (EMPLMNT EQ 2 AND RELIGION EQ 4) THEN CELL = 8.
/GROUP       CODES(CELL) ARE 1 TO 8.
             NAMES(CELL) ARE A,B,C,D,E,F,G,H.
/EST         METHOD=REGR.
/PRINT       MATR=DIS.
/END
```

UNIVARIATE SUMMARY STATISTICS

GROUP IS B   SIZE IS   63

VARIABLE	SAMPLE SIZE	MEAN	STANDARD DEVIATION	COEFFICIENT OF VARIATION	SMALLEST VALUE	LARGEST VALUE	SMALLEST STANDARD SCORE	LARGEST STANDARD SCORE	SKEWNESS	KURTOSIS
2 ATTDRUG	63	7.66667	0.98374	0.128314	6.00000	10.00000	-1.69	2.37	0.29	-0.54
3 PHYHEAL	63	4.61905	2.47848	0.536578	2.00000	14.00000	-1.06	3.78	1.36	2.18
4 MENHEAL	63	5.84127	4.75261	0.813626	0.00000	18.00000	-1.23	2.56	0.79	-0.40
5 PSYDRUG	63	2.34921	3.41325	1.452938	0.00000	14.00000	-0.69	3.41	1.39	1.11

356

ESTIMATES OF MISSING DATA, MAHALANOBIS D-SQUARED (CHI-SQUARED) AND SQUARED MULTIPLE CORRELATIONS WITH AVAILABLE VARIABLES

CASE LABEL	CASE NUMBER	MISSING VARIABLE	ESTIMATE	R-SQUARED	GROUP	CHI-SQ	CHISQ/DF	D.F.	SIGNIFICANCE
127	107				B	8.943	2.236	4	0.0625
128	108				G	25.968	6.492	4	0.0000 *
129	109				C	1.889	0.472	4	0.7562
130	110				F	5.043	1.261	4	0.2829
131	111				F	3.051	0.763	4	0.5494
132	112				A	1.553	0.388	4	0.8172
133	113				B	12.571	3.143	4	0.0136
134	114				H	3.871	0.968	4	0.4238
135	115				E	2.090	0.522	4	0.7193
136	116				C	0.212	0.053	4	0.9948
137	117				C	12.940	3.235	4	0.0116
138	118				F	1.841	0.460	4	0.7650
139	119				H	4.073	1.018	4	0.3962
140	120				G	6.660	1.665	4	0.1550
141	121				C	1.782	0.445	4	0.7759
142	122				B	2.517	0.629	4	0.6417
143	123				B	1.639	0.410	4	0.8017
144	124				D	1.119	0.280	4	0.8913
145	125				D	11.153	2.788	4	0.0249
146	126				H	20.657	5.164	4	0.0004 *
148	127				F	5.246	1.311	4	0.2630

In addition, five cases are multivariate outliers in their respective groups as identified by extreme Mahalanobis distance from group means. The Mahalanobis distance is evaluated as χ^2 with degrees of freedom equal to the number of variables (in this case the three covariates plus the DV). BMDPAM prints significance values for each case, and places an asterisk for each case that exceeds $\alpha = .001$ for Mahalanobis distance. Notice the two outliers in Table 8.16. The first, with $\chi^2 = 25.968$ is from the seventh group (unemployed Protestant women) and the second, with $\chi^2 = 20.657$ departs significantly from the eighth group (unemployed Jewish women). As it turns out, all five outliers are from these two groups—three in group 7 and two in group 8—so that about 4% of the cases in each group are outliers.

This is a borderline case in terms of whether to transform variables or delete outliers—somewhere between "few" and "many." A log transform of the two skewed variables is undertaken to see if outliers remain after transformation. LPSYDRUG is

TABLE 8.17 CHECK FOR MULTIVARIATE OUTLIERS WITH
TRANSFORMED VARIABLES. SETUP AND PARTIAL
OUTPUT FROM BMDPAM

```
          /PROBLEM   TITLE IS 'LARGE SAMPLE ANCOVA, MULTIVARIATE OUTLIERS'.
          /INPUT     VARIABLES ARE 7.  FORMAT IS '(A4,6F4.0)'.
                     FILE = ANCOVA.
          /VARIABLE  NAMES ARE SUBJNO,ATTDRUG,PHYHEAL,MENHEAL,PSYDRUG,
                     EMPLMNT,RELIGION.
                     MISSING = (7)9.
                     LABEL = SUBJNO.
                     GROUPING = CELL.
                     USE=2,4,LPHYHEAL,LPSYDRUG,CELL.
          /TRANSFORM IF (EMPLMNT EQ 1 AND RELIGION EQ 1) THEN CELL = 1.
                     IF (EMPLMNT EQ 1 AND RELIGION EQ 2) THEN CELL = 2.
                     IF (EMPLMNT EQ 1 AND RELIGION EQ 3) THEN CELL = 3.
                     IF (EMPLMNT EQ 1 AND RELIGION EQ 4) THEN CELL = 4.
                     IF (EMPLMNT EQ 2 AND RELIGION EQ 1) THEN CELL = 5.
                     IF (EMPLMNT EQ 2 AND RELIGION EQ 2) THEN CELL = 6.
                     IF (EMPLMNT EQ 2 AND RELIGION EQ 3) THEN CELL = 7.
                     IF (EMPLMNT EQ 2 AND RELIGION EQ 4) THEN CELL = 8.
                     LPHYHEAL=LOG(PHYHEAL).
                     LPSYDRUG=LOG(PSYDRUG+1).
          /GROUP     CODES(CELL) ARE 1 TO 8.
                     NAMES(CELL) ARE A,B,C,D,E,F,G,H.
          /EOT       METHOD=REGR.
          /PRINT     MATR=CORR,DIS.
          /END
```

CASE LABEL	CASE NUMBER	MISSING VARIABLE	ESTIMATE	R-SQUARED	GROUP	CHI-SQ	CHISQ/DF	D.F.	SIGNIFICANCE
127	107				B	8.169	2.042	4	0.0856
128	108				G	11.375	2.844	4	0.0227
129	109				C	2.408	0.602	4	0.6612
130	110				F	5.401	1.350	4	0.2486
131	111				F	3.199	0.800	4	0.5251
132	112				A	2.196	0.549	4	0.6997
133	113				B	13.731	3.433	4	0.0082
134	114				H	4.181	1.045	4	0.3821
135	115				E	2.510	0.628	4	0.6428
136	116				C	0.743	0.186	4	0.9460
137	117				C	9.712	2.428	4	0.0456
138	118				F	2.711	0.678	4	0.6072
139	119				H	2.851	0.713	4	0.5831
140	120				G	4.771	1.193	4	0.3116
141	121				C	2.696	0.674	4	0.6100
142	122				B	3.647	0.912	4	0.4558
143	123				B	2.429	0.607	4	0.6575
144	124				D	1.649	0.412	4	0.7999
145	125				D	11.776	2.944	4	0.0191
146	126				H	13.582	3.395	4	0.0088
148	127				F	4.340	1.085	4	0.3619

created as the logarithm of PSYDRUG (incremented by 1 since many of the values are at zero) and LPHYHEAL as the logarithm of PHYHEAL.

A second run of BMDPAM with the four variables (the DV plus three covariates, two of them transformed) reveals no outliers at $\alpha = .001$ (Table 8.17). All five former outliers are within acceptable distance from their groups once the two covariates are transformed. The decision is to proceed with the analysis using the two transformed covariates rather than to delete cases, although the alternative decision is also acceptable in this situation.

8.7.1.5 Multicollinearity and Singularity

BMDPAM provides squared multiple correlations for each variable as a DV with the remaining variables acting as IVs. This is helpful for detecting the presence of multicollinearity and singularity among the DV-covariate set, as seen in Table 8.18 for the transformed variables. Because the largest $R^2 = .30$, there is no danger of multicollinearity or singularity.

8.7.1.6 Homogeneity of Variance

Sample variances are available from the full BMDPAM run in the section on UNIVARIATE SUMMARY STATISTICS (cf. Table 8.16). For each variable, find the largest and smallest standard deviations over the groups and square them to get variances. For example, the variance for PSYDRUG in the EMPLYD CATHOLIC group is $(3.413)^2 = 11.648569$. The sample variances in the full BMDPAM run with transformed covariates are very similar in the eight groups. The ratio of largest to smallest variance is less than 2 to 1 after transformation. With the use of two-tailed tests, with ratio of largest to smallest sample size less than 4 to 1, and with absence of outliers, robustness is expected and no formal test of homogeneity of variance is necessary.

8.7.1.7 Homogeneity of Regression

Because this is a between-subjects design, the test of homogeneity of regression is readily available through BMDP1V when the two-way factorial is recoded into a one-way design with 8 groups (Table 8.19). The obtained $F(21, 430) = 1.0367$ for equality of slopes indicates that regression coefficients for the eight groups are homogeneous at the .05 level of significance. In addition, none of the specific tests for equality of slopes shows significant deviation from equality.

8.7.1.8 Reliability of Covariates

The three covariates, MENHEAL, PHYHEAL, and PSYDRUG, were measured as counts of symptoms or drug use—"have you ever

TABLE 8.18 CHECK FOR MULTICOLLINEARITY THROUGH BMDPAM.
SETUP APPEARS IN TABLE 8.17

```
SQUARED MULTIPLE CORRELATIONS OF EACH VARIABLE WITH ALL OTHER VARIABLES
-----------------------------------------------------------------------
(MEASURES OF MULTICOLLINEARITY OF VARIABLES)
AND TESTS OF SIGNIFICANCE OF MULTIPLE REGRESSION
DEGREES OF FREEDOM FOR F-STATISTICS ARE      3 AND     451
```

VARIABLE NO.	NAME	R-SQUARED	F-STATISTIC	SIGNIFICANCE (P LESS THAN)
2	ATTDRUG	0.093267	15.46	0.00000
4	MENHEAL	0.286702	60.42	0.00000
9	LPHYHEAL	0.303041	65.37	0.00000
10	LPSYDRUG	0.229175	44.70	0.00000

TABLE 8.19 SETUP AND SELECTED OUTPUT FOR BMDP1V TEST OF HOMOGENEITY OF REGRESSION

```
/PROBLEM     TITLE IS 'LARGE SAMPLE ANCOVA, BMDP1V'.
/INPUT       VARIABLES ARE 7.  FORMAT IS '(A4,GF4.0)'.
             FILE = ANCOVA.
/VARIABLE    NAMES ARE SUBJNO,ATTDRUG,PHYHEAL,MENHEAL,PSYDRUG,
             EMPLMNT,RELIGION.
             MISSING = (7)9.
             LABEL = SUBJNO.
             GROUPING = CELL.
             USE=2,4,LPHYHEAL,LPSYDRUG,CELL.
/TRANSFORM   IF (EMPLMNT EQ 1 AND RELIGION EQ 1) THEN CELL = 1.
             IF (EMPLMNT EQ 1 AND RELIGION EQ 2) THEN CELL = 2.
             IF (EMPLMNT EQ 1 AND RELIGION EQ 3) THEN CELL = 3.
             IF (EMPLMNT EQ 1 AND RELIGION EQ 4) THEN CELL = 4.
             IF (EMPLMNT EQ 2 AND RELIGION EQ 1) THEN CELL = 5.
             IF (EMPLMNT EQ 2 AND RELIGION EQ 2) THEN CELL = 6.
             IF (EMPLMNT EQ 2 AND RELIGION EQ 3) THEN CELL = 7.
             IF (EMPLMNT EQ 2 AND RELIGION EQ 4) THEN CELL = 8.
             LPHYHEAL=LOG(PHYHEAL).
             LPSYDRUG=LOG(PSYDRUG+1).
/GROUP       CODES(CELL) ARE 1 TO 8.
             NAMES(CELL) ARE A,B,C,D,E,F,G,H.
/DESIGN      DEPENDENT IS ATTDRUG.
             INDEPENDENT ARE LPHYHEAL,MENHEAL,LPSYDRUG.
/END
```

ANALYSIS OF COVARIANCE TABLE
==============================

SOURCE OF VARIATION	D.F.	SUM OF SQ.	MEAN SQ.	F-VALUE	PROB(TAIL)
EQUALITY OF ADJ. MEANS	7	22.9147	3.2735	2.7382	0.009

ZERO SLOPES

ALL COVARIATES	3	55.4604	18.4868	15.4634	0.000
VAR. 4 MENHEAL	1	1.4289	1.4289	1.1952	0.275
VAR. 9 LPHYHEAL	1	0.6300	0.6300	0.5270	0.468
VAR. 10 LPSYDRUG	1	46.7374	46.7374	39.0939	0.000
ERROR	451	539.1786	1.1955		

EQUALITY OF SLOPES

ALL COVARIATES, ALL GROUPS	21	25.9839	1.2373	1.0367	0.417
ERROR	430	513.1948	1.1935		

FOR EACH VARIABLE SEPARATELY

VAR. 4 MENHEAL , ALL GROUPS	7	7.4122	1.0589	0.8841	0.519
ERROR	444	531.7664	1.1977		
VAR. 9 LPHYHEAL , ALL GROUPS	7	6.5384	0.9341	0.7786	0.606
ERROR	444	532.6403	1.1996		
VAR. 10 LPSYDRUG , ALL GROUPS	7	10.6053	1.5150	1.2726	0.262
ERROR	444	528.5733	1.1905		

TABLE 8.19 (Continued)

```
BETWEEN EACH GROUP AND ALL OTHERS

    GROUP    1 A       ,
    ALL COVARIATES         3      0.1874      0.0625     0.0519     0.984
    ERROR                448    538.9913      1.2031

    GROUP    2 B       ,
    ALL COVARIATES         3      8.2559      2.7520     2.3222     0.074
    ERROR                448    530.9227      1.1851

    GROUP    3 C       ,
    ALL COVARIATES         3      5.2829      1.7610     1.4776     0.220
    ERROR                448    533.8958      1.1917

    GROUP    4 D       ,
    ALL COVARIATES         3      0.6694      0.2231     0.1856     0.906
    ERROR                448    538.5093      1.2020

    GROUP    5 E       ,
    ALL COVARIATES         3      3.6113      1.2038     1.0069     0.389
    ERROR                448    535.5674      1.1955

    GROUP    6 F       ,
    ALL COVARIATES         3      1.9523      0.6508     0.5427     0.653
    ERROR                448    537.2264      1.1992
    GROUP    7 G       ,
    ALL COVARIATES         3      3.1080      1.0360     0.8658     0.459
    ERROR                448    536.0707      1.1966

    GROUP    8 H       ,
    ALL COVARIATES         3      6.2889      2.0963     1.7624     0.154
    ERROR                448    532.8898      1.1895
```

. . . . ?'' It is assumed that people are reasonably consistent in reporting the presence or absence of symptoms and that high reliability is likely. Therefore no adjustment in ANCOVA is made for unreliability of covariates.

8.7.2 Analysis of Covariance

The program chosen for the major two-way analysis of covariance is BMDP4V. The cell size weights (Table 8.11, number 4, Method 3) approach to adjustment of unequal-n is chosen for this set of survey data. Ease of use, then, makes BMDP4V a convenient program for this unequal-n data set.

Selected output from application of BMDP4V to these data appears in Table 8.20 that shows unadjusted marginal and cell means and sample sizes, and the source table. Sources of variance are the combined covariates; the main effects for RELIGION and EMPLMNT, unadjusted (R and E, respectively) and adjusted for each other (R¦E and E¦R, respectively); the interaction adjusted for both main effects, RE; and the error term, ERROR.

BMDP4V provides the researcher with results of all options for adjustment and the researcher chooses among them. In this example, the researcher can choose the main effect of RELIGION adjusted for the effect of EMPLYMNT (R¦E) or the main effect of EMPLYMNT adjusted for RELIGION (E¦R). In this case the decision is made to give RELIGION priority over EMPLYMNT because for most people religious affiliation precedes employment status. The sources used, then, are RE-LIGION (R) adjusted for covariates, EMPLYMENT adjusted for RELIGION (E¦R) and covariates, and the interaction RE adjusted for all lower-order effects and co-variates.

```
/PROBLEM      TITLE IS 'P4V WITH TRANSFORMS'.
/INPUT        VARIABLES ARE 7.  FORMAT IS '(A4,6F4.0)'.
              FILE = ANCOVA.
/VARIABLE     NAMES ARE SUBJNO, ATTDRUG, PHYHEAL, MENHEAL,
                      PSYDRUG, EMPLMNT, RELIGION.
              MISSING = (7)9.
              LABEL = SUBJNO.
              USE=ATTDRUG,MENHEAL,EMPLMNT,RELIGION,LPHYHEAL,LPSYDRUG.
/TRANSFORM    LPHYHEAL = LOG(PHYHEAL).
              LPSYDRUG = LOG(PSYDRUG + 1).
/BETWEEN      FACTORS ARE RELIGION,EMPLMNT.
              CODES(EMPLMNT) ARE 1,2.
              NAMES(EMPLMNT) ARE EMPLYD, UNEMPLYD.
              CODES(RELIGION) ARE 1 TO 4.
              NAMES(RELIGION) ARE NONOTHER,CATHOLIC,PROTEST,JEWISH.
/WEIGHTS      BETWEEN ARE SIZES.
/END
ANALYSIS      PROCEDURE IS FACTORIAL.
              COVARIATES ARE LPHYHEAL,MENHEAL,LPSYDRUG./
```

SUMMARY STATISTICS FOR VARIATE(S):

VARIATE	COUNT	MEAN	STDERROR	STD-DEV	WTD-MEAN	MAXIMUM	MINIMUM
ATTDRUG	462	7.684	0.5386E-01	1.158	7.684	10.00	5.000
MENHEAL	462	6.136	0.1954	4.199	6.136	18.00	0.0000
LPHYHEAL	462	0.6496	0.9587E-02	0.2061	0.6496	1.176	0.3010
LPSYDRUG	462	0.4159	0.2331E-01	0.5009	0.4159	1.643	0.0000

LEVEL 1 MARGINALS

FACTOR	LEVEL	VARIATE	COUNT	MEAN	STDERROR	STD-DEV	WTD-MEAN	MAXIMUM	MINIMUM
RELIGION	NONOTHER	ATTDRUG	76	7.4474	0.1503	1.3104	7.4474	10.0000	5.0000
		MENHEAL	76	5.9342	0.4582	3.9944	5.9342	17.0000	0.0000
		LPHYHEAL	76	0.6564	0.0192	0.1675	0.6564	1.0000	0.3010
		LPSYDRUG	76	0.4011	0.0571	0.4975	0.4011	1.5185	0.0000
	CATHOLIC	ATTDRUG	119	7.8403	0.0962	1.0495	7.8403	10.0000	6.0000
		MENHEAL	119	6.1345	0.4270	4.6576	6.1345	18.0000	0.0000
		LPHYHEAL	119	0.6303	0.0190	0.2067	0.6303	1.1461	0.3010
		LPSYDRUG	119	0.3271	0.0403	0.4395	0.3271	1.4150	0.0000
	PROTEST	ATTDRUG	175	7.6686	0.0867	1.1467	7.6686	10.0000	5.0000
		MENHEAL	175	6.0857	0.3100	4.1005	6.0857	16.0000	0.0000
		LPHYHEAL	175	0.6618	0.0187	0.2209	0.6618	1.1139	0.3010
		LPSYDRUG	175	0.4348	0.0390	0.5164	0.4348	1.6435	0.0000
	JEWISH	ATTDRUG	92	7.7065	0.1212	1.1630	7.7065	10.0000	5.0000
		MENHEAL	92	6.4022	0.4142	3.9726	6.4022	18.0000	0.0000
		LPHYHEAL	92	0.6450	0.0215	0.2000	0.6450	1.1781	0.3010
		LPSYDRUG	92	0.5070	0.0559	0.5359	0.5070	1.4472	0.0000
EMPLMNT	EMPLYD	ATTDRUG	245	7.5959	0.0712	1.1144	7.5959	10.0000	5.0000
		MENHEAL	245	6.1020	0.2622	4.1044	6.1020	18.0000	0.0000
		LPHYHEAL	245	0.6396	0.0130	0.2039	0.6396	1.1461	0.3010
		LPSYDRUG	245	0.4206	0.0314	0.4913	0.4206	1.5185	0.0000
	UNEMPLYD	ATTDRUG	217	7.7834	0.0814	1.1996	7.7834	10.0000	5.0000
		MENHEAL	217	6.1751	0.2928	4.3126	6.1751	18.0000	0.0000
		LPHYHEAL	217	0.6608	0.0141	0.2083	0.6608	1.1781	0.3010
		LPSYDRUG	217	0.4105	0.0348	0.5127	0.4105	1.6435	0.0000

CELL STATISTICS

```
==================================================
FACTOR    LEVEL
RELIGION  NONOTHER
==>
```

FACTOR	LEVEL	VARIATE	COUNT	MEAN	STDERROR	STD-DEV	WTD-MEAN	MAXIMUM	MINIMUM
EMPLMNT	EMPLYD	ATTDRUG	46	7.6739	0.1992	1.3508	7.6739	10.0000	5.0000
		MENHEAL	46	6.5435	0.5977	4.0536	6.5435	17.0000	0.0000
		LPHYHEAL	46	0.6733	0.0251	0.1702	0.6733	0.9542	0.3010
		LPSYDRUG	46	0.4882	0.0789	0.5355	0.4882	1.5185	0.0000
	UNEMPLYD	ATTDRUG	30	7.1000	0.2163	1.1847	7.1000	10.0000	5.0000
		MENHEAL	30	5.0000	0.6898	3.7783	5.0000	12.0000	0.0000
		LPHYHEAL	30	0.6304	0.0297	0.1626	0.6304	1.0000	0.3010
		LPSYDRUG	30	0.2676	0.0741	0.4060	0.2676	1.1461	0.0000

```
==================================================
```

TABLE 8.20 (Continued)

```
FACTOR     LEVEL
RELIGION   CATHOLIC
==>

  FACTOR     LEVEL    VARIATE   COUNT  MEAN     STDERROR   STD-DEV   WTD-MEAN   MAXIMUM    MINIMUM

  EMPLMNT    EMPLYD   ATTDRUG    63    7.6667   0.1239     0.9837    7.6667     10.0000    6.0000
                      MENHEAL    63    5.8413   0.5988     4.7526    5.8413     18.0000    0.0000
                      LPHYHEAL   63    0.6095   0.0275     0.2186    0.6095      1.1461    0.3010
                      LPSYDRUG   63    0.3275   0.0509     0.4039    0.3275      1.1761    0.0000
             UNEMPLYD ATTDRUG    56    8.0357   0.1463     1.0949    8.0357     10.0000    6.0000
                      MENHEAL    56    6.4643   0.6105     4.5685    6.4643     18.0000    0.0000
                      LPHYHEAL   56    0.6537   0.0256     0.1918    0.6537      1.0414    0.3010
                      LPSYDRUG   56    0.3267   0.0642     0.4802    0.3267      1.4150    0.0000
========================================================
FACTOR     LEVEL
RELIGION   PROTEST
==>

  FACTOR     LEVEL    VARIATE   COUNT  MEAN     STDERROR   STD-DEV   WTD-MEAN   MAXIMUM    MINIMUM

  EMPLMNT    EMPLYD   ATTDRUG    92    7.5109   0.1130     1.0843    7.5109     10.0000    5.0000
                      MENHEAL    92    5.8804   0.3879     3.7206    5.8804     16.0000    0.0000
                      LPHYHEAL   92    0.6521   0.0223     0.2139    0.6521      1.0792    0.3010
                      LPSYDRUG   92    0.4333   0.0522     0.5010    0.4333      1.3802    0.0000
             UNEMPLYD ATTDRUG    83    7.8434   0.1311     1.1943    7.8434     10.0000    5.0000
                      MENHEAL    83    6.3133   0.4935     4.4964    6.3133     16.0000    0.0000
                      LPHYHEAL   83    0.6724   0.0252     0.2292    0.6724      1.1139    0.3010
                      LPSYDRUG   83    0.4364   0.0588     0.5361    0.4364      1.6435    0.0000

========================================================
FACTOR     LEVEL
RELIGION   JEWISH
==>

  FACTOR     LEVEL    VARIATE   COUNT  MEAN     STDERROR   STD-DEV   WTD-MEAN   MAXIMUM    MINIMUM

  EMPLMNT    EMPLYD   ATTDRUG    44    7.5909   0.1668     1.1064    7.5909      9.0000    5.0000
                      MENHEAL    44    6.4773   0.6026     3.9970    6.4773     18.0000    0.0000
                      LPHYHEAL   44    0.6214   0.0289     0.1917    0.6214      0.9542    0.3010
                      LPSYDRUG   44    0.4568   0.0803     0.5329    0.4568      1.4472    0.0000
             UNEMPLYD ATTDRUG    48    7.8125   0.1753     1.2144    7.8125     10.0000    6.0000
                      MENHEAL    48    6.3333   0.5761     3.9911    6.3333     15.0000    0.0000
                      LPHYHEAL   48    0.6680   0.0315     0.2179    0.6680      1.1761    0.3010
                      LPSYDRUG   48    0.5530   0.0780     0.5402    0.5530      1.4150    0.0000


  --- ANALYSIS SUMMARY ---


THE FOLLOWING EFFECTS ARE COMPONENTS OF THE SPECIFIED
LINEAR MODEL FOR THE BETWEEN DESIGN.  ESTIMATES AND TESTS
OF HYPOTHESES FOR THESE EFFECTS CONCERN PARAMETERS OF THAT MODEL.

          COVARIATES
          OVALL: GRAND MEAN
          RE
          R:E
          E:R

THE FOLLOWING EFFECTS INVOLVE WEIGHTED COMBINATIONS OF
REGRESSION INTERCEPTS (ADJUSTED MEANS FOR CONTRAST EFFECTS).
THEY ARE NOT COMPONENTS OF THE LINEAR MODEL FOR THE BETWEEN DESIGN,
SO ESTIMATES AND TESTS MAY NOT BE MEANINGFUL.
          R: RELIGION
          E: EMPLMNT
========================================================
```

TABLE 8.20   (Continued)

```
==================================================================================

EFFECT    VARIATE         STATISTIC              F          DF          P
----------------------------------------------------------------------------------
COVARIATES
          ATTDRUG
                    SS=         55.460423
                    MS=         18.486808         15.46     3,    451    0.0000
OVALL: GRAND MEAN
          ATTDRUG
                    SS=       2225.713856
                    MS=       2225.713856       1861.71     1,    451    0.0000
R: RELIGION
          ATTDRUG
                    SS=         10.178336
                    MS=          3.392779          2.84     3,    451    0.0377
E: EMPLMNT
          ATTDRUG
                    SS=          4.189173
                    MS=          4.189173          3.50     1,    451    0.0619
RE
          ATTDRUG
                    SS=          8.870919
                    MS=          2.956973          2.47     3,    451    0.0611
R:E
          ATTDRUG
                    SS=          9.774202
                    MS=          3.258067          2.73     3,    451    0.0438
E:R
          ATTDRUG
                    SS=          3.787008
                    MS=          3.787008          3.17     1,    451    0.0758


ERROR
          ATTDRUG
                    SS=        539.17869323
                    MS=          1.19551817
```

The significance level is provided for each computed F, so one need not consult a table of critical F; any p less than .05 is statistically significant with $\alpha = .05$. In this example, only the main effect of RELIGION reaches statistical significance.

Entries in the sum of squares column of the source table are used to calculate η^2 as a measure of strength of association for any significant main effect and interaction (Sections 8.4 and 8.5.2.4). Because there is only one significant effect, the more standard equation (Equation 8.7) is preferable.

$$\eta^2 = \frac{10.178}{(10.178 + 3.787 + 8.871) + 539.179} = .02$$

Significance tests of individual covariates assess the utility of covariates, as described in Section 8.5.3. Tests of individual covariates are unavailable in BMDP4V but are in the BMDP1V run of Table 8.19 and reproduced in Table 8.21. In this run, covariates are adjusted for each other and for main effects and interactions. The table of pooled within-cell correlations between the DV ATTDRUG and the covariates, as provided in BMDPAM through the setup shown in Table 8.17 and in Table 8.22, is also needed for a full assessment of the covariates. Pooled within-cell correlations adjust the correlations for main effects and interactions, but not for the DV and other covariates.

The correlations provide a measure of the total relationship between DV and covariates (independent of factors in the design) and the significance tests provide a

TABLE 8.21 ANALYSIS OF COVARIANCE OF ATTITUDE TOWARD DRUGS

Source of variance	Adjusted SS	df	MS	F
Religion	10.178	3	3.393	2.84*
Employment status (adjusted for Religion)	3.787	1	3.787	3.17
Interaction	8.871	3	2.957	2.47
Covariates	55.460	3	18.487	15.46
Physical health (LOG)	0.630	1	0.630	0.53
Mental health	1.429	1	1.429	1.20
Drug uses (LOG)	46.737	1	46.737	39.09**
Error	539.179	451	1.196	

* $p < .05$
** $p < .01$

TABLE 8.22 POOL WITHIN-CELL CORRELATIONS AMONG THREE COVARIATES AND THE DEPENDENT VARIABLE, ATTITUDE TOWARD DRUGS

	Physical health (LOG)	Mental health	Drug uses (LOG)
Attitude toward drugs	.121*	.064	.301*
Physical health (LOG)		.510*	.365*
Mental health			.333*

* $p < .01$

measure of the unique contribution of each covariate, adjusted for the others, to adjustment of the DV. As seen in Table 8.22, both LPHYHEAL and LPSYDRUG are reliably related to the DV, but, as seen in Table 8.21, only LPSYDRUG provides reliable unique adjustment of the DV after all other effects are considered. According to the criteria of Section 8.5.4, then, use of MENHEAL as a covariate in future research is not warranted (it has, in fact, lowered the power of this analysis) and use of LPHYHEAL is questionable.

Adjusted marginal and cell means are likewise not available through BMDP4V, but again are provided by BMDP1V. Output showing adjusted cell means from the setup in Table 8.19 appears in Table 8.23. The adjusted cell means are appropriate to report for a significant interaction.

Because only the main effect of RELIGION is significant here, adjusted marginal

TABLE 8.23 OUTPUT FROM BMDP1V SHOWING CELL SIZES AND ADJUSTED CELL MEANS. SETUP APPEARS IN TABLE 8.19

GROUP	N	GRP.MEAN	ADJ.GRP.MEAN	STD.ERR.
A	46.	7.67391	7.62434	0.16143
B	63.	7.66667	7.73281	0.13827
C	92.	7.51087	7.49404	0.11409
D	44.	7.59091	7.57361	0.16533
E	30.	7.10000	7.19055	0.20050
F	56.	8.03571	8.10272	0.14660
G	83.	7.84337	7.82679	0.12015
H	48.	7.81250	7.71522	0.15847

means for the four religious groups are needed. Adjusted marginal means are found by weighting adjusted cell means by sample sizes, summing across appropriate cells, and then dividing by total sample size on the margin. Groups 1 and 5 (A and E) comprise the none-or-other RELIGION group, as seen from the coding in the setup portion of Table 8.19. Multiply each ADJ.GRP.MEAN by its cell size to find the adjusted cell sum:

$$\text{Group A:} \quad (46)(7.62434) = 350.72$$
$$\text{Group E:} \quad (30)(7.19055) = 215.72$$

Add the adjusted cell sums and divide by the sum of sample sizes to get the adjusted marginal mean:

$$\text{Weighted mean} = \frac{350.72 + 215.72}{46 + 30} = 7.453$$

TABLE 8.24 ADJUSTED AND UNADJUSTED MEAN AT-
TITUDE TOWARD DRUGS FOR FOUR LEV-
ELS OF RELIGION

Religion	Adjusted mean	Unadjusted mean
None-or-other	7.45	7.45
Catholic	7.91	7.84
Protestant	7.65	7.67
Jewish	7.65	7.71

TABLE 8.25 CHECKLIST FOR ANALYSIS OF COVARIANCE

1. Issues
    a. Unequal sample size and missing data
    b. Within-cell outliers
    c. Normality of sampling distributions
    d. Homogeneity of variance
    e. Within-cell linearity
    f. Homogeneity of regression
    g. Reliability of covariates
2. Major analyses
    a. Main effect(s) or planned comparison. If significant:
        (1) Adjusted marginal means
        (2) Strength of association
    b. Interactions or planned comparisons. If significant:
        (1) Adjusted cell means (in table or interaction graph)
        (2) Strength of association
3. Additional analyses
    a. Evaluation of covariate effects
    b. Evaluation of DV-covariate, covariate-covariate correlations
    c. Post hoc comparisons (if appropriate)
    d. Unadjusted marginal and/or cell means (if significant main
       effect and/or interaction) if nonexperimental application

Application of this procedure to the other three religious groups results in the adjusted marginal means shown in Table 8.24. The table includes the unadjusted marginal means from Table 8.20 for comparison.

Because no a priori hypotheses about differences among religious groups were generated, planned comparisons are not appropriate. A glance at the four adjusted means in Table 8.24 suggests a straightforward interpretation; the None-or-other group has the least favorable attitude toward use of psychotropic drugs, the Catholic group the most favorable attitude, and the Protestant and Jewish groups intermediate attitude. In the absence of specific questions about differences between means, there is no compelling reason to evaluate post hoc the significance of these differences, although they certainly provide a rich source of speculation for future research.

A checklist for analysis of covariance appears as Table 8.25. An example of a Results section, in journal format, follows for the analysis described above.

Results

A 2 × 4 between-groups analysis of covariance was performed on attitude toward drugs. Independent variables consisted of current employment status (employed and unemployed) and religious identification (None-or-other, Catholic, Protestant, and Jewish), factorially combined. Covariates were physical health, mental health, and the sum of psychotropic drug uses. Analyses were performed by BMDP4V, weighting cells by their sample sizes to adjust for unequal n.

Results of evaluation of the assumptions of normality of sampling distributions, linearity, homogeneity of variance, homogeneity of regression, and reliability of covariates were satisfactory. Presence of outliers led to transformation of two of the covariates. Logarithmic transforms were made of physical health and the sum of psychotropic drug uses. After transformation, no outliers remained. The original sample of 465 was reduced to 462 by three women who did not provide information as to religious affiliation.

After adjustment by covariates, attitude toward drugs varied significantly with religious affiliation, as summarized in Table 8.21, with $\underline{F}(3, 451) = 2.84$, $\underline{p} <$ .05. The strength of the relationship between adjusted attitudes toward drugs and religion was weak, however, with $\eta^2 = .02$. The adjusted marginal means, as displayed in Table 8.24, show that the most favorable attitudes toward drugs were held by Catholic women, and least favorable attitudes by women who either were unaffiliated with a religion or identified with some religion other than the major three. Attitudes among Protestant and Jewish women were almost identical, on average, for this sample, and fell between those of the two other groups.

Insert Tables 8.21 and 8.24 about here

No statistically significant main effect of current employment status was found. Nor was there a significant interaction between employment status and religion after adjustment for covariates.

Pooled within–group correlations among covariates and attitude toward drugs are shown in Table 8.22. Two of the covariates, logarithm of physical health and logarithm of drug use, were significantly associated with the dependent variable. However, only logarithm of drug use uniquely adjusted the attitude scores, $\underline{F}(1, 451) = 39.09$, $\underline{p} < .01$, after covariates were adjusted for other covariates, main effects, and interaction. The remaining two covariates, mental health and logarithm of physical health, provided no reliable unique adjustment.

Insert Table 8.22 about here

8.8 SOME EXAMPLES FROM THE LITERATURE

A straightforward, classic experimental application of ANCOVA is demonstrated in a study of Hall and colleagues (1977). Obese members of a weight reduction club were randomly assigned to five experimental treatments: self-management, external management, self-management plus external management, psychotherapy, and no-treatment control. Using pretreatment weight as a covariate, a one-way ANCOVA was done on posttreatment weight. After a 10-week treatment period, the groups differed significantly in weight after adjustment for the covariate. A Tukey HSD test showed that the three behavioral conditions did not differ significantly from one another but did differ from the two remaining conditions. The remaining conditions, psychotherapy and no-treatment control, did not differ from each other. ANCOVAs performed on weight at 3- and 6-month follow-ups showed that differences among groups disappeared.

Merrill and Towle (1976) investigated the effects of the availability of objectives in a graduate course in programmed instruction. Analyses of covariance were used on both student performance and anxiety level (as DVs) to evaluate the effect of presence versus absence of course objectives. The course consisted of 12 cognitive units. For the first half of the course, consisting of six units, half of the 32 students received objectives and half did not.

Pretest scores of knowledge of the material served as the covariate for posttest performance (scores on tests after the first six units and the final); no significant effect of course objectives was found. For anxiety as DV, the two covariates were A-Trait and A-State scales of the State-Trait Anxiety Inventory administered to graduate students during the first class session. The DV was the average of the short-form A-State scale administered after each unit test for the first six units. Availability of objectives significantly decreased A-State scores after adjustment for covariates. Six separate analyses of covariance were then done on the A-State scores, one for each of the first six units.[11] These analyses showed that anxiety was significantly reduced only for the first three units when objectives were given.

O'Kane and coworkers (1977) studied the relationship between anticipated social mobility and social political ideology. A total of 307 male Catholic adolescents were divided into four groups. They were first separated into either middle or working class on the basis of the Duncan Index score of the heads of households. These two groups were then further subdivided on the basis of predicted furture class, either same as class of origin or different—upward or downward depending on class of origin. The four groups, then, were middle-middle, working-middle, working-working, and middle-working. IQ served as the covariate for analyses of covariance on three measures of ideology: economic liberalism, noneconomic liberalism, and ethnocentrism.[12]

For economic liberalism, no significant effects were found for class of origin, class of destination, or the interaction between origin and destination classes. For both noneconomic liberalism and ethnocentrism, significant main effects were found for class of destination. Those who expected to join (or remain in) the working class

---

[11] This is a questionable practice because of inflated Type I error rates with multiple testing, especially when the DVs are correlated. Stepdown analysis as described in Chapter 9, or profile analysis as described in Chapter 10 is more appropriate. Or, if assumptions are met, "unit" could act as a six-level within-subjects IV in ANCOVA with both between- and within-subjects IVs.

[12] Because these three measures are probably correlated, multivariate analysis of covariance followed by stepdown analysis as described in Chapter 9 is a better analytic strategy.

showed more conservative scores on the scale of noneconomic liberalism, and scored higher in ethnocentrism, than those who expected to join (or remain in) the middle class. For these two measures, no significant effects of class of origin were found, nor was there a significant interaction between class of origin and class of destination. Additional t tests, presumably on the basis of planned comparisons, showed that the two stable groups differed from each other, in the expected direction, on measures of noneconomic liberalism and ethnocentrism, but not on economic liberalism.

Chapter 9

Multivariate Analysis of Variance and Covariance

9.1 GENERAL PURPOSE AND DESCRIPTION

Multivariate analysis of variance is a generalization of analysis of variance to a situation in which there are several DVs. For example, suppose a researcher is interested in the effect of different types of treatment on several types of anxiety: test anxiety, anxiety in reaction to minor life stresses, and so-called free-floating anxiety. The IV is different treatment with three levels (desensitization, relaxation training, and a waiting-list control). After random assignment of subjects to treatments and a subsequent period of treatment, subjects are measured for test anxiety, stress anxiety, and free-floating anxiety; scores on all three measures for each subject serve as DVs. MANOVA is used to ask whether a combination of the three anxiety measures varies as a function of treatment.

ANOVA tests whether mean differences among groups on a single DV are likely to have occurred by chance. MANOVA tests whether mean differences among groups on a combination of DVs are likely to have occurred by chance. In MANOVA, a new DV that maximizes group differences is created from the set of DVs. The new DV is a linear combination of measured DVs, combined so as to separate the groups as much as possible. ANOVA is then performed on the newly created DV. As in ANOVA, hypotheses about means are tested by comparing variances—hence multivariate analysis of variance.

In factorial or more complicated MANOVA, a different linear combination of DVs is formed for each main effect and interaction. If sex of subject is added to the example as a second IV, one combination of the three DVs maximizes the separation of the three treatment groups, a second combination maximizes separation of women and men, and a third combination maximizes separation of the cells of the interaction.

Further, if the treatment IV has more than two levels, the DVs can be recombined in yet other ways to maximize the separation of groups formed by comparisons.

MANOVA has a number of advantages over ANOVA. First, by measuring several DVs instead of only one, the researcher improves the chance of discovering what it is that changes as a result of different treatments and their interactions. For instance, desensitization may have an advantage over relaxation training or waiting-list control, but only on test anxiety. A second advantage of MANOVA over a series of ANOVAs when there are several DVs is protection against inflated Type I error due to multiple tests of (likely) correlated DVs.

Another advantage of MANOVA is that, under certain conditions, it may reveal differences not shown in separate ANOVAs. Such a situation is shown in Figure 9.1 for a one-way design with two levels. In this figure, the axes represent frequency distributions for each of two DVs, Y_1 and Y_2. Notice that from the point of view of either axis, the distributions are sufficiently overlapping that a mean difference might not be found in ANOVA. The ellipses in the quadrant, however, represent the distributions of Y_1 and Y_2 for each group separately. When responses to two DVs are considered in combination, group differences become apparent. Thus, MANOVA, which considers DVs in combination, may sometimes be more powerful than separate ANOVAs.

But there are no free lunches in statistics, just as there are none in life. MANOVA is a substantially more complicated analysis than ANOVA. There are several important assumptions to consider, and there is often some ambiguity in interpretation of the effects of IVs on any single DV. Further, the situations in which MANOVA is more powerful than ANOVA are quite limited; often MANOVA is considerably less powerful

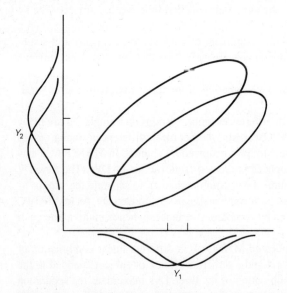

Figure 9.1 Advantage of MANOVA, which combines DVs, over ANOVA. Each axis represents a DV; frequency distributions projected to axes show considerable overlap, while ellipses, showing DVs in combination, do not.

than ANOVA. Thus, our recommendation is to avoid MANOVA except when there is compelling need to measure several DVs.[1]

Multivariate analysis of covariance (MANCOVA) is the multivariate extension of ANCOVA (Chapter 8). MANCOVA asks if there are statistically reliable mean differences among groups after adjusting the newly created DV for differences on one or more covariates. For the example, suppose that before treatment subjects are pretested on test anxiety, minor stress anxiety, and free-floating anxiety. When pretest scores are used as covariates, MANCOVA asks if mean anxiety in the composite score differs in the three treatment groups, after adjusting for preexisting differences in the three types of anxiety.

MANCOVA is useful in the same ways as ANCOVA. First, in experimental work, it serves as a noise-reducing device where variance associated with the covariate(s) is removed from error variance; smaller error variance provides a more powerful test of mean differences among groups. Second, in nonexperimental work, MANCOVA provides statistical matching of groups when random assignment to groups is not possible. Prior differences among groups are accounted for by adjusting DVs as if all subjects scored the same on the covariate(s). (But review Chapter 8 for a discussion of the logical difficulties of using covariates this way.)

ANCOVA is used after MANOVA (or MANCOVA) in stepdown analysis where the goal is to assess the contributions of the various DVs to a significant effect. One asks whether, after adjusting for differences on higher-priority DVs serving as covariates, there is any significant mean difference among groups on a lower-priority DV. That is, does a lower-priority DV provide additional separation of groups beyond that of the DVs already used? In this sense, ANCOVA is used as a tool in interpreting MANOVA results.

Although computing procedures and programs for MANOVA and MANCOVA are not as well developed as for ANOVA and ANCOVA, there is in theory no limit to the generalization of the model, despite complications that arise. There is no reason why all types of designs—one-way, factorial, repeated measures, nonorthogonal, and so on—cannot be extended to research with several DVs. Questions of strength of association, and specific comparisons and trend analysis are equally interesting with MANOVA. In addition, there is the question of importance of DVs—that is, which DVs are affected by the IVs and which are not.

MANOVA developed in the tradition of ANOVA. Typically, MANOVA is applied to experimental situations where all, or at least some, IVs are manipulated and subjects are randomly assigned to groups, usually with equal cell sizes. Discriminant function analysis (Chapter 11) developed in the context of nonexperimental research where groups are formed naturally and are not usually the same size. MANOVA asks if mean differences among groups on the combined DV are larger than expected by chance; discriminant function analysis asks if there is some combination of variables that reliably separates groups. But it should be noted that there is no mathematical distinction between MANOVA and discriminant function analysis. At a practical level, computer programs for discriminant analysis are more informative but are also, for

---

[1] In several years of working with students, we have found that, once the possibility of measuring several DVs is raised, they each and every one become necessary. That is, we have been royally unsuccessful at talking our students out of MANOVA, despite its disadvantage.

the most part, limited to one-way designs. Therefore, analysis of one-way MANOVA is deferred to Chapter 11 and the present chapter covers factorial MANOVA and MANCOVA.

9.2 KINDS OF RESEARCH QUESTIONS

The goal of research using MANOVA is to discover whether behavior, as reflected by the DVs, is changed by manipulation (or other action) of the IVs. Statistical techniques are currently available for answering the types of questions posed in Sections 9.2.1 through 9.2.8.

9.2.1 Main Effects of IVs

Holding all else constant, are mean differences in the composite DV among groups at different levels of an IV larger than expected by chance? The statistical procedures described in Sections 9.4.1 and 9.4.3 are designed to answer this question, by testing the null hypothesis that the IV has no systematic effect on the optimal linear combination of DVs.

As in ANOVA "holding all else constant" refers to a variety of procedures: (1) controlling the effects of other IVs by "crossing over" them in a factorial arrangement, (2) controlling extraneous variables by holding them constant (e.g., running females only as subjects), counterbalancing their effects, or randomizing their effects, or (3) using covariates to produce an "as if constant" state by statistically adjusting for differences on covariates.

In the anxiety-reduction example, the test of main effect asks, Are there mean differences in anxiety—measured by test anxiety, stress anxiety, and free-floating anxiety—associated with differences in treatment? With addition of covariates, the question is: Are there differences in anxiety associated with treatment, after adjustment for individual differences in anxiety prior to treatment?

When there are two or more IVs, separate tests are made for each IV. Further, when sample sizes are equal in all cells, the separate tests are independent of one another (except for use of a common error term) so that the test of one IV in no way predicts the outcome of the test of another IV. If the example is extended to include sex of subject as an IV, and if there are equal numbers of subjects in all cells, the design produces tests of the main effect of treatment and of sex of subject, the two tests independent of each other.

9.2.2 Interactions among IVs

Holding all else constant, does change in the DV over levels of one IV depend on the level of another IV? The test of interaction is similar to the test of main effect, but interpreted differently, as discussed more fully in Chapter 3 and in Sections 9.4.1 and 9.4.3. In the example, the test of interaction asks, Is the pattern of response to the three types of treatment the same for men as it is for women? If the interaction is significant, it indicates that one type of treatment "works best" for women while another type "works best" for men.

With more than two IVs, there are multiple interactions. Each interaction is tested separately from tests of other main effects and interactions, and these tests (but for a common error term) are independent when sample sizes in all cells are equal.

9.2.3 Importance of DVs

If there are significant differences for one or more main effects or interactions, the researcher usually asks which of the DVs are changed and which are unaffected by the IVs. If the main effect of treatment is significant, it may be that only test anxiety is changed while stress anxiety and free-floating anxiety do not differ with treatment. As mentioned in Section 9.1, stepdown analysis is often used where each DV is assessed in ANCOVA with higher-priority DVs serving as covariates. Stepdown analysis and other procedures for assessing importance of DVs appear in Section 9.5.2.

9.2.4 Adjusted Marginal and Cell Means

Ordinarily, marginal means are the best estimates of population parameters for main effects and cell means are the best estimates of population parameters for interactions. But when stepdown analysis is used to test the importance of the DVs, the means that are tested are adjusted means rather than sample means. In the example, suppose free-floating anxiety is given first, stress anxiety second, and test anxiety third priority. Now suppose that a stepdown analysis shows that only test anxiety is affected by differential treatment. The means that are tested for test anxiety are not sample means, but sample means adjusted for stress anxiety and free-floating anxiety. In MANCOVA, additional adjustment is made for covariates. Interpretation and reporting of results is based on both adjusted and sample means, as illustrated in Section 9.7.

9.2.5 Specific Comparisons and Trend Analysis

If an interaction or a main effect for an IV with more than two levels is significant, you probably want to ask which levels of main effect or cells of interaction are different from which others. If, in the example, treatment, with three levels, is significant, the researcher would be likely to want to ask if the pooled average for the two treated groups is different from the average for the waiting-list control, and if the average for relaxation training is different from the average for desensitization. Indeed, the researcher may have planned to ask these questions instead of the omnibus F question for treatment. Similarly, if the interaction of sex of subject and treatment is significant, you may want to ask if there is a significant difference in the average response of women and men to, for instance, desensitization.

Specific comparisons and trend analysis are discussed more fully in Sections 9.5.3, 3.2.6, 8.5.2.3, and 10.5.1.

9.2.6 Strength of Association

If a main effect or interaction reliably affects behavior, the next logical question is, How much? What proportion of variance of the linear combination of DV scores is attributable to the effect? If, for example, there is a reliable main effect of treatment, you can compute the proportion of variance in the composite DV score that is associated with differences in treatment, as described in Section 9.4.1. Procedures are also available for finding the strength of association between the IV and an individually significant DV, as demonstrated in Section 9.7.

9.2.7 Effects of Covariates

When covariates are used, the researcher normally wants to assess their utility. Do the covariates provide reliable adjustment and what is the nature of the DV-covariate relationship? For example, when pretests of test, stress, and free-floating anxiety are used as covariates, to what degree does each covariate adjust the composite DV? Assessment of covariates is demonstrated in Section 9.7.3.1.

9.2.8 Repeated-Measures Analysis of Variance

MANOVA is an alternative to repeated-measures ANOVA in which responses to the levels of the within-subjects IV are simply viewed as separate DVs. Suppose, in the example, that measures of test anxiety are taken four times (instead of measuring three different kinds of anxiety once), before, immediately after, 3 months after, and 6 months after treatment. Results could be analyzed as a two-way ANOVA, with treatment as a between-subjects IV and tests as a within-subject IV or as a one-way MANOVA with treatment as a between-subjects IV and the four testing occasions as four DVs.

As discussed in Sections 8.5.2.1 and 3.2.3, repeated measures ANOVA has the often-violated assumption of homogeneity of covariance.[2] When the assumption is violated, significance tests are too liberal and some alternative to ANOVA is necessary. Other alternatives are adjusted tests of the significance of the within-subjects IV (e.g., Huynh-Feldt), decomposition of the repeated-measures IV into an orthogonal series of single degree of freedom tests (e.g., trend analysis), or profile analysis (Chapter 10).

9.3 LIMITATIONS TO MULTIVARIATE ANALYSIS OF VARIANCE AND COVARIANCE

9.3.1 Theoretical Issues

As with all other procedures, attribution of causality to IVs is in no way assured by the statistical test. This caution is especially relevant because MANOVA, as an

---

[2] An additional assumption of repeated-measures ANOVA is homogeneity of variance over within-subjects IVs, tested in SPSS MANOVA as the FMAX test of "homogeneity of variance over measures." If the test fails (i.e., is significant), MANOVA is preferred over repeated-measures ANOVA.

extension of ANOVA, stems from experimental research where IVs are typically manipulated by the experimenter and desire for causal inference provides the reason behind elaborate controls. But the statistical test is available whether or not IVs are manipulated, subjects randomly assigned, and controls instituted. Therefore the inference that significant changes in the DVs are caused by concomitant changes in the IVs is a logical exercise, not a statistical one; care must be taken in interpreting causality from a significant IV-DV relationship.

Choice of variables is also a question of logic and research design rather than of statistics. Skill is required in choosing IVs and levels of IVs as well as DVs that have some chance of showing effects of the IVs. A further consideration in choice of DVs is the extent of likely correlation among them. The best choice is a set of DVs that are uncorrelated with each other because they each measure a separate aspect of the influence of the IVs. When DVs are correlated, they measure the same or similar facets of behavior in slightly different ways. What is gained by inclusion of several measures of the same thing? Might there be some way of combining DVs or deleting some of them so that the analysis is simpler?

In addition to choice of number and type of DVs is choice of the order in which DVs enter a stepdown analysis if stepdown F is the method chosen to assess the importance of DVs (see Section 9.5.2.2). Priority is usually given to more important DVs or to DVs that are considered causally prior to others in theory. The choice is not trivial because the significance of a DV may well depend on how high a priority it is given, just as in hierarchical multiple regression the significance of an IV is likely to depend on its position in the hierarchy.

When MANCOVA is used, the same limitations apply as in ANCOVA. Consult Sections 8.3.1, 8.5.3, and 8.5.4 for a review of some of the hazards associated with interpretation of designs that include covariates.

Finally, the usual limits to generalizability apply. The results of MANOVA and MANCOVA generalize only to those populations from which the researcher has randomly sampled. And, although MANCOVA may, in some very limited situations, adjust for failure to randomly assign subjects to groups, MANCOVA does not adjust for failure to sample from segments of the population to which one wishes to generalize.

9.3.2 Practical Issues

In addition to the theoretical and logical issues discussed above, the statistical procedure demands consideration of some practical matters.

9.3.2.1 Unequal Sample Sizes and Missing Data
Problems associated with unequal cell sizes are discussed in Section 9.5.4.2. Problems caused by incomplete data (and solutions to them) are discussed in Chapters 4 and 8 (particularly Section 8.3.2.1). The discussion applies to MANOVA and, in fact, may be even more relevant because, as experiments are complicated by numerous DVs and, perhaps, covariates, the probability of missing data increases.

In addition, when using MANOVA, it is necessary to have more cases than DVs in every cell. With numerous DVs this requirement can become burdensome, especially when the design is complicated and there are a lot of cells. There are two reasons for the requirement. First, the power of the analysis is lowered unless there are more cases than DVs in every cell because of reduced degrees of freedom for error. One likely outcome of reduced power is a nonsignificant multivariate F, but one or more significant univariate F's (and a very unhappy researcher).

The second reason for having more cases than DVs in every cell is associated with the assumption of homogeneity of variance-covariance matrices (see Section 9.3.2.4). If a cell has more DVs than cases, the cell becomes singular and the assumption is untestable. If the cell has only one or two more cases than DVs, the assumption is likely to be rejected. Thus MANOVA as an analytic strategy may be discarded because of a failed assumption when the assumption failed as an artifact of a cases-to-DVs ratio that is too low.

9.3.2.2 Multivariate Normality

Significance tests for MANOVA, MANCOVA, and other multivariate techniques are based on the multivariate normal distribution. Multivariate normality implies that the sampling distributions of means of the various DVs in each cell and all linear combinations of them are normally distributed. With univariate F and large samples, the central limit theorem suggests that the sampling distribution of means approaches normality even when raw scores do not. Univariate F is robust to modest violations of normality as long as the violations are not due to outliers.

Mardia (1971) shows that MANOVA is also robust to modest violation of normality if the violation is created by skewness rather than by outliers. *A sample size that produces 20 degrees of freedom[3] for error in the univariate case should ensure robustness of the test, as long as sample sizes are equal and two-tailed tests are used.* Even with unequal n and only a few DVs, a sample size of about 20 in the smallest cell should ensure robustness. With small, unequal samples, normality is assessed by reliance on judgment. Are the individual DVs expected to be fairly normally distributed in the population? If not, would some transformation be expected to produce normality (cf. Chapter 4)?

9.3.2.3 Outliers

One of the more serious limitations of MANOVA (and ANOVA) is its sensitivity to outliers. Especially worrisome is that an outlier can produce either a Type I or a Type II error, with no clue in the analysis as to which is occurring. Therefore, it is highly recommended that a test for outliers accompany any use of MANOVA.

Several programs are available for screening for univariate and multivariate outliers (cf. Chapter 4). *Run tests for univariate and multivariate outliers separately for each cell of the design and transform or eliminate them.* Report the transformation or deletion of outlying cases. Screening runs for within-cell univariate and multivariate outliers are shown in Sections 8.7.1.4 and 9.7.1.4.

9.3.2.4 Homogeneity of Variance-Covariance Matrices

The multivariate generalization of homogeneity of variance for individual DVs is homogeneity of variance-

---

[3] See Chapter 3 for a review of calculation of degrees of freedom for univariate F.

covariance matrices.[4] The assumption is that variance-covariance matrices within each cell of the design are sampled from the same population variance-covariance matrix and can reasonably be pooled to create a single estimate of error.[5] If the within-cell error matrices are heterogeneous, the pooled matrix is misleading as an estimate of error variance.

The following guidelines for testing this assumption in MANOVA are based on a generalization of a Monte Carlo test of robustness in T^2 (Hakstian, Roed, and Lind, 1979). If sample sizes are equal, robustness of significance tests is expected; disregard the outcome of Box's M test, a notoriously sensitive test of homogeneity of variance-covariance matrices available through SPSS MANOVA.

However, if sample sizes are unequal and Box's M test is significant at $p <$ .001, then robustness is not guaranteed. The more numerous the DVs and the greater the discrepancy in cell sample sizes, the greater the potential distortion of α levels. Look at both sample sizes and the sizes of the variances and covariances for the cells. If cells with larger samples produce larger variances and covariances, the α level is conservative so that null hypotheses can be rejected with confidence. If, however, cells with smaller samples produce larger variances and covariances, the significance test is too liberal. Null hypotheses may be retained with confidence but indications of mean differences are suspect. *Use Pillai's criterion instead of Wilks' Lambda (see Section 9.5.1) to evaluate multivariate significance* (Olson, 1979). Or equalize sample sizes by random deletion of cases, if power can be maintained at reasonable levels.

9.3.2.5 Linearity MANOVA and MANCOVA assume linear relationships among all pairs of DVs, all pairs of covariates, and all DV-covariate pairs in each cell. Deviations from linearity reduce the power of the statistical tests because (1) the linear combinations of DVs do not maximize the separation of groups for the IVs, and (2) covariates do not maximize adjustment for error. *With a small number of DVs and covariates, examine all within-cell scatterplots between pairs of DVs, pairs of covariates, and pairs of DV-covariate combinations* through SPSS SCATTERGRAM, SPSS[x] PLOT, or BMDP6D.

With a large number of DVs and/or covariates, however, a complete check is unwieldly; spot check the bivariate relationships for which nonlinearity is likely. Or, use the procedure for checking linearity with large numbers of variables as available through BMDP1V and described in Section 8.3.2.6. One DV (preferably the one with the lowest entry priority) is designated DEPENDENT, and the remainder are designated INDEPENDENT (covariates) in an ANOVA run and residuals plots are examined as described in Section 5.3.2.4. If serious curvilinearity is found with a covariate, consider deletion; if curvilinearity is found with a DV, consider transformation—provided, of course, that increased difficulty in interpretation of a transformed DV is worth the increase in power.

---

[4] In MANOVA, homogeneity of variance for each of the DVs is also assumed. See Section 8.3.2.5 for discussion and recommendations.

[5] Don't confuse this assumption with the assumption of homogeneity of covariance that is relevant to repeated measures ANOVA or MANOVA, as discussed in Section 8.5.2.1.

9.3.2.6 Homogeneity of Regression   In stepdown analysis (Section 9.5.2.2) and in MANCOVA (Section 9.4.3) it is assumed that the relationship between covariates and DVs in one group is the same as the relationship in other groups and that using the average regression to adjust for covariates in all groups is reasonable.

In MANCOVA (like ANCOVA) heterogeneity of regression implies that there is interaction between the IV(s) and the covariates and that a different adjustment of DVs for covariates is needed in different groups. If interaction between IVs and covariates is suspected, MANCOVA is an inappropriate analytic strategy, both statistically and logically. Consult Sections 8.3.2.7 and 8.5.5 for alternatives to MANCOVA where heterogeneity of regression is found.

In stepdown analysis, the importance of a DV in a hierarchy of DVs is assessed in ANCOVA with higher-priority DVs serving as covariates. Homogeneity of regression is required for each step of the analysis, as each DV, in turn, joins the list of covariates. If heterogeneity of regression is found at a step, the rest of the stepdown analysis is uninterpretable. Once violation occurs, the IV-"covariate" interaction is itself interpreted and the DV causing violation is eliminated from further steps.

For MANOVA, test for stepdown homogeneity of regression and for MANCOVA, test for overall and stepdown homogeneity of regression. These procedures are demonstrated in Section 9.7.1.6.

9.3.2.7 Reliability of Covariates   In MANCOVA as in ANCOVA, the F test for mean differences is more powerful if covariates are reliable. If covariates are not reliable, either increased Type I or Type II errors can occur. Reliability of covariates is discussed more fully in Section 8.3.2.8.

In stepdown analysis where all but the lowest-priority DV act as covariates in assessing other DVs, unreliability of any of the DVs (say $r_{yy} < .8$) raises questions about stepdown analysis as well as about the rest of the research effort. When DVs are unreliable, use another method for assessing the importance of DVs (Section 9.5.2) and report known or suspected unreliability of covariates and high-priority DVs in your Results section.

9.3.2.8 Multicollinearity and Singularity   When correlations among DVs are high, one DV is a near-linear combination of other DVs; the DV provides information that is redundant to the information available in one or more of the other DVs. It is both statistically and logically suspect to include all the DVs in analysis and the usual solution is deletion of the redundant DV. But if there is some compelling theoretical reason to retain all DVs, a principal components analysis (cf. Chapter 12) is done on the pooled within-cell correlation matrix, and component scores are entered as an alternative set of DVs.

BMDP4V and BMDP7M protect against multicollinearity and singularity through computation of pooled within-cell tolerance $(1 - \text{SMC})$ for each DV; DVs with insufficient tolerance are deleted from analysis. In SPSS MANOVA, singularity or multicollinearity may be present when the determinant of the within-cell correlation matrix is near zero (say, less than .0001). The redundant DV is identified by using the within-cell correlation matrix (from MANOVA) as input to the REGRESSION program and then performing several regressions, with each DV in turn serving as

DV with all other DVs serving as IVs in analysis. If R^2 approaches .99 for some DV, that DV is redundant. Alternatively, the within-cell correlation matrix is input to BMDP4M or to the SPSS^x FACTOR program where SMCs (called EST COMMU-NALITY in SPSS) are reported.

9.4 FUNDAMENTAL EQUATIONS FOR MULTIVARIATE ANALYSIS OF VARIANCE AND COVARIANCE

9.4.1 Multivariate Analysis of Variance

A minimum data set for MANOVA has one or more IVs, each with two or more levels, and two or more DVs for each subject within each combination of IVs. A fictitious small sample with two DVs and two IVs is illustrated in Table 9.1. The first IV is degree of disability with three levels—mild, moderate, and severe—and the second is treatment with two levels—treatment and no treatment. These two IVs in factorial arrangement produce six cells; three children are assigned to each cell so there are 3×6 or 18 children in the study. Each child produces two DVs, score on the reading subtest of the Wide Range Achievement Test (WRAT-R) and score on the arithmetic subtest (WRAT-A). In addition, an IQ score is given in parentheses for each child to be used as a covariate in Section 9.4.3.

The test of the main effect of treatment asks, Disregarding degree of disability, does treatment affect the composite score created from the two subtests of the WRAT? The test of interaction asks, Does the effect of treatment on the two subtests differ as a function of degree of disability?

The test of the main effect of disability is automatically provided in the analysis but is trivial in this example. The question is, Are scores on the WRAT affected by degree of disability? Because degree of disability is at least partially defined by difficulty in reading and/or arithmetic, a significant effect provides no useful information. On the other hand, absence of the effect, at least among the no treatment group, would lead us to question the adequacy of classification.

The sample size of three children per cell is probably inadequate for a realistic test but serves to illustrate the techniques of MANOVA. Additionally, if causal inference is intended, the researcher should randomly assign children to the levels of treatment. The reader is encouraged to analyze these data by hand and by computer. Setup and selected output for this example appear in Section 9.4.2 for several appropriate programs.

MANOVA follows the model of ANOVA where variance in scores is partitioned into variance attributable to difference among scores within groups and to differences among groups. Squared differences between scores and various means are summed (see Chapter 3); these sums of squares, when divided by appropriate degrees of freedom, provide estimates of variance attributable to different sources (main effects of IVs, interactions among IVs, and error). Ratios of variances provide tests of hypotheses about the effects of IVs on the DV.

In MANOVA, however, each subject has a score on each of several DVs. When

TABLE 9.1 SMALL SAMPLE DATA FOR ILLUSTRATION OF MULTIVARIATE ANALYSIS OF VARIANCE

	Mild			Moderate			Severe		
	WRAT-R	WRAT-A	(IQ)	WRAT-R	WRAT-A	(IQ)	WRAT-R	WRAT-A	(IQ)
Treatment	115	108	(110)	100	105	(115)	89	78	(99)
	98	105	(102)	105	95	(98)	100	85	(102)
	107	98	(100)	95	98	(100)	90	95	(100)
Control	90	92	(108)	70	80	(100)	65	62	(101)
	85	95	(115)	85	68	(99)	80	70	(95)
	80	81	(95)	78	82	(105)	72	73	(102)

several DVs for each subject are measured, there is a matrix of scores (subjects by DVs) rather than a simple set of DVs within each group. Matrices of difference scores are formed by subtracting from each score an appropriate mean; then the matrix of differences is squared. When the squared differences are summed, a sum-of-squares-and-cross-products matrix, an **S** matrix, is formed, analogous to a sum of squares in ANOVA. Determinants[6] of the various **S** matrices are found, and ratios between them provide tests of hypotheses about the effects of the IVs on the linear combination of DVs. In MANCOVA, the sums of squares and cross products in the **S** matrix are adjusted for covariates, just as sums of squares are adjusted in ANCOVA (Chapter 8).

The MANOVA equation for equal n is developed below through extension from ANOVA. The simplest partition apportions variance to systematic sources (variance attributable to differences between groups) and to unknown sources of error (variance attributable to differences in scores within groups). To do this, differences between scores and various means are squared and summed.

$$\sum_i \sum_j (Y_{ij} - \text{GM})^2 = n \sum_j (\overline{Y}_j - \text{GM})^2 + \sum_i \sum_j (Y_{ij} - \overline{Y}_j)^2 \qquad (9.1)$$

The total sum of squared differences between scores on Y (the DV) and the grand mean (GM) is partitioned into sum of squared differences between group means ($\overline{Y}_j$) and the grand mean (i.e., systematic or between-groups variability), and sum of squared difference between individual scores (Y_{ij}) and their respective group means.

or
$$\text{SS}_{\text{total}} = \text{SS}_{bg} + \text{SS}_{wg}$$

For factorial designs, or those with more than one IV, SS_{bg} is further partitioned into variance associated with the first IV (e.g, degree of disability, abbreviated D), variance associated with the second IV (treatment, or T), and variance associated with the interaction between degree of disability and treatment (or DT).

$$n_{km} \sum_k \sum_m (\text{DT}_{km} - \text{GM})^2 = n_k \sum_k (\text{D}_k - \text{GM})^2 + n_m \sum_m (T_m - \text{GM})^2$$

$$+ \left[n_{km} \sum_k \sum_m (\text{DT}_{km} - \text{GM})^2 - n_k \sum_k (\text{D}_k - \text{GM})^2 \right.$$

$$\left. - n_m \sum_m (T_m - \text{GM})^2 \right] \qquad (9.2)$$

The sum of squared differences between cell (DT_{km}) means and the grand mean is partitioned into (1) sum of squared differences between means associated with different levels of disability (D_k) and the grand mean; (2) sum of squared differences between means associated with different levels of treatment (T_m) and

---

[6] A determinant, as described in Appendix A, can be viewed as a measure of generalized variance for a matrix.

the grand mean; and (3) sum of squared differences associated with combinations of treatment and disability (DT_{km}) and the grand mean, from which differences associated with D_k and T_m are subtracted. Each n is the number of scores composing the relevant marginal or cell mean.

or
$$SS_{bg} = SS_D + SS_T + SS_{DT}$$

The full partition for this factorial between-subjects design is

$$\sum_i \sum_k \sum_m (Y_{ikm} - GM)^2 = n_k \sum_k (D_k - GM)^2 + n_m \sum_m (T_m - GM)^2$$

$$+ \left[n_{km} \sum_k \sum_m (DT_{km} - GM)^2 - n_k \sum_k (D_k - GM)^2 \right.$$

$$\left. - n_m \sum_m (T_m - GM)^2 \right] + \sum_i \sum_k \sum_m (Y_{ikm} - DT_{km})^2 \quad (9.3)$$

For MANOVA, there is no single DV but rather a column matrix (or vector) of Y_{ikm} values, of scores on each DV. For the example in Table 9.1, column matrices of Y scores for the three children in the first cell of the design (mild disability with treatment) are

$$\mathbf{Y}_{i11} = \begin{bmatrix} 115 \\ 108 \end{bmatrix} \begin{bmatrix} 98 \\ 105 \end{bmatrix} \begin{bmatrix} 107 \\ 98 \end{bmatrix}$$

Similarly, there is a column matrix of disability—D_k—means for mild, moderate, and severe levels of D, with one mean in each matrix for each DV.

$$\mathbf{D}_1 = \begin{bmatrix} 95.83 \\ 96.50 \end{bmatrix} \quad \mathbf{D}_2 = \begin{bmatrix} 88.83 \\ 88.00 \end{bmatrix} \quad \mathbf{D}_3 = \begin{bmatrix} 82.67 \\ 77.17 \end{bmatrix}$$

where 95.83 is the mean on WRAT-R and 96.50 is the mean on WRAT-A for children with mild disability, averaged over treatment and control groups.

Matrices for treatment—T_m—means, averaged over children with all levels of disability are

$$\mathbf{T}_1 = \begin{bmatrix} 99.89 \\ 96.33 \end{bmatrix} \quad \mathbf{T}_2 = \begin{bmatrix} 78.33 \\ 78.11 \end{bmatrix}$$

Similarly, there are six matrices of cell means (DT_{km}) averaged over the three children in each group. Finally, there is a single matrix of grand means (GM), one for each DV, averaged over all children in the experiment.

$$\mathbf{GM} = \begin{bmatrix} 89.11 \\ 87.22 \end{bmatrix}$$

As illustrated in Appendix A, differences are found by simply subtracting one matrix from another, to produce difference matrices. The matrix counterpart of a difference score, then, is a difference matrix. To produce the error term for this example, the matrix of grand means (GM) is subtracted from each of the matrixes of individual scores (Y_{ikm}). Thus for the first child in the example:

$$(Y_{111} - GM) = \begin{bmatrix} 115 \\ 108 \end{bmatrix} - \begin{bmatrix} 89.11 \\ 87.22 \end{bmatrix} = \begin{bmatrix} 25.89 \\ 20.78 \end{bmatrix}$$

In ANOVA, difference scores are squared. The matrix counterpart of squaring is multiplication by a transpose. That is, each column matrix is multiplied by its corresponding row matrix (see Appendix A for matrix transposition and multiplication) to produce a sum-of-squares and cross-products matrix. For example, for the first child in the first group of the design:

$$(Y_{111} - GM)(Y_{111} - GM)' = \begin{bmatrix} 25.89 \\ 20.78 \end{bmatrix} [25.89 \quad 20.78]$$

$$= \begin{bmatrix} 670.29 & 537.99 \\ 537.99 & 431.81 \end{bmatrix}$$

These matrices are then summed over subjects and over groups, just as squared differences are summed in univariate ANOVA. The order of summing and squaring is the same in MANOVA as in ANOVA for a comparable design. The resulting matrix (S) is called by various names: sum-of-squares and cross-products, cross-products, or sum-of-products. The MANOVA partition of sums-of-squares and cross-products for our factorial example is represented below in a matrix form of Equation 9.3:

$$\sum_i \sum_k \sum_m (Y_{ikm} - GM)(Y_{ikm} - GM)' = n_k \sum_k (D_k - GM)(D_k - GM)'$$

$$+ n_m \sum_m (T_m - GM)(T_m - GM)'$$

$$+ \left[n_{km} \sum_k \sum_m (DT_{km} - GM)(DT_{km} - GM)' \right.$$

$$- n_k \sum_k (D_k - GM)(D_k - GM)'$$

$$\left. - n_m \sum_m (T_m - GM)(T_m - GM)' \right]$$

$$+ \sum_i \sum_k \sum_m (Y_{ikm} - DT_{km})(Y_{ikm} - DT_{km})'$$

or $\quad S_{total} = S_D + S_T + S_{DT} + S_{S(DT)}$

The total cross-products matrix (S_{total}) is partitioned into cross-products matrices for differences associated with degree of disability, with treatment, with the interaction between disability and treatment, and for error—subjects within groups ($S_{S(DT)}$).

For the example in Table 9.1, the four resulting cross-products matrices[7] are

$$S_D = \begin{bmatrix} 520.78 & 761.72 \\ 761.72 & 1126.78 \end{bmatrix} \qquad S_T = \begin{bmatrix} 2090.89 & 1767.56 \\ 1767.56 & 1494.22 \end{bmatrix}$$

$$S_{DT} = \begin{bmatrix} 2.11 & 5.28 \\ 5.28 & 52.78 \end{bmatrix} \qquad SS_{S(DT)} = \begin{bmatrix} 544.00 & 31.00 \\ 31.00 & 539.33 \end{bmatrix}$$

Notice that all these matrices are symmetric, with the elements top left to bottom right representing sums of squares (that, when divided by degrees of freedom, produce variances), and with the other elements representing sums of cross products (that, when divided by degrees of freedom, produce covariances). That is, the first element in the major diagonal (top left to bottom right) is the sum of squares for the first DV, WRAT-R, and the second element is the sum of squares for the second DV, WRAT-A. The off-diagonal elements are the sums of cross products between WRAT-R and WRAT-A.

In ANOVA, sums of squares are divided by degrees of freedom to produce variances, or mean squares. In MANOVA, the matrix analog of variance is a determinant (see Appendix A); the determinant is found for each cross-products matrix. In ANOVA, ratios of variances are formed to test main effects and interactions. In MANOVA, ratios of determinants are formed to test main effects and interactions, when Wilks' Lambda criterion is used (see Section 9.5.1 for alternative criteria). These ratios follow the general form

$$\Lambda = \frac{|S_{error}|}{|S_{effect} + S_{error}|} \tag{9.4}$$

Wilks' Lambda (Λ) is the ratio of the determinant of the error cross-products matrix to the determinant of the sum of the error and effect cross-products matrices.

To find Wilks' Λ, the within-groups matrix is added to matrices corresponding to main effects and interactions before determinants are found. For example, the matrix produced by adding the S_{DT} matrix for interaction to the $S_{S(DT)}$ matrix for subjects within groups (error) is

$$S_{DT} + S_{S(DT)} = \begin{bmatrix} 2.11 & 5.28 \\ 5.28 & 52.78 \end{bmatrix} + \begin{bmatrix} 544.00 & 31.00 \\ 31.00 & 539.33 \end{bmatrix}$$

$$= \begin{bmatrix} 546.11 & 36.28 \\ 36.28 & 592.11 \end{bmatrix}$$

[7] Numbers producing these matrices were carried to 8 digits before rounding.

TABLE 9.2 MULTIVARIATE ANALYSIS OF VARIANCE OF WRAT-R AND WRAT-A SCORES

Source of variance	Wilks' Lambda	Hypoth. df	Error df	Multivariate F
Treatment	.13772	2.00	11.00	34.43570**
Disability	.25526	4.00	22.00	5.38602*
Treatment by disability	.90807	4.00	22.00	.27170

* $p < .01$.
** $p < .001$.

For the four matrices needed to test main effect of disability, main effect of treatment, and the treatment-disability interaction, the determinants are

$$|S_{S(DT)}| = 292436.34$$
$$|S_D + S_{S(DT)}| = 1145629.58$$
$$|S_T + S_{S(DT)}| = 2123390.88$$
$$|S_{DT} + S_{S(DT)}| = 322042.38$$

At this point a source table, similar to the source table for ANOVA, is useful, as presented in Table 9.2. The first column lists sources of variance, in this case the two main effects and the interaction. The error term does not appear. The second column contains the value of Wilks' Lambda.

Wilks' Lambda is a ratio of determinants, as described in Equation 9.4. For example, for the interaction between disability and treatment, Wilks' Lambda is

$$\Lambda = \frac{|S_{S(DT)}|}{|S_{DT} + S_{S(DT)}|} = \frac{292436.34}{322042.38} = .908068$$

Tables for evaluating Wilks' Lambda directly are not common, though they do appear in Harris (1975). However, an approximation to F has been derived that closely fits Lambda. The last three columns of Table 9.2, then, represent the approximate F values and their associated degrees of freedom.

The following procedure for calculating approximate F is based on Wilks' Lambda and the various degrees of freedom associated with it.

$$\text{Approximate } F(df_1, df_2) = \left(\frac{1-y}{y}\right)\left(\frac{df_2}{df_1}\right)$$

where df_1 and df_2 are defined below as the degrees of freedom for testing the F ratio, and y is

$$y = \Lambda^{1/s}$$

Λ is defined in Equation 9.4, and s is[8]

$$s = \sqrt{\frac{p^2(df_{effect})^2 - 4}{p^2 + (df_{effect})^2 - 5}} \tag{9.5}$$

where p is the number of DVs, and df_{effect} is the degrees of freedom for the effect being tested. And

$$df_1 = p(df_{effect})$$

and $$df_2 = s\left[(df_{error}) - \frac{p - df_{effect} + 1}{2}\right] - \left[\frac{p(df_{effect}) - 2}{2}\right]$$

where df_{error} is the degrees of freedom associated with the error term.

For the test of interaction in the sample problem, we have

$p = 2$ — the number of DVs

$df_{effect} = 2$ — the number of treatment levels minus 1 times the number of disability levels minus 1 or $(t - 1(d - 1)$

$df_{error} = 12$ — the number of treatment levels times the number of disability levels times the quantity $n - 1$ (where n is the number of scores per cell for each DV)—that is, $df_{error} = dt(n - 1)$

Thus

$$s = \sqrt{\frac{(2)^2(2)^2 - 4}{(2)^2 + (2)^2 - 5}} = 2$$

$$y = .908068^{1/2} = .952926$$

$$df_1 = 2(2) = 4$$

$$df_2 = 2\left[12 - \frac{2 - 2 + 1}{2}\right] - \left[\frac{2(2) - 2}{2}\right] = 22$$

$$\text{Approximate } F(4, 22) = \left(\frac{.047074}{.952926}\right)\left(\frac{22}{4}\right) = 0.2717$$

This approximate F value is tested for significance by using the usual tables of F at selected α. In this example, the interaction between disability and treatment

[8] When $df_{effect} = 1$, $S = 1$. When $p = 1$, we have univariate ANOVA.

is not statistically significant with 4 and 22 df, because the observed value of 0.2717 does not exceed the critical value of 2.82 at $\alpha = .05$.

Following the same procedures, the effect of treatment is statistically significant, with the observed value of 34.44 exceeding the critical value of 3.98 with 2 and 11 df, $\alpha = .05$. The effect of degree of disability is also statistically significant, with the observed value of 5.39 exceeding the critical value of 2.82 with 4 and 22 df, $\alpha = .05$. (As noted previously, this main effect is not of research interest but does serve to validate the classification procedure.) In Table 9.2, significance is indicated at the highest level of alpha reached, following standard practice.

A measure of strength of association is readily available from Wilks' Lambda.[9] For MANOVA:

$$\eta^2 = 1 - \Lambda \qquad (9.6)$$

This equation represents the variance accounted for by the best linear combination of DVs as explained below.

In a one-way analysis, according to Equation 9.4, Wilks' Lambda is the ratio of (the determinant of) the error matrix and (the determinant of) the total sum-of-squares and cross-products matrix. The determinant of the error matrix—Λ—is the variance not accounted for by the combined DVs so $1 - \Lambda$ is the variance that is accounted for.

Thus for each statistically significant effect, the proportion of variance accounted for is easily calculated using Equation 9.6. For example, the main effect of treatment:

$$\eta_T^2 = 1 - \Lambda_T = 1 - .137721 = .862279$$

In the example, 86% of the variance in the best linear combination of WRAT-R and WRAT-A scores is accounted for by assignment to levels of treatment.[10] The square root of η^2 ($\eta = .93$) is a form of correlation between WRAT scores and assignment to treatment.

9.4.2 Computer Analyses of Small Sample Example

Tables 9.3 through 9.6 show setup and selected minimal output for SPSS[x] MANOVA, BMDP4V, SYSTAT MGLH, and SAS GLM, respectively.

In SPSS[x] MANOVA (Table 9.3) simple MANOVA source tables, resembling those of ANOVA, are printed out when PRINT = SIGNIF(BRIEF) is requested. After interpretive material including the design matrix is printed, the TESTS OF

---

[9] An alternative measure of strength of association is canonical correlation, printed out by some computer programs. Canonical correlation is the correlation between the optimal linear combination of IV levels and the optimal linear combination of DVs where optimal is chosen to maximize the correlation between combined IVs and DVs. Canonical correlation as a general procedure is discussed in Chapter 6, and the relation between canonical correlation and MANOVA is discussed briefly in Chapter 13.

[10] However, unlike η^2 in the analogous ANOVA design, the sum of η^2 for all effects in MANOVA may be greater than 1.0 because DVs are recombined for each effect. This lessens the appeal of an interpretation in terms of proportion of variance accounted for, although the size of η^2 is a measure of the relative importance of an effect.

**TABLE 9.3   MANOVA ON SMALL SAMPLE EXAMPLE THROUGH SPSS'
MANOVA (SETUP AND OUTPUT)**

```
TITLE        SMAL_  SAMPLE MANOVA
FILE HANDLE  TAPE31
DATA LIST    FILE=TAPE81         FREE
             /SUBJNO,TREATMNT,DISABILITY,WRATR,WRATA,IQ,CELLNO
VALUE LABELS TREATMNT 1 'TREATMENT' 2 'CONTROL'/
             DISABILITY 1 'MILD' 2 'MODERATE' 3 'SEVERE'
MANOVA       WRATR, WRATA BY TREATMNT(1,2), DISABLTY(1,3)/
             PRINT=SIGNIF(BRIEF)/
             NOPRINT=PARAMETERS(ESTIM)/
             DESIGN/
```

```
* * * * * * * * * * * * * * A N A L Y S I S   O F   V A R I A N C E * * * * * * * * * * * * * * * * * * * * * * * * * *
```

```
18 CASES ACCEPTED.
0 CASES REJECTED BECAUSE OF OUT-OF-RANGE FACTOR VALUES.
0 CASES REJECTED BECAUSE OF MISSING DATA.
6 NON-EMPTY CELLS,
1 DESIGN WILL BE PROCESSED.
```

```
- - - - - - - - - - - - - - - - - - - - - - - - - - - - - - - - - - - - - - - - - - - - - - - - - - - -
CORRESPONDENCE BETWEEN EFFECTS AND COLUMNS OF BETWEEN-SUBJECTS DESIGN 1
```

```
STARTING   ENDING
COLUMN     COLUMN   EFFECT NAME

   1         1      CONSTANT
   2         2      TREATMNT
   3         4      DISABLT
   5         6      TREATMNT BY DISABLTY
```

```
- - - - - - - - - - - - - - - - - - - - - - - - - - - - - - - - - - - - - - - - - - - - - - - - - - - -
```

```
* * * * * * * * * * * * * * A N A L Y S I S   O F   V A R I A N C E * * * * * * * * * * * * * * * * * * * * * * * * * *
```

```
TESTS OF SIGNIFICANCE FOR WITHIN CELLS     USING SEQUENTIAL SUMS OF SQUARES     62
```

SOURCE OF VARIATION	WILKS LAMBDA	MULT. F	HYPOTH. DF	ERROR DF	SIG. OF F
CONSTANT	.00204	2687.77918	2.00	11.00	0
TREATMNT	.13772	34.43570	2.00	11.00	.000
DISABLTY	.25526	5.38602	4.00	22.00	.004
TREATMNT BY DISABLTY	.90807	.27170	4.00	22.00	.893

TABLE 9.4 MANOVA ON SMALL SAMPLE EXAMPLE THROUGH
BMDP4V (SETUP AND SELECTED OUTPUT)

```
        PROBLEM        TITLE IS 'SMALL SAMPLE MANOVA'.
        /INPUT         VARIABLES ARE 7.  FORMAT IS FREE.   FILE='TAPE81'.
        /VARIABLE      NAMES ARE SUBJNO,TREATMNT,DISABLTY,WRATR,WRATA,IQ,GROUPNO.
                       USE = TREATMNT,DISABLTY,WRATR,WRATA.
        /BETWEEN       FACTORS ARE TREATMNT, DISABLTY.
                       CODES(1) ARE 1,2.
                       NAMES(1) ARE TREATMENT, CONTROL.
                       CODES(2) ARE 1 TO 3.
                       NAMES(2) ARE MILD, MODERATE, SEVERE.
        /WEIGHTS       BETWEEN ARE EQUAL.
        /END
        ANALYSIS PROC IS FACTORIAL./
        /END


 --- ANALYSIS SUMMARY ---


THE FOLLOWING EFFECTS ARE COMPONENTS OF THE SPECIFIED
LINEAR MODEL FOR THE BETWEEN DESIGN.  ESTIMATES AND TESTS
OF HYPOTHESES FOR THESE EFFECTS CONCERN PARAMETERS OF THAT MODEL.


        OVALL: GRAND MEAN
        T: TREATMNT
        D: DISABLTY
        TD

=========================================================================

=========================================================================


EFFECT   VARIATE     STATISTIC          F         DF        P
-------------------------------------------------------------------------
 OVALL: GRAND MEAN
        -ALL----
                    TSQ=     5864.25    2687.78    2,   11  0.0000
        WRATR
                    SS=    142934.222222
                    MS=    142934.222222  3152.96   1,   12  0.0000
        WRATA
                    SS=    136938.888889
                    MS=    136938.888889  3046.85   1,   12  0.0000
 T: TREATMNT
        -ALL----
                    TSQ=      75.1324      34.44    2,   11  0.0000
        WRATR
                    SS=      2090.888889
                    MS=      2090.888889    46.12   1,   12  0.0000
        WRATA
                    SS=      1494.222222
                    MS=      1494.222222    33.25   1,   12  0.0001
 D: DISABLTY
        -ALL----
                    LRATIO=   0.255263      5.39    4,   22.00 0.0035
                    TRACE=    2.89503
                    TZSQ=     34.7404
                    P APPROXIMATION FOR ABOVE STATISTIC UNAVAILABLE - DF IS ZERO
                    MXROOT=   0.742748
                    P APPROXIMATION FOR ABOVE STATISTIC UNAVAILABLE
        WRATR
                    SS=       520.777778
                    MS=       260.388889     5.74    2,   12  0.0178
        WRATA
                    SS=      1126.777778
                    MS=       563.388889    12.54    2,   12  0.0012

 TD
        -ALL----
                    LRATIO=   0.908068      0.27    4,   22.00 0.8930
                    TRACE=    0.100954
                    TZSQ=     1.21144
                    P APPROXIMATION FOR ABOVE STATISTIC UNAVAILABLE - DF IS ZERO
                    MXROOT=   0.892854E-01
                    P APPROXIMATION FOR ABOVE STATISTIC UNAVAILABLE
        WRATR
                    SS=         2.111111
                    MS=         1.055556     0.02    2,   12  0.9770
        WRATA
                    SS=        52.777778
                    MS=        26.388889     0.59    2,   12  0.5711
```

TABLE 9.4 (Continued)

```
ERROR
     WRATR
            SS=        544.00000000
            MS=         45.33333333
     WRATA
            SS=        539.33333333
            MS=         44.94444444
```

TABLE 9.5 MANOVA ON SMALL SAMPLE EXAMPLE THROUGH
SYSTAT MGLH (SETUP AND SELECTED OUTPUT)

```
            USE TAPE81
            CATEGORY TREATMNT=2,DISABLTY=3
            MODEL WRATR,WRATA=CONSTANT+TREATMNT+DISABLTY+TREATMNT*DISABLTY
            ESTIMATE
            HYPOTHESIS EFFECT=TREATMNT*DISABLTY
            TEST
            HYPOTHESIS EFFECT=DISABLTY
            TEST
            HYPOTHESIS EFFECT=TREATMNT
            TEST

NUMBER OF CASES PROCESSED:     18

TEST FOR EFFECT CALLED:
                  TREATMNT
               BY
                  DISABLTY
```

UNIVARIATE F TESTS

VARIABLE	SS	DF	MS	F	P
WRATR	2.111	2	1.056	0.023	0.977
ERROR	544.000	12	45.333		
WRATA	52.778	2	26.389	0.587	0.571
ERROR	539.333	12	44.944		

MULTIVARIATE TEST STATISTICS

```
          WILKS' LAMBDA =      0.908
            F-STATISTIC =      0.272    DF =   4,  22    PROB =        0.893

          PILLAI TRACE =       0.092
            F-STATISTIC =      0.290    DF =   4,  24    PROB =        0.882

HOTELLING-LAWLEY TRACE =       0.101
            F-STATISTIC =      0.252    DF =   4,  20    PROB =        0.905

              THETA =   0.089   S = 2, M = -.5, N = 4.5 PROB =        0.766


TEST FOR EFFECT CALLED:
               DISABLTY
```

UNIVARIATE F TESTS

VARIABLE	SS	DF	MS	F	P
WRATR	520.778	2	260.389	5.744	0.018
ERROR	544.000	12	45.333		
WRATA	1126.778	2	563.389	12.535	0.001
ERROR	539.333	12	44.944		

MULTIVARIATE TEST STATISTICS

```
          WILKS' LAMBDA =      0.255
            F-STATISTIC =      5.386    DF =   4,  22    PROB =        0.004

          PILLAI TRACE =       0.750
            F-STATISTIC =      3.604    DF =   4,  24    PROB =        0.019

HOTELLING-LAWLEY TRACE =       2.895
            F-STATISTIC =      7.238    DF =   4,  20    PROB =        0.001

              THETA =   0.743   S = 2, M = -.5, N = 4.5 PROB =        0.002
```

TABLE 9.5 (Continued)

TEST FOR EFFECT CALLED:
                TREATMNT

UNIVARIATE F TESTS

VARIABLE	SS	DF	MS	F	P
WRATR	2090.889	1	2090.889	46.123	0.000
ERROR	544.000	12	45.333		
WRATA	1494.222	1	1494.222	33.246	0.000
ERROR	539.333	12	44.944		

MULTIVARIATE TEST STATISTICS

WILKS' LAMBDA =	0.138				
F-STATISTIC =	34.436	DF = 2, 11	PROB =	0.000	
PILLAI TRACE =	0.862				
F-STATISTIC =	34.436	DF = 2, 11	PROB =	0.000	
HOTELLING-LAWLEY TRACE =	6.261				
F-STATISTIC =	34.436	DF = 2, 11	PROB =	0.000	

TABLE 9.6 MANOVA ON SMALL SAMPLE EXAMPLE THROUGH SAS
          GLM (SETUP AND OUTPUT)

```
DATA SSAMPLE;
INFILE TAPE81;
INPUT SUBJNO TREATMNT DISABLTY WRATR WRATA IQ CELLNO;
PROC GLM;
  CLASS TREATMNT DISABLTY;
  MODEL WRATR WRATA = TREATMNT DISABLTY TREATMNT*DISABLTY/NOUNI;
  MANOVA H=-ALL-/SHORT;
```

SAS

GENERAL LINEAR MODELS PROCEDURE

CLASS LEVEL INFORMATION

CLASS	LEVELS	VALUES
TREATMNT	2	1 2
DISABLTY	3	1 2 3

NUMBER OF OBSERVATIONS IN DATA SET = 18

CHARACTERISTIC ROOTS AND VECTORS OF: E INVERSE * H,
WHERE   H = TYPE III SS&CP MATRIX FOR: TREATMNT   E = ERROR SS&CP MATRIX

CHARACTERISTIC ROOT	PERCENT	CHARACTERISTIC VECTOR V'EV=1	
		WRATR	WRATA
6.26103637	100.00	0.03206535	0.02680051
0.00000000	0.00	-0.02856728	0.03379300

MANOVA TEST CRITERIA AND EXACT F STATISTICS FOR THE HYPOTHESIS OF NO OVERALL TREATMNT EFFECT

S=1   M=0   N=5

STATISTIC	VALUE	F	NUM DF	DEN DF	PR > F
WILKS' LAMBDA	0.1377214	34.436	2	11	0.0001
PILLAI'S TRACE	0.8622786	34.436	2	11	0.0001
HOTELLING-LAWLEY TRACE	6.261036	34.436	2	11	0.0001
ROY'S GREATEST ROOT	6.261036	34.436	2	11	0.0001

TABLE 9.6 (Continued)

```
                  CHARACTERISTIC ROOTS AND VECTORS OF: E INVERSE * H,
    WHERE  H = TYPE III SS&CP MATRIX FOR: DISABLTY      E = ERROR SS&CP MATRIX

       CHARACTERISTIC    PERCENT     CHARACTERISTIC VECTOR     V'EV=1
            ROOT
                                      WRATR              WRATA

          2.88724085      99.73      0.02260839        0.03531017
          0.00779322       0.27     -0.03651215        0.02476743
```

MANOVA TEST CRITERIA AND F APPROXIMATIONS FOR THE HYPOTHESIS OF NO OVERALL DISABLTY EFFECT

```
                         S=2    M=-0.5    N=5
```

STATISTIC	VALUE	F	NUM DF	DEN DF	PR > F
WILKS' LAMBDA	0.2552626	5.386	4	22	0.0035
PILLAI'S TRACE	0.7504811	3.604	4	24	0.0195
HOTELLING-LAWLEY TRACE	2.895034	7.238	4	20	0.0009
ROY'S GREATEST ROOT	2.887241	17.323	2	12	0.0003

```
         NOTE: F STATISTIC FOR ROY'S GREATEST ROOT IS AN UPPER BOUND
               F STATISTIC FOR WILKS' LAMBDA IS EXACT

              CHARACTERISTIC ROOTS AND VECTORS OF: E INVERSE * H, WHERE
    H= TYPE III SS&CP MATRIX FOR: TREATMNT*DISABLTY     E = ERROR SS&CP MATRIX

       CHARACTERISTIC    PERCENT     CHARACTERISTIC VECTOR     V'EV=1
            ROOT
                                      WRATR              WRATA

          0.09803883      97.11      0.00187535        0.04291087
          0.00291470       2.89      0.04290407       -0.00434641
```

MANOVA TEST CRITERIA AND F APPROXIMATIONS FOR THE HYPOTHESIS OF NO OVERALL TREATMNT*DISABLTY EFFECT

```
                         S=2    M=-0.5    N=5
```

STATISTIC	VALUE	F	NUM DF	DEN DF	PR > F
WILKS' LAMBDA	0.9080679	0.272	4	22	0.8930
PILLAI'S TRACE	0.09219163	0.290	4	24	0.8816
HOTELLING-LAWLEY TRACE	0.1009535	0.252	4	20	0.9048
ROY'S GREATEST ROOT	0.09803883	0.588	2	12	0.5706

```
         NOTE: F STATISTIC FOR ROY'S GREATEST ROOT IS AN UPPER BOUND
               F STATISTIC FOR WILKS' LAMBDA IS EXACT
```

SIGNIFICANCE FOR WITHIN CELLS USING SEQUENTIAL SUMS OF SQUARES source table is shown.[11] WITHIN CELLS refers to the pooled within-cell error SSCP matrix (Section 9.4.1), the error term chosen by default for MANOVA. For the example, the two-way MANOVA source table consists of CONSTANT (test of the grand mean) followed by the two main effects and the interaction. For each source, you are given Wilks' Lambda, multivariate F with numerator and denominator degrees of freedom (HYPOTH. DF and ERROR DF, respectively), and the probability level achieved for the significance test.

BMDP4V (Table 9.4) requires a USE= instruction if there are more variables in your data set than are used in the MANOVA. After the IVs are listed in the BETWEEN instruction (and in the WITHIN instruction, if applicable) the remaining variables serve as DVs. There is a great deal of output showing the program's interpretation of the data and the analysis, and the cell and marginal means, most of which has been deleted from Table 9.4. Shown is a detailed source table starting

---

[11] Note that later versions of SPSS^x MANOVA use UNIQUE rather than SEQUENTIAL SUMS OF SQUARES as default to correct for unequal-n, cf. Section 8.5.2.2.

with the tests for the grand mean. Results for each effect are reported in a common format: the multivariate test for -ALL---- (the DVs combined) is followed by univariate tests for each DV. For the example, the multivariate test of T: TREATMNT is a form of Hotelling's T^2 (because TREATMNT has only two levels), for which you are given the value of TSQ, the associated F ratio (34.44), numerator and denominator degrees of freedom (2, 11), and the significance level attained (rounded to four digits). For each DV, BMDP4V prints out sum of squares and mean square (SS and MS) for the effect, the univariate F ratio, numerator and denominator df, and significance level.

For an effect with more than one df (e.g., D: DISABLTY and TD, the interaction), several multivariate tests of significance are provided: Wilks' Lambda (LRATIO), Hotelling-Lawley TRACE, a form of Hotelling's T^2 labeled TZSQ, and Roy's gcr (MXROOT), described in Section 9.5.1. Because of the small n in this example, significance values for TZSQ and MXROOT are not printed out. Note that, unlike other programs reviewed here, the same F ratio is estimated for Wilks' Lambda and Hotelling-Lawley TRACE.

The last source of variance in the source table is ERROR. This section contains error SS and MS for each DV, in turn. These mean square values form the denominators in the univariate F tests above.

In SYSTAT MGLH (Table 9.5) the CATEGORY instruction defines the IVs. An equation is formed in the MODEL statement with DVs on the left of the equation and effects on the right. The program then requires a separate HYPOTHESIS instruction for each main effect and interaction to be tested.

Separate source tables are provided for each effect in the output. Each source table begins with univariate F tests for each of the DVs (giving the usual SS, DF, MS, F, and P level) followed by four fully labeled multivariate tests. The last, THETA, is Roy's gcr, with its three df parameters.

In SAS GLM (Table 9.6) IVs are defined in a CLASS instruction while the MODEL instruction defines the DVs and the effects to be considered. The NOUNI instruction suppresses printing of descriptive statistics and univariate F tests. MANOVA H = _ALL_ requests tests of all main effects and interactions listed in the MODEL instruction, and SHORT condenses the printout.

The output begins with some interpretative information followed by separate sections for TREATMNT, DISABLTY, and TREATMNT*DISABLTY. Each source table is preceded by information about characteristic roots and vectors of the error SSCP matrix (as discussed in Chapters 6, 12, and 13), and the three df parameters (Section 9.4.1). Each source table shows results of four multivariate tests, fully labeled.

9.4.3 Multivariate Analysis of Covariance

In MANCOVA, the linear combination of DVs is adjusted for differences in the covariates. The adjusted linear combination of DVs is the combination that would be obtained if all participants had the same scores on the covariates. For this example, preexperimental IQ scores (listed in parentheses in Table 9.1) are used as covariates.

In MANCOVA the basic partition of variance is the same as in MANOVA. However, all the matrices—$\mathbf{Y}_{ikm}$, $\mathbf{D}_k$, $\mathbf{T}_m$, $\mathbf{DT}_{km}$, and $\mathbf{GM}$—have three entries in our example; the first entry is the covariate (IQ score) and the second two entries are the two DV scores (WRAT-R and WRAT-A). For example, for the first child with mild disability and treatment, the column matrix of covariate and DV scores is

$$\mathbf{Y}_{111} = \begin{bmatrix} 110 \\ 115 \\ 108 \end{bmatrix} \begin{matrix} \text{(IQ)} \\ \text{(WRAT-R)} \\ \text{(WRAT-A)} \end{matrix}$$

As in MANOVA, difference matrices are found by subtraction, and then the squares and cross-products matrices are found by multiplying each difference matrix by its transpose to form the S matrices.

At this point another departure from MANOVA occurs. The S matrices are partitioned into sections corresponding to the covariates, the DVs, and the cross products of covariates and DVs. For the example, the cross-products matrix for the main effect of treatment is

$$\mathbf{S}_T = \begin{bmatrix} [2.00] & [64.67 \quad 54.67] \\ \begin{bmatrix} 64.67 \\ 54.67 \end{bmatrix} & \begin{bmatrix} 2090.89 & 1767.56 \\ 1767.56 & 1494.22 \end{bmatrix} \end{bmatrix}$$

The lower right-hand partition is the $\mathbf{S}_T$ matrix for the DVs (or $\mathbf{S}_T^{(Y)}$), and is the same as the $\mathbf{S}_T$ matrix developed in Section 9.4.1. The upper left matrix is the sum of squares for the covariate (or $\mathbf{S}_T^{(X)}$). (With additional covariates, this segment becomes a full sum-of-squares and cross-products matrix.) Finally, the two off-diagonal segments contain cross products of covariates and DVs (or $\mathbf{S}_T^{(XY)}$).

Adjusted or S* matrices are formed from these segments. The S* matrix is the sums-of-squares and the cross-products of DVs adjusted for effects of covariates. Each sum of squares and each cross product is adjusted by a value that reflects variance due to differences in the covariate.

In matrix terms, the adjustment is

$$\mathbf{S}^* = \mathbf{S}^{(Y)} - \mathbf{S}^{(YX)}(\mathbf{S}^{(X)})^{-1}\mathbf{S}^{(XY)} \tag{9.7}$$

The adjusted cross-products matrix S* is found by subtracting from the unadjusted cross-products matrix of DVs ($\mathbf{S}^{(Y)}$) a product based on the cross-products matrix for covariate(s) ($\mathbf{S}^{(X)}$) and cross-products matrices for the relation between the covariates and the DVs ($\mathbf{S}^{(YX)}$ and $\mathbf{S}^{(XY)}$).

The adjustment is made for the regression of the DVs (Y) on the covariates (X). Because $\mathbf{S}^{(XY)}$ is the transpose of $\mathbf{S}^{(YX)}$, their multiplication is analogous to a squaring operation. Multiplying by the inverse of $\mathbf{S}^{(X)}$ is analogous to division. As shown in Chapter 3 for simple scalar numbers, the regression coefficient is the sum of cross products between X and Y, divided by the sum of squares for X.

An adjustment is made to each S matrix to produce S* matrices. The S* matrices are 2×2 matrices, but their entries are usually smaller than those in the original MANOVA S matrices. For the example, the reduced S* matrices are

$$\mathbf{S}_D^* = \begin{bmatrix} 388.18 & 500.49 \\ 500.49 & 654.57 \end{bmatrix} \qquad \mathbf{S}_T^* = \begin{bmatrix} 2059.50 & 1708.24 \\ 1708.24 & 1416.88 \end{bmatrix}$$

$$\mathbf{S}_{DT}^* = \begin{bmatrix} 2.06 & 0.87 \\ 0.87 & 19.61 \end{bmatrix} \qquad \mathbf{S}_{S(DT)}^* = \begin{bmatrix} 528.41 & -26.62 \\ -26.62 & 324.95 \end{bmatrix}$$

Note that, as in the lower right-hand partition, cross-products matrices may have negative values for entries other than the major diagonal which contains sums of squares.

Tests appropriate for MANOVA are applied to the adjusted S* matrices. Ratios of determinants are formed to test hypotheses about main effects and interactions by using Wilks' Lambda criterion (Equation 9.4). For the example, the determinants of the four matrices needed to test the three hypotheses (two main effects and the interaction) are

$$|\mathbf{S}_{S(DT)}^*| = 171032.69$$

$$|\mathbf{S}_D^* + \mathbf{S}_{S(DT)}^*| = 673383.31$$

$$|\mathbf{S}_T^* + \mathbf{S}_{S(DT)}^*| = 1680076.69$$

$$|\mathbf{S}_{DT}^* + \mathbf{S}_{S(DT)}^*| = 182152.59$$

The source table for MANCOVA, analogous to that produced for MANOVA, for the sample data is in Table 9.7.

One new item in this source table that is not in the MANOVA table of Section 9.4.1 is the variance in the DVs due to the covariate. (With more than one covariate, there is a line for combined covariates and a line for each of the individual covariates.) As in ANCOVA, one degree of freedom for error is used for each covariate so that df_2 and s of Equation 9.5 are modified. For MANCOVA, then,

$$s = \sqrt{\frac{(p + q)^2 (df_{effect})^2 - 4}{(p + q)^2 + (df_{effect})^2 - 5}} \tag{9.8}$$

TABLE 9.7 MULTIVARIATE ANALYSIS OF COVARIANCE OF WRAT-R AND WRAT-A SCORES

Source of variance	Wilks' Lambda	Hypoth. df	Error df	Multivariate F
Covariate	.58485	2.00	10.00	3.54913
Treatment	.10180	2.00	10.00	44.11554**
Disability	.25399	4.00	20.00	4.92112*
Treatment by disability	.93896	4.00	20.00	0.15997

* $p < .01$.
** $p < .001$.

where q is the number of covariates and all other terms are defined as in Equation 9.5. And

$$df_2 = s\left[(df_{error}) - \frac{(p + q) - df_{effect} + 1}{2}\right] - \left[\frac{(p + q)(df_{effect}) - 2}{2}\right]$$

Approximate F is used to test the significance of the covariate-DV relationship as well as main effects and interactions. If a significant relationship is found, Wilks' Lambda is used to find the strength of association as shown in Equation 9.6.

9.5 SOME IMPORTANT ISSUES

9.5.1 Criteria for Statistical Inference

Several multivariate statistics are available in MANOVA programs to test significance of main effects and interactions: Wilks' Lambda, Hotelling's trace criterion, Pillai's criterion, as well as Roy's gcr criterion. When an effect has only two levels (1 df), the F tests for Wilks' Lambda, Hotelling's trace, and Pillai's criterion are identical. And usually when an effect has more than two levels (df $>$ 1), the F values are often different but all three statistics are either significant or all are nonsignificant. Occasionally, however, some of the statistics are significant while others are not, and the researcher is left wondering which result to believe.

When there is only one degree of freedom for effect there is only one way to combine the DVs to separate the two groups from each other. However, when there is more than one degree of freedom for effect, there is more than one way to combine DVs to separate groups. For example, with three groups, one way of combining DVs may separate the first group from the other two while the second way of combining DVs separates the second group from the third. Each way of combining DVs is a dimension along which groups differ (as described in gory detail in Chapter 11) and each generates a statistic.

When there is more than one degree of freedom for effect, Wilks' Lambda, Hotelling's trace criterion, and Pillai's criterion pool the statistics from each dimension to test the effect; Roy's gcr criterion uses only the first dimension (in our example, the way of combining DVs that separates the first group from the other two) and is the preferred test statistic for a few researchers (Harris, 1975). Most researchers, however, use one of the pooled statistics to test the effect (Olson, 1976).

Wilks' Lambda, defined in Equation 9.4 and Section 9.4.1, is a likelihood ratio statistic that tests the likelihood of the data under the assumption of equal population mean vectors for all groups against the likelihood under the assumption that population mean vectors are identical to those of the sample mean vectors for the different groups. Wilks' Lambda is the pooled ratio of error variance to effect variance plus error variance. Hotelling's trace is the pooled ratio of effect variance to error variance. Pillai's criterion is simply the pooled effect variances. Equations describing these criteria are available in the SPSS Update 7–9 manual (Hull and Nie, 1981) and the SPSS MANOVA manual (Cohen and Burns, 1977).

Wilks' Lambda, Hotelling's trace, and Roy's gcr criterion are often more powerful than Pillai's criterion when there is more than one dimension but the first dimension provides most of the separation of groups; they are less powerful when separation of groups is distributed over dimensions. But Pillai's criterion is said to be more robust than the other three (Olson, 1979). As sample size decreases, unequal n's appear, and the assumption of homogeneity of variance-covariance matrices is violated (Section 9.3.2.4), the advantage of Pillai's criterion in terms of robustness is more important. When the research design is less than ideal, then, Pillai's criterion is the criterion of choice.

In terms of availability, all the MANOVA programs reviewed here provide Wilks' Lambda, as do most research reports, so that Wilks' Lambda is the criterion of choice unless there is reason to use Pillai's criterion. Programs differ in the other statistics provided (see Section 9.6).

In addition to potentially conflicting significance tests for multivariate F is the irritation of a nonsignificant multivariate F but a significant univariate F for one of the DVs. If the researcher measures only one DV—the right one—the effect is significant, but because more DVs are measured, it is not. Why doesn't MANOVA combine DVs with a weight of 1 for the significant DV and a weight of zero for the rest? In fact, MANOVA comes close to doing just that, but multivariate F is often not as powerful as univariate or stepdown F, and significance can be lost. If this happens, about the best one can do is report the nonsignificant multivariate F and offer the univariate and/or stepdown result as a guide to future research, with only tentative interpretation.

9.5.2 Assessing DVs

When a main effect or interaction is significant in MANOVA, the researcher has usually planned to pursue the finding to discover which DVs are affected. But the problems of assessing DVs for significant multivariate effects are similar to the problems of assigning importance to IVs in multiple regression (Chapter 5). First, there are multiple significance tests so some adjustment is necessary for inflated Type I error. Second, if DVs are uncorrelated, there is no ambiguity in assignment of variance to them, but if DVs are correlated, assignment of overlapping variance to DVs is problematical.

9.5.2.1 Univariate F If pooled within-group correlations among DVs are zero, univariate ANOVAs, one per DV, give the relevant information about their importance. Using ANOVA for uncorrelated DVs is analogous to assessing importance of IVs in multiple regression by the magnitude of their individual correlations with the DV. The DVs that have significant univariate F's are the important ones, and they can be ranked in importance by strength of association. However, because of inflated Type I error rate due to multiple testing, more stringent α levels are required.

Because there are multiple ANOVA's, a Bonferroni type adjustment is made for inflated Type I error. The researcher assigns alpha for each DV so that alpha for the set of DVs does not exceed some critical value.

$$\alpha = 1 - (1 - \alpha_1)(1 - \alpha_2) \cdots (1 - \alpha_p) \qquad (9.9)$$

The Type I error rate (α) is based on the error rate for testing the first DV (α_1), the second DV (α_2), and all other DVs to the pth, or last, DV (α_p).

All the alphas can be set at the same level, or more important DVs can be given more liberal alphas. For example, if there are four DVs and α for each DV is set at .01, the overall α level according to Equation 9.9 is .039, acceptably below .05 overall. Or, if alpha is set at .02 for 2 DVs, and at .001 for the other 2 DVs, overall alpha is .042, also below .05.

When DVs are correlated, there are two problems with univariate F's. First, correlated DVs measure overlapping aspects of the same behavior. To say that two of them are both "significant" mistakenly suggests that the IV affects two different behaviors. For example, if the two DVs are Stanford-Binet IQ and WISC IQ, they are so highly correlated that an IV that affects one surely affects the other. The second problem with reporting univariate F's for correlated DVs is inflation of Type I error rate; with correlated DVs, the univariate F's are not independent and no straightforward adjustment of the error rate is possible.

Although reporting univariate F for each DV is a simple tactic, the report should also contain the pooled within-group correlations among DVs so the reader can make necessary interpretive adjustments. The pooled within-group correlation matrix is provided by SPSS MANOVA, SAS GLM, and SYSTAT MGLH; the pooled within-group variance-covariance matrix is available through BMDP7M, and entries can be converted to correlations as described in Chapter 1.

In the example of Tables 9.1 and 9.2, there is a significant multivariate effect of treatment (and of disability, although, as previously noted, the disability effect is not interesting in this example). It is appropriate to ask which of the two DVs is affected by treatment. Univariate ANOVAs for WRAT-R and WRAT-A are in Tables 9.8 and 9.9, respectively. The pooled within-group correlation between WRAT-R and WRAT-A is .057 with 12 df. Because the DVs are uncorrelated, univariate F with adjustment of α for multiple tests is appropriate. There are two DVs, so each is set at α at .025.[12] With 2 and 12 df, critical F is 5.10; with 1 and 12 df, critical F is 6.55. There is a main effect of treatment (and disability) for both WRAT-R and WRAT-A.

9.5.2.2 Stepdown Analysis[13]   The problem of correlated univariate F tests with correlated DVs is resolved by stepdown analysis (Bock, 1966; Bock and Haggard, 1968). Stepdown analysis of DVs is analogous to testing the importance of IVs in multiple regression by hierarchical analysis. Priorities are assigned to DVs according to theoretical or practical considerations.[14] The highest-priority DV is tested in univariate ANOVA, with appropriate adjustment of alpha. The rest of the DVs are tested

---

[12] When the design is very complicated and generates many main effects and interactions, further adjustment of α is necessary in order to keep overall α under .15 or so, across the ANOVAs for the DVs.

[13] Stepdown analysis can be run in lieu of MANOVA where a significant stepdown F is interpreted as a significant multivariate effect for the main effect or interaction.

[14] It is also possible to assign priority on the basis of statistical criteria such as univariate F, but the analysis suffers all the problems inherent in stepwise regression, discussed in Chapter 5.

TABLE 9.8 UNIVARIATE ANALYSIS OF VARIANCE
OF WRAT-R SCORES

Source	SS	df	MS	F
D	520.7778	2	260.3889	5.7439
T	2090.8889	1	2090.8889	46.1225
DT	2.1111	2	1.0556	0.0233
S(DT)	544.0000	12	45.3333	

TABLE 9.9 ANALYSIS OF VARIANCE OF WRAT-A
SCORES

Source	SS	df	MS	F
D	1126.7778	2	563.3889	12.5352
T	1494.2222	1	1494.2222	33.2460
DT	52.7778	2	26.3889	0.5871
S(DT)	539.5668	12	44.9444	

in a series of ANCOVAs; each successive DV is tested with higher-priority DVs as covariates to see what, if anything, it adds to the combination of DVs already tested. Because successive ANCOVAs are independent, adjustment for inflated Type I error due to multiple testing is the same as in Section 9.5.2.1.

For the example, let's assign WRAT-R scores higher priority since reading problems represent the most common presenting symptoms for learning disabled children. To keep overall alpha below .05, individual alpha levels are set at .025 for each of the two DVs. WRAT-R scores are analyzed through univariate ANOVA, as displayed in Table 9.8. Because the main effect of disability is not interesting and the interaction is not statistically significant in MANOVA (Table 9.2), the only effect of interest is treatment. The critical value for testing the treatment effect (6.55 with 1 and 12 df at α = .025) is clearly exceeded by the obtained F of 46.1225.

WRAT-A scores are analyzed in ANCOVA with WRAT-R scores as covariate. The results of this analysis appear in Table 9.10.[15] For the treatment effect, critical F with 1 and 11 df at α = .025 is 6.72. This exceeds the obtained F of 5.49. Thus, according to stepdown analysis, the significant effect of treatment is represented in WRAT-R scores, with nothing added by WRAT-A scores.

Note that WRAT-A scores show significant univariate but not stepdown F. Because WRAT-A scores are not significant in stepdown analysis does not mean they are unaffected by treatment but rather that no unique variability is shared with treatment after adjustment for WRAT-R.

This procedure can be extended to sets of DVs through MANCOVA. If the DVs fall into categories, such as scholastic variables and attitudinal variables, one can ask whether there is any change in attitudinal variables as a result of an IV, after adjustment for the effects of scholastic variables. The attitudinal variables serve as

---

[15] A full stepdown analysis is produced as an option through SPSS[x] MANOVA. For illustration, however, it is helpful to show how the analysis develops.

TABLE 9.10 ANALYSIS OF COVARIANCE OF
WRAT-A SCORES, WITH WRAT-R
SCORES AS THE COVARIATE

Source	SS	df	MS	F
Covariate	1.7665	1	1.7665	0.0361
D	538.3662	2	269.1831	5.5082
T	268.3081	1	268.3081	5.4903
DT	52.1344	2	26.0672	0.5334
S(DT)	537.5668	11	48.8679	

DVs in MANCOVA while the scholastic variables serve as covariates. This type of analysis is demonstrated in the hierarchical discriminant function analysis of Chapter 11.

9.5.2.3 Choosing among Strategies for Assessing DVs You may find the procedures of Sections 11.6.3 and 11.6.4 more useful than univariate or stepdown F for assessing DVs when you have a significant multivariate main effect with more than two levels. Similarly, you may find the procedures described in Section 10.5.1 helpful for assessment of DVs if you have a significant multivariate interaction.

The choice between univariate and stepdown F is not always easy, and often you want to use both. When there is very little correlation among the DVs, univariate F with adjustment for Type I error is acceptable. When DVs are correlated, stepdown F is preferable on grounds of statistical purity, but you have to prioritize the DVs and the results can be difficult to interpret.

If DVs are correlated and there is some compelling priority ordering of them, stepdown analysis is clearly called for, with univariate F's and pooled within-cell correlations reported simply as supplemental information. For significant lower-priority DVs, marginal and/or cell means adjusted for higher-priority DVs are reported and interpreted.

If the DVs are correlated but the ordering is somewhat arbitrary, an initial decision in favor of stepdown analysis is made. If the pattern of results from stepdown analysis makes sense in the light of the pattern from univariate analysis, interpretation takes both patterns into account with emphasis on DVs that are significant in stepdown analysis. If, for example, a DV has a significant univariate F but a nonsignificant stepdown F, interpretation is straightforward: the variance the DV shares with the IV is already accounted for through overlapping variance with one or more higher-priority DVs. This is the interpretation of WRAT-A in the preceding section and the strategy followed in Section 9.7.

But if a DV has a nonsignificant univariate F and a significant stepdown F, interpretation is much more difficult. In the presence of higher-order DVs as covariates, the DV suddenly takes on "importance." In this case, interpretation is tied to the context in which the DVs entered the stepdown analysis. It may be worthwhile at this point, especially if there is only a weak basis for ordering DVs, to forgo evaluation of statistical significance of DVs and resort to simple description. After finding a significant multivariate effect, unadjusted marginal and/or cell means are reported for DVs with high univariate F's but significance levels are not given.

An alternative to attempting interpretation of either univariate or stepdown F is interpretation of discriminant functions, as discussed in Section 11.6.3. This process is facilitated when SPSS^x MANOVA, SAS GLM, or SYSTAT MGLH is used because information about the discriminant functions is provided as a routine part of the output.

9.5.3 Specific Comparisons and Trend Analysis

When there are more than two levels in a significant multivariate main effect and when a DV is important to the main effect, the researcher often wants to perform specific comparisions or trend analysis of the DV to pinpoint the source of the significant difference. Similarly, when there is a significant multivariate interaction and a DV is important to the interaction, the researcher follows up the finding with comparisons on the DV.[16] Review Sections 3.2.6, 8.5.2.3, and 10.5.1 for examples and discussions of comparisons. The issues and procedures are the same for individual DVs in MANOVA as in ANOVA.

Comparisons are either planned (performed in lieu of omnibus F) or post hoc (performed after omnibus F to snoop the data). When comparisons are post hoc, an extension of the Scheffé procedure is used to protect against inflated Type I error due to multiple tests. The procedure is very conservative but allows for an unlimited number of comparisons. Following Scheffé for ANOVA (see Section 3.2.6), the tabled critical value of F is multiplied by the degrees of freedom for the effect being tested to produce an adjusted, and much more stringent, F. If marginal means for a main effect are being contrasted, the degrees of freedom are those associated with the main effect. If cell means are being contrasted, the degrees of freedom are those associated with the interaction.

9.5.4 Design Complexity

In between-subjects designs with more than two IVs extension of MANOVA is straight-forward as long as sample sizes are equal within each cell of the design. The partition of variance continues to follow ANOVA, with a variance component computed for each main effect and interaction. The pooled variance-covariance matrix due to differences among subjects within cells serves as the single error term. Assessment of DVs and comparisons proceed as described in Sections 9.5.2 and 9.5.3.

Two major design complexities that arise, however, are inclusion of within-subjects IVs and unequal sample sizes in cells.

9.5.4.1 Within-Subjects and Between-Within Designs The simplest design with repeated measures is a one-way within-subjects design where the same subjects are measured on a single DV on several different occasions. The design can be complicated by addition of between-subjects IVs or more within-subjects IVs. Consult Chapters

---

[16] Comparisons and trend analysis can also be performed on a composite DV as illustrated in Tabachnick and Fidell (1983) Section 8.5.2 but the results for a composite DV are less interpretable.

3 and 8 for discussion of some of the problems that arise in ANOVA with repeated measures.

Repeated measures is extended to MANOVA when the researcher measures several DVs on several different occasions. The occasions can be viewed in two ways. In the traditional sense, occasions produce a within-subjects IV with as many levels as occasions (Chapter 3). Alternatively, occasions can be treated as separate DVs—one DV per occasion (Section 9.2.8). In this latter view, if there is more than one DV measured on each occasion, the design is said to be doubly multivariate. (There is no distinction between the two views when there are only two levels of the within-subjects IV.) Section 10.5.3 has an example of a doubly multivariate analysis of a small data set with a between-subjects IV (PROGRAM), a within-subjects IV (MONTH), and two DVs (WTLOSS and ESTEEM), both measured three times. Generalizations from the example are reasonably straightforward.

9.5.4.2 Unequal Sample Sizes When cells in a factorial ANOVA have an unequal number of scores, the sum of squares for effect plus error no longer equals the total sum of squares, and tests of main effects and interactions are not independent. There are a number of ways to adjust for overlap in sums of squares (cf. Woodward and Overall, 1975), as discussed in some detail in Section 8.5.2.2, particularly Table 8.11. Both the problem and the solutions generalize to MANOVA.

All the MANOVA programs described in Section 9.6 adjust for unequal n. SPSS MANOVA offers both Method 1 adjustment (METHOD = SSTYPE(UNIQUE)) and Method 3 adjustment (METHOD = SSTYPE(SEQUENTIAL)). Method 3 adjustment with survey data through SPSS[x] MANOVA is shown in Section 9.7.2. BMDP4V allows very flexible specification of method through choice of cell-weighting procedure. In SAS GLM Method 1 (called TYPE III or TYPE IV) is the default among four options available. For SYSTAT MGLH, default is also Method 1; alternative methods require specification of custom sums of squares for error terms.

9.6 COMPARISON OF PROGRAMS

SPSS[x], BMDP, SAS, and SYSTAT all have highly flexible and full-featured MANOVA programs, as seen in Table 9.11. One-way between-subjects MANOVA is also available through discriminant function programs, as discussed in Chapter 11.

9.6.1 SPSS Package

Within the original SPSS series of programs (Nie et al., 1975), no program was specifically designed for MANOVA (although one-way between-subjects MANOVA could be analyzed through DISCRIMINANT). The new MANOVA, however, is a comprehensive, flexible program that gives SPSS and SPSS[x] extended capability for both MANOVA and discriminant function analyses, as seen in Table 9.11.

The program offers several methods of adjustment for unequal n and is the only program that performs stepdown analysis as an option (Section 9.5.2.2). Should Pillai's criterion be desired as the multivariate significance test (Section 9.5.1), it and several other alternatives are available through SPSS MANOVA. The program is also reasonably simple to use for comparisons on both between- and within-subjects IVs (Section 10.5.1).

Except for outliers,[17] tests of assumptions are readily available in MANOVA. Multicollinearity and homogeneity of variance-covariance matrices are readily tested through within-cell correlations and homogeneity of dispersion matrices, respectively.

For between-subjects designs, Bartlett's test of sphericity tests the null hypothesis that correlations among DVs are zero; if they are, univariate F (with Bonferroni adjustment) is used instead of stepdown F to test the importance of DVs (Section 9.5.2.1). In repeated measures designs, the sphericity test evaluates the homogeneity of covariance assumption; if the assumption is rejected (that is, if the test is significant), one of the alternatives to repeated measures ANOVA—MANOVA, for instance—is appropriate (Section 8.5.5). Although this test of homogeneity of covariance is provided, adjustment for violation of the assumption is available only in the CDC CYBER version of the program. The FMAX test for homogeneity of variance among DVs also facilitates the choice between repeated-measures univariate ANOVA and MANOVA.

For designs with covariates or when stepdown F is used, homogeneity of regression is readily tested through control language illustrated in Table 9.14. If the assumption is violated, the manuals describe procedures for ANCOVA with separate regression estimates. Adjusted cell and marginal means can be obtained, although in different ways. The PMEANS procedure provides adjusted cell means. For adjusted marginal means, however, the CONSPLUS procedure is used, illustrated in Sections 9.7.2 and 9.7.3.

Finally, a principal components analysis can be performed on the DVs, as described in the manuals. In the case of multicollinearity or singularity among DVs (see Chapter 4), principal components analysis can be used to produce composite variables that are orthogonal to one another. However, the program still performs MANOVA on the raw DV scores, not the component scores. If MANOVA for component scores is desired, use the results of PCA and the COMPUTE facility to generate component scores for use as DVs.

9.6.2 BMD Series

The MANOVA program in the BMDP series is P4V. The program is also known as URWAS (University of Rochester Weighted Analysis of Variance System). It is unique among the programs reviewed here in that it uses a cell weighting system to adjust for unequal n and to specify hypotheses to be tested. Cell weighting can be used for situations where population sizes are unequal even though sample sizes are equal.

For repeated measures designs, both the Huynh-Feldt and the Greenhouse-Geisser-Imhof corrections for degrees of freedom are available to adjust for violation of the assumption of homogeneity of covariance.

Statistics appropriate for doubly multivariate analysis (in which levels of a

---

[17] The CYBER implentation of MANOVA provides graphic screening for multivariate outliers, but no formal statistical evaluation of them. Further, outliers are not assessed with respect to their own group.

TABLE 9.11 COMPARISON OF PROGRAMS FOR MULTIVARIATE ANALYSIS OF VARIANCE AND COVARIANCE

Feature	BMDP4V	SPSS MANOVA	SPSS^x MANOVA	SYSTAT MGLH	SAS GLM
Input					
Post hoc comparisons	No	No	No	No	Yes
Variety of strategies for unequal n	Yes	Yes	Yes	Yes	Yes
Alternative languages or procedures to specify design	Yes	Yes	Yes	Yes	Yes
Specific comparisons	Yes	Yes	Yes	Yes	Yes
Tests of simple effects (complete)	Yes	No[a]	Yes	No	No
Output					
Standard source table	Yes	PRINT = SIGNIF(BRIEF)	PRINT = SIGNIF(BRIEF)	No	No
F matrix of cell comparisons	No	No	No	No	No
Design matrix	Yes	Yes	Yes	Yes[b]	Yes[c]
Cell covariance matrices	No	Yes	Yes	No	No
Cell covariance matrix determinants	No	Yes	Yes	No	No
Cell correlation matrices	No	Yes	Yes	No	No
Cell SSCP matrices	No	Yes	Yes	No	No
Cell SSCP determinants	No	No	No	No	No
Marginal standard deviation for factorial design	Yes	Yes	Yes	No	No
Marginal means for factorial design	Yes	Yes	Yes	No	No
Weighted marginal means	Yes	Yes	Yes	No	No
Cell means	Yes	Yes	Yes	No	Yes
Cell standard deviations	Yes	Yes	Yes	No	No
Cell and marginal variance	Yes	No	No	No	No
Confidence interval around cell means	No	Yes	Yes	No	No
Cell and marginal minimum and maximum values	Yes	No	No	No	No
Adjusted cell means	No	PMEANS	PMEANS	No	LS MEANS
Adjusted marginal means	No	Yes[d]	Yes[d]	No	LS MEANS
Wilks' Lambda (U statistic) with approx. F statistic	LRATIO	Yes	Yes	Yes	LRATIO
Test for multivariate outliers	No[e]	No[a]	No	Yes	No
Canonical (discriminant function) statistics[f]	Eigenvalues and eigenvectors	Yes	Yes	Yes	Yes

Univariate F tests	Yes	Yes	Yes	Yes	Yes
Averaged univariate F tests	No	Yes	Yes	No	No
Stepdown F tests (DVs)	No[g]	Yes	Yes	No[g]	No[g]
Criteria other than Wilks'	Yes	Yes	Yes	Yes	Yes
Greenhouse-Geisser-Imhof and Huynh-Feldt statistic	Yes	No[a]	No	No	Yes
Bartlett test of sphericity for homogeneity of covariance	No	Yes	Yes	No	No
Tests for univariate homogeneity of variance	No	Box's M	Yes	No	No
Test for homogeneity of covariance matrices	No	Box's M	Box's M	No	No
Principal components analysis of residuals	No	Yes	Yes	Yes	No
Hypothesis SSCP matrices	Yes	No	No	Yes	Yes
Inverse of hypothesis SSCP matrices	No	No	No	No	Yes
Pooled with-cell error SSCP matrix	Yes	Yes	Yes	Yes	Yes
Pooled within-cell covariance matrix	No	Yes	Yes	Yes	Yes
Pooled within-cell correlation matrix	No	Yes	Yes	Yes	Yes
Determinants of pooled within-cell correlation matrix	No	Yes	Yes	No	No
Homogeneity of regression	No	Yes	Yes	Yes	No
ANCOVA with separate regression estimates	No	Yes	Yes	No	No
Tukey's test for nonadditivity	No	Yes	Yes	No	No
Effect sizes	No	No[a]	No	No	No
Observed power values	No	No[a]	No	No	No
Tolerance	No	No	No	No	Yes
R^2 for model	No	No	No	No	Yes
R^2 for each DV	No	No	No	Yes	No
Coefficient of variation	No	No	No	No	Yes
Normalized plots for each DV and covariate	No	Yes	Yes	No[h]	No
Predicted values and residuals for each case	No	Yes	No	No	Yes
Confidence limits for predicted values	No	No	No	No[h]	Yes
Residuals plots	No	Yes	Yes	No[h]	No[i]

[a] Available in CDC CYBER SPSS-6000 Version 9.0.
[b] Available by saving MODEL.
[c] Prints transformations matrices that define contrasts.
[d] Available through CONSPLUS procedure; see Section 9.7.
[e] Available through BMDPAM or BMDP7M.
[f] Discussed more fully in Chapter 11.
[g] Can be done through successive ANCOVAs as subproblems.
[h] Available through GRAPH program.
[i] Available through PLOT procedure.

within-subjects IV are treated as separate DVs) are automatically provided when multiple DVs are specified (cf. Chapter 10). Weighted marginal means are given that are useful if weights other than sample sizes are used (see Section 8.5.2.2). Adjusted (for covariates) cell and marginal means are planned but have not been implemented at the time of this writing. Three multivariate test criteria, as well as univariate F tests, are automatically provided. Specific comparisons include a simple effects procedure as described in Section 10.5.1.

For a factorial design in which tests of all main effects and interactions are desired, a simple statement so indicates. For complex and/or nonorthogonal designs, Model Description Language is available, in which the user specifies the design matrix of contrast coefficients and/or other design goodies. Finally, a structural-equation procedure allows the user to write out an equation that specifies the design and desired tests.

Neither stepdown analysis nor tests for homogeneity of regression are available in BMDP4V, although tests for MANOVA stepdown analysis can be done through BMDP1V, as described in Chapter 8. No direct evidence is available to evaluate multicollinearity, but the program does deal with the problem. If a sum-of-squares and products matrix is multicollinear, it is conditioned and a pseudoinverse calculated. And if there is any problem with the accuracy of a statistic, warning messages regarding precision are printed out.

9.6.3 SYSTAT System

In SYSTAT, the MGLH program serves for all varieties of linear modeling—multiple regression, ANOVA and ANCOVA, MANOVA and MANCOVA, and discriminant function analysis. Among the SYSTAT programs, MGLH is by far the most flexible and complex.

Model 1 adjustment for unequal-n is provided by default, along with a strong argument as to its benefits. Other options are available, however, by specification of error terms. Several criteria are provided for tests of multivariate hypotheses, along with a great deal of flexibility in specifying these hypotheses. Mahalanobis distance is provided for evaluation of multivariate outliers, but the values are the square root of those typically given. To apply a χ^2 criterion to evaluate significance of outliers, you must first square the distance value produced by SYSTAT.

For MANCOVA or stepdown F, tests are provided for homogeneity of regression. However, there is no provision in MGLH or elsewhere in the package for adjusted cell or marginal means. Other univariate statistics are not provided in the program, but they can be obtained through the STATS module.

For repeated-measures designs, there is no test for homogeneity of covariance. Instead, the manual (Wilkinson, 1986) recommends the doubly multivariate approach (cf. Chapter 10) that does not require satisfaction of this assumption.

Like SPSS^x MANOVA, principal components analysis can be done on the pooled within-cell correlation matrix. But also like the SPSS program, the MANOVA is performed on the original scores.

9.6.4 SAS System

MANOVA in SAS is done through the PROC GLM. This general linear model program, like that of SYSTAT MGLH, has great flexibility in testing models and specific comparisons. Four types of adjustment for unequal-n are available, called TYPE I through TYPE IV estimable functions (cf. 8.5.2.2); this program is considered by some to have provided the archetypes of the choices available for unequal-n adjustment. Cell means, adjusted cell means, and marginal means are printed out with the LSMEANS instruction. SAS has no test for multivariate outliers, however.

SAS GLM provides Greenhouse-Geisser and Huynh-Feldt adjustments to degrees of freedom and corresponding significance values for violation of homogeneity of covariance in repeated measures designs. There is no explicit test for homogeneity of regression, but because this program can be used for any form of multiple regression, the assumption can be tested as a regression problem where the interaction between the covariate(s) and IV(s) is an explicit term in the regression equation (Section 8.5.1).

Of the programs reviewed, GLM has the most extensive selection of options for testing contrasts. In addition to a CONTRAST procedure for specifying comparisons, numerous post hoc procedures are available.

There is abundant information about residuals, as expected from a program that can be used for multiple regression. Should you want to plot residuals, however, a run through the PLOT procedure is required. As with most SAS programs, the output requires a fair amount of effort to decode until you become accustomed to the style.

9.7 COMPLETE EXAMPLES OF MULTIVARIATE ANALYSIS OF VARIANCE AND COVARIANCE

In the research described in Appendix B, Section B.1, there is interest in whether the means of several of the variables differ as a function of sex role identification. Are there differences in self-esteem, introversion-extraversion, neuroticism, and so on, associated with a woman's masculinity and femininity?

Sex role identification is defined by the masculinity and femininity scales of the Bem Sex Role Inventory (Bem, 1974). Each scale is divided at its median to produce two levels of masculinity (high and low), two levels of femininity (high and low), and four groups: Undifferentiated (low femininity, low masculinity), Feminine (high femininity, low masculinity), Masculine (low femininity, high masculinity), and Androgynous (high femininity, high masculinity). The design produces a main effect of masculinity, a main effect of femininity, and a masculinity-femininity interaction.

DVs for this analysis are self-esteem (ESTEEM), internal versus external locus of control (CONTROL), attitudes toward women's role (ATTROLE), socioeconomic level (SEL2), introversion-extraversion (INTEXT), and neuroticism (NEUROTIC).

Omnibus MANOVA (Section 9.7.2) asks whether these DVs are associated with the two IVs (femininity and masculinity) or their interaction. Then stepdown analysis, in conjunction with the univariate F values, allows us to examine the pattern of relationships between DVs and each IV.

In a second example (Section 9.7.3), MANCOVA is performed with SEL2, CONTROL, and ATTROLE used as covariates and ESTEEM, INTEXT, and NEU-ROTIC used as DVs. The research question is whether the three personality DVs vary as a function of sex role identification (the two IVs and their interaction) after adjusting for differences in socioeconomic status, attitudes toward women's role, and beliefs regarding locus of control of reinforcements.

9.7.1 Evaluation of Assumptions

Before proceeding with MANOVA and MANCOVA, we assess the variables with respect to practical limitations of the techniques.

9.7.1.1 Unequal Sample Sizes and Missing Data
SPSS^X FREQUENCIES is run with SORT and SPLIT FILE to divide cases into the four groups. Data and distributions for each DV within each group are inspected for missing values, shape, and variance (see Table 9.12 for output on the CONTROL variable). The run reveals the presence of a case for which the CONTROL score is missing. No datum is missing on any

TABLE 9.12 SETUP AND SELECTED SPSS' FREQUENCIES OUTPUT
FOR MANOVA VARIABLES SPLIT BY GROUPS

```
            FILE HANDLE    MANOVA
            DATA LIST      FILE=MANOVA
                           /CASENO,ANDRM,FEM,MASC,ESTEEM,CONTROL,ATTROLE,
                           SEL2,INTEXT,NEUROTIC (A4,6F3.0,F8.5,F4.1,F3.0)
            VALUE LABELS   ANDRM 1 'UNDIFF' 2 'FEMNINE' 3 'MASCLNE' 4'ANDROGNS'/
                           FEM,MASC 1 'LOW' 2 'HIGH'
            MISSING VALUES  CONTROL(0)
            SORT CASES BY ANDRM
            SPLIT FILE BY ANDRM
            FREQUENCIES    VARIABLES=ESTEEM TO NEUROTIC/
                           FORMAT=NOTABLE/
                           HISTOGRAM=NORMAL/
                           STATISTICS=ALL/
```

CONTROL

```
      COUNT      VALUE     ONE SYMBOL EQUALS APPROXIMATELY  1.20 OCCURRENCES

       26        5.00    ******************:*****
       54        6.00    *****************************************:********
       51        7.00    *****************************************:
       18        8.00    ***************               .
       20        9.00    *********:*******
        3       10.00    *:*
                         I.........I.........I.........I.........I.........I
                         0        12        24        36        48        60
                                   HISTOGRAM FREQUENCY
```

```
   ANDRM:    2   FEMNINE
   CONTROL
```

```
   MEAN         6.773     STD ERR      .097     MEDIAN       7.000
   MODE         6.000     STD DEV     1.266     VARIANCE     1.603
   KURTOSIS     -.381     S E KURT    1.989     SKEWNESS      .541
   S E SKEW      .185     RANGE       5.000     MINIMUM      5.000
   MAXIMUM     10.000     SUM      1165.000
```

```
   VALID CASES    172     MISSING CASES     1
```

of the other DVs for the 369 women who were administered the Bem Sex Role Inventory. Deletion of the case with the missing value, then, reduces the available sample size to 368.

Sample sizes are quite different in the four groups: there are 71 Undifferentiated, 172 Feminine, 36 Masculine, and 89 Androgynous women in the sample. Because it is assumed that these differences in sample size reflect real processes in the population, the hierarchical approach to adjustment for unequal n is used with FEM (femininity) given priority over MASC (masculinity), and FEM by MASC (interaction between femininity and masculinity).

9.7.1.2 Multivariate Normality   The sample size of 368 includes over 35 cases for each cell of the 2×2 between-subjects design, far more than the 20 df for error suggested to assure multivariate normality of the sampling distribution of means, even with unequal sample sizes. Further, the distributions for the full run (of which CONTROL in Table 9.12 is a part) produce no cause for alarm. Skewness is not extreme and, when present, is roughly the same for the DVs.

Two-tailed tests are automatically performed by the computer programs used. That is, the F test looks for differences between means in either direction.

9.7.1.3 Linearity   The full output for the run of Table 9.12 reveals no cause for worry about linearity. All DVs in each group have reasonably balanced distributions so there is no need to examine scatterplots for each pair of DVs within each group. Had scatterplots been necessary, SPSS^x PLOT would have been used with the SORT and SPLIT FILE setup in Table 9.12.

9.7.1.4 Outliers   BMDPAM is used to check for univariate and multivariate outliers (cf. Table 8.16) within each of the four groups. Using a cutoff criterion of $p < .001$, no outliers are found.

9.7.1.5 Homogeneity of Variance-Covariance Matrices   As a preliminary check for robustness, sample variances (in the full run of Table 9.12) for each DV are compared across the four groups. For no DV does the ratio of largest to smallest variance approach $20:1$. As a matter of fact, the largest ratio is about $1.5:1$ for the Undifferentiated versus Androgynous groups on CONTROL.

Sample sizes are widely discrepant, with a ratio of almost $5:1$ for the Feminine to Masculine groups. However, with very small differences in variance and two-tailed tests, the discrepancy in sample sizes does not invalidate use of MANOVA. The very sensitive Box's M test for homogeneity of dispersion matrices (performed through SPSS MANOVA as part of the major analysis in Table 9.15) produces $F(63, 63020) = 1.07$, $p > .05$, confirming homogeneity of variance-covariance matrices.

9.7.1.6 Homogeneity of Regression   Because stepdown analysis is planned to assess the importance of DVs after MANOVA, a test of homogeneity of regression is necessary for each step of the stepdown analysis. Table 9.13 shows the SPSS MANOVA deck setup for tests of homogeneity of regression where each DV, in turn, serves as DV on one step and then becomes a covariate on the next step.

```
TITLE          HOMOGENEITY OF REGRESSION FOR LARGE SAMPLE MANOVA
FILE HANDLE    MANOVA
DATA LIST      FILE=MANOVA
               /CASENO,ANDRM,FEM,MASC,ESTEEM,CONTROL,ATTROLE,
               SEL2,INTEXT,NEUROTIC (A4,6F3.0,F8.5,F4.1,F3.0)
VALUE LABELS   ANDRM 1 'UNDIFF' 2 'FEMNINE' 3 'MASCLNE' 4 'ANDROGNS'/
               FEM,MASC 1 'LOW' 2 'HIGH'
MISSING VALUES  CONTROL(0)
MANOVA  ESTEEM,ATTROLE,NEUROTIC,INTEXT,CONTROL,SEL2 BY FEM(1,2) MASC(1,2)/
        PRINT=SIGNIF(BRIEF)/ NOPRINT=PARAMETERS(ESTIM)/
   ANALYSIS=ATTROLE/
   DESIGN=ESTEEM,FEM,MASC,FEM BY MASC, ESTEEM BY FEM + ESTEEM
          BY MASC + ESTEEM BY FEM BY MASC/
   ANALYSIS=NEUROTIC/
   DESIGN=ESTEEM,ATTROLE,FEM,MASC,FEM BY MASC,CONTIN(ESTEEM,ATTROLE)
          BY FEM + CONTIN(ESTEEM,ATTROLE) BY MASC + CONTIN
          (ESTEEM,ATTROLE) BY FEM BY MASC/
   ANALYSIS=INTEXT/
   DESIGN=ESTEEM,ATTROLE,NEUROTIC,FEM,MASC,FEM BY MASC,CONTIN(ESTEEM,
          ATTROLE,NEUROTIC) BY FEM + CONTIN(ESTEEM,ATTROLE,NEUROTIC)
          BY MASC + CONTIN(ESTEEM,ATTROLE,NEUROTIC) BY FEM BY MASC/
   ANALYSIS=CONTROL/
   DESIGN=ESTEEM,ATTROLE,NEUROTIC,INTEXT,FEM,MASC,FEM BY MASC,
          CONTIN(ESTEEM,ATTROLE,NEUROTIC,INTEXT) BY FEM +
          CONTIN(ESTEEM,ATTROLE,NEUROTIC,INTEXT) BY MASC +
          CONTIN(ESTEEM,ATTROLE,NEUROTIC,INTEXT) BY FEM BY MASC/
   ANALYSIS=SEL2/
   DESIGN=ESTEEM,ATTROLE,NEUROTIC,INTEXT,CONTROL,FEM,MASC,FEM BY MASC,
          CONTIN(ESTEEM,ATTROLE,NEUROTIC,INTEXT,CONTROL) BY FEM +
          CONTIN(ESTEEM,ATTROLE,NEUROTIC,INTEXT,CONTROL) BY MASC +
          CONTIN(ESTEEM,ATTROLE,NEUROTIC,INTEXT,CONTROL) BY FEM BY MASC/
```

TESTS OF SIGNIFICANCE FOR CONTROL USING SEQUENTIAL SUMS OF SQUARES

SOURCE OF VARIATION	SUM OF SQUARES	DF	MEAN SQUARE	F	SIG. OF F
WITHIN+RESIDUAL	442.61116	348	1.27187		
CONSTANT	16632.27174	1	16632.27174	13077.00992	0
ESTEEM	85.72041	1	85.72041	67.39708	0
ATTROLE	4.22285	1	4.22285	3.32019	.069
NEUROTIC	46.07474	1	46.07474	36.22595	0
INTEXT	.85481	1	.85481	.67209	.413
FEM	.05797	1	.05797	.04558	.831
MASC	.00005	1	.00005	.00004	.995
FEM BY MASC	.41040	1	.41040	.32267	.570
CONTIN(ESTEEM ATTROLE NEUROTIC INTEXT) B Y FEM + CONTIN(ESTEEM ATTROLE NEUROTIC I NTEXT) BY MASC + CONTIN(ESTEEM ATTROLE NEUROTIC INTEXT) BY FEM BY MASC	19.77587	12	1.64799	1.29572	.219

TESTS OF SIGNIFICANCE FOR SEL2 USING SEQUENTIAL SUMS OF SQUARES

SOURCE OF VARIATION	SUM OF SQUARES	DF	MEAN SQUARE	F	SIG. OF F
WITHIN+RESIDUAL	220340.09519	344	640.52353		
CONSTANT	617835.10457	1	617835.10457	964.57831	0
ESTEEM	938.10858	1	938.10858	1.46460	.227
ATTROLE	17.20327	1	17.20327	.02686	.870
NEUROTIC	.12133	1	.12133	.00019	.989
INTEXT	755.15901	1	755.15901	1.17897	.278
CONTROL	1407.82269	1	1407.82269	2.19792	.139
FEM	20.02234	1	20.02234	.03126	.860
MASC	384.09808	1	384.09808	.59966	.439
FEM BY MASC	297.09000	1	297.09000	.46382	.496
CONTIN(ESTEEM ATTROLE NEUROTIC INTEXT CO NTROL) BY FEM + CONTIN(ESTEEM ATTROLE NE UROTIC INTEXT CONTROL) BY MASC + CONTIN(ESTEEM ATTROLE NEUROTIC INTEXT CONTROL) BY FEM BY MASC	14017.21614	15	934.48108	1.45893	.118

Table 9.13 also contains output for the last two steps where CONTROL serves as DV with ESTEEM, ATTROLE, NEUROTIC, and INTEXT as covariates, and then SEL2 is the DV with ESTEEM, ATTROLE, NEUROTIC, INTEXT, and CONTROL as covariates. At each step, the relevant effect is the one appearing last in SOURCE OF VARIATION, so that for SEL2 the F value for homogeneity of regression is $F(15, 344) = 1.45893$, $p > .01$. (The more stringent cutoff is used here because robustness is expected.) Homogeneity of regression is established for all steps.

For MANCOVA, an overall test of homogeneity of regression is required, in addition to stepdown tests. Deck setups for all tests are shown in Table 9.14. The ANALYSIS sentence with three DVs specifies the overall test, while the ANALYSIS sentences with one DV each are for stepdown analysis. Output for the overall test and the last stepdown test are also shown in Table 9.14. Multivariate output is printed for the overall test because there are three DVs; univariate results are given for the stepdown tests. All runs show sufficient homogeneity of regression for this analysis.

9.7.1.7 Reliability of Covariates For the stepdown analysis in MANOVA, all DVs except ESTEEM must be reliable because all act as covariates. Based on the nature of scale development and data collection procedures, there is no reason to expect unreliability of a magnitude harmful to covariance analysis for ATTROLE, NEUROTIC, INTEXT, CONTROL, and SEL2. These same variables act as true or stepdown covariates in the MANCOVA analysis.

9.7.1.8 Multicollinearity and Singularity The determinant of the pooled within-cells correlation matrix is found (through SPSS^x MANOVA setup in Table 9.15) to be 0.81. This is sufficiently different from zero that neither multicollinearity nor singularity is judged to be a problem.

9.7.2 Multivariate Analysis of Variance

Deck setup and partial output of omnibus MANOVA produced by SPSS MANOVA appear in Table 9.15. The order of IVs listed in the MANOVA statement together with METHOD = SSTYPE(SEQUENTIAL) sets up the priority for testing FEM before MASC in this unequal-n design.[18] Results are reported for FEM by MASC, MASC, and FEM, in turn. Tests are reported out in order of adjustment where FEM by MASC is adjusted for both MASC and FEM, and MASC is adjusted for FEM.

Four multivariate statistics are reported for each effect. Because there is only one degree of freedom for each effect, three of the tests—Pillai's, Hotelling's, and Wilks'—produce the same F.[19] Both main effects are highly significant, but there is no statistically reliable interaction. If desired, strength of association for the composite DV for each main effect is found using Equation 9.6.

---

[18] Check carefully with your computer center to see what the default is for adjustment for unequal n in MANOVA; in some versions the sequential method (Method 3) is default. On most versions, particularly more recent ones, Method 1 is default. We recommend that you take positive control of unequal n adjustment by using appropriate control language rather than relying on a changing default method.

[19] For more complex designs, a single source table containing all effects can be obtained through PRINT = SIGNIF(BRIEF) but the table displays only Wilks' Lambda.

```
            TITLE        HOMOGENEITY OF REGRESSION FOR LARGE SAMPLE MANCOVA
            FILE HANDLE  MANOVA
            DATA LIST    FILE=MANOVA
                         /CASENO,ANDRM,FEM,MASC,ESTEEM,CONTROL,ATTROLE,
                         SEL2,INTEXT,NEUROTIC (A4,6F3.0,F8.5,F4.1,F3.0)
            VALUE LABELS ANDRM 1 'UNDIFF' 2 'FEMNINE' 3 'MASCLNE' 4 'ANDROGNS'/
                         FEM,MASC 1 'LOW' 2 'HIGH'
            MISSING VALUES  CONTROL(0)
            MANOVA      ESTEEM,ATTROLE,NEUROTIC,INTEXT,CONTROL,SEL2 BY FEM(1,2) MASC(1,2)/
                   PRINT=SIGNIF(BRIEF)/
                   NOPRINT=PARAMETERS(ESTIM)/
               ANALYSIS=ESTEEM,INTEXT,NEUROTIC/
               DESIGN=CONTROL,ATTROLE,SEL2,FEM,MASC,FEM BY MASC,
                      CONTIN(CONTROL,ATTROLE,SEL2) BY FEM +
                      CONTIN(CONTROL,ATTROLE,SEL2) BY MASC +
                      CONTIN(CONTROL,ATTROLE,SEL2) BY FEM BY MASC/
               ANALYSIS=ESTEEM/
               DESIGN=CONTROL,ATTROLE,SEL2,FEM,MASC,FEM BY MASC,
                      CONTIN(CONTROL,ATTROLE,SEL2) BY FEM +
                      CONTIN(CONTROL,ATTROLE,SEL2) BY MASC +
                      CONTIN(CONTROL,ATTROLE,SEL2) BY FEM BY MASC/
               ANALYSIS=INTEXT/
               DESIGN=CONTROL,ATTROLE,SEL2,ESTEEM,FEM,MASC,FEM BY MASC,
                      CONTIN(CONTROL,ATTROLE,SEL2,ESTEEM) BY FEM +
                      CONTIN(CONTROL,ATTROLE,SEL2,ESTEEM) BY MASC +
                      CONTIN(CONTROL,ATTROLE,SEL2,ESTEEN,INTEXT) BY FEM BY MASC/
               ANALYSIS=NEUROTIC/
               DESIGN=CONTROL,ATTROLE,SEL2,ESTEEM,INTEXT,FEM,MASC,FEM BY MASC,
                      CONTIN(CONTROL,ATTROLE,SEL2,ESTEEM,INTEXT) BY FEM +
                      CONTIN(CONTROL,ATTROLE,SEL2,ESTEEM,INTEXT) BY MASC +
                      CONTIN(CONTROL,ATTROLE,SEL2,ESTEEM,INTEXT) BY FEM BY MASC/
```

TESTS OF SIGNIFICANCE FOR WITHIN+RESIDUAL USING SEQUENTIAL SUMS OF SQUARES

SOURCE OF VARIATION	WILKS LAMBDA	MULT. F	HYPOTH. DF	ERROR DF	SIG. OF F
CONSTANT	.02360	4826.62570	3.00	350.00	0
CONTROL	.74523	39.88542	3.00	350.00	0
ATTROLE	.93181	8.53755	3.00	350.00	.000
SEL2	.99517	.56566	3.00	350.00	.638
FEM	.95317	5.73190	3.00	350.00	.001
MASC	.85120	20.39438	3.00	350.00	0
FEM BY MASC	.99734	.31119	3.00	350.00	.817
CONTIN(CONTROL ATTROLE SEL2) BY FEM + CONTIN(CONTROL ATTROLE SEL2) BY MASC + CONTIN(CONTROL ATTROLE SEL2) BY FEM BY MASC	.93292	.91150	27.00	1022.82	.596

TESTS OF SIGNIFICANCE FOR NEUROTIC USING SEQUENTIAL SUMS OF SQUARES

SOURCE OF VARIATION	SUM OF SQUARES	DF	MEAN SQUARE	F	SIG. OF F
WITHIN+RESIDUAL	6662.67324	344	19.36824		
CONSTANT	27930.53261	1	27930.53261	1442.07931	0
CONTROL	1487.85072	1	1487.85072	76.81911	0
ATTROLE	50.58323	1	50.58323	2.61166	.107
SEL2	1.67428	1	1.67428	.08644	.769
ESTEEM	535.03736	1	535.03736	27.62447	0
INTEXT	29.26997	1	29.26997	1.51124	.220
FEM	4.32678	1	4.32678	.22340	.637
MASC	.92184	1	.92184	.04760	.827
FEM BY MASC	4.94321	1	4.94321	.25522	.614
CONTIN(CONTROL ATTROLE SEL2 ESTEEM INTEXT) BY FEM + CONTIN(CONTROL ATTROLE SEL2 ESTEEM INTEXT) BY MASC + CONTIN(CONTROL ATTROLE SEL2 ESTEEM INTEXT) BY FEM BY MASC	420.18677	15	28.01245	1.44631	.124

TABLE 9.15 MULTIVARIATE ANALYSIS OF VARIANCE OF ESTEEM, CONTROL, ATTROLE, SEL2, INTEXT, AND NEUROTIC, AS A FUNCTION OF (TOP TO BOTTOM) FEMINITY BY MASCULINITY INTERACTION, MASCULINITY, AND FEMININITY (SETUP AND SELECTED OUTPUT FROM SPSS' MANOVA)

```
          TITLE          LARGE SAMPLE MANOVA
          FILE HANDLE    MANOVA
          DATA LIST      FILE=MANOVA
                         /CASENO,ANDRM,FEM,MASC,ESTEEM,CONTROL,ATTROLE,
                         SEL2,INTEXT,NEUROTIC (A4,6F3.0,F8.5,F4.1,F3.0)
          VALUE LABELS   ANDRM 1 'UNDIFF' 2 'FEMNINE' 3 'MASCLNE' 4 'ANDROGNS'/
                         FEM,MASC 1 'LOW' 2 'HIGH'
          MISSING VALUES CONTROL(0)
          MANOVA         ESTEEM,ATTROLE,NEUROTIC,INTEXT,CONTROL,SEL2 BY FEM(1,2) MASC(1,2)/
                         NOPRINT=PARAMETERS(ESTIM)/
                         PRINT=SIGNIF(STEPDOWN), ERROR(COR),
                              HOMOGENEITY(BARTLETT,COCHRAN,BOXM)/
                         METHOD=SSTYPE(SEQUENTIAL)/
                         DESIGN/


EFFECT .. FEM BY MASC

MULTIVARIATE TESTS OF SIGNIFICANCE (S = 1, M = 2 , N = 178 1/2)

TEST NAME        VALUE      APPROX. F    HYPOTH. DF      ERROR DF     SIG. OF F

PILLAIS          .00816       .49230        6.00          359.00        .814
HOTELLINGS       .00823       .49230        6.00          359.00        .814
WILKS            .99184       .49230        6.00          359.00        .814
ROYS             .00816


EFFECT .. MASC

MULTIVARIATE TESTS OF SIGNIFICANCE (S = 1, M = 2 , N = 178 1/2)

TEST NAME        VALUE      APPROX. F    HYPOTH. DF      ERROR DF     SIG. OF F

PILLAIS          .24363     19.27301        6.00          359.00         0
HOTELLINGS       .32211     19.27301        6.00          359.00         0
WILKS            .75637     19.27301        6.00          359.00         0
ROYS             .24363


EFFECT .. FEM

MULTIVARIATE TESTS OF SIGNIFICANCE (S = 1, M = 2 , N = 178 1/2)

TEST NAME        VALUE      APPROX. F    HYPOTH. DF      ERROR DF     SIG. OF F

PILLAIS          .08101      5.27423        6.00          359.00        .000
HOTELLINGS       .08815      5.27423        6.00          359.00        .000
WILKS            .91899      5.27423        6.00          359.00        .000
ROYS             .08101
```

Because omnibus MANOVA shows significant main effects, it is appropriate to investigate further the nature of the relationships among the IVs and DVs. Correlations, univariate F's, and stepdown F's help clarify the relationships.

The degree to which DVs are correlated provides information as to the independence of behaviors. Pooled within-cell correlations, adjusted for IVs, as produced by SPSS MANOVA through PRINT = ERROR(COR), appear in Table 9.16. (Diagonal elements are pooled standard deviations.) Correlations among ESTEEM, NEUROTIC, and CONTROL are in excess of .30 so stepdown analysis is appropriate.

Even if stepdown analysis is the primary procedure, knowledge of univariate F's is required to correctly interpret the pattern of stepdown F's. And, although the statistical significance of these F values is misleading, investigators frequently are interested in the ANOVA that would have been produced if each DV had been investigated in isolation. These univariate analyses are produced automatically by SPSS MANOVA and shown in Table 9.17 for the three effects in turn: FEM by MASC,

TABLE 9.16 POOLED WITHIN-CELL CORRELATIONS AMONG SIX DVs
(SELECTED OUTPUT FROM SPSS' MANOVA—SEE
TABLE 9.15 FOR SETUP)

WITHIN CELLS   CORRELATIONS WITH STD. DEVS. ON DIAGONAL

	ESTEEM	ATTROLE	NEUROTIC	INTEXT	CONTROL	SEL2
ESTEEM	3.53338					
ATTROLE	-.14529	6.22718				
NEUROTIC	.35772	.05070	4.96517			
INTEXT	-.16444	.01111	-.00923	3.58729		
CONTROL	.34789	-.03051	.38737	-.08315	1.26680	
SEL2	-.03510	.01635	-.01501	.05526	-.08417	25.50097

TABLE 9.17 UNIVARIATE ANAYSES OF VARIANCE OF SIX DVs
FOR EFFECTS OF (TOP TO BOTTOM) FEM BY MASC
INTERACTION, MASCULINITY, AND FEMININITY
(SELECTED OUTPUT FROM SPSS' MANOVA—SEE TABLE
9.15 FOR SETUP)

UNIVARIATE F-TESTS WITH (1,364) D. F.

VARIABLE	HYPOTH. SS	ERROR SS	HYPOTH. MS	ERROR MS	F	SIG. OF F
ESTEEM	17.48685	4544.44694	17.48685	12.48474	1.40066	.237
ATTROLE	36.79594	14115.12120	36.79594	38.77781	.94889	.331
NEUROTIC	.20239	8973.67662	.20239	24.65296	.00821	.928
INTEXT	.02264	4684.17900	.02264	12.86862	.00176	.967
CONTROL	.89539	584.14258	.89539	1.60479	.55795	.456
SEL2	353.58143	236708.96634	353.58143	650.29936	.54372	.461

UNIVARIATE F-TESTS WITH (1,364) D. F.

VARIABLE	HYPOTH. SS	ERROR SS	HYPOTH. MS	ERROR MS	F	SIG. OF F
ESTEEM	979.60086	4544.44694	979.60086	12.48474	78.46383	0
ATTROLE	1426.75675	14115.12120	1426.75675	38.77781	36.79313	0
NEUROTIC	179.53396	8973.67662	179.53396	24.65296	7.28245	.007
INTEXT	327.40797	4684.17900	327.40797	12.86862	25.44235	.000
CONTROL	11.85923	584.14258	11.85923	1.60479	7.38991	.007
SEL2	1105.38196	236708.96634	1105.38196	650.29936	1.69980	.193

UNIVARIATE F-TESTS WITH (1,364) D. F.

VARIABLE	HYPOTH. SS	ERROR SS	HYPOTH. MS	ERROR MS	F	SIG. OF F
ESTEEM	101.46536	4544.44694	101.46536	12.48474	8.12715	.005
ATTROLE	610.88860	14115.12120	610.88860	38.77781	15.75356	.000
NEUROTIC	44.05442	8973.67662	44.05442	24.65296	1.78698	.182
INTEXT	87.75996	4684.17900	87.75996	12.86862	6.81968	.009
CONTROL	2.83106	584.14258	2.83106	1.60479	1.76414	.185
SEL2	9.00691	236708.96634	9.00691	650.29936	.01385	.906

MASC, and FEM. *F* values are substantial for all DVs but SEL2 for MASC and for ESTEEM, ATTROLE, and INTEXT for FEM.

Finally, stepdown analysis, produced by PRINT = SIGNIF(STEPDOWN), allows a statistically pure look at the significance of DVs, in context, with Type I error rate controlled. For this study, the following priority order of DVs is developed, from most to least important: ESTEEM, ATTROLE, NEUROTIC, INTEXT, CONTROL, SEL2. Following the procedures for stepdown analysis (Section 9.5.2.2), the highest-priority DV, ESTEEM, is tested in univariate ANOVA. The second-priority DV, ATTROLE, is assessed in ANCOVA with ESTEEM as the covariate. The third-priority DV, NEUROTIC, is tested with ESTEEM and ATTROLE as covariates, and so on, until all DVs are analyzed. Stepdown analyses for the interaction and both main effects are in Table 9.18.

TABLE 9.18 STEPDOWN ANALYSES OF SIX ORDERED DVs FOR (TOP TO BOTTOM) FEM BY MASC INTERACTION, MASCULINITY, AND FEMININITY (SELECTED OUTPUT FROM SPSS' MANOVA—SEE TABLE 9.15 FOR SETUP)

EFFECT .. FEM BY MASC (CONT.)

ROY-BARGMAN STEPDOWN F - TESTS

VARIABLE	HYPOTH. MS	ERROR MS	STEP-DOWN F	HYPOTH. DF	ERROR DF	SIG. OF F
ESTEEM	17.48685	12.48474	1.40066	1	364	.237
ATTROLE	24.85653	38.06383	.65302	1	363	.420
NEUROTIC	2.69735	21.61699	.12478	1	362	.724
INTEXT	.26110	12.57182	.02077	1	361	.885
CONTROL	.41040	1.28441	.31952	1	360	.572
SEL2	297.09000	652.80588	.45510	1	359	.500

EFFECT .. MASC (CONT.)

ROY-BARGMAN STEPDOWN F - TESTS

VARIABLE	HYPOTH. MS	ERROR MS	STEP-DOWN F	HYPOTH. DF	ERROR DF	SIG. OF F
ESTEEM	979.60086	12.48474	78.46383	1	364	0
ATTROLE	728.51682	38.06383	19.13935	1	363	.000
NEUROTIC	4.14529	21.61699	.19176	1	362	.662
INTEXT	139.98354	12.57182	11.13471	1	361	.001
CONTROL	.00082	1.28441	.00064	1	360	.980
SEL2	406.59619	652.80588	.62284	1	359	.431

EFFECT .. FEM (CONT.)

ROY-BARGMAN STEPDOWN F - TESTS

VARIABLE	HYPOTH. MS	ERROR MS	STEP-DOWN F	HYPOTH. DF	ERROR DF	SIG. OF F
ESTEEM	101.46536	12.48474	8.12715	1	364	.005
ATTROLE	728.76735	38.06383	19.14593	1	363	.000
NEUROTIC	2.21946	21.61699	.10267	1	362	.749
INTEXT	47.98941	12.57182	3.81722	1	361	.052
CONTROL	.05836	1.28441	.04543	1	360	.831
SEL2	15.94930	652.80588	.02443	1	359	.876

For purposes of journal reporting, critical information from Tables 9.17 and 9.18 is consolidated into a single table with both univariate and stepdown analyses, as shown in Table 9.19. The α level established for each DV is reported along with the significance levels for stepdown F. For the main effect of FEM, ESTEEM and ATTROLE are significant. (INTEXT would be significant in ANOVA but its variance is already accounted for through overlap with ESTEEM, as noted in the pooled within-cell correlation matrix.) For the main effect of MASC, ESTEEM, ATTROLE, and INTEXT are significant. (NEUROTIC and CONTROL would be significant in ANOVA, but their variance is also already accounted for through overlap with ESTEEM.)

For the DVs significant in stepdown analysis, the relevant adjusted marginal means are needed for interpretation. Marginal means are needed for ESTEEM for FEM, and for MASC adjusted for FEM. Also needed are marginal means for ATTROLE with ESTEEM as a covariate for both FEM, and MASC adjusted for FEM; lastly, marginal means are needed for INTEXT with ESTEEM, ATTROLE, and NEUROTIC as covariates for MASC adjusted for FEM. Table 9.20 contains deck setup and control language for these marginal means as produced through SPSS[x] MANOVA. In the table, level of effect is identified under PARAMETER and mean

TABLE 9.19 TESTS OF FEMININITY, MASCULINITY, AND
THEIR INTERACTION

IV	DV	Univariate F	df	Stepdown F	df	α
Femininity	ESTEEM	8.13[a]	1/364	8.13**	1/364	.01
	ATTROLE	15.75[a]	1/364	19.15**	1/363	.01
	NEUROTIC	1.79	1/364	0.10	1/362	.01
	INTEXT	6.82[a]	1/364	3.82	1/361	.01
	CONTROL	1.76	1/364	0.05	1/360	.01
	SEL2	0.01	1/364	0.02	1/359	.001
Masculinity	ESTEEM	78.46[a]	1/364	78.46**	1/364	.01
	ATTROLE	36.79[a]	1/364	19.14**	1/363	.01
	NEUROTIC	7.28[a]	1/364	0.19	1/362	.01
	INTEXT	25.44[a]	1/364	11.13**	1/361	.01
	CONTROL	7.39[a]	1/364	0.00	1/360	.01
	SEL2	1.70	1/364	0.62	1/359	.001
Femininity by	ESTEEM	1.40	1/364	1.40	1/364	.01
masculinity	ATTROLE	0.95	1/364	0.65	1/363	.01
interaction	NEUROTIC	0.01	1/364	0.12	1/362	.01
	INTEXT	0.00	1/364	0.02	1/361	.01
	CONTROL	0.56	1/364	0.32	1/360	.01
	SEL2	0.54	1/364	0.46	1/359	.001

[a] Significance level cannot be evaluated but would reach $p < .01$ in univariate context.
** $p < .01$

TABLE 9.20 ADJUSTED MARGINAL MEANS FOR ESTEEM; ATTROLE
WITH ESTEEM AS A COVARIATE; AND INTEXT WITH
ESTEEM, ATTROLE, AND NEUROTIC AS COVARIATES
(SETUP AND SELECTED OUTPUT FROM SPSS' MANOVA)

```
         TITLE          ADJUSTED MEANS FOR LARGE SAMPLE MANOVA
         FILE HANDLE    MANOVA
         DATA LIST      FILE=MANOVA
                        /CASENO,ANDRM,FEM,MASC,ESTEEM,CONTROL,ATTROLE,
                         SEL2,INTEXT,NEUROTIC (A4,6F3.0,F8.5,F4.1,F3.0)
         VALUE LABELS   ANDRM 1 'UNDIFF' 2 'FEMNINE' 3 'MASCLNE' 4 'ANDROGNS'/
                        FEM,MASC 1 'LOW' 2 'HIGH'
         MISSING VALUES  CONTROL(0)
         MANOVA     ESTEEM,ATTROLE,NEUROTIC,INTEXT,CONTROL,SEL2 BY FEM(1,2) MASC(1,2)/
                        PRINT=PARAMETERS (ESTIM)/
                    ANALYSIS=ESTEEM/DESIGN=CONSPLUS FEM/
                                    DESIGN=FEM,CONSPLUS MASC/
                    ANALYSIS=ATTROLE WITH ESTEEM/DESIGN=CONSPLUS FEM/
                                            DESIGN=FEM, CONSPLUS MASC/
                    ANALYSIS=INTEXT WITH ESTEEM,ATTROLE,NEUROTIC/
                                            DESIGN=FEM, CONSPLUS MASC/
```

ESTIMATES FOR ESTEEM

CONSPLUS FEM   CONSPLUS MASC

PARAMETER   COEFF.   PARAMETER   COEFF.

    1   16.5700934579   2   17.1588559923
    2   15.4137931034   3   13.7138143624

ESTIMATES FOR ATTROLE ADJUSTED FOR 1 COVARIATE

CONSPLUS FEM   CONSPLUS MASC

PARAMETER   COEFF.   PARAMETER   COEFF.

    1   32.7152449841   2   35.3913780733
    2   35.8485394127   3   32.1158759823
```
```

TABLE 9.20   (Continued)

```
ESTIMATES FOR INTEXT ADJUSTED FOR 3 COVARIATES

CONSPLUS MASC

  PARAMETER            COEFF.

       2        11.0021422402
       3        12.4757067316
```

Note: COEFF. = adjusted marginal mean; first parameter=low, second parameter=high

is under COEFF. Thus, the mean for ESTEEM at level 1 of FEM is 16.57. Marginal means for effects with univariate, but not stepdown, differences are shown in Table 9.21 where means for NEUROTIC and CONTROL are found for the main effect of MASC adjusted for FEM.

Strength of association between the main effect and each DV with which it has a significant stepdown relationship is evaluated as η^2 (Equations 3.25, 3.26, 8.7, 8.8, or 8.9). The information you need for calculation of η^2 is available in SPSS MANOVA stepdown tables (see Table 9.18) but not in a convenient form; mean squares are given in the tables but you need sums of squares for calculation of η^2. Sums of squares are found by multiplying a mean square by its associated degrees of freedom. For this example, appropriate information is summarized in Table 9.22. Once sums of squares are computed, total sum of squares for a DV is calculated by adding the sums of squares for all effects and error.

TABLE 9.21   UNADJUSTED MARGINAL MEANS FOR NEUROTIC AND CONTROL (SETUP AND SELECTED OUTPUT FROM SPSS˟ MANOVA)

```
       TITLE          UNADJUSTED MEANS FOR LARGE SAMPLE MANOVA
       FILE HANDLE    MANOVA
       DATA LIST      FILE=MANOVA
                      /CASENO,ANDRM,FEM,MASC,ESTEEM,CONTROL,ATTROLE,
                      SEL2,INTEXT,NEUROTIC (A4,6F3.0,F8.5,F4.1,F3.0)
       VALUE LABELS   ANDRM 1 'UNDIFF' 2 'FEMNINE' 3 'MASCLNE' 4 'ANDROGNS'/
                      FEM,MASC 1 'LOW' 2 'HIGH'
       MISSING VALUES    CONTROL(0)
       MANOVA   ESTEEM,ATTROLE,NEUROTIC,INTEXT,CONTROL,SEL2 BY FEM(1,2) MASC(1,2)/
                PRINT=PARAMETERS (ESTIM)/
                ANALYSIS=NEUROTIC/DESIGN=FEM, CONSPLUS MASC/
                ANALYSIS=CONTROL/ DESIGN=FEM, CONSPLUS MASC/


ESTIMATES FOR NEUROTIC

CONSPLUS MASC

  PARAMETER          COEFF.

       2        9.3709383047
       3        7.8961041130



ESTIMATES FOR CONTROL

CONSPLUS MASC

  PARAMETER          COEFF.

       2        6.8916331030
       3        6.5125816027
```

Note: COEFF. = marginal means; first parameter = low, second parameter = high

TABLE 9.22 SUMMARY OF ADJUSTED SUMS OF SQUARES
AND η^2 FOR EFFECTS OF FEMININITY,
MASCULINITY, AND THEIR INTERACTION
ON ESTEEM; ATTROLE WITH ESTEEM
AS A COVARIATE; AND INTEXT WITH ESTEEM,
ATTROLE, AND NEUROTIC AS COVARIATES

Source of variance	ESTEEM		ATTROLE		INTEXT	
	SS'	η^2	SS'	η^2	SS'	η^2
F	101.47	.02	728.77	.05	47.99	—
M	979.60	.17	728.52	.05	139.98	.03
F by M	17.49	—	24.86	—	0.26	—
Error	4544.45		13817.17		4538.43	
Total	5643.01		15299.32		4726.67	

Note: See Table 9.20 for MS values. SS = (MS)(DF).

For example, ATTROLE is significantly related to FEM (F). For ATTROLE, the adjusted sum of squares is the mean square for hypothesis (HYPOTH. MS) times the degrees of freedom (HYPOTH. DF) from the FEM section of Table 9.18.

$$SS'_F = (728.76735)(1) = 728.77$$

Adjusted sums of squares for the remaining effects and error are necessary for computation of sum of squares total; these values are also available, but scattered around through Table 9.18. The adjusted sum of squares for MASC (middle part of Table 9.18) is

$$SS'_M = (728.51682)(1) = 728.52$$

and for FEM BY MASC (top part of Table 9.18) is

$$SS'_{F \text{ by } M} = (24.85653)(1) = 24.86$$

Error mean square and degrees of freedom are duplicated in all three segments of Table 9.18. Adjusted sum of squares for error, then, is

$$SS'_{\text{error}} = (38.06383)(363) = 13817.17$$

Finally, η^2 for ATTROLE is found by computing

$$\eta^2 = \frac{SS'_F}{SS'_{\text{total}}} = \frac{SS'_F}{SS'_F + SS'_M + SS'_{F \text{ by M}} + SS'_{\text{error}}}$$

$$= \frac{728.77}{728.77 + 728.52 + 24.86 + 13817.17} = \frac{728.77}{15299.32} = .05$$

A checklist for MANOVA appears in Table 9.23. An example of a Results section, in journal format, follows for the study just described.

TABLE 9.23 CHECKLIST FOR MULTIVARIATE ANALYSIS OF VARIANCE

1. Issues
   a. Unequal sample sizes and mising data
   b. Normality of sampling distributions
   c. Outliers
   d. Homogeneity of variance-covariance matrices
   e. Linearity
   f. In stepdown, when DVs act as covariates
      (1) Homogeneity of regression
      (2) Reliability of DVs
   g. Multicollinearity and singularity
2. Major analyses: Planned comparisons or omnibus F, when significant
   a. Multivariate strength of association
   b. Importance of DVs
      (1) Within-cell correlations, stepdown F, univariate F
      (2) Strength of association for significant F
      (3) Means or adjusted marginal and/or cell means for significant F
3. Additional analyses
   a. Post hoc comparisons
   b. Interpretation of IV-covariates interaction (if homogeneity of regression violated)

Results

A 2 × 2 between-subjects multivariate analysis of variance was performed on six dependent variables: self-esteem, attitude toward the role of women, neuroticism, introversion-extraversion, locus of control, and socioeconomic level. Independent variables were masculinity (low and high) and femininity (low and high).

SPSS[x] MANOVA was used for the analyses with the sequential adjustment for nonorthogonality. Order of entry of IVs was femininity, then masculinity. Total $\underline{N}$ of 369 was reduced to 368 with the deletion of a case missing a score on locus of control. There were no univariate or multivariate within-cell outliers at $\alpha = .001$. Results of evaluation of assumptions of normality, homogeneity of variance-covariance matrices, linearity, and multicollinearity were satisfactory.

With the use of Wilks' criterion, the combined DVs were significantly affected by both masculinity, $\underline{F}(6, 359) = 19.27$, $\underline{p} <$

.001, and femininity, $\underline{F}(6, 359) = 5.27$, $\underline{p} <$ .001, but not by their interaction, $\underline{F}(6,359)$ $= 0.49$, $\underline{p} > .05$. The results reflected a moderate association between masculinity scores (low vs. high) and the combined DVs, $\eta^2 = .24$. The association was less substantial between femininity and the DVs, $\eta^2 = .08$. [*F and lambda are from Table 9.15; η^2 is calculated according to Equation 9.6.*]

To investigate the impact of each main effect on the individual DVs, a stepdown analysis was performed on the prioritized DVs. All DVs were judged to be sufficiently reliable to warrant stepdown analysis. In stepdown analysis each DV was analyzed, in turn, with higher-priority DVs treated as covariates and with the highest-priority DV tested in a univariate ANOVA. Homogeneity of regression was achieved for all components of the stepdown analysis. Results of this analysis are summarized in Table 9.19. An experimentwise error rate of 5 percent was achieved by the apportionment of alpha as shown in the last column of Table 9.19 for each of the DVs.

Insert Table 9.19 about here

A unique contribution to predicting differences between those low and high on femininity was made by self-esteem, stepdown $\underline{F}(1, 364) = 8.13$, $\underline{p} < .01$, $\eta^2 = .02$. Self-esteem was scored inversely, so women with higher femininity scores showed greater self-esteem (mean self-esteem = 15.41) than those with lower femininity (mean self-esteem = 16.57). After the pattern of differences measured by self-esteem was entered, a difference was also found on attitude toward role of women, stepdown $\underline{F}(1, 363) = 19.15$, $\underline{p} < .01$, $\eta^2 = .05$. Women with higher femininity scores had more conservative attitudes toward women's role (adjusted mean attitude = 35.85) than those lower in femininity (adjusted mean attitude = 32.72). Although a univariate comparison revealed that those higher in femininity also were more extraverted, univariate $F(1, 364) = 6.82$, this difference was already represented in the stepdown analysis by higher-priority DVs.

Three DVs—self-esteem, attitude toward role of women, and introvert-extravert—made unique contributions to the composite DV that best distinguished between those high and low in masculinity. The greatest contribution was made by self-esteem, the highest-priority DV, stepdown $F(1, 364) = 78.46$, $p < .01$, $\eta^2 = 17$. Women scoring high in masculinity had higher self-esteem (mean self-esteem = 13.71) than those scoring low (mean self-esteem = 17.16). With differences due to self-esteem already entered, ATTROLE made a unique contribution, stepdown $F(1, 363) = 19.14$, $p < .01$, $\eta^2 = .05$. Women scoring lower in masculinity had more conservative attitudes toward the proper role of women (adjusted mean attitude = 35.39) than those scoring higher (adjusted mean attitude = 32.12). Introversion-extraversion, adjusted by self-esteem, attitudes toward women's role, and neuroticism also made a unique contribution to the composite DV, stepdown $F(1, 361) = 11.13$, $p < .01$, $\eta^2 = .03$. Women with higher masculinity were more extraverted (mean adjusted introversion-extraversion score = 12.48) than lower masculinity women (mean adjusted introversion-extraversion score = 11.00). Univariate analyses revealed that women with higher masculinity scores were also less neurotic, univariate $F(1, 364) = 7.28$, and had a more internal locus of control, univariate $F(1, 364) = 7.39$, differences that were already accounted for in the

composite DV by higher-priority DVs. [*Means
adjusted for main effects and for other DVs
for stepdown interpretation are from Table
9.20,* η^2 *values are from Table 9.22. Means
adjusted for main effects but not other DVs
for univariate interpretation are in Table
9.21.*]

High-masculinity women, then, have
greater self-esteem, less conservative
attitudes toward the role of women, and more
extraversion than women scoring low on
masculinity. High femininity is associated
with greater self-esteem and more
conservative attitudes toward women's role
than low femininity. Of the five effects,
however, only the association between
masculinity and self-esteem shows even a
moderate proportion of shared variance.

Pooled within-cell correlations among
DVs are shown in Table 9.16.

Insert Table 9.16 about here

9.7.3 Multivariate Analysis of Covariance

For MANCOVA the same six variables are used as for MANOVA but ESTEEM,
INTEXT, and NEUROTIC are used as DVs and CONTROL, ATTROLE, and SEL2
are used as covariates. The research question is whether there are personality differences associated with femininity, masculinity, and their interaction after adjustment
for differences in attitudes and socioeconomic status.

Deck setup and partial output of omnibus MANCOVA as produced by SPSS^x
MANOVA appear in Table 9.24. As in MANOVA, Method 3 adjustment for unequal
n is used with MASC adjusted for FEM. And, as in MANOVA, both main effects
are highly significant but there is no interaction.

9.7.3.1 Assessing Covariates
Under EFFECT..WITHIN CELLS REGRESSION
is the multivariate significance test for the relationship between the set of DVs (ES-
TEEM, INTEXT, and NEUROTIC) and the set of covariates (CONTROL, ATTROLE,
and SEL2). Because there is multivariate significance, it is useful to look at the three
multiple regression analyses of each DV in turn, with covariates acting as IVs (see
Chapter 5). The setup of Table 9.24 automatically produces these regressions. They
are done on the pooled within-cell correlation matrix, so that effects of the IVs are
eliminated.

The results of the DV-covariate multiple regressions are shown in Table 9.25.

TABLE 9.24 MULTIVARIATE ANALYSIS OF COVARIANCE OF ESTEEM, INTEXT, AND NEUROTIC AS A FUNCTION OF (TOP TO BOTTOM) FEM BY MASC INTERACTION, MASCULINITY, FEMININITY; COVARIATES ARE ATTROLE, CONTROL, AND SEL2 (SETUP AND SELECTED OUTPUT FROM SPSS* MANOVA)

```
        TITLE       LARGE SAMPLE MANCOVA
        FILE HANDLE MANOVA
        DATA LIST   FILE=MANOVA
                    /CASENO,ANDRM,FEM,MASC,ESTEEM,CONTROL,ATTROLE,
                      SEL2,INTEXT,NEUROTIC (A4,6F3.0,F8.5,F4.1,F3.0)
        VALUE LABELS ANDRM 1 'UNDIFF' 2 'FEMNINE' 3 'MASCLNE' 4 'ANDROGNS'/
                      FEM,MASC 1 'LOW' 2 'HIGH'
        MISSING VALUES CONTROL(0)
        MANOVA   ESTEEM,ATTROLE,NEUROTIC,INTEXT,CONTROL,SEL2 BY FEM(1,2) MASC(1,2)/
                  ANALYSIS=ESTEEM,INTEXT,NEUROTIC WITH CONTROL,ATTROLE,SEL2/
                  NOPRINT=PARAMETERS(ESTIM)/
                  PRINT=SIGNIF(STEPDOWN), ERROR(COR),
                      HOMOGENEITY(BARTLETT,COCHRAN,BOXM)/
                  METHOD=SSTYPE(SEQUENTIAL)/
                  DESIGN/
```

EFFECT .. WITHIN CELLS   REGRESSION

MULTIVARIATE TESTS OF SIGNIFICANCE (S = 3, M = -1/2, N = 178 1/2)

TEST NAME	VALUE	APPROX. F	HYPOTH. DF	ERROR DF	SIG. OF F
PILLAIS	.23026	10.00372	9.00	1083.00	0
HOTELLINGS	.29094	11.56236	9.00	1073.00	0
WILKS	.77250	10.86414	9.00	873.86	0
ROYS	.21770				

EFFECT .. FEM BY MASC

MULTIVARIATE TESTS OF SIGNIFICANCE (S = 1, M = 1/2, N = 178 1/2)

TEST NAME	VALUE	APPROX. F	HYPOTH. DF	ERROR DF	SIG. OF F
PILLAIS	.00263	.31551	3.00	359.00	.814
HOTELLINGS	.00264	.31551	3.00	359.00	.814
WILKS	.99737	.31551	3.00	359.00	.814
ROYS	.00263				

EFFECT .. MASC

MULTIVARIATE TESTS OF SIGNIFICANCE (S = 1, M = 1/2, N = 178 1/2)

TEST NAME	VALUE	APPROX. F	HYPOTH. DF	ERROR DF	SIG. OF F
PILLAIS	.14683	20.59478	3.00	359.00	0
HOTELLINGS	.17210	20.59478	3.00	359.00	0
WILKS	.85317	20.59478	3.00	359.00	0
ROYS	.14683				

EFFECT .. FEM

MULTIVARIATE TESTS OF SIGNIFICANCE (S = 1, M = 1/2, N = 178 1/2)

TEST NAME	VALUE	APPROX. F	HYPOTH. DF	ERROR DF	SIG. OF F
PILLAIS	.03755	4.66837	3.00	359.00	.003
HOTELLINGS	.03901	4.66837	3.00	359.00	.003
WILKS	.96245	4.66837	3.00	359.00	.003
ROYS	.03755				

At the top of Table 9.25 are the results of the univariate and stepdown analysis, summarizing the results of multiple regressions for the three DVs independently and then in priority order (see Section 9.7.3.2). At the bottom of Table 9.25 under REGRESSION ANALYSIS FOR WITHIN CELLS ERROR TERM are the separate regressions for each DV with covariates as IVs. For ESTEEM, two covariates, CONTROL and ATTROLE, are significantly related but SEL2 is not. None of the three

TABLE 9.25 UNIVARIATE, STEPDOWN, AND MULTIPLE REGRESSION
ANALYSES FOR THREE DVs WITH THREE COVARIATES
(SELECTED OUTPUT FROM SPSS' MANOVA—SEE TABLE 9.24
FOR SETUP)

UNIVARIATE F-TESTS WITH (3,361) D. F.

VARIABLE	SQ. MUL. R	MUL. R	ADJ. R-SQ.	HYPOTH. MS	ERROR MS	F	SIG. OF F
ESTEEM	.14542	.38134	.13832	220.28068	10.75791	20.47616	0
INTEXT	.00932	.09655	.00109	14.55535	12.85461	1.13231	.336
NEUROTIC	.15425	.39274	.14722	461.38686	21.02359	21.94615	0

EFFECT .. WITHIN CELLS   REGRESSION (CONT.)

ROY-BARGMAN STEPDOWN F - TESTS

VARIABLE	HYPOTH. MS	ERROR MS	STEP-DOWN F	HYPOTH. DF	ERROR DF	SIG. OF F
ESTEEM	220.28068	10.75791	20.47616	3	361	0
INTEXT	6.35936	12.60679	.50444	3	360	.679
NEUROTIC	239.94209	19.72942	12.16164	3	359	.000

REGRESSION ANALYSIS FOR WITHIN CELLS   ERROR TERM

DEPENDENT VARIABLE .. ESTEEM

COVARIATE	B	BETA	STD. ERR.	T-VALUE	SIG. OF T	LOWER 95 CL	UPPER 95 CL
CONTROL	.9817325741	.3519752398	.13625	7.20542	0	.71379	1.24967
ATTROLE	.0886059312	.1561581235	.02762	3.20773	.001	.03428	.14293
SEL2	-.0011127953	-.0080312308	.00677	-.16446	.869	-.01442	.01219

DEPENDENT VARIABLE .. INTEXT

COVARIATE	B	BETA	STD. ERR.	T-VALUE	SIG. OF T	LOWER 95 CL	UPPER 95 CL
CONTROL	-.2232208679	-.0788274507	.14894	-1.49877	.135	-.51611	.06967
ATTROLE	.0045581881	.0079125744	.03019	.15096	.880	-.05482	.06394
SEL2	.0068216006	.0484927597	.00740	.92231	.357	-.00772	.02137

DEPENDENT VARIABLE .. NEUROTIC

COVARIATE	B	BETA	STD. ERR.	T-VALUE	SIG. OF T	LOWER 95 CL	UPPER 95 CL
CONTROL	1.5312818992	.3906873702	.19047	8.03955	0	1.15671	1.90585
ATTROLE	.0497076488	.0623419410	.03861	1.28727	.199	-.02623	.12565
SEL2	.0032824106	.0168583493	.00946	.34703	.729	-.01532	.02188

covariates is related to INTEXT. Finally, for NEUROTIC, only CONTROL is significantly related. Because SEL2 provides no adjustment to any of the DVs, it could be omitted from future analyses.

9.7.3.2 Assessing DVs Procedures for evaluating DVs, now adjusted for covariates, follow those specified in Section 9.7.2 for MANOVA. Correlations among all DVs, among covariates, and between DVs and covariates are informative so all the correlations in Table 9.16 are still relevant.[20]

Univariate F's are now adjusted for covariates. The univariate ANCOVAs produced by the SPSS[x] MANOVA run specified in Table 9.24 are shown in Table 9.26. Although significance levels are misleading, there are substantial F values for ESTEEM and INTEXT for MASC (adjusted for FEM) and for FEM.

---

[20] For MANCOVA, SPSS[x] MANOVA prints pooled within-cell correlations among DVs (called criteria) adjusted for covariates. To get a pooled within-cell correlation matrix for covariates as well as DVs, you need a run in which covariates are included in the set of DVs.

TABLE 9.26 UNIVARIATE ANALYSES OF COVARIANCE OF THREE DVs ADJUSTED FOR THREE COVARIATES FOR (TOP TO BOTTOM) FEM BY MASC INTERACTION, MASCULINITY, AND FEMININITY (SELECTED OUTPUT FROM SPSS^x MANOVA—SEE TABLE 9.24 FOR SETUP)

UNIVARIATE F-TESTS WITH (1,361) D. F.

VARIABLE	HYPOTH. SS	ERROR SS	HYPOTH. MS	ERROR MS	F	SIG. OF F
ESTEEM	7.21931	3883.60490	7.21931	10.75791	.67107	.413
INTEXT	.02590	4640.51295	.02590	12.85461	.00202	.964
NEUROTIC	1.52636	7589.51604	1.52636	21.02359	.07260	.788

UNIVARIATE F-TESTS WITH (1,361) D. F.

VARIABLE	HYPOTH. SS	ERROR SS	HYPOTH. MS	ERROR MS	F	SIG. OF F
ESTEEM	533.61774	3883.60490	533.61774	10.75791	49.60237	0
INTEXT	264.44545	4640.51295	264.44545	12.85461	20.57204	.000
NEUROTIC	35.82929	7589.51604	35.82929	21.02359	1.70424	.193

UNIVARIATE F-TESTS WITH (1,361) D. F.

VARIABLE	HYPOTH. SS	ERROR SS	HYPOTH. MS	ERROR MS	F	SIG. OF F
ESTEEM	107.44454	3883.60490	107.44454	10.75791	9.98749	.002
INTEXT	74.94312	4640.51295	74.94312	12.85461	5.83006	.016
NEUROTIC	26.81431	7589.51604	26.81431	21.02359	1.27544	.259

TABLE 9.27 STEPDOWN ANALYSES OF THREE ORDERED DVs ADJUSTED FOR THREE COVARIATES FOR (TOP TO BOTTOM) FEM BY MASC INTERACTION, MASCULINITY, AND FEMININITY (SELECTED OUTPUT FROM SPSS^x MANOVA—SEE TABLE 9.24 FOR SETUP)

EFFECT .. FEM BY MASC (CONT.)

ROY-BARGMAN STEPDOWN F - TESTS

VARIABLE	HYPOTH. MS	ERROR MS	STEP-DOWN F	HYPOTH. DF	ERROR DF	SIG. OF F
ESTEEM	7.21931	10.75791	.67107	1	361	.413
INTEXT	.35520	12.60679	.02817	1	360	.867
NEUROTIC	4.94321	19.72942	.25055	1	359	.617

EFFECT .. MASC (CONT.)

ROY-BARGMAN STEPDOWN F - TESTS

VARIABLE	HYPOTH. MS	ERROR MS	STEP-DOWN F	HYPOTH. DF	ERROR DF	SIG. OF F
ESTEEM	533.61774	10.75791	49.60237	1	361	0
INTEXT	137.74436	12.60679	10.92621	1	360	.001
NEUROTIC	1.07421	19.72942	.05445	1	359	.816

EFFECT .. FEM (CONT.)

ROY-BARGMAN STEPDOWN F - TESTS

VARIABLE	HYPOTH. MS	ERROR MS	STEP-DOWN F	HYPOTH. DF	ERROR DF	SIG. OF F
ESTEEM	107.44454	10.75791	9.98749	1	361	.002
INTEXT	47.36159	12.60679	3.75683	1	360	.053
NEUROTIC	4.23502	19.72942	.21466	1	359	.643

TABLE 9.28   TESTS OF COVARIATES, FEMININITY, MASCULINITY, AND INTERACTION

Effect	DV	Univariate		Stepdown		
		F	df	F	df	α
Covariates	ESTEEM	20.48[a]	3/361	20.48**	3/361	.02
	INTEXT	1.13	3/361	0.50	3/360	.02
	NEUROTIC	21.95[a]	3/361	12.16**	3/359	.01
Femininity	ESTEEM	9.99[a]	1/361	9.99**	1/361	.02
	INTEXT	5.83[a]	1/361	3.76	1/360	.02
	NEUROTIC	1.28	1/361	0.21	1/359	.01
Masculinity	ESTEEM	49.60[a]	1/361	49.60**	1/361	.02
	INTEXT	20.57[a]	1/361	10.93**	1/360	.02
	NEUROTIC	1.70	1/361	0.05	1/359	.01
Femininity by	ESTEEM	0.67	1/361	0.67	1/361	.02
masculinity	INTEXT	0.00	1/361	0.03	1/360	.02
interaction	NEUROTIC	0.07	1/361	0.25	1/359	.01

[a] Significance level cannot be evaluated but would reach $p < .01$ in univariate context.
** $p < .01$.

TABLE 9.29   ADJUSTED MARGINAL MEANS FOR ESTEEM
ADJUSTED FOR THREE COVARIATES, AND INTEXT
ADJUSTED FOR ESTEEM PLUS THREE COVARIATES
(SETUP AND SELECTED OUTPUT FROM SPSS' MANOVA)

```
TITLE          ADJUSTED MEANS FOR LARGE SAMPLE MANOVA
FILE HANDLE    MANOVA
DATA LIST      FILE=MANOVA
               /CASENO,ANDRM,FEM,MASC,ESTEEM,CONTROL,ATTROLE,
               SEL2,INTEXT,NEUROTIC (A4,6F3.0,F8.5,F4.1,F3.0)
VALUE LABELS   ANDRM 1 'UNDIFF' 2 'FEMNINE' 3 'MASCLNE' 4 'ANDROGNS'/
               FEM,MASC 1 'LOW' 2 'HIGH'
MISSING VALUES CONTROL(0)
MANOVA         ESTEEM,ATTROLE,NEUROTIC,INTEXT,CONTROL,SEL2 BY FEM(1,2) MASC(1,2)/
               PRINT=PARAMETERS (ESTIM)/
               ANALYSIS=ESTEEM WITH CONTROL,ATTROLE,SEL2/
               DESIGN=CONSPLUS FEM/DESIGN=FEM, CONSPLUS MASC/
               ANALYSIS=INTEXT WITH CONTROL,ATTROLE,SEL2,ESTEEM/
               DESIGN=FEM, CONSPLUS MASC/


ESTIMATES FOR ESTEEM ADJUSTED FOR 3 COVARIATES

CONSPLUS FEM                         CONSPLUS MASC

PARAMETER      COEFF.                PARAMETER      COEFF.

    1      16.6136356957                 2      16.9194927002
    2      15.3959424542                 3      14.2190404415


ESTIMATES FOR INTEXT ADJUSTED FOR 4 COVARIATES

CONSPLUS MASC

PARAMETER      COEFF.

    2      11.0067954086
    3      12.4700349085
```

Note: COEFF. = adjusted marginal mean; first parameter = low, second parameter = high

TABLE 9.30 MARGINAL MEANS FOR INTEXT ADJUSTED FOR THREE COVARIATES ONLY (SETUP AND SELECTED OUTPUT FROM SPSS' MANOVA)

```
    TITLE           ADJUSTED MEANS FOR LARGE SAMPLE MANOVA
    FILE HANDLE     MANOVA
    DATA LIST       FILE=MANOVA
                    /CASENO,ANDRM,FEM,MASC,ESTEEM,CONTROL,ATTROLE,

                    SEL2,INTEXT,NEUROTIC (A4,6F3.0,F8.5,F4.1,F3.0)
    VALUE LABELS    ANDRM 1 'UNDIFF' 2 'FEMNINE' 3 'MASCLNE' 4 'ANDROGNS'/
                    FEM,MASC 1 'LOW' 2 'HIGH'
    MISSING VALUES  CASENO TO ATTROLE(0)
    MANOVA     ESTEEM,ATTROLE,NEUROTIC,INTEXT,CONTROL,SEL2 BY FEM(1,2) MASC(1,2)/
               PRINT=PARAMETERS (ESTIM)/
               ANALYSIS=INTEXT WITH CONTROL,ATTROLE,SEL2,ESTEEM/
               DESIGN=CONSPLUS FEM/


    ESTIMATES FOR INTEXT ADJUSTED FOR 4 COVARIATES

    CONSPLUS FEM

      PARAMETER            COEFF.

           1       11.0926434549
           2       11.9122113039
```

Note: COEFF. = adjusted marginal mean; first parameter = low, second parameter = high

For interpretation of effects of IVs on DVs adjusted for covariates, comparison of stepdown F's with univariate F's again provides the best information. The priority order of DVs for this analysis is ESTEEM, INTEXT, and NEUROTIC. ESTEEM is evaluated after adjustment only for the three covariates. INTEXT is adjusted for effects of ESTEEM and the three covariates; NEUROTIC is adjusted for ESTEEM

TABLE 9.31 CHECKLIST FOR MULTIVARIATE ANALYSIS OF COVARIANCE

1. Issues
   a. Unequal sample sizes and missing data
   b. Normality of sampling distributions
   c. Outliers
   d. Homogeneity of variance-covariance matrices
   e. Linearity
   f. Homogenity of regression
      (1) Covariates
      (2) DVs for stepdown analysis
   g. Reliability of covariates (and DVs for stepdown)
   h. Multicollinearity and singularity
2. Major analyses: Planned comparisons or omnibus F, when significant
   a. Multivariate strength of association
   b. Importance of DVs
      (1) Within-cell correlations, stepdown F, univariate F
      (2) Strength of association for significant F
      (3) Adjusted marginal and/or cell means for significant F
3. Additional analyses
   a. Assessment of covariates
   b. Interpretation of IV-covariates interaction (if homogeneity of regression violated for stepdown analysis)
   c. Post hoc comparisons

and INTEXT and the three covariates. In effect, then, INTEXT is adjusted for four covariates and NEUROTIC is adjusted for five.

Stepdown analysis for the interaction and two main effects is in Table 9.27. The results are the same as those in MANOVA except that there is no longer a main effect of FEM on INTEXT after adjustment for four covariates. The relationship between FEM and INTEXT is already represented by the relationship between FEM and ESTEEM. Consolidation of information from Tables 9.26 and 9.27, as well as some information from Table 9.25, appears in Table 9.28, along with apportionment of the .05 α error to the various tests.

For the DVs associated with significant main effects, interpretation requires associated marginal means. Table 9.29 contains setup and adjusted marginal means for ESTEEM and for INTEXT (which is adjusted for ESTEEM as well as covariates) for FEM and for MASC adjusted for FEM. Setup and marginal means for the main effect of FEM on INTEXT (univariate but not stepdown effect) appear in Table 9.30. A checklist for MANCOVA appears in Table 9.31. An example of a Results section, as might be appropriate for journal presentation, follows.

Results

    A 2 × 2 between–subjects multivariate
analysis of covariance was performed on
three dependent variables associated with
personality of respondents: self–esteem,
introvert–extravert, and neuroticism.
Adjustment was made for three covariates:
attitude toward role of women, locus of
control, and socioeconomic status.
Independent variables were masculinity
(high and Low) and femininity (high and
low).

    SPSS MANOVA was used for the analyses
with the hierarchical adjustment for
nonorthogonality. Order of entry of IVs was
femininity, then masculinity. Total $\underline{N}$ = 369
was reduced to 368 with the deletion of a
case missing a score on locus of control.
There were no univariate or multivariate
within–cell outliers at α = .001. Results of
evaluation of assumptions of normality,
homogeneity of variance–covariance
matrices, linearity, and multicollinearity
were satisfactory. Covariates were judged
to be adequately reliable for covariance
analysis.

With the use of Wilks' criterion, the combined DVs were significantly related to the combined covariates, approximate $F(9, 873) = 10.86$, $p < .01$, to femininity, $F(3, 359) = 4.67$, $p < .01$, and to masculinity, $F(3, 359) = 20.59$, $p < .001$ but not to the interaction, $F(3, 359) = 0.31$, $p > .05$. There was a moderate association between DVs and covariates, with $\eta^2 = .23$. A somewhat smaller association was found between combined DVs and the main effect of masculinity, $\eta^2 = .15$, and the association between the main effect of femininity and the combined DVs was smaller yet, $\eta^2 = .04$. [F and lambda are from Table 9.24; η^2 is calculated from Equation 9.6.]

To investigate more specifically the power of the covariates to adjust dependent variables, multiple regressions were run for each DV in turn, with covariates acting as multiple predictors. Two of the three covariates, locus of control and attitudes toward women's role, provided significant adjustment to self-esteem. The β value of .35 for locus of control was significantly different from zero, $t(361) = 7.21$, $p < .001$, as was the β value of .16 for attitudes toward women's role, $t(361) = 3.21$, $p < .01$. None of the covariates provided adjustment to the introversion-extraversion scale. For neuroticism, only locus of control reached statistical significance, with $\beta = .39$, $t(361) = 8.04$, $p < .001$. For none of the DVs did socioeconomic status provide significant adjustment.

Effects of masculinity and femininity on the DVs after adjustment for covariates were investigated in univariate and stepdown analysis, in which self-esteem was given the highest priority, introversion-extraversion second priority (so that adjustment was made for self-esteem as well as for the three covariates), and

neuroticism third priority (so that
adjustment was made for self—esteem and
introversion—extraversion as well as for
the three covariates). Homogeneity of
regression was satisfactory for this
analysis, and DVs were judged to be
sufficiently reliable to act as covariates.
Results of this analysis are summarized in
Table 9.28. An experimentwise error rate of
5% for each effect was achieved by
apportioning alpha according to the values
shown in the last column of the table.

Insert Table 9.28 about here

After adjusting for differences on the
covariates, self—esteem made a significant
contribution to the composite of the DVs
that best distinguishes between women who
were high or low in femininity, stepdown
$F(1, 361) = 9.99$, $p < .01$, $\eta^2 = .02$. With
self—esteem scored inversely, women with
higher femininity scores showed greater
self—esteem after adjustment for covariates
(adjusted mean self—esteem = 15.40) than
those scoring lower on femininity (adjusted
mean self—esteem = 16.61). Univariate
analysis revealed that a reliable
difference was also present on the
introversion—extraversion measure, with
higher—femininity women more extraverted,
univariate $F(1, 361) = 5.83$, a difference
already accounted for by covariates and the
higher—priority DV. [*Adjusted means are
from Tables 9.29 and 9.30; η^2 is calculated
as in Section 9.7.2.*]

Lower— versus higher—masculinity women
differed in self—esteem, the highest—
priority DV, after adjustment for
covariates, stepdown $F(1, 361) = 49.60$, $p <
.01$, $\eta^2 = .10$. Greater self—esteem was found
among higher—masculinity women (adjusted
mean = 14.22) than among lower—masculinity

women (adjusted mean = 16.92). The measure
of introversion and extraversion, adjusted
for covariates and self-esteem, was also
related to differences in masculinity,
stepdown $\underline{F}(1, 360) = 10.93$, $\underline{p} < .01$, $\eta^2 =$
.03. Women scoring higher on the masculinity
scale were more extraverted (adjusted mean
extraversion 12.47) than those showing
lower masculinity (adjusted mean
extraversion = 11.01).

    High-masculinity women, then, are
characterized by greater self-esteem and
extraversion than low-masculinity women
when adjustments are made for differences in
socioeconomic status, attitudes toward
women's role, and locus of control. High-
femininity women show greater self-esteem
than low-femininity women with adjustment
for those covariates.

    Pooled within-cell correlations among
dependent variables and covariates are
shown in Table 9.16.

Insert Table 9.16 about here

9.8 SOME EXAMPLES FROM THE LITERATURE

9.8.1 Examples of MANOVA

Wade and Baker (1977) surveyed clinical psychologists on their reasons for decisions
to use tests. Psychologists were divided into low and high test users, as the two levels
of the IV. Six "reasons," each rated on a 5-point scale of importance, served as the
DVs. Separate MANOVAs were performed on users of objective and projective tests.
After finding significant multivariate effects—that is, that high and low test users
rated importance of reasons differently—for both objective and projective users, the
researchers performed univariate ANOVAs on each of the reasons.

    They found that for objective test users, the distinguishing reasons were reli-
ability and validity, and agency requirements, for which high users gave higher ratings
of importance than did low users. For the projective test users, high users rated as
more important than low users graduate training experience, previous experience with
tests, and reliability and validity. A MANOVA testing reasons for use as a function
of eight major therapeutic orientations was not statistically significant.

In a study of romantic attraction, Giarrusso (1977) investigated the effects of physical attractiveness (low vs. high) and similarity of attitudes (similar vs. different) of a target date in a 2 × 2 between-groups MANOVA. Male students rated the target date on degree of liking, desire to date, comparability with previous dates, goodness of personality, and whether the target date liked them. These five ratings served as DVs. A significant multivariate effect of physical attractiveness was found, but not of similarity of attitudes or of the interaction between attractiveness and attitudes. After an a priori ordering of DVs, it was found that only the highest-priority variable, degree of liking, was significantly affected by physical attractiveness. That is, the more attractive the target date, the more she was liked. The remaining variables showed no statistically significant differences once the effect on degree of liking was partialed out.

Junior high school students were evaluated in terms of their behavioral changes after being exposed in class for one week to materials on "responsibility," in a study by Singh, Greer, and Hammond (1977). IVs were (1) whether or not materials were presented; (2) grade level: 7, 8, and 9; and (3) ability levels, three levels based on the School and College Ability Tests. The three DVs consisted of an attitude test designed to accompany the "responsibility" materials, an experimenter-designed essay, and objective tests, which tapped comprehension of various aspects of responsibility. With the use of a 2 × 3 × 3 between-groups multivariate analysis of variance on gain scores, only the main effect of program was found to be statistically significant. Bonferroni-type confidence intervals constructed for the three DVs, used instead of stepdown analysis, indicated that the attitude test alone accounted for the significant multivariate effect, but the authors felt that the magnitude of change was insufficient to support the program as implemented in the study.

Concerned that policy and decision makers seldom use evaluation information, Brown, Braskamp, and Newman (1978) investigated effects on readers' reactions to use of jargon and to objective vs. subjective statements in evaluation reports. After reading one of four statements (with presence vs. absence of jargon, and subjective vs. objective statements factorially combined), readers judged the reports on two dependent measures, difficulty and technicality, and judged the writer of the evaluation on nine measures (including such variables as thoroughness and believability). Separate MANOVAs were performed on two types of DVs—judgment of report and judgment of writer. Significant multivariate effects were interpreted in terms of cell means for DVs with significant univariate F's, rather than with the combination of univariate and stepdown analysis recommended here.

The only significant multivariate effect was that of jargon on ratings of the report. Both report DVs, technicality and difficulty, produced significant univariate F's, indicating that reports high in jargon were considered to be more technical and more difficult. Ratings of the report writer were not affected by jargon, objectivity, or the interaction. Further, a univariate ANOVA revealed that extent of agreement with the recommendations of the evaluation report was independent of use of jargon, objectivity, or the interaction.

Jakubczak (1977) was interested in age differences in regulation of caloric intake of rats. For the first experiment, he used as IVs three age levels, three levels of dilution of food with cellulose, two levels of previous food deprivation experience, and six daysets, a within-subjects factor. DVs were caloric intake and body weight. Significant multivariate main effects were found on all factors, in addition to a number of two- and three-way interactions. Univariate ANOVAs, rather than stepdown and univariate analyses, were then used for interpreting effects on individual DVs. An age-related decrement in caloric intake and body weight was found in response to dilution with cellulose. Previous deprivation experience did not affect this relationship.

In a second experiment, two age levels, two levels of dilution by water, and nine daysets were used as factors. As an additional IV, rats were or were not given quinine in their diets. Older rats behaved like younger rats in response to dilution of the diet by water, maintaining caloric intake and weight. With addition of quinine, however, older rats decreased caloric intake and body weight to a greater degree than younger rats.

For both experiments, MANCOVAs run with initial levels of caloric intake and body weight as covariates revealed no difference in decisions with respect to within-subjects effects, as could be expected (cf. Section 8.5.2.1). Jakubczak made no mention of changes in between-subjects effects with the use of covariates.

Willis and Wortman (1976) studied public acceptance of randomized control-group designs in medical experimentation. Factors were sex of subject, three levels of science emphasis in describing the medical experiment, and three levels of treatment scarcity situations. For the last IV, descriptions of the medical experiment explicitly mentioned nonscarcity (i.e., the proposed treatment was available to all), mentioned scarcity (the treatment was expensive and scarce), or failed to mention anything about scarcity. Dependent measures were nine opinion questions, answered on bipolar scales. MANOVA revealed significant multivariate effects for two of the three IVs, scarcity of treatment and sex of subject. None of the interactions was statistically significant. DVs were analyzed in terms of cell means for scarcity and sex associated with those DVs with significant univariate F's. For post hoc comparisons, among the three scarcity levels, multivariate F's were appropriately adjusted for inflated α error. It was concluded that the most favorable reactions occurred in situations in which scarcity was not mentioned at all, and that women tended to be less accepting than men of placebos and withholding of treatment.

9.8.2 Examples of MANCOVA

Cornbleth (1977) studied the effect of a protected hospital ward on geriatric patients. IVs were ward assignment (protected or unprotected) and identification of the patient as a wanderer or nonwanderer. These two IVs were factorially combined, forming a 2×2 between-subjects design. Separate analyses were run for two periods in time (rather than including time period as a within-subjects factor). DVs were two physical measures, five cognitive measures, and six psychosocial measures. Separate MANOVAs were done for each of these three categories of functioning. In additional analyses, preassessment measures on the appropriate DVs were used as covariates. Therefore, there were three MANOVAs and three MANCOVAs. Contribution of the DVs to multivariate effects were evaluated through examination of standardized discriminant function coefficients (cf. Chapter 11). In addition, means adjusted for covariates were reported for all DVs.

On physical variables, a significant multivariate interaction was found (marginal at time 1, $p < .05$ at time 2). The protected ward produced greater range of motion for wanderers, while the regular nonprotected ward facilitated performance for non-wanderers. In addition, independent of ward assignment, wanderers showed a lower level of psychosocial functioning than nonwanderers at both time points.

Modification of perceived locus of control in high school students was studied by Bradley and Gaa (1977). Subjects were blocked on sex and on two levels of prior achievement, and then randomly assigned to one of three treatment groups, goal-setting, conference (serving as a placebo), and control. Five variables measuring aspects of perceived locus of control were used as DVs. The covariate was a teacher-developed achievement pretest. In the report of results, no mention was made of main effects of the blocking IVs or any interactions between them and the manipulated IV.

Three separate analyses were reported, reflecting specific comparisons among groups rather than a single omnibus analysis of the three groups. These three comparisons were not mutually orthogonal, and they required more degrees of freedom than the omnibus test. However, no mention was made of adjusting (by Bonferroni procedures or others) for inflated α produced by multiple testing. Further, although it was reported that three MANCOVAs were performed, no multivariate statistics were presented. Instead, univariate F's for each DV were reported for each of the three analyses. It was concluded that goal-setting treatment was effective in promoting more internal orientation among students, but only on measures related to locus of control in an academic situation.

Chapter 10

Profile Analysis of Repeated Measures

10.1 GENERAL PURPOSE AND DESCRIPTION

Profile analysis is an application of multivariate analysis of variance (MANOVA) in which several DVs are measured and they are all measured on the same scale.[1] A classic example of profile analysis is the comparison of personality profiles on, say, the Profile of Mood States (POMS), for different groups of people—psychoanalysts and behavior therapists. The DVs are the various scales of the POMS, tension-anxiety, vigor, anger-hostility, and the like.

Profile analysis is also used in research where subjects are measured repeatedly on the same DV. In this case, profile analysis is an alternative to univariate repeated measures ANOVA (see Chapter 3). For example, math achievement tests are given at various points during a semester to test the effects of alternative educational programs such as traditional classroom vs. computer assisted instruction. The choice between profile analysis (called the multivariate approach) and univariate repeated measures ANOVA depends on sample size, power, and whether statistical assumptions of repeated measured ANOVA are met. These issues are discussed fully in Section 10.5.2.

Current computer programs allow application of profile analysis to complex designs where, for instance, groups are classified along more than one dimension to create multiple IVs as in ANOVA. For example, POMS profiles are examined for both male and female psychoanalysts and behavior therapists. Or changes in math achievement during a semester as a result of either traditional education or computer assisted instruction are evaluated for both elementary vs. junior high school students. In addition, "multivariate-multivariate" tests are available, in which multiple DVs

---

[1] The term profile analysis is also applied to techniques for measuring resemblance among profile patterns through cluster analysis rather than the MANOVA strategy described in this chapter.

are measured repeatedly. For example, the POMS is administered to psychoanalysts and behavior therapists at the beginning of professional training, at the end of training, and after some period of time in the profession. A discussion of types of doubly multivariate analysis and an example of one type appear in Section 10.5.3.

10.2 KINDS OF RESEARCH QUESTIONS

The major question to be answered by profile analysis is whether or not profiles of groups differ on a set of measures. To apply profile analysis, all measures must have the same range of possible scores, with the same score value having the same meaning on all the measures. The restriction on scaling of the measures is present because in two of the major tests of profile analysis (parallelism and flatness) the numbers that are actually tested are difference scores between DVs measured on adjacent occasions. Difference scores are called segments in profile analysis, and some of the limitations to use of profile analysis apply to segments rather than to DVs as originally measured.

10.2.1 Parallelism of Profiles

Do different groups have parallel profiles? This is commonly known as the test of parallelism and is the primary question addressed by profile analysis. Using the therapist example, do psychoanalysts and behavior therapists have the same pattern of highs and lows on the various mood states measured by the POMS?

When using profile analysis as a substitute for univariate repeated measures ANOVA, the parallelism test is the test of interaction. For example, do traditional and computer assisted instruction lead to the same pattern of gains in achievement over the course of testing? Put another way, do changes in achievement depend on which method of instruction is used?

10.2.2 Overall Difference among Groups

Whether or not groups produce parallel profiles, does one group, on average, score higher on the collected set of measures than another? For example, does one type of therapist have reliably higher scores on the set of states measured by the POMS than the other? Or, using the repeated measures example, does one method of instruction lead to greater overall achievement than the other method?

In profile analysis, this question is answered by test of the "groups" hypothesis or, in profile analysis jargon, the "levels" hypothesis. It addresses the same question as the between-subjects main effect in repeated-measures ANOVA.

10.2.3 Flatness of Profiles

The third question addressed by profile analysis concerns the similarity of response to all DVs, independent of groups. Do the DVs all elicit the same average response? This question is typically relevant only if the profiles are parallel. If the profiles are not parallel, then at least one of them is necessarily not flat. Although it is conceivable that nonflat profiles from two or more groups could cancel each other out to produce, on average, a flat profile, this result is usually not of research interest.

Using the therapist example, if psychoanalysts and behavior therapists have the same pattern of mood states on the POMS (that is, they have parallel profiles), one might ask whether the groups combined are notably high or low on any of the states. In the instructional example, the flatness test evaluates whether achievement changes over the period of testing. In this context the flatness test evaluates the same hypothesis as the within-subjects main effect in repeated-measures ANOVA.

10.2.4 Contrasts Following Profile Analysis

With more than two groups or more than two measures, differences in parallelism, flatness, and/or level can result from a variety of sources. For example, if a group of client-centered therapists is added to the therapist study, and the parallelism or levels hypothesis is rejected, it is not obvious whether it is the behavior therapists who differ from the other two groups, the psychoanalysts who differ from the client-centered therapists, or exactly which group differs from which other group or groups. Contrasts following profile analysis are demonstrated in Sections 10.5.1 and 10.7.2.

10.2.5 Marginal/Cell Means and Plots

Whenever statistically significant differences are found between groups or measures, parameters are estimated. For profile analysis, the major description of results is typically a plot of profiles where means for each of the DVs are plotted for each of the groups. In addition, if the null hypothesis regarding levels is rejected, assessment of group means and group standard deviations is helpful. And if the null hypothesis regarding flatness is rejected and the finding is of interest, a plot of scores combined over groups is instructive. Profile plots are demonstrated in Sections 10.4 and 10.7.

10.2.6 Strength of Association

As with all statistical techniques, a finding of statistically significant differences is accompanied by description of the strength of association between the relevant IVs and DVs. Such measures are demonstrated in Sections 10.4 and 10.7.

10.2.7 Treatment Effects in Multiple Time-Series Designs

In multiple time-series designs, two groups—experimental and control—are tested on several pretreatment and several posttreatment occasions to provide multiple DVs. In these types of designs, profile analysis is available as a substitute for univariate repeated-measures ANOVA.

Because random assignment to groups is rare in this design, the levels test of overall group difference is usually not of interest because the two groups may well differ prior to treatment. However, as Algina and Swaminathan (1979) suggest, the test of parallelism, or interaction, sometimes reflects treatment effects. Acceptance of parallelism suggests no treatment effects because the course of the time series is the same whether or not treatment occurs. Rejection of parallelism suggests that there may be treatment effects but further analysis, as described by Algina and Swaminathan, is required to reveal whether the deviation from parallelism supports treatment effects.

10.3 LIMITATIONS TO PROFILE ANALYSIS

10.3.1 Theoretical Issues

Choice of DVs is more limited in profile analysis than in usual applications of multivariate statistics because DVs must be commensurate. That is, they must all have been subjected to the same scaling techniques.

In applications where profile analysis is used as an alternative to univariate repeated-measures ANOVA, this requirement is met because all DVs are literally the same measure. In other applications of profile analysis, however, careful consideration of measurement of the DVs is required to assure that the units of measure are the same. One way of assuring commensurability is to use standardized scores such as z scores, instead of raw scores for the DVs. In this case, each DV is standardized using the pooled within-groups standard deviation (the square root of the error mean square for the DV) provided by univariate one-way between-subjects ANOVA for the DV.

If, and only if, groups are formed by random assignment, levels of IVs are manipulated, and proper experimental controls are maintained, are differences among profiles causally attributed to differences in treatments among groups. Causality, as usual, is not addressed by the statistical test. Generalizability, as well, is influenced by sampling strategy, not choice of statistical tests. That is, the results of profile analysis generalize only to the populations from which cases are randomly sampled.

10.3.2 Practical Issues

10.3.2.1 Sample Size and Missing Data   The sample size in each group is an important issue in profile analysis as in MANOVA, because there must be more research units in the smallest group than there are DVs. This is required both because of consideration of power and for evaluation of the assumption of homogeneity of variance-covariance matrices (cf. Section 9.3.2.1). In the choice between univariate repeated-measures ANOVA and profile analysis, sample size per group is often the deciding factor.

Unequal sample sizes typically provide no special difficulty in profile analysis because each hypothesis is tested as if in a one-way design and, as discussed in Section 8.5.2.2, unequal n creates difficulties in interpretation only in designs with more than one IV. However, unequal n sometimes has implications for evaluating homogeneity of variance-covariance matrices, as discussed in Section 10.3.2.4.

If some measures are missing from some cases, the usual problems and solutions discussed in Chapter 4 apply.

10.3.2.2 Multivariate Normality   Because the DVs tested in parallelism and flatness, the multivariate components of profile analysis, are actually differences between DVs or segments, assumptions of normality apply to segments rather than original DVs. However, profile analysis is as robust to violation of normality as other forms of MANOVA (cf. Section 9.3.2.2) because the central limit theorem also applies to means formed by differences between DVs. So unless there are under 20 cases in the smallest group and highly unequal n, deviation from normality is not expected. In the unhappy event of small, unequal samples, however, a look at the distributions of segments for each group is in order. If distributions of segments show marked, highly significant skewness, some normalizing transformations might be investigated (cf. Chapter 4). Conversion of DVs into segments and assessment of segments are shown in Section 10.7.1.

10.3.2.3 Outliers   As in all MANOVA, profile analysis is extremely sensitive to outliers. Tests for univariate and multivariate outliers, detailed in Chapter 4, are applied to segments. In addition, the original DVs, averaged over measures for each group, are examined for univariate outliers. These tests are demonstrated in Section 10.7.1.

10.3.2.4 Homogeneity of Variance-Covariance Matrices   If sample sizes are equal, evaluation of homogeneity of variance-covariance matrices is not necessary. However, if sample sizes are notably discrepant, Box's M test is available through SPSS MANOVA as a test of the homogeneity of the variance-covariance matrices of segments. Box's M is too sensitive for use at routine α levels but if the test of homogeneity is rejected at highly significant levels, the guidelines in Section 9.3.2.4 are appropriate.

Univariate homogeneity of variance is also assumed but the robustness of ANOVA generalizes to profile analysis. Unless sample sizes are highly divergent or there is evidence of strong heterogeneity of the original DVs (cf. Section 8.3.2.5) this assumption is probably safely ignored.

10.3.2.5 Linearity   For the parallelism and flatness tests, linearity of the relationships among segments is assumed. This assumption is evaluated by examining scatterplots between all pairs of segments through SPSS SCATTERGRAM, SPSS^x PLOT, BMDP6D, SYSTAT GRAPH, or SAS CORR or PLOT. Because the major consequence of failure of linearity is loss of power in the parallelism test, violation is somewhat mitigated by large sample sizes. Therefore with many DVs and large sample sizes, check just a few scatterplots to assure that the assumption is not too badly violated.

10.3.2.6 Multicollinearity and Singularity   Although correlations among segments are at issue, it is the original DVs that are discarded if correlations are excessive. If segments pass the test of nonredundancy, proceed with the profile analysis. If segments fail the test, then examine the SMCs among the original DVs to determine the source of difficulty.

BMDP4V protects against multicollinearity and singularity by calculating pooled within-cell tolerance and dealing with tolerances that are too low. SPSS[x] MANOVA provides the determinant of the within-cell correlation matrix. If the determinant is less than .0001, then SMCs for all original DVs are examined, most easily through BMDPAM or a factor analysis program (cf. Section 9.3.2.8).

10.4 FUNDAMENTAL EQUATIONS FOR PROFILE ANALYSIS

In Table 10.1 is a hypothetical data set appropriate for profile analysis used as an alternative to repeated measures ANOVA. The three groups (the IV) whose profiles are compared are belly dancers, politicians, and administrators (or substitute your favorite scapegoat). The five respondents in each of these occupational groups participate in four leisure activities (the DVs) and, during each, are asked to rate their satisfaction on a 10-point scale. The leisure activities are reading, dancing, watching TV, and skiing. The profiles are illustrated in Figure 10.1, where mean ratings of each group for each activity are plotted.

Profile analysis tests of parallelism and flatness are multivariate and involve

TABLE 10.1   SMALL SAMPLE OF HYPOTHETICAL DATA FOR ILLUSTRATION OF PROFILE ANALYSIS

Group		Case No.	Activity				Combined activities
			Read	Dance	TV	Ski	
Belly dancers		1	7	10	6	5	7.00
		2	8	9	5	7	7.25
		3	5	10	5	8	7.00
		4	6	10	6	8	7.50
		5	7	8	7	9	7.75
	Mean		6.60	9.40	5.80	7.40	7.30
Politicians		6	4	4	4	4	4.00
		7	6	4	5	3	4.50
		8	5	5	5	6	5.25
		9	6	6	6	7	6.25
		10	4	5	6	5	5.00
	Mean		5.00	4.80	5.20	5.00	5.00
Administrators		11	3	1	1	2	1.75
		12	5	3	1	5	3.50
		13	4	2	2	5	3.25
		14	7	1	2	4	3.50
		15	6	3	3	3	3.75
	Mean		5.00	2.00	1.80	3.80	3.15
	Grand mean		5.53	5.40	4.27	5.40	5.15

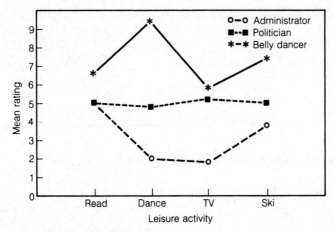

Figure 10.1 Profiles of leisure-time ratings for three occupations.

sum-of-squares and cross-products matrices. But the levels test is a univariate test, equivalent to the between-subjects main effect in repeated-measures ANOVA.[2]

10.4.1 Differences in Levels

For the example, the levels test examines differences between the means of the three occupational groups combined over the four activities. Are the group means of 7.30, 5.00, and 3.15 significantly different from each other?

The relevant equation for partitioning variance is adapted from Equation 3.8 as follows:

$$\sum_i \sum_j (Y_{ij} - \text{GM})^2 = p \sum_i \sum_j (Y_{ij} - \overline{Y}_j)^2 + np \sum_i \sum_j (\overline{Y}_j - \text{GM})^2 \qquad (10.1)$$

where n is the number of subjects in each group and p is the number of measures, in this case the number of ratings made by each respondent.

The partition of variance in Equation 10.1 produces the total, within-groups, and between-groups sums of squares, respectively, as in Equation 3.9. Because the score for each subject in the levels test is the subject's average score over the four activities, degrees of freedom follow Equations 3.10 through 3.13, with N equal to the total number of subjects and k equal to the number of groups.

For the hypothetical data of Table 10.1:

$$\text{SS}_{bg} = (5)(4)[(7.30 - 5.15)^2 + (5.00 - 5.15)^2 + (3.15 - 5.15)^2]$$

and

---

[2] Timm (1975) offers a multivariate form of this test, which he recommends when the parallelism hypothesis is rejected. It has been omitted here because it is not readily available through computer programs.

TABLE 10.2   ANOVA SUMMARY TABLE FOR TEST
OF LEVELS EFFECT FOR SMALL SAM-
PLE EXAMPLE OF TABLE 10.1

Source of variance	SS	df	MS	F
Between groups	172.90	2	86.45	44.14*
Within groups	23.50	12	1.96	

* $p < .001$

$$SS_{wg} = (4)[(7.00 - 7.30)^2 + (7.25 - 7.30)^2 + \cdots + (3.75 - 3.15)^2]$$

$$df_{bg} = k - 1 = 2$$

$$df_{wg} = N - k = 12$$

The levels test for the example produces a standard ANOVA source table for a one-way univariate test, as summarized in Table 10.2. There is a statistically significant difference between occupational groups in average rating of satisfaction during the four leisure activities.

Standard η^2 is used to evaluate the strength of association between occupational groups and averaged satisfaction ratings:

$$\eta^2 = \frac{SS_{bg}}{SS_{bg} + SS_{wg}} = \frac{172.90}{172.90 + 23.50} = .88 \qquad (10.2)$$

10.4.2 Parallelism

Tests of parallelism and flatness are conducted through hypotheses about adjacent segments of the profiles. The test of parallelism, for example, asks if the difference (segment) between reading and dancing is the same for belly dancers, politicians, and administrators? How about the difference between dancing and watching TV?

The most straightforward demonstration of the parallelism test begins by converting the data matrix into difference scores. For the example, the four DVs are turned into three differences, as shown in Table 10.3. The difference scores are created from adjacent pairs of activities but in this example, as in many uses of profile analysis, the order of the DVs is arbitrary. In profile analysis it is often true that segments are formed from arbitrarily ordered DVs and have no intrinsic meaning. As discussed later, this sometimes creates difficulty in interpreting statistical findings.

In Table 10.3 the first entry for the first case is the difference between READ and DANCE scores, that is, $7 - 10 = -3$. Her second score is the difference in ratings between DANCE and TV: $10 - 6 = 4$, and so on.

A one-way MANOVA on the segments tests the parallelism hypothesis. Because each segment represents a slope between two original DVs, if there is a multivariate difference between groups then one or more slopes are different and the profiles are not parallel.

Using procedures developed in Chapter 9, the total sum-of-squares and cross-products matrix (S_{tot}) is partitioned into the between-groups matrix S_{bg} and the within-

TABLE 10.3 SCORES FOR ADJACENT SEGMENTS FOR SMALL SAMPLE HYPOTHETICAL DATA

Group	Case no.	Read vs. dance	Dance vs. TV	TV vs. ski
Belly dancers	1	−3	4	1
	2	−1	4	−2
	3	−5	5	−3
	4	−4	4	−2
	5	−1	1	−2
Mean		−2.8	3.6	−1.6
Politicians	6	0	0	0
	7	2	−1	2
	8	0	0	−1
	9	0	0	−1
	10	−1	−1	1
Mean		0.2	−0.4	0.2
Administrators	11	2	0	−1
	12	2	2	−4
	13	2	0	−3
	14	6	−1	−2
	15	3	0	0
Mean		3.0	0.2	−2.0
Grand mean		0.13	1.13	−1.13

groups or error matrix $\mathbf{S}_{wg}$.[3] To produce the within-groups matrix, each person's score matrix, $\mathbf{Y}_{ikm}$, has subtracted from it the mean matrix for that group, $\mathbf{M}_m$. The resulting difference matrix is multiplied by its transpose to create the sum-of-squares and cross-products matrix. For the first belly dancer:

$$(\mathbf{Y}_{111} - \mathbf{M}_1) = \begin{bmatrix} -3 \\ 4 \\ 1 \end{bmatrix} - \begin{bmatrix} -2.8 \\ 3.6 \\ -1.6 \end{bmatrix} = \begin{bmatrix} -0.2 \\ 0.4 \\ 2.6 \end{bmatrix}$$

and

$$(\mathbf{Y}_{111} - \mathbf{M}_1)(\mathbf{Y}_{111} - \mathbf{M}_1)' = \begin{bmatrix} -0.2 \\ 0.4 \\ 2.6 \end{bmatrix} [-0.2 \quad 0.4 \quad 2.6]$$

$$= \begin{bmatrix} 0.04 & -0.08 & -0.52 \\ -0.08 & 0.16 & 1.04 \\ -0.52 & 1.04 & 6.76 \end{bmatrix}$$

[3] Other methods of forming $\mathbf{S}$ matrices can be used to produce the same result.

This is the sum-of-squares and cross-products matrix for the first case. When these matrices are added over all cases and groups, the result is the error matrix $\mathbf{S}_{wg}$:

$$\mathbf{S}_{wg} = \begin{bmatrix} 29.6 & -13.2 & 6.4 \\ -13.2 & 15.2 & -6.8 \\ 6.4 & -6.8 & 26.0 \end{bmatrix}$$

To produce the between-groups matrix, $\mathbf{S}_{bg}$, the grand matrix, $\mathbf{GM}$, is subtracted from each mean matrix $\mathbf{M}_k$, to form a difference matrix for each group. The mean matrix for each group in the example is

$$\mathbf{M}_1 = \begin{bmatrix} -2.8 \\ 3.6 \\ -1.6 \end{bmatrix} \quad \mathbf{M}_2 = \begin{bmatrix} 0.2 \\ -0.4 \\ 0.2 \end{bmatrix} \quad \mathbf{M}_3 = \begin{bmatrix} 3.0 \\ 0.2 \\ -2.0 \end{bmatrix}$$

and the grand mean matrix is

$$\mathbf{GM} = \begin{bmatrix} 0.13 \\ 1.13 \\ -1.13 \end{bmatrix}$$

The between-groups sum-of-squares and cross-products matrix $\mathbf{S}_{bg}$ is formed by multiplying each group difference matrix by its transpose, and then adding the three resulting matrices. After multiplying each entry by $n = 5$, to provide for summation over subjects,

$$\mathbf{S}_{bg} = \begin{bmatrix} 84.133 & -50.067 & -5.133 \\ -50.067 & 46.533 & -11.933 \\ -5.133 & -11.933 & 13.733 \end{bmatrix}$$

Wilks' Lambda (Λ) tests the hypothesis of parallelism by evaluating the ratio of the determinant of the within-groups cross-products matrix to the determinant of the sum of the within- and between-groups cross-products matrices:

$$\Lambda = \frac{|\mathbf{S}_{wg}|}{|\mathbf{S}_{wg} + \mathbf{S}_{bg}|} \tag{10.3}$$

For the example, Wilks' Lambda for testing parallelism is

$$\Lambda = \frac{6325.2826}{6325.2826 + 76598.7334} = .076279$$

By applying the procedures of Section 9.4.1, one finds an approximate $F(6, 20) = 8.74$, $p < .001$, leading to rejection of the hypothesis of parallelism. That is, the three profiles of Figure 10.1 are not parallel.

Strength of association is measured as eta square:[4]

$$\eta^2 = 1 - \qquad \qquad (10.4)$$

For this example, then,

$$\eta^2 = 1 - .076279 = .92$$

A whopping 92% of the variance in the segments as combined for this test is accounted for by the difference in shape of the profiles for the three groups. Recall from Chapter 9, however, that segments are combined here to maximize group differences for parallelism. A different combination of segments is used for the test of flatness.

10.4.3 Flatness

Because the hypothesis of parallelism is rejected for this example, the test of flatness is irrelevant; the question of flatness of combined profiles of Figure 10.1 makes no sense because at least one of them (and in this case probably two) is not flat. The flatness test is computed here to conclude the demonstration of this example.

Statistically, the test is whether, with groups combined, the three segments of Table 10.3 deviate from zero. That is, if segments are interpreted as slopes in Figure 10.1, are any of the slopes for the combined groups different from zero (nonhorizontal)? The test subtracts a set of hypothesized grand means representing the null hypothesis from the matrix of actual grand means:

$$(\mathbf{GM} - 0) = \begin{bmatrix} 0.13 \\ 1.13 \\ -1.13 \end{bmatrix} - \begin{bmatrix} 0 \\ 0 \\ 0 \end{bmatrix} = \begin{bmatrix} 0.13 \\ 1.13 \\ -1.13 \end{bmatrix}$$

The test of flatness is a multivariate generalization of the one-sample t-test demonstrated in Chapter 3. Because it is a one-sample test, it is most conveniently evaluated through Hotelling's T^2, or trace:[5]

$$T^2 = N (\mathbf{GM} - 0)' \ \mathbf{S}_{wg}^{-1} (\mathbf{GM} - 0) \qquad (10.5)$$

where N is the total number of cases and $\mathbf{S}_{wg}^{-1}$ is the inverse of the within-groups sum-of-squares and cross-products matrix developed in Section 10.4.2. For the example:

$$T^2 = (15)[0.13 \quad 1.13 \quad -1.13] \begin{bmatrix} .05517 & .04738 & -.00119 \\ .04738 & .11520 & .01847 \\ -.00119 & .01847 & .04358 \end{bmatrix} \begin{bmatrix} 0.13 \\ 1.13 \\ -1.13 \end{bmatrix}$$

$$= 2.5825$$

[4] Partial η^2 is available through some versions of SPSS MANOVA, and s is reported in all versions. If unavailable, the simpler $\eta^2 = 1 - \Lambda$ can be used.

[5] This is sometimes referred to as Hotelling's T.

From this is found F, with $p - 1$ and $N - k - p + 2$ degrees of freedom, where p is the number of original DVs (in this case 4), and k is the number of groups (3).

$$F = \frac{N - k - p + 2}{p - 1} (T^2) \tag{10.6}$$

so that

$$F = \frac{15 - 3 - 4 + 2}{4 - 1} (2.5825) = 8 \ 608$$

with 3 and 10 degrees of freedom, $p < .01$ and the test shows significant deviation from flatness.

A measure of strength of association is found through Hotelling's T^2 that bears a simple relationship to lambda.

$$\Lambda = \frac{1}{1 + T^2} = \frac{1}{1 + 2.5825} = .27913$$

Lambda, in turn, is used to find eta square:

$$\eta^2 = 1 - \Lambda = 1 - .27913 = .72$$

showing that 72% of the variance in this combination of segments is accounted for by nonflatness of the profile collapsed over groups.

10.4.4 Computer Analyses of Small Sample Example

Tables 10.4 through 10.7 show setup and selected output for computer analyses of the data in Table 10.1. Table 10.4 illustrates SPSS^x MANOVA using a repeated-measures setup (cf. Section 10.6.1). BMDP4V is illustrated in Table 10.5. Table 10.6 demonstrates profile analysis through SAS GLM, with SHORT printout requested. SYSTAT MGLH, using the REPEAT procedure, is illustrated in Table 10.7.

The four programs differ substantially in setup and presentation of the three tests. To set up SPSS^x MANOVA for profile analysis, the DVs (levels of the within-subject effect) READ TO SKI are followed in the MANOVA statement by the keyword BY and the grouping variable with its levels—OCCUP(1,3). The DVs are combined for profile analysis in the WSFACTOR instruction and labeled ACTIVITY(4) to indicate four levels for the within-subjects factor. The RENAME instruction makes the output easier to read: OVERALL labels the test of the combined DVs, and the remaining three variables specify the segments. For the example, RVSD labels the READ vs. DANCE segment.

In the SPSS^x MANOVA output, the levels test for differences among groups is the test of OCCUP in the section labeled TESTS OF SIGNIFICANCE FOR OVER-

ALL... The flatness and parallelism tests appear in the section labeled TESTS OF SIGNIFICANCE FOR WITHIN CELLS... as ACTIVITY and OCCUP BY ACTIVITY, respectively. Output is limited in this example by the PRINT = SIGNIF(BRIEF) instruction. Without this statement, separate source tables are printed for flatness and parallelism, each containing several multivariate tests.

In BMDP4V, the BETWEEN instruction specifies the grouping variable and the WITHIN instruction indicates the multiple DVs (repeated measures). There are two separate source tables in the output, one for the levels test and another for the flatness and parallelism tests. The first source table is labeled OBS: WITHIN CASE

TABLE 10.4 PROFILE ANALYSIS OF SMALL SAMPLE EXAMPLE THROUGH SPSS' MANOVA (SETUP AND SELECTED OUTPUT)

```
        TITLE            SMALL SAMPLE PROFILE ANALYSIS
        FILE HANDLE      TAPE42
        DATA LIST        FILE=TAPE42
                         /RESPOND,OCCUP,READ,DANCE,TV,SKI  (6F3.0)
        VAR LABELS       RESPOND 'RESPONDENT'  OCCUP 'OCCUPATION'
        VALUE LABELS     OCCUP 1 'BELLY DANCER' 2 'POLITICIAN' 3 'ADMINISTRATOR'
        MANOVA           READ TO SKI BY OCCUP(1,3)/
                         WSFACTOR=ACTIVITY(4)/
                         WSDESIGN=ACTIVITY/
                         ANALYSIS(REPEATED)/
                         TRANSFORM=REPEATED/
                         RENAME=OVERALL,RVSD,DVST,TVSS/
                         NOPRINT=PARAMETER(ESTIM)/
                         PRINT=SIGNIF(BRIEF)/



* * * * * * * * * * * A N A L Y S I S   O F   V A R I A N C E * * * * * * * * * * * *

* * * * * * * * * * * * * * * * * * * * * * * * * * * * * * * * * * * *
*                   *                                                 *
*  W A R N I N G    * THE LAST COMMAND IS NOT A "DESIGN"              *
*                   * SPECIFICATION.  A FULL FACTORIAL MODEL          *
*                   * IS GENERATED FOR THIS PROBLEM.                  *
*                   *                                                 *
* * * * * * * * * * * * * * * * * * * * * * * * * * * * * * * * * * * *


- - - - - - - - - - - - - - - - - - - - - - - - - - - - - - - - - - - - -


      15 CASES ACCEPTED.
       0 CASES REJECTED BECAUSE OF OUT-OF-RANGE FACTOR VALUES.
       0 CASES REJECTED BECAUSE OF MISSING DATA.
       3 NON-EMPTY CELLS.
       1 DESIGN WILL BE PROCESSED.

- - - - - - - - - - - - - - - - - - - - - - - - - - - - - - - - - - - - -

CORRESPONDENCE BETWEEN EFFECTS AND COLUMNS OF WITHIN-SUBJECTS DESIGN 1

STARTING  ENDING
 COLUMN   COLUMN   EFFECT NAME


    1        1     CONSTANT
    2        4     ACTIVITY

* * * * * * * * * * * A N A L Y S I S   O F   V A R I A N C E * * * * * * * * * * * *


CORRSPONDENCE BETWEEN EFFECTS AND COLUMNS OF BETWEEN-SUBJECTS DESIGN 1

STARTING  ENDING
 COLUMN   COLUMN   EFFECT NAME

    1        1     CONSTANT
    2        3     OCCUP


- - - - - - - - - - - - - - - - - - - - - - - - - - - - - - - - - - - - -
```

TABLE 10.4 (Continued)

```
* * * * * * * * * * * A N A L Y S I S   O F   V A R I A N C E * * * * * * * * * * *


ORDER OF VARIABLES FOR ANALYSIS


    VARIATES      COVARIATES      NOT USED

   *OVERALL                       RSSD
                                  DVST
                                  TVSS

    1 DEPENDENT VARIABLE
    0 COVARIATES
    3 VARIABLES NOT USED

- - - - - - - - - - - - - - - - - - - - - - - - - - - - - - - - - - - -


    NOTE..  "*" MARKS TRANSFORMED VARIABLES.

         THESE TRANSFORMED VARIABLES CORRESPOND TO THE
         'CONSTANT' WITHIN-SUBJECT EFFECT.

- - - - - - - - - - - - - - - - - - - - - - - - - - - - - - - - - - - - - -
* * * * * * * * * * * * * A N A L Y S I S   O F   V A R I A N C E * * * * * * * * * * * * * *


TESTS OF SIGNIFICANCE FOR OVERALL USING SEQUENTIAL SUMS OF SQUARES
```

SOURCE OF VARIATION	SUM OF SQUARES	DF	MEAN SQUARE	F	SIG. OF F
WITHIN CELLS	23.50000	12	1.95833		
CONSTANT	1591.35000	1	1591.35000	812.60426	0
OCCUP	172.90000	2	86.45000	44.14468	0

```
- - - - - - - - - - - - - - - - - - - - - - - - - - - - - - - - - - - -
* * * * * * * * * * * * * A N A L Y S I S   O F   V A R I A N C E * * * * * * * * * * * * * *


ORDER OF VARIABLES FOR ANALYSIS


    VARIATES      COVARIATES      NOT USED

   *RVSD                          OVERALL
   *DVST
   *TVSS

    3 DEPENDENT VARIABLES
    0 COVARIATES
    1 VARIABLE NOT USED

- - - - - - - - - - - - - - - - - - - - - - - - - - - - - - - - - - - -


    NOTE..  "*" MARKS TRANSFORMED VARIABLES.

         THESE TRANSFORMED VARIABLES CORRESPOND TO THE
         'ACTIVITY' WITHIN-SUBJECT EFFECT.

- - - - - - - - - - - - - - - - - - - - - - - - - - - - - - - - - - - -


STATISTICS FOR WITHIN CELLS      CORRELATIONS

DETERMINANT =                     .89580
BARTLETT TEST OF SPHERICITY =    1.11877 WITH 3 D. F.
SIGNIFICANCE =                    .773


F(MAX) CRITERION =               3.03723 WITH (3,12) D. F.

- - - - - - - - - - - - - - - - - - - - - - - - - - - - - - - - - - - -
```

TABLE 10.4 (Continued)

```
* * * * * * * * * * * * * * A N A L Y S I S   O F   V A R I A N C E * * * * * * * * * * * * * *
```

TESTS OF SIGNIFICANCE FOR WITHIN CELLS   USING SEQUENTIAL SUMS OF SQUARES

SOURCE OF VARIATION	WILKS LAMBDA	MULT. F	HYPOTH. DF	ERROR DF	SIG. OF F
ACTIVITY	.27914	8.60822	3.00	10.00	.004
OCCUP BY ACTIVITY	.07628	8.73584	6.00	20.00	.000

MEAN and the levels effect is O: OCCUP. In the second table, labeled A: ACTIVITY, the flatness test is simply called A. The parallelism test is called (A) X (O: OCCUP). For both these tests, both univariate and multivariate tests are printed out. For flatness, TSQ, a form of Hotelling's T^2, is used. For parallelism, there are four multivariate tests, Wilks' Lambda (LRATIO), Hotelling's TRACE, TZSQ, and MXROOT (Roy's greatest characteristic root, cf. Section 9.5.1). The F ratios for WCP are the univariate within-subjects tests with Greenhouse-Geisser and Huynh-Feldt adjustments to univariate F for violation of homogeneity of covariance, if necessary. All cell and marginal means are also printed out (not shown).

TABLE 10.5 SMALL SAMPLE PROFILE ANALYSIS THROUGH BMDP4V
(SETUP AND SELECTED OUTPUT)

```
          /PROBLEM       TITLE IS 'SMALL SAMPLE PROFILE ANALYSIS'.
          /INPUT         VARIABLES=6.  FORMAT IS '(6F3.0)'.
                         FILE = TAPE42.
          /VARIABLE      NAMES ARE RESPOND,OCCUP,READ,DANCE,TV,SKI.
                         LABEL IS RESPOND.
          /BETWEEN       FACTOR IS OCCUP.  CODES ARE 1,2,3.
                         NAMES ARE 'BELLY DANCER','POLITICIAN',
                                   'ADMINISTRATOR'.
          /WITHIN        FACTOR IS ACTIVITY.  CODES ARE 1 TO 4.
                         NAMES ARE READ,DANCE,TV,SKI.
          /WEIGHTS       WITHIN ARE EQUAL.  BETWEEN ARE EQUAL.
          /END
          ANALYSIS       PROC = FACTORIAL./
          END/



ANALYSIS CONTROL LANGUAGE

ANALYSIS      PROC = FACTORIAL./


FACTORIAL PROCEDURE .....

THE FOLLOWING STATEMENTS HAVE BEEN GENERATED:

DESIGN  TYPE IS BETWEEN, CONTRAST.  MODEL.
   CODE IS CONST.  NAME IS 'OVALL: GRAND MEAN'./

DESIGN  FACTOR IS OCCUP.
   CODE IS EFFECT.  NAME IS 'O: OCCUP'./

DESIGN  TYPE IS WITHIN, CONTRAST.  MODEL.
   CODE IS CONST.  NAME IS 'OBS: WITHIN CASE MEAN'./

DESIGN  FACTOR IS ACTIVITY.
   CODE IS EFFECT.  NAME IS 'A: ACTIVITY'./

ANALYSIS  /


END OF PROCEDURE-GENERATED STATEMENTS.
USER-SUPPLIED ANALYSIS CONTROL STATEMENTS RESUME FOLLOWING ANALYSIS OUTPUT.

=========================================================================
```

TABLE 10.5 (Continued)

```
--- ANALYSIS SUMMARY ---


THE FOLLOWING EFFECTS ARE COMPONENTS OF THE SPECIFIED
LINEAR MODEL FOR THE BETWEEN DESIGN.  ESTIMATES AND TESTS
OF HYPOTHESES FOR THESE EFFECTS CONCERN PARAMETERS OF THAT MODEL.

          OVALL: GRAND MEAN
          O: OCCUP

THE FOLLOWING EFFECTS ARE COMPONENTS OF THE SPECIFIED
LINEAR MODEL FOR THE WITHIN  DESIGN.  ESTIMATES AND TESTS
OF HYPOTHESES FOR THESE EFFECTS CONCERN PARAMETERS OF THAT MODEL.

          OBS: WITHIN CASE MEAN
          A: ACTIVITY

EFFECTS CONCERNING PARAMETERS OF THE COMBINED BETWEEN AND
WITHIN MODELS ARE THE COMBINATIONS (INTERACTIONS) OF EFFECTS
IN BOTH MODELS.


=====================================================================

=====================================================================
WITHIN EFFECT:
+          OBS: WITHIN CASE MEAN

EFFECT    VARIATE      STATISTIC        F      DF       P
---------------------------------------------------------------------
 OVALL: GRAND MEAN
        DEP-VAR
              SS=      1591.35
              MS=      1591.35       812.60   1,   12   .0000
 O: OCCUP
        DEP-VAR
              SS=      172.900
              MS=      86.4500        44.14   2,   12   .0000

 ERROR
        DEP-VAR
              SS=      23.500000
              MS=       1.9583333


---------------------------------------------------------------------

=====================================================================
WITHIN EFECT:
+          A: ACTIVITY

EFFECT    VARIATE      STATISTIC        F      DF       P
---------------------------------------------------------------------
  A
        DEP-VAR
              TSQ=      30.9896         8.61   3,   10   .0040
          WCP SS=       15.7833
          WCP MS=        5.26111       4.72   3,   36   .0070
          GREEHHOUSE-GEISSER ADJ. DF   4.72  2.40, 28.75 .0126
          HUYNH-FELDT ADJUSTED DF      4.72  3.00, 36.00 .0070
  (A) X (O: OCCUP)
        DEP-VAR
              LRATIO=  .762785E-01      8.74   6,   20.00 .0001
              TRACE=    5.42785
              TZSQ=    59.7063
                (P APPROXIMATION FOR ABOVE STATISTIC NOT AVAILABLE -
                 DF NOT IN RANGE OF ACCURACY.)
              MXROOT=   .779765
                (P APPROXIMATION FOR ABOVE STATISTIC NOT AVAILABLE -
                 DF NOT IN RANGE OF ACCURACY.)
          WCP SS=      55.3667
          WCP MS=       9.22778        8.28   6,   36   .0000
          GREENHOUSE-GEISSER ADJ. DF   8.28  4.79, 28.75 .0001
          HUYNH-FELDT ADJUSTED DF      8.28  6.00, 36.00 .0000

  ERROR
        DEP-VAR
              WCP SS=  40.100000
              WCP MS=   1.1138889

              GGI EPSILON=   .79856
              H-F EPSILON=  1.00000


---------------------------------------------------------------------

=====================================================================
```

```
DATA SSAMPLE;
INFILE TAPE42;
INPUT RESPOND OCCUP $ READ DANCE TV SKI;
PROC GLM;
   CLASS OCCUP;
   MODEL READ DANCE TV SKI = OCCUP/NOUNI;
   REPEATED ACTIVITY 4 PROFILE/SHORT;
```

GENERAL LINEAR MODELS PROCEDURE

CLASS LEVEL INFORMATION

CLASS	LEVELS	VALUES
OCCUP	3	ADMIN BELLY POLITIC

NUMBER OF OBSERVATIONS IN DATA SET = 15

GENERAL LINEAR MODELS PROCEDURE

REPEATED MEASURES ANALYSIS OF VARIANCE

REPEATED MEASURES LEVEL INFORMATION

DEPENDENT VARIABLE	READ	DANCE	TV	SKI

MANOVA TEST CRITERIA AND EXACT F STATISTICS FOR THE HYPOTHESIS OF NO ACTIVITY EFFECT[a]
H = TYPE III SS&CP MATRIX FOR: ACTIVITY   E = ERROR SS&CP MATRIX
S=1   M=0.5   N=4.5

STATISTIC	VALUE	F	NUM DF	DEN DF	PR > F
WILKS' LAMBDA	0.2791374	8.608	3	10	0.0040
PILLAI'S TRACE	0.7208626	8.608	3	10	0.0040
HOTELLING-LAWLEY TRACE	2.582466	8.608	3	10	0.0040
ROY'S GREATEST ROOT	2.582466	8.608	3	10	0.0040

F STATISTIC FOR WILKS' LAMBDA IS EXACT

MANOVA TEST CRITERIA AND F APPROXIMATIONS FOR THE HYPOTHESIS OF NO ACTIVITY*OCCUP EFFECT
H = TYPE III SS&CP MATRIX FOR: ACTIVITY*OCCUP   E = ERROR SS&CP MATRIX

S=2   M=0   N=4.5

STATISTIC	VALUE	F	NUM DF	DEN DF	PR > F
WILKS' LAMBDA	0.07627855	8.736	6	20	0.0001
PILLAI'S TRACE	1.433414	9.276	6	22	0.0001
HOTELLING-LAWLEY TRACE	5.42785	8.142	6	18	0.0002
ROY'S GREATEST ROOT	3.5406	12.982	3	11	0.0006

NOTE: F STATISTIC FOR ROY'S GREATEST ROOT IS AN UPPER BOUND
F STATISTIC FOR WILKS' LAMBDA IS EXACT

GENERAL LINEAR MODELS PROCEDURE

TESTS OF HYPOTHESES FOR BETWEEN SUBJECTS EFFECTS

SOURCE	DF	TYPE III SS	MEAN SQUARE	F VALUE	PR > F
OCCUP	2	172.90000000	86.45000000	44.14	0.0001
ERROR	12	23.50000000	1.95833333		

GENERAL LINEAR MODELS PROCEDURE

TABLE 10.6 (Continued)

SOURCE	DF	TYPE III SS	MEAN SQUARE	F VALUE	PR > F	ADJUSTED PR > F G - G	ADJUSTED PR > F H - F
		UNIVARIATE TESTS OF HYPOTHESES FOR WITHIN SUBJECT EFFECTS					
ACTIVITY	3	15.78333333	5.26111111	4.72	0.0070	0.0126	0.0070
ACTIVITY*OCCUP	6	55.36666667	9.22777778	8.28	0.0001	0.0001	0.0001
ERROR(ACTIVITY)	36	40.10000000	1.11388889				

GREENHOUSE-GEISSER EPSILON = 0.7986
HUYNH-FELDT EPSILON = 1.1778

In SAS GLM, the CLASS instruction identifies OCCUP as the grouping variable. The MODEL instruction shows the DVs on the left of an equation and the IV on the right. Profile analysis is distinguished from ordinary MANOVA by the instructions in the line beginning REPEATED.

The results are presented in two multivariate tables and two univariate tables. The first table, labeled ...HYPOTHESIS OF NO ACTIVITY EFFECT, shows four fully-labeled multivariate tests of flatness. The second table shows the same four multivariate tests of parallelism, labeled ...HYPOTHESIS OF NO ACTIVITY*OCCUP EFFECT. The test of levels is the test for OCCUP in the third table, labeled TEST FOR HYPOTHESES OF BETWEEN SUBJECTS EFFECTS. The UNIVARIATE TESTS OF HYPOTHESES FOR WITHIN SUBJECTS EFFECTS table provides the univariate ANOVA for the within-subjects main effect (ACTIVITY) and the within-between interaction (ACTIVITY*OCCUP). Greenhouse-Geisser (G-G) and Huynh-Feldt (H-F) adjustments for violation of homogeneity of covariance are also given for these univariate repeated-measures tests. More extensive output is available if the SHORT instruction is omitted.

TABLE 10.7 PROFILE ANALYSIS OF SMALL SAMPLE EXAMPLE THROUGH SYSTAT MGLH (SETUP AND OUTPUT)

```
        USE TAPE42
        CATEGORY OCCUP=3
        MODEL ACTIVITY(1-4)=CONSTANT+OCCUP/REPEAT
        ESTIMATE


NUMBER OF CASES PROCESSED:    15

DEPENDENT VARIABLE MEANS

        ACTIVITY(1) ACTIVITY(2) ACTIVITY(3) ACTIVITY(4)

            5.533       5.400       4.267       5.400
    UNIVARIATE AND MULTIVARIATE REPEATED MEASURES ANALYSIS


*****************************
* BETWEEN SUBJECTS EFFECTS *
*****************************

TEST FOR EFFECT CALLED:
                OCCUP

TEST OF HYPOTHESIS

        SOURCE        SS      DF      MS          F          P

    HYPOTHESIS     172.900      2    86.450    44.145      0.000
         ERROR      23.500     12     1.958
```

TABLE 10.7 (Continued)

```
****************************
* WITHIN SUBJECTS EFFECTS *
****************************
```

TEST FOR EFFECT CALLED:
                CONSTANT

SINGLE DEGREE-OF-FREEDOM POLYNOMIAL CONTRASTS

DEGREE	SS	DF	MS	F	P
1	1.763	1	1.763	1.117	0.311
ERROR	18.940	12	1.578		
2	6.017	1	6.017	6.748	0.023
ERROR	10.700	12	0.892		
3	8.003	1	8.003	9.182	0.010
ERROR	10.460	12	0.872		

UNIVARIATE REPEATED MEASURES F-TEST

SOURCE	SS	DF	MS	F	P
HYPOTHESIS	15.783	3	5.261	4.723	0.007
ERROR	40.100	36	1.114		

MULTIVARIATE TEST STATISTICS

WILKS' LAMBDA =	0.279				
F-STATISTIC =	8.608	DF =	3, 10	PROB =	0.004
PILLAI TRACE =	0.721				
F-STATISTIC =	8.608	DF =	3, 10	PROB =	0.004
HOTELLING-LAWLEY TRACE =	2.582				
F-STATISTIC =	8.608	DF =	3, 10	PROB =	0.004

TEST FOR EFFECT CALLED:
                OCCUP

SINGLE DEGREE-OF-FREEDOM POLYNOMIAL CONTRASTS

DEGREE	SS	DF	MS	F	P
1	2.247	2	1.123	0.712	0.510
ERROR	18.940	12	1.578		
2	27.033	2	13.517	15.159	0.001
ERROR	10.700	12	0.892		
3	26.087	2	13.043	14.964	0.001
ERROR	10.460	12	0.872		

UNIVARIATE REPEATED MEASURES F-TEST

SOURCE	SS	DF	MS	F	P
HYPOTHESIS	55.367	6	9.228	8.284	0.000
ERROR	40.100	36	1.114		

MULTIVARIATE TEST STATISTICS

WILKS' LAMBDA =	0.076				
F-STATISTIC =	8.736	DF =	6, 20	PROB =	0.000
PILLAI TRACE =	1.433				
F-STATISTIC =	9.276	DF =	6, 22	PROB =	0.000
HOTELLING-LAWLEY TRACE =	5.428				
F-STATISTIC =	8.142	DF =	6, 18	PROB =	0.000
THETA =	0.780	S =	2, M = .0, N = 4.0	PROB =	0.004

SYSTAT MGLH setup is similar to SAS except that CONSTANT is added to the IV side of the MODEL equation. The program first prints out the means for each leisure time activity (but not means for each group). The levels test follows in the BETWEEN SUBJECT EFFECTS section. The WITHIN SUBJECTS EFFECTS section begins with the flatness test, labeled CONSTANT but actually ACTIVITY.

A trend analysis is printed out (SINGLE DEGREE-OF-FREEDOM POLYNOMIAL CONTRASTS) before the univariate test of the within-subjects effect. Finally, three multivariate tests are given. The same tests are given for the parallelism test, labeled OCCUP but actually OCCUP by ACTIVITY. Roy's gcr, labeled THETA, is an additional test given for parallelism.

10.5 SOME IMPORTANT ISSUES

Issues discussed here are unique to profile analysis or affect profile analysis differently from traditional MANOVA. Issues such as choice among statistical criteria, for instance, are identical whether the DVs are analyzed directly (as in MANOVA) or converted into segments (as in profile analysis). Therefore, the reader is referred to Section 9.5.1 for consideration of these matters.

10.5.1 Contrasts in Profile Analysis

When there are more than two levels of a significant effect in profile analysis, it is often desirable to perform contrasts to pinpoint sources of variability. For instance, because there is an overall difference between administrators, belly dancers, and politicians in their ratings of satisfaction with leisure time activities (see Section 10.4), contrasts are needed to discover which groups differ from which other groups. Are belly dancers the same as administrators? Politicians? Neither?

It is probably easiest to think of contrasts following profile analysis as coming from a regular ANOVA design with (at least) one grouping variable and one repeated measure, even when the application of the technique is to multiple, commensurate DVs. That is, the most interpretable contrasts following profile analysis are likely to be the ones that would also be appropriate after a within- and between-subjects ANOVA.

There are, of course, numerous contrast procedures and the choice among them depends on what makes most sense in a given experimental setting. With a single control group, Dunnett's procedure often makes most sense. Or, if all pairwise comparisons are desired, the Tukey test is most appropriate. Or, if there are numerous repeated measures, and/or normative data are available, a confidence interval procedure such as that used in Section 10.7.2 may make most sense. With relatively few repeated measures, a Scheffé type procedure is probably the most general (if also the most conservative) and is the procedure illustrated in this section.

It is important to remember that the contrasts recommended here explore differences in original DV scores while the significance tests in profile analysis for parallelism and flatness evaluate segments. Although there is a logical problem with following up a significance test based on segments with a contrast based on the original scores, performing contrasts on segments seems even worse because of difficulty interpreting the results.

Contrasts in repeated-measures ANOVA with both grouping variables and repeated measures is not the easiest of topics, as you probably recall. First, when parallelism (interaction) is significant, there is the choice between a simple effects analysis and an interaction contrasts analysis. Second, there is a need in some cases to develop separate error terms for some of the contrasts. Third is the need to apply an adjustment such as Scheffé to the F test to avoid too liberal rejection of the null hypothesis. The researcher who is fascinated by these topics is referred to Keppel (1982, especially Chapters 11, 14, and 18) for a detailed and reasonably clear discussion of them. The present effort is to illustrate several possible approaches through BMDP and SPSS[x] and to recommend guidelines for when each is likely to be appropriate.

The most appropriate contrast to perform depends on which effect or combination of effects—levels, flatness, or parallelism—is significant. If either levels or flatness is significant, but parallelism (interaction) is not, contrasts are performed on marginal means. If the test for levels is significant, contrasts are formed on marginal values for the grouping variable. If the test for flatness is significant, contrasts are formed on the repeated-measures marginal values. Because contrasts formed on marginal values "fall out" of computer runs for interaction contrasts, they are illustrated in Section 10.5.1.3.

Sections 10.5.1.1 and 10.5.1.2 describe simple effects analyses, appropriate if parallelism is significant. In simple effects analysis one variable is held constant at some value while mean differences are examined on the levels of the other variable, as seen in Figure 10.2. For instance, the level of group is held constant at belly dancer while mean differences are examined among the leisure time activities [Figure 10.2(a)]. The researcher asks if belly dancers have mean differences in satisfaction with different leisure activities. Or, leisure activity is held constant at dance while mean differences are explored between administrators, politicians, and belly dancers [Figure 10.2(c)]. The researcher asks whether the three groups have different mean satisfaction while dancing.

Section 10.5.1.1 illustrates a simple effects analysis followed by simple contrasts [Figures 10.2(a) and (b)] for the case where parallelism and flatness effects are both significant, but the levels effect is not. Section 10.5.1.2 illustrates a simple effects analysis followed by simple contrasts [Figures 10.2(c) and (d)] for the case where parallelism and levels are both significant, but the flatness effect is not. This particular pattern of simple effects analysis is recommended because of the confounding inherent in analyzing simple effects.

The analysis is confounded because when the groups (levels) effect is held constant to analyze the repeated measure in a one-way within-subjects ANOVA, both the sum of squares for interaction and the sum of squares for the repeated measure are partitioned. When the repeated measure is held constant so the groups (levels) effect is analyzed in a one-way between-subjects ANOVA, both the sum of squares for interaction and the sum of squares for the group effect are partitioned. Because in simple effects analyses the interaction sum of squares is confounded with one or the other of the main effects, it seems best to confound it with a nonsignificant main effect where possible. This recommendation is followed in Sections 10.5.1.1 and 10.5.1.2.

Section 10.5.1.3 describes an interaction contrasts analysis. In such an analysis, an interaction between two IVs is examined through one or more smaller interactions (Figure 10.3). For instance, the significant interaction between the three groups on four leisure activities in the example might be reduced to examination of the difference

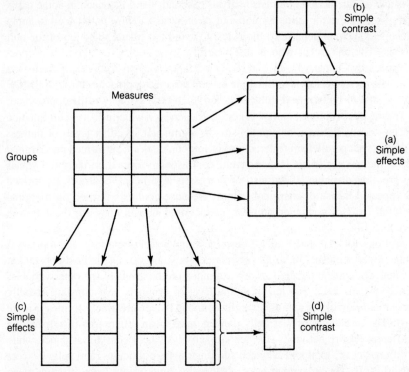

Figure 10.2 Simple effects analysis exploring: (a) differences among measures for each group, followed by (b) a simple contrast between measures for one group; and (c) differences among groups for each measure, followed by (d) a simple contrast between groups for one measure.

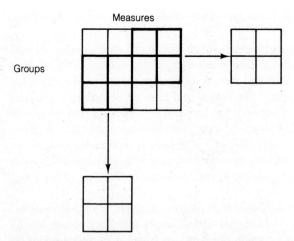

Figure 10.3 Interaction contrasts analyses exploring small (2 × 2) interactions formed by partitioning a large (3 × 4) interaction.

between two groups on only two of the activities. One could, for instance, ask if there is a significant interaction in satisfaction between belly dancers and administrators while watching TV vs. dancing. Or, one could pool the results for administrator and politician and contrast them with belly dancer for one side of the interaction, while pooling the results for the two sedentary activities (watching TV and reading) against the results for the two active activities (dancing and skiing) as the other side of the interaction. The researcher asks whether there is an interaction between dancers and the other professionals in their satisfaction while engaged in sedentary vs. active leisure time activities.

An interaction contrasts analysis is not a confounded analysis; only the sum of squares for interaction is partitioned. Thus, it is appropriate whenever the interaction is significant and regardless of the significance of the other two effects. However, because simple effects are generally easier to understand and explain, it seems better to perform them when possible. For this reason, we recommend an interaction contrasts analysis to explore the parallelism effect only when both the levels and flatness effects are also significant.

10.5.1.1 Parallelism and Flatness Significant, Levels Not Significant (Simple Effects Analysis)
When parallelism and flatness are both significant, a simple effects analysis is recommended where differences among means for groups are examined separately at each level of the repeated measure [Figure 10.2(c)]. For the example, differences in means among politicians, administrators, and belly dancers are sought first in the reading variable, then in dance, then in TV, and finally in skiing. (Not all these effects need to be examined, of course, if they are not of interest.)

Both BMDP and SPSS^x have command language to request simple effects analyses. In BMDP one specifies PROCEDURE IS SIMPLE and HOLD IS (variable name) in the ANALYSIS paragraph. In SPSS^x the keyword MWITHIN is used in the MANOVA paragraph. Table 10.8 shows control language and partial output from both BMDP and SPSS^x for performing simple effects analysis on groups with repeated measures held constant.

Table 10.8(a) shows the control language and output produced by BMDP. The simple effect of groups at READ is identified as (OBS.R) X (O: OCCUP) (sum of squares = 8.53333, mean square = 4.26667, and F = 2.67) in the output. The simple effect of groups at DANCE is (OBS.D) X (O: OCCUP) in the output, while simple effects of TV is (OBS.T) X (O: OCCUP) and SKI is (OBS.S) X (O: OCCUP).

Table 10.8(b) shows the control language and output produced by SPSS^x. The simple effect of groups at READ is identified as OCCUP BY MWITHIN ACTIVITY(1) in the output. Simple effects of DANCE, TV, and SKI are identified as OCCUP BY MWITHIN ACTIVITY(2), ACTIVITY(3), and ACTIVITY(4), respectively.

The same separate error terms are developed by both programs for each of the simple effects. The error term formed by subjects nested within groups at READ is used to test for differences between groups when the activity is READ. Similar error terms are developed for DANCE, TV, and SKI. In the BMDP output, the error terms are called ERROR, in SPSS^x, WITHIN CELLS. For example, both BMDP and SPSS^x compute the sum of squares for subjects nested within groups at READ as 19.2000 (mean square = 1.6000).

To evaluate the significance of the simple effects, a Scheffé adjustment (see Section 3.2.6.1) is applied to unadjusted critical F under the assumptions that these tests are performed post hoc and that the researcher wants to control for familywise Type I error. For these contrasts the Scheffé adjustment is

$$F_s = (k - 1) F_{(k-1),\ k(n-1)} \tag{10.7}$$

TABLE 10.8 SIMPLE EFFECTS ANALYSIS OF OCCUPATION, HOLDING ACTIVITY CONSTANT, BMDP (10.8a) AND SPSS[x] (10.8b)

```
                    (a)  SETUP AND SELECTED OUTPUT FROM BMDP4V

         /PROBLEM   TITLE IS 'SMALL SAMPLE EXAMPLE FOR PROFILE CHAPTER'.
         /INPUT     VARIABLES ARE 5.
                    FORMAT IS '(F1,4F2)'.
         /VARIABLES NAMES ARE OCCUP, READ,DANCE,TV,SKI.
         /BETWEEN   FACTOR IS OCCUP.
                    CODES ARE 1,2,3.
                    NAMES ARE BELLY, POL, ADMIN.
         /WITHIN    FACTOR IS ACTIV.
                    CODES ARE 1 TO 4.
                    NAMES ARE READ, DANCE, TV, SKI.
         /WEIGHTS   BETWEEN ARE EQUAL.
                    WITHIN ARE EQUAL.
         /END
         ANALYSIS   PROCEDURE IS SIMPLE.
                    HOLD IS ACTIV./
         END/
```

```
WITHIN EFFECT:
+           OBS.R: MEAN AT READ

EFFECT   VARIATE      STATISTIC           F         DF        P
------------------------------------------------------------------
  OBS.R
        DEP-VAR
                 SS=     459.267
                 MS=     459.267        287.04     1,   12    .0000
 (OBS.R) X (O: OCCUP)
        DEP-VAR
                 SS=     8.53333
                 MS=     4.26667          2.67     2,   12    .1101

 ERROR
        DEP-VAR
                 SS=     19.200000
                 MS=     1.6000000

------------------------------------------------------------------

==================================================================
WITHIN EFFECT:
+           OBS.D: MEAN AT DANCE

EFFECT   VARIATE      STATISTIC           F         DF        P
------------------------------------------------------------------
  OBS.D
        DEP-VAR
                 SS=     437.400
                 MS=     437.400        524.88     1,   12    .0000
 (OBS.D) X (O: OCCUP)
        DEP-VAR
                 SS=     139.600
                 MS=     69.8000         83.76     2,   12    .0000

 ERROR
        DEP-VAR
                 SS=     10.000000
                 MS=     .83333333

------------------------------------------------------------------

==================================================================
```

TABLE 10.8 (Continued)

```
WITHIN EFFECT:
+                 OBS.T: MEAN AT TV

EFFECT    VARIATE      STATISTIC          F        DF        P
-----------------------------------------------------------------
  OBS.T
        DEP-VAR
                 SS=     273.067
                 MS=     273.067        390.10    1,    12    .0000
  (OBS.T) X (O: OCCUP)
        DEP-VAR
                 SS=     46.5333
                 MS=     23.2667         33.24    2,    12    .0000

  ERROR
        DEP-VAR
                 SS=     8.4000000
                 MS=     .70000000

-----------------------------------------------------------------

=================================================================
WITHIN EFFECT:
+                 OBS.S: MEAN AT SKI

EFFECT    VARIATE      STATISTIC          F        DF        P
-----------------------------------------------------------------
  OBS.S
        DEP-VAR
                 SS=     437.400
                 MS=     437.400        201.88    1,    12    .0000
  (OBS.S) X (O: OCCUP)
        DEP-VAR
                 SS=     33.6000
                 MS=     16.8000          7.75    2,    12    .0069

  ERROR
        DEP-VAR
                 SS=     26.000000
                 MS=     2.1666667

-----------------------------------------------------------------

=================================================================
```

(b) SETUP AND SELECTED OUTPUT FROM SPSS'

```
        TITLE          SMALL SAMPLE PROFILE ANALYSIS.
        FILE HANDLE    TAPE42
        DATA LIST      FILE=TAPE42
                       /RESPOND,OCCUP,READ,DANCE,TV,SKI (6F3.0)
        VAR LABELS     RESPOND 'RESPONDENT'  OCCUP 'OCCUPATION'
        VALUE LABELS   OCCUP 1 'BELLY DANCER' 2 'POLITICIAN' 3 'ADMINISTRATOR'
        MANOVA         READ TO SKI BY OCCUP(1,3)/
                       WSFACTOR=ACTIVITY(4)/
                       WSDESIGN=MWITHIN ACTIVITY(1) MWITHIN ACTIVITY(2)
                                MWITHIN ACTIVITY(3) MWITHIN ACTIVITY(4)/
                       ANALYSIS(REPEATED)/
                       NOPRINT=PARAMETER(ESTIM)/
```

* * * * * * * * * * * * * * A N A L Y S I S   O F   V A R I A N C E * * * * * * * * * * * * * * *

TESTS OF SIGNIFICANCE FOR READ USING SEQUENTIAL SUMS OF SQUARES

SOURCE OF VARIATION	SUM OF SQUARES	DF	MEAN SQUARE	F	SIG. OF F
WITHIN CELLS	19.20000	12	1.60000		
MWITHIN ACTIVITY(1)	459.26667	1	459.26667	287.04167	0
OCCUP BY MWITHIN ACTIVITY(1)	8.53333	2	4.26667	2.66667	.110

TESTS OF SIGNIFICANCE FOR DANCE USING SEQUENTIAL SUMS OF SQUARES

SOURCE OF VARIATION	SUM OF SQUARES	DF	MEAN SQUARE	F	SIG. OF F
WITHIN CELLS	10.00000	12	.83333		
MWITHIN ACTIVITY(2)	437.40000	1	437.40000	524.88000	0
OCCUP BY MWITHIN ACTIVITY(2)	139.60000	2	69.80000	83.76000	0

TABLE 10.8   (Continued)

```
TESTS OF SIGNIFICANCE FOR TV USING SEQUENTIAL SUMS OF SQUARES

SOURCE OF VARIATION            SUM OF SQUARES      DF    MEAN SQUARE              F     SIG. OF F

WITHIN CELLS                         8.40000      12        .70000
MWITHIN ACTIVITY(3)                273.06667       1     273.06667       390.09524            0
OCCUP BY MWITHIN ACTIVITY(3)        46.53333       2      23.26667        33.23810         .000
```

```
TESTS OF SIGNIFICANCE FOR SKI USING SEQUENTIAL SUMS OF SQUARES

SOURCE OF VARIATION            SUM OF SQUARES      DF    MEAN SQUARE              F     SIG. OF F

WITHIN CELLS                        26.00000      12       2.16667
MWITHIN ACTIVITY(4)                437.40000       1     437.40000       201.87692            0
OCCUP BY MWITHIN ACTIVITY(4)        33.60000       2      16.80000         7.75385         .007
```

where k is the number of groups and n is the number of subjects in each group. For the example, using alpha $= 0.05$

$$F_s = (3 - 1) F_{(2, 12)}$$
$$= 7.76$$

By this criterion, there are not reliable mean differences between the groups when the DV is READ or SKI, but there are reliable differences when the DV is DANCE or TV.

Because there are three groups, these findings are still ambiguous. Which group or groups are different from which other group or groups? To pursue the analysis, simple contrasts are performed [Figure 10.2(d)]. Contrast coefficients are applied to the levels of the grouping variable to determine the source of the difference. For the example, contrast coefficients compare the mean for belly dancers with the mean for the other two groups combined for DANCE. Control language and partial output from both BMDP and SPSS^x are shown in Table 10.9.

In BMDP [Table 10.9(a)], the control language is appropriate to an ANOVA with a between-subjects IV. DANCE is identified as the DV in the ANALYSIS paragraph. The sum of squares for the contrast is identified by its name as specified in the DESIGN paragraph (BDVSOTHR).

In SPSS^x [Table 10.9(b)], the CONTRAST procedure is used.[6] The sum of squares for the contrast is labeled OCCUP(1). Both programs use the error term formed by subjects nested within groups for the particular DV. This error term is called ERROR in BMDP and WITHIN CELLS in SPSS^x.

For this analysis, the sum of squares and mean square for the contrast is 120.000, error mean square is .83333, and F is 144.00. This F exceeds the F_s adjusted critical value of 7.76; it is no surprise to find there is a reliable difference between belly dancers and others in their satisfaction while engaging in DANCE.

10.5.1.2 Parallelism and Levels Significant, Flatness Not Significant (Simple Effects Analysis)   This combination of findings occurs rarely because if parallelism

[6] This could also be done in conjunction with the MWITHIN procedure, but the setup is more complicated and the labeling in the output misleading.

TABLE 10.9 SIMPLE COMPARISONS ANALYSIS ON OCCUPATIONS,
HOLDING ACTIVITY CONSTANT, BMDP (10.9a)
AND SPSS* (10.9b)

(a) SETUP AND SELECTED OUTPUT FROM BMDP4V

```
/PROBLEM   TITLE IS 'SMALL SAMPLE EXAMPLE FOR PROFILE CHAPTER'.
/INPUT     VARIABLES ARE 5.
           FORMAT IS '(F1,4F2)'.
/VARIABLES NAMES ARE OCCUP, READ,DANCE,TV,SKI.
/BETWEEN   FACTOR IS OCCUP.
           CODES ARE 1,2,3.
           NAMES ARE BELLY, POL, ADMIN.
/WEIGHTS   BETWEEN ARE EQUAL.
/END
DESIGN  FACTOR IS OCCUP.
        TYPE = BETWEEN, REGRESSION.
        CODE = READ.
        VALUES = 2,-1,-1.
        NAME IS BDVSOTHR./
ANALYSIS  DEPENDENT IS DANCE.
          PROC = FACT./
END/
```

```
==================================================================

EFFECT   VARIATE     STATISTIC         F       DF       P
------------------------------------------------------------------
 BDVSOTHR
      DANCE
              SS=    120.000
              MS=    120.000      144.00    1,    12    .0000
 OVALL: GRAND MEAN
      DANCE
              SS=    437.400
              MS=    437.400      524.88    1,    12    .0000
 O: OCCUP
      DANCE
              SS=    139.600
              MS=     69.8000      83.76    2,    12    .0000

 ERROR
      DANCE
              SS=     10.000000
              MS=      .83333333
```

(b) SETUP AND SELECTED OUTPUT FROM SPSS*

```
TITLE        SMALL SAMPLE PROFILE ANALYSIS
FILE HANDLE  TAPE42
DATA LIST    FILE=TAPE42
             /RESPOND,OCCUP,READ,DANCE,TV,SKI (6F3.0)
VAR LABELS   RESPOND 'RESPONDENT'  OCCUP 'OCCUPATION'
VALUE LABELS OCCUP 1 'BELLY DAR' 2 'POLITICIAN' 3 'ADMINISTRATOR'
MANOVA       DANCE BY OCCUP(1,3)/
             PARTITION(OCCUP)/
             CONTRAST(OCCUP)=SPECIAL(1 1 1, 2 -1 -1, 0 1 -1)/
             NOPRINT=PARAMETER(ESTIM)/
             DESIGN=OCCUP(1)/
```

TESTS OF SIGNIFICANCE FOR DANCE USING SEQUENTIAL SUMS OF SQUARES

SOURCE OF VARIATION	SUM OF SQUARES	DF	MEAN SQUARE	F	SIG. OF F
WITHIN CELLS	10.00000	12	.83333		
CONSTANT	437.40000	1	437.40000	524.88000	0
OCCUP(1)	120.00000	1	120.00000	144.00000	0

and levels are significant, flatness is nonsignificant only if profiles for different groups are mirror images that cancel each other out.

The simple effects analysis recommended here examines mean differences among the various DVs in series of one-way within-subjects ANOVAs with each group in turn held constant [Figure 10.2(a)]. For the example, mean differences between READ, DANCE, TV, and SKI are sought first for belly dancers, then for politicians, and then for administrators. The researcher inquires whether or not each group, in turn, is more satisfied during some activities than during others.

The same setups that were used in Section 10.5.1.1 are appropriate here except that the other effect is specified. Table 10.10(a) contains control language and partial output for BMDP; Table 10.10(b) shows the same analyses through SPSS^x.

For BMDP the simple effect of repeated measures for belly dancers is labeled

TABLE 10.10 SIMPLE EFFECTS OF ACTIVITY, HOLDING OCCUPATION CONSTANT, BMDP (10.10a) AND SPSS· (10.10b)

```
                      (a)   SETUP AND SELECTED FROM BMDP4V

        /PROBLEM    TITLE IS 'SMALL SAMPLE EXAMPLE FOR PROFILE CHAPTER'.
        /INPUT      VARIABLES ARE 5.
                    FORMAT IS '(F1,4F2)'.
        /VARIABLES  NAMES ARE OCCUP, READ,DANCE,TV,SKI.
        /BETWEEN    FACTOR IS OCCUP.
                    CODES ARE 1,2,3.
                    NAMES ARE BELLY, POL, ADMIN.
        /WITHIN     FACTOR IS ACTIV.
                    CODES ARE 1 TO 4.
                    NAMES ARE READ, DANCE, TV, SKI.
        /WEIGHTS    BETWEEN ARE EQUAL.
                    WITHIN ARE EQUAL.
        /END
        ANALYSIS    PROCEDURE IS SIMPLE.
                    HOLD IS OCCUP./
        END/
```

```
WITHIN EFFECT: A: ACTIV

EFFECT   VARIATE      STATISTIC              F         DF          P
-----------------------------------------------------------------------
(A) X (OVALL.B: MEAN AT BELLY)
        DEP-VAR
                TSQ=      51.5102          14.31      3,     10  0.0006
                WCP SS=      35.800000
                WCP MS=      11.933333      10.71      3,     36  0.0000
                GREENHOUSE-GEISSER ADJ. DF  10.71   2.40,  28.75  0.0002
                HUYNH-FELDT ADJUSTED DF     10.71   3.00,  36.00  0.0000

   ERROR
        DEP-VAR
                WCP SS=      40.10000000
                WCP MS=       1.11388889

                GGI EPSILON    0.79856
                H-F EPSILON    1.00000
-----------------------------------------------------------------------

WITHIN EFFECT: A: ACTIV

EFFECT   VARIATE      STATISTIC              F         DF          P
-----------------------------------------------------------------------
(A) X (OVALL.P: MEAN AT POL)
        DEP-VAR
                TSQ=       0.705068          0.20      3,     10  0.8968
                WCP SS=       0.400000
                WCP MS=       0.133333       0.12      3,     36  0.9479
                GREENHOUSE-GEISSER ADJ. DF   0.12   2.40,  28.75  0.9177
                HUYNH-FELDT ADJUSTED DF      0.12   3.00,  36.00  0.9479

   ERROR
        DEP-VAR
                WCP SS=      40.10000000
                WCP MS=       1.11388889

                GGI EPSILON    0.79856
                H-F EPSILON    1.00000
-----------------------------------------------------------------------
```

TABLE 10.10   (Continued)

```
WITHIN EFFECT: A: ACTIV

EFFECT     VARIATE       STATISTIC           F          DF          P
------------------------------------------------------------------------
(A) X (OVALL.B: MEAN AT ADMIN)
        DEP-VAR
                 TSQ=         43.9086      12.20      3,     10  0.0011
            WCP SS=          34.950000
            WCP MS=          11.650000     10.46      3,     36  0.0000
            GREENHOUSE-GEISSER ADJ. DF     10.46  2.40,  28.75  0.0002
            HUYNH-FELDT ADJUSTED DF        10.46  3.00,  36.00  0.0000

ERROR
       DEP-VAR
            WCP SS=          40.10000000
            WCP MS=           1.11388889

            GGI EPSILON      0.79856
            H-F EPSILON      1.00000

------------------------------------------------------------------------

            (b)  SETUP AND SELECTED OUTPUT FROM SPSSˣ


          TITLE          SMALL SAMPLE PROFILE ANALYSIS.
          FILE HANDLE    TAPE42
          DATA LIST      FILE=TAPE42
                         /RESPOND,OCCUP,READ,DANCE,TV,SKI (6F3.0)
          VAR LABELS     RESPOND 'RESPONDENT'  OCCUP 'OCCUPATION'
          VALUE LABELS   OCCUP 1 'BELLY DANCER' 2 'POLITICIAN' 3 'ADMINISTRATOR'
          MANOVA         READ TO SKI BY OCCUP(1,3)/
                         WSFACTOR=ACTIVITY(4)/
                         WSDESIGN=ACTIVITY/
                         ANALYSIS(REPEATED)/
                         NOPRINT=PARAMETER(ESTIM)/
                         PRINT=SIGNIF(BRIEF)/
                         PRINT=SIGNIF(AVERF)/
                         DESIGN=MWITHIN OCCUP(1) MWITHIN OCCUP(2) MWITHIN OCCUP(3)/


AVERAGED TESTS OF SIGNIFICANCE FOR MEAS.1 USING SEQUENTIAL SUMS OF SQUARES

SOURCE OF VARIATION           SUM OF SQUARES     DF    MEAN SQUARE        F   SIG. OF F

WITHIN CELLS                     40.10000        36     1.11389
MWITHIN OCCUP(1) BY ACTIVITY     35.80000         3    11.93333    10.71322      .000
MWITHIN OCCUP(2) BY ACTIVITY       .40000         3      .13333      .11970      .948
MWITHIN OCCUP(3) BY ACTIVITY     34.95000         3    11.65000    10.45885      .000
```

(A) X (OVALL.B: MEAN AT BELLY); the simple effect of repeated measures for politicians is (A) X (OVALL.P: MEAN AT POL) and for administrators (A) X (OVALL.B: MEAN AT ADMIN). For SPSSˣ the simple effect for belly dancers is MWITHIN OCCUP(1) BY ACTIVITY, for politicians MWITHIN OCCUP(2) BY ACTIVITY, and for administrators MWITHIN OCCUP(3) BY ACTIVITY. The sum of squares for the simple effect at belly dancers is 35.8000.

Both BMDP and SPSSˣ use the error term from the full analysis (interaction between repeated measures and subjects nested within groups) as the error term for each of the contrasts. BMDP calls the term ERROR and SPSSˣ calls it WITHIN CELLS; the sum of squares is 40.1000.

Some feel that it is best to develop a separate error term for each segment of an analysis, that is, that the error term for the simple effect for belly dancers should be the repeated measures by subjects interaction for that group alone. If you are of this opinion, judicious use of the USE = (OCCUP EQ 1) feature in BMDP or the

SELECT IF (OCCUP EQ 1) procedure in SPSS^x is used to separate out the group on whose data a one-way within-subject ANOVA is then performed.

For these simple effects, the Scheffé adjustment to critical F is

$$F_s = (p - 1) F_{(p-1), k(p-1)(n-1)} \tag{10.8}$$

where p is the number of repeated measures, n is the number of subjects in each group, and k is the number of groups. For the example

$$F_s = (4 - 1) F_{(3, 36)}$$
$$= 8.76$$

Two of the F values for simple effects (10.71322 and 10.45885) exceed adjusted critical F; the researcher concludes that there are reliable differences in mean satisfaction during different activities among belly dancers and among administrators.

These findings are also ambiguous because there are more than two activities. Contrast coefficients are therefore applied to the levels of the repeated measure to examine the pattern of differences in greater detail [Figure 10.2(b)].

Table 10.11 shows setups and partial outputs from BMDP and SPSS^x for a simple contrast. The contrast that is illustrated compares the pooled mean for the two sedentary activities (READ and TV) against the pooled mean for the two active activities (DANCE and SKI) for (you guessed it) belly dancers.

For BMDP [Table 10.11(a)] the contrast is defined and named in the DESIGN paragraph. The contrast is calculated for only the first group because of the USE = (OCCUP EQ 1) sentence in the TRANSFORM paragraph. The error term that is developed by BMDP for this contrast is the interaction of subjects with the contrast on the repeated measure for the group. The F value of 15.37 exceeds F_s of 8.76 and indicates that belly dancers have a reliable mean difference in their satisfaction while engaging in active vs. sedentary activities.

For SPSS^x [Table 10.11(b)] the contrast is specified in the WSFACTOR segment of the MANOVA paragraph while the group is selected by DESIGN = MWITHIN OCCUP(1). SPSS^x differs from BMDP in computation of the error term. SPSS^x uses the interaction of subjects with the contrast on the repeated measure for all groups combined as the error term. In this case the F value of 16.98246 also exceeds F_s of 8.76 and indicates that belly dancers have reliable mean differences in their satisfaction during active vs. sedentary activities.

10.5.1.3 Parallelism, Levels, and Flatness Significant (Interaction Contrasts
When all three effects are significant, an interaction contrasts analysis is often most appropriate. This analysis partitions the sum of squares for interaction into a series of smaller interactions (Figure 10.3). Smaller interactions are obtained by deleting or combining groups or measures with use of appropriate contrast coefficients.

For the example, illustrated in Table 10.12, means for administrators and politicians are combined and compared with the mean of belly dancers, while means for TV and READ are combined and compared with the combined mean of DANCE

TABLE 10.11 SIMPLE COMPARISONS ANALYSIS ON ACTIVITY,
HOLDING OCCUPATION CONSTANT, BMDP
(10.11a) AND SPSS* (10.11b)

(a) SETUP AND SELECTED OUTPUT FROM BMDP4V

```
/PROBLEM   TITLE IS 'SMALL SAMPLE EXAMPLE FOR PROFILE CHAPTER'.
/INPUT     VARIABLES ARE 5.
           FORMAT IS '(F1,4F2)'.
/VARIABLES  NAMES ARE OCCUP, READ,DANCE,TV,SKI.
/TRANSFORM USE = OCCUP EQ 1.
/BETWEEN   FACTOR IS OCCUP.
           CODES ARE 1,2,3.
           NAMES ARE BELLY, POL, ADMIN.
/WITHIN    FACTOR IS ACTIV.
           CODES ARE 1 TO 4.
           NAMES ARE READ, DANCE, TV, SKI.
/WEIGHTS   BETWEEN ARE EQUAL.
           WITHIN ARE EQUAL.
/END
DESIGN     FACTOR IS ACTIV.
           TYPE = WITHIN, REGRESSION.
           CODE = READ.
           VALUES = -1,1,-1,1.
           NAME IS SEDVSACT./
ANALYSIS  PROC = FACT./
END/
```

WITHIN EFFECT:
+   SEDVSACT

EFFECT	VARIATE	STATISTIC		F	DF	P
SEDVSACT						
	DEP-VAR					
		SS=	24.2000			
		MS=	24.2000	15.37	1, 4	.0173
ERROR						
	DEP-VAR					
		SS=	6.3000000			
		MS=	1.5750000			

(b) SETUP AND SELECTED OUTPUT FROM SPSS*

```
TITLE        SMALL SAMPLE PROFILE ANALYSIS.
FILE HANDLE  TAPE42
DATA LIST    FILE=TAPE42
             /RESPOND,OCCUP,READ,DANCE,TV,SKI (6F3.0)
VAR LABELS   RESPOND 'RESPONDENT'  OCCUP 'OCCUPATION'
VALUE LABELS OCCUP 1 'BELLY DANCER' 2 'POLITICIAN' 3 'ADMINISTRATOR'
MANOVA       READ TO SKI BY OCCUP(1,3)/
             WSFACTOR=ACTIVITY(4)/
             PARTITION(ACTIVITY)/
             CONTRAST(ACTIVITY)=SPECIAL (1 1 1 1, -1 1 -1 1,
                                         -1 0 1 0, 0 -1 0 1)/
             WSDESIGN=ACTIVITY(1)/
             ANALYSIS(REPEATED)/
             NOPRINT=PARAMETER(ESTIM)/
             PRINT=SIGNIF(BRIEF)/
             PRINT=SIGNIF(AVERF)/
             DESIGN=MWITHIN OCCUP(1)/
```

TESTS OF SIGNIFICANCE FOR DANCE USING SEQUENTIAL SUMS OF SQUARES

SOURCE OF VARIATION	SUM OF SQUARES	DF	MEAN SQUARE	F	SIG. OF F
WITHIN CELLS	17.10000	12	1.42500		
MWITHIN OCCUP(1) BY ACTIVITY(1)	24.20000	1	24.20000	16.98246	.001

and SKI. The researcher asks whether belly dancers and others have the same pattern of satisfaction during sedentary vs. active leisure activities.

In BMDP [Table 10.12(a)] two DESIGN paragraphs are used, one to specify the contrast on the within-subjects segment of the design (the repeated measures) and the other to specify the contrast on the between-subjects segment (the grouping effect). In the heavily edited output, the interaction contrast is labeled (SEDVSACT)

TABLE 10.12 INTERACTION CONTRASTS, BELLY DANCERS VS. OTHERS AND ACTIVE VS. SEDENTARY ACTIVITIES, BMDP (10.12a) AND SPSS· (10.12b)

(a) SETUP AND SELECTED OUTPUT FROM BMDP4V

```
/PROBLEM   TITLE IS 'SMALL SAMPLE EXAMPLE FOR PROFILE CHAPTER'.
/INPUT     VARIABLES ARE 5.
           FORMAT IS '(F1,4F2)'.
/VARIABLES  NAMES ARE OCCUP, READ,DANCE,TV,SKI.
/BETWEEN   FACTOR IS OCCUP.
           CODES ARE 1,2,3.
           NAMES ARE BELLY, POL, ADMIN.
/WITHIN    FACTOR IS ACTIV.
           CODES ARE 1 TO 4.
           NAMES ARE READ, DANCE, TV, SKI.
/WEIGHTS   BETWEEN ARE EQUAL.
           WITHIN ARE EQUAL.
/END
DESIGN   TYPE=WITHIN,REGRESSION.
         CODE=READ.
         VALUES=-1,1,-1,1.
         NAME=SEDVSACT./

DESIGN   TYPE=BETWEEN,REGRESSION.
         CODE=READ.
         VALUES=2,-1,-1.
         NAME=BDVSOTHR./

ANALYSIS  PROCEDURE IS FACTORIAL./
END/
```

```
WITHIN EFFECT:
+          SEDVSACT

EFFECT   VARIATE      STATISTIC          F        DF        P
-------------------------------------------------------------------
  (SEDVSACT) X (BDVSOTHR)
         DEP-VAR
                 SS=    21.6750
                 MS=    21.6750        15.21     1,    12    .0021
  SEDVSACT
         DEP-VAR
                 SS=    3.75000
                 MS=    3.75000         2.63     1,    12    .1307
  ERROR
         DEP-VAR
                 SS=    17.100000
                 MS=    1.4250000

====================================================================
WITHIN EFFECT:
+          OBS: WITHIN CASE MEAN

EFFECT   VARIATE      STATISTIC          F        DF        P
-------------------------------------------------------------------
  BDVSOTHR
         DEP-VAR
                 SS=    138.675
                 MS=    138.675        70.81     1,    12    .0000
  ERROR
         DEP-VAR
                 SS=    23.500000
                 MS=    1.9583333

-------------------------------------------------------------------
```

TABLE 10.12 (Continued)

(b) SETUP AND SELECTED OUTPUT SPSS*

```
TITLE          SMALL SAMPLE PROFILE ANALYSIS.
FILE HANDLE    TAPE42
DATA LIST      FILE=TAPE42
               /RESPOND ,OCCUP ,READ ,DANCE ,TV ,SKI (6F3.0)
VAR LABELS     RESPOND 'RESPONDENT'  OCCUP 'OCCUPATION'
VALUE LABELS   OCCUP 1 'BELLY DANCER' 2 'POLITICIAN' 3 'ADMINISTRATOR'
MANOVA         READ TO SKI BY OCCUP(1,3)/
               WSFACTOR=ACTIVITY(4)/
               PARTITION(ACTIVITY)/
               CONTRAST(ACTIVITY)=SPECIAL (1 1 1 1, -1 1 -1 1,
                                           -1 0 1 0, 0 -1 0 1)/
               WSDESIGN=ACTIVITY(1)/
               ANALYSIS(REPEATED)/
               PARTITION(OCCUP)/
               CONTRAST(OCCUP)=SPECIAL (1 1 1, 2 -1 -1, 0 1 -1)/
               NOPRINT=PARAMETER(ESTIM)/
               PRINT=SIGNIF(BRIEF)/
               DESIGN=OCCUP(1)/
```

TESTS OF SIGNIFICANCE FOR READ USING SEQUENTIAL SUMS OF SQUARES

SOURCE OF VARIATION	SUM OF SQUARES	DF	MEAN SQUARE	F	SIG. OF F
WITHIN CELLS	23.50000	12	1.95833		
CONSTANT	1591.35000	1	1591.35000	812.60426	0
OCCUP(1)	138.67500	1	138.67500	70.81277	0

TESTS OF SIGNIFICANCE FOR DANCE USING SEQUENTIAL SUMS OF SQUARES

SOURCE OF VARIATION	SUM OF SQUARES	DF	MEAN SQUARE	F	SIG. OF F
WITHIN CELLS	17.10000	12	1.42500		
ACTIVITY(1)	3.75000	1	3.75000	2.63158	.131
OCCUP(1) BY ACTIVITY(1)	21.67500	1	21.67500	15.21053	.002

X (BDVSOTHR), sum of squares = 21.6750. The marginal test of the repeated measure is labeled SEDVSACT, sum of squares = 3.75000. Both of these are tested against ERROR, sum of squares = 17.100000. The error term is the interaction of subjects with the contrast applied to the repeated measure nested with groups.

The marginal test of the groupings variable is labeled BDVSOTHR, sum of squares = 138.675. It is tested against subjects nested within groups. The two marginal tests are themselves of interest in examination of significant levels or flatness effects. The marginal value of the coefficients applied to the repeated measures is a contrast appropriate as follow up of a significant flatness effect; the marginal value of the coefficients applied to groups is a contrast appropriate as a follow-up of a significant levels effect. Both marginal contrasts are automatically provided during a run for interaction contrasts.

SPSS* gives less voluminous output, as seen in Table 10.12(b). The interaction contrast is identified as OCCUP(1) BY ACTIVITY(1) (sum of squares = 21.675) with its corresponding WITHIN CELLS error term (sum of squares = 17.100). The second WITHIN CELLS error term is the appropriate one for the interaction contrast, and it is the one used by SPSS*. It is the interaction of subjects with contrast applied to the repeated measure nested in groups.

The first WITHIN CELLS error term (sum of squares = 23.500) is subjects nested in groups, the error term that accompanies OCCUP(1) (sum of squares = 138.675), the marginal test of contrast applied to groups and the test appropriate as follow-up of a significant levels effect. The effect labeled ACTIVITY(1) (sum of squares = 3.75) is the marginal test of contrast applied to repeated measures and is the test appropriate as follow- up of a significant flatness effect. This test is evaluated against the second error term.

　　　Interaction contrasts also need Scheffé adjustment to critical F to hold down the rate of familywise error. For an interaction, the Scheffé adjustment is

$$F_s = (p - 1)(k - 1) F_{(p-1)(k-1),\ k(p-1)(n-1)} \tag{10.9}$$

where p is the number of repeated measures, k is the number of groups, and n is the number of subjects in each group. For the example

$$F_s = (4 - 1)(3 - 1) F_{(6,\ 36)}$$
$$= 14.52$$

Because the F value for the interaction contrast is 15.21053, a value that exceeds F_s, there is an interaction between belly dancers vs. others in their satisfaction during sedentary vs. active leisure time activities. A look at the means in Table 10.1 reveals that belly dancers favor active leisure time activities to a greater extent than others.

10.5.1.4 Only Parallelism Significant　Do whatever makes the most sense—simple effects or interaction contrasts—and is the easiest to explain.[7]

10.5.2 Multivariate Approach to Repeated Measures

Research where the same cases are repeatedly measured with the same instrument is common in many sciences. Longitudinal or developmental studies, research that requires follow-up, studies where changes in time are of interest—all involve repeated measurement. Further, many studies of short-term phenomena have repeated measurement of the same subjects under several experimental conditions, resulting in an economical research design.

　　　When there are repeated measures, a variety of analytical strategies are available, all with advantages and disadvantages. Choice among the strategies depends upon details of research design and conformity between the data and the assumptions of analysis.

　　　Univariate repeated-measures ANOVA with more than 1 df for the repeated-measure IV requires homogeneity of covariance. Although the test for homogeneity of covariance is fairly complicated, the notion is conceptually simple. All pairs of levels of the within-subjects variable need to have equivalent correlations. For example, consider a longitudinal study in which children are measured yearly from ages 5 to 10. If there is homogeneity of covariance the correlation between scores on the DV for ages 5 and 6 should be about the same as the correlation between scores between

---

[7] If you got this far, go have a beer.

ages 5 and 7, or 5 and 8, or 6 and 10, etc. In applications like these, however, the assumption is almost surely violated. Things measured closer in time tend to be more highly correlated than things measured farther away in time; the correlation between scores measured at ages 5' and 6 is likely to be much higher than the correlation between scores measured at ages 5 and 10. Thus, whenever time is a within-subjects IV, the assumption of homogeneity of covariance is likely to be violated, leading to increased Type I error.

In the event of violation of the assumption of homogeneity of covariance, several strategies are available, as discussed in Section 8.5.2.1. One of these, available through BMDP2V, BMDP4V, and SAS, is a more stringent adjustment of the statistical criterion leading to a more honest Type I error rate, but lower power. This strategy has the advantage of simplicity of interpretation (because familiar main effects and interactions are evaluated) and simplicity of decision-making (you decide on one of the strategies before performing the analysis and then take your chances with respect to power). If BMDP2V is used, you are safe from temptation because both the Greenhouse-Geisser-Imhof and Huynh-Feldt adjustments of results for univariate, repeated-measures ANOVA are reported directly in the output.[8]

If you use BMDP4V, however, results of profile analysis are also printed out, and you have availed yourself of the second strategy, whether you meant to or not. Profile analysis, called the multivariate approach to repeated measures, is a statistically acceptable alternative to repeated-measures ANOVA because multiple DVs replace the within-subjects IV and the assumption of homogeneity of covariance is no longer required. Although other requirements such as homogeneity of variance-covariance matrices and absence of multicollinearity and singularity must be met, they are less likely to be violated.

Profile analysis requires more cases than univariate repeated-measures ANOVA—certainly more cases than DVs in the smallest group. If the sample is too small, the choice between multivariate and univariate approaches is automatically resolved in favor of the univariate approach, with adjustment for failure of homogeneity of covariance, as necessary.

Sometimes, however, the choice is not so simple, and you find yourself with two sets of results. If the conclusions from both sets of results are the same, it is often easier to report the univariate solution, while noting that the multivariate solution is similar. But if conclusions differ between the two sets of results, you have a dilemma. Choice between conflicting results requires attention to the details of the research design. Clean, counterbalanced experimental designs "fit" better within the univariate model, while nonexperimental or contaminated designs often require the multivariate model that is more forgiving statistically but more ambiguous to interpret.

The best solution is often to perform trend analysis (or some other set of single df contrasts) instead of either profile analysis or repeated-measures ANOVA if that makes conceptual sense within the context of the research design. Many longitudinal, follow-up, and other time-related studies lend themselves beautifully to interpretation

---

[8] See Keppel (1982, pp 468–472) for a discussion of the differences between the two types of adjustment. These adjustments are also available in the CDC SPSS-6000 Version 9.0.

in terms of trends. Because statistical tests of trends and other contrasts use single degrees of freedom of the within-subjects IV, there is no possibility of violation of homogeneity of covariance. Furthermore, none of the assumptions of the multivariate approach is relevant. BMDP2V offers simple, straightforward setup and output for polynomial decomposition of the within-subjects IV, in which results for each trend are printed out (1 = linear, 2 = quadratic, 3 = cubic, etc.)

A fourth alternative is straightforward MANOVA where DVs are treated directly (cf. Chapter 9), without conversion to segments. The design becomes a one-way between-subjects analysis of the grouping variable with the repeated measures used simply as multiple DVs. There are two problems with conversion of repeated measures to MANOVA. First, because the design is now one-way between-subjects, MANOVA does not produce the interaction (parallelism) test most often of interest in a repeated-measures design. Second, MANOVA allows a stepdown analysis, but not a trend analysis of DVs after finding a multivariate effect.

In summary then, if the levels of the IV differ along a single dimension such as time or dosage and trend analysis makes sense, use it. Or, if the design is a clean experiment where cases have been randomly assigned to treatment and there are expected to be no carry-over effects, the univariate repeated-measures approach is probably justified. (But just to be on the safe side, use a program that tests and adjusts for violation of homogeneity of covariance.) If, however, the levels of the IV do not vary along a single dimension but violation of homogeneity of covariance is likely, and if there are lots more cases than DVs, it is probably a good idea to choose either profile analysis or MANOVA.

10.5.3 Doubly Multivariate Designs

10.5.3.1 Kinds of Doubly Multivariate Analysis
There are two ways that an analysis can be doubly multivariate. In the first, commensurate DVs are administered to groups more than once, creating repeated measures over time. For example, the POMS is administered to psychoanalysts and behavior therapists before training, immediately after training, and 1 year later. This is analogous to a univariate between-within-within design with time as one within-subjects factor and the set of DVs as the other. The analysis produces tests of three main effects (levels, flatness over time, and flatness over DVs), three two-way interactions (parallel groups over time, parallel groups over DVs, and doubly parallel time by DVs) as well as the mind-boggling three-way interaction.

In the second and more common doubly multivariate design, noncommensurate DVs are repeatedly measured. For example, children in classrooms with either traditional or computer assisted instruction are measured at several points over the semester on reading achievement, general information, and math achievement. There are two ways to conceptualize the analysis. If treated in a singly multivariate fashion, this is a between-within design (groups by time) with multiple DVs. The time effect, however, has the assumption of homogeneity of covariance. To circumvent the assumption, the analysis becomes doubly multivariate where both the within-subjects part of the design and the multiple DVs are analyzed multivariately. The between-subject effect is singly multivariate; the within-subject effects and interactions are doubly multivariate.

For both commensurate and noncommensurate applications, the limiting factor is the large number of cases needed for analysis. The number of measures for each case is the number of DVs times the number of repetitions. With 10 DVs measured three times, there are 30 measures per case. To avoid singularity of variance-covariance matrices (cf. Section 10.3.2) more than 30 cases per group are needed.

BMDP4V is particularly well set up to handle analysis of noncommensurate DVs. The multiple DVs are designated VARIATES, and an appropriate name is given to the within-subjects IV treated multivariately. This is illustrated in the example in Section 10.5.3.2.

For SPSS MANOVA, the procedure is not much more difficult, but there was little guidance in the 1981 manual (Hull and Nie, 1981). This has been rectified with the introduction of SPSS^x (SPSS Inc., 1985) where a fully documented example with two between-subjects IVs is given. SAS GLM and SYSTAT MGLH have no provision for specification of doubly multivariate designs other than through design matrices that the user supplies.

For repeatedly measured commensurate DVs, neither SPSS nor BMDP4V provides explicit information for analysis. The procedures offered by BMDP4V and SPSS^x result in MANOVA where the tests of flatness and parallelism apply to the repeated-measures factor but there are no tests of the higher-order interaction or of parallelism or flatness of the commensurate DVs. Instead, special design matrices have to be devised, a topic beyond the scope of this book.

10.5.3.2 Example of a Doubly Multivariate Analysis of Variance   In this example, a small hypothetical data set with repeatedly measured noncommensurate DVs is used. The between-subjects IV is three weight-loss programs (PROGRAM): a control group (CONTROL), a group that diets (DIET), and a group that both diets and exercises (DIET + EX). Each group has 12 participants, so total $N = 36$. The major DV is weight loss (WTLOSS) and a secondary DV is self-esteem (ESTEEM). The DVs are measured at the end of the first, second, and third months of treatment. The within-subject IV treated multivariately, then, is MONTH that the measures are taken. The data set is in Table 10.13.

The setup and results from BMDP4V as applied to this data set appear in Table 10.14. Note that in the WITHIN paragraph, VARIATES[9] appears first, indicating that in the data set the three values for the first DV appear before the three values of the second DV. Months "change" fastest in this arrangement of data.

The first effect of interest is that of PROGRAM, the between-subject IV. Note that there is no significant difference between groups when the DVs are combined (–ALL—), with Wilks' Lambda (LRATIO) = .773006, $F(4, 64) = 2.20$, $p = .0791$. Therefore the results of the main effect of program for the individual DVs are ignored.

The second effect that appears is M (MONTH). The section labeled –ALL— is the one that tests combined DVs and is the repeated-measures effect treated doubly

---

[9] VARIATES is a BMDP4V key word which must always be used to specify the multiple DVs.

TABLE 10.13 HYPOTHETICAL DATA SET TO DEMONSTRATE DOUBLY MULTIVARIATE ANALYSIS

DV	Weight loss			Self-esteem		
Month	1	2	3	1	2	3
Group Control	4	3	3	14	13	15
	4	4	3	13	14	17
	4	3	1	17	12	16
	3	2	1	11	11	12
	5	3	2	16	15	14
	6	5	4	17	18	18
	6	5	4	17	16	19
	5	4	1	13	15	15
	5	4	1	14	14	15
	3	3	2	14	15	13
	4	2	2	16	16	11
	5	2	1	15	13	16
Diet	6	3	2	12	11	14
	5	4	1	13	14	15
	7	6	3	17	11	18
	6	4	2	16	15	18
	3	2	1	16	17	15
	5	5	4	13	11	15
	4	3	1	12	11	14
	4	2	1	12	11	11
	6	5	3	17	16	19
	7	6	4	19	19	19
	4	3	2	15	15	15
	7	4	3	16	14	18
Diet + exercise	8	4	2	16	12	16
	3	6	3	19	19	16
	7	7	4	15	11	19
	4	7	1	16	12	18
	9	7	3	13	12	17
	2	4	1	16	13	17
	3	5	1	13	13	16
	6	5	2	15	12	18
	6	6	3	15	13	18
	9	5	2	16	14	17
	7	9	4	16	16	19
	8	6	1	17	17	17

multivariately. Wilks' Lambda is also used to test this effect and $F(4, 130) = 40.40$. There is, then, a significant difference over the 3 months in the combination of weight loss and self-esteem measures.

Below this section appear the tests of differences over months for the individual DVs. For each DV there is both the multivariate (TSQ) and univariate (WCP) results, the latter with the two forms of adjustment for violation of homogeneity of covariance. According to the multivariate results both the DVs show highly significant effects, with $F(2, 32) = 120.62$ for weight loss and $F(2, 32) = 12.19$ for self-esteem.

BMDP4V offers no automatic stepdown analysis. To test changes in self-esteem over the 3 months after adjustment for differences in weight loss, a second run with self-esteem as the DV and weight loss as a covariate is needed.

The final doubly multivariate test is of the program by month interaction, highly significant with Wilks' criterion (LRATIO) $F(8, 60) = 4.43$. According to tests of the individual DVs, weight loss shows significant change as a joint function of program and months, with $F(4, 64) = 8.31$ for LRATIO, but there is no significant multivariate effect for self-esteem, $F(4, 64) = 1.71$. This finding also needs to be verified in a run in which self-esteem scores are adjusted for differences in weight loss.

Table 10.15 shows the setup and results of an ANCOVA run through BMDP2V with self-esteem as the DV and weight loss as a covariate. BMDP2V is more convenient

TABLE 10.14 SETUP AND RESULTS OF DOUBLY MULTIVARIATE ANALYSIS THROUGH BMDP4V

```
/PROBLEM      TITLE IS 'DOUBLY MULTIVARIATE ANAYSIS THROUG BMDP4V'.
/INPUT        VARIABLES ARE 7.  FORMAT IS FREE.  FILE=DBLDAT.
/VARIABLES    NAMES ARE PROGRAM, WTLOSS1, WTLOSS2, WTLOSS3,
                            ESTEEM1, ESTEEM2, ESTEEM3.
/BETWEEN      FACTOR IS PROGRAM.
              CODES(PROGRAM) ARE 1, 2, 3.
              NAMES(PROGRAM) ARE CONTROL, DIET, 'DIET+EX'.
/WITHIN       FACTORS = VARIATES, MONTH.
              CODES(VARIATES) = 1, 2.
              NAMES(VARIATES) = WTLOSS, ESTEEM.
              CODES(MONTH) = 1, 2, 3.
              NAMES(MONTH) = MONTH1, MONTH2, MONTH3.
/WEIGHTS      BETWEEN ARE EQUAL.  WITHIN ARE EQUAL.
/END
ANALYSIS      PROC = FACTORIAL./
END/
```

```
==============================================================================
WITHIN EFFECT:
+              OBS: WITHIN CASE MEAN
```

EFFECT	VARIATE	STATISTIC		F	DF		P
OVALL: GRAND MEAN							
	-ALL----						
		TSQ=	2637.25	1278.67	2,	32	.0000
	WTLOSS						
		SS=	1688.23				
		MS=	1688.23	429.01	1,	33	.0000
	ESTEEM						
		SS=	24390.1				
		MS=	24390.1	2608.75	1,	33	.0000
P: PROGRAM							
	-ALL----						
		LRATIO=	.773006	2.20	4,	64.00	.0791
		TRACE=	.292626				
		TZSQ=	9.65665				
		CHISQ =	9.68		17.481		.0704
		MXROOT=	.224252				.0582
	WTLOSS						
		SS=	36.9074				
		MS=	18.4537	4.69	2,	33	.0161
	ESTEEM						
		SS=	13.7222				
		MS=	6.86111	.73	2,	33	.4877
ERROR							
	WTLOSS						
		SS=	129.86111				
		MS=	3.9351852				
	ESTEEM						
		SS=	308.52778				
		MS=	9.3493266				

TABLE 10.14 (Continued)

```
=====================================================================
WITHIN EFFECT:
+          M: MONTH

EFFECT   VARIATE      STATISTIC          F         DF        P
---------------------------------------------------------------------
   M
        -ALL----
                TSQ=     260.614       59.23     4,    30    .0000
             WCP LRATIO=  .198756      40.40     4, 130.00   .0000
             WCP TRACE=  3.20173
             WCP TZSQ=   211.314
                  CHISQ =  172.06                   3.159    .0000
             WCP MXROOT=  .744727                             .0000
        WTLOSS
                TSQ=     248.783      120.62     2,    32    .0000
             WCP SS=     181.352
             WCP MS=      90.6759      88.37     2,    66    .0000
             GREENHOUSE-GEISSER ADJ. DF  88.37  1.56,  51.34  .0000
             HUYNH-FELDT ADJUSTED DF     88.37  1.72,  56.92  .0000
        ESTEEM
                TSQ=      25.1352      12.19     2,    32    .0001
             WCP SS=      86.7222
             WCP MS=      43.3611      18.78     2,    66    .0000
             GREENHOUSE-GEISSER ADJ. DF  18.78  1.58,  52.07  .0000
             HUYNH-FELDT ADJUSTED DF     18.78  1.75,  57.83  .0000
(M) X (P: PROGRAM)
        -ALL----
                LRATIO=  .395418        4.43     8,  60.00   .0003
                TRACE=  1.47509
                TZSQ=   45.7278
                  CHISQ =    6.48                  23.712    .0002
                MXROOT=  .589762                             .0001
             WCP LRATIO=  .679124       3.47     8, 130.00   .0012
             WCP TRACE=   .464989
             WCP TZSQ=   30.6893
                  CHISQ =   22.97                   5.800    .0007
             WCP MXROOT=  .309520                             .0737
        WTLOSS
                LRATIO=  .433300        8.31     4,  64.00   .0000
                TRACE=  1.28611
                TZSQ=   42.4415
                  CHISQ =    2.90                  17.481    .0000
                MXROOT=  .559268                             .0000
             WCP SS=     20.9259
             WCP MS=      5.23148       5.10     4,    66    .0012
             GREENHOUSE-GEISSER ADJ. DF   5.10  3.11,  51.34  .0033
             HUYNH-FELDT ADJUSTED DF      5.10  3.45,  56.92  .0023

        ESTEEM
                LRATIO=  .816616        1.71     4,  64.00   .1597
                TRACE=   .224443
                TZSQ=    7.40661
                  CHISQ =   11.53                  17.481    .1506
                MXROOT=  .182936                             .1206
             WCP SS=     25.5556
             WCP MS=      6.38889       2.77     4,    66    .0344
             GREENHOUSE-GEISSER ADJ. DF   2.77  3.16,  52.07  .0483
             HUYNH-FELDT ADJUSTED DF      2.77  3.50,  57.83  .0420

  ERROR
        WTLOSS
             WCP SS=     67.722222
             WCP MS=      1.0260943

             GGI EPSILON=   .77784
             H-F EPSILON=   .86244
        ESTEEM
             WCP SS=    152.38889
             WCP MS=      2.3089226

             GGI EPSILON=   .78896
             H-F EPSILON=   .87623

---------------------------------------------------------------------

=====================================================================
```

TABLE 10.15 SETUP AND RESULTS OF BMDP2V ANALYSIS OF SELF-ESTEEM ADJUSTED FOR WEIGHT LOSS

```
/PROBLEM      TITLE IS 'DOUBLY MULTIVARIATE ANAYSIS THROUG BMDP2V'.
/INPUT        VARIABLES ARE 7.  FORMAT IS FREE.
              FILE = DBLDAT.
/VARIABLES    NAMES ARE PROGRAM, WTLOSS1, WTLOSS2, WTLOSS3,
              ESTEEM1, ESTEEM2, ESTEEM3.
/DESIGN       GROUPING IS PROGRAM.
              DEPENDENT ARE ESTEEM1, ESTEEM2, ESTEEM3.
              COVARIATES ARE WTLOSS1, WTLOSS2, WTLOSS3.
              LEVEL IS 3.
              NAME IS MONTH.
/GROUP        CODES(PROGRAM) ARE 1, 2, 3.
              NAMES(PROGRAM) ARE CONTROL, DIET, 'DIET+EX'.
/END
```

ANALYSIS OF VARIANCE FOR 1-ST
DEPENDENT VARIABLE - ESTEEM1 ESTEEM2 ESTEEM3

	SOURCE	SUM OF SQUARES	DEGREES OF FREEDOM	MEAN SQUARE	F	TAIL PROB.	GREENHOUSE GEISSER PROB.	HUYNH FELDT PROB.	REGRESSION COEFFICIENTS
	PROGRAM	1.71025	2	.85512	.12	.8894			
	1-ST COVAR	75.98203	1	75.98203	10.46	.0028			.76492
1	ERROR	232.54575	32	7.26705					
MP	MONTH	56.03087	2	28.01544	11.96	.0000	.0002	.0001	
	1-ST COVAR	19.01707	4	4.75427	2.03	.1006	.1180	.1093	
2	ERROR	152.26327	65	2.34251	.05	.8176			-.04307

ERROR
TERM   EPSILON FACTORS FOR DEGREES OF FREEDOM ADJUSTMENT

         GREENHOUSE-GEISSER   HUYNH-FELDT
2   .7933   .8915

to use than BMDP4V because adjusted cell means for self-esteem are available. The results for the two programs are the same because in the case of a covariate changing with each level of the DV, BMDP4V prints only univariate ANCOVA.

As seen in Table 10.15, the ANCOVA results are consistent with those of MANOVA by all criteria. For adjusted self-esteem scores, there is a significant difference due to months, $F(2, 65) = 11.96$, but no significant month by program interaction effect, $F(4, 65) = 2.03$.

To interpret the significant effects, cell and marginal means for the DVs are needed. These are most easily found in the BMDP2V run, although means for weight loss are available through the BMDP4V run. Table 10.16 shows the cell and marginal means for weight loss (called the 1-ST COVARIATE) and adjusted cell means for self-esteem.

For weight loss with all three groups combined, there is a decrease over the 3 months, with an average of 5.28 pounds lost the first month, 4.39 pounds the second month, and 2.19 pounds the third month. A plot of the month by program interaction based on the cell means for weight loss is helpful to show how the weight loss over the 3 months differs for the three program groups.

The adjusted cell means for self-esteem need to be averaged in order to see the progress over the 3 months for the combined groups. Because all cells have equal sample sizes, a simple unweighted averaging is appropriate. For example, for the third month,

$$\text{Mean}_{\text{Month3}} = (15.58007 + 16.36914 + 17.78581)/3$$
$$= 16.58$$

For the second month, the adjusted average self-esteem score is 13.80 and for the first month it is 14.70.

Strength of association measures for multivariate effects are, as usual, based

---

TABLE 10.16 CELL AND MARGINAL MEANS FROM BMDP2V (SETUP APPEARS IN TABLE 10.15)

CELL MEANS FOR   1-ST COVARIATE

PROGRAM =		CONTROL	DIET	DIET+EX	MARGINAL
	MONTH				
WTLOSS1	1	4.50000	5.33333	6.00000	5.27778
WTLOSS2	2	3.33333	3.91667	5.91667	4.38889
WTLOSS3	3	2.08333	2.25000	2.25000	2.19444
MARGINAL		3.30556	3.83333	4.72222	3.95370
COUNT		12	12	12	36

ADJUSTED CELL MEANS FOR 1-ST DEPENDENT VARIABLE

PROGRAM =		CONTROL	DIET	DIET+EX
	MONTH			
ESTEEM1	1	14.60491	14.46693	15.03987
ESTEEM2	2	14.49809	13.75984	13.14534
ESTEEM3	3	15.58007	16.36914	17.78581

on Wilks' Lambda (Equation 10.3) and for univariate effects, eta square for adjusted scores (Equation 8.7).

10.5.4 Classifying Profiles

A procedure typically available in programs designed for discriminant function analysis is classification of cases into groups on the basis of a best-fit statistical function. Although classification is done on the basis of scores rather than segments, the principle of classification is often of interest in research where profile analysis is appropriate. If it is found that groups differ on their profiles, it could be useful to classify new cases into groups according to their profiles.

For example, given profile of scores for different groups on a standardized test such as the Illinois Test of Psycholinguistic Abilities, one might use the profile of a new child to see if that child more closely resembles a group of children who have difficulty reading or a group who does not show such difficulty. If reliable profile differences were available before the age at which children are taught to read, classification according to profiles could provide a powerful diagnostic tool.

Note that this is no different from using classification procedures in discriminant function analysis. It is simply mentioned here because choice of profile analysis as the initial vehicle for testing group differences does not preclude use of classification. To use a discriminant function program such as SPSS DISCRIMINANT or BMDP7M for classification, one simply defines the levels of the IV as "groups" and the DVs as "predictors."

10.6 COMPARISON OF PROGRAMS

Programs for MANOVA are covered in detail in Chapter 9. Therefore this section is limited to those features of particular relevance to profile analysis. SYSTAT, SPSS, SPSS[x], and BMDP each have one program useful for profile analysis. SAS has two programs that can be used for profile analysis, one limited to equal-*n* designs. The BMDP, SYSTAT, and SAS programs are easier to set up and more flexible in terms of choice between profile analysis and univariate repeated-measures ANOVA, but the SPSS programs provide more straightforward output. The BMDP and SPSS[x] manuals show by example how to set up doubly multivariate designs. No such help is available in the SPSS, SYSTAT, or SAS manuals. Features of the programs appear in Table 10.17.

10.6.1 SPSS Package

The SPSS and SPSS[x] MANOVA programs differ little in application. The primary difference between them is in the enhanced documentation of SPSS[x]. A few changes, however, ease the use of SPSS[x] MANOVA in profile analysis. The MWITHIN feature provides for testing simple effects when repeated measures are analyzed. And the MEASURES command allows the multiple DVs to be given a generic name in doubly multivariate designs, making the output more readable. Both these enhancements are

TABLE 10.17 COMPARISON OF PROGRAMS FOR PROFILE ANALYSIS

	SPSS MANOVA	SPSS^x MANOVA	BMDP4V	SAS GLM and ANOVA	SYSTAT MGLH
Input					
Variety of strategies for unequal n	Yes	Yes	Yes	Yes^c	Yes
Requires specification of segments	Yes	Yes	No	No	No
Multiple analyses in single run	No	Yes	No	Yes	No
Special specification for doubly multivariate analysis	Yes	Yes	Yes	No	No
Output					
Single source table	No	No	Yes	Option	No
Specific comparisons	Yes	Yes	Yes	Yes	Yes
Simple effects (complete)	No^a	Yes	Yes	No	No
Design matrix	Yes	Yes	Yes	No^d	Yes^f
Within-cells correlation matrix	Yes	Yes	No	No	Yes^g
Determinant of within-cells correlation matrix	Yes	Yes	No	No	No
Squared multiple correlations for DVs	No	No	No	No	Yes
Cell means	Yes	Yes	Yes	Yes	No
Cell standard deviations	Yes	Yes	Yes	No	Yes
Marginal means	Yes	Yes	Yes	No	Yes^h
Marginal standard deviations	No	No	Yes	No	No
Cell and marginal variances	No	No	Yes	No	No
Cell and marginal minimum and maximum values	No	No	Yes	No	No
Confidence intervals around cell means	Yes	Yes	No	No	No
Wilks' Lambda for parallelism	Yes	Yes	LRATIO	L	Yes
F for parallelism	Yes	Yes	Yes	Yes	Yes
Hotelling's T-square for flatness	Yes	No	TSQ	No	No
Additional statistical criteria	Yes	Yes	Yes	Yes	Yes
Test for homogeneity of covariance	Yes	Yes	Yes	No	No
Greenhouse-Geisser-Imhof statistic	No^a	No	Yes	Yes	No
Huynh-Feldt adjusted F	No^a	No	Yes	Yes	No
Predicted values and residuals for each case	Yes	No	No	Yes	Yes

Residuals plot	Yes	Yes	No	No[e]	No[i]
Allows generic labeling of DVs	No[a]	Yes	No	No	No
Homogeneity of variance-covariance matrices	Yes	Yes	No	No	No
Test for multivariate outliers	No[a]	No	No[b]	No	Yes
Effect sizes (strength of association)	No[a]	No	No	No	No
Observed power values	No[a]	No	No	No	No
Characteristic roots and vectors	Yes	Yes	No	Yes	Yes[g]

[a] Available in CDC CYBER SPSS-6000, Version 9.0.
[b] Available through BMDPAM or BMDP7M with recoded DVs.
[c] SAS ANOVA requires equal n.
[d] Prints transformation matrices that define contrasts.
[e] Available through PLOT procedure.
[f] Available by saving MODEL.
[g] Through PROFILE format only.
[h] For each DV across groups only.
[i] Available through GRAPH program.

available in the CDC CYBER SPSS-6000 Version 9.0. Also, several repeated-measures analyses can be specified within a single run.

The full output for both SPSS and SPSS^x MANOVA consists of three separate source tables, one each for parallelism, levels, and flatness. Specification of PRINT = SIGNIF(BRIEF) is used both to simplify the multivariate output and to provide results appropriate to univariate repeated-measures ANOVA in the column labeled AVER-AGED F.[10]

The two SPSS manuals (Hull and Nie, 1981; SPSS, Inc., 1985) show somewhat different setups depending on whether you want to specify profile analysis or repeated measures tested multivariately. Although major results of the two setups are identical, the repeated-measures setup is more easily interpreted because of more straightforward labeling of effects, as illustrated in Table 10.14.

The determinant of the within-cells correlation matrix is used as an aid to determining whether or not further investigation of multicollinearity is needed; the residuals plots provide evidence of skewness and outliers. And Box's M test is available as an ultrasensitive test of homogeneity of variance-covariance matrices. But tests of linearity and homoscedasticity are not directly available within SPSS or SPSS^x MANOVA and a direct test for outliers requires use of SPSS NEW REGRESSION or SPSS^x REGRESSION.[11]

10.6.2 BMD Series

BMDP4V provides a very simple setup for profile analysis—it is the same as the setup for univariate repeated-measures ANOVA and is fully described in the manual. The output automatically includes relevant statistics for both types of analysis. Furthermore, in case of violation of homogeneity of covariance, several modifications on the univariate tests are presented. If there is doubt about whether a univariate or a multivariate approach to a data set is more appropriate, then the elaborate BMDP4V output is especially useful, and is the progam of choice.

Although the BMDP manual (Dixon, 1985) is helpful in interpreting the output, a few additional hints might be in order. The first main effect shown is the test for levels, easily interpreted in any case because it is a univariate test. The second main effect is the test for flatness based on Hotelling's T^2 and labeled TSQ. Finally the parallelism test is shown as an interaction effect, and several multivariate statistics are given, including LRATIO for Wilks' Lambda.

There is no provision for evaluation of homogeneity of variance-covariance matrices directly within BMDP4V. Nor can multicollinearity, linearity, skewness, heteroscedasticity of variance, or outliers be evaluated within that program. Instead, it is necessary to apply other programs to both the regular data set and one in which DVs have been converted to segments (cf. Section 10.4.2).

---

[10] For the CDC CYBER implementation of SPSS MANOVA, univariate results are produced by PRINT = SIGNIF(AVONLY). PRINT = SIGNIF(AVERF) adds univariate results to the full multivariate profile output.

[11] The CDC CYBER SPSS-6000 Version 9.0 does include a method for screening for multivariate outliers but data need to be subdivided into groups and coded into segments because otherwise the screening is done on the raw data for the entire sample with all groups combined; further, there is no probability level provided in the output for case distance.

10.6.3 SAS System

Profile analysis is available through the GLM (general linear model) procedure of SAS (SAS Institute, Inc., 1985) or through ANOVA if the groups have equal sample sizes. The two procedures use very similar setup conventions and provide the same output. In GLM and ANOVA, profile analysis is treated as a special case of repeated-measures ANOVA. PROFILE is an explicit statement that generates the segments between adjacent levels of the within-subjects factor.

Both univariate and multivariate results are provided by default, with the multivariate output providing the tests of parallelism and flatness. However, the output is not particularly easy to read. Each multivariate effect appears separately but instead of a single table for each effect, there are multiple sections filled with cryptic symbols. These symbols are defined above the output for that effect. Only the test of levels appears in a familiar source table. Specification of SHORT within the REPEATED statement provides condensed output for the multivariate tests of flatness and parallelism, but the tests for each effect are still separated.

There is no explicit test for homogeneity of covariance, but the univariate output provides the same modifications on the univariate test as BMDP4V if the assumption is violated. By default, both univariate and multivariate results are output. The univariate results are printed in a separate ANOVA table so that there is no ambiguity in identifying the appropriate output.

SAS GLM and ANOVA have no provision for evaluating the remaining assumptions of profile analysis. The superior data-management capabilities of SAS are useful for creating matrices and plotting exotic functions, but the statistical procedures such as GLM and ANOVA have minimal built-in capabilities for testing assumptions.

10.6.4 SYSTAT System

The MGLH program in SYSTAT (Wilkinson, 1986) handles profile analysis through either the PROFILE format or the REPEAT format. The PROFILE format requires separate sets of instructions for tests of parallelism, flatness, and levels but is quite flexible. Coding of DVs is either by adjacent segments (DIFFERENCE) as in Section 10.4 or by polynomials, equally or unequally spaced. For the REPEAT format, coding is by polynomials, but only a single instruction is necessary to produce the three major tests.

The labeling of the output associated with the REPEAT format is somewhat confusing. There are separate subsections for the between-subjects effects and within-subjects effects. The between-subjects subsection includes the levels test labeled with the name of the grouping factor. So far so good. However, the within-subjects subsection reports the flatness test first, labeled CONSTANT, followed by the parallelism test, labeled with the name of the *grouping* factor.

In addition to the two formats for running profile analysis, there are two forms for printing output—long and short. The short form is the default option. The long form provides such extras as error correlation matrices and canonical analysis. The combination of PROFILE format and long form produces the most voluminous output.

The MGLH program is stingy with description statistics; only the marginal means for DVs, averaged across groups, are printed. Additional statistics are available through the STATS program but the data set must be sorted by groups. Multivariate outliers are found by applying the discriminant function procedure detailed in the SYSTAT manual. Mahalanobis distances (the square root of values given by other programs) for each group are saved to a file. Multivariate outliers are cases that are significantly distant from their own groups. Assumptions are not directly tested through the MGLH procedure although assumptions such as linearity, and homogeneity of variance can be evaluated through STATS and GRAPH. No tests for homogeneity of covariance or homogeneity of variance-covariance matrices are available.

10.7 COMPLETE EXAMPLE OF PROFILE ANALYSIS

To illustrate the application of profile analysis, variables are chosen from among those in the learning disabilities data bank described in Appendix B, Section B.2. Three groups are formed on the basis of the preference of learning-disabled children for age of playmates (AGEMATE): children whose parents report that they have (1) preference for playmates younger than themselves, (2) preference for playmates older than themselves, and (3) preference for playmates the same age as themselves or no preference.

DVs are the 11 subtests of the Wechsler Intelligence Scale for Children given either in its original or revised (WISC-R) form, depending on the date of administration of the test. The subtests are information (INFO), similarities (SIMIL), arithmetic (ARITH), comprehension (COMP), vocabulary (VOCAB), digit span (DIGIT), picture completion (PICTCOMP), picture arrangement (PARANG), block design (BLOCK), object assembly (OBJECT), and CODING.

The primary question is whether profiles of learning-disabled children on the WISC-R subscales differ if the children are grouped on the basis of their choice of age of playmates (the parallelism test). Secondary questions are whether preference for age of playmates is associated with overall IQ (the levels test), and whether the subtest pattern of the combined group of learning-disabled children is flat (the flatness test), as it is for the population on which the WISC was standardized.

10.7.1 Evaluation of Assumptions

Assumptions and limitations of profile analysis are evaluated as described in Section 10.3.2. Because the DVs in the parallelism and flatness tests are adjacent segments of original DVs, most of the formal tests of assumptions are performed on DVs converted to segments.

10.7.1.1 Unequal Sample Sizes and Missing Data   From the sample of 177 learning-disabled children given the WISC or WISC-R, 168 could be grouped according to preferred age of playmates. Among the 168 children who could be grouped, a preliminary run of BMDPAM (Table 10.18) is used to reveal the extent and pattern of missing data.[12]

---

[12] When transformed variables are used, it is safest to identify missing data through BMDPAM rather than to trust other programs to catch the cases.

```
          /PROBLEM   TITLE IS 'BMPAM ON SEGMENTS FOR PROFILE ANALYSIS'.
          /INPUT     VARIABLES=13. FORMAT IS '(13F5.0)'.
                     CASES ARE 177.
          /VARIABLE  NAMES ARE CLIENT,AGEMATE,INFO,COMP,ARITH,SIMIL,
                          VOCAB,DIGIT,PICTCOMP,PARANG,BLOCK,
                          OBJECT,CODING,AVSB,BVSC,CVSD,DVSE,EVSF,
                          FVSG,GVSH,HVSI,IVSJ,JVSK.   ADD=10.
                     GROUPING=AGEMATE.
                     USE=2, 14 TO 23.
                     MISSING=13*0.
          /TRANSFORM AVSB = INFO - SIMIL.
                     BVSC = SIMIL - ARITH.
                     CVSD = ARITH - COMP.
                     DVSE = COMP - VOCAB.
                     EVSF = VOCAB - DIGIT.
                     FVSG = DIGIT - PICTCOMP.
                     GVSH = PICTCOMP - PARANG.
                     HVSI = PARANG - BLOCK.
                     IVSJ = BLOCK - OBJECT.
                     JVSK = OBJECT - CODING.
          /GROUP     CODES(2) ARE 1 TO 3.
                     NAMES(2) ARE 'YOUNGER','OLDER','SAME-NO PREF'.
          /ESTIMATE  METHOD=REGR.
          /PRINT     MATR=DIS.
          /END
```

TABLE OF SAMPLE SIZES

(PERCENTAGES OF MISSING INCLUDE CASES WITH ANY VARIABLE
MISSING OR BEYOND MAXIMUM OR MINIMUM LIMITS)

GROUP	SIZE	COMPLETE CASES	PERCENT MISSING
YOUNGER	46	45	2.2
OLDER	55	54	1.8
SAME-NO	67	65	3.0

TABLE OF SAMPLE SIZES, NUMBER MISSING AND PERCENT
MISSING FOR ALL GROUPS TAKEN TOGETHER. NUMBER AND
PERCENT FOR MISSING INCLUDES VALUES OUTSIDE LIMITS.

VARIABLE NO.	NAME	SAMPLE SIZE	NUMBER MISSING	PERCENT MISSING
14	AVSB	168	0	0.0
15	BVSC	168	0	0.0
16	CVSD	166	2	1.2
17	DVSE	166	2	1.2
18	EVSF	167	1	0.6
19	FVSG	167	1	0.6
20	GVSH	168	0	0.0
21	HVSI	168	0	0.0
22	IVSJ	168	0	0.0
23	JVSK	167	1	0.6

ESTIMATES OF MISSING DATA, MAHALANOBIS D-SQUARED (CHI-SQUARED)
AND SQUARED MULTIPLE CORRELATIONS WITH AVAILABLE VARIABLES

CASE LABEL	CASE NUMBER	MISSING VARIABLE	ESTIMATE	R-SQUARED	GROUP	CHI-SQ	CHISQ/DF	D.F.	SIGNIFICANCE
	54				YOUNGER	6.407	0.641	10	0.7800
	55				SAME-NO	4.214	0.421	10	0.9372
	57				YOUNGER	7.325	0.732	10	0.6945
	59				OLDER	3.570	0.397	9	0.9374
	60				OLDER	10.885	1.088	10	0.3666
	61				SAME-NO	4.518	0.452	10	0.9210
	62				SAME-NO	3.524	0.352	10	0.9663
	63				OLDER	6.837	0.684	10	0.7408
	64				YOUNGER	7.474	0.747	10	0.6800
	65				OLDER	15.337	1.534	10	0.1203
	66				YOUNGER	5.986	0.599	10	0.8164
	67				SAME-NO	29.471	2.947	10	0.0010
	68				OLDER	6.778	0.678	10	0.7462
	69				OLDER	6.377	0.638	10	0.7826
	70				SAME-NO	6.505	0.650	10	0.7712

In the BMDPAM setup, segments are created by subtracting adjacent subtests from each other in the TRANSFORM paragraph. Missing data are sought among the segments in cases grouped by AGEMATE as indicated in the USE and GROUPING sentences of the VARIABLE paragraph. The 11 subtests of the WISC yield 10 segments, so cases with complete data have 10 df in the case by case listing of the MAHALANOBIS D-SQUARED test for outliers in the bottom portion of Table 10.18. Cases with fewer than 10 df are those with missing data. Four children are identified through the BMDPAM run, among them case 59 from the group preferring younger playmates. Because so few cases have missing data, and the missing vaiables are scattered over groups and DVs, as seen in the upper portion of Table 10.18, it is decided to delete them from analysis, leaving $N = 164$. Other strategies for dealing with missing data are discussed in Chapter 4.

Of the remaining 164 children, 45 are in the group preferring younger playmates, 54 older playmates, and 65 same age playmates or no preference. This leaves 4.5 times as many cases as DVs in the smallest group, posing no problems for multivariate analysis.

10.7.1.2 Multivariate Normality Groups are large and not notably discrepant in size. Therefore the central limit theorem should assure acceptably normal sampling distributions of means for use in profile analysis. Other portions of the BMDPAM output show all of the segments to be well-behaved; univariate summary statistics for the YOUNGER group, for example, are in Table 10.19. Although cases with missing data are not yet deleted in this run, resulting in slightly different sample sizes for different segments, skewness and kurtosis values are acceptable and unlikely to be much influenced by deletion of so few cases.

Normality for AVERAGE, the variable used in the univariate test for levels, is assessed through BMDP7D after cases with missing values are deleted. Setup and output appear in Table 10.20.

10.7.1.3 Linearity Considering the well-behaved nature of these DVs and the known linear relationship among the measures from which they are derived, no threats to linearity are anticipated.

10.7.1.4 Outliers A second run of BMDPAM, after deletion of cases with missing values, provides information for evaluation of several issues, including univariate and multivariate outliers among segments. As seen in the univariate summary statistics of Table 10.21 for the various groups, all the standard scores are within 3.5 standard deviations or so, suggesting no serious univariate outliers in a sample of this type. Similarly, the descriptive statistics for the average of all 11 subtests, the DV for the univariate test of levels, reveals no outliers, as seen in Table 10.20.

A later portion of the run in Table 10.21 (not shown), identical in format to the final portion of Table 10.18, also reveals no multivariate outliers with a criterion of $p = .001$.

10.7.1.5 Homogeneity of Variance-Covariance Matrices With relatively equal sample sizes and no gross discrepancy in within-cell variances, there is no need to

TABLE 10.19 SAMPLE UNIVARIATE STATISTICS FOR FIRST GROUP PRODUCED BY BMDPAM. SEE TABLE 10.18 FOR SETUP

UNIVARIATE SUMMARY STATISTICS

GROUP IS YOUNGER   SIZE IS   46

VARIABLE	SAMPLE SIZE	MEAN	STANDARD DEVIATION	COEFFICIENT OF VARIATION	SMALLEST VALUE	LARGEST VALUE	SMALLEST STANDARD SCORE	LARGEST STANDARD SCORE	SKEWNESS	KURTOSIS
14 AVSB	46	-0.78261	3.33275	-4.258518	-10.00000	6.00000	-2.77	2.04	-0.28	-0.07
15 BVSC	46	0.56522	3.10306	5.490030	-5.00000	8.00000	-1.79	2.40	0.40	-0.21
16 CVSD	46	-0.36957	2.68625	-7.268678	-7.00000	6.00000	-2.47	2.37	0.11	0.21
17 DVSE	46	-0.71739	2.67219	-3.724866	-10.00000	6.00000	-3.47	2.51	-0.83	2.02
18 EVSF	45	1.75556	3.19200	1.818230	-3.00000	12.00000	-1.49	3.21	1.14	1.27
19 FVSG	45	-2.66667	3.63068	-1.361504	-10.00000	4.00000	-2.02	1.84	0.16	-0.97
20 GVSH	46	1.13043	2.97104	2.628224	-6.00000	7.00000	-2.40	1.98	0.00	-0.32
21 HVSI	46	0.00000	3.24551	UNDEFINED	-9.00000	6.00000	-2.77	1.85	-0.32	0.16
22 IVSJ	46	-0.41304	2.71274	-6.567677	-7.00000	6.00000	-2.43	2.36	0.00	-0.23
23 JVSK	46	1.67391	3.59663	2.148636	-7.00000	9.00000	-2.41	2.04	-0.45	-0.36

487

TABLE 10.20 SETUP AND SELECTED BMDP7D OUTPUT TO
EVALUATE DISTRIBUTIONS FOR 'AVERAGE', THE
DV USED IN UNIVARIATE TEST FOR LEVELS

```
/PROBLEM    TITLE IS 'EVALUATION OF AVERAGE OF SUBTESTS'.
/INPUT      VARIABLES=13. FORMAT IS '(13F5.0)'.
            FILE = PROFILE.
/VARIABLE   NAMES ARE CLIENT,AGEMATE,INFO,COMP,ARITH,SIMIL,VOCAB,
                DIGIT,PICTCOMP,PARANG,BLOCK,OBJECT,CODING.
            LABEL IS CLIENT.
            MISSING=13*0.
/TRANSFORM  AVERAGE = MEAN(INFO,COMP,ARITH,SIMIL,VOCAB,DIGIT,
                PICTCOMP,PARANG,BLOCK,OBJECT,CODING).
            DELETE = 59, 123, 129, 130.
/GROUP      CODES(2) ARE 1 TO 3.
            NAMES(2) ARE 'YOUNGER', 'OLDER', 'SAME-NO PREF'.
/HISTOGRAM  GROUPING IS AGEMATE.
            VARIABLE = AVERAGE.
/END


          ************                    ************
HISTOGRAM OF * AVERAGE * ( 14)   GROUPED  BY  * AGEMATE * ( 2)
          ************                    ************                   CASES WITH
                                                                         UNUSED
                                                                         VALUES FOR
          YOUNGER             OLDER              SAME-NO                  AGEMATE
MIDPOINTS.................................+......................+.......................+.................+
  16.500)
  16.000)
  15.500)*
  15.000)
  14.500)                        *                   *
  14.000)
  13.500)
  13.000)*                       *                   *                   *
  12.500)**                     ***                 ****
  12.000)*                     *****                ****                 *
  11.500)***                    ***                *****
  11.000)**                  ********               *****                **
  10.500)*****                 ****               *******                M*
  10.000)**                   M****               M******                *
   9.500)M**********          *********           *******                *
   9.000)****                  ****              *********
   8.500)****                  ****               *******
   8.000)***                   ***                 **                    *
   7.500)****                  ***                *****
   7.000)*                                          *
   6.500)*                     *
   6.000)
        GROUP MEANS ARE DENOTED BY M'S IF THEY COINCIDE WITH *'S, N'S OTHERWISE
MEAN        9.721        10.146         9.996         10.535
STD.DEV.    1.791         1.624         1.611          1.477
S.E.M.      0.267         0.221         0.200          0.492
MAXIMUM    15.545        14.636        14.455         13.091
MINIMUM     6.364         6.545         7.091          7.909
CASES EXCL. (   0)       (   0)        (   0)         (   0)
CASES INCL.   45            54            65              9
```

consult the overly sensitive Box's M test available in SPSS^x MANOVA. Evidence for relatively equal variances is available from the BMDPAM run of Table 10.19, where standard deviations are given for each variable within each group. All the variances (squared deviations) are quite close in value; for no variable is there a between-group ratio of largest to smallest variance approaching 20:1.

10.7.1.6 Multicollinearity and Singularity BMDPAM provides evidence of multicollinearity and/or singularity by printing out squared multiple correlations of each variable with remaining variables, or in this case segments. As seen in the last portion of Table 10.21, none of the R-squared values approaches the .99 level which could pose danger to matrix inversion in performing profile analysis.

TABLE 10.21 EVALUATION OF OUTLIERS AND HOMOGENEITY OF VARIANCE (SETUP AND SELECTED OUTPUT FROM BMDPAM)

```
/PROBLEM      TITLE IS 'BMDPAM ON SEGMENTS FOR PROFILE ANALYSIS, PART 2'.
/INPUT        VARIABLES=13, FORMAT IS '(13F5.0)'.
              CASES ARE 177.
/VARIABLE     NAMES ARE CLIENT,AGEMATE,INFO,COMP,ARITH,SIMIL,
                  VOCAB,DIGIT,PICTCOMP,PARANG,BLOCK,
                  OBJECT,CODING,AVSB,BVSC,CVSD,DVSE,EVSF,
                  FVSG,GVSH,HVSI,IVSJ,JVSK.  ADD=10.

              GROUPING=AGEMATE.
              USE=2, 14 TO 23.
              MISSING=13*0.
                  DELETE = 59, 123, 129, 130.
/TRANSFORM    AVSB = INFO - SIMIL.
              BVSC = SIMIL - ARITH.
              CVSD = ARITH - COMP.
              DVSE = COMP - VOCAB.
              EVSF = VOCAB - DIGIT.
              FVSG = DIGIT - PICTCOMP.
              GVSH = PICTCOMP - PARANG.
              HVSI = PARANG - BLOCK.
              IVSJ = BLOCK - OBJECT.
              JVSK = OBJECT - CODING.
/GROUP        CODES(2) ARE 1 TO 3.
              NAMES(2) ARE 'YOUNGER','OLDER','SAME-NO PREF'.
/ESTIMATE     METHOD=REGR.
/PRINT        MATR=DIS.
/END
```

UNIVARIATE SUMMARY STATISTICS

GROUP IS YOUNGER   SIZE IS   45

VARIABLE	SAMPLE SIZE	MEAN	STANDARD DEVIATION	COEFFICIENT OF VARIATION	SMALLEST VALUE	LARGEST VALUE	SMALLEST STANDARD SCORE	LARGEST STANDARD SCORE	SKEWNESS	KURTOSIS
14 AVSB	45	-0.80000	3.36830	-4.210377	-10.00000	6.00000	-2.73	2.02	-0.26	-0.13
15 BVSC	45	0.64444	3.09072	4.795939	-5.00000	8.00000	-1.83	2.38	0.37	-0.18
16 CVSD	45	-0.28889	2.65985	-9.206480	-7.00000	6.00000	-2.52	2.36	0.08	0.32
17 DVSE	45	-0.77778	2.67045	-3.433438	-10.00000	6.00000	-3.45	2.54	-0.81	2.04
18 EVSF	45	1.75556	3.19200	1.818230	-3.00000	12.00000	-1.49	3.21	1.14	1.27
19 FVSG	45	-2.66667	3.63068	-1.361504	-10.00000	4.00000	-2.02	1.84	0.16	-0.97
20 GVSH	45	1.11111	3.00168	2.701515	-6.00000	7.00000	-2.37	1.96	0.01	-0.37
21 HVSI	45	0.04444	3.26800	73.529877	-9.00000	6.00000	-2.77	1.82	-0.36	0.16
22 IVSJ	45	-0.42222	2.74267	-6.495790	-7.00000	6.00000	-2.40	2.34	0.01	-0.29
23 JVSK	45	1.82222	3.49213	1.916411	-7.00000	9.00000	-2.53	2.06	-0.47	-0.19

TABLE 10.21 (Continued)

UNIVARIATE SUMMARY STATISTICS

GROUP IS OLDER   SIZE IS   54

VARIABLE	SAMPLE SIZE	MEAN	STANDARD DEVIATION	COEFFICIENT OF VARIATION	SMALLEST VALUE	LARGEST VALUE	SMALLEST STANDARD SCORE	LARGEST STANDARD SCORE	SKEWNESS	KURTOSIS
14 AVSB	54	-1.01852	3.20077	-3.142575	-8.00000	7.00000	-2.18	2.51	0.16	-0.29
15 BVSC	54	2.40741	3.35069	1.391824	-5.00000	11.00000	-2.21	2.56	0.16	-0.36
16 CVSD	54	-1.62963	2.81033	-1.724523	-8.00000	5.00000	-2.27	2.36	-0.22	-0.08
17 DVSE	54	-1.03704	2.57684	-2.484811	-9.00000	4.00000	-3.09	1.95	-0.48	0.55
18 EVSF	54	2.44444	3.02609	1.237947	-4.00000	10.00000	-2.13	2.50	0.01	-0.16
19 FVSG	54	-0.77778	3.61687	-4.650264	-9.00000	10.00000	-2.27	2.98	0.69	1.09
20 GVSH	54	-0.90741	3.91547	-4.315005	-10.00000	9.00000	-2.32	2.53	-0.12	0.03
21 HVSI	54	0.40741	3.11126	7.636731	-7.00000	8.00000	-2.38	2.44	-0.22	-0.13
22 IVSJ	54	-0.61111	3.17666	-5.198178	-8.00000	7.00000	-2.33	2.40	0.24	0.00
23 JVSK	54	2.09259	4.12662	1.972014	-6.00000	8.00000	-1.96	1.43	-0.27	-1.13

UNIVARIATE SUMMARY STATISTICS

GROUP IS SAME-NO   SIZE IS   E5

VARIABLE	SAMPLE SIZE	MEAN	STANDARD DEVIATION	COEFFICIENT OF VARIATION	SMALLEST VALUE	LARGEST VALUE	SMALLEST STANDARD SCORE	LARGEST STANDARD SCORE	SKEWNESS	KURTOSIS
14 AVSB	65	-1.38462	2.71392	-1.960056	-9.00000	4.00000	-2.81	1.98	-0.21	-0.49
15 BVSC	65	1.61538	3.07049	1.900777	-5.00000	7.00000	-2.15	1.75	-0.18	-0.72
16 CVSD	65	-0.98462	2.99735	-3.044188	-7.00000	8.00000	-2.01	3.00	0.25	0.35
17 DVSE	65	-0.24615	2.75585	-11.195643	-10.00000	6.00000	-3.54	2.27	-0.62	0.93
18 EVSF	65	1.63077	3.60802	2.212463	-6.00000	9.00000	-2.11	2.04	0.08	-0.83
19 FVSG	65	-2.41538	3.96426	-1.641256	-12.00000	6.00000	-2.42	2.12	-0.10	-0.26
20 GVSH	65	0.76923	3.26807	4.248492	-6.00000	10.00000	-2.07	2.82	0.49	0.17
21 HVSI	65	-0.26154	2.98578	-11.416231	-7.00000	7.00000	-2.26	2.43	0.13	-0.28
22 IVSJ	65	-0.43077	3.17199	-7.363555	-8.00000	5.00000	-2.39	1.71	-0.45	-0.41
23 JVSK	65	2.87692	3.70181	1.286724	-6.00000	13.00000	-2.40	2.73	-0.05	-0.05

SQUARED MULTIPLE CORRELATIONS OF EACH VARIABLE WITH ALL OTHER VARIABLES

(MEASURES OF MULTICOLLINEARITY OF VARIABLES)
AND TESTS OF SIGNIFICANCE OF MULTIPLE REGRESSION
DEGREES OF FREEDOM FOR F-STATISTICS ARE   9 AND   152

VARIABLE NO.	NAME	R-SQUARED	F-STATISTIC	SIGNIFICANCE (P LESS THAN)
14	AVSB	0.547211	20.41	0.00000
15	BVSC	0.724222	44.35	0.00000
16	CVSD	0.706265	40.61	0.00000
17	DVSE	0.696029	38.67	0.00000
18	EVSF	0.699417	39.30	0.00000
19	FVSG	0.746806	49.81	0.00000
20	GVSH	0.707653	40.88	0.00000
21	HVSI	0.670488	34.37	0.00000
22	IVSJ	0.539415	19.78	0.00000
23	JVSK	0.422214	12.44	0.00000

10.7.2 Profile Analysis

Deck setup and major output for profile analysis of the 11 WISC subtests for the three groups as produced by SPSS^x MANOVA appear in Table 10.22. After cell statistics, significance tests are shown, in turn, for levels, parallelism, and flatness.

The levels test for AGEMATE in the portion of output called the TEST OF SIGNIFICANCE FOR OVERALL (the average of all subtests) shows no significant univariate effect of the three AGEMATE groups on the combined subtests, F (2, 161) = .81235. That is, the groups do not differ on their combined WISC subtest scores.

The parallelism test, called the test of the AGEMATE BY SUBTEST effect, shows significantly different profiles for the three AGEMATE groups. The various multivariate tests of parallelism produce slightly different probability levels for alpha, all less than 0.05. Because there is no reason to doubt compliance with multivariate assumptions in this data set, Wilks' Lambda (.78398) is used for statistical evaluation and strength of association using Equation 10.4. The test shows that there are reliable differences among the three AGEMATE groups in their profiles on the WISC. The profiles are illustrated in Figure 10.4. Mean values for the plots are found in the cell means portion of the output in Table 10.20, produced by the statement PRINT = CELLINFO(MEANS).

For interpretation of the nonparallel profiles, a contrast procedure is needed to determine which WISC subtests separate the three groups of children. Because there are so many subtests, however, the procedure of Section 10.5.1 is unwieldy. The decision is made to evaluate profiles in terms of subtests on which group averages fall outside the confidence interval of the pooled profile. Cell statistics in Table 10.22 provide the 95% confidence intervals for the pooled profile (confidence interval for entire sample for each variable) and group means for each variable (cell means).

In order to compensate for multiple testing, a wider confidence interval is developed for each test to reflect an experimentwise 95% confidence interval. Alpha rate is set at .0015 for each test to account for the 33 comparisons available—3 groups at each of 11 subtests—generating a 99.85% confidence interval. Because an N of 164 produces a t distribution similar to z, it is appropriate to base the confidence interval on $z = 3.19$.

For the first subtest, INFO,

$$P(\overline{Y} - z\,s_m < \mu < \overline{Y} + z\,s_m) = 99.85 \qquad (10.10)$$

$$P[9.55488 - 3.19\,(3.03609/\sqrt{164}) < \mu$$
$$< 9.55488 + 3.19\,(3.03609/\sqrt{164})] = 99.85$$
$$P(8.79860 < \mu < 10.31116) = 99.85$$

Because none of the group means on INFO falls outside this interval for the INFO subtest, profiles are not differentiated on the basis of the information subtest of the WISC. It is not necessary to calculate intervals for any variable for which none of the groups deviate from the 95% confidence interval because they cannot deviate from a wider interval. Therefore intervals are calculated only for SIMIL, COMP, VOCAB, and PICTCOMP. Applying Equation 10.10 to these variables, significant

TABLE 10.22 SETUP AND SELECTED OUTPUT FROM SPSS'
MANOVA PROFILE ANALYSIS OF 11 WISC SUBTESTS

```
TITLE         LARGE SAMPLE PROFILE ANALYSIS THROUGH SPSSX
FILE HANDLE   TAPE39
DATA LIST     FILE=PROFILE.
              /CLIENT,AGEMATE,INFO,COMP,ARITH,SIMIL,VOCAB,DIGIT,
              PICTCOMP,PARANG,BLOCK,OBJECT,CODING
              (13F5.0)
VAR LABELS    AGEMATE 'PREFERRED AGE OF PLAYMATES'
VALUE LABELS  AGEMATE 1 'YOUNGER' 2 'OLDER' 3 'SAME OR NO PREFER.'
SELECT IF     ($CASENUM NE 59 OR $CASENUM NE 123 OR $CASENUM NE 129
              OR $CASENUM NE 130)
MANOVA        INFO,SIMIL,ARITH,COMP,VOCAB,DIGIT,PICTCOMP,PARANG,
              BLOCK,OBJECT,CODING BY AGEMATE(1,3)/
              WSFACTOR=SUBTEST(11)/
              WSDESIGN=SUBTEST/
              TRANSFORM=REPEATED/
              ANALYSIS(REPEATED)/
              RENAME=OVERALL,AVSB,BVSC,CVSD,DVSE,EVSF,FVSG,GVSH,HVSI,
              IVSJ,JVSK/
              NOPRINT=PARAMETERS(ESTIM)/
              PRINT=CELLINFO(MEANS),HOMOGENEITY(BOXM)/
              DESIGN=AGEMATE/
```

CELL MEANS AND STANDARD DEVIATIONS

VARIABLE .. INFO

FACTOR	CODE	MEAN	STD. DEV.	N	95 PERCENT CONF. INTERVAL	
AGEMATE	YOUNGER	9.06667	3.32620	45	8.06736	10.06597
AGEMATE	OLDER	10.18519	3.27411	54	9.29153	11.07884
AGEMATE	SAME OR	9.36923	2.54073	65	8.73967	9.99879
FOR ENTIRE SAMPLE		9.55488	3.03609	164	9.08674	10.02302

493

TABLE 10.22 (Continued)

VARIABLE .. SIMIL

FACTOR	CODE	MEAN	STD. DEV.	N	95 PERCENT CONF. INTERVAL	
AGEMATE	YOUNGER	9.66667	3.25856	45	8.88769	10.84565
AGEMATE	OLDER	11.20370	2.98031	54	10.39024	12.01717
AGEMATE	SAME OR	10.75385	3.44815	65	9.89993	11.60776
FOR ENTIRE SAMPLE		10.65854	3.26994	164	10.15434	11.16274

VARIABLE .. ARITH

FACTOR	CODE	MEAN	STD. DEV.	N	95 PERCENT CONF. INTERVAL	
AGEMATE	YOUNGER	9.22222	2.71267	45	8.40725	10.03720
AGEMATE	OLDER	8.79630	2.25187	54	8.18165	9.41094
AGEMATE	SAME OR	9.13846	2.49297	65	8.52073	9.75619
FOR ENTIRE SAMPLE		9.04878	2.47144	164	8.66770	9.42986

VARIABLE .. COMP

FACTOR	CODE	MEAN	STD. DEV.	N	95 PERCENT CONF. INTERVAL	
AGEMATE	YOUNGER	9.51111	2.89688	45	8.64079	10.38143
AGEMATE	OLDER	10.42593	2.87213	54	9.64199	11.20987
AGEMATE	SAME OR	10.12308	2.88047	65	9.40933	10.83682
FOR ENTIRE SAMPLE		10.05488	2.88693	164	9.60973	10.50002

VARIABLE .. VOCAB

FACTOR	CODE	MEAN	STD. DEV.	N	95 PERCENT CONF. INTERVAL	
AGEMATE	YOUNGER	10.28889	3.48126	45	9.24300	11.33478
AGEMATE	OLDER	11.46296	2.80641	54	10.69696	12.22897
AGEMATE	SAME OR	10.36923	2.75323	65	9.68701	11.05145
FOR ENTIRE SAMPLE		10.70732	3.01525	164	10.24239	11.17225

VARIABLE .. DIGIT

FACTOR	CODE	MEAN	STD. DEV.	N	95 PERCENT CONF. INTERVAL	
AGEMATE	YOUNGER	8.53333	2.68498	45	7.72668	9.33999
AGEMATE	OLDER	9.01852	2.53646	54	8.32620	9.71084
AGEMATE	SAME OR	8.73846	2.62367	65	8.08835	9.38858
FOR ENTIRE SAMPLE		8.77439	2.60327	164	8.37299	9.17579

VARIABLE .. PICTCOMP

FACTOR	CODE	MEAN	STD. DEV.	N	95 PERCENT CONF. INTERVAL	
AGEMATE	YOUNGER	11.20000	2.66799	45	10.39845	12.00155
AGEMATE	OLDER	9.79630	3.33297	54	8.88657	10.70602
AGEMATE	SAME OR	11.15385	2.77956	65	10.46510	11.84259
FOR ENTIRE SAMPLE		10.71951	2.99805	164	10.25724	11.18179

VARIABLE .. PARANG

FACTOR	CODE	MEAN	STD. DEV.	N	95 PERCENT CONF. INTERVAL	
AGEMATE	YOUNGER	10.08889	2.37240	45	9.37614	10.80164
AGEMATE	OLDER	10.70370	2.95009	54	9.89849	11.50892
AGEMATE	SAME OR	10.38462	2.59623	65	9.74130	11.02793
FOR ENTIRE SAMPLE		10.40854	2.65574	164	9.99904	10.81803

VARIABLE .. BLOCK

FACTOR	CODE	MEAN	STD. DEV.	N	95 PERCENT CONF. INTERVAL	
AGEMATE	YOUNGER	10.04444	2.97685	45	9.15010	10.93879
AGEMATE	OLDER	10.29630	2.89195	54	9.50695	11.08565
AGEMATE	SAME OR	10.64615	2.45860	65	10.03684	11.25536
FOR ENTIRE SAMPLE		10.36585	2.74706	164	9.94228	10.78943

TABLE 10.22 (Continued)

VARIABLE .. OBJECT

FACTOR	CODE	MEAN	STD. DEV.	N	95 PERCENT CONF. INTERVAL	
AGEMATE	YOUNGER	10.46667	2.60768	45	9.68323	11.25010
AGEMATE	OLDER	10.90741	2.87651	54	10.12227	11.69254
AGEMATE	SAME OR	11.07692	2.92782	65	10.35145	11.80240
FOR ENTIRE SAMPLE		10.85366	2.82027	164	10.41879	11.28852

VARIABLE .. CODING

FACTOR	CODE	MEAN	STD. DEV.	N	95 PERCENT CONF. INTERVAL	
AGEMATE	YOUNGER	8.64444	2.52403	45	7.88614	9.40275
AGEMATE	OLDER	8.81481	2.97203	54	8.00361	9.62602
AGEMATE	SAME OR	8.20000	2.80736	65	7.50437	8.89563
FOR ENTIRE SAMPLE		8.52439	2.78570	164	8.09486	8.95392

ORDER OF VARIABLES FOR ANALYSIS

VARIATES	COVARIATES	NOT USED
*OVERALL		AVSB
		BVSC
		CVSD
		DVSE
		EVSF
		FVSG
		GVSH
		HVSI
		IVSJ
		JVSK

1 DEPENDENT VARIABLE
0 COVARIATES
10 VARIABLES NOT USED

NOTE.. "*" MARKS TRANSFORMED VARIABLES.

THESE TRANSFORMED VARIABLES CORRESPOND TO THE
'CONSTANT' WITHIN-SUBJECT EFFECT.

* * * * * * * * * * * * * * A N A L Y S I S   O F   V A R I A N C E * * * * * * * * * * * * * * * * * *

TESTS OF SIGNIFICANCE FOR OVERALL USING SEQUENTIAL SUMS OF SQUARES

SOURCE OF VARIATION	SUM OF SQUARES	DF	MEAN SQUARE	F	SIG. OF F
WITHIN CELLS	4916.22681	161	30.53557		
CONSTANT	179321.61641	1	179321.61641	5872.54848	0
AGEMATE	49.61133	2	24.80567	.81235	.446

497

TABLE 10.22 (Continued)

ORDER OF VARIABLES FOR ANALYSIS

VARIATES	COVARIATES	NOT USED
*AVSB		OVERALL
*BVSC		
*CVSD		
*DVSE		
*EVSF		
*FVSG		
*GVSH		
*HVSI		
**IVSJ		
*JVSK		

10 DEPENDENT VARIABLES
 0 COVARIATES
 1 VARIABLE NOT USED

NOTE.. "*" MARKS TRANSFORMED VARIABLES.

     THESE TRANSFORMED VARIABLES CORRESPOND TO THE
     'SUBTEST' WITHIN-SUBJECT EFFECT.

STATISTICS FOR WITHIN CELLS   CORRELATIONS

DETERMINANT =   .42489
BARTLETT TEST OF SPHERICITY = 134.23767 WITH 45 D. F.
SIGNIFICANCE =.   .000

F(MAX) CRITERION =   2.51277 WITH (10,161) D. F.

```
* * * * * * * * * *  A N A L Y S I S   O F   V A R I A N C E  * * * * * * * * * * *

EFFECT .. AGEMATE BY SUBTEST

MULTIVARIATE TESTS OF SIGNIFICANCE (S = 2, M = 3 1/2, N = 75 )

TEST NAME        VALUE        APPROX. F    HYPOTH. DF    ERROR DF    SIG. OF F

PILLAIS          .22243       1.91452      20.00         306.00      .011
HOTELLINGS       .26735       2.01851      20.00         302.00      .007
WILKS            .78398       1.96682      20.00         304.00      .009
ROYS             .18838


EFFECT .. SUBTEST

MULTIVARIATE TESTS OF SIGNIFICANCE (S = 1, M = 4 , N = 75 )

TEST NAME        VALUE        APPROX. F    HYPOTH. DF    ERROR DF    SIG. OF F

PILLAIS          .47516       13.76103     10.00         152.00      0
HOTELLINGS       .90533       13.76103     10.00         152.00      0
WILKS            .52484       13.76103     10.00         152.00      0
ROYS             .47516
```

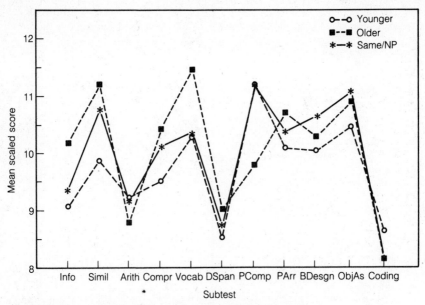

Figure 10.4 Profiles of WISC scores for three agemate groups.

profile deviation is found for vocabulary and picture completion. (The direction of differences is given in the Results section that follows.)

The final omnibus test produced by SPSS^x MANOVA is the SUBTEST effect, for which the flatness hypothesis is rejected. All multivariate criteria show essentially the same result, but Hotelling's criterion, with approximate F (10, 152) = 13.76, $p < .001$, is most appropriately reported because it is a test of a single group (all groups combined). Strength of association, however, is best computed through Wilks' Lambda, as shown in Section 10.4.3.

Although not usually of interest when the hypothesis of parallelism is rejected, the flatness test is interesting in this case because it reveals differences between learning-disabled children (our three groups combined) and the sample used for standardizing the WISC, for which the profile is necessarily flat. (The WISC was standardized so that all subtests produce the same mean value.) Any group that differs from a flat profile, that is, has different mean values on various subtests, diverges from the standard profile of the WISC.

Appropriate contrasts for the flatness test in this example are simple one-sample z tests (cf. Section 3.2.1) against the standardized population values for each subtest with mean = 10.0 and standard deviation = 3.0. In this case we are less interested in how the subtests differ from one another than in how they differ from the normative population. (Had we been interested in differences among subtests for this sample, the contrasts procedures of Section 10.5.1 could have been applied.)

As a correction for post hoc inflation of experimentwise Type I error rate, individual α for each of the 11 z tests is set at .0045, meeting the requirements of

$$\alpha_{ew} = 1 - (1 - .0045)^{11} < .05$$

as per Equation 9.9. Because most z tables (cf. Table C.1) are set up for testing one-sided hypotheses, tabled α is divided in half to find critical z for rejecting the hypothesis of no difference between our sample and the population; the resulting criterion z is ± 2.845.

TABLE 10.23 RESULTS OF Z TESTS COMPARING EACH SUB-TEST WITH WISC/WISC-R POPULATION MEAN (ALPHA = .0045, TWO-TAILED TEST)

Subtest	Mean for entire sample	z for comparison with population mean
Information	9.55488	−1.90
Similarities	10.65854	2.81
Arithmetic	9.04878	−4.06*
Comprehension	10.05488	0.23
Vocabulary	10.70732	3.02*
Digit span	8.77439	−5.23*
Picture completion	10.71951	3.07*
Picture arrangement	10.40854	1.74
Block design	10.36585	1.56
Object assembly	10.85366	3.64*
Coding	8.52439	−6.30*

* $p < .0045$

TABLE 10.24 CHECKLIST FOR PROFILE ANALYSIS

1. Issues
   a. Unequal sample sizes and missing data
   b. Normality of sampling distributions
   c. Outliers
   d. Homogeneity of variance-covariance matrices
   e. Linearity
   f. Multicollinearity and singularity
2. Major analysis
   a. Tests for parallelism. If significant:
      (1) Strength of association
      (2) Figure showing profile for deviation from parallelism
   b. Test for differences among levels, if appropriate. If significant:
      (1) Strength of association
      (2) Marginal means for groups
   c. Test for deviation from flatness, if appropriate. If significant:
      (1) Strength of association
      (2) Means for measures
3. Additional analyses
   a. Planned comparisons
   b. Post hoc comparisons appropriate for significant effect(s)
      (1) Comparisons among groups
      (2) Comparisons among measures
      (3) Comparisons among measures within groups
   c. Power analysis for nonsignificant effects

For the first subtest, INFO, the mean for our entire sample (from Table 10.22) is 9.55488. Application of the z test results in

$$z = \frac{\overline{Y} - \mu}{\sigma/\sqrt{N}} = \frac{9.55488 - 10.0}{3.0/\sqrt{164}} = -1.900$$

For INFO, then, there is no significant difference between the learning disabled group and the normative population. Results of these individual z tests appear in Table 10.23.

A checklist for profile analysis appears in Table 10.24. Following is an example of a Results section in APA journal format.

Results

A profile analysis was performed on 11 subtests of the Wechsler Intelligence Scale for Children (WISC): information, similarities, arithmetic, comprehension, vocabulary, digit span, picture completion, picture arrangement, block design, object assembly, and coding. The grouping variable was preference for age of playmates, divided into children who (1) prefer younger playmates, (2) prefer older playmates, and (3) those who have no preference or prefer playmates the same age as themselves.

BMDPAM and BMDP7D were used for data screening. Four children in the original sample had missing data on one or more subtest, reducing the sample size to 164. No univariate or multivariate outliers were detected among these children, with $\underline{p}$ = .001. After deletion of cases with missing data, assumptions regarding normality of sampling distributions, homogeneity of variance–covariance matrices, linearity, and multicollinearity were met.

SPSS^{x} MANOVA was used for the major analysis. Using Wilks' criterion, the profiles, seen in Figure 10.4, deviated significantly from parallelism, $\underline{F}(20, 304)$ = 1.97, $\underline{p}$ = .009, η^2 = .11 [cf. Equation

10.4]. For the levels test, no reliable
differences were found among groups when
scores were averaged over all subtests, $\underline{F}(2,$
$161) = 0.81$, $\underline{p} = .45$. When averaged over
groups, however, subtests were found by
Hotelling's criterion to deviate
significantly from flatness, $\underline{F}(10, 152) =$
13.76, $\underline{p} < .001$, $\eta^2 = .48$.

Insert Figure 10.4 about here

To evaluate deviation from parallelism of
the profiles, confidence limits were
calculated around the mean of the profile
for the three groups combined. Alpha error
for each confidence interval was set at
.0015 to achieve an experimentwise error
rate of 5%. Therefore, 99.85% limits were
evaluated for the pooled profile. For two of
the subtests, one or more groups had means
that fell outside these limits. Children who
preferred older playmates had a reliably
higher mean on the vocabulary subtest (mean
$= 11.46$) than that of the pooled groups
(where the 99.85% confidence limits were
9.956 to 11.458); children who preferred
older playmates had reliably lower scores on
the picture completion subtest (mean =
9.80) than that of the pooled groups (99.85%
confidence limits were 9.973 to 11.466).

Deviation from flatness was evaluated
by identifying which subtests differed from
those of the standardization population of
the WISC, with mean = 10 and standard
deviation = 3 for each subtest.
Experimentwise alpha = .05 was achieved by
setting alpha for each test at .0045. As seen
in Table 10.23, learning-disabled children
showed significantly lower scores than the
WISC normative population in arithmetic,
digit span, and coding. On the other hand,
these children showed significantly higher

```
than normal performance on vocabulary,
picture completion, and object assembly.
```
———————————————
```
Insert Table 10.23 about here
```

10.8 SOME EXAMPLES FROM THE LITERATURE

Gray-Toft (1980) used profile analysis to evaluate the effectiveness of a counseling program for hospice nurses. Two groups of nurses participated in the program at different times, with each group acting as both treatment (during participation) and control. One group consisted of day shift nurses, while the other group was made up of nurses on evening and night shifts. (Because groups participated as a whole, the use of individual subjects for calculation of error variance is questionable.) Variables were background characteristics, seven subscales of a nursing stress scale, and five subscales of a job satisfaction index.

Profile analysis in this study was limited to evaluation of pretest scores. For the nursing stress scale, profiles were parallel and the groups did not differ in level. There were, however, differences in subscales (flatness), which were evaluated through stepdown analysis. Because subscales were based on factor analysis of a larger set of questions, priority for stepdown analysis was presumably based on percentage of variance accounted for by successive factors. Nurses experienced significantly more stress associated with death and dying and with their workload, and significantly less stress associated with conflict with other nurses. Their means did not differ on conflict with physicians, inadequate preparation, need for support, and uncertainty concerning treatment.

On the job satisfaction index, profiles were also parallel. However, significant differences were found in both levels and subscales. Day shift nurses showed greater satisfaction than evening or night shift nurses, and stepdown analysis showed nurses to be more satisfied with supervision than with pay, promotions, and coworkers. (No rationale for ordering of subscales was provided for the satisfaction index.)

To study career-making decisions, Mitchell and Krumboltz (1984) performed separate profile analyses for several time periods, during and after university students participated in one of three groups: training in cognitive restructuring, decision skills intervention, or a test-only control. Students were administered, among other instruments, a Vocational Exploratory Behavior Inventory (VEBI), which measures four components of appropriate career decision-making behavior.

No differences in profiles (parallelism) were reported between groups, nor were differences between components noted (flatness). Group (levels) differences were, however, reported at a number of the time periods. These differences began to emerge after the third week of treatment with the treatment groups performing more appropriate behaviors than controls during the treatment period. After treatment had ended, the cognitive restructuring group generally performed more vocational exploratory behaviors than the other two groups. That is, the cognitive restructuring treatment was more effective in maintaining vocational exploratory behavior. Specific differences between groups were explored via contrast analysis.

Chapter 11

Discriminant Function Analysis

11.1 GENERAL PURPOSE AND DESCRIPTION

The major purpose of discriminant function analysis is to predict group membership from a set of predictors. For example, can a differential diagnosis between a group of normal children, a group of children with learning disability, and a group with emotional disorder be made reliably from a set of psychological test scores? The three groups are normal children, children with learning disability, and children with emotional disorder. The predictors are a set of psychological test scores such as the Illinois Test of Psycholinguistic Ability, subtests of the Wide Range Achievement Test, Figure Drawing tests, and the Wechsler Intelligence Scale for Children.

Discriminant function analysis (DISCRIM) is MANOVA turned around. In MANOVA, we ask whether group membership produces reliable differences on a combination of DVs. If the answer to that question is yes, then the combination of variables can be used to predict group membership—the DISCRIM perspective. In univariate terms, a significant difference between groups implies that, given a score, you can predict (imperfectly, no doubt) which group it comes from.

Semantically, however, confusion arises between MANOVA and DISCRIM because in MANOVA the IVs are the groups and the DVs predictors while in DISCRIM the IVs are the predictors and the DVs groups. We have tried to avoid confusion here by always referring to IVs as "predictors" and to DVs as "groups" or "grouping variables."[1]

Mathematically, MANOVA and DISCRIM are the same, although the emphases

---

[1] Many texts also refer to IVs or predictors as discriminating variables and to DVs or groups as classification variables. However, there are also discriminant functions and classification functions to contend with, so the terminology becomes quite confusing. We have tried to simplify it by use only of the terms predictors and groups.

often differ. The major question in MANOVA is whether group membership is associated with reliable differences in combined DV scores, analogous in DISCRIM to the question of whether predictors can be combined to predict group membership reliably. In many cases, DISCRIM is carried to the point of actually putting cases into groups in a process called classification.

Classification is a major extension of DISCRIM over MANOVA. The adequacy of classification is tested in most computer programs for DISCRIM. How well does the classification procedure do? How many learning-disabled kids in the original sample, or a cross-validation sample, are classified correctly? When errors occur, what is their nature? Are learning-disabled kids more often confused with normal kids or with kids suffering emotional disorder?

A second difference involves interpretation of differences among the predictors. In MANOVA, there is frequently an effort to decide which DVs are associated with group differences, but rarely an effort to interpret the pattern of differences among the DVs as a whole. In DISCRIM, there is often an effort to interpret the pattern of differences among the predictors as a whole in an attempt to understand the dimensions along which groups differ.

Complexity arises with this attempt, however, because with more than two groups there may be more than one way to combine the predictors to differentiate among groups. There may, in fact, be as many dimensions that discriminate among groups as there are degrees of freedom for groups. For instance, if there are three groups, there may be two dimensions that discriminate among groups: a dimension that separates the first group from the second and third groups, and a dimension that separates the second group from the third group, for example.

In our example of three groups of children (normal, learning-disabled, and emotionally disordered) given a variety of psychological measures, one way of combining the psychological test scores may separate the normal group from the two groups with disorders while a second way of combining the test scores may separate the group with learning disability from the group with emotional disorder. The researcher attempts to understand the "message" in the two ways of combining test scores to separate groups differently. What is the meaning of the combination of scores that separates normal from disordered kids, and what is the meaning of the different combination of scores that separates one kind of disorder from another? This attempt is facilitated by statistics available in many of the canned computer programs for DISCRIM that are not printed in some programs for MANOVA.

Thus there are two facets of DISCRIM, and one or both may be emphasized in any given research application. The researcher may simply be interested in a decision rule for classifying cases where the number of dimensions and their meaning is irrelevant. Or, the emphasis may be on interpreting the results of DISCRIM in terms of the combinations of predictors—called discriminant functions—that separate various groups from each other.

A DISCRIM analog to covariates analysis (MANCOVA) is available, because DISCRIM can be set up in a hierarchical manner. When hierarchical DISCRIM is used, the covariate is simply a predictor that is given top priority. For example, a researcher might consider the score on the Wechsler Intelligence Scale for Children a covariate and ask how well the Wide Range Achievement Test, the Illinois Test of

Psycholinguistic Ability, and Figure Drawings differentiate between normal, learning-disabled, and emotionally disordered children after differences in IQ are accounted for.

If groups are arranged in a factorial design, it is frequently best to rephrase research questions so that they are answered within the framework of MANOVA. (However, DISCRIM can in some circumstances be directly applied to factorial designs as discussed in Section 11.6.5.) Similarly, DISCRIM programs make no provision for within-subjects variables. If a within-subjects analysis is desired, the question is also rephrased in terms of MANOVA. For this reason the emphasis in this chapter is on one-way between-subjects DISCRIM.

11.2 KINDS OF RESEARCH QUESTIONS

The primary goals of DISCRIM are to find the dimension or dimensions along which groups differ and to find classification functions to predict group membership. The degree to which these goals are met depends, of course, on choice of predictors. Typically, the choice is made either on the basis of theory about which variables should provide information about group membership, or on the basis of pragmatic considerations such as expense or convenience. Sometimes, for instance, one seeks predictors that are easy to obtain unobtrusively.

It should be emphasized that the same data are profitably analyzed through either MANOVA or DISCRIM programs, and frequently both, depending on the kinds of questions you want to ask. In any event, statistical procedures are readily available within canned computer programs for answering the following types of questions generally associated with DISCRIM.

11.2.1 Significance of Prediction

Can group membership be predicted reliably from the set of predictors? For example, can we do better than chance in predicting whether children are learning-disabled, emotionally disordered, or normal on the basis of the set of psychological test scores? This is the major question of DISCRIM that the statistical procedures described in Section 11.6.1 are designed to answer. The question is identical to the question about "main effects of IVs" for a one-way MANOVA.

11.2.2 Number of Significant Discriminant Functions

Along how many dimensions do groups differ reliably? For the three groups of children in our example, two discriminant functions are possible. It is possible that both are statistically reliable. For example, the first function may separate the normal group from the other two while the second separates the group with learning disability from the group with emotional disorder. Or there may be only one reliable function where, for instance, the group with learning disability is midway between the normal group and the group with emotional disorder on a single discriminant function (highly unlikely with this example).

In DISCRIM, like canonical correlation (Chapter 6), the first discriminant function provides the best separation among groups. Then a second discriminant function, orthogonal to the first, is found that best separates groups on the basis of associations not used in the first discriminant function. This procedure of finding successive orthogonal discriminant functions continues until all possible dimensions are evaluated. The number of possible dimensions is either one fewer than the number of groups or equal to the number of predictor variables, whichever is smaller. Typically, only the first one or two discriminant functions reliably discriminate among groups; remaining functions provide no additional information about group membership and are better ignored. Tests of significance for discriminant functions are discussed in Section 11.6.2.

11.2.3 Dimensions of Discrimination

How can the dimensions along which groups are separated be interpreted? Where are groups located along the discriminant functions and how do predictors correlate with the discriminant functions? In our example, if two significant discriminant functions are found, which predictors correlate highly with each function? What pattern of test scores discriminates between normal children and the other two groups (first discriminant function)? And what pattern of scores discriminates between children with learning disability and children with emotional disorder (second discriminant function)? These questions are discussed in Section 11.6.3.

11.2.4 Classification Functions

What linear equation(s) can be used to classify new cases into groups? For example, suppose we have the battery of psychological test scores for a group of new, undiagnosed children. How can we combine (weight) their scores to achieve the most reliable diagnosis? Procedures for deriving and using classification functions are discussed in Sections 11.4.2 and 11.6.6.[2]

11.2.5 Adequacy of Classification

Given classification functions, what proportion of cases is correctly classified? When errors occur, into which groups are cases misclassified? For instance, what proportion of learning-disabled children is correctly classified as learning-disabled, and, among those who are incorrectly classified, are they more often put into the group of normal children or into the group of emotionally disordered children?

Classification functions are used to predict group membership for new cases and to check the adequacy of classification for cases in the same sample through cross-validation. If the researcher knows that some groups are more likely to occur,

---

[2] Discriminant function analysis provides classification of cases into groups where group membership is known, at least for the sample from whom the classification equations are derived. Cluster analysis is a similar procedure except that group membership is not known. Instead, the analysis develops groups on the basis of similarities among cases.

or if some kinds of misclassification are especially undesirable, the classification procedure can be modified. Procedures for deriving classification functions and modifying them are discussed in Section 11.4.2; procedures for testing them are discussed in Section 11.6.6.

11.2.6 Strength of Association

What is the degree of relationship between group membership and the set of predictors? If the first discriminant function separates the normal group from the other two groups, how much does the variance for groups overlap the variance in combined test scores? If the second discriminant function separates learning-disabled from emotionally disordered children, how much does the variance for these groups overlap the combined test scores for this discriminant function? This is basically a question of percent of variance accounted for and, as seen in Section 11.4.1, is answered through canonical correlation. A canonical correlation is found for each discriminant function that, when squared, indicates the proportion of variance shared between groups and predictors on that function.

11.2.7 Importance of Predictor Variables

Which predictors are most important in predicting group membership? Which test scores are helpful for separating normal children from children with disorders, and which are helpful for separating learning-disabled from emotionally disordered children?

Questions about importance of predictors are analogous to those of importance of DVs in MANOVA, to those of IVs in multiple regression, and to those of IVs and DVs in canonical correlation. One procedure in DISCRIM is to interpret the correlations between the predictors and the discriminant functions, as discussed in Section 11.6.3.2. A second procedure is to evaluate predictors by how well they separate each group from all the others, as discussed in Section 11.6.4. (Or importance can be evaluated as in MANOVA, Section 9.5.2.)

11.2.8 Significance of Prediction with Covariates

After statistically removing the effects of one or more covariates, can one reliably predict group membership from a set of predictors? In DISCRIM, as in MANOVA, the ability of some predictors to promote group separation can be assessed after adjustment for prior variables. If scores on the Wechsler Intelligence Scale for Children are considered the covariate and given first entry in DISCRIM, do scores on the Illinois Test of Psycholinguistic Ability, the Wide Range Achievement Test, and Figure Drawings contribute to prediction of group membership when they are added to the equation?

Rephrased in terms of hierarchical discriminant function analysis, the question becomes, Do scores on the Illinois Test of Psycholinguistic Ability, the Wide Range Achievement Test, and Figure Drawings provide significantly better classification among the three groups than that afforded by scores on the WISC alone? Hierarchical DISCRIM is discussed in Section 11.5.2 and demonstrated in Section 11.8.3. A test for contribution of added predictors is given in Section 11.6.6.3.

11.2.9 Estimation of Group Means

If predictors discriminate among groups, it is important to report just how the groups differ on those variables. The best estimate of central tendency in a population is the sample mean. If, for example, the Illinois Test of Psycholinguistic Ability (ITPA) discriminates between groups with learning disability and emotional disorder, it is worthwhile to compare and report the mean ITPA score for learning-disabled children and the mean ITPA score for emotionally disordered children.

11.3 LIMITATIONS TO DISCRIMINANT FUNCTION ANALYSIS

11.3.1 Theoretical Issues

Because DISCRIM is typically used to predict membership in naturally occurring groups rather than groups formed by random assignment, questions such as why we can reliably predict group membership, or what causes differential membership are not often asked. If, however, group membership has occurred by random assignment, inferences of causality are justifiable as long as proper experimental controls have been instituted. The DISCRIM question then becomes, Does treatment following random assignment to groups produce enough difference in the predictors that we can now reliably separate groups on the basis of those variables?

As implied, limitations to DISCRIM are the same as limitations to MANOVA. The usual difficulties of generalizability apply to DISCRIM. But the cross-validation procedure described in Section 11.6.6.1 gives some indication of the generalizability of a solution.

11.3.2 Practical Issues

Practical issues for DISCRIM are basically the same as for MANOVA. Therefore, they are discussed here only to the extent of identifying the similarities between MANOVA and DISCRIM and identifying the situations in which assumptions for MANOVA and DISCRIM differ.

Classification makes fewer statistical demands than does inference. If classification is the primary goal, then, most of the following requirements (except for outliers and homogeneity of variance-covariance matrices) are relaxed. If, for example, you achieve 95% accuracy in classification, you hardly worry about the shape of distributions. Nevertheless, DISCRIM is optimal under the same conditions where MANOVA is optimal; and, if the classification rate is unsatisfactory, it may be because of violation of assumptions or limitations.

11.3.2.1 Unequal Sample Sizes and Missing Data As DISCRIM is typically a one-way analysis, no special problems are posed by unequal sample sizes in the groups.[3] In classification, however, a decision is required as to whether or not you want the a priori probabilities of assignment to groups to be influenced by sample size. That is, do you want the probability with which a case is assigned to a group to reflect the fact that the group itself is more (or less) probable in the sample? Section 11.4.2 discusses this issue, and use of unequal a priori probabilities is demonstrated in Section 11.8.

Regarding missing data (absence of scores on predictors for some cases), consult Section 8.3.2.1 and Chapter 4 for a review of problems and potential solutions.

As discussed in Section 9.3.2.1, the sample size of the smallest group should exceed the number of predictor variables. Although hierarchical and stepwise DISCRIM avoid the problems of multicollinearity and singularity by a tolerance test at each step, overfitting (producing results so close to the sample they don't generalize to other samples) occurs with all forms of DISCRIM if the number of cases does not notably exceed the number of predictors in the smallest group.

11.3.2.2 Multivariate Normality When using statistical inference in DISCRIM, the assumption of multivariate normality is that scores on predictors are independently and randomly sampled from a population, and that the sampling distribution of any linear combination of predictors is normally distributed. No tests are currently feasible for testing the normality of all linear combinations of sampling distributions of means of predictors.

However, DISCRIM, like MANOVA, is robust to failures of normality if violation is caused by skewness rather than outliers. A sample size that would produce 20 df for error in the univariate case should ensure robustness with respect to multivariate normality, as long as sample sizes are equal and two-tailed tests are used. (Calculation of df for error in the univariate case is discussed in Chapter 3.)

Because tests for DISCRIM typically are two-tailed, this requirement poses no difficulty. Sample sizes, however, are often not equal for applications of DISCRIM because naturally occurring groups rarely occur or are sampled with equal numbers of cases in groups. As differences in sample size among groups increase, larger overall sample sizes are necessary to assure robustness. As a conservative recommendation, robustness is expected with 20 cases in the smallest group if there are only a few predictors.

If samples are both small and unequal in size, assessment of normality is a matter of judgment. Are predictors expected to have normal sampling distributions in the population being sampled? If not, transformation of one or more predictors (cf. Chapter 4) may be worthwhile.

11.3.2.3 Outliers DISCRIM, like MANOVA, is highly sensitive to inclusion of outliers. Therefore, for each group separately, run a test for univariate and multivariate outliers, and transform or eliminate significant outliers before DISCRIM (see Chapter 4).

---

[3] Actually a problem does occur because discriminant functions may be nonorthogonal with unequal *n* (cf. Chapter 12), but rotation of axes is uncommon in discriminant function analysis and therefore is not discussed in this book.

11.3.2.4 Homogeneity of Variance-Covariance Matrices   In inference, when sample sizes are equal or large, DISCRIM, like MANOVA (Section 9.3.2.4) is robust to violation of the assumption of equal within-group variance-covariance (dispersion) matrices. When sample sizes are unequal and small, however, results of significance testing may be misleading if there is heterogeneity of the variance-covariance matrices.

When sample sizes are unequal and small, homogeneity of variance-covariance matrices is assessed through procedures of Section 9.3.2.4 or by inspection of scatterplots of scores on the first two canonical discriminant functions produced separately for each group. These scatterplots are available through BMDP7M or SPSS^x DISCRIMINANT. Rough equality in overall size of the scatterplots is evidence of homogeneity of variance-covariance matrices.

Although inference is usually robust with respect to heterogeneity of variance-covariance matrices with decently sized samples, classification is not. Cases tend to be overclassified into groups with greater dispersion. If classification is an important goal of analysis, examine scatterplots or otherwise test for homogeneity of variance-covariance matrices. If the matrices are grossly different, transformation of predictors may equalize them. (Examine individual predictors for skewness and kurtosis to estimate the necessary type of transformation.) SPSS^x DISCRIMINANT and SAS DISCRIM allow classification on the basis of separate covariance matrices. Because this procedure often leads to overfitting, it should be used only when the sample is large enough to permit cross-validation (Section 11.6.6.1).

11.3.2.5 Linearity   The DISCRIM model assumes a linear relationship among all pairs of predictors within each group. The assumption is less serious (from some points of view) than others, however, in that violation leads to reduced power rather than increased Type I error. The procedures in Section 8.3.2.6 may be applied to test for and improve linearity and to increase power.

11.3.2.6 Multicollinearity and Singularity   With highly redundant predictors, multicollinearity or singularity may occur, making matrix inversion unreliable. Fortunately, most computer programs for DISCRIM protect against this possibility by testing tolerance. Predictors with insufficient tolerance are excluded.

Guidelines for assessing multicollinearity and singularity for programs that do not include tolerance tests, and for dealing with multicollinearity or singularity when it occurs, are in Section 9.3.2.8. Note that analysis is done on predictors, not "DVs."

11.4 FUNDAMENTAL EQUATIONS FOR DISCRIMINANT FUNCTION ANALYSIS

For demonstration of DISCRIM, hypothetical scores on four predictors are given for three groups of learning-disabled children. Scores for three cases in each of the three groups are shown in Table 11.1.

The three groups are MEMORY (children whose major difficulty seems to be with tasks related to memory), PERCEPTION (children who show difficulty in visual perception), and COMMUNICATION (children with language difficulty). The four

TABLE 11.1 HYPOTHETICAL SMALL SAMPLE DATA FOR ILLUSTRATION OF DISCRIMINANT FUNCTION ANALYSIS

Group	Predictors			
	PERF	INFO	VERBEXP	AGE
MEMORY	87	5	31	6.4
	97	7	36	8.3
	112	9	42	7.2
PERCEPTION	102	16	45	7.0
	85	10	38	7.6
	76	9	32	6.2
COMMUNICATION	120	12	30	8.4
	85	8	28	6.3
	99	9	27	8.2

predictors are PERF (Performance Scale IQ of the WISC-R [revised]), INFO (Information subtest of the WISC-R), VERBEXP (Verbal Expression subtest of the ITPA), and AGE (chronological age in years). The grouping variable, then, is type of learning disability, and the predictors are selected scores from psychodiagnostic instruments and age.

Fundamental equations are presented for two major parts of DISCRIM: discriminant functions and classification equations. Setup and selected output for this example appear in Section 11.4.3 for BMDP7M, SPSS^x DISCRIM, SAS DISCRIM, and SYSTAT MGLH.

11.4.1 Derivation and Test of Discriminant Functions

The fundamental equations for testing the significance of a set of discriminant functions are the same as for MANOVA, discussed in Chapter 9. Variance in the set of predictors is partitioned into two sources: variance attributable to differences between groups and variance attributable to differences within groups. Through procedures shown in Equations 9.1 to 9.3, cross-products matrices are formed.

$$S_{total} = S_{bg} + S_{wg} \tag{11.1}$$

The total cross-products matrix (S_{total}) is partitioned into a cross-products matrix associated with differences between groups (S_{bg}) and a cross-products matrix of differences within groups (S_{wg}).

For the example in Table 11.1, the resulting cross-products matrices are

$$S_{bg} = \begin{bmatrix} 314.89 & -71.56 & -180.00 & 14.49 \\ -71.56 & 32.89 & 8.00 & -2.22 \\ -180.00 & 8.00 & 168.00 & -10.40 \\ 14.49 & -2.22 & -10.40 & 0.74 \end{bmatrix}$$

$$S_{wg} = \begin{bmatrix} 1286.00 & 220.00 & 348.33 & 50.00 \\ 220.00 & 45.33 & 73.67 & 6.37 \\ 348.33 & 73.67 & 150.00 & 9.73 \\ 50.00 & 6.37 & 9.73 & 5.49 \end{bmatrix}$$

Determinants[4] for these matrices are

$$|S_{wg}| = 4.70034789 \times 10^{13}$$
$$|S_{bg} + S_{wg}| = 448.63489 \times 10^{13}$$

Following procedures in Equation 9.4, Wilks' Lambda[5] for these matrices is

$$\Lambda = \frac{|S_{wg}|}{|S_{bg} + S_{wg}|} = .010477$$

To find the approximate F ratio, as per Equation 9.5, the following values are used:

$p = 4$ the number of predictor variables

$df_{effect} = 2$ the number of groups minus one, or $k - 1$

$df_{error} = 6$ the number of groups times the quantity $n - 1$, where n is the number of cases per group. Because n is often not equal for all groups in DISCRIM, an alternative equation for df_{error} is $N - k$, where N is the total number of cases in all groups—9 in this case.

Thus we obtain

$$s = \sqrt{\frac{(4)^2(2)^2 - 4}{(4)^2 + (2)^2 - 5}} = 2$$

$$y = (.010477)^{1/2} = .102357$$

$$df_1 = 4(2) = 8$$

$$df_2 = (2)\left[6 - \frac{4 - 2 + 1}{2}\right] - \left[\frac{4(2) - 2}{2}\right] = 6$$

$$\text{Approximate } F(8, 6) = \left(\frac{1 - .102357}{.102357}\right)\left(\frac{6}{8}\right) = 6.58$$

---

[4] A determinant, as described in Appendix A, can be viewed as a measure of generalized variance of a matrix.

[5] Alternative statistical criteria are discussed in Section 11.6.1.1.

Critical F with 8 and 6 df at $\alpha = 0.05$ is 4.15. Because obtained F exceeds critical F, we conclude that the three groups of children can be distinguished on the basis of the combination of the four predictors.

This is a test of overall relationship between groups and predictors. It is the same as the overall test of main effect in MANOVA. In MANOVA, this result is followed by an assessment of the importance of the various DVs to the main effect. In DISCRIM, however, when an overall relationship is found between groups and predictors, the next step is to examine the discriminant functions that compose the overall relationship.

The maximum number of discriminant functions is limited by the number of groups or the number of predictors. It is either (1) the number of predictors or (2) the degrees of freedom for groups, whichever is smaller. Because there are three groups (and four predictors) in this example, there are potentially two discriminant functions contributing to the overall relationship. And, because the overall relationship is reliable, at least the first discriminant function is very likely to be reliable, and both may be reliable.

Discriminant functions are like regression equations; a discriminant function score is predicted from the sum of the series of predictors, each weighted by a coefficient. There is one set of discriminant function coefficients for the first discriminant function, a second set of coefficients for the second discriminant function, and so forth. Subjects get separate discriminant function scores for each discriminant function when their own scores on predictors are inserted into the equations.

To solve for the (standardized) discriminant function score for the ith function, Equation 11.2 is used.

$$D_i = d_{i1}z_1 + d_{i2}z_2 + \cdots + d_{ip}z_p \qquad (11.2)$$

A child's standardized score on the ith discriminant function (D_i) is found by multiplying the standardized score on each predictor (z) by its standardized discriminant function coefficient (d_i) and then adding the products for all predictors.

Discriminant function coefficients are found in the same manner as are coefficients for canonical variates (Section 6.4). In fact, DISCRIM is basically a problem in canonical correlation with group membership on one side of the equation and predictors on the other, where successive canonical variates (here called discriminant functions) are computed. In DISCRIM d_i are chosen to maximize differences between groups relative to differences within groups.

Just as in multiple regression, Equation 11.2 can be written either for raw scores or for standardized scores. A discriminant function score, then, can also be produced by multiplying the raw score on each predictor by its associated unstandardized discriminant function coefficient, adding the products over all predictors, and adding a constant to adjust for the means. The score produced in this way is the same D_i as produced in Equation 11.2.

The mean of each discriminant function over all cases is zero, because the mean of each predictor, when standardized, is zero. Further, the standard deviation of D_i is 1.

Just as D_i can be calculated for each case, a mean value of D_i can be calculated for each group. The members of each group considered together have a mean score on a discriminant function that is the distance of the group in standard deviation units from the zero mean of the discriminant function. Group means on D_i are typically called centroids in reduced space, the spacing having been reduced from that of the p predictors to a single dimension, or discriminant function.

A canonical correlation is found for each discriminant function following procedures in Chapter 6. Successive discriminant functions are evaluated for significance, as discussed in Section 11.6.2. Also discussed in subsequent sections are standardized and unstandardized discriminant coefficients, loading matrices, and group centroids.

If there are only two groups, discriminant function scores can be used to classify cases into groups. A case is classified into one group if its D_i score is above zero, and into the other group if the D_i score is below zero. With numerous groups, classification is possible from the discriminant functions, but it is simpler to use the procedure in the following section.

11.4.2 Classification

To assign cases into groups, a classification equation is developed for each group. Three classification equations are developed for the example of Table 11.1 where there are three groups. Data for each case are inserted into each classification equation to develop a classification score for each group for the case. The case is assigned to the group for which it has the highest classification score.

In its simplest form, the basic classification equation for the jth group ($j = 1$, $2, \ldots , k$) is

$$C_j = c_{j0} + c_{j1}X_1 + c_{j2}X_2 + \cdots + c_{jp}X_p \qquad (11.3)$$

A score on the classification function for group j (C_j) is found by multiplying the raw score on each predictor (X) by its associated classification function coefficient (c_j), summing over all predictors, and adding a constant c_{j0}.

The classification coefficients, c_j, are found from the means of the p predictors and the pooled within-group variance-covariance matrix, $\mathbf{W}$. The within-group covariance matrix is produced by dividing each element in the cross-products matrix, $\mathbf{S}_{wg}$, by the within-group degrees of freedom, $N - k$. In matrix form,

$$\mathbf{C}_j = \mathbf{W}^{-1}\mathbf{M}_j \qquad (11.4)$$

The row matrix of classification coefficients for group j ($\mathbf{C}_j = c_{j1}, c_{j2}, \ldots , c_{jp}$) is found by multiplying the inverse of the within-group variance-covariance matrix ($\mathbf{W}^{-1}$) by a column matrix of means for group j on the p variables ($\mathbf{M}_j = \overline{X}_{j1}, \overline{X}_{j2}, \ldots , \overline{X}_{jp},$).

The constant for group j, c_{j0}, is found as follows:

$$c_{j0} = (-\tfrac{1}{2}) \, \mathbf{C}_j\mathbf{M}_j \qquad (11.5)$$

The constant for the classification function for group j (c_{j0}) is formed by multiplying the row matrix of classification coefficients for group j $(\mathbf{C}_j)$ by the column matrix of means for group j $(\mathbf{M}_j)$.

For the sample data, each element in the $\mathbf{S}_{wg}$ matrix from Section 11.4.1 is divided by $df_{wg} = df_{error} = 6$ to produce the within-group variance-covariance matrix:

$$\mathbf{W} = \begin{bmatrix} 214.33 & 36.67 & 58.06 & 8.33 \\ 36.67 & 7.56 & 12.28 & 1.06 \\ 58.06 & 12.28 & 25.00 & 1.62 \\ 8.33 & 1.06 & 1.62 & 0.92 \end{bmatrix}$$

The inverse of the within-group variance-covariance matrix is

$$\mathbf{W}^{-1} = \begin{bmatrix} 0.04362 & -0.20195 & 0.00956 & -0.17990 \\ -0.20195 & 1.62970 & -0.37073 & 0.60623 \\ 0.00956 & -0.37073 & 0.20071 & -0.01299 \\ -0.17990 & 0.60623 & -0.01299 & 2.05006 \end{bmatrix}$$

Multiplying $\mathbf{W}^{-1}$ by the column matrix of means for the first group gives the matrix of classification coefficients for that group, as per Equation 11.4.

$$\mathbf{C}_1 = \mathbf{W}^{-1} \begin{bmatrix} 98.67 \\ 7.00 \\ 36.33 \\ 7.30 \end{bmatrix} = [1.92, \ -17.56, 5.55, 0.99]$$

The constant for group 1, then, according to Equation 11.5, is

$$c_{1,0} = (-\tfrac{1}{2}) [1.92, \ -17.56, 5.55, 0.99] \begin{bmatrix} 98.67 \\ 7.00 \\ 36.33 \\ 7.30 \end{bmatrix} = -137.83$$

(Values used in these calculations were carried to several decimal places before rounding.) When these procedures are repeated for groups 2 and 3, the full set of classification equations is produced, as shown in Table 11.2.

In its simplest form, classification proceeds as follows for the first case in group 1. Three classification scores, one for each group, are calculated for the case by applying Equation 11.3:

$C_1 = -137.83 + (1.92)(87) + (-17.56)(5) + (5.55)(31) + (0.99)(6.4) = 119.80$

$C_2 = -71.29 + (0.59)(87) + (-8.70)(5) + (4.12)(31) + (5.02)(6.4) = 96.39$

$C_3 = -71.24 + (1.37)(87) + (-10.59)(5) + (2.97)(31) + (2.91)(6.4) = 105.69$

TABLE 11.2 CLASSIFICATION FUNCTION COEFFICIENTS
FOR SAMPLE DATA OF TABLE 11.1

	Group 1: MEMORY	Group 2: PERCEP	Group 3: COMMUN
PERF	1.92420	0.58704	1.36552
INFO	− 17.56221	− 8.69921	− 10.58700
VERBEXP	5.54585	4.11679	2.97278
AGE	0.98723	5.01749	2.91135
(CONSTANT)	− 137.82892	− 71.28563	− 71.24188

Because this child has the highest classification score in group 1, the child is assigned to group 1.

This simple classification scheme is most appropriate when equal group sizes are expected in the population. If unequal group sizes are expected, the classification procedure can be modified by a priori probabilities set by group size. Although a number of highly sophisticated classification schemes have been suggested (e.g., Tatsuoka, 1975), the most straightforward involves adding to each classification equation a term that adjusts for group size.[6] The classification equation for group j (C_j) then becomes

$$C_j = c_{j0} + \sum_{i=1}^{p} c_{ji}X_i + \ln(n_j/N) \tag{11.6}$$

where n_j = size of group j and N = total sample size.

It should be reemphasized that the classification procedures are highly sensitive to heterogeneity of variance-covariance matrices. Cases are more likely to be classified into the group with the greatest dispersion—that is, into the group for which the determinant of the within-group covariance matrix is greatest. Section 11.3.2.4 provides suggestions for dealing with this problem.

Uses of classification procedures are discussed more fully in Section 11.6.6.

11.4.3 Computer Analyses of Small Sample Example

Setup and selected output for computer analyses of the data in Table 11.1, using the simplest methods, are in Tables 11.3 through 11.6, for BMDP7M, SPSS^x DISCRIMINANT, SAS DISCRIM, and SYSTAT MGLH, respectively. There is less consensus than usual in these programs regarding the statistics that are reported owing, in part, to the complexity of discriminant function analysis.

Table 11.3 shows a BMDP7M run with all but the most relevant output suppressed. Direct DISCRIM is specified in the LEVEL instruction, LEVEL = 0,0,1,1,1,1. The first two zeros delete the first two predictors named in the VARIABLE paragraph from the problem. The remaining four predictors, with level 1, *are* to be used as

---

[6] Output from computer programs reflects this adjustment (except for the earliest releases of SPSS DISCRIMINANT—prior to update 8.0.)

```
            /PROBLEM      TITLE IS 'SMALL SAMPLE DISCRIMINANT FUNCTION ANALYSIS'.
            /INPUT        VARIABLES=6.  FORMAT IS FREE.  FILE='TAPE41'.
            /VARIABLE     NAMES ARE SUBJNO, GROUP, PERF, INFO, VERBEXP, AGE.
                          LABEL = SUBJNO.
                          GROUPING IS GROUP.
            /GROUP        CODES(GROUP) ARE 1 TO 3.
            /DISCRIM      LEVEL=0,0,1,1,1,1.  FORCE=1.
            /END
```

PRIOR PROBABILITIES. . . .   0.33333   0.33333   0.33333

****

STEP NUMBER   0

```
  VARIABLE     F TO   FORCE TOLERNCE *   VARIABLE     F TO   FORCE TOLERNCE
              REMOVE LEVEL             *             ENTER  LEVEL
         DF =  2    7                  *        DF =  2    6
                                       *   3 PERF       0.73    1   1.00000
                                       *   4 INFO       2.18    1   1.00000
                                       *   5 VERBEXP    3.36    1   1.00000
                                       *   6 AGE        0.40    1   1.00000
```

****

STEP NUMBER   4
VARIABLE ENTERED   6 AGE

```
  VARIABLE     F TO   FORCE TOLERNCE *   VARIABLE     F TO   FORCE TOLERNCE
              REMOVE LEVEL             *             ENTER  LEVEL
         DF =  2    3                  *        DF =  2    2
  3 PERF       6.89    1   0.10696 *
  4 INFO      18.84    1   0.08121 *
  5 VERBEXP    9.74    1   0.19930 *
  6 AGE        0.27    1   0.53278 *
```

U-STATISTIC(WILKS' LAMBDA) 0.0104766   DEGREES OF FREEDOM   4   2   6
APPROXIMATE F-STATISTIC   6.577   DEGREES OF FREEDOM   8.00   6.00

```
  F - MATRIX         DEGREES OF FREEDOM =    4    3

            *1       *2
*2        9.70
*3        7.19     4.57
```

CLASSIFICATION FUNCTIONS

```
          GROUP =   *1          *2          *3
VARIABLE
   3 PERF          1.92420     0.58704     1.36552
   4 INFO        -17.56222    -8.69921   -10.58700
   5 VERBEXP       5.54585     4.11679     2.97278
   6 AGE           0.09872     0.50175     0.29114

CONSTANT        -138.91110   -72.38436   -72.34031
```

CLASSIFICATION MATRIX

```
GROUP     PERCENT   NUMBER OF CASES CLASSIFIED INTO GROUP -
          CORRECT
                    *1       *2       *3
*1        100.0     3        0        0
*2        100.0     0        3        0
*3        100.0     0        0        3

TOTAL     100.0     3        3        3
```

TABLE 11.3 (Continued)

```
JACKKNIFED CLASSIFICATION

GROUP      PERCENT    NUMBER OF CASES CLASSIFIED INTO GROUP -
           CORRECT
                        *1        *2        *3
  *1       100.0        3         0         0
  *2       100.0        0         3         0
  *3        66.7        0         1         2

TOTAL       88.9        3         4         2

EIGENVALUES

                  13.48590    5.58923

CUMULATIVE PROPORTION OF TOTAL DISPERSION

                   0.70699    1.00000

CANONICAL CORRELATIONS

                   0.96487    0.92100

VARIABLE      COEFFICIENTS FOR CANONICAL VARIABLES

  3 PERF         0.17100     -0.10069
  4 INFO        -1.26953     -0.10325
  5 VERBEXP      0.26493      0.35776
  6 AGE         -0.05254      0.02469

              STANDARDIZED   (BY POOLED WITHIN VARIANCES)
VARIABLE      COEFFICIENTS FOR CANONICAL VARIABLES

  3 PERF         0.85502     -0.50343
  4 INFO       -12.14749     -0.98793
  5 VERBEXP      0.00000      0.00000
  6 AGE          0.00000      0.00000

CONSTANT        -9.67374     -3.45293

GROUP       CANONICAL VARIABLES EVALUATED AT GROUP MEANS
  *1           4.10234      0.69097
  *2          -2.98068      1.94169
  *3          -1.12166     -2.63265
```

predictors. FORCE = 1 forces all predictors with level 1 into the equation. Because all predictors to be included have the same level, this is a direct rather than a stepwise or hierarchical analysis.

In the output, PRIOR PROBABILITIES of group membership are set equal by default. At STEP NUMBER 0 univariate ANOVAs for each predictor considered separately, with df = 2 and 6, are given; F ratios for each predictor appear in the column labeled F TO ENTER. TOLERNCE is 1-SMC for each predictor already in the equation. Because there are no predictors in the equation at this point, SMCs are zero.

At STEP NUMBER 4 all predictors have entered the equation. F TO REMOVE (with 2 and 3 df) tests the significance of the reduction in prediction if a predictor is removed from the equation. Both Wilks' Lambda and the APPROXIMATE F-STATISTIC for the multivariate effect (cf. Sections 9.4.1 and 11.4.1) are provided as a test of the discriminant functions considered together. The F − MATRIX, with 4 and 3 df, shows multivariate pairwise comparisons among the three groups. For example, for the difference between groups 1 and 2, multivariate $F(4, 3) = 9.70$.

In the CLASSIFICATION FUNCTIONS section, the CLASSIFICATION MA-TRIX shows all cases to be perfectly classified. Rows represent actual group membership, columns show the group into which cases are classified, following procedures in Section 11.4.2. With JACKKNIFED CLASSIFICATION, in which classification equations are developed for each case with that case deleted (cf. Section 11.6.6.2), one case in the third group is misclassified into the second group, reducing the PERCENT CORRECT for group 3 to 66.7 and the overall percentage to 88.9.

The remaining output describes the two discriminant functions (CANONICAL VARIABLES) produced for the three groups. The EIGENVALUES show the relative proportion of variance contributed by the two discriminant functions, followed by the CUMULATIVE PROPORTION OF TOTAL DISPERSION (variance) in the solution accounted for by the functions—the first discriminant function accounts for 70.699 percent of the variance in the solution while the second accounts for the remaining variance. CANONICAL CORRELATIONS are the multiple correlations between the predictors and the discriminant functions. COEFFICIENTS FOR CANONICAL VARIABLES are weighting coefficients for canonical correlation as described in Chapter 6, and STANDARDIZED .... COEFFICIENTS FOR CANONICAL VARIABLES are raw score discriminant function coefficients computed using the pooled within-cell variance-covariance matrix. Finally, the section labeled CANONICAL VARIABLES EVALUATED AT GROUP MEANS shows the average discriminant function score for each group for each function.

SPSS^x DISCRIMINANT (Table 11.4) also assigns equal PRIOR PROBABILITY FOR EACH GROUP by default. The output summarizing the CANONICAL DISCRIMINANT FUNCTIONS appears in two parallel tables separated by :'s. At the left are shown EIGENVALUE, PERCENT OF VARIANCE and CUMULATIVE PERCENT of variance accounted for by each function, and CANONICAL CORRELATION for each discriminant function. At the right are the ''peel off'' significance tests of successive discriminant functions. When AFTER FUNCTION is 0, (CHISQUARED = 20.514) all discriminant functions are tested together (0 removed). AFTER FUNCTION 1 is removed, CHI-SQUARED is still statistically significant at α = .05 because SIGNIFICANCE = .0370. This means that the second discriminant function is significant as well as the first. If not, the second would not have been ...USED IN THE REMAINING ANALYSIS.

STANDARDIZED CANONICAL DISCRIMINANT FUNCTION COEFFICIENTS (Equation 11.2) are given for deriving discriminant function scores from standardized predictors. Correlations (loadings) between predictors and discriminant functions are given in the STRUCTURE MATRIX. These are separated into chunks, so that predictors loading on the first discriminant function are listed first, and those loading on the second discriminant function second. Then GROUP CENTROIDS are shown, indicating the average discriminant score for each group on each function.

In the CLASSIFICATION RESULTS table, rows represent actual group membership and columns represent predicted group membership. Within each cell, the number and percent of cases correctly classified are shown. For this example, all of the diagonal cells show 100% correction classification (100.0P).

SAS DISCRIM output (Table 11.5) begins with a summary of GROUPS, their FREQUENCY (number of cases), and PRIOR PROBABILITY (set equal by default). Information is then provided about the POOLED COVARIANCE MATRIX, which can signal problems in multicollinearity/singularity if the rank of the matrix is not

TABLE 11.4 SETUP AND SELECTED SPSS' DISCRIMINANT OUTPUT FOR DIRECT DISCRIMINANT FUNCTION ANALYSIS ON SAMPLE DATA IN TABLE 11.1

```
TITLE       SMALL SAMPLE DISCRIMINANT FUNCTION ANALYSIS
FILE HANDLE TAPE41
DATA LIST   FILE=TAPE41      FREE
            /SUBJNO,GROUP,PERF,INFO,VERBEXP,AGE
DISCRIMINANT GROUPS=GROUP(1,3)/
             VARIABLES=PERF TO AGE/
             METHOD=DIRECT/
```

PRIOR PROBABILITY FOR EACH GROUP IS .33333

CANONICAL DISCRIMINANT FUNCTIONS

FUNCTION	EIGENVALUE	PERCENT OF VARIANCE	CUMULATIVE PERCENT	CANONICAL CORRELATION	: AFTER : FUNCTION	WILKS' LAMBDA	CHI-SQUARED	D.F.	SIGNIFICANCE
					: 0 :	.0104766	20.514	8	.0086
1*	13.48590	70.70	70.70	.9648665	: 1 :	.1517629	8.4845	3	.0370
2*	5.58923	29.30	100.00	.9209979	: :				

* MARKS THE 2 CANONICAL DISCRIMINANT FUNCTION(S) TO BE USED IN THE REMAINING ANALYSIS.

STANDARDIZED CANONICAL DISCRIMINANT FUNCTION COEFFICIENTS

	FUNC 1	FUNC 2
PERF	-2.50352	-1.47106
INFO	3.48961	-.28380
VERBEXP	-1.32466	1.78381
AGE	.50273	.23325

STRUCTURE MATRIX:

POOLED WITHIN-GROUPS CORRELATIONS BETWEEN CANONICAL DISCRIMINANT FUNCTIONS
AND DISCRIMINATING VARIABLES. VARIABLES ARE ORDERED BY THE FUNCTION WITH
LARGEST CORRELATION AND THE MAGNITUDE OF THAT CORRELATION.

	FUNC 1	FUNC 2
INFO	.22796*	.06642
VERBEXP	-.02233	.44630*
PERF	-.07546	-.17341*
AGE	-.02786	-.14861*

CANONICAL DISCRIMINANT FUNCTIONS EVALUATED AT GROUP MEANS (GROUP CENTROIDS)

GROUP	FUNC 1	FUNC 2
1	-4.10234	.69097
2	2.98068	1.94169
3	1.12166	-2.63285

CLASSIFICATION RESULTS -

ACTUAL GROUP	NO. OF CASES	PREDICTED GROUP MEMBERSHIP		
		1	2	3
GROUP 1	3	3 100.0P	0 .0P	0 .0P
GROUP 2	3	0 .0P	3 100.0P	0 .0P
GROUP 3	3	0 .0P	0 .0P	3 100.0P

PERCENT OF "GROUPED" CASES CORRECTLY CLASSIFIED: 100.00P

TABLE 11.5 SETUP AND SAS DISCRIM OUTPUT FOR DISCRIMINANT FUNCTION ANALYSIS OF SMALL SAMPLE DATA OF TABLE 11.1

```
DATA SSAMPLE;
INFILE TAPE41;
INPUT SUBJNO GROUP PERF INFO VERBEXP AGE;
PROC DISCRIM;
CLASS GROUP;
VAR PERF INFO VERBEXP AGE;
```

DISCRIMINANT ANALYSIS

GROUP	FREQUENCY	PRIOR PROBABILITY
1	3	0.33333333
2	3	0.33333333
3	3	0.33333333
----	-	----------
TOTAL	9	1.00000000

DISCRIMINANT ANALYSIS   POOLED COVARIANCE MATRIX INFORMATION

COVARIANCE MATRIX RANK	NATURAL LOG OF DETERMINANT OF THE COVARIANCE MATRIX
4	11.11026440

DISCRIMINANT ANALYSIS   PAIRWISE SQUARED GENERALIZED DISTANCES BETWEEN GROUPS

$$D^2(I!J) = (X - X)' \, COV^{-1} \, (X - X)$$
$$ \quad I \quad J \quad\quad\quad I \quad J$$

GENERALIZED SQUARED DISTANCE TO GROUP

FROM GROUP	1	2	3
1	0.00000000	51.73350724	38.33673132
2	51.73350724	0.00000000	24.38052536
3	38.33673132	24.38052536	0.00000000

DISCRIMINANT ANALYSIS   LINEAR DISCRIMINANT FUNCTION

$$\text{CONSTANT} = -.5\ \bar{X}_J'\ \text{COV}^{-1}\ \bar{X}_J \qquad \text{COEFFICIENT VECTOR} = \text{COV}^{-1}\ \bar{X}_J$$

GROUP

	1	2	3
CONSTANT	-137.81246658	-71.28575224	-71.24169825
PERF	1.92419912	0.58704261	1.36551640
INFO	-17.56221495	-8.69921249	-10.58699872
VERBEXP	5.54585425	4.11679027	2.97277918
AGE	0.09872266	0.50174934	0.29113509

DISCRIMINANT ANALYSIS   CLASSIFICATION SUMMARY FOR CALIBRATION DATA: WORK.SSAMPLE

GENERALIZED SQUARED DISTANCE FUNCTION: POSTERIOR PROBABILITY OF MEMBERSHIP IN EACH GROUP:

$$D_J^2(X) = (X - \bar{X}_J)'\ \text{COV}^{-1}\ (X - \bar{X}_J) \qquad PR(J|X) = EXP(-.5\ D_J^2(X)) / \underset{K}{\text{SUM}}\ EXP(-.5\ D_K^2(X))$$

NUMBER OF OBSERVATIONS AND PERCENTS CLASSIFIED INTO GROUP:

FROM GROUP	1	2	3	TOTAL
1	3	0	0	3
	100.00	0.00	0.00	100.00
2	0	3	0	3
	0.00	100.00	0.00	100.00
3	0	0	3	3
	0.00	0.00	100.00	100.00
TOTAL PERCENT	3 33.33	3 33.33	3 33.33	9 100.00
PRIORS	0.3333	0.3333	0.3333	

TABLE 11.6 SETUP AND SELECTED SYSTAT MGLH OUTPUT FOR
DISCRIMINANT FUNCTION ANALYSIS OF SMALL
SAMPLE DATA IN TABLE 11.1

```
                    USE TAPE41
                    CATEGORY GROUP=3
                    MODEL PERF INFO VERBEXP AGE=CONSTANT+GROUP
                    PRINT LONG
                    ESTIMATE
                    HYPOTHESIS
                    EFFECT=GROUP
                    TEST


TEST FOR EFFECT CALLED:
                    GROUP

MULTIVARIATE TEST STATISTICS

            WILKS' LAMBDA =        0.010
              F-STATISTIC =        6.577     DF =   8,   6      PROB =        0.017

            PILLAI TRACE =         1.779
              F-STATISTIC =        8.058     DF =   8,   8      PROB =        0.004

    HOTELLING-LAWLEY TRACE =      19.075
              F-STATISTIC =        4.769     DF =   8,   4      PROB =        0.074

                    THETA =   0.931   S =   2, M =   .5, N =   .5 PROB =      0.249

TEST OF RESIDUAL ROOTS

    ROOTS  1 THROUGH  2
      CHI-SQUARE STATISTIC =      20.514     DF =        8      PROB =        0.009

    ROOTS  2 THROUGH  2
      CHI-SQUARE STATISTIC =       8.484     DF =        3      PROB =        0.037


CANONICAL CORRELATIONS

                    1          2

                 0.965      0.921

DEPENDENT VARIABLE CANONICAL COEFFICIENTS
STANDARDIZED BY CONDITIONAL (WITHIN GROUPS) STANDARD DEVIATIONS

                    1          2

        PERF      2.504      1.474
        INFO     -3.490      0.284
     VERBEXP      1.325     -1.789
        AGE      -0.503     -0.236


CANONICAL LOADINGS (CORRELATIONS BETWEEN CONDITIONAL
DEPENDENT VARIABLES AND DEPENDENT CANONICAL FACTORS)

                    1          2

        PERF      0.075      0.173
        INFO     -0.228     -0.066
     VERBEXP      0.022     -0.446
        AGE       0.028      0.149
```

equal to the number of predictors. An equation is provided by which distances between groups are derived and then presented in the matrix labeled PAIRWISE SQUARED GENERALIZED DISTANCES BETWEEN GROUPS. For example the greatest distance is between groups 1 and 2 (51.73350724) while the smallest distance is between groups 2 and 3 (24.38052536). Equations for classification functions and classification coefficients (Equation 11.6) are given in the following matrix, labeled GROUP. Finally, results of classification are presented in the NUMBER OF OBSERVATIONS AND PERCENTS CLASSIFIED INTO GROUP table where, as usual, rows represent actual group and columns represent predicted group. Cell values show number of cases classified and percentage correct. Prior probabilities are repeated at the bottom of this table.

SYSTAT MGLH (Table 11.6) shows the results of four multivariate tests of the discriminant function solution in the MULTIVARIATE TEST STATISTICS table with Roy's gcr labeled THETA, as described in Section 9.5.1. TEST OF RESIDUAL ROOTS shows the "peel off" significance tests, using CHI-SQUARE, of successive discriminant functions. CANONICAL CORRELATIONS of predictors with discriminant functions are shown. The next matrix, DEPENDENT VARIABLE... shows standardized discriminant coefficients (d_i) for deriving discriminant function scores from standardized predictor scores, matching those produced by SPSS^x but for sign. The last table presents the CANONICAL LOADINGS.

11.5 TYPES OF DISCRIMINANT FUNCTION ANALYSIS

The three types of discriminant function analysis—standard (direct), hierarchical, and statistical (stepwise)—are analogous to the three types of multiple regression discussed in Section 5.5. Criteria for choosing among the three strategies are the same as those discussed in Section 5.5.4 for multiple regression.

11.5.1 Direct Discriminant Function Analysis

In standard (direct) DISCRIM, like standard multiple regression, all predictors enter the equations at once and each predictor is assigned only the unique association it has with groups. Variance shared among predictors contributes to the total relationship, but not to any one predictor.

The overall test of relationship between predictors and groups in direct DISCRIM is the same as the test of main effect in MANOVA where all discriminant functions are combined and DVs are considered simultaneously. Direct DISCRIM is the model demonstrated in Section 11.4.1.

All the computer programs described in Table 11.11 perform direct DISCRIM; the use of some of them for that purpose is shown in Tables 11.3 through 11.6.

11.5.2 Hierarchical Discriminant Function Analysis

Hierarchical (or, as some prefer to call it, sequential) DISCRIM is used to evaluate contributions to prediction of group membership by predictors as they enter the equations in an order determined by the researcher. The researcher assesses improvement in classification when a new predictor is added to a set of prior predictors. Does classification of cases into groups improve reliably when the new predictor or predictors are added (cf. Section 11.6.6.3)?

If predictors with early entry are viewed as covariates and an added predictor is viewed as a DV, DISCRIM is used for analysis of covariance. Indeed, hierarchical

DISCRIM can be used to perform stepdown analysis following MANOVA (cf. Section 9.5.2.2) because stepdown analysis is a sequence of ANCOVAs.

Hierarchical DISCRIM is useful when a reduced set of predictors is desired and there is some basis for establishing a priority order among them. If, for example, some predictors are easy or inexpensive to obtain and they are given early entry, a useful, cost-effective set of predictors may be found through the hierarchical procedure.

Hierarchical DISCRIM through BMDP7M for the data of Section 11.4 is shown in Table 11.7. Age is given first priority because it is one of the easiest pieces of information to obtain about a child, and, furthermore, may be useful as a covariate. The two WISC-R scores are given the same priority two entry level and they compete with each other in a stepwise fashion for entry (see Section 11.5.3). Third priority is assigned the VERBEXP score because it is the least likely to be available without special testing.

At each step, significance of discrimination is assessed and classification functions are provided. Adequacy of classification is shown automatically at the last step but can be requested after any step.

Hierarchical DISCRIM through SPSS^x DISCRIMINANT is shown in Section 11.8.3, Table 11.23. Although classification information is provided only at the last step, other useful information, unavailable in other programs, is given at each step. Particularly handy for stepdown analysis is the "change in Rao's V" at every step, as discussed in Section 11.6.1.1.[7]

In SPSS, hierarchical DISCRIM differs from hierarchical multiple regression in the way predictors with the same priority enter. In DISCRIM, predictors with the same priority compete with one another stepwise, and only one predictor enters at each step. In multiple regression, several IVs can enter in a single step. This makes DISCRIM slightly less flexible than multiple regression. If you have only two groups, you might consider performing DISCRIM through regression where the DV is a dichotomous variable representing group membership. If classification is desired, preliminary multiple regression analysis with fully flexible entry of predictors could be followed by DISCRIM to provide classification.

In BMDP, the less flexible method (where predictors with the same priority compete for entry stepwise and only one enters at a time) is used for both hierarchical regression and hierarchical DISCRIM. In SYSTAT MGLH and SAS DISCRIM, hierarchical DISCRIM is accomplished through a series of runs.

11.5.3 Stepwise Discriminant Function Analysis

When the researcher has no reasons for assigning some predictors higher priority than others, statistical criteria can be used to determine order of entry. That is, if a researcher wants a reduced set of predictors but has no preferences among them, stepwise DISCRIM can be used to produce the reduced set. Entry of predictors is

---

[7] In using DISCRIM for stepdown analysis following MANOVA, change in Rao's V can replace stepdown F as the criterion for evaluating significance of successive DVs. Rao's V statistics are available through an option in SPSS^x DISCRIMINANT (see Section 11.6.1).

```
            /PROBLEM    TITLE = 'SMALL SAMPLE HIERARCHICAL ANALYSIS'.
            /INPUT      VARIABLES=6.  FORMAT IS FREE.   FILE='TAPE41'.
            /VARIABLES  NAMES ARE CASENO,GROUP,PERF,INFO,VERBEXP,AGE.
                        LABEL IS CASENO.
                        GROUPING IS GROUP.
            /GROUP      CODES(GROUP) ARE 1,2,3.
                        NAMES(GROUP) ARE MEMORY, PERCEPT, COMMUN.
            /DISC       LEVEL=0,0,2,2,3,1.  FORCE=3.
            /END
```

STEP NUMBER   0

```
  VARIABLE    F TO    FORCE TOLERNCE *   VARIABLE    F TO    FORCE TOLERNCE
              REMOVE  LEVEL            *             ENTER   LEVEL
      DF =  2    7                     *      DF =  2  6
                                       *   3 PERF     0.73      2   1.00000
                                       *   4 INFO     2.18      2   1.00000
                                       *   5 VERBEXP  3.36      3   1.00000
                                       *   6 AGE      0.40      1   1.00000
```

****

STEP NUMBER   1
VARIABLE ENTERED   6 AGE

```
  VARIABLE    F TO    FORCE TOLERNCE *   VARIABLE    F TO    FORCE TOLERNCE
              REMOVE  LEVEL            *             ENTER   LEVEL
      DF =  2    6                     *      DF =  2  5
  6 AGE      0.40      1   1.00000     *   3 PERF     0.31      2   0.64611
                                       *   4 INFO     2.47      2   0.83723
                                       *   5 VERBEXP  3.49      3   0.88503
```

U-STATISTIC(WILKS' LAMBDA) 0.8819122   DEGREES OF FREEDOM   1   2   6
APPROXIMATE F-STATISTIC   0.402   DEGREES OF FREEDOM   2.00   6.00

  F - MATRIX   DEGREES OF FREEDOM =   1   6

```
          MEMORY    PERCEPT
PERCEPT    0.22
COMMUN     0.18     0.80
```

CLASSIFICATION FUNCTIONS

```
          GROUP =   MEMORY        PERCEPT        COMMUN
VARIABLE
  6 AGE            0.79733        0.75728        0.83374

CONSTANT        -30.20116      -27.35104      -32.91961
```

****

STEP NUMBER   2
VARIABLE ENTERED   4 INFO   1

```
  VARIABLE    F TO    FORCE TOLERNCE *   VARIABLE    F TO    FORCE TOLERNCE
              REMOVE  LEVEL            *             ENTER   LEVEL
      DF =  2    5                     *      DF =  2  4
  4 INFO     2.47      2   0.83723     *   3 PERF     9.30      2   0.10809
  6 AGE      0.77      1   0.83723     *   5 VERBEXP 13.13      3   0.20140
```

U-STATISTIC(WILKS' LAMBDA) 0.4435523   DEGREES OF FREEDOM   2   2   6
APPROXIMATE F-STATISTIC   1.254   DEGREES OF FREEDOM   4.00   10.00

  F - MATRIX   DEGREES OF FREEDOM =   2   5

```
          MEMORY    PERCEPT
PERCEPT    2.65
COMMUN     0.59     1.12
```

CLASSIFICATION FUNCTIONS

```
          GROUP =   MEMORY        PERCEPT        COMMUN
VARIABLE
  4 INFO           -0.23089       0.57401        0.12959
  6 AGE            0.82409        0.69075        0.81872

CONSTANT        -30.36978      -28.39318      -32.97273
```

****

TABLE 11.6   (Continued)

```
STEP NUMBER   3
VARIABLE ENTERED   3 PERF

  VARIABLE     F TO   FORCE TOLERNCE *   VARIABLE      F TO   FORCE TOLERNCE
             REMOVE  LEVEL            *                ENTER  LEVEL
        DF =  2    4                  *          DF =  2    3
  3 PERF       9.30    2   0.10809 *   5 VERBEXP     9.74    3   0.19930
  4 INFO      18.01    2   0.14007 *
  6 AGE        0.44    1   0.53300 *

U-STATISTIC(WILKS' LAMBDA) 0.0784812   DEGREES OF FREEDOM   3    2    6
APPROXIMATE F-STATISTIC        3.426   DEGREES OF FREEDOM   6.00    8.00

  F - MATRIX       DEGREES OF FREEDOM =    3    4

            MEMORY   PERCEPT
PERCEPT     13.85
COMMUN       1.78     5.95

CLASSIFICATION FUNCTIONS

         GROUP =   MEMORY      PERCEPT      COMMUN
VARIABLE
  3 PERF          1.65998      0.39091     1.22389
  4 INFO         -7.31838     -1.09502    -5.09593
  6 AGE           0.13461      0.52839     0.31037

CONSTANT        -62.29002    -30.16335   -50.32441
*****************************************************************************

STEP NUMBER   4
VARIABLE ENTERED   5 VERBEXP

  VARIABLE     F TO   FORCE TOLERNCE *   VARIABLE      F TO   FORCE TOLERNCE
             REMOVE  LEVEL            *                ENTER  LEVEL
        DF =  2    3                  *          DF =  2    2
  3 PERF       6.89    2   0.10696 *
  4 INFO      18.84    2   0.08121 *
  5 VERBEXP    9.74    3   0.19930 *
  6 AGE        0.27    1   0.53278 *

U-STATISTIC(WILKS' LAMBDA) 0.0104766   DEGREES OF FREEDOM   4    2    6
APPROXIMATE F-STATISTIC        6.577   DEGREES OF FREEDOM   8.00    6.00

  F - MATRIX       DEGREES OF FREEDOM =    4    3

            MEMORY   PERCEPT
PERCEPT      9.70
COMMUN       7.19     4.57

CLASSIFICATION FUNCTIONS

         GROUP =   MEMORY      PERCEPT      COMMUN
VARIABLE
  3 PERF          1.92420      0.58704     1.36552
  4 INFO        -17.56222     -8.69921   -10.58700
  5 VERBEXP      5.54585      4.11679     2.97278
  6 AGE          0.09872      0.50175     0.29114

CONSTANT       -138.91110    -72.38436   -72.34031

CLASSIFICATION MATRIX

GROUP     PERCENT    NUMBER OF CASES CLASSIFIED INTO GROUP -
          CORRECT
                     MEMORY   PERCEPT   COMMUN
MEMORY    100.0        3         0         0
PERCEPT   100.0        0         3         0
COMMUN    100.0        0         0         3

TOTAL     100.0        3         3         3
```

SUMMARY TABLE

STEP NO.	VARIABLE ENTERED REMOVED	F VALUE TO ENTER REMOVE	NO. OF VARIAB. INCLUDED	U-STATISTIC	APPROXIMATE F-STATISTIC	DEGREES OF FREEDOM	
1	6 AGE	0.402	1	0.8819	0.402	2.0	6.0
2	4 INFO	2.471	2	0.4436	1.254	4.0	10.0
3	3 PERF	9.303	3	0.0785	3.426	6.0	8.0
4	5 VERBEXP	9.737	4	0.0105	6.577	8.0	6.0

determined by user-specified statistical criteria, of which several are available as discussed in Section 11.6.1.2.

Stepwise DISCRIM has the same controversial aspects as stepwise multiple regression (see Section 5.5.3). Order of entry may be dependent on trivial differences in relationships among predictors in the sample that do not reflect population differences. However, if cross-validation is used (cf. Section 11.8) this bias is reduced.

Application of stepwise analysis to our small sample example through SPSS[x] DISCRIMINANT is illustrated in Table 11.8. One of five statistical criteria for entry of predictors was selected for this example. Notice that AGE is dropped as a predictor by the statistical criterion and the analysis has only three steps. Compared with output in Table 11.7 where AGE is included, F for discrimination has improved ($F = 6.58$ with AGE included, $F = 10.67$ without AGE). That is, AGE is worthless as a predictor in this example; its inclusion only decreases within-groups degrees of freedom while not increasing between-groups sums of squares. If this result replicates, AGE should be dropped from consideration when classifying future cases.

Progression of the stepwise analysis is summarized at the bottom of Table 11.8. The summary table is followed by discriminant function coefficients, the loading matrix, and group centroids. Moreover, a great deal of classification information is available, not shown in this segment of the output but discussed in Section 11.6.6.

In BMDP stepwise as well as hierarchical DISCRIM are produced by BMDP7M. Format of output for stepwise analysis is identical to that of hierarchial analysis, as shown in Table 11.7. The user can select among four statistical criteria for entry of predictors. In SAS, stepwise discriminant analysis is provided through a separate program—STEPDISC. Three entry methods are available (cf. Section 11.6.1.2), as well as several additional statistical criteria. SYSTAT does not provide for stepwise DISCRIM.

11.6 SOME IMPORTANT ISSUES

11.6.1 Statistical Inference

Section 11.6.1.1 contains a discussion of criteria for evaluating the overall statistical significance of a set of predictors for predicting group membership. Section 11.6.1.2 summarizes methods for directing the progression of stepwise DISCRIM and statistical criteria for entry of predictors.

11.6.1.1 Criteria for Overall Statistical Significance
Criteria for evaluating overall statistical reliability in DISCRIM are the same as those in MANOVA. The choice between Wilks' Lambda, Roy's gcr, Hotelling's trace, and Pillai's criterion is based on the same considerations as discussed in Section 9.5.1. Different statistics are available in different programs, as noted in Section 11.7.

Two additional statistical criteria, Mahalanobis' D^2 and Rao's V, are especially relevant to stepwise DISCRIM. Mahalanobis' D^2 is based on distance between pairs of group centroids which is then generalizable to distances over multiple pairs of

TABLE 11.8 SETUP AND SELECTED SPSS' DISCRIMINANT OUTPUT FOR STEPWISE DISCRIMINANT FUNCTION ANALYSIS OF DATA IN TABLE 11.1

```
TITLE        SMALL SAMPLE DISCRIMINANT FUNCTION ANALYSIS
FILE HANDLE  TAPE41
DATA LIST    FILE=TAPE41     FREE
             /SUBJNO,GROUP,PERF,INFO,VERBEXP,AGE
DISCRIMINANT GROUPS=GROUP(1,3)/
             VARIABLES=PERF TO AGE/
             METHOD=WILKS
```

PRIOR PROBABILITY FOR EACH GROUP IS   .33333

------------- VARIABLES NOT IN THE ANALYSIS AFTER STEP   0 -------------

| | | MINIMUM | | |
VARIABLE	TOLERANCE	TOLERANCE	F TO ENTER	WILKS' LAMBDA
PERF	1.0000000	1.0000000	.73458	.8033037
INFO	1.0000000	1.0000000	2.1765	.5795455
VERBEXP	1.0000000	1.0000000	3.3600	.4716981
AGE	1.0000000	1.0000000	.40170	.8819122

* *

AT STEP   1, VERBEXP WAS INCLUDED IN THE ANALYSIS.

		DEGREES OF FREEDOM		SIGNIFICANCE	BETWEEN GROUPS
WILKS' LAMBDA	.4716981	1	2	6.0	
EQUIVALENT F	3.3600000	2	2	6.0	.1050

------------- VARIABLES IN THE ANALYSIS AFTER STEP   1 -------------

VARIABLE	TOLERANCE	F TO REMOVE	WILKS' LAMBDA
VERBEXP	1.0000000	3.3600	

------------- VARIABLES NOT IN THE ANALYSIS AFTER STEP   1 -------------

| | | MINIMUM | | |
VARIABLE	TOLERANCE	TOLERANCE	F TO ENTER	WILKS' LAMBDA
PERF	.3709896	.3709896	5.4218	.1488599
INFO	.2019444	.2019444	13.134	.0754301
AGE	.8850270	.8850270	.70229	.3682506
```
```

```
* * * * * * * * * * * * * * * * * * * * * * * * * * * * * * * * * * * *

AT STEP   2, INFO       WAS INCLUDED IN THE ANALYSIS.

                             DEGREES OF FREEDOM   SIGNIFICANCE   BETWEEN GROUPS
WILKS' LAMBDA    .0754301     2     2     6.0        .0072
EQUIVALENT F     6.602645           4    10.0

----------- VARIABLES IN THE ANALYSIS AFTER STEP   2 -----------

VARIABLE   TOLERANCE   F TO REMOVE   WILKS' LAMBDA

INFO       .2019444     13.134        .4716981
VERBEXP    .2019444     16.708        .5795455

----------- VARIABLES NOT IN THE ANALYSIS AFTER STEP   2 -----------

                     MINIMUM
VARIABLE   TOLERANCE  TOLERANCE   F TO ENTER   WILKS' LAMBDA

PERF       .1676353   .0912506     10.231       .0123339
AGE        .8349769   .1905241     .57354       .0586198

* * * * * * * * * * * * * * * * * * * * * * * * * * * * * * * * * * * *

AT STEP   3, PERF       WAS INCLUDED IN THE ANALYSIS.

                             DEGREES OF FREEDOM   SIGNIFICANCE   BETWEEN GROUPS
WILKS' LAMBDA    .0123339     3     2     6.0        .0019
EQUIVALENT F     10.67242           6     8.0

----------- VARIABLES IN THE ANALYSIS AFTER STEP   3 -----------

VARIABLE   TOLERANCE   F TO REMOVE   WILKS' LAMBDA

PERF       .1676353     10.231        .0754301
INFO       .0912506     22.138        .1488599
VERBEXP    .1993789     13.555        .059272

----------- VARIABLES NOT IN THE ANALYSIS AFTER STEP   3 -----------

                     MINIMUM
VARIABLE   TOLERANCE  TOLERANCE   F TO ENTER   WILKS' LAMBDA

AGE        .5327805   .0812130     .26592       .0104766

F LEVEL OR TOLERANCE OR VIN INSUFFICIENT FOR FURTHER COMPUTATION.
```

TABLE 11.8 (Continued)

SUMMARY TABLE

	ACTION	VARS	WILKS'		
STEP	ENTERED REMOVED	IN	LAMBDA	SIG.	LABEL
1	VERBEX	1	.471698	.1050	
2	INFO	2	.075430	.0072	
3	PERF	3	.012334	.0019	

CANONICAL DISCRIMINANT FUNCTIONS

FUNCTION	EIGENVALUE	PERCENT OF VARIANCE	CUMULATIVE PERCENT	CANONICAL CORRELATION	: AFTER FUNCTION	WILKS' LAMBDA	CHI-SQUARED	D.F.	SIGNIFICANCE
					: 0	.0123339	21.977	6	.0012
1*	11.71790	68.55	68.55	.9598805	: 1	.1568607	9.2620	2	.0097
2*	5.37508	31.45	100.00	.9182262	:				

* MARKS THE 2 CANONICAL DISCRIMINANT FUNCTION(S) TO BE USED IN THE REMAINING ANALYSIS.

STANDARDIZED CANONICAL DISCRIMINANT FUNCTION COEFFICIENTS

	FUNC 1	FUNC 2
PERF	1.88474	-1.42704
INFO	-3.30213	-.07042
VERBEXP	1.50580	1.64500

STRUCTURE MATRIX:

POOLED WITHIN-GROUPS CORRELATIONS BETWEEN CANONICAL DISCRIMINANT FUNCTIONS AND DISCRIMINATING VARIABLES VARIABLES ARE ORDERED BY THE FUNCTION WITH LARGEST CORRELATION AND THE MAGNITUDE OF THAT CORRELATION.

	FUNC 1	FUNC 2
INFO	-.23964*	.09887
VERBEXP	.05067	.45030*
AGE	.29955	-.31955*
PERF	.07023	-.18655*

CANONICAL DISCRIMINANT FUNCTIONS EVALUATED AT GROUP MEANS (GROUP CENTROIDS)

GROUP	FUNC 1	FUNC 2
1	3.89650	.44988
2	-2.52348	2.06052
3	-1.37301	-2.51038

groups. Rao's V is a generalized distance measure that attains its largest value when there is greatest overall separation among groups.

These two criteria are available both to direct the progression of stepwise DISCRIM and to evaluate the reliability of a set of predictors to predict group membership. Like Wilks' Lambda, Mahalanobis' D^2 and Rao's V are based on all discriminant functions rather than one. Note that Lambda, D^2, and V are descriptive statistics; they are not, themselves, inferential statistics, although inferential statistics are applied to them.

SAS DISCRIM provides no overall statistical evaluation of DISCRIM. In an SAS STEPDISC run, this information is at the last step after all predictors are forced into the equation.

11.6.1.2 Stepping Methods Related to criteria for statistical inference is the choice among methods to direct the progression of entry of predictors in stepwise DISCRIM. Different methods of progression maximize group differences along different statistical criteria, as indicated in the Purpose column of Table 11.9. The variety of stepping methods, and the computer programs in which they are available, are presented in Table 11.9.

Choice among stepping methods depends on the availability of programs and choice of statistical criterion. If, for example, the statistical criterion is Wilks' Lambda, it is beneficial to choose the stepping method that minimizes Lambda. (In SPSS[x] DISCRIMINANT Lambda is the least expensive method, and is recommended in the absence of contrary reasons.) Or, if the statistical criterion is "change in Rao's V," the obvious choice of stepping method is RAO.

Statistical criteria can also be used to modify stepping. For example, the user

TABLE 11.9 METHODS FOR DIRECTING STEPWISE DISCRIMINANT FUNCTION ANALYSIS

Label	Purpose	Program and option
WILKS	Produces smallest values of Wilks' Lambda (therefore largest multivariate F)	SPSS:WILKS BMDP:METHOD = 2.
MAHAL	Produces largest distance (D^2) for two closest groups	SPSS:MAHAL
MAXMINF	Maximizes the smallest F between pairs of groups	SPSS:MAXMINF
MINRESID	Produces smallest average residual variance $(1 - R^2)$ between variables and pairs of groups	SPSS:MINRESI
RAO	Produces at each step largest increase in distance between groups as measured by Rao's V	SPSS:RAO
F TO ENTER	At each step picks variables with largest F TO ENTER	BMDP:METHOD = 2.
Partial R^2	Specifies the partial R^2 (p) to enter and/or stay at each step	SAS:PR2ENTRY = p, PR2STAY = p
Significance	Specified the significance level (p) to enter and/or stay at each step	SAS:SLENTRY = p, SLSTAY = p
CONTRAST	Stepping procedure based on groups as defined by contrasts	BMDP7M:CONTRAST

can modify minimum F for a predictor to enter, minimum F to avoid removal, and so on. Tolerance (the proportion of variance for a potential predictor that is not already accounted for by predictors in the equation) can also be modified in BMDP, SAS, and SPSS^x stepwise programs. Comparison of programs with respect to these stepwise statistical criteria is provided in Table 11.12.

11.6.2 Number of Discriminant Functions

In DISCRIM with more than two groups, a number of discriminant functions are extracted. The maximum number of functions is the lesser of either degrees of freedom for groups or, as in canonical correlation, principal components analysis and factor analysis, equal to the number of predictors. As in these other analyses, often not all the functions carry worthwhile information. It is frequently the case that the first one or two discriminant functions account for the lion's share of discriminating power, with no additional information forthcoming from the remaining functions.

Many of the programs evaluate successive discriminant functions. For the SPSS^x DISCRIMINANT example of Table 11.4, note that eigenvalues, percents of variance, and canonical correlations are given for each discriminant function for the small sample data of Table 11.1. To the right, with both functions included, the $\chi^2(8)$ of 20.514 indicates highly reliable relationship between groups and predictors. With the first discriminant function removed, there is still reliable relationship between groups and predictors as indicated by the $\chi^2(3) = 8.484$, $p = .037$. This finding indicates that the second discriminant function is also reliable.

How much between-group variability is accounted for by each discriminant function? The eigenvalues associated with discriminant functions indicate the relative proportion of between-group variability accounted for by each function. In the small sample example of Table 11.4, 70.70% of the between-group variability is accounted for by the first discriminant function and 29.30% by the second.

Only SPSS^x DISCRIMINANT offers flexibility with regard to number of discriminant functions. The user can choose the number of functions, critical value for proportion of variance accounted for (with succeeding discriminant functions dropped once that value is exceeded), or the significance level of additional functions.

11.6.3 Interpreting Discriminant Functions

If a primary goal of analysis is to discover and interpret the combinations of predictors (the discriminant functions) that separate groups in various ways, then the next two sections are relevant. Section 11.6.3.1 reveals how groups are spaced out along the various discriminant functions. Section 11.6.3.2 reveals correlations between predictors and the discriminant functions.

11.6.3.1 Discriminant Function Plots Groups are spaced along the various discriminant functions according to their centroids. Recall from Section 11.4.1 that centroids are mean discriminant scores for each group on a function. Discriminant functions form axes and the centroids of the groups are plotted along the axes. If there is a big difference between the centroid of one group and the centroid of another along a discriminant function axis, the discriminant function separates the two groups. If there is not a big distance, the discriminant function does not separate the two groups. Many groups can be plotted along a single axis.

An example of a discriminant function plot is illustrated in Figure 11.1 for the data of Section 11.4. Centroids are obtained from the section called canonical discriminant functions evaluated at group means in Table 11.4.

The plot emphasizes the utility of both discriminant functions in separating the three groups. On the first discriminant function (X axis), the MEMORY group is some distance from the other two groups, but the COMMUNICATION and PERCEPTION groups are close together. On the second function the COMMUNICATION group is far from the MEMORY and PERCEPTION groups. It takes both discriminant functions, then, to separate the three groups from each other.

If there are four or more groups and, therefore, more than two reliable discriminant functions, then pairwise plots of axes are used. One discriminant function is the X axis and another is the Y axis. Each group has a centroid for each discriminant function; paired centroids are plotted with respect to their values on the X and Y axes. Because centroids are only plotted pairwise, three signficiant discriminant functions require three plots (function 1 vs. function 2; function 1 vs. function 3; and function 2 vs. function 3) and so on.

BMDP7M and SPSS^x DISCRIMINANT provide a plot of group centroids for the first pair of discriminant functions (called canonical variates in BMDP7M). In SPSS^x DISCRIMINANT, cases as well as means are plotted, making separations among groups harder to see than with simpler plots, but facilitating evaluation of classification. In BMDP7M, simplified plots of group means as well as plots including cases are available.

Plots of centroids on additional pairs of reliable discriminant functions have to

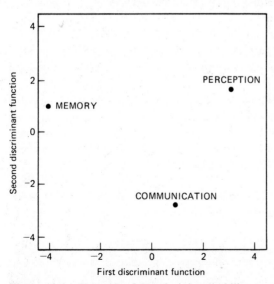

Figure 11.1 Centroids of three learning disability groups on the two discriminant functions derived from sample data of Table 11.1.

be prepared by hand, or discriminant scores can be passed to a "plotting" program such as BMDP6D. Passing discriminant scores to plotting programs is the tack taken by SAS and SYSTAT.

With factorial designs (Section 11.6.5) separate sets of plots are required for each significant main effect and interaction. Main effect plots have the same format as Figure 11.1, with one centroid per group per margin. Interaction plots have as many centroids as cells in the design.

11.6.3.2 Loading Matrices Plots of centroids tell you how groups are separated by a discriminant function, but they do not reveal the meaning of the discriminant function. The meaning of the function is inferred by a researcher from the pattern of correlations between the function and the predictors.[8] Correlations between predictors and functions are called loadings in both discriminant function analysis and factor analysis (see Chapter 12). If predictors X_1, X_2, and X_3 load (correlate) highly with the function but predictors X_4 and X_5 do not, the researcher attempts to understand what X_1, X_2, and X_3 have in common with each other that is different from X_4 and X_5; the meaning of the function is determined by this understanding. (Read Section 12.6.5 for further insights into the art of interpreting loadings.)

Mathematically, the matrix of loadings is the pooled within-group correlation matrix multiplied by the matrix of standardized discriminant function coefficients.[9]

$$A = R_w D \tag{11.7}$$

The loading matrix of correlations between predictors and discriminant functions, **A,** is found by multiplying the matrix of within-group correlations among predictors, R_w, by a matrix of standardized discriminant function coefficients, **D** (standardized using pooled within-group standard deviations).

For the example of Table 11.1, the loading matrix, called POOLED WITHIN-GROUPS CORRELATIONS BETWEEN CANONICAL DISCRIMINANT FUNCTIONS AND DISCRIMINATING VARIABLES by SPSS^x DISCRIMINANT, appears as the middle matrix in Table 11.4.

Loading matrices are read in columns; the column is the discriminant function (FUNC 1 and FUNC 2), the rows are predictors (INFO to AGE), and the entries in the column are correlations. For this example, the first discriminant function correlates most highly with INFO (WISC-R Information scores, $r = .22796$), while the second function correlates most highly with VERBEXP (ITPA Verbal Expression scale, $r = .44630$).

These findings are related to discriminant function plots (e.g., Figure 11.1) for full interpretation. The first discriminant function is largely a measure of INFOrmation,

---

[8] Some researchers interpret standardized discriminant function coefficients, however they suffer from the same difficulties in interpretation as standardized regression coefficients, discussed in Section 5.6.1.

[9] Some texts (e.g., Cooley & Lohnes, 1971) and early versions of SPSS use the total correlation matrix to find standardized coefficients rather than the within-group matrix.

and it separates the group with MEMORY problems from the groups with PER-
CEPTION and COMMUNICATION problems. The second discriminant function is
largely a measure of VERBEXP (verbal expression) and it separates the group with
COMMUNICATION problems from the groups with PERCEPTION and MEMORY
problems. (Interpretation in this example is reasonably straightforward because only
one predictor is highly correlated with each discriminant function; interpretation is
much more interesting when several predictors correlate with a discriminant function.)

Consensus is lacking regarding how high correlations in a loading matrix must
be to be interpreted. By convention, correlations in excess of .30 (9% of variance)
may be considered eligible while lower ones are not. Guidelines suggested by Comrey
(1973) are included in Section 12.6.5. However, the size of loadings depends both
on the value of the correlation in the population and on the homogenity of scores in
the sample taken from it. If the sample is unusually homogeneous with respect to a
predictor, the loadings for the predictor are lower and it may be wise to lower the
criterion for determining whether or not to interpret the predictor as part of a discriminant function.

Caution is always necessary in interpreting loadings, however, because they
are full, not partial or semipartial, correlations. The loading could be substantially
lower if correlations with other predictors were partialed out. For a review of this
material, read Section 5.6.1. Section 11.6.4 deals with methods for interpreting
predictors after variance associated with other predictors is removed, if that is desired.

In some cases, rotation of the loading matrix may facilitate interpretation, as
discussed in Chapter 12. SPSS^x DISCRIMINANT and SPSS^x MANOVA allow rotation
of discriminant functions. But rotation of discriminant function loading matrices is
still considered experimental and not recommended for the novice.

11.6.4 Evaluating Predictor Variables

Another tool for evaluating contribution of predictors to separation of groups is
available through the CONTRAST procedure of BMDP7M. Means for predictors for
each group are contrasted with means for other groups pooled. For instance, if there
are three groups, means on predictors for group 1 are contrasted with pooled means
from groups 2 and 3; then means for group 2 are contrasted with pooled means from
groups 1 and 3; finally means for group 3 are contrasted with pooled means from
groups 1 and 2. This procedure is used to determine which predictors are important
for isolating one group from the rest.

One BMDP7M run is required to isolate the means for each group and contrast
them with means for other groups. Because there are only two groups (one group
versus pooled other groups), there is only one discriminant function per run. F TO
ENTER for each predictor at step 0 is univariate F for testing the reliability of the
mean difference between the group singled out and the other groups. Therefore, at
step 0, F TO ENTER shows how important a predictor is, by itself, in separating
the members of a particular group.

At the last step of each contrast run, with all predictors forced into the analysis,
F TO REMOVE reflects the reduction in prediction that would result if a predictor
were removed from the equation. It is the unique contribution the predictor makes,

in this particular set of predictors, to separation of groups. F TO REMOVE is the importance of a predictor after adjustment for all other predictors in the set.

In order to avoid overinterpretation, it is probably best to consider only predictors with F ratios "significant" after adjusting error for the number of predictors in the set. The adjustment is made on the basis of

$$\alpha = 1 - (1 - \alpha_i)^p \qquad (11.8)$$

Type I error rate (α) for evaluating contribution of p predictors to between-group contrasts is based on the error rate for evaluating each predictor and the number of predictors, $i = 1, 2, \ldots, p$.

F TO REMOVE for each predictor at the last step, then, is evaluated at α_i. Predictors that meet this criterion contribute reliable unique variance to separation of one group from the others. Prediction of group membership is reliably reduced if the predictor is deleted from the equation.

Even with this adjustment, there is danger of inflation of Type I error rate because multiple nonorthogonal contrasts are performed. If there are numerous groups, further adjustment might be considered such as multiplication of critical F by $k - 1$, where $k =$ number of groups. Or interpretation can proceed very cautiously, deemphasizing statistical justification.

As an example of use of F TO ENTER and F TO REMOVE for interpretation, partial output from the three BMDP7M CONTRAST runs for the data of Table 11.1 is shown in Table 11.10. The contrast between the group with MEMORY problems and the other two groups appears in Table 11.10(a), between the group with PERCEPTION problems and the other groups in Table 11.10(b), and between the group with COMMUNICATION problems and the other groups in Table 11.10(c).

With 1 and 3 df at the last step, it is not reasonable to evaluate predictors with respect to statistical significance. However, the pattern of F TO ENTER and F TO REMOVE reveals the predictors most likely to separate each group from the other two. The group with MEMORY deficits is characterized by low scores on WISC-R INFOrmation, as corroborated by the group means in Table 11.10(a). When the group with PERCEPTUAL deficits is contrasted with the other two, there is no clear pattern among the predictors at step 0. At the last step, however, F TO REMOVE for WISC-R PERFormance scores is large and seems to indicate that the PERCEPTION group can be distinguished from the other two groups by this predictor, as shown in Table 11.10(b). For the group with disorders in COMMUNICATION, low ITPA VERBal EXPression scores stand out as predominant for prediction, as seen in Table 11.10(c).

A form of squared semipartial correlation is useful as a measure of strength of association between each predictor and dichotomized group membership. The percent of variance contributed at the last step by reliable predictors is

$$sr_i^2 = \frac{F_i}{df_{res}} (1 - r_c^2) \qquad (11.9)$$

The squared semipartial correlation (sr_i^2) between predictor i and the discriminant function for the difference between two groups (one group vs. all others) is

TABLE 11.10 SETUP AND PARTIAL OUTPUT FOR BMDP7M
CONTRAST RUNS ON SMALL SAMPLE EXAMPLE: (a) MEMORY
VS. OTHER TWO GROUPS; PERCEPTION VS. OTHER TWO
GROUPS; (c) COMMUNICATION VS. OTHER TWO GROUPS

```
                                      ( a )
         /PROBLEM       TITLE = 'SMALL SAMPLE CONTRASTS'.
         /INPUT         VARIABLES ARE 6.  FORMAT IS FREE.  FILE='TAPE41'.
         /VARIABLE      NAMES ARE CASENO,GROUP,PERF,INFO,VERBEXP,AGE.
                        LABEL IS CASENO.
                        GROUPING IS GROUP.
         /GROUP         CODES(GROUP) ARE 1 TO 3.
                        NAMES(GROUP) ARE MEMORY, PERCEPT, COMMUN.
         /DISCRIM       LEVEL=0,0,1,1,1,1.  FORCE=1.
                        CONTRAST=2,-1,-1.
         /PLOT          NO CANON.  CONTRAST.
         /PRINT         NO STEP.
         /END

   MEANS

        GROUP =    MEMORY      PERCEPT     COMMUN      ALL GPS.
VARIABLE
   3 PERF          98.66666    87.66666    101.33330   95.88889
   4 INFO           7.00000    11.66667      9.66667    9.44444
   5 VERBEXP       36.33333    38.33333     28.33333   34.33333
   6 AGE           73.00000    69.33334     76.33334   72.88889
COUNTS                  3.          3.           3.         9.

STEP NUMBER   0

   VARIABLE    F TO   FORCE TOLERNCE *  VARIABLE    F TO   FORCE TOLERNCE
               REMOVE LEVEL            *            ENTER  LEVEL
        DF =  1     7                  *     DF =  1    6
                                       *  3 PERF      0.16    1   1.00000
                                       *  4 INFO      3.56    1   1.00000
                                       *  5 VERBEXP   0.72    1   1.00000
                                       *  6 AGE       0.00    1   1.00000


STEP NUMBER   4
VARIABLE ENTERED   6 AGE

   VARIABLE    F TO   FORCE TOLERNCE *  VARIABLE    F TO   FORCE TOLERNCE
               REMOVE LEVEL            *            ENTER  LEVEL
        DF =  1     3                  *     DF =  1    2
   3 PERF       2.90    1   0.10696 *
   4 INFO      33.37    1   0.08121 *
   5 VERBEXP    2.72    1   0.19930 *
   6 AGE        0.34    1   0.53278 *

EIGENVALUES
             12.98000
CUMULATIVE PROPORTION OF TOTAL DISPERSION
              1.00000
CANONICAL CORRELATIONS
              0.96357



                                      ( b )
         /PROBLEM       TITLE IS 'SMALL SAMPLE CONTRASTS'.
         /INPUT         VARIABLES ARE 6.  FORMAT IS FREE.  FILE='TAPE41'.
         /VARIABLE      NAMES ARE CASENO,GROUP,PERF,INFO,VERBEXP,AGE.
                        LABEL IS CASENO.
                        GROUPING IS GROUP.
         /GROUP         CODES(GROUP) ARE 1 TO 3.
                        NAMES(GROUP) ARE MEMORY, PERCEPT, COMMUN.
         /DISCRIM       LEVEL=0,0,1,1,1,1.  FORCE=1.
                        CONTRAST=-1,2,-1.
         /PLOT          NO CANON.  CONTRAST.
         /PRINT         NO STEP.
         /END


STEP NUMBER   0


   VARIABLE    F TO   FORCE TOLERNCE *  VARIABLE    F TO   FORCE TOLERNCE
               REMOVE LEVEL            *            ENTER  LEVEL
        DF =  1     7                  *     DF =  1    6
                                       *  3 PERF      1.42    1   1.00000
                                       *  4 INFO      2.94    1   1.00000
                                       *  5 VERBEXP   2.88    1   1.00000
                                       *  6 AGE       0.62    1   1.00000
```

TABLE 11.10 (Continued)

```
STEP NUMBER     4
VARIABLE ENTERED    5 VERBEXP
```

VARIABLE	F TO REMOVE	FORCE LEVEL	TOLERNCE	*	VARIABLE	F TO ENTER	FORCE LEVEL	TOLERNCE
DF =	1	3		*	DF =	1	2	
3 PERF	13.23	1	0.10696	*				
4 INFO	3.87	1	0.08121	*				
5 VERBEXP	0.01	1	0.19930	*				
6 AGE	0.51	1	0.53278	*				

```
EIGENVALUES
          9.49094
CUMULATIVE PROPORTION OF TOTAL DISPERSION
          1.00000
CANONICAL CORRELATIONS
          0.95115
```

(c)

```
          /PROBLEM      TITLE = 'SMALL SAMPLE CONTRASTS'.
          /INPUT        VARIABLES ARE 6.  FORMAT IS FREE.  FILE='TAPE41'.
          /VARIABLE     NAMES ARE CASENO,GROUP,PERF,INFO,VERBEXP,AGE.
                        LABEL IS CASENO.
                        GROUPING IS GROUP.
          /GROUP        CODES(GROUP) ARE 1 TO 3.
                        NAMES(GROUP) ARE MEMORY, PERCEPT, COMMUN.
          /DISCRIM      LEVEL=0,0,1,1,1,1.  FORCE=1.
                        CONTRAST=-1,-1,2.
          /PLOT         NO CANON.  CONTRAST.
          /PRINT        NO STEP.
          /END
```

```
STEP NUMBER     0
```

VARIABLE	F TO REMOVE	FORCE LEVEL	TOLERNCE	*	VARIABLE	F TO ENTER	FORCE LEVEL	TOLERNCE
DF =	1	7		*	DF =	1	6	
				*	3 PERF	0.62	1	1.00000
				*	4 INFO	0.03	1	1.00000
				*	5 VERBEXP	6.48	1	1.00000
				*	6 AGE	0.58	1	1.00000

```
STEP NUMBER     4
VARIABLE ENTERED    6 AGE
```

VARIABLE	F TO REMOVE	FORCE LEVEL	TOLERNCE	*	VARIABLE	F TO ENTER	FORCE LEVEL	TOLERNCE
DF =	1	3		*	DF =	1	2	
3 PERF	0.04	1	0.10858	*				
4 INFO	0.68	1	0.08121	*				
5 VERBEXP	12.25	1	0.19930	*				
6 AGE	0.00	1	0.53278	*				

```
EIGENVALUES
          6.14175
CUMULATIVE PROPORTION OF TOTAL DISPERSION
          1.00000
CANONICAL CORRELATIONS
          0.92735
```

calculated from F TO REMOVE for the ith predictor at the last step with all predictors forced, degrees of freedom for error (df_{res}) at the last step, and r_c^2 (CANONICAL CORRELATIONS, squared).

For example, strength of association between INFO and the group with MEMORY problems vs. the other two groups (assuming statistical reliability) is

$$sr^2_{info} = \frac{33.37}{3}(1 - .96357^2) = .7957$$

Almost 80% of the variance in INFO scores overlaps that of the discriminant function that separates the MEMORY group from the PERCEP and COMMUN groups.

The procedures detailed in this section are most useful when the number of groups is small and the separations among groups are fairly uniform on the discriminant function plot for the first two functions. With numerous groups, some closely clustered, other kinds of contrasts might be suggested by the discriminant function plot (e.g., groups 1 and 2 might be pooled and contrasted with pooled groups 3, 4, and 5). Or, with a very large number of groups, the procedures of Section 11.6.3 may suffice.

If there is logical basis for assigning priorities to predictors, a hierarchical rather than standard approach to contrasts can be used. Instead of evaluating each predictor after adjustment for all other predictors, it is evaluated after adjustment by only higher priority predictors. This strategy is accomplished through a series of SPSS^x MANOVA runs, in which stepdown F's (cf. Chapter 9) are evaluated for each contrast.

All the procedures for evaluation of DVs in MANOVA apply to evaluation of predictor variables in DISCRIM. Interpretation of stepdown analysis, univariate F, pooled within-group correlations among predictors, or standardized discriminant function coefficients is as appropriate (or inappropriate) for DISCRIM as for MANOVA. These procedures are summarized in Section 9.5.2.

11.6.5 Design Complexity: Factorial Designs

The notion of placing cases into groups is easily extended to situations where groups are formed by differences on more than one dimension. An illustration of factorial arrangement of groups is the large sample example of Section 9.7, where women are classified by femininity (high or low) and also by masculinity (high or low) on the basis of scores on the Bem Sex Role Inventory (BSRI). Dimensions of femininity and masculinity (each with two levels) are factorially combined to form four groups: high-high, high-low, low-high, low-low. Unless you want to classify cases, factorial designs are best analyzed through MANOVA. If classification is your goal, however, some issues require attention.

As long as sample sizes are equal in all cells, factorial designs are fairly easily analyzed through a factorial DISCRIM program such as BMDP7M. A separate run is required for each main effect and interaction wherein the main effect or interaction is specified by contrast coding. Discriminant and classification functions are produced for the effect during the run.

When sample sizes are unequal in cells, a two-stage analysis is often best. First, questions about reliability of separation of groups by predictors are answered through MANOVA. Second, if classification is desired after MANOVA, it is found through DISCRIM programs.

Formation of groups for DISCRIM depends on the outcome of MANOVA. If the interaction is statistically significant, groups are formed for the cells of the design. That is, in a two-by-two design, four groups are formed and used as the grouping variable in DISCRIM. Note that main effects as well as interactions influence group means (cell means) in this procedure, but for most purposes classification of cases into cells seems reasonable.

If interaction(s) is not statistically reliable, classification is based on significant main effects. For example, in the data of Section 9.7, interaction is not reliable, but both main effect of masculinity and main effect of femininity are reliable. One DISCRIM run is used to produce the classification equations for main effect of masculinity and a second run is used to produce the classification equations for main effect of femininity. That is, classification of main effects is based on marginal groups.

11.6.6 Use of Classification Procedures

The basic technique for classifying cases into groups is outlined in Section 11.4.2. Results of classification are presented in tables such as the CLASSIFICATION MATRIX of BMDP7M (Table 11.3), the CLASSIFICATION RESULTS of SPSS^x (Table 11.4), or NUMBER OF OBSERVATIONS AND PERCENTS CLASSIFIED INTO GROUP of SAS (Table 11.5) where actual group membership is compared to predicted group membership. From these tables, one finds the percent of cases correctly classified and the number and nature of errors of classification.

But how good is the classification? When there are equal numbers of cases in every group, it is easy to determine the percent of cases that should be correctly classified by chance alone to compare to the percent correctly classified by the classification procedure. If there are two equally sized groups, 50% of the cases should be correctly classified by chance alone (cases are randomly assigned into two groups and half of the assignments in each group are correct), while three equally sized groups should produce 33% correct classification by chance, and so forth. However, when there are unequal numbers of cases in the groups, computation of the percent of cases that should be correctly classified by chance alone is a bit more complicated.

The easier way to find it[10] is to first compute the number of cases in each group that should be correct by chance alone and then add across the groups to find the overall expected percent correct. Consider an example where there are 60 cases, 10 in Group 1, 20 in Group 2, and 30 in Group 3. If prior probabilities are specified as .17, .33, and .50, respectively, the programs will assign 10, 20, and 30 cases to the groups. If 10 cases are assigned at random to Group 1, .17 of them (or 1.7) should be correct by chance alone. If 20 cases are randomly assigned to Group 2, .33 (or 6.6) of them should be correct by chance alone, and if 30 cases are assigned to Group 3, .50 of them (or 15) should be correct by chance alone. Adding together 1.7, 6.6, and 15 gives 23.3 cases correct by chance alone, 39% of the total. The percent correct using classification equations has to be substantially larger than the percent expected correct by chance alone if the equations are to be useful.

Some of the computer programs offer sophisticated additional features that are helpful in many classification situations.

11.6.6.1 Cross-Validation and New Cases Classification is based on classification coefficients derived from samples. Because of the coefficients are only estimates of

---

[10] The harder way to find it is to expand the multinomial distribution, a procedure that is more technically correct but produces identical results to those of the simpler method presented here.

population classification coefficients, it is often most desirable to know how well the coefficients generalize to a new sample of cases because they usually work too well for the sample from which they were derived. Testing the utility of coefficients on a sample from which they were not derived is called cross-validation. One form of cross-validation involves dividing a single large sample randomly in two parts, deriving classification functions on one part, and testing them on the other. A second form of cross-validation involves deriving classification functions from a sample measured at one time, and testing them on a sample measured at a later time. In either case, cross-validation techniques are especially well developed in the BMDP7M, SAS DISCRIM, and SPSS^x DISCRIMINANT programs.

For a large sample randomly divided into parts, you simply omit information about actual group membership for some cases (hide it in the program) as shown in Section 11.8. SPSS^x DISCRIMINANT and BMDP7M do not include these cases in the derivation of classification functions, but do include them in the classification phase. In SAS DISCRIM, the withheld cases are put in a separate data file. The accuracy with which the classification functions predict group membership for cases in this data file is then examined. The SAS manual (SAS Institute, 1985, pp. 322–323) provides details for classifying cases in the data file by including "calibration" information for their classification in yet another file.

When the new cases are measured at a later time classifying them is somewhat more complicated unless you use SAS DISCRIM (in the same way that you would for cross-validation). This is because none of the other computer programs for DIS-CRIM allows classification of new cases without repeated entry of the original cases to derive the classification functions. You "hide" the new cases, derive the classification functions from the old cases, and test classification on all cases. Or, you can input the classification coefficients along with raw data for the new cases and run the data only through the classification phase, Or, it may be easiest to write your own program based on the classification coefficients to classify cases as shown in Section 11.4.2.

11.6.6.2 Jackknifed Classification Bias enters classification if the coefficients used to assign a case to a group are derived, in part, from the case. In jackknifed classification, the data from the case are left out when the coefficients used to assign it to a group are computed. Each case has a set of coefficients that are developed from all other cases. Jackknifed classification gives a more realistic estimate of the ability of predictors to separate groups.

BMDP7M provides for jackknifed classification. When the procedure is used with all predictors forced into the equation, bias in classified is eliminated. When it is used with stepwise entry of predictors (where they may not all enter), bias is reduced. An application of jackknifed classification is shown in Section 11.8.

11.6.6.3 Evaluating Improvement in Classification In hierarchical DISCRIM, it is useful to determine if classification improves as a new set of predictors is added to the analysis. McNemar's repeated-measures chi square provides a simple, straight-forward test of improvement. Cases are tabulated one by one, by hand, as to whether

they are correctly or incorrectly classified before the step and after the step where
the predictors are added.

Early step classification

		Correct	Incorrect
Later step classification	Correct	(A)	B
	Incorrect	C	(D)

Cases that have the same result at both steps (either correctly classified—cell A—
or incorrectly classified—cell D) are ignored because they do not change. Therefore,
χ^2 for change is

$$\chi^2 = \frac{(|B - C| - 1)^2}{B + C} \qquad df = 1 \qquad (11.10)$$

Ordinarily, the researcher is only interested in improvement in χ^2, that is, in situations
where $B > C$ because more cases are correctly classified after addition of predictors.
When $B > C$ and χ^2 is greater than 3.84 (critical value of χ^2 with 1 df at $\alpha = .05$),
the added predictors reliably improve classification. This test is applied to the sample
data of Section 11.8.3.

　　With very large samples hand tabulation of cases is not reasonable. An alternative, but possibly less desirable, procedure is to test the significance of the difference
between two lambdas, as suggested by Frane (personal communication). Wilks'
Lambda from the step with the larger number of predictors (Λ_2) is divided by Lambda
from the step with fewer predictors (Λ_1) to produce (Λ_D).

$$\Lambda_D = \frac{\Lambda_2}{\Lambda_1} \qquad (11.11)$$

Wilks' Lambda for testing the significance of the difference between two Lambdas (Λ_D) is calculated by dividing the smaller Lambda (Λ_2) by the larger Lambda
(Λ_1).

　　Λ_D is evaluated with three degree of freedom parameters: p, the number of
predictors after addition of predictors; df_{effect}, the number of groups minus 1; and
the df_{error} at the step with the added predictors. Approximate F is found according
to procedures in Section 11.4.1.

　　For the hierarchical example of Table 11.7, one can test whether addition of
INFO, PERF, and VERBEXP at the last step ($\Lambda_2 = .0105$) reliably improves classification of cases over that achieved at step 1 with AGE in the equation ($\Lambda_1 = .8819$).

$$\Lambda_D = \frac{.0105}{.8819} = .0119$$

where
$$df_p = p = 4$$
$$df_{effect} = k - 1 = 2$$
$$df_{error} = N - 1 = 6$$

From Section 11.4.1,

$$s = \frac{(4)^2(2)^2 - 4}{(4)^2 + (2)^2 - 5} = 2$$

$$y = .0119^{1/2} = .1091$$

$$df_1 = (4)(2) = 8$$

$$df_2 = 2 \qquad 6 - \left[\frac{4 - 2 + 1}{2}\right] - \left[\frac{4(2) - 2}{2}\right] = 6$$

$$\text{Approximate } F(8, 6) = \left(\frac{1 - .1091}{.1091}\right) \left(\frac{6}{8}\right) = 6.124$$

Because critical $F(8, 6)$ is 4.15 at $\alpha = .05$, there is reliable improvement in classification into the three groups when INFO, PERF and VERBEXP scores are added to AGE scores.

11.7 COMPARISON OF PROGRAMS

There are numerous programs for discriminant function analysis in the four packages, some general and some special purpose. Both SPSS^x and BMDP have a general purpose discriminant function analysis program that performs direct, hierarchical, or stepwise DISCRIM with classification. In addition, SPSS^x MANOVA performs DISCRIM, but not classification. SYSTAT has a single, limited, program for direct discriminant analysis, but offers classification through another of its programs. SAS has three programs, one for inference, one for discriminant function coefficients and classification, and a third for stepwise analysis. Finally, if the only question is reliability of predictors to separate groups, any of the MANOVA programs discussed in Chapter 9 is appropriate. Table 11.11 compares features of direct discriminant function programs. Features for stepwise discriminant function are compared in Table 11.12.

11.7.1 SPSS Package

SPSS^x DISCRIMINANT, features of which are described in both Tables 11.11 and 11.12, is the basic program in this package for DISCRIM. The program provides direct (standard), hierarchical, or stepwise entry of predictors with numerous options. Strong points include several types of plots and plenty of information about classi-

TABLE 11.11 COMPARISON OF PROGRAMS FOR DIRECT DISCRIMINANT FUNCTION ANALYSIS

Feature	SAS DISCRIM	SAS CANDISC	SPSSx DISCRIMINANT	SPSSx MANOVA[a]	SYSTAT MGLH	BMDP7M
Input						
Optional matrix input	Yes	Yes	Yes	Yes	Yes	No
Optional prior probabilities	Yes	No	Yes	No	Yes	Yes
Missing data options	No	No	Yes	No	No	No
Threshold for classification	Yes	No	Yes	Yes	No	No
Restrict number of discriminant functions	No	Yes	Yes	Yes	No	CONTRASTS
Factorial arrangement of groups	No	No	No	Yes	No	Yes
Specify tolerance	No	Yes	Yes	No	No[b]	No
Rotation of discriminant functions	No	No	Yes	Yes	No	No
Specify separate covariance matrices for classification	Yes	No	Yes	No	No	No
Output						
Wilks' Lambda with approx. F	No	Yes	Yes	Yes	Yes	Yes
χ^2	No	No	Yes	No	No	No
Generalized distance between groups	Yes	No	No	No	No	No
Hotelling's trace criterion	No	Yes	No	Yes	Yes	No
Roy's gcr	No	Yes	No	Yes	Theta	No
Pillai's criterion	No	Yes	No	Yes	Yes	No
Tests of successive discriminant functions (roots)	No	Yes	Yes	Yes	Yes	No
Univariate F ratios	No	Yes	Yes	Yes	Yes	Yes[c]
Group means	Yes	Yes	Yes	Yes	Yes	Yes
Total and within-group standardized group means	No	Yes	No	No	No	No
Group standard deviations	Yes	Yes	Yes	Yes	No	Yes
Total, within-group and between-group standard deviations	No	Yes	No	No	No	No
Coefficient of variation	No	No	No	No	No	Yes

Feature						
Standardized discriminant function (canonical) coefficients	No	Yes	Yes	Yes	Yes	No
Unstandardized discriminant function (canonical) coefficients	Yes	Yes	Yes	Yes	No	Yes
Group centroids	No	Yes	Yes	No	Yes	Yes
Within-groups cross-products matrix	No	Yes	No	Yes	Yes	No
Between-groups cross-products matrix	No	Yes	No	No	Yes	No
Hypothesis cross-products matrix	No	No	No	Yes	Yes	No
Total cross-products matrix	No	Yes	No	No	Yes	No
Within-groups correlation matrix	No	Yes	Yes	Yes	Yes	Yes
Determinant of within-group correlation matrix	No	No	No	Yes	No	No
Between-groups correlation matrix	No	Yes	No	No	No	No
Group correlation matrices	Yes	No	No	No	No	No
Total correlation matrix	No	Yes	No	No	No	No
Partial correlation matrix	Yes	No	No	No	No	No
Total covariance matrix	No	Yes	Yes	Yes	Yes	No
Pooled within-groups covariance matrix	Yes	Yes	Yes	Yes	Yes	Yes
Group covariance matrices	Yes	No	Yes	No	No	No
Between-group covariance matrix	No	Yes	No	Yes	No	No
Determinants of group covariance matrices	Yes	No	No	Yes	No	No
Equality of group covariance matrices	Yes	No	Yes	Yes	No	No
F matrix, pairwise group comparison	No	Yes	Yes	No[d]	No	Yes
Canonical correlations	No	Yes	Yes	Yes	Yes	Yes
Eigenvalues	No	Yes	Yes	Yes	No	Yes
SMCs for each variable	No	Yes	No	No	Yes	No
Correlations between variables and discriminant functions	No	Yes	Yes	Yes	Yes	No
Classification features						
Classification function coefficient	No	No	Yes	No	No	Yes
Classification of cases	Yes	No	Yes	No	No[e]	Yes
Classification matrix	Yes	No	Yes	No	No[e]	Yes
Probability of cases classification	Yes	No	Yes	No	Yes	Yes
Individual discriminant scores	No	No	Yes	No	Yes	Yes

TABLE 11.11 (Continued)

Feature	SAS DISCRIM	SAS CANDISC	SPSS^x DISCRIMINANT	SPSS^x MANOVA[a]	SYSTAT MGLH	BMDP7M
Mahalanobis' D^2 for cases (outliers)	No	No	No	No	Yes	Yes
Jackknifed classification matrix	No	No	No	No	No	Yes
Classification with a cross-validation sample	Yes	No	Yes	No	No	Yes
All groups plot	No[f]	No	Yes	No	No[b]	Yes
Plot of group centroids alone	No	No	No	No	No	Yes
Separate plots by group	No[f]	No	Yes	No	No[b]	Yes
Territorial map	No	No	Yes	No	No	No

[a] Additional features reviewed in Section 9.6.
[b] Available through SYSTAT FACTOR.
[c] STEP NUMBER 0, F TO ENTER.
[d] Can be obtained through CONTRAST procedure.
[e] Available through SYSTAT TABLES.
[f] Available through SAS PLOT.

fication. Territorial maps are handy for classification using discriminant function scores if there are only a few cases to classify. In addition, a test of homogeneity of covariance is provided through plots and, should heterogeneity be found, appropriate adjustment is available in the classification phase. Other useful features are evaluation of successive discriminant functions and availability of loading matrices.

SPSS^x MANOVA can also be used for DISCRIM and has some features unobtainable in any of the other DISCRIM programs. SPSS^x MANOVA is described rather fully in Table 9.11, but some aspects especially pertinent to DISCRIM are featured in Table 11.11. MANOVA offers a variety of statistical criteria for testing the significance of the set of predictors (cf. Section 11.6.1) and routinely prints loading matrices. Many other matrices can be printed out, and these, along with determinants, are useful for the more sophisticated researcher. Successive discriminant functions (roots) are evaluated, as in SPSS^x DISCRIMINANT.

SPSS^x MANOVA provides discriminant functions for more complex designs such as factorial arrangements with unequal sample sizes. The program is limited, however, in that it includes no classification phase. Further, only standard DISCRIM is available, with no provision for stepwise or hierarchical analysis other than stepdown analysis as described in Chapter 9. Updated versions of SPSS programs (Hull and Nie, 1981) share features with SPSS^x DISCRIMINANT and MANOVA.

11.7.2 BMD Series

BMDP7M deals well with all varieties of DISCRIM. Although designed as a stepwise program, direct DISCRIM is easily produced by forcing entry of all predictors and suppressing all steps except the last.

BMDP7M has one very substantial advantage over most other DISCRIM programs, namely, identification of outliers through the case classification procedure. Further, the JACKKNIFED CLASSIFICATION feature in BMDP7M assigns a case to a group without using the case in developing the classification coefficients for it. This eliminates or reduces bias in classification, as discussed in Section 11.6.6.2.

The BMDP7M provision of stepping by contrasts allows analysis of factorial designs as long as sample sizes are equal in cells (Section 11.6.5). Because this program allows for classification as well as statistical inference, it it the program of choice for equal-n factorial designs. (Unequal-n factorial designs require contrast coefficients that reflect choice of adjustment procedure and some considerable sophistication on the part of the user.)

11.7.3 SYSTAT System

SYSTAT MGLH offers direct DISCRIM as well as ANOVA, MANOVA, regression, and canonical correlation. It is a rich program with respect to supplemental matrices but lacks some features for DISCRIM. For example, there is no provision for stepwise or hierarchical analysis, although these analyses can be simulated through separate runs at each step. And although classification is not provided directly, discriminant scores can be saved and tabulated or plotted through the TABLES or GRAPH programs to classify cases, as fully described in the SYSTAT manual. SYSTAT MGLH can

TABLE 11.12 COMPARISON OF PROGRAMS FOR STEPWISE AND HIERARCHICAL DISCRIMINANT FUNCTION ANALYSIS

Feature	SPSS[x] DISCRIMINANT	BMDP7M	SAS STEPDISC
Input			
Optional matrix input	Yes	No	Yes
Prior probabilities optional	Yes	Yes	No[a]
Missing data options	Yes	No	No
Specify contrast	No	Yes	No
Factorial arrangement of groups	No	CONTRASTS	No
Suppress intermediate steps	No	Yes	No
Suppress all but summary table	Yes	Yes	Yes
Optional methods for order of entry and removal	3	3	5
Forced entry by level (hierarchical)	Yes	Yes	No
Specify tolerance	Yes	Yes	Yes
Specify maximum number of steps	Yes	Yes	Yes
Specify F to enter and remove	Yes	Yes	No
Specify significance of F to enter and remove	Yes	No	Yes
Specify partial R^2 to enter and remove	No	No	Yes
Restrict number of discriminant functions	Yes	No	No[b]
Rotation of discriminant functions	Yes	No	No
Specify separate covariance matrices for classification	Yes	No	No[a]
Output			
Wilks' Lambda with approximate F	Yes	Yes	Yes
χ^2	Yes	No	No
Mahalanobis' D^2 (between groups)	Yes[c]	No	No[b]
Rao's V	Yes[c]	No	No
Pillai's criterion	No	No	Yes
Tests of successive discriminant functions (roots)	Yes	No	No[b]
Univariate F ratios	Yes	Yes[d]	Yes[e]
Group means	Yes	Yes	Yes
Within-group and total standardized group means	No	No	Yes
Group standard deviations	Yes	Yes	No
Total and pooled within-group standard deviations	No	No	Yes

Coefficients of variation	No	Yes	No
Classification function coefficients	Yes	Yes	No[a]
Standardized discriminant function (canonical) coefficients	Yes	No	No[b]
Unstandardized discriminant function (canonical) coefficients	Yes	Yes	No[a,b]
Group centroids	Yes	No	Yes
Within-group correlation matrix	Yes	Yes	Yes
Determinant of within-group correlation matrix	No	Yes	No
Total correlation matrix	No	No	Yes
Total covariance matrix	Yes	No	No[b]
Pooled within-group covariance matrix	Yes	Yes	No[a,b]
Group covariance matrices	Yes	No	No[a]
Equality of group covariance matrices	Yes	Yes	No[a]
F matrix, pairwise group comparison	Yes	Yes	No[b]
Canonical correlations, each discriminant function	Yes	Yes	No[b]
Canonical correlations, average	No	No	Yes
Eigenvalues	Yes	Yes	No[b]
Correlations between variables and discriminant functions	Yes	No	No[b]
Partial R^2 to enter and remove, each step	No	No	Yes
Classification features			
Classification of cases	Yes	Yes	No[a]
Classification matrix	Yes	Yes	No[a]
Probability of cases classification	Yes	Yes	No[a]
Individual discriminant (canonical variate) scores	Yes	Yes	No
Mahalanobis' D^2 for cases (outliers)	No	Yes	No
Jackknifed classification matrix	No	Yes	No
Classification with a cross-validation sample	Yes	Yes	No
Classification information at each step	No	Yes	No
All groups plot	Yes	Yes	No
Plot of group centroids alone	No	Yes	No
Separate plots by group	Yes	Yes	No
Territorial map	Yes	No	No

[a] Available through SAS DISCRIM.
[b] Available through SAS CANDISC.
[c] Available as a stepping option.
[d] STEP 0, F TO ENTER.
[e] STEP 1 F.

be used to assess outliers through Mahalanobis distance of each case to its group centroid; these distances are saved along with discriminant scores given the appropriate instructions.

11.7.4 SAS System

In SAS, there are three separate programs to deal with different aspects of discriminant analysis, with surprisingly little overlap among them. Basic discriminant function information is available through CANDISC, which analyzes roots sequentially and gives alternative inferential statistics for the analysis. This is the program that provides canonical correlations, eigenvalues, and the loading matrix.

For unstandardized discriminant function coefficients and for classification, one uses DISCRIM. This program shows relevant correlation and covariance matrices, but is devoid of inferential statistics. This program is, however, especially handy for classifying new cases or performing cross-validation (Section 11.6.6.1).

Finally, stepwise (but not hierarchical) analysis is accomplished through STEP-DISC. As seen in Table 11.12, very few additional amenities are available in this program. There is no classification, nor is there information about the discriminant functions. On the other hand, this program offers more options for entry and removal of predictors than any of the other programs reviewed.

11.8 COMPLETE EXAMPLES OF DISCRIMINANT FUNCTION ANALYSIS

Examples of discriminant function analysis in this section explore how role-dissatisfied housewives, role-satisfied housewives, and employed women differ in attitudes and demography. The sample of 465 women is described in Appendix B, Section B.1. The grouping variable is role-dissatisfied housewives (UNHOUSE), role-satisfied housewives (HAPHOUSE), and working women (WORKING).

Predictors are internal vs. external locus of control (CONTROL), satisfaction with current marital status (ATTMAR), attitude toward women's role (ATTROLE), and attitude toward homemaking (ATTHOUSE). A fifth attitudinal variable, attitude toward paid work, was dropped from analysis because data were available only for women who had been employed within the past five years and use of this predictor would have involved nonrandom missing values (cf. Chapter 4). The first example of DISCRIM, then, involves prediction of group membership from the four attitudinal variables.

The second example is a hierarchical discriminant function analysis in which demographic variables are entered first, followed by the four attitudinal variables. Demographic predictors are marital status (MARITAL), whether or not the women have had children (CHILDREN), religious affiliation (RELIGION), RACE, socio-economic level (SEL), AGE, and years of education (EDUC).

The first discriminant function analysis allows us to evaluate the distinctions among the three groups on the basis of attitudes alone, while the second explores the prediction available from demography alone, followed by demography plus attitudes. We explore the dimensions on which the groups differ, the predictors contributing to differences among groups on these dimensions, and the degree to which we can accurately classify members into their own groups. We also evaluate efficiency of classification with a cross-validation sample.

11.8.1 Evaluation of Assumptions

The data are first evaluated with respect to practical limitations of DISCRIM.

11.8.1.1 Unequal Sample Sizes and Missing Data In a screening run through BMDP7D (cf. Chapters 4 or 8), seven cases had missing values among the four attitudinal predictors and twenty-six cases showed data missing among the fourteen demographic and attitudinal predictors. For both sets of predictors, missing data were scattered over predictors and groups in apparently random fashion, so that deletion of the cases was deemed appropriate.[11] For the direct analysis of attitudinal variables, the full data set includes 458 cases once cases with missing values are deleted.

The BMDP7D run does not reveal whether any cases have data missing on more than one predictor but indicates that, at worst, there are 439 (465 − 26) cases available for further evaluation in the set with both demographic and attitudinal predictors. During the search for multivariate outliers, one case is found to have missing data on two predictors, so the sample size is 440 for the hierarchical analysis once cases with missing values are deleted.

During classification, unequal sample sizes are used to modify the probabilities with which cases are classified into groups. Because the sample is randomly drawn from the population of interest, sample sizes in groups are believed to represent some real process in the population that should be reflected in classification. For example, knowledge that over half the women are employed implies that greater weight should be given the WORKING group.

11.8.1.2 Multivariate Normality After deletion of cases with missing data, there are still over 70 cases per group. Although a BMDP7D run (not shown) reveals skewness in ATTMAR, sample sizes are large enough to suggest normality of sampling distributions of means. Therefore there is no reason to expect distortion of results due to failure of multivariate normality.

11.8.1.3 Linearity Some of the demographic predictors (e.g., RELIGION) initially had several categories. Because the coding schemes are arbitrary, there is no reason to expect the numbers representing the codes to be linearly related to other predictors. Therefore, dummy variable coding is used to render the demographic predictors dichotomous. MARITAL, for instance, initially coded into single, married, and broken, is recoded into two dummy variables: single vs. other (SINGLE), and married vs. other (MARRIED).[12] Similarly, RELIGION, initially coded as None-or-other, Catholic, Protestant, and Jewish, is recoded as None-or-other vs. all others (NON-

---

[11] Alternative strategies for dealing with missing data are discussed in Chapter 4.

[12] A variety of sources discuss dummy variable coding (e.g. Cohen & Cohen, 1975).

OTHER), Catholic vs. all others (CATHOLIC), and Protestant vs. all others (PRO-TEST). Recoding results in a total of 10 demographic predictors: CHILDREN, RACE (dichotomous), SEL, AGE, EDUC, SINGLE, MARRIED, CATHOLIC, PROTES-TANT, and NONOTHER.

Although ATTMAR and RACE are skewed, there is no expectation of cur-vilinearity between these two and the remaining predictors. As a dichotomous variable, RACE cannot cause curvilinearity. And, at worst, ATTMAR in conjunction with the remaining continuous, well-behaved predictors may contribute to a mild reduction in association.

11.8.1.4 Outliers To identify univariate outliers, z scores associated with minimum and maximum values on each of the 14 predictors are investigated through BMDP7D for each group separately. There are some questionable values on EDUC (one woman reports 24 years of schooling and another 4 years), and ATTHOUSE, with a few exceptionally positive scores. These values are about 4.5 standard deviations from their group means, but still plausible in a sample of this size. So, the cases are retained unless they prove to be multivariate outliers.

BMDP7M is used to search for multivariate outliers in each group separately. Two runs are necessary, one with all 14 predictors and a second with only the 4 attitudinal predictors. A portion of BMDP7M output for the WORKING group with 14 predictors is shown in Table 11.13. Outliers are identified as cases with too large a Mahalanobis D^2, evaluated as χ^2 with degrees of freedom equal to the number of predictors. Critical χ^2 with 14 df at $\alpha = .001$ is 36.12; any case with $D^2 > 36.12$ is an outlier. In Table 11.13, case 102 is identified as an outlier in the group of WORKING women. Four outliers are found in the run with 14 predictors. Two outliers are found in the run with 4 predictors.

The four outliers in the run with 14 predictors are examined through BMDP9R as discussed in Chapter 4. Two of the cases are the same cases that have extreme univariate scores on ATTHOUSE. (These are the two outliers in the run with four predictors, as well.) The other two multivariate outliers are nonwhite women, both single. Because transformation is inappropriate for RACE and questionable for ATTHOUSE (where it seems unreasonable to transform the predictor for only two cases) it is decided to delete multivariate outliers for both the direct and the hierarchical runs.

Therefore, for the direct discriminant function, of the original 465 cases, 7 are lost due to missing values and 2 are multivariate outliers, leaving a total of 456 cases for analysis. For the hierarchical analysis, of the original 465 cases, 25 are lost due to missing values and 4 are multivariate outliers, leaving a total of 436 cases for analysis.

11.8.1.5 Homogeneity of Variance-Covariance Matrices Examination of sample variances (through standard deviations produced by the BMDP7M run shown partially in Table 11.13) for the 14 predictors reveals no gross discrepancies among the groups. MARRIED shows the largest ratio of variances, where the ratio is 6.8 to 1 for the WORKING vs. HAPHOUSE groups. There is more variance in marital status for working women than for role-satisfied housewives. Sample sizes in the groups are

TABLE 11.13 IDENTIFICATION OF OUTLIERS AMONG SET OF 14 VARIABLES. SETUP AND SELECTED OUTPUT FROM BMDP7M

```
/PROBLEM    TITLE IS 'BMDP7M OUTLIER RUN, ALL 14 VARIABLES'.
/INPUT      VARIABLES ARE 13.  FORMAT IS '(A3,12F4.0)'.
            FILE=DISCRIM.
/VARIABLES  NAMES ARE CASESEQ,WORKSTAT,MARITAL,CHILDREN,RELIGION,RACE,
            CONTROL,ATTMAR,ATTROLE,SEL,ATTHOUSE,AGE,EDUC,
            NONOTHER,CATHOLIC,PROTEST,SINGLE,MARRIED.
            ADD=5.
            GROUPING IS WORKSTAT.
            MISSING = 2*0, 3*9, 5*0, 1, 2*0.
            LABEL IS CASESEQ.
/TRANSFORM  NONOTHER=RELIGION EQ 1.   CATHOLIC=RELIGION EQ 2.
            PROTEST=RELIGION EQ 3.
            SINGLE=MARITAL EQ 1.      MARRIED=MARITAL EQ 2.
/GROUP      CODES(WORKSTAT) ARE 1 TO 3.
            NAMES(WORKSTAT) ARE WORKING,HAPHOUSE,UNHOUSE.
/DISC       LEVEL=0,0,0,1,0,1,1,1,1,1,1,1,1,1,1,1,1,1,1.
            FORCE=1.  TOL=.001.
/PRINT      NO STEP.
/END
```

	INCORRECT CLASSIFICATIONS		MAHALANOBIS D-SQUARE FROM AND POSTERIOR PROBABILITY FOR GROUP -		
GROUP	WORKING		WORKING	HAPHOUSE	UNHOUSE
CASE					
87	107	HAPHOUSE	7.2 0.287	5.1 0.463	5.4 0.250
88	108		25.7 0.736	27.6 0.156	27.3 0.108
90	110		14.5 0.611	15.8 0.177	14.5 0.212
91	111		10.5 0.652	12.6 0.128	10.5 0.220
93	113		9.7 0.589	9.5 0.351	12.1 0.060
98	118		23.8 0.760	25.7 0.163	26.3 0.077
101	121		21.3 0.643	21.9 0.258	22.9 0.099
102	122		38.5 0.831	42.4 0.066	40.5 0.103
103	123		18.2 0.663	19.6 0.178	18.9 0.159
104	124		32.6 0.899	37.5 0.043	35.9 0.058
105	125		13.5 0.752	15.3 0.167	15.9 0.080
107	127	HAPHOUSE	9.1 0.307	6.9 0.497	7.8 0.196
109	129		14.7 0.724	16.3 0.178	16.5 0.097
112	132	HAPHOUSE	13.6 0.428	12.3 0.444	13.8 0.128

fairly discrepant, with almsot 3.5 times as many working women as role-dissatisfied housewives, but with two-tailed tests and reasonable homogeneity of variance, DIS-CRIM is robust enough to handle the discrepancies.

Homogeneity of variance-covariance matrices is also examined through plots of the first two discriminant functions (called canonical variables) produced by BMDP7M. As seen in Figure 11.2, the spread of cases for the three groups is relatively equal. Therefore no further test of homogeneity of variance-covariance matrices is necessary.

11.8.1.6 Multicollinearity and Singularity Because BMDP7M, used for the major analysis, protects against multicollinearity through checks of tolerance, no formal evaluation is necessary (cf. Chapter 4).

11.8.2 Direct Discriminant Function Analysis

Direct DISCRIM is performed through BMDP7M with the 4 attitudinal predictors all forced into the equation. The program instructions and a segment of the output

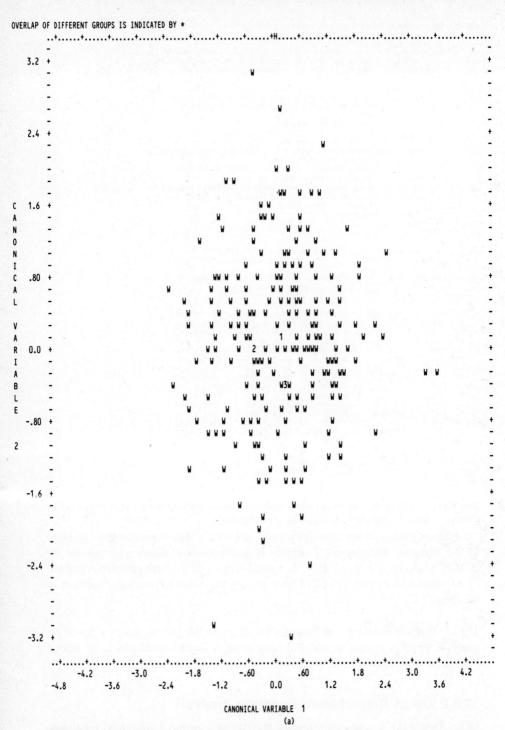

Figure 11.2 Scatterplots of cases on first two canonical variates for (a) working women, (b) role-satisfied housewives, and (c) role-dissatisfied housewives.

OVERLAP OF DIFFERENT GROUPS IS INDICATED BY *

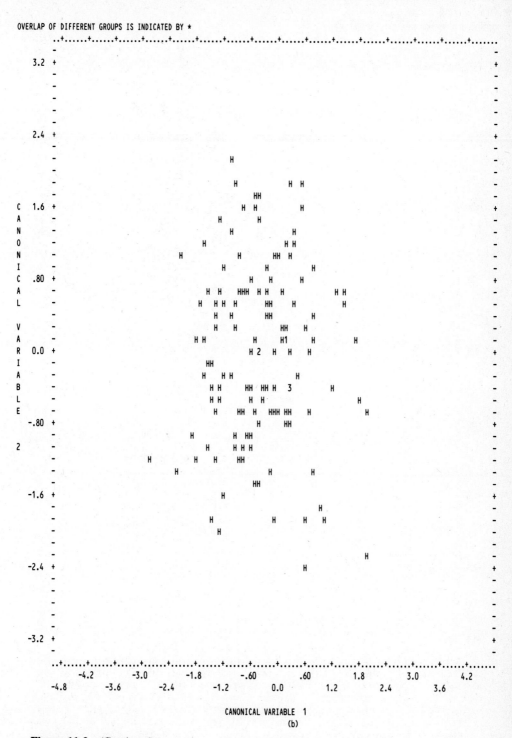

CANONICAL VARIABLE 1

(b)

Figure 11.2 (Continued)

OVERLAP OF DIFFERENT GROUPS IS INDICATED BY *

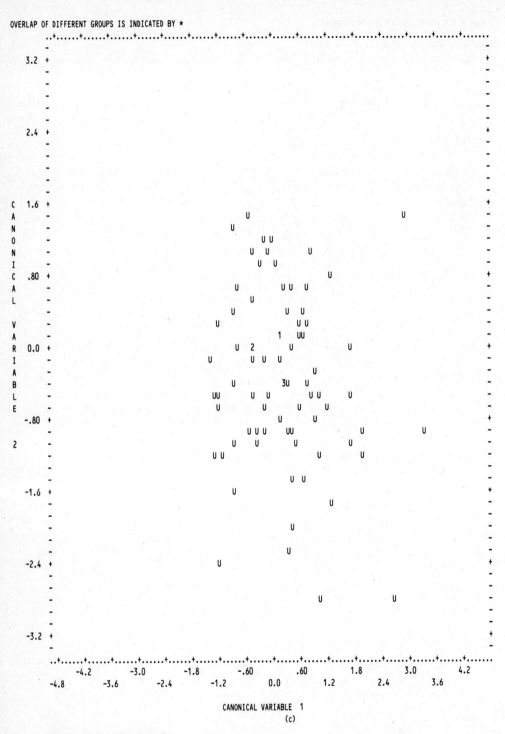

CANONICAL VARIABLE 1

(c)

Figure 11.2 (Continued)

appear in Table 11.14. Because this is a direct analysis, stepwise output is suppressed through NO STEP.

In Table 11.14 the F TO ENTER values at step 0 are univariate F ratios with 2 and 453 df for the individual predictors. Three of the four predictors, all except CONTROL, show univariate F ratios that are significant at the .01 level. When all 4 predictors are used, the approximate F of 6.274 (with 8 and 900 df based on Wilks' Lambda) is highly significant. That is, there is reliable separation of the three groups based on all four predictors combined, as discussed in Section 11.6.1.1.

Table 11.14 also shows the classification functions used to classify cases into the three groups (see Equation 11.3) and the results of that classification, with and without jackknifing (see Section 11.6.6). In this case, classification is made on the basis of a modified equation in which unequal prior probabilities are set to reflect unequal group sizes by use of the PRIOR convention in the control language.

A total of 54.4% of cases are correctly classified by normal procedures, 53.5% by jackknifed procedures. How do these compare with random assignment? Prior probabilities, specified as .52 (WORKING), .30 (HAPHOUSE), and .18 (UNHOUSE), put 237 cases (.52 × 456) in the WORKING group, 137 in the HAPHOUSE group, and 82 in the UNHOUSE group. Of those randomly assigned to the WORKING group, 123 (.52 × 237) should be correct, while 41.1 (.30 × 137) and 14.8 (.18 × 82) should be correct by chance in the HAPHOUSE and UNHOUSE groups, respectively. Over all three groups 178.9 out of the 456 cases or 39% should be correct by chance alone. Both classification procedures correctly classify substantially more than that.

Although BMDP7D does not provide tests of the discriminant functions, it does provide considerable information about them as seen near the end of Table 11.14. Canonical correlations for each discriminant function (.26699 and .18437), although small, are relatively equal for the two discriminant functions. CANONICAL VARIABLES EVALUATED AT GROUP MEANS are centroids on the discriminant functions for the groups, discussed in Sections 11.4.1 and 11.6.3.1.

An additional BMDP7M run for cross-validation is shown in Table 11.15. Approximately 25% of the original cases are randomly selected out in the TRANSFORM paragraph to use for cross-validation. The 111 cases randomly selected out are used to assess the results of classification. New group membership is designated in the GROUP paragraph for these cases: NEWWORK (for WORKING), NEWHAP (for HAPHOUSE), and NEWUN (for UNHOUSE). The other 345 cases are used, as specified in the GROUP paragraph, to develop discriminant functions and classification equations. Table 11.15 shows results of classification and jackknifed classification for original and cross-validation cases. Unexpectedly, classification improves for the cross-validation sample, although ordinarily it is worse.

SPSS^x DISCRIMINANT is used to test the reliability of the discriminant functions and to provide the loadings for them, as shown in Table 11.16. The total overlap between groups and predictors (AFTER FUNCTION 0) is tested as CHI-SQUARE with 8 df (49.002) and found to be reliable. After the first discriminant function is removed (AFTER FUNCTION 1) CHI-SQUARE with 3 df is 15.614, and still reliable at $p = 0.0014$. This second chi square is a test of the overlap between groups and predictors after the first function is removed. Because there are only two possible

```
            /PROBLEM    TITLE IS 'BMDP7M DISCRIM RUN, DIRECT WITH FOUR VARIABLES'.
            /INPUT      VARIABLES ARE 13.  FORMAT IS '(A3,12F4.0)'.
                        FILE=DISCRIM.
            /VARIABLES  NAMES ARE CASESEQ,WORKSTAT,MARITAL,CHILDREN,RELIGION,RACE,
                        CONTROL,ATTMAR,ATTROLE,SEL,ATTHOUSE,AGE,EDUC.
                        GROUPING IS WORKSTAT.
                        MISSING = 2*0, 3*9, 5*0, 1, 2*0.
                        LABEL IS CASESEQ.
                        USE = 2,7,8,9,11.
            /TRANSFORM  DELETE = 261, 299.
            /GROUP      CODES(WORKSTAT) ARE 1 TO 3.
                        NAMES(WORKSTAT) ARE WORKING,HAPHOUSE,UNHOUSE.
                        PRIOR = 0.52, 0.30, 0.18.
            /DISC       LEVEL = 6*0, 1,1,1,0,1, 2*0.
                        FORCE = 1.  TOL=0.001.
            /PLOT       GROUP=1. GROUP=2. GROUP=3.
            /PRINT      NO STEP.  CORR.
            /END


PRIOR PROBABILITIES. . . . .   0.52000   0.30000   0.18000

NUMBER OF CASES READ. . . . . . . . . . . .      465
   CASES WITH USE SET TO NEGATIVE VALUE . . . .     2
      REMAINING NUMBER OF CASES . . . . . . . .    463
   CASES WITH DATA MISSING OR BEYOND LIMITS . .      7
      REMAINING NUMBER OF CASES . . . . . . . .    456

   MEANS

        GROUP =    WORKING      HAPHOUSE     UNHOUSE      ALL GPS.
VARIABLE
   7 CONTROL       6.71548      6.63235      7.04938      6.75000
   8 ATTMAR       23.39749     20.60294     25.61728     22.95833
   9 ATTROLE      33.86192     37.19118     35.66667     35.17544
  11 ATTHOUSE     23.81172     22.50735     24.92593     23.62061

COUNTS             239.         136.          81.         456.

   STANDARD DEVIATIONS

        GROUP =    WORKING      HAPHOUSE     UNHOUSE      ALL GPS.
VARIABLE
   7 CONTROL       1.23780      1.30984      1.25401      1.26253
   8 ATTMAR        8.53003      6.62350     10.29753      8.36830
   9 ATTROLE       6.95618      6.45843      5.75977      6.61150
  11 ATTHOUSE      4.45544      3.88348      3.95846      4.20608

WITHIN CORRELATION MATRIX


                CONTROL      ATTMAR       ATTROLE      ATTHOUSE


                   7            8            9           11
CONTROL    7    1.00000
ATTMAR     8    0.17169      1.00000
ATTROLE    9    0.00912     -0.07010      1.00000
ATTHOUSE  11    0.15500      0.28229     -0.29145      1.00000

STEP NUMBER   0

  VARIABLE    F TO  FORCE TOLERNCE *   VARIABLE     F TO  FORCE TOLERNCE
            REMOVE LEVEL           *               ENTER LEVEL
     DF =  2  454                  *      DF =  2   453
                                   * .  7 CONTROL    2.96    1   1.00000
                                   * .  8 ATTMAR     9.81    1   1.00000
                                   *    9 ATTROLE   11.26    1   1.00000
                                   *   11 ATTHOUSE   8.91    1   1.00000

*************************************************************************
```

TABLE 11.14 (Continued)

```
STEP NUMBER   4
VARIABLE ENTERED   7 CONTROL

   VARIABLE    F TO   FORCE TOLERNCE *   VARIABLE    F TO   FORCE TOLERNCE
              REMOVE LEVEL            *             ENTER  LEVEL
        DF =  2  450                 *       DF =  2  449
   7 CONTROL   1.08   1   0.95517 *
   8 ATTMAR    4.90   1   0.90351 *
   9 ATTROLE   9.31   1   0.91201 *
  11 ATTHOUSE  3.22   1   0.83315 *

U-STATISTIC(WILKS' LAMBDA) 0.8971503  DEGREES OF FREEDOM   4   2   453
APPROXIMATE F-STATISTIC         6.274  DEGREES OF FREEDOM   8.00   900.00

 F - MATRIX        DEGREES OF FREEDOM =    4   450

            WORKING  HAPHOUSE
HAPHOUSE    7.57
UNHOUSE     4.12     7.30

CLASSIFICATION FUNCTIONS

         GROUP =  WORKING      HAPHOUSE     UNHOUSE
VARIABLE
   7 CONTROL     3.22293      3.21674      3.36966
   8 ATTMAR      0.07664      0.04406      0.09771
   9 ATTROLE     1.08148      1.15040      1.13735
  11 ATTHOUSE    1.64843      1.62485      1.71834

CONSTANT       -50.30873    -52.00294    -56.54160

CLASSIFICATION MATRIX
GROUP      PERCENT   NUMBER OF CASES CLASSIFIED INTO GROUP -
           CORRECT
                     WORKING  HAPHOUSE UNHOUSE
WORKING    86.2      206      31       2
HAPHOUSE   27.9      96       38       2
UNHOUSE    4.9       66       11       4

TOTAL      54.4      368      80       8


JACKKNIFED CLASSIFICATION

GROUP      PERCENT   NUMBER OF CASES CLASSIFIED INTO GROUP -
           CORRECT
                     WORKING  HAPHOUSE UNHOUSE
WORKING    84.9      203      33       3
HAPHOUSE   27.2      97       37       2
UNHOUSE    4.9       66       11       4

TOTAL      53.5      366      81       9



EIGENVALUES

              0.07675      0.03519

CUMULATIVE PROPORTION OF TOTAL DISPERSION

              0.68566      1.00000

CANONICAL CORRELATIONS

              0.26699      0.18437


GROUP      CANONICAL VARIABLES EVALUATED AT GROUP MEANS
   WORKING    0.14072      0.15053
   HAPHOUSE  -0.41601     -0.05393
   UNHOUSE    0.28328     -0.35361
```

TABLE 11.15 CROSS-VALIDATION OF CLASSIFICATION OF CASES BY FOUR ATTITUDINAL VARIABLES. SETUP AND SELECTED OUTPUT FROM BMDP7M

```
PROGRAM INSTRUCTIONS

/INPUT       VARIABLES ARE 13.  FORMAT IS FREE.
             FILE=DISCRIM.
/VARIABLES   NAMES ARE SUBNO, WORKSTAT, MARITAL, CHILDREN,
             RELIGION, RACE, CONTROL, ATTMAR, ATTROLE, SEL,
             ATTHOUSE, AGE, EDUC.
             USE = 2, 7, 8, 9, 11.
             GROUPING IS WORKSTAT.
             MISSING = 2*0, 3*9, 5*0, 1, 2*0.
             LABEL IS SUBNO.
/TRANSFORM   DELETE = 261, 299.
             IF(RNDU(11738) LE .25) THEN WORKSTAT = WORKSTAT+3.
/GROUP       CODES(WORKSTAT) ARE 1 TO 6.
             NAMES(WORKSTAT) ARE WORKING,HAPHOUSE,UNHOUSE,
                                 NEWWORK,NEWHAP,NEWUN.
             PRIOR = 0.39, 0.225, 0.135, 0.13, 0.075, 0.045.
             USE = 1 TO 3.
/DISC        LEVEL = 6*0, 3*1, 0, 1, 2*0.
             FORCE = 1.  TOL=.001.
/PLOT        NO CANON.
/PRINT       NO STEP.  NO POST.  NO POINT.
/END


CLASSIFICATION FUNCTIONS

          GROUP =   WORKING      HAPHOUSE      UNHOUSE
VARIABLE
   7 CONTROL       2.84787       2.80040       3.03103
   8 ATTMAR        0.09174       0.05442       0.11964
   9 ATTROLE       0.98078       1.04430       1.02213
  11 ATTHOUSE      1.42151       1.40166       1.47110


CONSTANT         -45.18960     -46.41790     -50.85888

CLASSIFICATION MATRIX

GROUP       PERCENT    NUMBER OF CASES CLASSIFIED INTO GROUP -
            CORRECT
                        WORKING   HAPHOUSE  UNHOUSE
WORKING      86.3        158        24         1
HAPHOUSE     25.0         70        24         2
UNHOUSE       7.6         54         7         5
NEWWORK       0.0         49         7         0
NEWHAP        0.0         28        12         0
NEWUN         0.0         12         3         0


TOTAL        54.2        371        77         8

JACKKNIFED CLASSIFICATION

GROUP       PERCENT    NUMBER OF CASES CLASSIFIED INTO GROUP -
            CORRECT
                        WORKING   HAPHOUSE  UNHOUSE
WORKING      85.8        157        25         1
HAPHOUSE     22.9         72        22         2
UNHOUSE       4.5         56         7         3
NEWWORK       0.0         49         7         0
NEWHAP        0.0         28        12         0
NEWUN         0.0         12         3         0


TOTAL        52.8        374        76         6
```

discriminant functions with three groups, this is a test of the second discriminant function.

SPSS^x DISCRIMINANT, unlike BMDP7M, also provides the loadings for the predictors on the discriminant functions. Labeled POOLED WITHIN-GROUP COR-RELATIONS BETWEEN CANONICAL DISCRIMINANT FUNCTIONS AND DIS-CRIMINATING VARIABLES, the first column is the correlations between the predictors and the first discriminant function; the second column is the correlations between the predictors and the second discriminant function. The loadings could be

TABLE 11.16 SETUP AND PARTIAL OUTPUT FROM SPSS' DISCRIMINANT FOR PREDICTION OF MEMBERSHIP IN THREE GROUPS ON THE BASIS OF FOUR ATTITUDINAL VARIABLES

```
TITLE        LARGE SAMPLE DIRECT DISCRIM THROUGH SPSSX
FILE HANDLE  DISCRIM
DATA LIST    FILE=DISCRIM    FREE
             /CASESEQ,WORKSTAT,MARITAL,CHILDREN,RELIGION,RACE,
             CONTROL,ATTMAR,ATTROLE,SEL,ATTHOUSE,AGE,EDUC
MISSING VALUES  CASESEQ,WORKSTAT,RACE,CONTROL,ATTMAR,ATTROLE,
                SEL,AGE,EDUC(0) ATTHOUSE(1) MARITAL TO RELIGION(9)
SELECT IF    (CASESEQ NE 346)
SELECT IF    (CASESEQ NE 407)
DISCRIMINANT GROUPS=WORKSTAT(1,3)/
             VARIABLES=CONTROL ATTMAR ATTROLE ATTHOUSE/
             ANALYSIS=CONTROL ATTMAR ATTROLE ATTHOUSE/
             PRIORS=SIZE
STATISTICS   4
```

PRIOR PROBABILITIES

GROUP	PRIOR	LABEL
1	.52412	
2	.29825	
3	.17763	
TOTAL	1.00000	

CANONICAL DISCRIMINANT FUNCTIONS

FUNCTION	EIGENVALUE	PERCENT OF VARIANCE	CUMULATIVE PERCENT	CANONICAL CORRELATION	: AFTER : FUNCTION	WILKS' LAMBDA	CHI-SQUARED	D.F.	SIGNIFICANCE
					: 0 :	.8971503	49.002	8	.0000
1*	.07675	68.57	68.57	.2699868	: 1 :	.9660094	15.614	3	.0014
2*	.03519	31.43	100.00	.1843655	:				

* MARKS THE 2 CANONICAL DISCRIMINANT FUNCTION(S) TO BE USED IN THE REMAINING ANALYSIS.

STRUCTURE MATRIX:

POOLED WITHIN-GROUPS CORRELATIONS BETWEEN CANONICAL DISCRIMINANT FUNCTIONS AND DISCRIMINATING VARIABLES VARIABLES ARE ORDERED BY THE FUNCTION WITH LARGEST CORRELATION AND THE MAGNITUDE OF THAT CORRELATION.

	FUNC 1	FUNC 2
ATTMAR	.71846*	.32299
ATTHOUSE	.67945*	.33332
ATTROLE	-.63925	.72223*
CONTROL	.28168	.44494*

calculated by hand according to Equation 11.7, but they are more conveniently found through SPSS^x, as shown in Table 11.16. For interpretation of this example, loadings less than .50 are not considered.

A plot of the placement of the centroids for the three group on the two discriminant functions (canonical variables) as axes appears in Figure 11.3. The points that are plotted are given in Table 11.14 as CANONICAL VARIABLES EVALUATED AT GROUP MEANS.

A summary of information appropriate for publication appears in Table 11.17. In the table are the loadings, univariate F for each predictor, and pooled within-group correlations among predictors (as found through SPSS^x DISCRIMINANT).

Contrasts run through BMDP7M are shown in Tables 11.18 to 11.20. In Table 11.18, means on predictors for WORKING women are contrasted with the pooled means for HAPHOUSE and UNHOUSE to determine which predictors distinguish WORKING women from others. Table 11.19 has the HAPHOUSE group contrasted with the other two groups; Table 11.20 shows the UNHOUSE group contrasted with the other two groups. F TO ENTER is the same univariate F ratio as in ANOVA where each predictor is evaluated separately. F TO REMOVE is the contribution each predictor makes to the contrast after adjustment for all other predictors. Critical F TO REMOVE for the four predictors at α_i = .01 (cf. Section 11.6.4) with 1 and 450 df is approximately 6.8.

On the basis of both F TO ENTER and F TO REMOVE, the predictor that most clearly distinguishes the WORKING group from the other two is ATTROLE. The HAPHOUSE group differs from the other two groups on the basis of ATTMAR after adjustment for the remaining predictors; ATTMAR, ATTROLE, and ATTHOUSE reliably separate this group from the other two in ANOVA. The UNHOUSE group

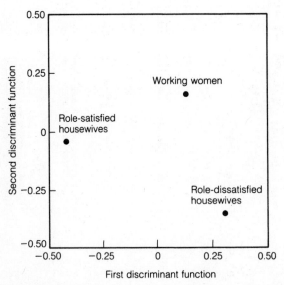

Figure 11.3 Plots of three group centroids on two discriminant functions derived from four attitudinal variables.

TABLE 11.17 RESULTS OF DISCRIMINANT FUNCTION ANALYSIS OF ATTITUDINAL VARIABLES

Predictor variable	Correlations of predictor variables with discriminant functions		Univariate $F(2, 453)$	Pooled within-group correlations among predictors		
	1	2		ATTMAR	ATTROLE	ATTHOUSE
CONTROL	.28	.44	2.96	.17	.01	.16
ATTMAR	.72	.32	9.81		−.07	.28
ATTROLE	−.64	.72	11.26			−.29
ATTHOUSE	.68	.33	8.91			
Canonical R	.27	.18				
Eigenvalue	.08	.04				

TABLE 11.18 SETUP AND PARTIAL OUTPUT OF BMDP7M
CONTRASTING THE WORKING GROUP WITH
THE OTHER TWO GROUPS FOR
ATTITUDINAL VARIABLES

```
          /PROBLEM    TITLE IS 'BMDP7M DIRECT DISCRIM, CONTRASTS'.
          /INPUT      VARIABLES ARE 13.  FORMAT IS '(A3,12F4.0)'.
                      FILE=DISCRIM.
          /VARIABLES NAMES ARE CASESEQ,WORKSTAT,MARITAL,CHILDREN,RELIGION,RACE,
                      CONTROL,ATTMAR,ATTROLE,SEL,ATTHOUSE,AGE,EDUC.
                      GROUPING IS WORKSTAT.
                      MISSING = 2*0, 3*9, 5*0, 1, 2*0.
                      LABEL IS CASESEQ.
                      USE = 2,7,8,9,11.
          /TRANSFORM DELETE = 261, 299.
          /GROUP      CODES(WORKSTAT) ARE 1 TO 3.
                      NAMES(WORKSTAT) ARE WORKING,HAPHOUSE,UNHOUSE.
                      PRIOR = 0.52, 0.30, 0.18.
          /DISC       LEVEL = 6*0, 1,1,1,0,1, 2*0.
                      FORCE = 1.  TOL=0.001.
                      CONTRAST = -2,1,1.
          /PLOT       NO CANON.  CONTRAST.
          /PRINT      NO STEP.  NO POST.
          /END
```

```
********************************************************************************

STEP NUMBER    0

   VARIABLE    F TO   FORCE TOLERNCE *   VARIABLE    F TO   FORCE TOLERNCE
              REMOVE  LEVEL           *              ENTER  LEVEL
      DF =  1  454                    *      DF =  1  453
                                      *    7 CONTROL     1.08     1   1.00000
                                      *    8 ATTMAR      0.13     1   1.00000
                                      *    9 ATTROLE    16.55     1   1.00000
                                      *   11 ATTHOUSE    0.06     1   1.00000

********************************************************************************

STEP NUMBER    4
VARIABLE ENTERED    8 ATTMAR

   VARIABLE    F TO   FORCE TOLERNCE *   VARIABLE    F TO   FORCE TOLERNCE
              REMOVE  LEVEL           *              ENTER  LEVEL
      DF =  1  450                    *      DF =  1  449
    7 CONTROL    0.79     1  0.95517  *
    8 ATTMAR     0.22     1  0.90351  *
    9 ATTROLE   16.87     1  0.91201  *
   11 ATTHOUSE   0.83     1  0.83315  *

EIGENVALUES
                0.04001

CANONICAL CORRELATIONS
                0.19803
```

TABLE 11.19 SETUP AND PARTIAL OUTPUT OF BMDP7M
CONTRASTING THE HAPHOUSE GROUP WITH
THE OTHER TWO GROUPS FOR
ATTITUDINAL VARIABLES

```
          /PROBLEM    TITLE IS 'BMDP7M DIRECT DISCRIM, CONTRASTS'.
          /INPUT      VARIABLES ARE 13.  FORMAT IS '(A3,12F4.0)'.
                      FILE=DISCRIM.
          /VARIABLES NAMES ARE CASESEQ,WORKSTAT,MARITAL,CHILDREN,RELIGION,RACE,
                      CONTROL,ATTMAR,ATTROLE,SEL,ATTHOUSE,AGE,EDUC.
                      GROUPING IS WORKSTAT.
                      MISSING = 2*0, 3*9, 5*0, 1, 2*0.
                      LABEL IS CASESEQ.
                      USE = 2,7,8,9,11.
          /TRANSFORM DELETE = 261, 299.
          /GROUP      CODES(WORKSTAT) ARE 1 TO 3.
                      NAMES(WORKSTAT) ARE WORKING,HAPHOUSE,UNHOUSE.
                      PRIOR = 0.52, 0.30, 0.18.
          /DISC       LEVEL = 6*0, 1,1,1,0,1, 2*0.
                      FORCE = 1.  TOL=0.001.
                      CONTRAST = 1,-2,1.
          /PLOT       NO CANON.  CONTRAST.
          /PRINT      NO STEP.  NO POST.
          /END
```

TABLE 11.19 (Continued)

```
********************************************************************************

STEP NUMBER   0

    VARIABLE     F TO    FORCE TOLERNCE *   VARIABLE     F TO   FORCE TOLERNCE
                REMOVE  LEVEL            *               ENTER  LEVEL
        DF =  1  454                     *       DF =  1  453
                                         *    7 CONTROL     3.42    1   1.00000
                                         *    8 ATTMAR     18.95    1   1.00000
                                         *    9 ATTROLE    11.73    1   1.00000
                                         *   11 ATTHOUSE   17.05    1   1.00000


********************************************************************************

STEP NUMBER   4
VARIABLE ENTERED    7 CONTROL

    VARIABLE     F TO    FORCE TOLERNCE *   VARIABLE     F TO   FORCE TOLERNCE
                REMOVE  LEVEL            *               ENTER  LEVEL
        DF =  1  450                     *       DF =  1  449
    7 CONTROL      0.78    1   0.95517  *
    8 ATTMAR       9.66    1   0.90351  *
    9 ATTROLE      5.45    1   0.91201  *
   11 ATTHOUSE     4.09    1   0.83315  *

EIGENVALUES
                0.07624

CANONICAL CORRELATIONS
                0.26615
```

TABLE 11.20 SETUP AND PARTIAL OUTPUT OF BMDP7M
            CONTRASTING THE UNHOUSE GROUP WITH THE
            OTHER TWO GROUPS FOR ATTITUDINAL VARIABLES

```
            /PROBLEM    TITLE IS 'BMDP7M DIRECT DISCRIM, CONTRASTS'.
            /INPUT      VARIABLES ARE 13.  FORMAT IS '(A3,12F4.0)'.
                        FILE=DISCRIM.
            /VARIABLES  NAMES ARE CASESEQ,WORKSTAT,MARITAL,CHILDREN,RELIGION,RACE,
                        CONTROL,ATTMAR,ATTROLE,SEL,ATTHOUSE,AGE,EDUC.
                        GROUPING IS WORKSTAT.
                        MISSING = 2*0, 3*9, 5*0, 1, 2*0.
                        LABEL IS CASESEQ.
                        USE = 2,7,8,9,11.
            /TRANSFORM  DELETE = 261, 299.
            /GROUP      CODES(WORKSTAT) ARE 1 TO 3.
                        NAMES(WORKSTAT) ARE WORKING,HAPHOUSE,UNHOUSE.
                        PRIOR = 0.52, 0.30, 0.18.
            /DISC       LEVEL = 6*0, 1,1,1,0,1, 2*0.
                        FORCE = 1.  TOL=0.001.
                        CONTRAST = 1,1,-2.
            /PLOT       NO CANON.  CONTRAST.
            /PRINT      NO STEP.  NO POST.
            /END

********************************************************************************

STEP NUMBER   0

    VARIABLE     F TO    FORCE TOLERNCE *   VARIABLE     F TO   FORCE TOLERNCE
                REMOVE  LEVEL            *               ENTER  LEVEL
        DF =  1  454                     *       DF =  1  453
                                         *    7 CONTROL     5.81    1   1.00000
                                         *    8 ATTMAR     12.27    1   1.00000
                                         *    9 ATTROLE     0.03    1   1.00000
                                         *   11 ATTHOUSE   11.58    1   1.00000


********************************************************************************

STEP NUMBER   4
VARIABLE ENTERED    9 ATTROLE

    VARIABLE     F TO    FORCE TOLERNCE *   VARIABLE     F TO   FORCE TOLERNCE
                REMOVE  LEVEL            *               ENTER  LEVEL
        DF =  1  450                     *       DF =  1  449
    7 CONTROL      2.13    1   0.95517  *
    8 ATTMAR       5.56    1   0.90351  *
    9 ATTROLE      1.14    1   0.91201  *
   11 ATTHOUSE     6.20    1   0.83315  *

EIGENVALUES
                0.04909

CANONICAL CORRELATIONS
                0.21633
```

**TABLE 11.21   CHECKLIST FOR DIRECT DISCRIMINANT
FUNCTION ANALYSIS**

1. Issues
   a. Unequal sample sizes and missing data
   b. Normality of sampling distributions
   c. Outliers
   d. Linearity
   e. Homogeneity of variance-covariance matrices
   f. Multicollinearity and singularity
2. Major analysis
   a. Significance of discriminant functions. If significant:
      (1) Variance accounted for
      (2) Plot(s) of discriminant functions
      (3) Loading matrix
   b. Variables separating each group
3. Additional analyses
   a. Group means for high-loading variables ($r > .50$)
   b. Pooled within-group correlations among predictor variables
   c. Classification results
      (1) Jackknifed classification
      (2) Cross-validation
   d. Change in Rao's V (or stepdown F) plus univariate F for predictors

does not differ from the other two when each predictor is adjusted for all others but does differ on ATTHOUSE and ATTMAR without adjustment.

A checklist for a direct discriminant function analysis appears in Table 11.21. It is followed by an example of a Results section, in journal format, for the study just described.

Results

A direct discriminant function analysis was performed using four attitudinal variables as predictors of membership in three groups. Predictors were locus of control, attitude toward marital status, attitude toward role of women, and attitude toward homemaking. Groups were working women, role-satisfied housewives, and role-dissatisfied housewives.

Of the original 465 cases, seven were dropped from analysis because of missing data. Missing data appeared to be randomly scattered throughout groups and predictors. Two additional cases were identified as multivariate outliers with $p < .001$, and were also deleted. Both of the outlying cases were in the working group; they were women with extraordinarily favorable attitudes toward housework. For the remaining 456 cases (239 working women, 136 role-satisfied housewives, and 81 role-dissatisfied housewives), evaluation of assumptions of linearity, normality, multicollinearity or singularity, and homogeneity of variance-covariance matrices revealed no threat to multivariate analysis.

Two discriminant functions were calculated, with a combined $\chi^2(8) = 49.00$, $p < .01$. After removal of the first function, there was still strong association between groups and predictors, $\chi^2(3) = 15.61$, $p < .01$. The two discriminant functions accounted for 69% and 31%, respectively, of the between-group variability. [χ^2 values and percent of variance are from Table 11.16; cf. Section 11.6.2.] As shown in Figure 11.3, the first discriminant function maximally separates role-satisfied housewives from the other two groups. The second discriminant function discriminates role-dissatisfied housewives from working women, with role satisfied housewives falling between these two groups.

Insert Figure 11.3 about here

The loading matrix of correlations between predictors and discriminant functions, as seen in Table 11.17, suggests

that the best predictors for distinguishing
between role-satisfied housewives and the
other two groups (first function) are
attitudes toward current marital status,
toward women's role, and toward homemaking.
Role-satisfied housewives have more
favorable attitudes toward marital status
(mean = 20.60) than working women (mean =
23.40) or role-dissatified housewives
(mean = 25.62), and more conservative
attitudes toward women's role (mean =
37.19) than working women (mean = 33.86) or
dissatisfied housewives (mean = 35.67).
Role-satisfied women are more favorable
toward homemaking (mean = 22.51) than
either working women (mean = 23.81) or role-
dissatisfied housewives (mean = 24.93).
[Group means are shown in Table 11.14.]
Loadings less than .50 are not interpreted.

Insert Table 11.17 about here

One predictor, attitudes toward women's
role, has a loading in excess of .50 on the
second discriminant function, which
separates role-dissatisfied housewives
from working women. Role-dissatisfied
housewives have more conservative attitudes
toward the role of women than working women
(means have already been cited).

Three contrasts were performed where
each group, in turn, was contrasted with the
other two groups, pooled, to determine which
predictors reliably separate each group
from the other two groups. When working
women were contrasted with the pooled groups
of housewives, after adjustment for all
other predictors, and keeping overall $\alpha <$
.05 for the four predictors, only attitude
toward women's role significantly separates
working women from the other two groups,
$F(1, 450) = 16.87$. The squared semipartial

correlation between working women vs. the pooled groups of housewives and attitudes toward women's role is .04. [*Equation 11.9 applied to output in Table 11.18.*] (Attitude toward women's role also separates working women from the other two groups without adjustment for the other predictors.)

Role—satisfied housewives differ from the other two groups on attitudes toward marital status, $F(1, 450) = 9.66$, $p < .05$. The squared semipartial correlation between this predictor and the role—satisfied group vs. the other two groups pooled is 0.02. (Attitudes toward marital status, toward the role of women, and toward housework separate role—satisfied housewives from the other two groups without adjustment for the other predictors.)

The group of role—dissatisfied housewives does not differ from the other two groups on any predictor after adjustment for all other predictors. Without adjustment, however, the group differs on attitudes toward marital status (less favorable) and attitudes toward homemaking (less favorable).

Pooled within—group correlations among the four predictors are shown in Table 11.17. [*Produced by run of Table 11.14 in section of output not shown called WITHIN CORRELATION MATRIX.*] Of the six correlations, four would show statistical significance at $\alpha = .01$ if tested individually. There is a small positive relationship beween locus of control and attitude toward marital status, with $r(454) = .17$, $p < .01$, indicating that women who are more satisfied with their current marital status are less likely to attribute control of reinforcements to external sources. Attitude toward homemaking is positively correlated with locus of control, $r(454) = .16$, $p < .01$, and attitude toward marital status, $r(454) = .28$, $p < .01$, and negatively correlated with attitude toward women's role, $r(454) = -.29$, $p < .01$. This indicates that women with negative attitudes toward homemaking are likely to attribute control to external sources, to be dissatisfied with their current marital status, and to have more liberal attitudes toward women's role.

With the use of a jackknifed classification procedure for the total usable sample of 456 women, 244 (53.5%) were classified correctly, compared to 178.9 (39%) that would be correctly classified by chance alone. The 53.5% classification rate was achieved by classifying a disproportionate number of cases as working women. Although 52% of the women actually were employed, the classification scheme, using sample proportions as prior probabilities, classified 80.3% of the women as employed [*366/456 from jackknifed classification matrix in Table 11.14*]. This means that the working women were more likely to be correctly classified (84.9% correct classifications) than either the role-satisfied housewives (27.2% correct classifications) or the role-dissatisfied housewives (4.9% correct classifications).

The stability of the classification procedure was checked by a cross-validation run. Approximately 25% of the cases were withheld from calculation of the classification functions in this run. For the 75% of the cases from whom the functions were derived, there was a 52.8% correct classification rate. For the cross-validation cases, classification actually improved to 55%. This indicates a high degree of consistency in the classification scheme, and an unusual random division of cases into the cross-validation sample.

11.8.3 Hierarchical Discriminant Function Analysis

Partial output of hierarchical DISCRIM produced by BMDP7M appears in Table 11.22. The 10 demographic predictors (after dummy variable coding) are assigned an entry level of 1; the 4 attitudinal predictors are assigned entry level 2. Both entry levels are then forced to include all predictors into the analysis. The two predictors that were recoded, MARITAL and RELIGION are omitted by assigning them an entry level of 0.

In DISCRIM, order of entry of predictors can be specified, but predictors enter individually on each step. Although 14 steps are produced by this analysis, the only steps shown in Table 11.22 are the two that include entry of all demographic predictors at step 10, and entry of all demographic plus attitudinal predictors at step 14.

Classification equations and results of classification with and without jackknifing are also shown in Table 11.22. These were requested only for steps 10 and 14, as

TABLE 11.22 SETUP AND PARTIAL OUTPUT FROM BMDP7M
HIERARCHICAL DISCRIMINANT FUNCTION
ANALYSIS OF 10 DEMOGRAPHIC AND 4
ATTITUDINAL VARIABLES

```
/PROBLEM     TITLE IS 'BMDP7M DISCRIM, HIERARCHICAL ANALYSIS'.
/INPUT       VARIABLES ARE 13.  FORMAT IS '(A3,12F4.0)'.
             FILE=DISCRIM.
/VARIABLES   NAMES ARE CASESEQ,WORKSTAT,MARITAL,CHILDREN,RELIGION,RACE,
             CONTROL,ATTMAR,ATTROLE,SEL,ATTHOUSE,AGE,EDUC,
             NONOTHER,CATHOLIC,PROTEST,SINGLE,MARRIED.
             ADD=5.   GROUPING IS WORKSTAT.
             MISSING = 2*0, 3*9, 5*0, 1, 2*0.   LABEL IS CASESEQ.
             USE=2, 4, 6 TO 18.
/TRANSFORM   NONOTHER=RELIGION EQ 1.   CATHOLIC=RELIGION EQ 2.
             PROTEST=RELIGION EQ 3.
             SINGLE=MARITAL EQ 1.       MARRIED=MARITAL EQ 2.
             DELETE=102,261,292,299.
/GROUP       CODES(WORKSTAT) ARE 1 TO 3.
             NAMES(WORKSTAT) ARE WORKING,HAPHOUSE,UNHOUSE.
             USE = 1 TO 3.
             PRIOR = 0.53, 0.30, 0.17.
/DISC        LEVEL=0,0,0,1,0,1,2,2,2,1,2,1,1,1,1,1,1,1.
             FORCE=2. TOL=0.001.
/PRINT       CLASS = 10,14.  NO POINT.
/PLOT        GROUP=1. GROUP=2. GROUP=3.
/END
```

MEANS

VARIABLE GROUP =	WORKING	HAPHOUSE	UNHOUSE	ALL GPS.
4 CHILDREN	0.78355	0.92248	0.86842	0.83945
6 RACE	1.10390	1.02326	1.10526	1.08028
7 CONTROL	6.70563	6.68217	7.02632	6.75459
8 ATTMAR	23.36364	20.47287	25.31579	22.84862
9 ATTROLE	33.90476	37.20155	35.61842	35.17890
10 SEL	51.51948	56.58139	49.52632	52.66972
11 ATTHOUSE	23.76623	22.57364	24.92105	23.61468
12 AGE	4.41558	4.52713	4.00000	4.37615
13 EDUC	13.57143	13.06977	12.63158	13.25917
14 NONOTHER	0.19048	0.16279	0.11842	0.16972
15 CATHOLIC	0.25541	0.20930	0.34211	0.25688
16 PROTEST	0.38095	0.37209	0.34211	0.37156
17 SINGLE	0.07792	0.01550	0.05263	0.05505
18 MARRIED	0.70563	0.96899	0.81579	0.80275
COUNTS	231.	129.	76.	436.

TABLE 11.22 (Continued)

STANDARD DEVIATIONS

	GROUP =	WORKING	HAPHOUSE	UNHOUSE	ALL GPS.
VARIABLE					
4	CHILDREN	0.41272	0.26846	0.34028	0.36310
6	RACE	0.30579	0.15130	0.30893	0.27012
7	CONTROL	1.24758	1.31092	1.21077	1.26043
8	ATTMAR	8.51736	6.42149	10.29590	8.31177
9	ATTROLE	6.96198	6.38526	5.84002	6.61103
10	SEL	22.64372	25.01396	23.54824	23.52376
11	ATTHOUSE	4.45039	3.87051	4.01252	4.21162
12	AGE	2.16905	2.09940	2.33238	2.17813
13	EDUC	2.47472	1.95331	1.77270	2.21929
14	NONOTHER	0.39353	0.37061	0.32525	0.37575
15	CATHOLIC	0.43704	0.40840	0.47757	0.43619
16	PROTEST	0.48668	0.48525	0.47757	0.48469
17	SINGLE	0.26863	0.12403	0.22478	0.22722
18	MARRIED	0.45675	0.17401	0.39023	0.38229

****

STEP NUMBER   0

VARIABLE	F TO REMOVE	FORCE LEVEL	TOLERNCE	*	VARIABLE	F TO ENTER	FORCE LEVEL	TOLERNCE
DF =	2	434		*	DF =	2	433	
				*	4 CHILDREN	6.35	1	1.00000
				*	6 RACE	4.08	1	1.00000
				*	7 CONTROL	2.15	2	1.00000
				*	8 ATTMAR	9.06	2	1.00000
				*	9 ATTROLE	10.50	2	1.00000
				*	10 SEL	2.74	1	1.00000
				*	11 ATTHOUSE	7.75	2	1.00000
				*	12 AGE	1.48	1	1.00000
				*	13 EDUC	5.80	1	1.00000
				*	14 NONOTHER	1.08	1	1.00000
				*	15 CATHOLIC	2.22	1	1.00000
				*	16 PROTEST	0.18	1	1.00000
				*	17 SINGLE	3.13	1	1.00000
				*	18 MARRIED	19.70	1	1.00000

****

STEP NUMBER 10
VARIABLE ENTERED 14 NONOTHER

VARIABLE	F TO REMOVE	FORCE LEVEL	TOLERNCE	*	VARIABLE	F TO ENTER	FORCE LEVEL	TOLERNCE
DF =	2	424		*	DF =	2	423	
4 CHILDREN	3.98	1	0.65289	*	7 CONTROL	1.26	2	0.94429
6 RACE	5.08	1	0.89144	*	8 ATTMAR	4.63	2	0.78131
10 SEL	1.76	1	0.81127	*	9 ATTROLE	12.33	2	0.74594
12 AGE	1.89	1	0.81312	*	11 ATTHOUSE	9.33	2	0.91176
13 EDUC	6.92	1	0.81090	*				
14 NONOTHER	0.54	1	0.63267	*				
15 CATHOLIC	1.11	1	0.54203	*				
16 PROTEST	0.63	1	0.53967	*				
17 SINGLE	1.11	1	0.59238	*				
18 MARRIED	12.19	1	0.72781	*				

U-STATISTIC(WILKS' LAMBDA) 0.8346487   DEGREES OF FREEDOM 10   2   433
APPROXIMATE F-STATISTIC   4.010   DEGREES OF FREEDOM 20.00   848.00

F - MATRIX   DEGREES OF FREEDOM =   10 424

	WORKING	HAPHOUSE
HAPHOUSE	6.77	
UNHOUSE	2.33	2.03

TABLE 11.22   (Continued)

CLASSIFICATION FUNCTIONS

GROUP =	WORKING	HAPHOUSE	UNHOUSE
VARIABLE			
4 CHILDREN	5.42126	6.47025	6.29802
6 RACE	20.35967	18.88255	19.74751
10 SEL	0.00778	0.01833	0.01226
12 AGE	0.86404	0.81395	0.72963
13 EDUC	3.31711	3.14008	3.11036
14 NONOTHER	6.14624	5.94448	5.67628
15 CATHOLIC	5.33587	4.84825	5.43543
16 PROTEST	7.47213	7.10998	7.22779
17 SINGLE	11.35422	12.32452	12.04162
18 MARRIED	6.59238	8.35262	7.19469
CONSTANT	-44.07142	-43.18656	-42.58067

CLASSIFICATION MATRIX

GROUP	PERCENT CORRECT	NUMBER OF CASES CLASSIFIED INTO GROUP -		
		WORKING	HAPHOUSE	UNHOUSE
WORKING	78.8	182	45	4
HAPHOUSE	43.4	72	56	1
UNHOUSE	0.0	53	23	0
TOTAL	54.6	307	124	5

JACKKNIFED CLASSIFICATION

GROUP	PERCENT CORRECT	NUMBER OF CASES CLASSIFIED INTO GROUP -		
		WORKING	HAPHOUSE	UNHOUSE
WORKING	75.8	175	51	5
HAPHOUSE	38.8	78	50	1
UNHOUSE	0.0	53	23	0
TOTAL	51.6	306	124	6

****

STEP NUMBER 14
VARIABLE ENTERED   7 CONTROL

VARIABLE	F TO REMOVE	FORCE LEVEL	TOLERNCE	*	VARIABLE	F TO ENTER	FORCE LEVEL	TOLERNCE
DF =	2	420		*	DF =	2	419	
4 CHILDREN	3.54	1	0.63687	*				
6 RACE	6.81	1	0.87555	*				
7 CONTROL	0.34	2	0.90388	*				
8 ATTMAR	1.39	2	0.68524	*				
9 ATTROLE	9.88	2	0.70182	*				
10 SEL	2.80	1	0.76717	*				
11 ATTHOUSE	4.24	2	0.77251	*				
12 AGE	3.45	1	0.73946	*				
13 EDUC	2.84	1	0.72671	*				
14 NONOTHER	0.30	1	0.61914	*				
15 CATHOLIC	2.68	1	0.51352	*				
16 PROTEST	1.82	1	0.51068	*				
17 SINGLE	0.46	1	0.58111	*				
18 MARRIED	8.40	1	0.60400	*				

U-STATISTIC(WILKS' LAMBDA) 0.7561023   DEGREES OF FREEDOM 14   2   433
APPROXIMATE F-STATISTIC   4.501   DEGREES OF FREEDOM 28.00   840.00

F - MATRIX   DEGREES OF FREEDOM =   14   420

	WORKING	HAPHOUSE
HAPHOUSE	7.03	
UNHOUSE	2.62	3.24

TABLE 11.22 (Continued)

CLASSIFICATION FUNCTIONS

	GROUP =	WORKING	HAPHOUSE	UNHOUSE
VARIABLE				
4 CHILDREN		0.86735	1.97436	1.50241
6 RACE		16.54148	14.76095	15.82294
7 CONTROL		4.06134	4.06747	4.15490
8 ATTMAR		0.33115	0.32520	0.36037
9 ATTROLE		1.46804	1.56382	1.52083
10 SEL		0.09652	0.11056	0.10353
11 ATTHOUSE		1.58535	1.54003	1.66019
12 AGE		0.27823	0.13321	0.12847
13 EDUC		4.35369	4.26303	4.18638
14 NONOTHER		10.12893	9.87199	9.82004
15 CATHOLIC		8.89079	8.12568	9.14709
16 PROTEST		9.50582	8.83322	9.36534
17 SINGLE		5.17305	5.83870	5.59402
18 MARRIED		9.94245	11.59567	10.84075
CONSTANT		-111.58170	-113.30400	-115.67680

CLASSIFICATION MATRIX

GROUP	PERCENT CORRECT	NUMBER OF CASES CLASSIFIED INTO GROUP -		
		WORKING	HAPHOUSE	UNHOUSE
WORKING	80.1	185	39	7
HAPHOUSE	54.3	56	70	3
UNHOUSE	11.8	52	15	9
TOTAL	60.6	293	124	19

JACKKNIFED CLASSIFICATION

GROUP	PERCENT CORRECT	NUMBER OF CASES CLASSIFIED INTO GROUP -		
		WORKING	HAPHOUSE	UNHOUSE
WORKING	78.8	182	41	8
HAPHOUSE	51.9	59	67	3
UNHOUSE	7.9	54	16	6
TOTAL	58.5	295	124	17

SUMMARY TABLE

STEP NO.	VARIABLE ENTERED REMOVED	F VALUE TO ENTER REMOVE	NO. OF VARIAB. INCLUDED	U-STATISTIC	APPROXIMATE F-STATISTIC	DEGREES OF FREEDOM	
1	18 MARRIED	19.696	1	0.9166	19.696	2.0	433.0
2	13 EDUC	5.019	2	0.8958	12.218	4.0	864.0
3	6 RACE	5.703	3	0.8727	10.122	6.0	862.0
4	10 SEL	1.699	4	0.8659	8.027	8.0	860.0
5	4 CHILDREN	1.895	5	0.8583	6.813	10.0	858.0
6	12 AGE	2.352	6	0.8489	6.087	12.0	856.0
7	15 CATHOLIC	1.364	7	0.8436	5.416	14.0	854.0
8	17 SINGLE	1.234	8	0.8387	4.896	16.0	852.0
9	16 PROTEST	0.488	9	0.8368	4.401	18.0	850.0
10	14 NONOTHER	0.541	10	0.8346	4.010	20.0	848.0
11	9 ATTROLE	12.326	11	0.7887	4.846	22.0	846.0
12	11 ATTHOUSE	7.022	12	0.7633	5.085	24.0	844.0
13	8 ATTMAR	1.654	13	0.7573	4.828	26.0	842.0
14	7 CONTROL	0.343	14	0.7561	4.501	28.0	840.0

EIGENVALUES

0.23454   0.07131

CUMULATIVE PROPORTION OF TOTAL DISPERSION

0.76685   1.00000

CANONICAL CORRELATIONS

0.43587   0.25800

GROUP	CANONICAL VARIABLES EVALUATED AT GROUP MEANS	
WORKING	0.40039	0.11877
HAPHOUSE	-0.70712	0.12850
UNHOUSE	-0.01673	-0.57911

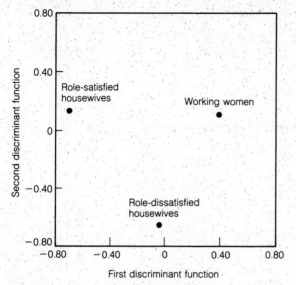

Figure 11.4 Plot of three group centroids on two discriminant functions derived from ten demographic and four attitudinal variables.

specified in the PRINT paragraph. Figure 11.4 shows the centroids for the three groups plotted on the two discriminant functions as axes. The centroids are called CANONICAL VARIABLES EVALUATED AT GROUP MEANS in the BMDP7M run of Table 11.22.

SPSS^x DISCRIMINANT is used to provide inference about the reliability of the two discriminant functions and to provide the loadings of the predictors on them. These two segments of output are shown in Table 11.23.

Finally, in a form that might be suitable for publication, Table 11.24 summarizes information from Tables 11.22 and 11.23; pooled within-group correlations among predictors (produced by SPSS^x DISCRIMINANT) are given in Table 11.25.

Results of contrasts of each group with the other two groups pooled are shown in Tables 11.26, 11.27, and 11.28. Critical F TO REMOVE for evaluating the association between each predictor (adjusted for all other predictors) and groups at α_i = .003 (cf. Equation 11.8) with 1 and 420 df is approximately 9.5.

McNemar's change test (Section 11.6.6.3, Equation 11.10) is used to see if there is improvement in classification when the 4 attitudinal predictors are added to the 10 demographic predictors. A portion of the output used for construction of the contingency table appears in Table 11.29. Cases 26, 40, and 44 are among those tabulated into cell B, incorrectly classified at step 10 but correctly classified at step 14. Case 36, correctly classified at step 10 but incorrectly classified at step 14, is

TABLE 11.23 SETUP AND PARTIAL OUTPUT FROM SPSS' DISCRIMINANT FOR PREDICTION OF MEMBERSHIP IN 3 GROUPS ON THE BASIS OF 10 DEMOGRAPHIC AND 4 ATTITUDINAL VARIABLES

```
TITLE         LARGE SAMPLE DIRECT DISCRIM THROUGH SPSSX
FILE HANDLE   DISCRIM
DATA LIST     FILE=DISCRIM       FREE
              /CASESEQ,WORKSTAT,MARITAL,CHILDREN,RELIGION,RACE,
              CONTROL,ATTMAR,ATTROLE,SEL,ATTHOUSE,AGE,EDUC
MISSING VALUES  CASESEQ,WORKSTAT,RACE TO SEL,AGE,EDUC(0)
                ATTHOUSE(1) MARITAL TO RELIGION(9)

SELECT IF  (CASESEQ NE 122)
SELECT IF  (CASESEQ NE 348)
SELECT IF  (CASESEQ NE 400)
SELECT IF  (CASESEQ NE 407)
COMPUTE  NONOTHER=0
COMPUTE  CATHOLIC=0
COMPUTE  PROTEST=0
COMPUTE  SINGLE=0
COMPUTE  MARRIED=0
IF  (RELIGION EQ 1) NONOTHER=1
IF  (RELIGION EQ 2) CATHOLIC=1
IF  (RELIGION EQ 3) PROTEST=1
IF  (MARITAL EQ 1) SINGLE=1
IF  (MARITAL EQ 2) MARRIED=1
DISCRIMINANT  GROUPS=WORKSTAT(1,3)/
              VARIABLES=MARITAL TO MARRIED/
              ANALYSIS=CHILDREN, RACE TO MARRIED/
              PRIORS=SIZE
```

CANONICAL DISCRIMINANT FUNCTIONS

FUNCTION	EIGENVALUE	PERCENT OF VARIANCE	CUMULATIVE PERCENT	CANONICAL CORRELATION	: : :	AFTER FUNCTION	WILKS' LAMBDA	CHI-SQUARED	D.F.	SIGNIFICANCE
					:	0	.7561023	119.24	28	.0000
1*	.23454	76.69	76.69	.4358687	:	1	.9334384	29.377	13	.0058
2*	.07131	23.31	100.00	.2579954	:					

* MARKS THE   2 CANONICAL DISCRIMINANT FUNCTION(S) TO BE USED IN THE REMAINING ANALYSIS.

STRUCTURE MATRIX:

POOLED WITHIN-GROUPS CORRELATIONS BETWEEN CANONICAL DISCRIMINANT FUNCTIONS AND DISCRIMINATING VARIABLES
VARIABLES ARE ORDERED BY THE FUNCTION WITH LARGEST CORRELATION AND THE MAGNITUDE OF THAT CORRELATION.

	FUNC 1	FUNC 2
MARRIED	.62240*	.04091
ATTROLE	.45116*	.10177
CHILDREN	.34661*	.12777
RACE	-.26808*	.16748
SINGLE	-.24813*	-.01118
ATTHOUSE	-.25091	.54287*
ATTMAR	-.30945	.52150*
EDUC	-.20834	-.48231*
CONTROL	-.01353	.37266*
CATHOLIC	-.09246	.34006*
AGE	.04361	-.29946*
SEL	.19222	-.23630*
NONOTHER	-.06858	-.23379*
PROTEST	-.01742	-.10443*

TABLE 11.24   RESULTS   OF   DISCRIMINANT   FUNCTION
ANALYSIS OF DEMOGRAPHIC PLUS
ATTITUDINAL VARIABLES

Predictor variable	Correlation of predictor variables with discriminant functions		Univariate $F(2, 433)$
	1	2	
Demographic			
CHILDREN	.35	.13	6.35
RACE	−.27	.17	4.08
SEL	.19	−.24	2.74
AGE	.04	−.30	1.48
EDUC	−.21	−.48	5.80
NONOTHER	−.07	−.23	1.08
CATHOLIC	−.09	.34	2.22
PROTEST	−.02	−.10	0.18
SINGLE	−.25	−.01	3.13
MARRIED	.62	.04	19.70
Attitudinal			
CONTROL	−.01	.37	2.15
ATTMAR	−.31	.52	9.06
ATTROLE	.45	.10	10.50
ATTHOUSE	−.25	.54	7.75
Canonical R	.44	.26	
Eigenvalue	.24	.07	

tabulated into cell C. Tabulation of all cases from the two runs produces the following contingency table:

		Step 10 classification	
		Correct	Incorrect
Step 14 classification	Correct	(A)	$B = 60$
	Incorrect	$C = 31$	(D)

Applying Equation 11.10, then, we obtain

$$\chi^2 = \frac{(|60 - 31| - 1)^2}{60 + 31} = 8.62$$

The obtained value of χ^2 (8.62) exceeds the critical value of χ^2 with 1 df at $\alpha = 0.01$. Therefore, addition of attitudinal predictors reliably improves classification over that provided by demographic predictors alone.

TABLE 11.25 POOLED WITHIN-GROUP CORRELATIONS AMONG PREDICTORS FOR DATA OF TABLE 11.24

	Demographic variables									Attitudinal variables			
	RACE	SEL	AGE	EDUC	NON-OTHER	CATHOLIC	PROTEST	SINGLE	MARRIED	CON-TROL	ATTMAR	ATTROLE	ATTHOUSE
Demographic													
CHILDREN	.13	−.07	.36	−.10	−.05	.12	−.02	−.52	.30	−.01	−.05	.13	.05
RACE		−.11	−.06	−.21	−.02	.25	−.12	−.09	.06	.06	−.01	.17	−.07
SEL			.11	.36	.04	−.16	−.02	−.02	.11	−.12	−.01	−.24	.00
AGE				.02	−.08	−.02	.10	−.28	.08	−.11	−.05	.25	−.10
EDUC					.08	−.14	−.05	.16	−.05	−.07	−.06	−.35	.14
NONOTHER						−.26	−.35	.10	−.17	−.05	.03	−.13	.08
CATHOLIC							−.45	−.10	.10	.01	.00	.11	−.15
PROTEST								−.04	.00	−.09	−.06	.14	−.07
SINGLE									−.48	.05	.23	−.06	.02
MARRIED										−.01	−.43	.01	−.02
Attitudinal													
CONTROL											.18	.00	.15
ATTMAR												−.06	.28
ATTROLE													−.30

TABLE 11.26 SETUP AND PARTIAL OUTPUT OF
BMDP7M CONTRASTING THE WORKING GROUP
WITH THE OTHER TWO GROUPS FOR
DEMOGRAPHIC AND ATTITUDINAL VARIABLES

```
          /PROBLEM    TITLE IS 'BMDP7M DISCRIM, CONTRASTS'.
          /INPUT      VARIABLES ARE 13.  FORMAT IS '(A3,12F4.0)'.
                      FILE=DISCRIM.
          /VARIABLES  NAMES ARE CASESEQ,WORKSTAT,MARITAL,CHILDREN,RELIGION,RACE,
                      CONTROL,ATTMAR,ATTROLE,SEL,ATTHOUSE,AGE,EDUC,
                      NONOTHER,CATHOLIC,PROTEST,SINGLE,MARRIED.
                      ADD=5.    GROUPING IS WORKSTAT.
                      MISSING = 2*0, 3*9, 5*0, 1, 2*0.   LABEL IS CASESEQ.
                      USE=2, 4, 6 TO 18.
          /TRANSFORM  NONOTHER=RELIGION EQ 1.  CATHOLIC=RELIGION EQ 2.
                      PROTEST=RELIGION EQ 3.
                      SINGLE=MARITAL EQ 1.     MARRIED=MARITAL EQ 2.
                      DELETE=102,261,292,299.
          /GROUP      CODES(WORKSTAT) ARE 1 TO 3.
                      NAMES(WORKSTAT) ARE WORKING,HAPHOUSE,UNHOUSE.
                      USE = 1 TO 3.
                      PRIOR = 0.53, 0.30, 0.17.
          /DISC       LEVEL=0,0,0,1,0,1,2,2,2,1,2,1,1,1,1,1,1,1.
                      FORCE=2. TOL=0.001.
                      CONTRAST = 2,1,1.
          /PRINT      NO STEP.  NO POST.  NO POINT.
          /PLOT       NO CANON.  CONTRAST.
          /END
```

STEP NUMBER   0

VARIABLE	F TO REMOVE	FORCE LEVEL	TOLERNCE	*	VARIABLE	F TO ENTER	FORCE LEVEL	TOLERNCE
DF = 1 434				*	DF = 1 433			
				*	4 CHILDREN	9.94	1	1.00000
				*	6 RACE	2.25	1	1.00000
				*	7 CONTROL	1.45	2	1.00000
				*	8 ATTMAR	0.33	2	1.00000
				*	9 ATTROLE	15.03	2	1.00000
				*	10 SEL	0.45	1	1.00000
				*	11 ATTHOUSE	0.00	2	1.00000
				*	12 AGE	0.51	1	1.00000
				*	13 EDUC	11.04	1	1.00000
				*	14 NONOTHER	1.84	1	1.00000
				*	15 CATHOLIC	0.23	1	1.00000
				*	16 PROTEST	0.25	1	1.00000
				*	17 SINGLE	3.90	1	1.00000
				*	18 MARRIED	24.98	1	1.00000

****

STEP NUMBER 14
VARIABLE ENTERED   11 ATTHOUSE

VARIABLE	F TO REMOVE	FORCE LEVEL	TOLERNCE	*	VARIABLE	F TO ENTER	FORCE LEVEL	TOLERNCE
DF = 1 420				*	DF = 1 419			
4 CHILDREN	5.60	1	0.63687	*				
6 RACE	8.84	1	0.87555	*				
7 CONTROL	0.31	2	0.90388	*				
8 ATTMAR	0.56	2	0.68524	*				
9 ATTROLE	15.23	2	0.70182	*				
10 SEL	4.12	1	0.76717	*				
11 ATTHOUSE	0.26	2	0.77251	*				
12 AGE	6.82	1	0.73946	*				
13 EDUC	5.23	1	0.72671	*				
14 NONOTHER	0.61	1	0.61914	*				
15 CATHOLIC	0.55	1	0.51352	*				
16 PROTEST	1.73	1	0.51068	*				
17 SINGLE	0.77	1	0.58111	*				
18 MARRIED	12.86	1	0.60400	*				

EIGENVALUES
            0.16905
CUMULATIVE PROPORTION OF TOTAL DISPERSION
            1.00000
CANONICAL CORRELATIONS
            0.38027

```
          /PROBLEM     TITLE IS 'BMDP7M DISCRIM, CONTRASTS'.
          /INPUT       VARIABLES ARE 13.  FORMAT IS '(A3,12F4.0)'.
                       FILE=DISCRIM.
          /VARIABLES   NAMES ARE CASESEQ,WORKSTAT,MARITAL,CHILDREN,RELIGION,RACE,
                       CONTROL,ATTMAR,ATTROLE,SEL,ATTHOUSE,AGE,EDUC,
                       NONOTHER,CATHOLIC,PROTEST,SINGLE,MARRIED.
                       ADD=5.   GROUPING IS WORKSTAT.
                       MISSING = 2*0, 3*9, 5*0, 1, 2*0.   LABEL IS CASESEQ.
                       USE=2, 4, 6 TO 18.
          /TRANSFORM   NONOTHER=RELIGION EQ 1.  CATHOLIC=RELIGION EQ 2.
                       PROTEST=RELIGION EQ 3.
                       SINGLE=MARITAL EQ 1.     MARRIED=MARITAL EQ 2.
                       DELETE=102,261,292,299.
          /GROUP       CODES(WORKSTAT) ARE 1 TO 3.
                       NAMES(WORKSTAT) ARE WORKING,HAPHOUSE,UNHOUSE.
                       USE = 1 TO 3.
                       PRIOR = 0.53, 0.30, 0.17.
          /DISC        LEVEL=0,0,0,1,0,1,2,2,2,1,2,1,1,1,1,1,1,1.
                       FORCE=2. TOL=0.001.
                       CONTRAST = 1,-2,1.
          /PRINT       NO STEP.  NO POST.  NO POINT.
          /PLOT        NO CANON.  CONTRAST.
          /END
************************************************************************

STEP NUMBER   0

   VARIABLE    F TO    FORCE TOLERNCE *   VARIABLE    F TO   FORCE TOLERNCE
               REMOVE  LEVEL           *               ENTER  LEVEL
       DF =  1   434                   *       DF =  1   433
                                       *    4 CHILDREN    5.83    1   1.00000
                                       *    6 RACE        7.48    1   1.00000
                                       *    7 CONTROL     1.75    2   1.00000
                                       *    8 ATTMAR     17.85    2   1.00000
                                       *    9 ATTROLE    11.24    2   1.00000
                                       *   10 SEL         5.47    1   1.00000
                                       *   11 ATTHOUSE   14.57    2   1.00000
                                       *   12 AGE         1.77    1   1.00000
                                       *   13 EDUC        0.02    1   1.00000
                                       *   14 NONOTHER    0.04    1   1.00000
                                       *   15 CATHOLIC    3.47    1   1.00000
                                       *   16 PROTEST     0.04    1   1.00000
                                       *   17 SINGLE      3.96    1   1.00000
                                       *   18 MARRIED    24.48    1   1.00000
************************************************************************

STEP NUMBER  14
VARIABLE ENTERED   7 CONTROL

   VARIABLE    F TO    FORCE TOLERNCE *   VARIABLE    F TO   FORCE TOLERNCE
               REMOVE  LEVEL           *               ENTER  LEVEL
       DF =  1   420                   *       DF =  1   419
    4 CHILDREN    3.58    1   0.63687 *
    6 RACE        8.95    1   0.87555 *
    7 CONTROL     0.16    2   0.90388 *
    8 ATTMAR      1.36    2   0.68524 *
    9 ATTROLE    10.27    2   0.70182 *
   10 SEL         3.22    1   0.76717 *
   11 ATTHOUSE    6.47    2   0.77251 *
   12 AGE         1.19    1   0.73946 *
   13 EDUC        0.01    1   0.72671 *
   14 NONOTHER    0.06    1   0.61914 *
   15 CATHOLIC    5.36    1   0.51352 *
   16 PROTEST     2.98    1   0.51068 *
   17 SINGLE      0.42    1   0.58111 *
   18 MARRIED     8.87    1   0.60400 *

EIGENVALUES
              0.17845
CUMULATIVE PROPORTION OF TOTAL DISPERSION
              1.00000
CANONICAL CORRELATIONS
              0.38913
```

```
/PROBLEM     TITLE IS 'BMDP7M DISCRIM, CONTRASTS'.
/INPUT       VARIABLES ARE 13.  FORMAT IS '(A3,12F4.0)'.
             FILE=DISCRIM.
/VARIABLES   NAMES ARE CASESEQ,WORKSTAT,MARITAL,CHILDREN,RELIGION,RACE,
             CONTROL,ATTMAR,ATTROLE,SEL,ATTHOUSE,AGE,EDUC,
             NONOTHER,CATHOLIC,PROTEST,SINGLE,MARRIED.
             ADD=5.   GROUPING IS WORKSTAT.
             MISSING = 2*0, 3*9, 5*0, 1, 2*0.   LABEL IS CASESEQ.
             USE=2, 4, 6 TO 18.
/TRANSFORM   NONOTHER=RELIGION EQ 1.  CATHOLIC=RELIGION EQ 2.
             PROTEST=RELIGION EQ 3.
             SINGLE=MARITAL EQ 1.      MARRIED=MARITAL EQ 2.
             DELETE=102,261,292,299.
/GROUP       CODES(WORKSTAT) ARE 1 TO 3.
             NAMES(WORKSTAT) ARE WORKING,HAPHOUSE,UNHOUSE.
             USE = 1 TO 3.
             PRIOR = 0.53, 0.30, 0.17.
/DISC        LEVEL=0,0,0,1,0,1,2,2,2,1,2,1,1,1,1,1,1,1.
             FORCE=2. TOL=0.001.
             CONTRAST = 1,1,-2.
/PRINT       NO STEP.  NO POST.  NO POINT.
/PLOT        NO CANON.  CONTRAST.
/END
```

```
*******************************************************************************

STEP NUMBER   0

  VARIABLE    F TO   FORCE TOLERNCE *   VARIABLE     F TO   FORCE TOLERNCE
             REMOVE  LEVEL           *              ENTER  LEVEL
      DF =  1   434                  *       DF =  1   433
                                     *   4 CHILDREN    0.11   1   1.00000
                                     *   6 RACE        1.47   1   1.00000
                                     *   7 CONTROL     4.30   2   1.00000
                                     *   8 ATTMAR     10.33   2   1.00000
                                     *   9 ATTROLE     0.01   2   1.00000
                                     *  10 SEL         2.29   1   1.00000
                                     *  11 ATTHOUSE   10.69   2   1.00000
                                     *  12 AGE         2.89   1   1.00000
                                     *  13 EDUC        5.96   1   1.00000
                                     *  14 NONOTHER    1.48   1   1.00000
                                     *  15 CATHOLIC    3.91   1   1.00000
                                     *  16 PROTEST     0.31   1   1.00000
                                     *  17 SINGLE      0.04   1   1.00000
                                     *  18 MARRIED     0.20   1   1.00000
```

```
*******************************************************************************

STEP NUMBER  14
VARIABLE ENTERED   9 ATTROLE

  VARIABLE    F TO   FORCE TOLERNCE *   VARIABLE     F TO   FORCE TOLERNCE
             REMOVE  LEVEL           *              ENTER  LEVEL
      DF =  1   420                  *       DF =  1   419
 4 CHILDREN    0.03   1   0.63687 *
 6 RACE        0.11   1   0.87555 *
 7 CONTROL     0.66   2   0.90388 *
 8 ATTMAR      2.76   2   0.68524 *
 9 ATTROLE     0.04   2   0.70182 *
10 SEL         0.00   1   0.76717 *
11 ATTHOUSE    7.41   2   0.77251 *
12 AGE         1.19   1   0.73946 *
13 EDUC        3.00   1   0.72671 *
14 NONOTHER    0.16   1   0.61914 *
15 CATHOLIC    2.24   1   0.51352 *
16 PROTEST     0.26   1   0.51068 *
17 SINGLE      0.01   1   0.58111 *
18 MARRIED     0.03   1   0.60400 *


EIGENVALUES
           0.07316
CUMULATIVE PROPORTION OF TOTAL DISPERSION
           1.00000
CANONICAL CORRELATIONS
           0.26110
```

TABLE 11.29 PARTIAL OUTPUT FROM BMDP7M FOR CLASSIFICATION OF CASES ON THE BASIS OF (a) DEMOGRAPHY ALONE AND (b) DEMOGRAPHY PLUS ATTITUDES

(a) Demography alone

		INCORRECT CLASSIFICATIONS		MAHALANOBIS D-SQUARE FROM AND POSTERIOR PROBABILITY FOR GROUP -		
GROUP	WORKING		WORKING	HAPHOUSE	UNHOUSE
CASE						
2	2	HAPHOUSE	3.6 0.383	2.2 0.438	2.8 0.179
3	3		6.6 0.480	6.4 0.292	5.8 0.228
6	6		8.1 0.501	8.2 0.272	7.4 0.227
10	10		11.6 0.677	13.3 0.164	12.2 0.158
16	16		13.2 0.702	16.0 0.098	13.5 0.199
18	22		8.2 0.715	10.6 0.116	8.8 0.169
20	24	UNHOUSE	8.0 0.335	6.9 0.324	5.7 0.341
21	25		6.8 0.421	6.4 0.286	5.2 0.294
22	26		9.9 0.427	8.8 0.424	9.7 0.149
23	27		3.7 0.503	3.1 0.377	4.2 0.120
25	29	HAPHOUSE	10.3 0.316	8.1 0.552	9.8 0.132
26	30	HAPHOUSE	5.8 0.325	3.8 0.491	4.7 0.184
29	33		11.9 0.893	16.5 0.051	15.2 0.056
30	34		4.1 0.583	4.2 0.313	5.3 0.104
31	35		8.7 0.586	8.8 0.316	10.0 0.098
32	36		3.5 0.523	3.3 0.333	3.8 0.144
33	37	HAPHOUSE	7.5 0.310	5.5 0.474	6.0 0.217
36	40		7.9 0.481	8.0 0.254	6.8 0.265
38	46		13.5 0.581	15.1 0.148	12.7 0.271
39	47		5.6 0.436	4.5 0.429	5.6 0.136
40	48	HAPHOUSE	6.5 0.433	5.3 0.442	6.7 0.125
41	49		7.7 0.670	8.5 0.258	9.9 0.071
42	50		5.4 0.429	4.6 0.352	4.5 0.218
44	52	HAPHOUSE	8.8 0.358	7.0 0.500	8.4 0.142

(b) Demography plus attitudes

		INCORRECT CLASSIFICATIONS		MAHALANOBIS D-SQUARE FROM AND POSTERIOR PROBABILITY FOR GROUP -		
GROUP	WORKING		WORKING	HAPHOUSE	UNHOUSE
CASE						
2	2	HAPHOUSE	5.7 0.380	4.0 0.501	5.8 0.119
3	3		8.3 0.411	7.4 0.366	7.3 0.223
6	6		13.5 0.698	16.6 0.083	13.5 0.218
10	10		22.0 0.690	26.5 0.042	21.6 0.267
16	16		15.5 0.808	18.8 0.089	17.4 0.103
18	22		17.7 0.632	18.3 0.272	19.2 0.096
20	24	HAPHOUSE	14.0 0.295	12.4 0.356	11.4 0.349
21	25		11.9 0.389	11.1 0.337	10.3 0.274
22	26		11.6 0.459	11.3 0.304	10.6 0.236
23	27		4.1 0.593	4.4 0.278	4.8 0.129
25	29	HAPHOUSE	14.3 0.360	12.7 0.452	13.3 0.188
26	30		10.7 0.498	11.3 0.209	9.5 0.293
29	33		15.0 0.831	18.9 0.065	16.9 0.104
30	34		11.2 0.596	12.1 0.212	11.1 0.192
31	35		13.4 0.515	13.1 0.332	13.6 0.153
32	36		6.5 0.739	8.4 0.161	8.3 0.100
33	37	HAPHOUSE	18.5 0.158	15.0 0.524	14.8 0.318
36	40	UNHOUSE	23.5 0.279	22.5 0.257	20.2 0.464
38	46		18.7 0.601	23.0 0.041	17.5 0.358
39	47		10.3 0.550	9.8 0.391	12.5 0.058
40	48		8.6 0.539	8.2 0.361	9.7 0.100
41	49		9.6 0.589	9.4 0.368	12.5 0.043
42	50		9.6 0.365	8.5 0.360	7.9 0.275
44	52		14.9 0.563	15.6 0.222	14.5 0.215

TABLE 11.30   CHECKLIST FOR HIERARCHICAL
DISCRIMINANT FUNCTION ANALYSIS

1. Issues
   a. Unequal sample sizes and missing data
   b. Normality of sampling distributions
   c. Outliers
   d. Linearity
   e. Homogeneity of variance-covariance matrices
   f. Multicollinearity and singularity
2. Major analysis
   a. Significance of discriminant at each major step
   b. Significance of change at each major step
   c. Number of significant functions at last significant step. If any:
      (1) Variance accounted for
      (2) Plot(s) of discriminant functions
      (3) Loading matrix
   d. Variables separating each group
3. Additional analyses
   a. Group means for high-loading variables ($r > .50$)
   b. Pooled within-group correlations among predictor variables
   c. Classification results
      (1) Jackknifed results
      (2) Cross-validation
   d. Change in Rao's V (or stepdown F) plus univariate F for predictors

Table 11.30 is a checklist for hierarchical discriminant function analysis. An example of a Results section, in journal format, follows.

Results
A hierarchical discriminant function
analysis was performed to assess prediction
of membership in the three groups first from
ten demographic predictors and then after
addition of four attitudinal predictors.
Groups were working women, role-satisfied
housewives, and role-dissatisfied
housewives. Demographic predictors were
children (presence or absence), race
(Caucasian vs. other), socioeconomic
status, age, three dichotomous predictors
of religious affiliation, and two
dichotomous predictors of marital status.
Attitudinal predictors were locus of
control, attitude toward marital status,

attitude toward role of women, and attitude toward homemaking.

Of the original 465 cases, 25 were deleted due to missing data; missing data appeared to be randomly scattered over groups and predictors. An additional four cases were identified as multivariate outliers with $p < .001$, and were also deleted. All outlying cases were in the working group. Among the outliers, two were nonwhite and single. The remaining two outlying cases were women with extraordinarily favorable attitudes toward homemaking. For the 436 cases retained (231 working women, 129 role-satisfied housewives, and 76 role-dissatisfied housewives), evaluation of assumptions of linearity, normality, multicollinearity, and homogeneity of variance-covariance matrices revealed no threat to multivariate analysis.

There was statistically significant separation among the three groups from the ten demographic predictors alone, $F(20, 848) = 4.010$, $p < .01$. At this point, based on a jackknife procedure that reduces bias in classification, 51.6% of the women were correctly classified. After addition of the four attitudinal predictors, $F(28, 840) = 4.501$, $p < .01$, indicating again reliability in separation of groups. With all 14 predictors in the equation, the jackknife procedure correctly classified 58.5% of the women. McNemar's χ^2 test for change indicated reliable improvement in classification with addition of the attitudinal predictors to the demographic predictors, $\chi^2(1) = 8.62$, $p < .01$.

For classification, sample sizes were used to estimate prior probabilities of group membership. However, a disproportionate number of women were assigned into the employed group that actually constituted only 53% of the sample. On the basis of the ten demographic predictors, 70% of all the women were classified as employed. With all 14 predictors, 68% were so classified. With the use of demographic predictors alone, correct classification rates were 75.8%, 38.8%, and 0% for working women, role-satisfied housewives, and role-dissatisfied housewives, respectively. With 14 predictors, correct classification rates were 78.8%, 51.9%, and 7.9%, respectively. The gain with additional predictors, then, was strongest in properly classifying the two groups of housewives.

On the basis of all 14 predictors, there was reliable association between groups and predictors, $\chi^2(28) = 119.24$, $\underline{p} < .01$. After removal of the first function, reliable association remained, $\chi^2(13) = 29.38$, $\underline{p} < .01$. The two discriminant functions accounted for 77% and 23%, respectively, of the between-group variability [χ^2 *and percent of variance are from Table 11.23; cf. Section 11.6.2*] in discriminating among groups. As shown in Figure 11.4, the first discriminant function separates working women from role-satisfied housewives, with role-dissatisfied housewives falling between these two groups. The second discriminant function separates role-dissatisfied housewives from the other two groups.

Insert Figure 11.4 about here

The loading matrix of correlations between the 14 predictor variables and the two discriminant functions, as seen in Table 11.24, shows that the primary predictor (loadings of .50 and above) for the first discriminant function (separation of

working women and role—satisfied
housewives) is whether or not the woman is
married. Working women are less likely to be
currently married (70.6%) than role—
satisfied housewives (96.9%).

Insert Table 11.24 about here

The primary predictors on the second
discriminant function (separation of role—
dissatisfied housewives from the other two
groups) are attitudes toward housework and
current marital status. Attitudes toward
housework are less favorable among role—
dissatisfied housewives (mean = 24.92) than
among working women (mean = 23.77) or role—
satisfied women (22.57). Similarly, role—
dissatisfied women have less favorable
attitudes toward current marital status
(mean = 25.32) than working women (mean =
23.36) or role—satisfied housewives (mean =
20.47).

The predictors that separated working
women from the other two groups, after
adjustment for all other predictors and
adjustment of Type I error rate for 14
variables, are attitude toward women's
role, $F(1, 420) = 15.23$, and whether or not
the woman is currently married, $F(1, 420) =
12.86$. Squared semipartial correlations
between the groups (working woman vs.
housewives) and attitude toward women's
role, and between groups and whether or not
the woman is married are both .03. [*Equation
11.9 applied to results of Table 11.26.*] In
univariate ANOVA, without adjustment for
the other predictors, these two groups
differed on whether or not there were
children (working women have fewer),
attitudes toward women's role (more liberal
for working women), education (higher for
working women), and marital status (fewer
working women are married).

After adjustment for all other predictors, role-satisfied housewives were separated from the other two groups on the basis of attitudes toward women's role, $\underline{F}(1, 420) = 10.27$, with squared semipartial correlation = .02. [*Equation 11.9 applied to results of Table 11.27.*] In univariate ANOVA, the groups differed on attitudes toward marital status (role-satisfied housewives were more satisfied), attitudes toward women's role (role-satisfied housewives were more conservative), attitudes toward homemaking (role-satisfied housewives were more satisfied by it), and marital status (role-satisfied housewives were more likely to be married).

After adjusting all predictors for one another and adjusting for inflated Type I error rate with 14 predictors, none of the predictors significantly separated role-dissatisfied housewives from the two other groups. However, in univariate ANOVA, without adjustment, role-dissatisfied housewives were different on attitudes toward marital status (they were less satisfied) and attitudes toward homemaking (also less satisfied).

Many of the pooled within-group correlations (all $\underline{r} > .124$) in Table 11.25 would reach statistical significance at the .01 level, tested a priori. Because the emphasis here was on the combination of demography and attitude, only those correlations involving attitudinal predictors or both attitudinal and demographic predictors are discussed. Among attitudinal predictors, four of the six within-group correlations show statistical significance when tested individually. There is a small positive relationship between locus of control and attitude toward marital status, $\underline{r}(434) = .18$, indicating that women who are satisfied with their

current marital status are less likely to
attribute control to external sources.
Attitude toward homemaking is positively
correlated with locus of control, $r(434)$ =
.15, and attitude toward current marital
status, $r(434)$ = .28, and negatively
correlated with attitude toward women's
role, $r(434)$ = $-.30$. This indicates that
women with negative attitudes toward
homemaking are likely to attribute control
to external sources, to be dissatisfied with
their current marital status, and to have
more liberal attitudes toward women's role.

Insert Table 11.25 about here

Eleven of the within-group correlations
between attitudinal and demographic
variables would reach statistical
significance tested individually. Attitude
toward women's role is positively
correlated with children, $r(434)$ = .13, and
race, $r(434)$ = .17. This indicates that
women with conservative attitudes toward
women's role are more likely to have
children and less likely to be white.
Attitude toward women's role is negatively
correlated with socioeconomic level, $r(434)$
= $-.24$, educational level, $r(434)$ = $-.35$,
and positively correlated with age, $r(434)$
= .25. Women with traditional attitudes
toward women's role, then, are
characterized by lower socioeconomic level,
less education, and greater age. These women
with traditional attitudes also are more
likely to be Protestant, $r(434)$ = .14, and
less likely to report other or no religious
affiliation, $r(434)$ = $-.13$.
      Attitude toward homemaking is
correlated with whether or not a woman is
Catholic, $r(434)$ = $-.15$, with Catholic
women showing more positive attitudes
toward homemaking than women claiming other

```
religious affiliation. And there is a
significant correlation between education
and attitude toward homemaking, r(422) =
.14. The more education a woman has, the less
satisfied she is likely to be with
housework. Finally, attitude toward marital
status is related to marital status. Women
who are currently married are likely to be
satisfied with that status, r(434) = -.43;
single women are not so likely to be
satisfied with their marital status, r(434)
= .23.
```

11.9 SOME EXAMPLES FROM THE LITERATURE

Curtis and Simpson (1977) explored predictors for three types of drug users: daily opioid users, less-than-daily opioid users, and nonusers of opioids. A preliminary discriminant function analysis used 31 predictors related to demographic character-istics, alcohol use, and drug history to predict group membership. Both discriminant functions were statistically reliable and virtually all the predictors showed significant univariate F's for group differences, not surprising with the sample size greater than 23,000. For the second analysis, only the 13 predictors with univariate $F > 100$ were selected.

The first discriminant function accounted for 20% of the variance between groups, the second for 1% of the variance. The first discriminant function separated daily opioid users from the other two groups, presumably based on comparison of group centroids, although they were not reported. Correlations between predictors and the discriminant function were used to interpret the function. Daily opioid users were likely to be black, to be older, to have used illicit drugs for a longer period of time, to be responsible for slightly more family members, and to have begun daily use of illicit drugs with a drug other than marijuana.

A series of two-group stepwise discriminant functions was run by Strober and Weinberg (1977) on families that did or did not purchase various merchandise. Each item (e.g., dishwasher, hobby and recreational items, college education) was analyzed in a separate DISCRIM. The hypothesis was that predictors based on wife's em-ployment status would be unrelated to purchase of the items considered in the study. Other predictors were family income, net family assets, stage of family life cycle, whether or not the family recently moved, and whether or not the family was in the market for the item being considered. In none of the analyses did predictors related to wife's employment status show significant F to enter the stepwise discriminant function. Different predictors were important for various expenditures. For example, for college education, color TV, and washer, life-cycle stage of the family was the first predictor to enter. For purchases of furniture, net family assets was the first predictor to enter.

For those analyses in which sample size was large enough, part of the sample was reserved for cross-validation. Percent correct classification for original sample and cross-validation sample ranged from 53% to 73%. In all except one analysis in which cross-validation was possible, the difference in correct prediction between the two samples was less than 2%. For hobby and recreational items, however, the classification coefficients did not generalize to the cross-validation sample.

Men involved in midcareer changes were studied by Wiener and Vaitenas (1977). Two personality inventories were used to distinguish midcareer changers from vocationally stable men. Separate discriminant function analyses were run for each inventory. The Edwards Personal Preference Schedule generated 15 predictors. The Gordon Personal Profile and Gordon Personal Inventory generated 8 predictors. For each analysis, Wilks' Lambda showed statistically significant association between groups and predictors, with discriminant functions accounting for 28% and 42% of the variance in the Edwards and Gordon analyses, respectively.

Intepretation was based on standardized discriminant function coefficients (rather than the less biased correlations of predictors with discriminant functions). Midcareer changers scored lower on endurance, dominance, and order scales on the Edwards inventory, and showed less responsibility and ascendancy as measured by the Gordon inventory.

Caffrey and Lile (1976) explored differences between psychology and physics students in attitudes toward science. Predictors were based on a scientific attitude questionnaire, in which the respondent expressed degree of agreement with 15 quotations from various writers on science. Stepwise discriminant function analysis on upper-division psychology vs. physics majors followed preliminary analyses on a random sample of humanities, social science, and nature science majors.

Three items entered the stepwise analysis. Entering first was a statement suggesting that the results of scientific knowledge about behavior should be used to change behavior. Psychology majors were more likely to agree with this item than physics majors. Second, physics majors were more likely to agree with a statement suggesting that humans are free agents, not amenable to scientific prediction, and third, more likely to agree with a statement that Shakespeare conveys more truth about human nature than results of questionnaires.

Distinction between menopausal and nonmenopausal women in Hawaii was studied by Goodman, Steward, and Gilbert (1977). Predictors included medical, gynecological, and obstetrical history; age and age squared; physical measurements; and blood test results. Separate stepwise discriminant function analyses were run for Caucasian women, Japanese women, and then both groups pooled. Because only two groups were compared at a time, analyses were run through a stepwise multiple regression program. Preliminary runs with 35 predictors allowed elimination of nonsignificant predictors; then the stepwise analyses were rerun. In all analyses, age and age squared were forced into the stepwise series first because of known relationships with the remaining predictors.

Percent of variance accounted for in separating menopausal from nonmenopausal women was 45% and 36% for Caucasian and Japanese women, respectively. After adjustment for effects of age, the only predictor retained in the stepwise analysis for both races was ''surgery related to a female disorder.'' For Caucasian women only, one additional predictor, medication, was selected in the stepwise analysis. For age, age squared, surgery, and medication, discriminant function coefficients and their standard errors were given.[13] Finally, predictors common to both races were selected and heterogeneity of coefficients between Japanese and Caucasian samples were tested.[14] No significant heterogeneity was found—that is, the two races did not produce significantly different discriminant functions. An alternative strategy for this test is a factorial discriminant function analysis, with race as one IV and menopausal vs. nonmenopausal as the other.

---

[13] Because this two-group analysis was run through a multiple regression program, discriminant function coefficients were produced as β weights (cf. Chapter 5) and standard errors were available.

[14] The test for heterogeneity reported was that of Rao (1952) as applied by Goodman et al. (1974).

Chapter 12

Principal Components and Factor Analysis

12.1 GENERAL PURPOSE AND DESCRIPTION

Principal components analysis (PCA) and factor analysis (FA) are statistical techniques applied to a single set of variables where the researcher is interested in discovering which variables in the set form coherent subsets that are relatively independent of one another. Variables that are correlated with one another but largely independent of other subsets of variables are combined into factors.[1] Factors are thought to reflect underlying processes that have created the correlations among variables.

Suppose, for instance, a researcher is interested in studying characteristics of graduate students. The researcher measures a large sample of graduate students on personality characteristics, motivation, intellectual ability, scholastic history, familial history, health and physical characteristics, etc. Each of these areas is assessed by numerous variables; the variables all enter the analysis individually at one time and correlations among them are studied. The analysis reveals patterns of correlation among the variables that are thought to reflect underlying processes affecting the behavior of graduate students. For instance, several individual variables from the personality measures combine with some variables from motivation and scholastic history to suggest a person who prefers to work independently, an independence factor. Several variables from the intellectual ability measures combine with some others from scholastic history to suggest an intelligence factor.

A major use of PCA and FA is in development of objective tests for measurement of personality and intelligence and the like. The researcher starts out with a very large number of items reflecting a first guess about the items that may eventually

---

[1] PCA produces components while FA produces factors, but it is less confusing here to call the results of both analyses factors.

prove useful. The items are given to randomly selected subjects and factors are derived. As a result of the first factor analysis, items are added and deleted, a second test is devised, and that test is given to other randomly selected subjects. The process continues until the researcher has a test with numerous items forming several factors that represent the area to be measured. The validity of the factors is tested in research where predictions are made regarding differences in behavior of persons who score high or low on a factor.

The specific goals of PCA or FA are to summarize patterns of correlations among observed variables, to reduce a large number of observed variables to a smaller number of factors, to provide an operational definition (a regression equation) for an underlying process by using observed variables, or to test a theory about the nature of underlying processes. Some or all of these goals may be the focus of a particular research project.

PCA and FA have considerable utility in reducing numerous variables down to a few factors. Mathematically, PCA and FA produce several linear combinations of observed variables, each linear combination a factor. The factors summarize the patterns of the correlations in the observed correlation matrix and can, in fact, be used to reproduce the observed correlation matrix. But the number of factors is usually far fewer than the number of observed variables so there is considerable parsimony in a factor analysis. Further, when scores on factors are estimated for each subject, they are often more reliable than scores on individual observed variables.

Steps in PCA or FA include selecting and measuring a set of variables, preparing the correlation matrix (to perform either PCA or FA), extracting a set of factors from the correlation matrix, determining the number of factors, (probably) rotating the factors to increase interpretability, and, finally, interpreting the results. Although there are relevant statistical considerations to most of these steps, the final test of the analysis is usually its interpretability.

A good PCA or FA "makes sense"; a bad one does not. Interpretation and naming of factors depend on the meaning of the particular combination of observed variables that correlate highly with each factor. A factor is more easily interpreted when several observed variables correlate highly with it and those variables do not correlate with other factors.

One of the problems with PCA and FA is that there is no criterion beyond interpretability against which to test the solution. In regression analysis, for instance, the DV is a criterion and the correlation between observed and predicted DV scores serves as a test of the solution. Similarly for the two sets of variables in canonical correlation. In discriminant function analysis, profile analysis, and multivariate analysis of variance, the solution is judged by how well it predicts group membership. But in PCA and FA there is no external criterion such as group membership against which to test the solution.

A second problem with FA or PCA is that, after extraction, there are an infinite number of rotations available, all accounting for the same amount of variance in the original data, but with factors defined slightly differently. The final choice among alternatives depends on the researcher's assessment of its interpretability and scientific utility. In the presence of an infinite number of mathematically identical solutions, researchers are bound to differ regarding which is best. Because the differences cannot be resolved by appeal to objective criteria, arguments over the best solution sometimes become vociferous. However, those who expect a certain amount of ambiguity with respect to the best FA solution will not be surprised when other researchers select a different one. Nor will they be surprised when results are not replicated if different decisions are made at one, or more, of the steps in performing FA.

A third problem is that FA is frequently used in an attempt to ''save'' poorly conceived research. If no other statistical procedure is applicable, at least data can usually be factor analyzed. Thus in the minds of many, FA is associated with sloppy research. The very power of PCA and FA to create apparent order from real chaos contributes to their somewhat tarnished reputations as scientific tools.

There are two major types of FA: exploratory and confirmatory. In exploratory FA, one seeks to describe and summarize data by grouping together variables that are correlated. The variables themselves may or may not have been chosen with potential underlying processes in mind. Exploratory FA is usually performed in the early stages of research, when it provides a tool for consolidating variables and for generating hypotheses about underlying processes.

Confirmatory FA is a much more sophisticated technique used in the advanced stages of the research process to test a theory about latent processes or to investigate hypothesized differences in latent processes between groups of subjects. Variables are carefully and specifically chosen to reveal underlying processes.

Before we go on, it is helpful to define a few terms. The first terms involve correlation matrices. The correlation matrix produced by the observed variables is called the *observed correlation matrix*. The correlation matrix produced from factors is called the *reproduced correlation matrix*. The difference between observed and reproduced correlation matrices is the *residual correlation matrix*. In a good FA, correlations in the residual matrix are small, indicating a close fit between observed and reproduced matrices.

A second set of terms refers to matrices produced and interpreted as part of the solution. Rotation of factors is a process by which the solution is made more interpretable without changing its underlying mathematical properties. There are two general classes of rotation: orthogonal and oblique. If rotation is *orthogonal* (so that all the factors are uncorrelated with each other), a *loading* matrix is produced. The loading matrix is a matrix of correlations between observed variables and factors. The sizes of the loadings reflect the extent of relationship between each observed variable and each factor. Orthogonal FA is interpreted from the loading matrix by looking at which observed variables correlate with each factor.

If rotation is *oblique* (so that the factors themselves are correlated), several additional matrices are produced. The *factor correlation* matrix contains the correlations among the factors. The loading matrix from orthogonal rotation splits into two matrices: a *structure* matrix of correlations between factors and variables and a *pattern* matrix of unique relationships (uncontaminated by overlap among factors) between each factor and each observed variable. Following oblique rotation, the meaning of factors is ascertained from the pattern matrix.

Lastly, for both types of rotation, there is a *factor-score* coefficients matrix, a matrix of coefficients used to estimate scores on factors from scores on observed variables for each individual.

FA produces *factors,* while PCA produces *components.* However, the processes are similar except in preparation of the observed correlation matrix for extraction. The difference between PCA and FA is in the variance that is analyzed. In PCA, all the variance in the observed variables is analyzed. In FA, only shared variance is analyzed; attempts are made to estimate and eliminate variance due to error and variance that is unique to each variable. The term *factor* is used here to refer to both components and factors unless the distinction is critical, in which case the appropriate term is used.

12.2 KINDS OF RESEARCH QUESTIONS

The goal of research using PCA or FA is to reduce a large number of variables to a smaller number of factors, to concisely describe (and perhaps understand) the relationships among observed variables, or to test theory about underlying processes. Some of the specific questions that are frequently asked are presented in Sections 12.2.1 through 12.2.6.

12.2.1 Number of Factors

How many reliable and interpretable factors are there in the data set? How many factors are needed to summarize the pattern of correlations in the correlation matrix? In the graduate student example, two factors are discussed; are these both reliable? Are any more reliable factors present? Strategies for choosing an appropriate number of factors and for assessing the correspondence between observed and reproduced correlation matrices are discussed in Section 12.6.2.

12.2.2 Nature of Factors

What is the meaning of the factors? How are the factors to be interpreted? Factors are interpreted by the variables that correlate with them. Rotation to improve inter-pretability is discussed in Section 12.6.3; interpretation itself is discussed in Section 12.6.5.

12.2.3 Importance of Solutions and Factors

How much variance in a data set is accounted for by the factors? Which factors account for the most variance? In a good factor analysis, a high percentage of the variance in the observed variables is accounted for by the first few factors. And, because factors are computed in descending order of magnitude, the first factor accounts for the most variance, with later factors accounting for less and less of the variance until they are no longer reliable. Methods for assessing the importance of solutions and factors are in Section 12.6.4.

12.2.4 Testing Theory in FA

How well does the obtained factor solution fit an expected factor solution? The researcher generates hypotheses regarding both the number and the nature of the factors expected of graduate students. Comparisons between the hypothesized factors and the factor solution provide a test of the hypotheses. Tests of theory in FA are addressed, in preliminary form, in Sections 12.6.2 and 12.6.7.

12.2.5 Comparing Factor Solutions for Different Groups

How similar are the factors for persons with different characteristics or different experiences? For instance, are the factors for graduate students the same as the factors for undergraduate business majors or other groups that have not gone on to graduate school? Similarity in factors between two groups is assessed, in preliminary fashion, using techniques described in Section 12.6.7.

12.2.6 Estimating Scores on Factors

Had factors been measured directly, what scores would subjects have received on each of them? For instance, if each graduate student were measured directly on independence and intelligence, what scores would each student receive for each of them? Estimation of factor scores is the topic of Section 12.6.6.

12.3 LIMITATIONS

12.3.1 Theoretical Issues

Most applications of PCA or FA are exploratory in nature; FA is used as a tool for reducing the number of variables or examining patterns of correlations among variables without a serious intent to test theory. Under these circumstances, both the theoretical and the practical limitations to FA are relaxed in favor of a frank exploration of the data. Decisions about number of factors and rotational scheme are based on pragmatic rather than theoretical criteria.

The research project that is designed specifically to be factor analyzed, however, differs from other projects in several important respects. Among the best detailed discussions of the differences is the one found in Comrey (1973, pp. 189–211), from which some of the following discussion is taken.

The first task of the researcher is to generate hypotheses about factors believed to underlie the domain of interest. Statistically, it is important to make the research inquiry broad enough to include five or six hypothesized factors so that the solution is stable. Logically, in order to reveal the processes underlying a research area, all relevant factors have to be included. Failure to measure some important factor may distort the apparent relationships among measured factors. Inclusion of all relevant factors poses a logical, but not statistical, problem to the researcher.

Next, one selects variables to observe. For each hypothesized factor, five or six variables, each thought to be a relatively pure measure of the factor, are included. Pure measures are called marker variables. Marker variables are highly correlated with one and only one factor, and load on it regardless of extractional or rotational technique. Marker variables are useful because they define clearly the nature of a factor; adding potential variables to a factor to round it out is much more meaningful if the factor is unambiguously defined by marker variables to begin with.

The complexity of the variables is also considered. Complexity is indicated by the number of factors with which a variable correlates. A pure variable is correlated with only one factor, whereas a complex variable is correlated with several. If variables differing in complexity are all included in an analysis, those with similar complexity levels may "catch" each other in factors that have little to do with underlying processes. Variables with similar complexity may correlate with each other *because of their complexity* and not because they relate to the same factor. Estimating the complexity of variables is part of generating hypotheses about factors and selecting variables to measure them.

Several other considerations are required of the researcher planning a factor analytic study. It is important, for instance, that the sample chosen exhibit spread in scores with respect to the variables and the factors they measure. If all subjects achieve about the same score on some factor, correlations among the observed variables are low and the factor may not emerge in analysis. Selection of subjects expected to differ on the observed variables and underlying factors is an important design consideration.

One should also be leery about pooling the results of several samples, or the same sample with measures repeated in time, for factor analytic purposes. First, samples that are known to be different with respect to some criterion (e.g., socioeconomic status) may also have different factors. Examination of group differences is often quite revealing. Second, underlying factor structure may shift in time for the same subjects with learning or with experience in an experimental setting and these differences may also be quite revealing. Pooling results from diverse groups in FA may obscure differences rather than illuminate them.

On the other hand, if different samples do produce the same factors, pooling them is desirable because of increase in sample size. For example, if men and women produce the same factors, the samples should be combined and the results of the single FA reported. Strategies for evaluating differences in factors among groups are discussed in Section 12.6.7.

12.3.2 Practical Issues

Because FA and PCA are exquisitely sensitive to the sizes of correlations, it is critical that honest, reliable correlations be employed. Sensitivity to outlying cases, problems created by missing data, and degradation of correlations between poorly distributed variables all plague FA and PCA. A review of these issues in Chapter 4 is important to FA and PCA. Thoughtful solutions to some of the problems, including variable transformations, may markedly enhance FA, whether performed for exploratory or confirmatory purposes. However, the limitations apply with greater force to confirmatory FA.

12.3.2.1 Sample Size and Missing Data Correlation coefficients tend to be less reliable when estimated from small samples. Therefore, it is important that sample size be large enough that correlations are reliably estimated. Comrey (1973) gives as a guide sample sizes of 50 as very poor, 100 as poor, 200 as fair, 300 as good, 500 as very good, and 1000 as excellent. Others suggest that a sample size of 100 to 200 is good enough for most purposes, particularly when factors are strong and distinct and number of variables is not too large. *As a general rule of thumb, it is comforting to have at least five cases for each observed variable.*

The required sample size depends also on magnitude of population correlations and number of factors. If there are strong, reliable correlations and a few, distinct factors, a sample size of 50 may even be adequate, as long as there are notably more cases than factors.

If cases have missing data, either the missing values are estimated or the cases are deleted. Consult Chapter 4 for methods of finding and estimating missing values. Consider the distribution of missing values (is it random?) and remaining sample size when deciding between estimation and deletion. If cases are missing values in a nonrandom pattern or if sample size becomes too small, estimation is in order. However, beware of using estimation procedures (such as regression) that are likely to overfit the data and cause correlations to be too high. These procedures may "create" factors.

12.3.2.2 Normality As long as PCA and FA are used descriptively as convenient ways to summarize the relationships in a large set of observed variables, assumptions regarding the distributions of variables are not in force. If variables are normally distributed, the solution is enhanced. To the extent that normality fails, the solution is degraded but may still be worthwhile.

However, when statistical inference is used to determine the number of factors, multivariate normality is assumed. Multivariate normality is the assumption that all variables, and all linear combinations of variables, are normally distributed. Although normality of all linear combinations of variables is not testable, *normality among single variables is assessed by skewness and kurtosis* (see Chapter 4 and Section 12.8.1.2). If a variable has substantial skewness and kurtosis, variable transformation is considered.

12.3.2.3 Linearity Multivariate normality also implies that relationships among pairs of variables are linear. Because correlation measures linear relationship and does not reflect nonlinear relationship, the analysis is degraded when linearity fails. *Linearity among pairs of variables is assessed through inspection of scatterplots.* Consult Chapter 4 and Section 12.8.1.3 for methods of screening for linearity. If nonlinearity is found, transformation of variables is considered.

12.3.2.4 Outliers among Cases As in all multivariate techniques, cases may be outliers either on individual variables (univariate) or on combinations of variables (multivariate). Such cases have more influence on the factor solution than other cases. Consult Chapter 4 and Section 12.8.1.4 for methods of detecting and reducing the influence of both univariate and multivariate outliers.

12.3.2.5 Multicollinearity and Singularity In PCA, multicollinearity is not a problem because there is no need to invert a matrix. For most forms of FA and for estimation of factor scores in any form of FA, singularity or extreme multicollinearity is a problem. *For FA, if the determinant of R and eigenvalues associated with some factors approach 0, multicollinearity or singularity may be present.*

To investigate further, look at the SMCs for each variable where it serves as DV with all other variables as IVs. If any of the SMCs is one, singularity is present; if any of the SMCs is very large (near one), multicollinearity is present. Delete the variable with multicollinearity or singularity. Chapter 4 and Section 12.8.1.5 provide examples of screening for and dealing with multicollinearity and singularity.

12.3.2.6 Factorability of R A matrix that is factorable should include several sizable correlations. The expected size depends, to some extent, on N (larger sample sizes tend to produce smaller correlations), but if no correlation exceeds .30, use of FA is questionable because there is probably nothing to factor analyze. *Inspect R for correlations in excess of .30 and, if none is found, reconsider use of FA,* except in its most exploratory and pragmatic sense.

High bivariate correlations are, however, not ironclad proof that the correlation matrix contains factors. It is possible that the correlations are between only two variables and do not reflect underlying processes that are simultaneously affecting several variables. For this reason, it is helpful to examine matrices of partial correlations where pairwise correlations are adjusted for effects of all other variables. If there are factors present, then high bivariate correlations become very low partial correlations. BMDP, SPSS^x, and SAS produce partial correlation matrices.

Bartlett's test of sphericity (1954) is a notoriously sensitive test of the hypothesis that the correlations in a correlation matrix are zero. The test is available in SPSS^x FACTOR but because of its sensitivity and its dependence on N, the test is likely to be significant with samples of substantial size even if correlations are very low. Therefore, use of the test is recommended only if there are fewer than, say, five cases per variable.

Several more sophisticated tests of the factorability of **R** are available through SPSS^x and SAS. Both programs give significance tests of correlations, the anti-image correlation matrix, and the Kaiser's (1970, 1974) measure of sampling adequacy. Significance tests of pairs of correlations in the correlation matrix provide an indication of the reliability of the relationships between pairs of variables. If **R** is factorable, numerous pairs are significant. The anti-image correlation matrix contains the negatives of partial correlations between pairs of variables with effects of other variables removed. If **R** is factorable, there are mostly small values among the off-diagonal elements of the anti-image matrix. Finally, Kaiser's measure of sampling adequacy is a ratio of the sum of squared correlations to the sum of squared correlations plus sum of squared partial correlations. The value approaches 1 if partial correlations are small.[2] Values of .6 and above are required for good FA.

---

[2] BMDP4M prints partial correlations between pairs of variables with effects of other variables removed through the PARTIAL option so Kaiser's measure of sampling adequacy could (with some pain) be hand-calculated.

12.3.2.7 Outliers among Variables After FA, in both exploratory and confirmatory FA, variables that are unrelated to others in the set are identified. These variables are usually not correlated with the first few factors although they often correlate with factors extracted later. These factors are usually unreliable, both because they account for very little variance and because factors that are defined by just one or two variables are not stable. Therefore one never knows whether or not these factors are "real." Suggestions for determining reliability of factors defined by one or two variables are in Section 12.6.2.

In exploratory FA, if the variance accounted for by a factor defined by only one or two variables is high enough, the factor is interpreted with great caution or ignored, as pragmatic considerations dictate. In confirmatory FA, the factor represents either a promising lead for future work or (probably) error variance, but its interpretation awaits clarification by more research.

A variable with a low squared multiple correlation with all other variables and low correlations with all important factors is an outlier among the variables. The variable is usually ignored in the current FA and either deleted or given friends in future research. Screening for outliers among variables is illustrated in Section 12.8.1.7.

12.3.2.8 Outlying Cases among the Factors In FA and PCA, cases may be unusual with respect to their scores on the factors. These are cases that have unusually large or small scores on the factors as estimated from factor score coefficients. The deviant scores are from cases for which the factor solution is inadequate. Examination of these cases for consistency is informative if it reveals the kinds of cases for which the FA is *not* appropriate.

If BMDP4M is used, *outlying cases among the factors are cases with large Mahalanobis distances, estimated as chi square values, from the location of the case in the space defined by the factors to the centroid of all cases in the same space.* If scatterplots between pairs of factors are requested, these cases appear along the borders. Screening for these outlying cases is in Section 12.8.1.8.

12.4 FUNDAMENTAL EQUATIONS FOR FACTOR ANALYSIS

Because of the variety and complexity of the calculations involved in preparing the correlation matrix, extracting factors, and rotating them, and because, in our judgment, little insight is produced by demonstrations of some of these procedures, this section does not show them all. Instead, the relationships between some of the more important matrices are shown, with an assist from SPSS^x FACTOR for underlying calculations.

Table 12.1 lists many of the important matrices in FA and PCA. Although the list is lengthy, it is composed mostly of *matrices of correlations* (between variables, between factors, and between variables and factors), *matrices of standard scores* (on variables and on factors), *matrices of regression weights* (for producing scores on factors from scores on variables), and the *pattern matrix* of unique relationships between factors and variables after oblique rotation.

**TABLE 12.1   COMMONLY ENCOUNTERED MATRICES IN
FACTOR ANALYSES**

Label	Name	Rotation	Size[a]	Description
R	Correlation matrix	Both orthogonal and oblique	$p \times p$	Matrix of correlations between variables
Z	Variable matrix	Both orthogonal and oblique	$N \times p$	Matrix of standardized observed variable scores
F	Factor-score matrix	Both orthogonal and oblique	$N \times m$	Matrix of standard scores on factors or components
A	Factor loading matrix Pattern matrix	Orthogonal Oblique	$p \times m$	Matrix of regressionlike weights used to estimate the unique contribution of each factor to the variance in a variable. If orthogonal, also correlations between variables and factors
B	Factor-score coefficients matrix	Both orthogonal and oblique	$p \times m$	Matrix of regression weights used to generate factor scores from variables
C	Structure matrix[b]	Oblique	$p \times m$	Matrix of correlations between variables and (correlated) factors
Φ	Factor correlation matrix	Oblique	$m \times m$	Matrix of correlations among factors
L	Eigenvalue matrix[c]	Both orthogonal and oblique	$m \times m$	Diagonal matrix of eigenvalues, one per factor[e]
V	Eigenvector matrix[d]	Both orthogonal and oblique	$p \times m$	Matrix of eigenvectors, one vector per eigenvalue

[a] Row by column dimensions where
  p = number of variables
  N = number of subjects
  m = number of factors or components

[b] In most textbooks, the structure matrix is labeled **S.** However, we have used **S** to represent the sum-of-squares and cross-products matrix elsewhere and will use **C** for the structure matrix here.

[c] Also called characteristic roots or latent roots.

[d] Also called characteristic vectors.

[e] If the matrix is of full rank, there are actually p rather than m eigenvalues and eigenvectors. Only m are of interest, however, so the remaining $p - m$ are not displayed.

Also in the table are the matrix of eigenvalues and the matrix of their corresponding eigenvectors. Eigenvalues and eigenvectors are discussed here and in Appendix A, albeit scantily, because of their importance in factor extraction, the frequency with which one encounters the terminology, and the close association between eigenvalues and variance in statistical applications.

A data set appropriate for FA consists of numerous subjects each measured on several variables. A grossly inadequate data set appropriate for FA is in Table 12.2. Five subjects who were trying on ski boots late on a Friday night in January were asked about the importance of each of four variables to their selection of a ski resort.

TABLE 12.2 SMALL SAMPLE OF HYPOTHETICAL DATA
FOR ILLUSTRATION OF FACTOR ANALYSIS

		Variables		
Skiers	COST	LIFT	DEPTH	POWDER
S_1	32	64	65	67
S_2	61	37	62	65
S_3	59	40	45	43
S_4	36	62	34	35
S_5	62	46	43	40

Correlation matrix

	COST	LIFT	DEPTH	POWDER
COST	1.000	−.953	−.055	−.130
LIFT	−.953	1.000	−.091	−.036
DEPTH	−.055	−.091	1.000	.990
POWDER	−.130	−.036	.990	1.000

The variables were cost of ski ticket (COST), kind of ski lift (LIFT), depth of snow (DEPTH), and kind of snow (POWDER). Larger numbers indicate greater importance. The researcher wanted to investigate the pattern of relationships among the variables in an effort to understand better the dimensions underlying choice of ski area.

Notice the pattern of correlations in the correlation matrix as set off by the vertical and horizontal lines. The strong correlations in the upper left and lower right quadrants show that scores on COST and LIFT are related, as are scores on DEPTH and POWDER. The other two quadrants show that scores on DEPTH and LIFT are unrelated, as are scores on POWDER and LIFT, and so on. With luck, FA will find this pattern of correlations, easy to see in a small correlation matrix but not in a very large one.

An important theorem from matrix algebra indicates that, under certain conditions, matrices can be diagonalized. Correlation and covariance matrices are among those that often can be diagonalized. When a matrix is diagonalized, it is transformed into a matrix with numbers in the positive diagonal[3] and zeros everywhere else. In this application, the numbers in the positive diagonal represent variance from the correlation matrix that has been repackaged as follows:

$$\mathbf{L} = \mathbf{V'RV} \tag{12.1}$$

[3] The positive diagonal runs from upper left to lower right in a matrix.

TABLE 12.3 EIGENVECTORS AND
CORRESPONDING
EIGENVALUES FOR
THE EXAMPLE

Eigenvector 1	Eigenvector 2
− .283	.651
.177	− .685
.658	.252
.675	.207
Eigenvalue 1	**Eigenvalue 2**
2.00	1.91

Diagonalization of **R** is accomplished by post- and premultiplying it by the matrix **V** and its transpose.

The columns in **V** are called eigenvectors, and the values in the main diagonal of **L** are called eigenvalues. The first eigenvector corresponds to the first eigenvalue, and so forth.

Because there are four variables in the example, there are four eigenvalues with their corresponding eigenvectors. However, because the goal of FA is to summarize a pattern of correlations with as few factors as possible, and because each eigenvalue corresponds to a different potential factor, usually only factors with large eigenvalues are retained. These few factors duplicate the correlation matrix as faithfully as possible.

In this example, when no limit is placed on the number of factors, eigenvalues of 2.02, 1.94, .04, and .00 are computed for each of the four possible factors. Only the first two factors, with values over 1.00, are large enough to be retained in subsequent analyses. FA is rerun specifying extraction of just the first two factors; they have eigenvalues of 2.00 and 1.91, respectively, as indicated in the Table 12.3.

Using Equation 12.1 and inserting the values from the example, we obtain

$$\mathbf{L} = \begin{bmatrix} -.283 & .177 & .658 & .675 \\ .651 & -.685 & .252 & .207 \end{bmatrix} \begin{bmatrix} 1.000 & -.953 & -.055 & -.130 \\ -.953 & 1.000 & -.091 & -.036 \\ -.055 & -.091 & 1.000 & .990 \\ -.130 & -.036 & .990 & 1.000 \end{bmatrix} \begin{bmatrix} -.283 & .651 \\ .177 & -.685 \\ .658 & .252 \\ .675 & .207 \end{bmatrix}$$

$$= \begin{bmatrix} 2.00 & .00 \\ .00 & 1.91 \end{bmatrix}$$

(All values agree with computer output. Hand calculation may produce discrepancies due to rounding error.)

The matrix of eigenvectors premultiplied by its transpose produces the identity matrix with ones in the positive diagonal and zeros elsewhere. Therefore, pre- and postmultiplying the correlation matrix by eigenvectors does not change it so much as repackage it.

$$\mathbf{V'V} = \mathbf{I} \tag{12.2}$$

For the example:

$$\begin{bmatrix} -.283 & .177 & .658 & .675 \\ .651 & -.685 & .252 & .207 \end{bmatrix} \begin{bmatrix} -.283 & .651 \\ .177 & -.685 \\ .658 & .252 \\ .675 & .207 \end{bmatrix} = \begin{bmatrix} 1.000 & .000 \\ .000 & 1.000 \end{bmatrix}$$

The important point is that because correlation matrices often meet requirements for diagonalizability, it is possible to use on them the matrix algebra of eigenvectors and eigenvalues with FA as the result. When a matrix is diagonalized, the information contained in it is repackaged. In FA, the variance in the correlation matrix is condensed into eigenvalues. The factor with the largest eigenvalue has the most variance and so on, down to factors with small or negative eigenvalues that are usually omitted from solutions.

Calculations for eigenvectors and eigenvalues are extremely laborious and not particularly enlightening (although they are illustrated in Appendix A for a small matrix). They require solving p equations in p unknowns with additional side constraints and are rarely performed by hand. Once the eigenvalues and eigenvectors are known, however, the rest of FA (or PCA) more or less "falls out," as is seen from Equations 12.3 to 12.6.

Equation 12.1 can be reorganized as follows:

$$\mathbf{R} = \mathbf{VLV'} \tag{12.3}$$

The correlation matrix can be considered a product of three matrices—the matrices of eigenvalues and corresponding eigenvectors.

After reorganization, the square root is taken of the matrix of eigenvalues.

$$\mathbf{R} = \mathbf{V}\sqrt{\mathbf{L}}\sqrt{\mathbf{L}}\mathbf{V'} \tag{12.4}$$

or

$$\mathbf{R} = (\mathbf{V}\sqrt{\mathbf{L}})(\sqrt{\mathbf{L}}\mathbf{V'})$$

If $\mathbf{V}\sqrt{\mathbf{L}}$ is called $\mathbf{A}$, and $\sqrt{\mathbf{L}}\mathbf{V'}$ is $\mathbf{A'}$, then

$$\mathbf{R} = \mathbf{AA'} \tag{12.5}$$

The correlation matrix can also be considered a product of two matrices, each a combination of eigenvectors and the square root of eigenvalues.

Equation 12.5 is frequently called the fundamental equation for FA.[4] It represents the assertion that the correlation matrix is a product of the factor loading matrix, $\mathbf{A}$, and its transpose.

---

[4] In order to reproduce the correlation matrix exactly, as indicated in Equations 12.4 and 12.5, all eigenvalues and eigenvectors are necessary, not just the first few of them.

Equations 12.4 and 12.5 also reveal that the major work of FA (and PCA) is calculation of eigenvalues and eigenvectors. Once they are known, the (unrotated) factor loading matrix is found by straightforward matrix multiplication, as follows.

$$\mathbf{A} = \mathbf{V}\sqrt{\mathbf{L}} \tag{12.6}$$

For the example:

$$\mathbf{A} = \begin{bmatrix} -.283 & .651 \\ .177 & -.685 \\ .658 & .252 \\ .675 & .207 \end{bmatrix} \begin{bmatrix} \sqrt{2.00} & 0 \\ 0 & \sqrt{1.91} \end{bmatrix} = \begin{bmatrix} -.400 & .900 \\ .251 & -.947 \\ .932 & .348 \\ .956 & .286 \end{bmatrix}$$

The factor loading matrix is a matrix of correlations between factors and variables. The first column is correlations between the first factor and each variable in turn, COST ($-.400$), LIFT ($.251$), DEPTH ($.932$), and POWDER ($.956$). The second column is correlations between the second factor and each variable in turn, COST ($.900$), LIFT ($-.947$), DEPTH ($.348$), and POWDER ($.286$). A factor is interpreted from the variables that are highly correlated with it—that have high loadings on it. Thus the first factor is primarily a snow conditions factor (DEPTH and POWDER), while the second reflects resort conditions (COST and LIFT). (The negative correlation indicates that more attractive lifts are also more costly.)

Notice, however, that all the variables are correlated with both factors to a considerable extent. Interpretation is fairly clear for this hypothetical example, but most likely would not be for real data. Usually a factor is most interpretable when a few variables are highly correlated with it and the rest are not.

Rotation is ordinarily used after extraction to maximize high correlations and minimize low ones. Numerous methods of rotation are available (see Section 12.5.2) but the most commonly used, and the one illustrated here, is varimax. Varimax is a *vari*ance *max*imizing procedure. The goal of varimax rotation is to maximize the variance of factor loadings by making high loadings higher and low ones lower for each factor.

This goal is accomplished by means of a transformation matrix $\mathbf{\Lambda}$ (as defined in Equation 12.8), where

$$\mathbf{A}_{\text{unrotated}}\,\mathbf{\Lambda} = \mathbf{A}_{\text{rotated}} \tag{12.7}$$

The unrotated factor loading matrix is multiplied by the transformation matrix to produce the rotated loading matrix.

For the example:

$$\mathbf{A}_{\text{rotated}} = \begin{bmatrix} -.400 & .900 \\ .251 & -.947 \\ .932 & .348 \\ .956 & .286 \end{bmatrix} \begin{bmatrix} .946 & -.325 \\ .325 & .946 \end{bmatrix} = \begin{bmatrix} -.086 & .981 \\ -.071 & -.977 \\ .994 & .026 \\ .997 & -.040 \end{bmatrix}$$

TABLE 12.4 RELATIONSHIPS AMONG LOADINGS, COMMUNALITIES, SSLs, VARIANCE, AND COVARIANCE OF ORTHOGONALLY ROTATED FACTORS

	Factor 1	Factor 2	Communalities (h^2)
COST	$-.086$	.981	$\Sigma a^2 = .970$
LIFT	$-.071$	$-.977$	$\Sigma a^2 = .960$
DEPTH	.994	.026	$\Sigma a^2 = .989$
POWDER	.997	$-.040$	$\Sigma a^2 = .996$
SSLs	$\Sigma a^2 = 1.994$	$\Sigma a^2 = 1.919$	3.915
Proportion of variance	.50	.48	.98
Proportion of covariance	.51	.49	

Compare the rotated and unrotated loading matrices. Notice that in the rotated matrix the low correlations are lower and the high ones higher than in the unrotated loading matrix. Emphasizing differences in loadings facilitates interpretation of factors by making unambiguous the variables that correlate with a factor.

The numbers in the transformation matrix have a spatial interpretation.

$$\Lambda = \begin{bmatrix} \cos \psi & -\sin \psi \\ \sin \psi & \cos \psi \end{bmatrix} \qquad (12.8)$$

The transformation matrix is a matrix of sines and cosines of an angle ψ.

For the example, the angle is approximately 19°. That is, cos 19 ≈ .946 and sin 19 ≈ .325. Geometrically, this corresponds to a 19° swivel of the factor axes about the origin. Greater detail regarding the geometric meaning of rotation is in Section 12.5.2.3.

Once the rotated loading matrix is available, other relationships are found, as in Table 12.4. The communality for a variable is the variance accounted for by the factors. It is the squared multiple correlation of the variable as predicted from the factors. Communality is the sum of squared loadings (SSL) for a variable across factors. In Table 12.4, the communality for COST is $(-.086)^2 + .981^2 = .970$. That is, 97% of the variance in COST is accounted for by Factor 1 plus Factor 2.

The proportion of variance *in the set of variables* accounted for by a factor is the SSL for the factor divided by the number of variables (if rotation is orthogonal).[5] For the first factor, the proportion of variance is $[(-.086)^2 + (-.071)^2 + .994^2 + .997^2]/4 = 1.994/4 = .50$. Fifty percent of the variance in the variables is accounted for by the first factor. The second factor accounts for 48% of the variance

---

[5] For unrotated factors only, the sum of the squared loadings for a factor is equal to the eigenvalue. Once loadings are rotated, the sum of squared loadings is called SSL and is no longer equal to the eigenvalue.

in the variables and, because rotation is orthogonal, the two factors together account for 98% of the variance in the variables.

The proportion of variance *in the solution* accounted for by a factor—the proportion of covariance—is the SSL for the factor divided by the sum of communalities (or, equivalently, the sum of the SSLs). The first factor accounts for 51% of the variance in the solution (1.994/3.915) while the second factor accounts for 49% of the variance in the solution (1.919/3.915). The two factors together account for all of the covariance.

The reproduced correlation matrix for the example is generated using Equation 12.5:

$$\bar{\mathbf{R}} = \begin{bmatrix} -.086 & .981 \\ -.071 & -.977 \\ .994 & .026 \\ .997 & -.040 \end{bmatrix} \begin{bmatrix} -.086 & -.071 & .994 & .997 \\ .981 & -.977 & .026 & -.040 \end{bmatrix}$$

$$= \begin{bmatrix} .970 & -.953 & -.059 & -.125 \\ -.953 & .962 & -.098 & -.033 \\ -.059 & -.098 & .989 & .990 \\ -.125 & -.033 & .990 & .996 \end{bmatrix}$$

Notice that the reproduced correlation matrix differs slightly from the original correlation matrix. The difference between the original and reproduced correlation matrices is the residual correlation matrix:

$$\mathbf{R}_{res} = \mathbf{R} - \bar{\mathbf{R}} \tag{12.9}$$

The residual correlation matrix is the difference between the observed correlation matrix and the reproduced correlation matrix.

For the example, with communalities inserted in the positive diagonal of **R**:

$$\mathbf{R}_{res} = \begin{bmatrix} .970 & -.953 & -.055 & -.130 \\ -.953 & .960 & -.091 & -.036 \\ -.055 & -.091 & .989 & .990 \\ -.130 & -.036 & .990 & .996 \end{bmatrix} - \begin{bmatrix} .000 & .000 & .004 & -.005 \\ .000 & -.002 & .007 & -.003 \\ .004 & .007 & .000 & .000 \\ -.005 & -.003 & .000 & .000 \end{bmatrix}$$

$$= \begin{bmatrix} .000 & .000 & .004 & -.005 \\ .000 & -.002 & .007 & -.003 \\ .004 & .007 & .000 & .000 \\ -.005 & -.003 & .000 & .000 \end{bmatrix}$$

In a "good" FA, the numbers in the residual correlation matrix are small because there is little difference between the original correlation matrix and the correlation matrix generated from factor loadings.

Scores on factors can be predicted for each case once the loading matrix is available. Regression-like coefficients are computed for weighting variable scores to produce factor scores. Because $\mathbf{R}^{-1}$ is the inverse of the matrix of correlations among variables and $\mathbf{A}$ is the matrix of correlations between factors and variables, Equation 12.10 for factor score coefficients is similar to Eq. 5.6 for regression coefficients in multiple regression.

$$\mathbf{B} = \mathbf{R}^{-1}\mathbf{A} \qquad (12.10)$$

Factor score coefficients for estimating factor scores from variable scores are a product of the inverse of the correlation matrix and the factor loading matrix.

For the example:[6]

$$\mathbf{B} = \begin{bmatrix} 25.485 & 22.689 & -31.655 & 35.479 \\ 22.689 & 21.386 & -24.831 & 28.312 \\ -31.655 & -24.831 & 99.917 & -103.950 \\ 35.479 & 28.312 & -103.950 & 109.567 \end{bmatrix} \begin{bmatrix} -.086 & .981 \\ -.072 & -.978 \\ .994 & .027 \\ .997 & -.040 \end{bmatrix}$$

$$= \begin{bmatrix} 0.082 & 0.537 \\ 0.054 & -0.461 \\ 0.190 & 0.087 \\ 0.822 & -0.074 \end{bmatrix}$$

To estimate a subject's score for the first factor, then, all of the subject's scores on variables are standardized and then the standardized score on COST is weighted by 0.082, LIFT by 0.054, DEPTH by 0.190, and POWDER by 0.822 and the results are added. In matrix form,

$$\mathbf{F} = \mathbf{ZB} \qquad (12.11)$$

Factor scores are a product of standardized scores on variables and factor score coefficients.

For the example:

$$\mathbf{F} = \begin{bmatrix} -1.22 & 1.14 & 1.15 & 1.14 \\ 0.75 & -1.02 & 0.92 & 1.01 \\ 0.61 & -0.78 & -0.36 & -0.47 \\ -0.95 & 0.98 & -1.20 & -1.01 \\ 0.82 & -0.30 & -0.51 & -0.67 \end{bmatrix} \begin{bmatrix} 0.082 & 0.537 \\ 0.054 & -0.461 \\ 0.190 & 0.087 \\ 0.822 & -0.074 \end{bmatrix}$$

$$= \begin{bmatrix} 1.12 & -1.16 \\ 1.01 & 0.88 \\ -0.45 & 0.69 \\ -1.08 & -0.99 \\ -0.60 & 0.58 \end{bmatrix}$$

[6] The numbers in **B** are different from the factor score coefficients generated by computer for the small data set. The difference is due to rounding error following inversion of a multicollinear correlation matrix.

The first subject has an estimated standard score of 1.12 on the first factor and -1.16 on the second factor, and so on for the other four subjects. The first subject values the snow factor highly and the cost factor low. The second subject values both the snow factor and the cost factor; the third subject values the cost factor more than the snow factor, and so forth. The sum of standardized factor scores across subjects for a single factor is zero.

Predicting scores on variables from scores on factors is also possible. The equation for doing so is

$$\mathbf{Z} = \mathbf{FA'} \tag{12.12}$$

Predicted standardized scores on variables are a product of scores on factors weighted by factor loadings.

For example:

$$\mathbf{Z} = \begin{bmatrix} 1.12 & -1.16 \\ 1.01 & 0.88 \\ -0.45 & 0.69 \\ -1.08 & -0.99 \\ -0.60 & 0.58 \end{bmatrix} \begin{bmatrix} -.086 & -.072 & .994 & .997 \\ .981 & -.978 & .027 & -.040 \end{bmatrix}$$

$$= \begin{bmatrix} -1.23 & 1.05 & 1.08 & 1.16 \\ 0.78 & -0.93 & 1.03 & 0.97 \\ 0.72 & -0.64 & -0.43 & -0.48 \\ -0.88 & 1.05 & -1.10 & -1.04 \\ 0.62 & -0.52 & -0.58 & -0.62 \end{bmatrix}$$

That is, the first subject (the first row of $\mathbf{Z}$) is predicted to have a standardized score of -1.23 on COST, 1.05 on LIFT, 1.08 on DEPTH, and 1.16 on POWDER. Like the reproduced correlation matrix, these values are similar to the observed values if the FA captures the relationship among the variables.

It is helpful to see these values written out because they provide an insight into how scores on variables are conceptualized in factor analysis. For example, for the first subject,

$$-1.23 = -.086(1.12) + .981(-1.16)$$
$$1.05 = -.072(1.12) - .978(-1.16)$$
$$1.08 = .994(1.12) + .027(-1.16)$$
$$1.16 = .997(1.12) - .040(-1.16)$$

Or, in algebraic form,

$$z_{\text{COST}} = a_{11}F_1 + a_{12}F_2$$
$$z_{\text{LIFT}} = a_{21}F_1 + a_{22}F_2$$

$$z_{\text{DEPTH}} = a_{31}F_1 + a_{32}F_2$$

$$z_{\text{POWDER}} = a_{41}F_1 + a_{42}F_2$$

A score on an observed variable is conceptualized as a properly weighted and summed combination of the scores on factors that underlie it. The underlying processes are factors; the outward manifestations are scores on variables, driven by underlying processes.

All the relationships mentioned thus far are for orthogonal rotation. Most of the complexities of orthogonal rotation remain and several others are added when oblique (correlated) rotation is used. Consult Table 12.1 for a listing of additional matrices and a hint of the discussion to follow.

SPSS^{x} FACTOR is run on the data from Table 12.2 using the default option for oblique rotation (cf. Section 12.5.2.2) to get values for the pattern matrix, **A**, and factor-score coefficients, **B**.

In oblique rotation, the loading matrix becomes the pattern matrix. Values in the pattern matrix, when squared, represent the unique contribution of each factor to the variance of each variable but do not include segments of variance that come from overlap between correlated factors. For the example the pattern matrix following oblique rotation is

$$\mathbf{A} = \begin{bmatrix} -.079 & .981 \\ -.078 & -.978 \\ .994 & .033 \\ .997 & -.033 \end{bmatrix}$$

The first factor makes a unique contribution of $(-.079)^2$ to the variance in COST, $(-.078)^2$ to LIFT, $.994^2$ to DEPTH and $.997^2$ to POWDER.

Factor-score coefficients following oblique rotation are also found:

$$\mathbf{B} = \begin{bmatrix} 0.104 & 0.584 \\ 0.081 & -0.421 \\ 0.159 & -0.020 \\ 0.856 & 0.034 \end{bmatrix}$$

Applying Equation 12.11 to produce factor scores results in the following values:

$$\mathbf{F} = \begin{bmatrix} 1.12 & 1.14 & 1.15 & 1.14 \\ 0.75 & -1.02 & 0.92 & 1.01 \\ 0.61 & -0.78 & -0.36 & -0.47 \\ -0.95 & 0.98 & -1.20 & -1.01 \\ 0.82 & -0.30 & -0.51 & -0.67 \end{bmatrix} \begin{bmatrix} 0.104 & 0.584 \\ 0.081 & -0.421 \\ 0.159 & -0.020 \\ 0.856 & 0.034 \end{bmatrix}$$

$$= \begin{bmatrix} 1.12 & -1.18 \\ 1.01 & 0.88 \\ -0.46 & 0.68 \\ -1.07 & -0.98 \\ -0.59 & 0.59 \end{bmatrix}$$

Once the factor scores are determined, correlations among factors can be obtained. Among the equations used for the purpose is

$$\mathbf{\Phi} = \left(\frac{1}{N-1}\right)\mathbf{F'F} \qquad (12.13)$$

One way to compute correlations among factors is from cross products of standardized factor scores divided by the number of cases minus one.

The factor correlation matrix is a standard part of computer output following oblique rotation. For the example:

$$\mathbf{\Phi} = \frac{1}{4}\begin{bmatrix} 1.12 & 1.01 & -0.46 & -1.07 & -0.59 \\ -1.18 & 0.88 & 0.68 & -0.98 & 0.59 \end{bmatrix}\begin{bmatrix} 1.12 & -1.18 \\ 1.01 & 0.88 \\ -0.46 & 0.68 \\ -1.07 & -0.98 \\ -0.59 & 0.59 \end{bmatrix}$$

$$= \begin{bmatrix} 1.00 & -0.01 \\ -0.01 & 1.00 \end{bmatrix}$$

The correlation between the first and second factor is quite low, $-.01$. For this example, there is almost no relationship between the two factors, although considerable correlation could have been produced had it been warranted. Ordinarily one uses orthogonal rotation in a case like this because complexities introduced by oblique rotation are not warranted by such a low correlation among factors.

However, if oblique rotation is used, the structure matrix, **C**, is the correlations between variables and factors. These correlations represent both the unique relationship between the variable and the factor (shown in the pattern matrix) and the relationship between the variable and the variance a factor shares with other factors. The equation for the structure matrix is

$$\mathbf{C} = \mathbf{A\Phi} \qquad (12.14)$$

The structure matrix is a product of the pattern matrix and the factor correlation matrix.

For example:

$$\mathbf{C} = \begin{bmatrix} -.079 & .981 \\ -.078 & -.978 \\ .994 & .033 \\ .997 & -.033 \end{bmatrix}\begin{bmatrix} 1.00 & -.01 \\ -.01 & 1.00 \end{bmatrix} = \begin{bmatrix} -.069 & .982 \\ -.088 & -.977 \\ .994 & .023 \\ .997 & -.043 \end{bmatrix}$$

COST, LIFT, DEPTH, and POWDER correlate $-.069$, $-.088$, $.994$, and $.997$ with the first factor and $.982$, $-.977$, $.023$, and $-.043$ with the second factor, respectively.

There is some debate as to whether one should interpret the pattern matrix or the structure matrix following oblique rotation. The structure matrix is appealing because it is readily understood. However, the correlations between variables and factors are inflated by any overlap between factors. The problem becomes more severe as the correlations among factors increase and it may be hard to determine which variables are related to a factor. On the other hand, the pattern matrix contains values representing the unique contributions of each factor to the variance in the variables. Shared variance is omitted (as it is with standard multiple regression), but the set of variables that composes a factor is usually easier to see. If factors are very highly correlated, it may appear that no variables are related to them because there is almost no unique variance once overlap is omitted.

Most researchers interpret and report the pattern matrix rather than the structure matrix. However, if the researcher reports either the structure or the pattern matrix and also Φ, then the interested reader can generate the other using Equation 12.14 as desired.

In oblique rotation, $\overline{\mathbf{R}}$ is produced as follows:

$$\overline{\mathbf{R}} = \mathbf{CA'} \qquad (12.15)$$

The reproduced correlation matrix is a product of the structure matrix and the transpose of the pattern matrix.

Once the reproduced correlation matrix is available, Equation 12.9 is used to generate the residual correlation matrix to diagnose adequacy of fit in FA.

Examples of setup and output for the example using SPSS^x, BMDP, SAS, and SYSTAT are shown in Tables 12.5 through 12.8. Principal factor analysis with varimax rotation is shown for SPSS^x FACTOR, BMDP4M, and SAS FACTOR. For SYSTAT FACTOR, version 3.0, principal components is the only available method of extraction.

SPSS^x FACTOR (Table 12.5) begins by printing out SMCs for each variable, labeled COMMUNALITY in the INITIAL STATISTICS portion of the output. In a parallel but unrelated table, EIGENVALUEs, PCT OF VARiance, and percent of variance cumulated over the four factors (CUM PCT) are printed out for the four initial factors. (Be careful not to confuse factors with variables.) The program then indicates the number of factors extracted with eigenvalues greater than 1 (the default value).

For the two extracted factors, an unrotated FACTOR loading MATRIX is then printed. In the output labeled FINAL STATISTICS are final COMMUNALITY values for each variable (h^2 in Table 12.4), and, parallel with them, EIGENVALUEs for the two retained factors (see Table 12.3), the PCT OF VARiance for each factor and CUM PCT of variance accounted for by successive factors. The ROTATED FACTOR loading MATRIX, which matches loadings in Table 12.4 is given along with the FACTOR TRANSFORMATION MATRIX (Equation 12.8) for orthogonal varimax rotation with Kaiser normalization.

BMDP4M (Table 12.6) first prints out the CORRELATION MATRIX for the variables, and then the COMMUNALITIES (h^2) for each variable, OBTAINED FROM 2 FACTORS with eigenvalues greater than 1. In the next table, eigenvalues are given for all 4 factors along with two types of cumulative proportion of variance, neither

TABLE 12.5 SETUP AND SPSS' FACTOR OUTPUT FOR FACTOR
ANALYSIS ON SAMPLE DATA OF TABLE 12.2

```
        TITLE       SMALL SAMPLE FACTOR ANALYSIS
        FILE HANDLE TAPE17
        DATA LIST   FILE=TAPE17        FREE
                    /SUBJNO,COST,LIFT,DEPTH,POWDER
        FACTOR      VARIABLES=COST,LIFT,DEPTH,POWDER/
                    EXTRACTION=PAF/
```

ANALYSIS NUMBER 1 LISTWISE DELETION OF CASES WITH MISSING VALUES

EXTRACTION 1 FOR ANALYSIS 1, PRINCIPAL AXIS FACTORING (PAF)

INITIAL STATISTICS:

VARIABLE	COMMUNALITY	FACTOR	EIGENVALUE	PCT OF VAR	CUM PCT
COST	.96076	1	2.01631	50.4	50.4
LIFT	.95324	2	1.94151	48.5	98.9
DEPTH	.98999	3	.03781	.9	99.9
POWDER	.99087	4	.00437	.1	100.0

 PAF EXTRACTED 2 FACTORS.   4 ITERATIONS REQUIRED.

FACTOR MATRIX:

	FACTOR 1	FACTOR 2
COST	-.40027	.89978
LIFT	.25060	-.94706
DEPTH	.93159	.34773
POWDER	.95596	.28615

FINAL STATISTICS:

VARIABLE	COMMUNALITY	FACTOR	EIGENVALUE	PCT OF VAR	CUM PCT
COST	.96983	1	2.00473	50.1	50.1
LIFT	.95973	2	1.90933	47.7	97.9
DEPTH	.98877				
POWDER	.99574				

- - - - - - - - - - - F A C T O R   A N A L Y S I S - - - - - - - - - - -

VARIMAX ROTATION 1 FOR EXTRACTION 1 IN ANALYSIS 1 - KAISER NORMALIZATION.

 VARIMAX CONVERGED IN   3 ITERATIONS.

ROTATED FACTOR MATRIX:

	FACTOR 1	FACTOR 2
COST	-.08591	-.98104
LIFT	-.07100	.97708
DEPTH	.99403	-.02588
POWDER	.99706	.04028

FACTOR TRANSFORMATION MATRIX:

	FACTOR 1	FACTOR 2
FACTOR 1	.94565	.32519
FACTOR 2	.32519	-.94565

TABLE 12.6 SETUP AND SELECTED BMDP4M OUTPUT FOR FACTOR
ANALYSIS ON SAMPLE DATA OF TABLE 12.2.

```
              /PROBLEM      TITLE IS 'SMALL SAMPLE FACTOR ANALYSIS THROUGH BMDP4M'.
              /INPUT        VARIABLES ARE 5.  FORMAT IS '(A2,4F4.0)'.
              /VARIABLE     NAMES ARE SUBJNO, COST, LIFT, DEPTH, POWDER.
                            LABEL IS SUBJNO.
              /FACTOR       METHOD=PFA.
              /END
```

CORRELATION MATRIX

```
              COST      LIFT      DEPTH     POWDER

              2         3         4         5
COST      2   1.000
LIFT      3   -0.953    1.000
DEPTH     4   -0.055    -0.091    1.000
POWDER    5   -0.130    -0.036    0.990     1.000
```

COMMUNALITIES OBTAINED FROM 2 FACTORS AFTER   4 ITERATIONS.

THE COMMUNALITY OF A VARIABLE IS ITS SQUARED MULTIPLE
CORRELATION WITH THE FACTORS.

```
    2 COST      0.9698
    3 LIFT      0.9597
    4 DEPTH     0.9888
    5 POWDER    0.9957
```

FACTOR	VARIANCE EXPLAINED	CUMULATIVE PROPORTION OF VARIANCE		CARMINES THETA
		IN DATA SPACE	IN FACTOR SPACE	
1	2.0047	0.5110	0.5122	0.6810
2	1.9093	0.9977	1.0000	
3	0.0090	1.0000		
4	-0.0103			

TOTAL VARIANCE IS DEFINED AS THE SUM OF THE POSITIVE EIGEN VALUES OF THE
CORRELATION MATRIX.

NEGATIVE VALUES FOR VARIANCE EXPLAINED INDICATE THE DEGREE
TO WHICH THE COMMUNALITIES HAVE BEEN UNDERESTIMATED.
LARGE NEGATIVE VALUES FOR VARIANCE EXPLAINED INDICATE THAT
THE COVARIANCE OR CORRELATION MATRIX IS POORLY ESTIMATED.

UNROTATED FACTOR LOADINGS (PATTERN)

```
              FACTOR    FACTOR
              1         2

COST      2   -0.400    -0.900
LIFT      3   0.251     0.947
DEPTH     4   0.932     -0.348
POWDER    5   0.956     -0.286

          VP  2.005     1.909
```

THE VP FOR EACH FACTOR IS THE SUM OF THE SQUARES OF THE
ELEMENTS OF THE COLUMN OF THE FACTOR LOADING MATRIX
CORRESPONDING TO THAT FACTOR. THE VP IS THE VARIANCE
EXPLAINED BY THE FACTOR.

ORTHOGONAL ROTATION, GAMMA =   1.0000

TABLE 12.6 (Continued)

```
ROTATED FACTOR LOADINGS (PATTERN)
--------------------------------

                  FACTOR      FACTOR
                    1           2

COST       2      -0.086      0.981
LIFT       3      -0.071     -0.977
DEPTH      4       0.994      0.026
POWDER     5       0.997     -0.040

           VP      1.995      1.919

THE VP FOR EACH FACTOR IS THE SUM OF THE SQUARES OF THE
ELEMENTS OF THE COLUMN OF THE FACTOR PATTERN MATRIX
CORRESPONDING TO THAT FACTOR.  WHEN THE ROTATION IS
ORTHOGONAL, THE VP IS THE VARIANCE EXPLAINED BY THE FACTOR.


FACTOR SCORE COVARIANCE (COMPUTED FROM FACTOR SCORES)
----------------------------------------------------

                  FACTOR     FACTOR

                    1          2
        FACTOR   1   0.997
        FACTOR   2  -0.000     0.982
```

of which exactly matches that of SPSS[x]. CARMINES THETA is a measure of the reliability of the first principal component that is closely related to Cronbach's alpha.

UNROTATED FACTOR LOADINGS are given for each variable on the two factors, along with the SSLs for each factor, labeled VP. This is followed by ROTATED FACTOR LOADINGS, with SSLs again shown as VP. Finally, BMDP4M prints out the FACTOR SCORE COVARIANCE matrix, in which the off-diagonal elements are set to zero by the nature of orthogonal rotation. The diagonal elements are the SMCs of each factor with the variables.

SAS FACTOR (Table 12.7) requires a bit more instruction to produce a principal factor analysis with orthogonal rotation for two factors. You specify the initial communalities (PRIOR = SMC), number of factors to be extracted (NFACTOR = 2) and the type of rotation (ROTATE = V). PRIOR COMMUNALITY ESTIMATES— SMCs—are given, followed by EIGENVALUES for all four factors; also given is the TOTAL of the eigenvalues and their AVERAGE. The next row shows DIFFERENCEs between successive eigenvalues. For example, there is a small difference between the first and second eigenvalues (0.099606) and between the third and fourth eigenvalues (0.020622), but a large difference between the second and third eigenvalues (1.897534). PROPORTION and CUMULATIVE proportion of variance are then printed for each factor.

The FACTOR PATTERN matrix contains unrotated factor loadings for the first two factors. SSLs for each factor are in the table labeled VARIANCE EXPLAINED BY EACH FACTOR. Both FINAL COMMUNALITY ESTIMATES (h^2) and the TOTAL h^2 are then given. The transformation matrix for orthogonal rotation (Equation 12.8) is followed by the rotated factor loadings in the ROTATED FACTOR PATTERN matrix. SSLs for rotated factors—VARIANCE EXPLAINED BY EACH FACTOR— appear below the loadings. FINAL COMMUNALITY ESTIMATES are then repeated.

SYSTAT FACTOR (Table 12.8) currently provides only principal components extraction and requires instructions for both number of factors (EIGEN = 2) and type

```
DATA SSAMPLE;
INFILE TAPE17;
INPUT SUBJNO COST LIFT DEPTH POWDER;
PROC FACTOR
     PRIOR=SMC
     NFACTORS=2
     ROTATE=V;
VAR   COST LIFT DEPTH POWDER;
```

INITIAL FACTOR METHOD: PRINCIPAL FACTORS

PRIOR COMMUNALITY ESTIMATES: SMC

	COST	LIFT	DEPTH	POWDER
	0.960761	0.953241	0.989992	0.990873

EIGENVALUES OF THE REDUCED CORRELATION MATRIX: TOTAL = 3.89487   AVERAGE = 0.973717

	1	2	3	4
EIGENVALUE	2.002343	1.902738	0.005204	-0.015418
DIFFERENCE	0.099606	1.897534	0.020622	
PROPORTION	0.5141	0.4885	0.0013	-0.0040
CUMULATIVE	0.5141	1.0026	1.0040	1.0000

2 FACTORS WILL BE RETAINED BY THE NFACTOR CRITERION

FACTOR PATTERN

	FACTOR1	FACTOR2
COST	-0.38290	0.90476
LIFT	0.23317	-0.95009
DEPTH	0.93844	0.33099
POWDER	0.95953	0.26820

VARIANCE EXPLAINED BY EACH FACTOR

FACTOR1	FACTOR2
2.002343	1.902738

FINAL COMMUNALITY ESTIMATES: TOTAL = 3.905081

COST	LIFT	DEPTH	POWDER
0.965198	0.957031	0.990217	0.992634

ROTATION METHOD: VARIMAX

ORTHOGONAL TRANSFORMATION MATRIX

	1	2
1	0.95140	-0.30795
2	0.30795	0.95140

ROTATED FACTOR PATTERN

	FACTOR1	FACTOR2
COST	-0.08566	0.97870
LIFT	-0.07075	-0.97572
DEPTH	0.99476	0.02591
POWDER	0.99549	-0.04033

VARIANCE EXPLAINED BY EACH FACTOR

FACTOR1	FACTOR2
1.992897	1.912184

FINAL COMMUNALITY ESTIMATES: TOTAL = 3.905081

COST	LIFT	DEPTH	POWDER
0.965198	0.957031	0.990217	0.992634

TABLE 12.8 SETUP AND SYSTAT FACTOR OUTPUT FOR PRINCIPAL COMPONENTS ANALYSIS OF SAMPLE DATA OF TABLE 12.2.

```
USE TAPE17
OUTPUT=OUTFILE
EIGEN=2
ROTATE=VARIMAX
FACTOR COST LIFT DEPTH POWDER
```

MATRIX TO BE FACTORED

	COST	LIFT	DEPTH	POWDER
COST	1.000			
LIFT	-0.953	1.000		
DEPTH	-0.055	-0.091	1.000	
POWDER	-0.130	-0.036	0.990	1.000

LATENT ROOTS (EIGENVALUES)

1	2	3	4
2.016	1.942	0.038	0.004

COMPONENT LOADINGS

	1	2
COST	-0.500	0.856
LIFT	0.357	-0.925
DEPTH	0.891	0.449
POWDER	0.919	0.390

VARIANCE EXPLAINED BY COMPONENTS

1	2
2.016	1.942

PERCENT OF TOTAL VARIANCE EXPLAINED

1	2
50.408	48.538

ROTATED LOADINGS

	1	2
COST	-0.087	0.988
LIFT	-0.072	-0.989
DEPTH	0.997	0.026
POWDER	0.998	-0.040

VARIANCE EXPLAINED BY ROTATED COMPONENTS

1	2
2.003	1.955

PERCENT OF TOTAL VARIANCE EXPLAINED

1	2
50.067	48.879

of rotation. The output includes the correlation matrix—MATRIX TO BE FAC-TORED—and the eigenvalues for all four factors. The unrotated loading matrix (COMPONENT LOADINGS) is printed, followed by the SSLs (VARIANCE EX-PLAINED BY COMPONENTS) and the percent of variance for each component. That information is repeated for the rotated solution after the matrix of ROTATED LOADINGS.

12.5 MAJOR TYPES OF FACTOR ANALYSIS

Numerous procedures for factor extraction and rotation are available. However, only those procedures available in the BMDP, SPSS^x, SAS, and SYSTAT packages are summarized here. Other extraction and rotational techniques are described in Mulaik (1972), Harman (1976, 1983), and Rummel (1970).

12.5.1 Factor Extraction Techniques

Among the extraction techniques available in the four packages are principal com-ponents (PCA), principal factors, maximum likelihood factoring, Rao's canonical factoring, image factoring, alpha factoring, and unweighted and generalized least squares factoring (see Table 12.9). Of these, PCA and principal factors are the most commonly used.

All the extraction techniques calculate a set of orthogonal components or factors that, in combination, reproduce **R.** Criteria used to establish the solution, such as maximizing variance or minimizing residual correlations, differ from technique to technique. But differences in solutions are small for a data set with a large sample, numerous variables and similar communality estimates. In fact, one test of the stability of a FA solution is that it appears regardless of which extraction technique is employed. Table 12.10 shows solutions for the same data set after extraction with several different techniques, followed by varimax rotation. Similarities among the solutions are ob-vious.

None of the extraction techniques routinely provides an interpretable solution without rotation. All types of extraction may be rotated by any of the procedures described in Section 12.5.2, except Kaiser's Second Little Jiffy Extraction, which has its own rotational procedure.

Lastly, when using FA the researcher should hold in abeyance well-learned proscriptions against data snooping. It is quite common to use PCA as a preliminary extraction technique, followed by one or more of the other procedures, perhaps varying number of factors, communality estimates, and rotational methods with each run. Analysis terminates when the researcher decides on the preferred solution.

12.5.1.1 PCA vs. FA One of the most important decisions is the choice between PCA and FA. Mathematically, the difference involves the contents of the positive diagonal in the correlation matrix (the diagonal that contains the correlation between a variable and itself). In either PCA or FA, the variance that is analyzed is the sum of the values in the positive diagonal. In PCA ones are in the diagonal and there is

TABLE 12.9 SUMMARY OF EXTRACTION PROCEDURES

Extraction technique	Program	Goal of analysis	Special features
Principal components	SPSS^x BMDP4M SAS SYSTAT	Maximize variance extracted by orthogonal components	Mathematically determined, empirical solution with common, unique, and error variance mixed into components
Principal factors	SPSS^x BMDP4M SAS	Maximize variance extracted by orthogonal factors	Estimates communalities to attempt to eliminate unique and error variance from factors
Image factoring	SPSS^x BMDP4M (Second Little Jiffy) SAS (Image and Harris)		Uses SMCs between each variable and all others as communalities to generate a mathematically determined solution with error variance and unique variance eliminated
Maximum likelihood factoring	BMDP4M SPSS^x SAS	Estimate factor loadings for population that maximize the likelihood of sampling the observed correlation matrix	Has significance test for factors; especially useful for confirmatory factor analysis
Alpha factoring	SPSS^x SAS	Maximize the generalizability of orthogonal factors	Somewhat likely to produce communalities larger than 1
Unweighted least squares	SPSS^x SAS	Minimize squared residual correlations	
Generalized least squares	SPSS^x SAS	Weights variables by shared variance before minimizing squared residual correlations	

as much variance to be analyzed as there are observed variables; each variable contributes a unit of variance by contributing a 1 to the positive diagonal of the correlation matrix. All the variance is distributed to components, including error and unique variance for each observed variable. So if all components are retained, PCA duplicates exactly the observed correlation matrix and the standard scores of the observed variables.

In FA, on the other hand, only the variance that each observed variable shares with other observed variables is available for analysis. Exclusion of error and unique variance from FA is based on the belief that such variance only confuses the picture of underlying processes. Shared variance is estimated by *communalities,* values between 0 and 1 that are inserted in the positive diagonal of the correlation matrix.[7] The solution in FA concentrates on variables with high communality values. The sum of the communalities (sum of the SSLs) is the variance that is distributed among

---

[7] Maximum likelihood extraction manipulates off-diagonal elements rather than values in the diagonal.

TABLE 12.10 RESULTS OF DIFFERENT EXTRACTION METHODS ON SAME DATA SET

Variables	Factor 1				Factor 2			
	PCA	PFA	Rao	Alpha	PCA	PFA	Rao	Alpha
				Unrotated factor loadings				
1	.58	.63	.70	.54	.68	.68	− .54	.76
2	.51	.48	.56	.42	.66	.53	− .47	.60
3	.40	.38	.48	.29	.71	.55	− .50	.59
4	.69	.63	.55	.69	− .44	− .43	.54	− .33
5	.64	.54	.48	.59	− .37	− .31	.40	− .24
6	.72	.71	.63	.74	− .47	− .49	.59	− .40
7	.63	.51	.50	.53	− .14	− .12	.17	− .07
8	.61	.49	.47	.50	− .09	− .09	.15	− .03
				Rotated factor loadings (varimax)				
1	.15	.15	.15	.16	.89	.91	.87	.92
2	.11	.11	.10	.12	.83	.71	.72	.73
3	− .02	.01	.02	.00	.81	.67	.69	.66
4	.82	.76	.78	.76	− .02	− .01	− .03	.01
5	.74	.62	.62	.63	.01	.04	.03	.04
6	.86	.86	.87	.84	.04	− .02	− .01	− .03
7	.61	.49	.48	.50	.20	.18	.21	.17
8	.57	.46	.45	.46	.23	.20	.20	.19

Note: The largest difference in communality estimates for a single variable between extraction techniques was 0.08.

factors and is less than the total variance in the set of observed variables. Because unique and error variances are omitted, a linear combination of factors approximates, but does not duplicate, the observed correlation matrix and scores on observed variables.

PCA analyzes variance, FA analyzes covariance (communality). The goal of PCA is to extract maximum variance from a data set with a few orthogonal components. The goal of FA is to reproduce the correlation matrix with a few orthogonal factors. PCA is a unique mathematical solution, whereas most forms of FA are not unique.

The choice between PCA and FA depends on your assessment of the fit between the models, the data set, and the goals of the research. If you are interested in a theoretical solution uncontaminated by unique and error variability, FA is your choice. If, on the other hand, you want an empirical summary of the data set, PCA is the better choice.

12.5.1.2 Principal Components The goal of PCA is to extract maximum variance from the data set with each component. The first principal component is the linear combination of observed variables that maximally separates subjects by maximizing the variance of their component scores. The second component is formed from residual correlations; it is the linear combination of observed variables that extracts maximum variability uncorrelated with the first component. Subsequent components also extract maximum variability from residual correlations and are orthogonal to all previously extracted components.

The principal components are ordered, with the first component extracting the most variance and the last component the least variance. The solution is mathematically unique and, if all components are retained, exactly reproduces the observed correlation matrix. Further, since the components are orthogonal, their use in other analyses (e.g., as DVs in MANOVA) may greatly facilitate interpretation of results.

PCA is the solution of choice for the researcher who is primarily interested in reducing a large number of variables down to a smaller number of components. PCA is also recommended as the first step in FA where it reveals a great deal about probable number and nature of factors. PCA is available through all four computer packages.

12.5.1.3 Principal Factors Principal factors extraction differs from PCA in that estimates of communality, instead of ones, are in the positive diagonal of the observed correlation matrix. These estimates are derived through an iterative procedure, with SMCs (squared multiple correlations of each variable with all other variables) used as the starting values in the iteration. The goal of analysis, like that for PCA, is to extract maximum orthogonal variance from the data set with each succeeding factor. Advantages to principal factors extraction are that it is widely used (and understood) and that it conforms to the factor analytic model in which common variance is analyzed with unique and error variance removed. Because the goal is to maximize variance extracted, however, principal factors is sometimes not as good as other extraction techniques in reproducing the correlation matrix. Also, communalities must be estimated and the solution is, to some extent, determined by those estimates. Principal factor analysis is available through SPSS, BMDP, and SAS.

12.5.1.4 Image Factor Extraction The technique is called image factoring because the analysis distributes among factors the variance of an observed variable that is *reflected* by the other variables, the SMC. Image factor extraction provides an interesting compromise between PCA and principal factors. Like PCA, image extraction provides a mathematically unique solution because there are fixed values in the positive diagonal of **R**. Like principal factors, the values in the diagonal are communalities with unique and error variability excluded. The compromise is struck by using the squared multiple correlation (SMC or R^2) of each variable as DV with all others serving as IVs as the communality for that variable. With a hefty sample size and more than 10 to 15 observed variables, SMCs are stable and, depending on the adequacy of sampling of *variables,* provide a decent estimate of communality.

Image factoring is available through SPSS[x] FACTOR, BMDP4M (as Kaiser's Second Little Jiffy), and SAS FACTOR (with two types—"image" and Harris component analysis).

12.5.1.5 Maximum Likelihood Factor Extraction The maximum likelihood method of factor extraction was developed originally by Lawley in the 1940s (see Lawley and Maxwell, 1963). Maximum likelihood extraction estimates population values for factor loadings by calculating loadings that maximize the probability of sampling the observed correlation matrix from a population. Within constraints imposed by the correlations among variables, population estimates for factor loadings are calculated that have the greatest probability of yielding a sample with the observed correlation matrix. This method of extraction also maximizes the canonical correlations between the variables and the factors (see Chapter 6).

Maximum likelihood extraction is available through BMDP4M, SPSS^x FACTOR, and SAS FACTOR. It is the extraction procedure recommended by BMDP when the common factor model is appropriate, the number of variables is fewer than 60, and the correlation matrix is not singular.

12.5.1.6 Unweighted Least Squares (Minres) Factoring
The goal of unweighted least squares (minimum residual) factor extraction is to minimize squared differences between the observed and reproduced correlation matrices. Only off-diagonal differences are considered; communalities are derived from the solution rather than estimated as part of the solution. This procedure gives the same results as principal factors if communalities are the same. The procedure was developed by Comrey (1962) and Harman and Jones (1966) and is available through SPSS^x FACTOR and SAS FACTOR.

12.5.1.7 Generalized Least Squares Factoring
Generalized least squares extraction also seeks to minimize (off-diagonal) squared differences between observed and reproduced correlation matrices but in this case weights are applied to the variables. Differences for variables that have substantial shared variance with other variables are weighted more heavily than differences for variables that have substantial unique variance. In other words, differences for variables that are not as strongly related to other variables in the set are not as important to the solution. This relatively new method of extraction is available through SPSS^x FACTOR and SAS FACTOR.

12.5.1.8 Alpha Factoring
Alpha factor extraction, available through SPSS^x FACTOR and SAS FACTOR, grew out of psychometric research where the interest is in discovering which common factors are found consistently when repeated samples of *variables* are taken from a population of *variables*. The problem is the same as identifying mean differences that are found consistently among samples of subjects taken from a population of subjects—a question at the heart of most univariate and multivariate statistics.

In alpha factoring, however, the concern is with the reliability of the common factors rather than with the reliability of group differences. Coefficient alpha is a measure derived in psychometrics for the reliability (also called generalizability) of a score taken in a variety of situations. In alpha factoring, communalities are estimated, using iterative procedures, that maximize coefficient alpha for the factors.

Probably the greatest advantage to the procedure is that it focuses the researcher's attention squarely on the problem of sampling variables from the domain of variables of interest. Disadvantages stem from the relative unfamiliarity of most researchers with the procedure and the reason for it.

12.5.2 Rotation

The results of factor extraction, unaccompanied by rotation, are likely to be hard to interpret regardless of which method of extraction is used. After extraction, rotation is used to improve the interpretability and scientific utility of the solution. It is *not* used to improve the quality of the mathematical fit between the observed and reproduced correlation matrices because all orthogonally rotated solutions are mathematically equivalent to one another and to the solution before rotation.

Just as the different methods of extraction tend to give similar results with a good data set, so also the different methods of rotation tend to give similar results if the pattern of correlations in the data is fairly clear. In other words, a stable solution tends to appear regardless of the method of rotation used.

A decision is required between orthogonal and oblique rotation. In orthogonal rotation, the factors are uncorrelated. Orthogonal solutions offer ease of interpreting, describing, and reporting results; yet they strain ''reality'' unless the researcher is convinced that underlying processes are almost independent. The researcher who believes that underlying processes are correlated uses an oblique rotation. In oblique rotation the factors may be correlated, with conceptual advantages but practical disadvantages in interpreting, describing, and reporting results.

Among the dozens of rotational techniques that have been proposed, only those available in the four reviewed packages are included in this discussion (see Table 12.11). The reader who wishes to know more about these or other techniques is referred to Gorsuch (1983), Harman (1976), or Mulaik (1972). For the industrious, a presentation of rotation by hand is in Comrey (1973, pp. 109—145).

12.5.2.1 Orthogonal Rotation   Varimax, quartimax, and equamax—three orthogonal techniques—are available in all four packages. Varimax is easily the most commonly used of all the rotations available.

Just as the extraction procedures have slightly different statistical goals, so also the rotational procedures maximize or minimize different statistics. The goal of varimax rotation is to simplify factors by maximizing the variance of the loadings within factors, across variables. The spread in loadings is maximized—loadings that are high after extraction become higher and loadings that are low become lower. Interpreting a factor is easier because it is obvious which variables correlate with it. Varimax also tends to reapportion variance among factors so that they become relatively equal in importance; variance is taken from the first factors extracted and distributed among the later ones.

Quartimax does for variables what varimax does for factors. It simplifies variables by increasing the dispersion of the loadings within variables, across factors. Varimax operates on the columns of the loading matrix, quartimax operates on the rows. Quartimax is not nearly as popular as varimax because one is usually more interested in simple factors than in simple variables.

Equamax is a hybrid between varimax and quartimax that tries simultaneously to simplify the factors and the variables. Mulaik (1972) reports that equamax tends to behave erratically unless the researcher can specify the number of factors with confidence.

TABLE 12.11 SUMMARY OF ROTATIONAL TECHNIQUES

Rotational technique	Program	Type	Goals of analysis	Comments
Varimax	BMDP4M SAS SPSS[x] SYSTAT	Orthogonal	Minimize complexity of factors (simplify columns of loading matrix) by maximizing variance of loadings on each factor	Most commonly used rotation; recommended as default option (in BMDP, Γ [gamma] = 1)
Quartimax	BMDP4M SAS SPSS[x] SYSTAT	Orthogonal	Minimize complexity of variables (simplify rows of loading matrix) by maximizing variance of loadings on each variable	First factor tends to be general with others subclusters of variables (in BMDP, Γ = 0)
Equamax	BMDP4M SAS SPSS[x] SYSTAT	Orthogonal	Simplify both variables and factors (rows and columns); compromise between quartimax and varimax	May behave erratically (Γ = 1/2)
Orthogonal with gamma	BMDP4M SAS	Orthogonal	Simplify either factors or variables, depending on the value of gamma (Γ)	Gamma (Γ) continuously variable
Direct oblimin	BMDP4M SPSS[x]	Oblique	Simplify factors by minimizing cross products of loadings	Continuous values of gamma, Γ (BMDP) or delta, δ (SPSS[x]), available; allows wide range of factor intercorrelations
(Direct) quartimin	BMDP4M SPSS[x]	Oblique	Simplify factors by minimizing sum of cross products of squared loadings in pattern matrix	Permits fairly high correlations among factors. Recommended oblique rotation by BMDP series; in BMDP, Γ = 0. Achieved in SPSS[x] by setting δ = 0
Orthoblique	BMDP4M SAS (HK)	Both orthogonal and oblique	Rescale factor loadings to yield orthogonal solution; nonscaled loadings may be correlated	Accompanies Kaiser's Second Little Jiffy (image) extraction in BMDP4M
Promax	SAS	Oblique	Orthogonal factors rotated to oblique positions	Fast and inexpensive
Procrustes	SAS	Oblique	Rotate to target matrix	Useful in confirmatory FA

Although varimax rotation simplifies the factors, quartimax the variables, and equamax both, they do so in BMDP4M and SAS FACTOR by setting levels on a simplicity criterion—Γ (gamma)—of 1, 0, and 1/2, respectively. Gamma can also be continuously varied between 0 (variables simplified) and 1 (factors simplified) by using the orthogonal rotation that allows the user to specify Γ level.

For many applications, varimax is the rotation of choice; it is the default option of all four packages.

12.5.2.2 Oblique Rotation An *embarrasse de richesse* awaits the researcher who uses oblique rotation (see Table 12.11). Oblique rotations offer a continuous range of correlations between factors. The amount of correlation permitted between factors is determined by a variable called delta (δ) by SPSS[x] FACTOR and gamma (Γ) by BMDP4M.[8] The values of delta and gamma determine the maximum amount of correlation permitted among factors. When the value is less than zero, solutions are increasingly orthogonal; at about -4 the solution is orthogonal. When the value is zero, solutions can be fairly highly correlated. Values near 1 can produce factors that are very highly correlated. Although there is a relationship between values of delta or gamma and size of correlation, the maximum correlation at a given size of gamma or delta depends on the data set.

It should be stressed that factors do not necessarily correlate when an oblique rotation is used. Often, in fact, they do not correlate and the researcher reports the simpler orthogonal rotation.

The family of procedures used for oblique rotation with varying degrees of correlation in SPSS and BMDP is direct oblimin. In the special case where Γ or δ = 0 (the default option for both SPSS and BMDP), the procedure is called direct quartimin. Values of gamma or delta greater than zero permit high correlations among factors, and the researcher should take care that the correct number of factors is chosen. Otherwise highly correlated factors may be indistinguishable one from the other. Some trial and error, coupled with inspection of the scatterplots of relationships between pairs of factors, may be required to determine the most useful size of gamma or delta. Or, one might simply trust to the default value.

Orthoblique rotation is designed to accompany Kaiser's Second Little Jiffy (Image) Factor extraction, and does so automatically through BMDP4M. Orthoblique rotation uses the quartimax algorithm to produce an orthogonal solution *on rescaled factor loadings;* therefore the solution may be oblique with respect to the original factor loadings.

Promax and Procrustes are available through SAS. In promax rotation, an orthogonally rotated solution (usually varimax) is rotated again to allow correlations among factors. The orthogonal loadings are raised to powers (usually powers of 2, 4, or 6) to drive small and moderate loadings to zero while larger loadings are reduced, but not to zero. Even though factors correlate, simple structure is maximized by clarifying which variables do and do not correlate with each factor. Promax has the additional advantage of being fast and inexpensive.

---

[8] In BMDP, gamma is used to indicate the nature of simplicity in orthogonal rotation and the amount of obliqueness in oblique rotation.

In Procrustes rotation, a target matrix of loadings (usually zeros and ones) is specified by the researcher and a transformation matrix is sought to rotate extracted factors to the target, if possible. If the solution can be rotated to the target, then the hypothesized factor structure is said to be confirmed. Unfortunately, as Gorsuch (1983) reports, with Procrustean rotation, factors are often extremely highly correlated and sometimes a correlation matrix generated by random processes is rotated to a target with apparent ease.

12.5.2.3 Geometric Interpretation A geometric interpretation of rotation is in Figure 12.1 where 12.1(a) is the unrotated and 12.1(b) the rotated solution to the example in Table 12.2. Points are represented in two-dimensional space by listing their co-ordinates with respect to X and Y axes. With the first two unrotated factors as axes, unrotated loadings are COST $(-.400, .900)$, LIFT $(.251, -.947)$, DEPTH $(.932, .348)$, and POWDER $(.956, .286)$.

The points for these variables are also located with respect to the first two rotated factors as axes in Figure 12.1(b). The position of points does not change, but their coordinates change in the new axis system. COST is now $(-0.86, .981)$, LIFT $(-.071, -.977)$, DEPTH $(.994, .026)$, and POWDER $(.997, -.040)$. Statistically, the effect of rotation is to amplify high loadings and reduce low ones. Spatially, the effect is to rotate the axes so that they "shoot through" the variable clusters more closely.

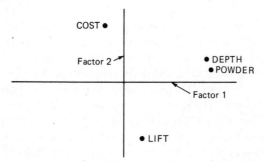

(a) Location of COST, LIFT, DEPTH, and POWDER after extraction (before rotation)

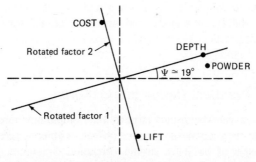

(b) Location of COST, LIFT, DEPTH, and POWDER vis-à-vis rotated axes

Figure 12.1 Illustration of rotation of axes to provide a better definition of factors vis à vis the variables with which they correlate.

Factor extraction yields a solution in which observed variables are vectors that terminate at the points indicated by the coordinate system. The factors serve as axes for the system. The coordinates of each point are the entries from the loading matrix for the variable. If there are three factors, then the space has three axes and three dimensions, and each observed variable is positioned by three coordinates. The length of the vector for each variable is the communality of the variable.

If the factors are orthogonal, the factor axes are all at right angles to one another and the coordinates of the variable points are correlations between the common factors and the observed variables. Correlations (factor loadings) are read directly from these graphs by projecting perpendicular lines from each point to each of the factor axes.

One of the primary goals of PCA or FA, and the motivation behind extraction, is to discover the minimum number of factor axes needed to reliably position variables. A second major goal, and the motivation behind rotation, is to discover the meaning of the factors that underlie responses to observed variables. This goal is met by interpreting the factor axes that are used to define the space. Factor rotation repositions factor axes so as to make them maximally interpretable. Repositioning the axes changes the coordinates of the variable points but not the positions of the points with respect to each other.

Factors are usually interpretable when some observed variables load highly on them and the rest do not. And, ideally, each variable loads on one, and only one, factor. In graphic terms this means that the point representing each variable lies far out along one axis but near the origin on the other axes, that is, that coordinates of the point are large for one axis and near zero for the other axes.

If you have only one observed variable, it is trivial to position the factor axis— variable point and axis overlap in a space of one dimension. However, with many variables and several factor axes, compromises are required in positioning the axes. The variables form a 'swarm'' in which variables that are correlated with one another form a cluster of points. The goal is to shoot an axis to the swarm of points. With luck, the swarms are about 90° away from one another so that an orthogonal solution is indicated. And with lots of luck, the variables cluster in just a few swarms with empty spaces between them so that the factor axes are nicely defined.

In oblique rotation the situation is slightly more complicated. Because factors may correlate with one another, factor axes are not necessarily at right angles. And, though it is easier to position each axis near a cluster of points, axes may be very near each other (highly correlated), making the solution harder to interpret. See Section 12.6.3 for practical suggestions of ways to use graphic techniques to judge the adequacy of rotation.

12.5.3 Some Practical Recommendations

Although an almost overwhelmingly large number of combinations of extraction and rotation techniques is available, in practice differences among them are often slight. With a large number of variables, strong correlations among them, with the same, well-chosen number of factors, and with similar values for communality, the results of extraction are similar regardless of which method is used. Further, differences that are apparent after extraction tend to disappear after rotation.

Most researchers begin their FA by using principal components extraction and varimax rotation. From the results, one estimates the factorability of the correlation matrix (Sections 12.3.2.6 and 12.8.1.6), the rank of the observed correlation matrix (Sections 12.3.2.5 and 12.8.1.5), the number of factors (Sections 12.6.2), and variables that might be excluded from subsequent analyses (Sections 12.3.2.7 and 12.8.1.7).

During the next few runs, researchers experiment with different numbers of factors, different extraction techniques, and both orthogonal and oblique rotations. Some number of factors with some combination of extraction and rotation produces the solution with the greatest scientific utility, consistency, and meaning; this is the solution that is interpreted.

12.6 SOME IMPORTANT ISSUES

Some of the issues raised in this section can be resolved through several different methods. Usually different methods lead to the same conclusion; occasionally they do not. When they do not, results are judged by the interpretability and scientific utility of the solutions.

12.6.1 Estimates of Communalities

FA differs from PCA in that communality values (numbers between 0 and 1) replace ones in the positive diagonal of **R** before factor extraction. Communality values are used instead of ones to remove the unique and error variance of each observed variable; only the variance a variable shares with the factors is used in the solution. But communality values are estimated, and there is some dispute regarding how that should be done.

The SMC of each variable as DV with the others in the sample as IVs is usually the starting estimate of communality. As the solution develops, communality estimates are adjusted by iterative procedures (which can be directed by the researcher) to fit the reproduced to the observed correlation matrix with the smallest number of factors. Iteration stops when successive communality estimates are very similar.

Final estimates of communality are also SMCs, but now between each variable as DV and the factors as IVs. Final communality values represent the proportion of variance in a variable that is predictable from the factors underlying it. Communality estimates do not change with orthogonal rotation.

Image extraction and maximum likelihood extraction are slightly different. In image extraction, SMCs of each variable with all other variables are used as the communality values throughout. Image extraction produces a mathematically unique solution because communality values are not changed. In maximum likelihood extraction, number of factors instead of communality values are estimated and off-diagonal correlations are "rigged" to produce the best fit between observed and reproduced matrices.

BMDP, SPSS^x, and SAS provide several different starting statistics for communality estimation. BMDP4M offers SMCs, user-specified values, or maximum absolute correlation with any other variable as initial communality estimates. SPSS[x] FACTOR permits user supplied values for principal factor extraction only, but otherwise uses SMCs. SAS FACTOR offers, for each variable, a choice of SMC, SMC adjusted so that the sum of the communalities is equal to the sum of the maximum absolute correlations, maximum absolute correlation with any other variable, user-specified values, or random numbers between 0 and 1. Fewer iterations are usually required when starting from SMCs.

The seriousness with which estimates of communality should be regarded depends on the number of observed variables. If the number of variables exceeds, say, 20, sample SMCs probably provide reasonable estimates of communality. Furthermore, with 20 or more variables, the elements in the positive diagonal are few compared with the total number of elements in **R**, and their sizes do not influence the solution very much. Actually, if the communality values for all variables in FA are of approximately the same magnitude, results of PCA and FA are very similar.

If communality values equal or exceed 1, problems with the solution are indicated. There are too few data, or starting communality values are wrong, or the number of factors extracted is wrong; addition or deletion of factors may reduce the communality below 1. Very low communality values, on the other hand, indicate that the variables with them are outliers (Sections 12.3.2.7 and 12.8.1.7).

12.6.2 Adequacy of Extraction and Number of Factors

Because inclusion of more factors in a solution improves the fit between observed and reproduced correlation matrices, adequacy of extraction is tied to number of factors. The more factors extracted, the better the fit and the greater the percent of variance in the data "explained" by the factor solution. However, the more factors extracted, the less parsimonious the solution. To account for all the variance (PCA) or covariance (FA) in a data set, one would normally have to have as many factors as observed variables. It is clear, then, that a trade-off is required: One wants to retain enough factors for an adequate fit, but not so many that parsimony is lost.

Selection of the number of factors is probably more critical than selection of extraction and rotational techniques or communality values. In confirmatory FA, selection of the number of factors is really selection of the number of theoretical processes underlying a research area. You can partially confirm a hypothesized factor structure by asking if the theoretical number of factors adequately fits the data.

There are several ways to assess adequacy of extraction and number of factors. For a highly readable summary of these methods, not all currently available through the statistical packages, see Gorsuch (1983). Reviewed below are methods available through SPSS[x], BMDP, SAS, and SYSTAT. SYSTAT provides the first two and SPSS[x], BMDP, and SAS all the methods described here.

A first quick estimate of the number of factors is obtained from the sizes of the eigenvalues reported as part of an initial run with principal components extraction. Eigenvalues represent variance. Because the variance that each standardized variable contributes to a principal components extraction is 1, a component with an eigenvalue less than 1 is not as important, from a variance perspective, as an observed variable.

The number of components with eigenvalues greater than 1 is usually somewhere between the number of variables divided by 3 and the number of variables divided by 5 (e.g., 20 variables should produce between 7 and 4 components with eigenvalues greater than 1). If this is a reasonable number of factors for the data, if the number of variables is 40 or fewer, and if sample size is large, the number of factors indicated by this criterion is probably about right. In other situations, this criterion may either over- or underestimate the number of factors in the data set.

A second criterion is the scree test (Cattell, 1966) of eigenvalues plotted against factors. Factors, in descending order, are arranged along the abscissa with eigenvalue as the ordinate. The plot is appropriately used with principal components or factor analysis at initial and later runs to find the number of factors. The scree plot is available through SPSS[x] FACTOR and SAS FACTOR, but it is easy to produce from the list of factors and eigenvalues available through BMDP and SYSTAT.

Usually the scree plot is negatively decreasing—the eigenvalue is highest for the first factor and moderate but decreasing for the next few factors before reaching small values for the last several factors, as illustrated for real data through SPSS[x] in Figure 12.2. What you are looking for is the point where a line drawn through the points changes direction. In the example, a single straight line can comfortably fit the first four eigenvalues. After that, another line, with a noticeably different slope, best fits the remaining eight points. Therefore, there appear to be about four factors in the data of Figure 12.2.

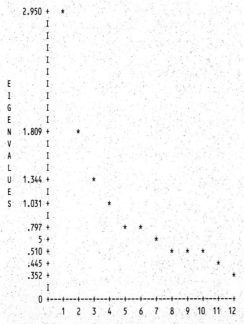

Figure 12.2 Screeoutput for sample data produced by SPSS[x] FACTOR. Note break in size of eigenvalues between the fourth and fifth factors.

Unfortunately, the scree test is not exact; it involves judgment of where the discontinuity in eigenvalues occurs and researchers are not perfectly reliable judges. As Gorsuch (1983) reports, results of the scree test are more obvious (and reliable) when sample size is large, communality values are high, and each factor has several variables with high loadings. Under less than optimal conditions, the scree test is still usually accurate to within one or two factors. If you are unsure of the number of factors, perform several factor analyses, each time specifying a different number of factors, repeating the scree test, and examining the residual correlation matrix.

The residual correlation matrix is available through SPSS[x], BMDP, and SAS. As discussed in Section 12.4, the residual correlation matrix is obtained by subtracting the reproduced correlation matrix from the observed correlation matrix. The numbers in the residual matrix are actually partial correlations between pairs of variables with effects of factors removed. If the analysis is good, the residuals are small. Several moderate residuals (say, .05 to .10) or a few large residuals (say $>.10$) suggest the presence of another factor.

Once you have determined the number of factors by these criteria, it is important to look at the rotated loading matrix to determine the number of variables that load on each factor (see Section 12.6.5). If only one variable loads highly on a factor, the factor is poorly defined. If two variables load on a factor, then whether or not it is reliable depends on the pattern of correlations of these two variables with each other and with other variables in **R**. If the two variables are highly correlated with each other (say, $r > .70$) and relatively uncorrelated with other variables, the factor may be reliable. Interpretation of factors defined by only one or two variables is cautious, however, under even the most exploratory factor analysis.

For principal components extraction and maximum likelihood extraction in confirmatory factor analysis there are significance tests for number of factors. Bartlett's test evaluates all factors together and each factor separately against the hypothesis that there are no factors. However, there is some dispute regarding use of these tests. The interested reader is referred to Gorsuch (1983) or one of the other newer factor analysis texts for discussion of significance testing in FA.

There is debate about whether it is better to retain too many or too few factors if the number is ambiguous. Sometimes a researcher wants to rotate, but not interpret, marginal factors for statistical purposes (e.g., to keep all communality values < 1). Other times the last few factors represent the most interesting and unexpected findings in a research area. These are good reasons for retaining factors of marginal reliability. However, if the researcher is interested in using only demonstrably reliable factors, the fewest possible factors are retained.

12.6.3 Adequacy of Rotation and Simple Structure

The decision between orthogonal and oblique rotation is made as soon as the number of reliable factors is apparent. In many factor analytic situations, oblique rotation seems more reasonable on the face of it than orthogonal rotation because it seems more likely that factors are correlated than that they are not. However, reporting the results of oblique rotation requires reporting the elements of the pattern matrix (**A**) and the factor correlation matrix (**φ**), whereas reporting orthogonal rotation requires only the loading matrix **(A)**. Thus simplicity of reporting results favors orthogonal rotation. Further, if factor scores or factorlike scores (Section 12.6.6) are to be used as IVs or DVs in other analyses, or if a goal of analysis is comparison of factor structure in groups (Section 12.6.7), then orthogonal rotation has distinct advantages.

Perhaps the best way to decide between orthogonal and oblique rotation is to request oblique rotation with the desired number of factors and look at the correlations among factors. The oblique rotations available by default in SPSS^x, BMDP, and SAS calculate factors that are fairly highly correlated if necessary to fit the data. However, if factor correlations are not driven by the data, the solution remains nearly orthogonal.

Look at the factor correlation matrix for correlations of .30 and above. If correlations exceed .30, then there is 10% (or more) overlap in variance among factors, enough variance to warrant oblique rotation unless there are compelling reasons for orthogonal rotation. Compelling reasons include a desire to compare structure in groups, a need for orthogonal factors in other analyses, or a theoretical need for orthogonal rotation.

Once the decision is made between orthogonal and oblique rotation, the *adequacy* of rotation is assessed several ways. Perhaps the simplest way is to compare the pattern of correlations in the correlation matrix with the factors. Are the patterns represented in the rotated solution? Do highly correlated variables tend to load on the same factor? If you included marker variables, do they load on the predicted factors?

Another criterion is simple structure (Thurstone, 1947). If simple structure is present (and factors are not too highly correlated), several variables correlate highly with each factor and only one factor correlates highly with each variable. In other words, the columns of **A,** which define factors, have several high and many low values while the rows of **A,** which define variables vis-à-vis factors, have only one high value. Rows with more than one high correlation correspond to variables that are said to be complex because they reflect the influence of more than one factor. It is usually best to avoid complex variables because they make interpretation of factors more ambiguous.

Adequacy of rotation is also ascertained through the PLOT commands of SPSS^x FACTOR, SAS FACTOR, and BMDP4M. In the figures, factors are considered two at a time with a different pair of factors as axes for each plot. Look at the *distance, clustering,* and *direction* of the points representing variables relative to the factor axes in the figures.

The *distance* of a variable point from the origin reflects the size of factor loadings; variables highly correlated with a factor are far out on that factor's axis. Ideally, each variable point is far out on one axis and near the origin on all others. *Clustering* of variable points reveals how clearly defined a factor is. One likes to see a cluster of several points near the end of each axis and all other points near the origin. A smattering of points at various distances along the axis indicates a factor that is not clearly defined, while a cluster of points midway between two axes reflects the presence of another factor or the need for oblique rotation. The *direction* of clusters after orthogonal rotation may also indicate the need for oblique rotation. If clusters of points fall between factor axes after orthogonal rotation, if the angle between clusters with the respect to the origin is not 90°, then a better fit to the clusters is

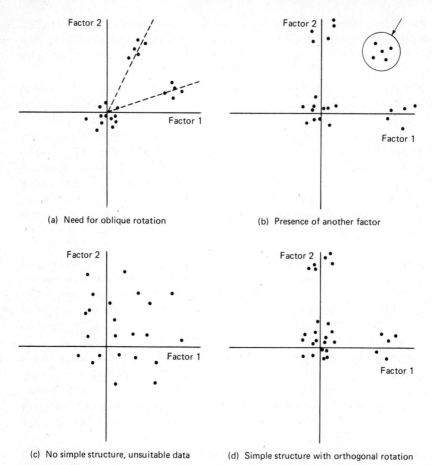

Figure 12.3   Pairwise plots of factor loadings following orthogonal rotation and indicating: (a) need for oblique rotation; (b) presence of another factor; (c) unsuitable data; and (d) simple structure.

provided by axes that are not orthogonal. Oblique rotation may reveal substantial correlations among factors. Several of these relationships are depicted in Figure 12.3.

12.6.4 Importance and Internal Consistency of Factors

The importance of a factor (or a set of factors) is evaluated by the proportion of variance or covariance associated with the factor after rotation. The proportion of variance attributable to individual factors differs before and after rotation because rotation tends to redistribute variance among factors somewhat. Ease of ascertaining proportions of variance for factors depends on whether rotation was orthogonal or oblique.

After orthogonal rotation, the importance of individual factors is related to the sizes of their SSLs (*S*um of *S*quared *L*oadings from **A** after rotation). SSLs are converted to proportion of variance for a factor by dividing its SSL by p, the number of variables. SSLs are converted to proportion of covariance for a factor by dividing its SSL by the sum of SSLs or, equivalently, sum of communalities. These computations are illustrated in Table 12.4 and Section 12.5 for the example.

The proportion of variance accounted for by a factor is the amount of variance in the original variables (where each has contributed one unit of variance) that has been condensed into the factor. Proportion of *variance* is the variance of a factor relative to the variance in the variables. The proportion of covariance accounted for by a factor indicates the relative importance of the factor to the total covariance accounted for by all factors. Proportion of *covariance* is the variance of a factor relative to the variance in the solution. The variance in the solution is likely to account for only a fraction of the variance in the original variables.

In oblique rotation, proportions of variance and covariance can be obtained from **A** *before* rotation by the methods just described, but they are only rough indicators of the proportions of variance and covariance of factors after rotation. Because factors are correlated, they share overlapping variability, and assignment of variance to individual factors is ambiguous. After oblique rotation the size of the SSL associated with a factor is a rough approximation of its importance—factors with bigger SSLs are more important—but proportions of variance and covariance cannot be specified.

An estimate of the internal consistency of the solution—the certainty with which factor axes are fixed in the variable space—is given by the squared multiple correlations of factor scores predicted from scores on observed variables. In a good solution, SMCs range between 0 and 1; the larger the SMCs, the more stable the factors. A high SMC (say, .70 or better) means that the observed variables account for substantial variance in the factor scores. A low SMC means the factors are poorly defined by the observed variables. If an SMC is negative, too many factors have been retained. If an SMC is above 1, the entire solution needs to be reevaluated.

BMDP4M prints these SMCs as the positive diagonal of the factor-score covariance matrix. SPSS[x] FACTOR prints them as the diagonal of the covariance matrix for estimated regression factor scores. In SAS FACTOR, SMCs are printed along with factor score coefficients by the SCORE option.

12.6.5 Interpretation of Factors

To interpret a factor, one tries to understand the underlying dimension that unifies the group of variables loading on it. In both orthogonal and oblique rotations, loadings are obtained from the loading matrix, **A,** but the meaning of the loadings is different for the two rotations.

After orthogonal rotation, the values in the loading matrix are correlations between variables and factors. The researcher decides on a criterion for meaningful correlation (usually .30 or larger), collects together the variables with loadings in excess of the criterion, and searches for a concept that unifies them.

After oblique rotation, the process is the same, but the interpretation of the values in **A,** the pattern matrix, is no longer straightforward. The loading is not a correlation but is a measure of the unique relationship between the factor and the variable. Because factors correlate, the correlations between variables and factors (available in the structure matrix, **C**) are inflated by overlap between factors. A variable may correlate with one factor through its correlation with another factor rather than directly. The elements in the pattern matrix have overlapping variance among factors "partialled out," but at the expense of conceptual simplicity.

Actually, the reason for interpretation of the pattern matrix rather than the structure matrix is pragmatic—it's easier. The difference between high and low loadings is more apparent in the pattern matrix than in the structure matrix.

As a rule of thumb, only variables with loadings of .30 and above are interpreted. The greater the loading, the more the variable is a pure measure of the factor. Comrey (1973) suggests that loadings in excess of .71 (50% overlapping variance) are considered excellent, .63 (40% overlapping variance) very good, .55 (30% overlapping variance) good, .45 (20% overlapping variance) fair, and .32 (10% overlapping variance) poor. Choice of the cutoff for size of loading to be interpreted is a matter of researcher preference. Sometimes there is a gap in loadings across the factors and, if the cutoff is in the gap, it is easy to specify which variables load and which do not. Other times, the cutoff is selected because one can interpret factors with that cutoff but not with a lower cutoff.

The size of loadings is influenced by the homogeneity of scores in the sample. If homogeneity is suspected, interpretation of lower loadings is warranted. That is, if the sample produces similar scores on observed variables, a lower cutoff is used for interpretation of factors.

At some point, a researcher usually tries to characterize a factor by assigning it a name or a label, a process that involves art as well as science. Rummel (1970) provides numerous helpful hints on interpreting and naming factors. Interpretation of factors is facilitated by output of the matrix of sorted loadings where variables are grouped by their correlations with factors. Sorted loadings are produced routinely by BMDP4M, by REORDER in SAS FACTOR, and SORT in SPSS[x] FACTOR.

The replicability, utility, and complexity of factors are also considered in interpretation. Is the solution replicable in time and/or with different groups? Is it trivial or is it a useful addition to scientific thinking in a research area? Where do the factors fit in the hierarchy of "explanations" about a phenomenon? Are they complex enough to be intriguing without being so complex that they are uninterpretable?

12.6.6 Factor Scores

Among the potentially more useful outcomes of PCA or FA are factor scores. Factor scores are estimates of the scores subjects would have received on each of the factors had they been measured directly.

Because there are normally fewer factors than observed variables, and because factor scores are nearly uncorrelated if factors are orthogonal, use of factor scores in other analyses may be very helpful. Multicollinear matrices can be reduced to orthogonal components using PCA, for instance. Or, one could use FA to reduce a large number of DVs to a smaller number of factors for use as DVs in MANOVA. Alternatively, once could reduce a large number of IVs to a small number of factors for purposes of predicting a DV in multiple regression or group membership in discriminant analysis. If factors are few in number, stable, and interpretable, their use enhances subsequent analyses.

Procedures for estimating factor scores range between simple-minded (but frequently adequate) and sophisticated. Comrey (1973) describes several rather simple-minded techniques for estimating factor scores. Perhaps the simplest is to sum scores on variables that load highly on each factor. Variables with bigger standard deviations contribute more heavily to the factor scores produced by this procedure, a problem that is alleviated if variable scores are standardized first or if the variables have roughly equal standard deviations to begin with. For many research purposes, this "quick and dirty" estimate of factor scores is entirely adequate.

There are several sophisticated statistical approaches to estimating factors. All produce factor scores that are correlated, but not perfectly, with the factors. The correlations between factors and factor scores are higher when communalities are higher and when the ratio of variables to factors is higher. But as long as communalities are estimated, factor scores suffer from indeterminacy because there is an infinite number of possible factor scores that all have the same mathematical characteristics and there is no way to decide among them. As long as factor scores are considered only estimates, however, the researcher is not overly beguiled by them.

The method described in Section 12.4 (especially Equations 12.10 and 12.11) is the regression approach to estimating factor scores. This approach results in the highest correlations between factors and factor scores. The distribution of each factor's scores has a mean of zero and a standard deviation of 1 (after PCA) or equal to the SMC between factors and variables (after FA). However, this regression method, like all others (see Chapter 5), capitalizes on chance relationships among variables so that factor-score estimates are biased (too close to "true" factor scores). Further, there are often correlations among scores for factors even if factors are orthogonal and factor scores sometimes correlate with other factors (in addition to the one they are estimating).

The regression approach to estimating factor scores is available through SYSTAT and the other three packages. All four packages write factor scores to files for use in other analyses and all four print standardized factor score coefficients—routinely through SYSTAT and BMDP, and through FSCORE in SPSS^x and SCORE in SAS.

SPSS^x FACTOR provides two additional methods of estimating factor scores. In the Bartlett method, factor scores correlate only with their own factors and the factor scores are unbiased (that is, neither systematically too close nor too far away from "true" factor scores). The factor scores correlate with the factors almost as well as in the regression approach and have the same mean and standard deviation as in the regression approach. However, factor scores may still be correlated with each other.

The Anderson-Rubin approach (discussed by Gorsuch, 1983) produces factor scores that are uncorrelated with each other even if factors are correlated. Factor scores have mean zero, standard deviation 1. Factor scores correlate with their own factors almost as well as in the regression approach, but they sometimes also correlate with other factors (in addition to the one they are estimating) and they are somewhat biased. If you need uncorrelated scores, the Anderson-Rubin approach is best; otherwise the regression approach is probably best simply because it is best understood and most widely available.

12.6.7 Comparisons among Solutions and Groups

Frequently a researcher is interested in deciding whether or not two groups that differ in experience or characteristics have the same factors. Comparisons among factor solutions involve the *pattern* of the correlations between variables and factors, or both the *pattern and magnitude* of the correlations between them. Rummel (1970), Levine (1977), and Gorsuch (1983) have excellent summaries of several comparisons that might be of interest. Only the easier comparison techniques are mentioned here.

It is important to note that theory can be tested in FA using these procedures. Theory regarding factor structure is used to generate one set of loadings that is compared with loadings derived from a sample. Estimation of the magnitude of factor loadings for variables from theory does not have to be very precise: ones can be used as loadings for variables that are expected to load on a factor, while zeros are used as loadings for the other variables. Comparisons between the *pattern* of loadings from theory and loadings from sample data are then conducted in confirmatory FA.

The first step in comparing factors from two different samples is to generate them. When comparison is the goal, similar procedures are employed at the various stages of analysis with the two data sets. Similar variables and, if possible, similar marker variables are measured during data collection. Similar procedures for handling missing data and outliers are employed. The same considerations are used to produce transformations of variables, if desired. Extractional and rotational techniques are the same, as is the criterion for determining number of factors. If factor scores are to be compared, they are generated by the same procedures.

Once data sets are factor analyzed, one decides which pairs of factors to compare. Comparing all possible pairs of factors can result in spuriously significant results by capitalizing on chance relationships. Presence of marker variables simplifies choosing pairs.

Before going on, careful inspection of the loading matrices for both groups may reveal similarities or differences in factor structure sufficiently clear as to obviate the need for more formal procedures. Did both groups generate the same number of factors? Do almost the same variables load highly on the different factors for the two groups? Could you reasonably use the same labels to name factors for both groups? If all three questions are answered in the affirmative, it is unnecessary to proceed to statistical comparisons.

If formal procedures are needed, an important decision is whether to compare just the pattern of loadings or both the pattern and magnitude of loadings in the data sets. Comparisons involving both pattern and magnitude are more stringent than those involving just pattern. Magnitude of loadings is influenced by extraneous features of data collection such as homogeneity of a sample for factors being compared, so magnitude is considered when the researcher believes that these influences are absent.

Cattell's salient similarly index, s (Cattell and Baggaley, 1960; Cattell, 1957), is used to compare patterns of loadings. The Pearson product-moment correlation coefficient, r, is used to compare both pattern and magnitude of loadings.

To illustrate these two methods, the two loading matrices in Table 12.12 are used. Both are products of overactive imagination but do illustrate a typical problem in factor comparison, namely, that factor 1 in Set 1 is similar to factor 2 (rather than

TABLE 12.12 LOADING MATRICES FROM TWO HYPOTHETICAL DATA SETS.

	Set 1		Set 2	
	Factor 1	Factor 2	Factor 1	Factor 2
COST	−.086	.981	.732	.265
LIFT	−.072	−.978	.649	.537
DEPTH	.994	.027	.211	.874
POWDER	.997	−.040	.189	.796

to factor 1) in Set 2. Sometimes it is hard to decide which factors to compare. But if the decision is that difficult, perhaps you have your answer.

In calculating s, the first step is to construct a two-way frequency table, like that in Table 12.13, with pairs of loadings for each variable on each factor contributing a single tally to the table according to whether the loadings are positively salient (PS), negatively salient (NS), or neither (hyperplane or HP) on the two factors being compared. Cattell used a cut of .10 for determining salience; loadings at or above .10 were salient while lower ones were not. But a cut of .32 or higher is better for this example and is employed here.

For the COST variable, the loading of −.086 in Set 1 is in the hyperplane as

TABLE 12.13 CALCULATION OF CATTELL'S SALIENT SIMILARITY INDEX s

		Set 1		
		PS	HP	NS
Set 2	PS	c_{11}	c_{12}	c_{13}
	HP	c_{21}	c_{22}	c_{23}
	NS	c_{31}	c_{32}	c_{33}

For the example:

		Set 1		
		PS	HP	NS
Set 2	PS	2	1	
	HP		1	
	NS			

$$s = \frac{2 + 0 - 0 - 0}{2 + 0 + 0 + 0 + .5(1 + 0 + 0 + 0)}$$

$$= \frac{2}{2.5}$$

$$= .80$$

is the loading of .265 in Set 2. Therefore, a tally is placed in the c_{22} cell of the table. For the LIFT variable, $-.072$ is HP while .537 is PS, resulting in a tally in cell c_{12}. For DEPTH, .994 and .874 are both PS as, for POWDER, are both .997 and .796. These give the two tallies in cell c_{11}.

Once the frequency table is constructed, s is calculated as follows:

$$s = \frac{c_{11} + c_{33} - c_{13} - c_{31}}{c_{11} + c_{33} + c_{13} + c_{31} + .5(c_{12} + c_{21} + c_{23} + c_{32})} \tag{12.16}$$

The c values in the equation are replaced by frequency counts from cells in the frequency table. Application of the equation for comparison of factor 1 in Set 1 and factor 2 in Set 2 results in an s value of .80.

Estimates of probability values for s are provided by Cattell and colleagues (1969) and reproduced in Appendix C, Table C.7. Probabilities are assessed considering both the number of variables, p, and the percentage of cases that fall into the hyperplane for the pair of factors being compared: 60%, 70%, 80%, or 90%. If a value of s exceeds that of v_s for some hyperplane percentage and number of variables, then the factors are reliably similar. For instance, if the hyperplane count is 60% and 40 variables are compared, an s value in excess of .26 indicates similarity of factors at the .041 significance level. (Significance for the example cannot be determined because only 25% of the loadings are in the hyperplane and only four variables are included.)

The pattern and magnitude of loadings are compared for two factors and two groups by computing Pearson's r (see Equation 3.29). The loadings for factor 1 in Set 1 and factor 2 in Set 2 from Table 12.12 correlate .91. Although calculating r for loadings is a straightforward procedure and the meaning of r is widely understood, this method of comparing factors has drawbacks. If there are numerous variables, it is possible for r to be large even though no variables with large loadings are the same for the two factors. The correlation is large because of the numerous variables with small loadings that are not loaded on either factor. Thus, caution is urged in interpretation of r used to compare factors.

Another method, experimental at this stage, involves generating pairs of factor scores for a group by using the factor-score coefficients for that group and then generating factor scores by using factor-score coefficients for the other group. The pairs of factor scores are correlated. If correlation is high, it implies that there is good correspondence between factor scores generated by the two different groups and that factor structure is therefore similar.

12.7 COMPARISON OF PROGRAMS

BMDP, SPSS^x, and SAS each have a single program to handle both FA and PCA. All three programs have numerous options for extraction and rotation and give the user considerable latitude in directing the progress of the analysis. The programs are all flexible and rich in information. The SYSTAT FACTOR 3.0 program is limited to PCA and orthogonal rotation. Features of all four programs are described in Table 12.14.

TABLE 12.14 COMPARISON OF FACTOR ANALYSIS PROGRAMS

Feature	SPSS[x] FACTOR	BMDP4M	SAS FACTOR	SYSTAT FACTOR
Input				
Correlation matrix	Yes	Yes	Yes	Yes
About origin	No	Yes	Yes	No
Covariance matrix	No	Yes	Yes	Yes
About origin	No	Yes	Yes	No
Factor loadings (unrotated pattern)	Yes	Yes	Yes	No
Factor-score coefficients	No	Yes	Yes	No
Factor loadings (rotated pattern) and factor correlations	Yes	No	Yes	No
Specify maximum number of factors	Yes	Yes	Yes	Yes
Extraction method (see Table 12.9)				
PCA	Yes	Yes	Yes	Yes
PFA	Yes	Yes	Yes	No
Image (Little Jiffy, Harris)	Yes	Yes	Yes[a]	No
Maximum likelihood (Rao's canonical)	Yes	Yes	Yes	No
Alpha	Yes	No	Yes	No
Unweighted least squares	Yes[b]	No	Yes	No
Generalized least squares	Yes[b]	No	Yes	No
Specify communalities	Yes	Yes	Yes	N.A.
Specify minimum eigenvalue	Yes	Yes	Yes	Yes
Specify proportion of variance to be accounted for	No	No	Yes	No
Specify maximum number of iterations	Yes	Yes	Yes	N.A.
Option to allow communalities > 1	No	No	Yes	N.A.
Specify tolerance	No	Yes	Yes	No
Specify convergence criterion	Yes	Yes	Yes	N.A.
Rotation method (see Table 12.11)				
Varimax	Yes	Yes	Yes	Yes
Quartimax	Yes	Yes	Yes	Yes
Equamax	Yes	Yes	Yes	Yes
Direct oblimin	Yes	Yes	No	No
Direct quartimin	$\delta = 0$	Yes	No	No
Indirect oblimin	No	Yes	No	No
Orthoblique	No	Yes	HK	No
Promax	No	No	Yes	No
Procrustes	No	No	Yes	No
Prerotation criteria	No	No	Yes	No
Optional Kaiser's normalization	Yes[c]	Yes	Yes	Normalized only
Optional weighting by Cureton-Mulaik technique	No	No	Yes	No
Optional rescaling of pattern matrix to covariances	No	No	Yes	No
Maximum numbers of factors	No limit	10[d]	No limit	No limit
Differential case weighting	No	Yes	No	No
Differential variable weighting	No	No	Yes	No
Output				
Means and standard deviations	Yes	Yes	Yes	No
Number of cases per variable (missing data)	Yes	No	No	No

TABLE 12.14   (Continued)

Feature	SPSS^x FACTOR	BMDP4M	SAS FACTOR	SYSTAT FACTOR
Coefficients of variation	No	Yes	No	No
Minimums and maximums	No	Yes	No	No
z for minimums and maximums	No	Yes	No	No
First case for minimums and maximums	No	Yes	No	No
Correlation matrix	Yes	Yes	Yes	Yes
Significance of correlations	Yes[b]	No	No	No
Covariance matrix	No	Yes	Yes	Yes
Initial communalities	Yes	Yes	Yes	N.A.
Final communalities	Yes	Yes	Yes	N.A.
Eigenvalues	Yes	Yes	Yes	Yes
Difference between successive eigenvalues	No	No	Yes	No
Carmine's theta	No	Yes	No	No
Percent of variance explained by factors	Yes	No	No	No
Cumulative percent of variance	Yes	No	No	No
Percent of covariance	No	Yes	Yes	Yes
Cumulative percent of covariance	No	Yes[e]	Yes	Yes
Unrotated factor loadings	Yes	Yes[e]	Yes	Yes
Variance explained by factors for all loading matrices	No	Yes	Yes	Yes
Simplicity criterion, each iteration	δ^f	γ	No	No
Rotated factor loadings (pattern)	Yes	Yes	Yes	Yes[9]
Rotated factor loadings (structure)	Yes	Yes	Yes	Yes[9]
Transformation matrix	Yes	No	Yes	No
Factor-score coefficients	Yes	Yes	Yes	Data file
Factor scores	Data file	Yes	Data file	Data file
Scree plot	Yes[b]	No	Yes	No
Plots of unrotated factor loadings	No	Yes	Yes	No
Plots of rotated factors loadings	Yes[i]	Yes	Yes	Yes
Plots of factor scores	No	Yes	No[j]	No[h]
Sorted rotated factor loadings	Yes	Yes	Yes	Yes
Inverse of correlation matrix	Yes	Yes	Yes	Yes
Determinant of correlation matrix	Yes	No	No	No
Factor-score covariance matrices	No	Yes	No	No
Mahalanobis distance of cases	No	Yes	No	No
Standard scores—each variable for each case	No	Yes	No	No
Partial correlations (anti-image matrix)	Yes[b]	Yes	Yes	No
Measure of sampling adequacy	Yes[b]	No	Yes	No
Anti-image covariance matrix	Yes[b]	No	No	No
Bartlett's test of sphericity	Yes[b]	No	No	No
Residual correlation matrix	Yes[b]	Yes	Yes	No
Reproduced correlation matrix	Yes[b]	No	No	No
Shaded correlations	No	Yes	No	No
Correlations among factors	Yes	Yes	Yes	N.A.
Print matrix entries × 100 and rounded	No	No	ROUND	No

TABLE 12.14 (Continued)

Feature	SPSS[x] FACTOR	BMDP4M	SAS FACTOR	SYSTAT FACTOR
Print low matrix values as missing values	No	No	FUZZ	No

[a] Two types.
[b] Not available in original SPSS FACTOR.
[c] Normalization only in original SPSS FACTOR.
[d] Can be increased with reduced number of variables.
[e] Labeled proportion of variance.
[f] Oblique only.
[g] Orthogonal rotation only, therefore pattern = structure.
[h] Available through SYSTAT GRAPH.
[i] For orthogonal rotation only in original SPSS FACTOR.
[j] Available through SAS PLOT.

12.7.1 SPSS Package

The program in the SPSS[x] package developed for PCA and FA is FACTOR. A number of features have been added since the original SPSS FACTOR program, as footnoted in Table 12.14. SPSS[x] FACTOR is a much more appealing program than SPSS FACTOR. In fact, this program is now one of the most attractive of those available for FA.

SPSS[x] FACTOR does a PCA or FA on a correlation matrix or a factor loading matrix, helpful to the researcher who is interested in higher-order factoring (extracting factors from previous FAs). Several extraction methods and a variety of orthogonal rotation methods are available. Oblique rotation is done using direct oblimin, one of the best methods currently available (see Section 12.5.2.2).

Univariate output is limited to means, standard deviations, and number of cases per variable, so that the search for univariate outliers must be conducted through other programs. Similarly, there is no provision for screening for multivariate outliers among cases. But the program is very helpful in assessing factorability of **R**, as discussed in Section 12.3.2.6.

Output of extraction and rotation information is extensive. The residual and reproduced correlation matrices are provided as an aid to diagnosing adequacy of extraction and rotation. SPSS[x] FACTOR is the only program reviewed that, under conditions requiring matrix inversion, prints out the determinant of the correlation matrix, helpful in signaling the need to check for multicollinearity and singularity (Sections 12.3.2.5 and 4.1.7). Determination of number of factors is aided by an optional printout of a scree plot (Section 12.6.2).

Several estimation procedures for factor scores (Section 12.6.6) are available as output to a file; they are not given as part of regular printed output. Information about outliers among the factors is not available.

12.7.2 BMD Series

BMDP4M is the program for FA and PCA in the BMDP package. BMDP4M is flexible in its input, performing PCA and FA on either correlation or covariance matrices, with an option for analysis to proceed around the origin (setting all variable means to zero).

Several varieties of factor extraction and both orthogonal and oblique rotation are available. Information is given for all varieties of outlier detection—variables, univariate and multivariate outliers among cases, and outliers in the solution. About the only features missing are measures of the factorability of **R,** the determinant and the transformation matrix, generally of limited interest, and a scree plot, easily done by hand.

A large variety of plots can be obtained. Both rotated and unrotated factor loadings can be plotted, as well as factor scores. The table of sorted rotated factor loadings, where variables that load highly on each factor can be easily identified visually, is helpful. (Loadings below a particular cutoff, specified by the user, are printed out as 0.0.) The residual correlation matrix is printed (see Section 12.6.2) as an aid to evaluating adequacy extraction and rotation. The factor-score covariance matrix is useful for assessing internal consistency (Section 12.6.5).

12.7.3 SAS System

SAS FACTOR is another highly flexible, full-featured program for FA and PCA. About the only weakness is in screening for outliers. In addition to all the input options of BMDP4M, SAS FACTOR accepts rotated loading matrices, as long as factor correlations are provided. There are several options for extraction, as well as orthogonal and oblique rotation. And a target pattern matrix can be specified as a criterion for oblique rotation in confirmatory FA. Additional options include specification of proportion of variance to be accounted for in determining the number of factors to retain, and the option to allow communalities to be greater than 1.0. Variables in the data set can be weighted differentially to allow the generalized least squares method of extraction.

Factor scores can be written to a data file, but information based on those scores, such as Mahalanobis distance for detection of outliers in the factor solution, is not available. A factor score covariance matrix is not available but SMCs of factors as DVs with variables as IVs are given. Two options unavailable in any of the other programs are the ability to print whole-number matrices (in which entry is multiplied by 100 and rounded to the nearest whole number) and to substitute low matrix values as missing values.

12.7.4 SYSTAT System

The SYSTAT FACTOR program is much more limited than those of the other three packages. Wilkinson (1986) advocates the use of PCA rather than FA because of the indeterminacy problem (Section 12.6.6). Therefore PCA is the only extraction method now available. Although three methods of orthogonal rotation are provided, there is no provision for oblique rotation. SYSTAT FACTOR can accept correlation or covariance matrices as well as raw data.

The SYSTAT FACTOR program provides plots of factor loadings and will optionally sort the loading matrix by size of loading to aid interpretation. Additional information is available by requesting that factor scores, their coefficients, and loadings be sent to a data file. You can also request that the factor scores not be standardized if rotation has not been performed.

12.8 COMPLETE EXAMPLE OF FA

During the second year of the panel study described in Appendix B Section B.1, men and women completed the Bem Sex Role Inventory (BSRI; Bem, 1974). The women were 369 middle-class, English-speaking women between the ages of 21 and 60 who were interviewed in person. One hundred sixty-two men who were "close to" the women also filled out and returned a written version of the BSRI when requested to do so by letter several months after the women were interviewed. Most of the men were husbands; a few were fathers, sons, boyfriends, and so forth.

The BSRI is a 45-item inventory where 20 items measure femininity, 20 masculinity,[9] and 5 social desirability. Respondents attribute traits (e.g., "gentle," "shy," "dominant") to themselves by assigning numbers between 1 ("never or almost never true of me") and 7 ("always or almost always true of me") to each of the items. Responses are summed to produce separate masculine and feminine scores. Masculinity and femininity are conceived as orthogonal dimensions of personality with both, one, or neither descriptive of any given individual.

Previous factor analytic work had indicated the presence of between three and five factors underlying the items of the BSRI, with potentially different factors for men and women. Comparison of factor structures between women and men is a goal of this analysis.

12.8.1 Evaluation of Limitations

Because the BSRI was neither developed through nor designed for factor analytic work, it meets only marginally the requirements listed in Section 12.3.1. For instance, marker variables are not included and variables from the feminine scale differ in social desirability as well as in meaning (e.g., "tender" and "gullible"), so some of these variables are likely to be complex.

Nonetheless, the widespread use of the instrument and its convergent validity with some behavioral measures make factor structure, and the comparison of structure for men and women, important research questions.

12.8.1.1 Sample Size and Missing Data   Missing and out-of-range values are replaced with the mean response for the variable in question for the appropriate sex.

---

[9] Due to clerical error, one of the masculine items, "aggression," was omitted from the questionnaires.

That is, if a woman is missing a score on TENDER, the mean for women on TENDER is inserted.

Data are available initially from 369 women and 162 men. With outlying cases deleted (see below), FAs are conducted on responses of 344 women and 158 men. Using the guidelines of Section 12.3.2.1, the ratio of cases to variables is 7.6 to 1 for women and 3.5 to 1 for men. We retain a good sample size for women but a marginal one for men.

12.8.1.2 Normality Distributions of the 44 variables are examined for skewness through BMDP2D (cf. Chapter 4). Many of the variables are negatively skewed and a few are positively skewed. However, because the BSRI is already published and in use, and because variable transformations would differ for the male and female samples, no deletion of variables or transformations of them is performed.

Because the variables fail in normality, significance tests are inappropriate. And because the direction of skewness is different for different variables, we also anticipate a weakened analysis due to lowering of correlations in **R**.

12.8.1.3 Linearity The differences in skewness for variables suggest the possibility of curvilinearity for some pairs of variables. With 44 variables, however, examination of all pairwise scatterplots for two samples (about 2000 plots) is impractical. Therefore, a spot check is run on a few plots. Figure 12.4 shows the plot expected to be among the worst—between LOYAL (with strong negative skewness) and MASCULIN (with strong positive skewness) for women. Although the plot is far from pleasing, and shows departure from linearity as well as the possibility of outliers, there is no evidence of true curvilinearity (Section 4.1.5.2). And again, transformations are viewed with disfavor considering the variable set and the goals of analysis.

12.8.1.4 Outliers Among Cases Multivariate outliers among men and women are identified separately and deleted from subsequent analysis using BMDPAM (cf. Chapter 4). Using a criterion of $\alpha = .001$, 25 women and 4 men are deleted.

Because of the large number of outliers and variables, a case-by-case analysis (cf. Chapter 4) is not feasible. Instead, for each gender, a discriminant function analysis (BMDP7M) is used to identify variables that significantly discriminate between outliers and nonoutliers. On the last step of the discriminant analysis for women, five variables (RELIANT, FLATTER, FEMININE, TRUTHFUL, and LEADACT) discriminate outliers as a group with $p < .01$. For men, outliers differ on four variables (DOMINANT, HAPPY, WARM, TRUTHFUL). Table 12.15 shows a portion of the BMDP7M run for men. Included in the table are means for outliers and nonoutliers (COMPLETE cases) on the four relevant variables. Also included is an edited part of the last step in the analysis, showing F TO REMOVE for the two groups on those four variables.

12.8.1.5 Multicollinearity and Singularity The original nonrotated PCA runs through BMDP4M for women and men reveal that the smallest eigenvalues (the ones associated with the 44th factor) are 0.126 for women and 0.089 for men, neither dangerously close to zero. Simultaneously it is observed that SMCs between variables where each, in turn, serves as DV for the others, do not approach 1. The largest

```
/PROBLEM    TITLE IS 'LINEARITY CHECK FOR WOMEN'.
/INPUT VARIABLES ARE 45. FORMAT IS '(A3,44F1.0)'.  FILE='FACTORW.DAT'.
/VARIABLE NAMES ARE SUBNO,HELPFUL,RELIANT,DEFBEL,YIELDING,CHEERFUL,INDPT,
           ATHLET,SHY,ASSERT,STRPERS,FORCEFUL,AFFECT,FLATTER,LOYAL,
           ANALYT,FEMININE,SYMPATHY,MOODY,SENSITIV,UNDSTAND,COMPASS,
           LEADERAB,SOOTHE,RISK,DECIDE,SELFSUFF,CONSCIEN,DOMINANT,
           MASCULINE,STAND,HAPPY,SOFTSPOK,WARM,TRUTHFUL,TENDER,
           GULLIBLE,LEADACT,CHILDLIK,INDIVID,FOULLANG,LOVECHIL,
           COMPETE,AMBITIOU,GENTLE.
           LABEL = SUBNO.
           USE = 2 TO 45.
/PLOT      XVAR = LOYAL.  YVAR = MASCULIN.
           SIZE = 40,25.
/END
```

```
                    ...+.....+.....+.....+.....+.....+.....+...
                     -                                         -
                     -                          1    3 -
                     -                                         -
                 8   +                                         +
                     -                                         -
                     -                                         -
                     -                                         -
               M     -                                         -
               A     -                                         -
               S   6 +                     1    1    1 +
               C     -                                         -
               U     -                                         -
               L     -                          1    5 -
               I     -                                         -
               N     -                                         -
                   4 +                               6    H +
                     -                                         -
                     -                               F    L -
                     -                                         -
                     -                                         -
                   2 +                     1    4    N    V +
                     Y                                    Y
                     -                                         -
                     - 1          3          A    *    * -
                    ...+.....+.....+.....+.....+.....+.....+...
                     1          3          5          7
                          2          4          6
```

```
               N = 369
               R = .003            LOYAL
               P = .949
                                        MEAN    S.D.
               --REGRESSION LINE--  -RES.MS-  X  6.5935  .72048
               Y= 1.6943 +.00609*X    1.7383  Y  1.7344  1.3167
```

Figure 12.4 Spot check for linearity among variables for women. Setup and output from BMDP6D.

SMC among the variables for women is .76 (Table 12.16) and for men is .81. Multicollinearity and singularity are not a threat in these data sets.

12.8.1.6 Factorability of R   Correlation matrices among the 44 items for both women and men produced by BMDP4M reveal numerous correlations in excess of .30 and some considerably higher. Patterns in responses to variables are therefore anticipated.

TABLE 12.15 DESCRIPTION OF OUTLIERS USING BMDP7M. SETUP
AND SELECTED OUTPUT FOR MEN.

```
/PROBLEM          TITLE IS 'MEN FA, DESCRIPTION OF OUTLIERS'.
/INPUT VARIABLES ARE 45. FORMAT IS '(A3,44F1.0)'. FILE = MFACTOR.
/VARIABLE NAMES ARE SUBNO,HELPFUL,RELIANT,DEFBEL,YIELDING,CHEERFUL,INDPT,
                    ATHLET,SHY,ASSERT,STRPERS,FORCEFUL,AFFECT,FLATTER,LOYAL,
                    ANALYT,FEMININE,SYMPATHY,MOODY,SENSITIV,UNDSTAND,COMPASS,
                    LEADERAB,SOOTHE,RISK,DECIDE,SELFSUFF,CONSCIEN,DOMINANT,
                    MASCULINE,STAND,HAPPY,SOFTSPOK,WARM,TRUTHFUL,TENDER,
                    GULLIBLE,LEADACT,CHILDLIK,INDIVID,FOULLANG,LOVECHIL,
                    COMPETE,AMBITIOU,GENTLE,DUMMY.  ADD=1.
                    GROUPING IS DUMMY.
                    LABEL = SUBNO.  USE = 2 TO 46.
/TRANSFORM DUMMY = 0.
           IF(KASE EQ 22 OR KASE EQ 110 OR KASE EQ 140
           OR KASE EQ 153) THEN DUMMY = 1.0.
/GROUP     CODES(DUMMY) ARE 0, 1.
           NAMES(DUMMY) ARE COMPLETE, OUTLIERS.
/PRINT     NO STEP.  NO POST.  NO POINT.
/PLOT      NO CANON.
/END
```

```
     MEANS

         GROUP =  COMPLETE    OUTLIERS    ALL GPS.
VARIABLE

      .
      .
29 DOMINANT      4.78481     3.00000     4.74074
30 MASCULIN      6.21519     6.50000     6.22222
31 STAND         5.77215     4.00000     5.72839
32 HAPPY         5.50000     3.25000     5.44444
33 SOFTSPOK      4.37975     3.75000     4.36420
34 WARM          5.13291     5.25000     5.13580
35 TRUTHFUL      6.19620     3.50000     6.12963
      .
      .

COUNTS            158.         4.         162.

STEP NUMBER   8
VARIABLE ENTERED   45 GENTLE

   VARIABLE    F TO    FORCE  TOLERNCE *  VARIABLE    F TO   FORCE  TOLERNCE
               REMOVE  LEVEL           *             ENTER  LEVEL
        DF =  1  153                   *     DF =   1  152
29 DOMINANT    9.81    1    0.75975 *   2 HELPFUL    0.13    1    0.65867
32 HAPPY       9.90    1    0.70491 *   4 DEFBEL     0.02    1    0.76609
34 WARM       15.11    1    0.51159 *   5 YIELDING   0.03    1    0.66214
35 TRUTHFUL   34.12    1    0.87038 *   6 CHEERFUL   0.07    1    0.42922
```

(More sophisticated measures of factorability would have been available had SPSS[x] or SAS FACTOR been used for the analysis.)

12.8.1.7 Outliers among Variables SMCs among variables (Table 12.16) are also used to screen for outliers among variables, as discussed in Section 12.3.2.7. The lowest SMC among variables is .11 for women and .33 for men. However, these SMCs are for different variables. It is decided to retain all 44 variables in both FAs, although many are largely unrelated to others in the set, particularly for women. (In fact, 45% of the 44 variables in the analysis for women and 25% of the 44 variables in the analysis for men have loadings too low on all the factors to assist interpretation in the final solution.)

12.8.1.8 Outlying Cases among the Factors The ESTIMATED FACTOR SCORES table produced by the final PFA run is used to find cases that are not well fit by the solution. The second column of CHISQ/DF (see Table 12.17) gives Mahalanobis

TABLE 12.16 SETUP AND SELECTED BMDP4M OUTPUT TO ASSESS MULTICOLLINEARITY AMONG WOMEN.

```
                    /PROBLEM  TITLE IS 'PRINCIPAL COMPONENTS ANALYSIS FOR WOMEN'.
                    /INPUT  VARIABLES ARE 45.  FORMAT IS '(A3,44F1.0)'.  FILE=WFACTOR.
                    /VARIABLE NAMES ARE SUBNO,HELPFUL,RELIANT,DEFBEL,YIELDING,CHEERFUL,INDPT,
                             ATHLET,SHY,ASSERT,STRPERS,FORCEFUL,AFFECT,FLATTER,LOYAL,
                             ANALYT,FEMININE,SYMPATHY,MOODY,SENSITIV,UNDSTAND,COMPASS,
                             LEADERAB,SOOTHE,RISK,DECIDE,SELFSUFF,CONSCIEN,DOMINANT,
                             MASCULINE,STAND,HAPPY,SOFTSPOK,WARM,TRUTHFUL,TENDER,
                             GULLIBLE,LEADACT,CHILDLIK,INDIVID,FOULLANG,LOVECHIL,
                             COMPETE,AMBITIOU,GENTLE.
                             USE = 2 TO 45.   LABEL = SUBNO.
                    /TRANSFORM   DELETE = 15,20,42,58,64,88,89,109,126,140,147,153,169,
                                          199,201,203,214,220,225,284,310,312,341,343,359.
                    /PRINT      FSCORE = 0.
                    /PLOT       FINAL=0.  FSCORE=0.
                    /END
```

```
SQUARED MULTIPLE CORRELATIONS (SMC) OF
EACH VARIABLE WITH ALL OTHER VARIABLES
--------------------------------------

  2 HELPFUL    0.37427
  3 RELIANT    0.46116
  4 DEFBEL     0.41691
  5 YIELDING   0.22995
  6 CHEERFUL   0.49202
  7 INDPT      0.53847
  8 ATHLET     0.25761
  9 SHY        0.32498
 10 ASSERT     0.53767
 11 STRPERS    0.59340
 12 FORCEFUL   0.56565
 13 AFFECT     0.55264
 14 FLATTER    0.29586
 15 LOYAL      0.39073
 16 ANALYT     0.24184
 17 FEMININE   0.35791
 18 SYMPATHY   0.45290
 19 MOODY      0.38091
 20 SENSITIV   0.48604
 21 UNDSTAND   0.61663
 22 COMPASS    0.64934
 23 LEADERAB   0.76269
 24 SOOTHE     0.43514
 25 RISK       0.42237
 26 DECIDE     0.48930
 27 SELFSUFF   0.63268
 28 CONSCIEN   0.39916
 29 DOMINANT   0.56213
 30 MASCULIN   0.31595
 31 STAND      0.57284
 32 HAPPY      0.53577
 33 SOFTSPOK   0.40179
 34 WARM       0.61523
 35 TRUTHFUL   0.35627
 36 TENDER     0.60454
 37 GULLIBLE   0.29683
 38 LEADACT    0.76136
 39 CHILDLIK   0.29604
 40 INDIVID    0.37905
 41 FOULLANG   0.11346
 42 LOVECHIL   0.28419
 43 COMPETE    0.46468
 44 AMBITIOU   0.45870
 45 GENTLE     0.57954
```

distance of each case from the centroid of the factor scores evaluated as χ^2/df. Cases with large values are deviant cases in the space of the factor solution. Critical χ^2 (cf. Table C.4) at the chosen probability level is divided by number of factors as degrees of freedom to provide the cutoff value.

At $\alpha = .001$, the cutoff for four factors (women) is 4.62 (18.47/4) and for

TABLE 12.17 SELECTED BMDP4M PFA OUTPUT USED TO EVALUATE
OUTLYING CASES IN SOLUTION FOR MEN. SETUP
SHOWN IN TABLE 12.19.

ESTIMATED FACTOR SCORES AND MAHALANOBIS DISTANCES (CHI-SQUARE S) FROM
EACH CASE TO THE CENTROID OF ALL CASES FOR THE ORIGINAL DATA
(44 D.F.) FACTOR SCORES (4 D.F.) AND THEIR DIFFERENCE (40 D.F.).
EACH CHI-SQUARE HAS BEEN DIVIDED BY ITS DEGREES OF FREEDOM.

CASE LABEL	NO.	CHISQ/DF 44	CHISQ/DF 4	CHISQ/DF 40	FACTOR 1	FACTOR 2	FACTOR 3	FACTOR 4	
149	43	1.043	1.464	0.989	-1.231	1.785	0.594	-1.245	-0.225
150	44	1.446	0.481	1.570	-0.173	-1.152	-0.356	-0.075	-0.848
152	45	1.772	7.588	1.026	-3.109	0.791	-2.686	1.240	-3.655
153	46	0.755	0.860	0.742	0.490	1.348	0.934	-0.864	0.241
154	47	0.546	0.340	0.572	-0.612	-0.478	0.010	0.707	-0.533
158	48	0.825	0.525	0.864	0.391	0.929	-0.539	0.895	-0.292
160	49	1.253	1.403	1.234	0.435	1.759	-0.186	-1.554	-0.408
183	50	1.126	0.741	1.175	0.071	0.733	-0.163	0.915	-1.244
190	51	0.747	0.666	0.757	0.447	1.397	-0.018	0.804	-0.343
192	52	1.064	0.712	1.110	0.649	-0.625	0.334	0.686	-1.219
203	53	0.637	0.486	0.656	-0.669	0.265	-0.263	1.101	-0.285
206	54	0.915	1.708	0.813	0.648	-1.113	-0.131	1.542	1.746
208	55	1.233	1.143	1.245	1.071	-0.183	-0.051	0.054	-1.857
210	56	0.776	0.724	0.783	0.284	0.456	-0.026	-1.033	1.230
230	57	1.813	6.510	1.210	-3.802	-3.099	-1.151	-1.052	-2.087
232	58	0.745	1.113	0.698	0.621	-0.378	0.706	-1.326	1.282

five factors (men) is 4.103. Cases with values larger than this in the column are outliers. Three cases among the men are outliers in the solution space, two of them in Table 12.17 with χ^2 values of 7.588 and 6.510. No women are outliers in the space of the solution. No attempt is made to discover unifying characteristics, if any, between men well fit and poorly fit by the factor analysis, but in some research such an attempt might prove fruitful.

12.8.2 Principal Factors Extraction with Varimax Rotation: Comparison of Two Groups

Principal components extraction with varimax rotation through BMDP4M is used in two initial runs, one for women and one for men, to evaluate assumptions and limitations as discussed in Section 12.8.1 and to estimate the number of factors from eigenvalues. The first 13 eigenvalues for women and for men are shown in Table 12.18. The maximum number of factors (eigenvalues larger than 1) is 11 for women and 12 for men. However, retention of 11 or 12 factors seems unreasonable so sharp breaks in size of eigenvalues are sought using the scree test (Section 12.6.2).

For men, a clear break in size occurs between factors 3 and 4 (slightly more than one unit). The differences between the first four factors are large, but there is little difference in variance explained between factors 4 and 5 and thereafter. This is taken as evidence that there are probably 4 or 5 factors for men. Far less distinction in size of successive eigenvalues is seen for women but differences seem to level off after factor 4. That is, factor 6 adds little to factor 5, and so on. This is taken as evidence that there are probably between 4 and 6 factors for women. Earlier research suggests 3 to 5 factors for the BSRI, so these results are consistent.

A common factor extraction model that removes unique and error variability from each variable is used for the next several runs and the final solution. Principal factors is chosen from among methods for common factor extractions because it is

TABLE 12.18 EIGENVALUES AND PROPORTIONS OF VARIANCE FOR FIRST 13 COMPONENTS, MEN AND WOMEN. BMDP4M PCA SETUP AND SELECTED OUTPUT

(a) Women

```
/PROBLEM    TITLE IS 'PRINCIPAL COMPONENTS ANALYSIS FOR WOMEN'.
/INPUT      VARIABLES ARE 45.  FORMAT IS '(A3,44F1.0)'.  FILE=WFACTOR.
/VARIABLE   NAMES ARE SUBNO,HELPFUL,RELIANT,DEFBEL,YIELDING,CHEERFUL,INDPT,
            ATHLET,SHY,ASSERT,STRPERS,FORCEFUL,AFFECT,FLATTER,LOYAL,
            ANALYT,FEMININE,SYMPATHY,MOODY,SENSITIV,UNDSTAND,COMPASS,
            LEADERAB,SOOTHE,RISK,DECIDE,SELFSUFF,CONSCIEN,DOMINANT,
            MASCULINE,STAND,HAPPY,SOFTSPOK,WARM,TRUTHFUL,TENDER,
            GULLIBLE,LEADACT,CHILDLIK,INDIVID,FOULLANG,LOVECHIL,
            COMPETE,AMBITIOU,GENTLE.
            USE = 2 TO 45.    LABEL = SUBNO.
/TRANSFORM  DELETE = 15,20,42,58,64,88,89,109,126,140,147,153,169,
            199,201,203,214,220,225,284,310,312,341,343,359.
/PRINT      FSCORE=0.
/PLOT       FSCORE=0.  FINAL=0.
/END
```

FACTOR	VARIANCE EXPLAINED	CUMULATIVE PROPORTION OF VARIANCE IN DATA SPACE	IN FACTOR SPACE	CARMINES THETA
1	8.1940	0.1862	0.3057	0.8984
2	5.1535	0.3034	0.4980	
3	2.5905	0.3622	0.5946	
4	2.0730	0.4093	0.6719	
5	1.6476	0.4468	0.7334	
6	1.4152	0.4789	0.7862	
7	1.2907	0.5083	0.8344	
8	1.2212	0.5360	0.8799	
9	1.1095	0.5613	0.9213	
10	1.0776	0.5857	0.9615	
11	1.0317	0.6092	1.0000	
12	0.9513	0.6308		
13	0.9417	0.6522		

(b) Men

```
/PROBLEM    TITLE IS 'PRINCIPAL COMPONENTS ANALYSIS FOR MEN'.
/INPUT      VARIABLES ARE 45.  FORMAT IS '(A3,44F1.0)'.  FILE=MFACTOR.
/VARIABLE   NAMES ARE SUBNO,HELPFUL,RELIANT,DEFBEL,YIELDING,CHEERFUL,INDPT,

            ATHLET,SHY,ASSERT,STRPERS,FORCEFUL,AFFECT,FLATTER,LOYAL,
            ANALYT,FEMININE,SYMPATHY,MOODY,SENSITIV,UNDSTAND,COMPASS,
            LEADERAB,SOOTHE,RISK,DECIDE,SELFSUFF,CONSCIEN,DOMINANT,
            MASCULINE,STAND,HAPPY,SOFTSPOK,WARM,TRUTHFUL,TENDER,
            GULLIBLE,LEADACT,CHILDLIK,INDIVID,FOULLANG,LOVECHIL,
            COMPETE,AMBITIOU,GENTLE.
            USE = 2 TO 45.    LABEL = SUBNO.
/TRANSFORM  DELETE = 22,110,140,153.
/PRINT      FSCORE = 0.
/PLOT       FINAL=0.  FSCORE=0.
/END
```

WFACTOR	VARIANCE EXPLAINED	CUMULATIVE PROPORTION OF VARIANCE IN DATA SPACE	IN FACTOR SPACE	CARMINES THETA
1	8.8490	0.2011	0.2863	0.9076
2	6.5099	0.3491	0.4969	
3	2.8968	0.4149	0.5906	
4	1.8794	0.4576	0.6514	
5	1.7008	0.4963	0.7064	
6	1.6331	0.5334	0.7593	
7	1.5280	0.5681	0.8087	
8	1.4856	0.6019	0.8568	
9	1.2206	0.6296	0.8962	
10	1.1349	0.6554	0.9330	
11	1.0570	0.6794	0.9672	
12	1.0153	0.7025	1.0000	
13	0.9581	0.7243		

most widely used. Several PFA runs are performed through BMDP4M to find the optimal number of factors; runs specifying 4 to 6 factors are planned for women and men.

For women, the trial PFA run with 5 factors has 5 eigenvalues larger than 1 among unrotated factors. But after rotation, the eigenvalue for the fifth factor is below 1 and it has no loadings larger than .45, the criterion for interpretation chosen for this research. The solution with four factors, on the other hand, meets the goals of interpretability and is chosen for follow-up runs. The first six eigenvalues from the four-factor solution for women are shown in Table 12.19(a).

For the men, the results of PFA runs are quite different from the PCA run. In all the trial runs, prerotation eigenvalues are larger than 1 for the first five factors. In addition, the fifth factor is interpretable in terms of size of loadings. The 6-factor solution revealed no additional interpretable factor. The 5-factor solution, therefore, is chosen. The first six eigenvalues for this solution are in Table 12.19(b).

As another test of adequacy of extraction and number of factors, it is noted (but not shown) that most values in the residual correlation matrices for the four-factor orthogonal solution for women and the five-factor solution for men are near zero. This is further confirmation that a reasonable number of factors is 4 for women and 5 for men.

Principal factors extraction with oblique rotation is employed, as discussed below, as a check on adequacy of rotation and simple structure. Altogether, 10 analyses are performed, 5 for women and 5 for men, before the final solutions are selected.

The decision between oblique and orthogonal rotation is made by requesting principal factor extraction with oblique rotation of four factors for women and five for men. Direct quartimin (DQUART) is the oblique method employed. The highest correlations are between factors 2 and 3, .333 for women and .389 for men (see Table 12.20). The generally oblong shape of the scatterplot of factor scores between these two factors (see Figure 12.5) confirms the correlation. Correlations between .2 and .3 are found between factor 4 and all other factors for men. If not for the desire to compare factors for the two sexes, oblique rotations are probably warranted. But because of complexities added in reporting results with oblique rotation, and because of the desire to compare factors for the two sexes, orthogonal rotation is considered adequate.

The solutions that are evaluated, interpreted and reported are the runs with principal factors extraction, varimax rotation, and 4 factors for women and 5 factors for men. In other words, after "trying out" oblique rotation, the decision is made to interpret the earlier runs with orthogonal rotation. Setups for these two runs are in Table 12.19.

Communalities are inspected to see if the variables are well defined by the solutions. Communalities indicate the percent of variance in a variable that overlaps variance in the factors. As seen in Table 12.21, communality values for a number of variables are quite low (e.g., FOULLANG for women, ATHLET for men). Ten of the variables for women and five of those for men have communality values lower than .2. One is unable to predict scores on many of the variables from scores on factors by using the solutions chosen here. There is considerable heterogeneity among

TABLE 12.19 EIGENVALUES AND PROPORTIONS OF VARIANCE FOR FIRST SIX FACTORS, WOMEN AND MEN. PRINCIPAL FACTORS EXTRACTION AND VARIMAX ROTATION. BMDP4M SETUP AND SELECTED OUTPUT

(a) Women

```
/PROBLEM   TITLE IS 'FACTOR ANALYSIS FOR WOMEN'.
/INPUT     VARIABLES ARE 45.  FORMAT IS '(A3,44F1.0)'.  FILE=WFACTOR.
/VARIABLE  NAMES ARE SUBNO,HELPFUL,RELIANT,DEFBEL,YIELDING,CHEERFUL,INDPT,
           ATHLET,SHY,ASSERT,STRPERS,FORCEFUL,AFFECT,FLATTER,LOYAL,
           ANALYT,FEMININE,SYMPATHY,MOODY,SENSITIV,UNDSTAND,COMPASS,
           LEADERAB,SOOTHE,RISK,DECIDE,SELFSUFF,CONSCIEN,DOMINANT,
           MASCULINE,STAND,HAPPY,SOFTSPOK,WARM,TRUTHFUL,TENDER,
           GULLIBLE,LEADACT,CHILDLIK,INDIVID,FOULLANG,LOVECHIL,
           COMPETE,AMBITIOU,GENTLE.
           USE = 2 TO 45.   LABEL = SUBNO.
/TRANSFORM DELETE = 15,20,42,58,64,88,89,109,126,140,147,153,169,
                    199,201,203,214,220,225,284,310,312,341,343,359.
/FACTOR    METHOD = PFA.  NUMBER = 4.
/PRINT     NO CORR.  LOLEV = .45   RESI.
/END
```

FACTOR	VARIANCE EXPLAINED	CUMULATIVE PROPORTION OF VARIANCE IN DATA SPACE	IN FACTOR SPACE	CARMINES THETA
1	7.6193	0.3669	0.4871	0.9599
2	4.6011	0.5885	0.7812	
3	1.9576	0.6828	0.9064	
4	1.4646	0.7533	1.0000	
5	0.9360	0.7984		
6	0.7715	0.8355		

(b) Men

```
/PROBLEM   TITLE IS 'FACTOR ANALYSIS FOR MEN'.
/INPUT     VARIABLES ARE 45.  FORMAT IS '(A3,44F1.0)'.  FILE=MFACTOR.
/VARIABLE  NAMES ARE SUBNO,HELPFUL,RELIANT,DEFBEL,YIELDING,CHEERFUL,INDPT,
           ATHLET,SHY,ASSERT,STRPERS,FORCEFUL,AFFECT,FLATTER,LOYAL,
           ANALYT,FEMININE,SYMPATHY,MOODY,SENSITIV,UNDSTAND,COMPASS,
           LEADERAB,SOOTHE,RISK,DECIDE,SELFSUFF,CONSCIEN,DOMINANT,
           MASCULINE,STAND,HAPPY,SOFTSPOK,WARM,TRUTHFUL,TENDER,
           GULLIBLE,LEADACT,CHILDLIK,INDIVID,FOULLANG,LOVECHIL,
           COMPETE,AMBITIOU,GENTLE.
           USE = 2 TO 45.   LABEL = SUBNO.
/TRANSFORM DELETE = 22,110,140,153.
/FACTOR    METHOD = PFA.  NUMBER = 5.
/PRINT     NO CORR.  LOLEV = .45   RESI.
/END
```

FACTOR	VARIANCE EXPLAINED	CUMULATIVE PROPORTION OF VARIANCE IN DATA SPACE	IN FACTOR SPACE	CARMINES THETA
1	8.3917	0.3331	0.4350	0.9534
2	6.0018	0.5713	0.7461	
3	2.3391	0.6641	0.8673	
4	1.3786	0.7188	0.9388	
5	1.1814	0.7657	1.0000	
6	0.9903	0.8050		

the variables. It should be recalled, however, that factorial purity was not a criterion used to devise the BSRI.

Adequacy of rotation (Section 12.6.3) is assessed, in part, by scatterplots with pairs of rotated factors as axes and variables as points, as shown in Figure 12.6. Ideally, variable points are at the origin (the unmarked middle of figures) or in clusters at the ends of factor axes. Scatterplots between factor 1 and factor 2 for both sexes,

TABLE 12.20 SETUP AND SELECTED BMDP4M PFA OUTPUT OF
CORRELATIONS AMONG FACTORS FOLLOWING DIRECT
QUARTIMIN ROTATION FOR WOMEN AND MEN.

(a) Women

```
/PROBLEM   TITLE IS 'FACTOR ANALYSIS FOR WOMEN, OBLIQUE'.
/INPUT     VARIABLES ARE 45.   FORMAT IS '(A3,44F1.0)'.   FILE=WFACTOR.
/VARIABLE  NAMES ARE SUBNO,HELPFUL,RELIANT,DEFBEL,YIELDING,CHEERFUL,INDPT,
           ATHLET,SHY,ASSERT,STRPERS,FORCEFUL,AFFECT,FLATTER,LOYAL,
           ANALYT,FEMININE,SYMPATHY,MOODY,SENSITIV,UNDSTAND,COMPASS,
           LEADERAB,SOOTHE,RISK,DECIDE,SELFSUFF,CONSCIEN,DOMINANT,
           MASCULINE,STAND,HAPPY,SOFTSPOK,WARM,TRUTHFUL,TENDER,
           GULLIBLE,LEADACT,CHILDLIK,INDIVID,FOULLANG,LOVECHIL,
           COMPETE,AMBITIOU,GENTLE.
           USE = 2 TO 45.   LABEL = SUBNO.
/TRANSFORM DELETE = 15,20,42,58,64,88,89,109,126,140,147,153,169,
           199,201,203,214,220,225,284,310,312,341,343,359.
/FACTOR    METHOD = PFA.  NUMBER = 4.
/ROTATE    METHOD = DQUART.
/PRINT     NO CORR.   LOLEV = .45 FSCORE=0.
/PLOT      FINAL = 0.  FSCORE = 0.
/END
```

FACTOR CORRELATIONS FOR ROTATED FACTORS

		FACTOR	FACTOR	FACTOR	FACTOR
		1	2	3	4
FACTOR	1	1.000			
FACTOR	2	0.173	1.000		
FACTOR	3	0.089	0.333	1.000	
FACTOR	4	0.113	0.013	0.034	1.000

(b) Men

```
/PROBLEM   TITLE IS 'FACTOR ANALYSIS FOR MEN, OBLIQUE'.
/INPUT     VARIABLES ARE 45.   FORMAT IS '(A3,44F1.0)'.   FILE=MFACTOR.
/VARIABLE  NAMES ARE SUBNO,HELPFUL,RELIANT,DEFBEL,YIELDING,CHEERFUL,INDPT,
           ATHLET,SHY,ASSERT,STRPERS,FORCEFUL,AFFECT,FLATTER,LOYAL,
           ANALYT,FEMININE,SYMPATHY,MOODY,SENSITIV,UNDSTAND,COMPASS,
           LEADERAB,SOOTHE,RISK,DECIDE,SELFSUFF,CONSCIEN,DOMINANT,
           MASCULINE,STAND,HAPPY,SOFTSPOK,WARM,TRUTHFUL,TENDER,
           GULLIBLE,LEADACT,CHILDLIK,INDIVID,FOULLANG,LOVECHIL,
           COMPETE,AMBITIOU,GENTLE.
           USE = 2 TO 45.   LABEL = SUBNO.
/TRANSFORM DELETE = 22,110,140,153.
/FACTOR    METHOD = PFA.  NUMBER = 5.
/ROTATE    METHOD = DQUART.
/PRINT     NO CORR.   LOLEV = .45  FSCORE=0.
/PLOT      FINAL=0.  FSCORE=0.
/END
```

FACTOR CORRELATIONS FOR ROTATED FACTORS

		FACTOR	FACTOR	FACTOR	FACTOR	FACTOR
		1	2	3	4	5
FACTOR	1	1.000				
FACTOR	2	0.113	1.000			
FACTOR	3	-0.053	0.389	1.000		
FACTOR	4	0.225	0.291	0.254	1.000	
FACTOR	5	-0.108	0.089	0.084	-0.151	1.000

and between factor 2 and factor 4 for women, seem reasonably clear. The scatterplots
between some pairs of factors show evidence of correlation among factors as found
during oblique rotation. Otherwise, the scatterplots are disappointing but consistent
with other evidence of heterogeneity among the variables in the BSRI.

Simplicity of structure (Section 12.6.3) in factor loadings following orthogonal
rotation is assessed from the table of ROTATED FACTOR LOADINGS (see Table

FACTOR SCORES

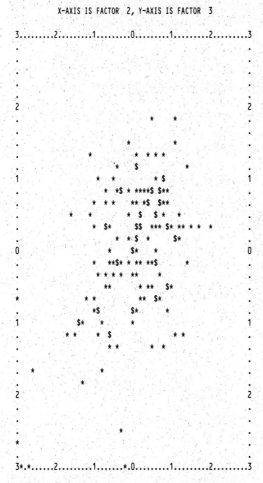

Figure 12.5 Scatterplot of factor scores for men with pairs of factor (2 and 3) as axes following oblique rotation. (Setup in Table 12.21.)

12.22). In each column there are a few high and many low correlations between variables and factors. There are also numerous moderate loadings so several variables will be complex (load on more than one factor) unless a fairly high cutoff for interpreting loadings is established. Although not as clear as one might hope, the factors are, however, mostly correlated with different variables.

Complexity of variables (Section 12.6.5) is assessed by examining loadings for a variable across factors. With a loading cut of .45 only two variables for women, WARM and INDPT, and one for men, CHEERFUL, load on more than one factor.

The importance of each of factor (Sections 12.4 and 12.6.4) is assessed by the percent of variance and covariance it represents. SSLs, called VP at the bottom of each column of loadings shown in Table 12.22, are used in the calculations. It is important to use SSLs from rotated factors, because the variance is redistributed during rotation. Proportion of variance for a factor is SSL for the factor divided by number of variables. Proportion of covariance is SSL divided by sum of SSLs. Results, converted to percent, are shown in Table 12.23. Each of the factors for both sexes accounts for between 4 and 16% of the variance in the set of variables, not an outstanding performance. For men, the first two factors account for most of the covariance; for women, only the first factor accounts for substantial covariance.

Internal consistency of the factors (Section 12.6.4) is shown as SMCs in the diagonal of the FACTOR SCORE COVARIANCE matrix. Factors serve as DVs with variables as IVs. Factors that are well defined by the variables have high SMCs, whereas poorly defined factors have low SMCs. As can be seen in Table 12.24, all factors are internally consistent for both women and men. (The off-diagonal elements in these matrices are correlations among factor scores. Although uniformly low, the

TABLE 12.21 COMMUNALITY VALUES FOR WOMEN (FOUR FACTORS) AND MEN (FIVE FACTORS), SELECTED OUTPUT FROM BMDP4M PFA. SEE TABLE 12.19 FOR SETUP.

(a) Women

COMMUNALITIES OBTAINED FROM   4 FACTORS AFTER   6 ITERATIONS.

THE COMMUNALITY OF A VARIABLE IS ITS SQUARED MULTIPLE CORRELATION WITH THE FACTORS.

```
 2 HELPFUL     0.2825
 3 RELIANT     0.3979
 4 DEFBEL      0.2490
 5 YIELDING    0.1511
 6 CHEERFUL    0.3598
 7 INDPT       0.4541
 8 ATHLET      0.1837
 9 SHY         0.1568
10 ASSERT      0.4403
11 STRPERS     0.5074
12 FORCEFUL    0.4635
13 AFFECT      0.4796
14 FLATTER     0.2002
15 LOYAL       0.2938
16 ANALYT      0.1514
17 FEMININE    0.1562
18 SYMPATHY    0.4405
19 MOODY       0.2713
20 SENSITIV    0.4440
21 UNDSTAND    0.5813
22 COMPASS     0.6846
23 LEADERAB    0.5771
24 SOOTHE      0.3877
25 RISK        0.2763
26 DECIDE      0.3765
27 SELFSUFF    0.6365
28 CONSCIEN    0.3502
29 DOMINANT    0.5400
30 MASCULIN    0.1894
31 STAND       0.4385
32 HAPPY       0.4421
33 SOFTSPOK    0.2775
34 WARM        0.6316
35 TRUTHFUL    0.1683
36 TENDER      0.5346
37 GULLIBLE    0.2214
38 LEADACT     0.5407
39 CHILDLIK    0.1920
40 INDIVID     0.2362
41 FOULLANG    0.0248
42 LOVECHIL    0.1366
43 COMPETE     0.3376
44 AMBITIOU    0.2645
45 GENTLE      0.5139
```

TABLE 12.21 (Continued)

(b) Men

COMMUNALITIES OBTAINED FROM 5 FACTORS AFTER 10 ITERATIONS.

THE COMMUNALITY OF A VARIABLE IS ITS SQUARED MULTIPLE
CORRELATION WITH THE FACTORS.

2	HELPFUL	0.4784
3	RELIANT	0.3424
4	DEFBEL	0.3835
5	YIELDING	0.4613
6	CHEERFUL	0.6696
7	INDPT	0.7119
8	ATHLET	0.0293
9	SHY	0.2242
10	ASSERT	0.4950
11	STRPERS	0.6005
12	FORCEFUL	0.6111
13	AFFECT	0.5033
14	FLATTER	0.0703
15	LOYAL	0.3339
16	ANALYT	0.2489
17	FEMININE	0.2929
18	SYMPATHY	0.5524
19	MOODY	0.3774
20	SENSITIV	0.5495
21	UNDSTAND	0.6828
22	COMPASS	0.6132
23	LEADERAB	0.5450
24	SOOTHE	0.5872
25	RISK	0.3456
26	DECIDE	0.4478
27	SELFSUFF	0.6616
28	CONSCIEN	0.4240
29	DOMINANT	0.5428
30	MASCULIN	0.3361
31	STAND	0.5699
32	HAPPY	0.6594
33	SOFTSPOK	0.2850
34	WARM	0.6804
35	TRUTHFUL	0.3995
36	TENDER	0.6842
37	GULLIBLE	0.2057
38	LEADACT	0.5449
39	CHILDLIK	0.1981
40	INDIVID	0.3675
41	FOULLANG	0.1198
42	LOVECHIL	0.2337
43	COMPETE	0.1933
44	AMBITIOU	0.2230
45	GENTLE	0.6064

values are not zero. As discussed in Section 12.6.6, low correlations among scores on factors are often obtained even with orthogonal rotation.)

Factors are interpreted through their factor loadings (Section 12.6.5). The process is facilitated with BMDP4M by requesting output of SORTED ROTATED FACTOR LOADINGS through the PRINT paragraph (see setup, Table 12.19). In the sorted, rotated loading matrix, variables are grouped by factors and reordered by size of loading. Examples from the data for women and men are shown in Table 12.25. Interpretation of factors comes from meaning of variables that cluster together for each factor.

Although a default cut of .25 is used for inclusion of variables in factors by BMDP4M, from a variance perspective the cutoff values proposed by Comrey (1973) and reported in Section 12.6.5 make better sense. It is decided to use a loading of .45 (20% variance overlap between variable and factor) for inclusion of a variable

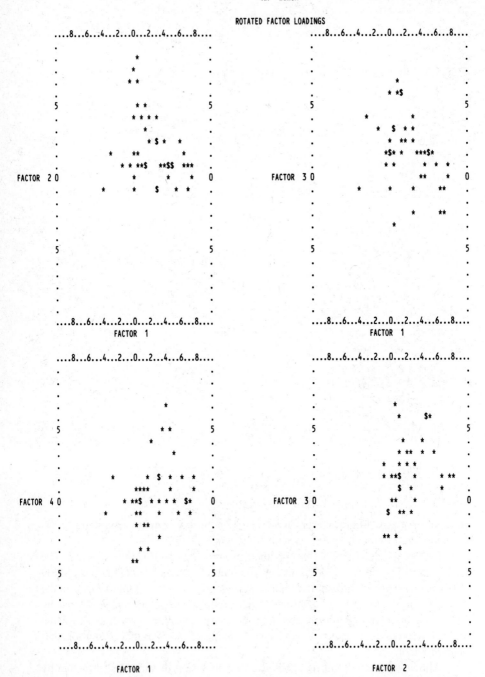

Figure 12.6 Selected BMDP4M PFA output showing scatterplots of variable loadings with factors as axes for (a) all four factors for women and (b) first four factors for men (setup in Table 12.19).

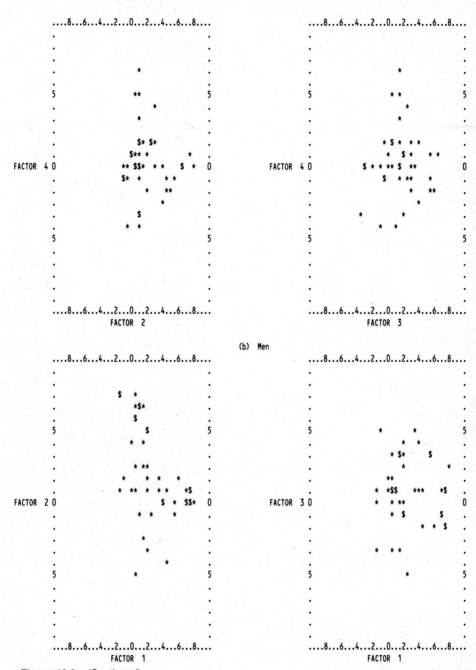

Figure 12.6 *(Continued)*

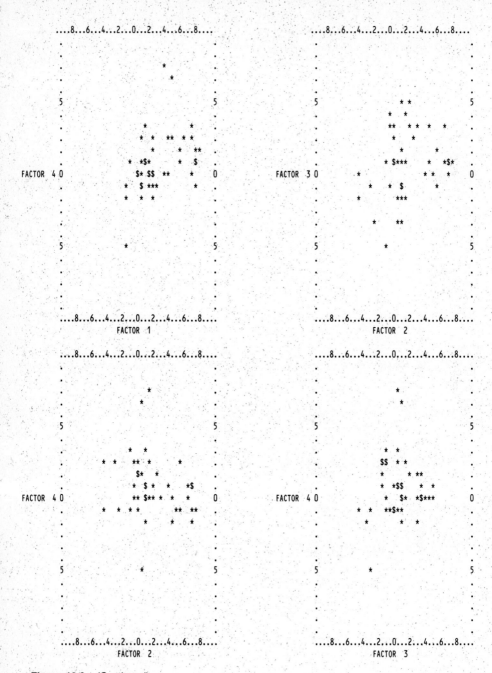

Figure 12.6 *(Continued)*

TABLE 12.22 FACTOR LOADINGS FOR PRINCIPAL FACTORS
EXTRACTION AND VARIMAX ROTATION OF FOUR
FACTORS FOR WOMEN AND FIVE FACTORS FOR
MEN. SELECTED BMDP4M OUTPUT. SETUP APPEARS
IN TABLE 12.19.

(a) Women

ROTATED FACTOR LOADINGS (PATTERN)

		FACTOR 1	FACTOR 2	FACTOR 3	FACTOR 4
HELPFUL	2	0.311	0.269	0.296	0.160
RELIANT	3	0.365	0.083	0.167	0.480
DEFBEL	4	0.413	0.280	0.001	0.017
YIELDING	5	-0.139	0.113	0.345	-0.009
CHEERFUL	6	0.167	0.088	0.559	0.109
INDPT	7	0.466	0.043	0.036	0.484
ATHLET	8	0.324	-0.122	0.247	-0.054
SHY	9	-0.383	-0.074	-0.050	-0.042
ASSERT	10	0.643	0.142	-0.084	0.014
STRPERS	11	0.701	0.102	-0.052	-0.054
FORCEFUL	12	0.645	0.055	-0.210	-0.013
AFFECT	13	0.300	0.392	0.391	-0.289
FLATTER	14	0.165	0.095	0.215	-0.343
LOYAL	15	0.200	0.388	0.319	-0.038
ANALYT	16	0.277	0.233	-0.055	0.131
FEMININE	17	0.057	0.189	0.324	0.111
SYMPATHY	18	-0.042	0.649	0.129	-0.024
MOODY	19	0.030	0.103	-0.374	-0.346
SENSITIV	20	0.056	0.660	0.069	0.028
UNDSTAND	21	0.024	0.731	0.177	0.124
COMPASS	22	0.052	0.811	0.151	0.036
LEADERAB	23	0.739	0.086	0.046	0.146
SOOTHE	24	0.070	0.540	0.297	-0.060
RISK	25	0.496	0.082	0.154	0.016
DECIDE	26	0.483	0.085	0.130	0.345
SELFSUFF	27	0.418	0.110	0.136	0.657
CONSCIEN	28	0.203	0.285	0.235	0.416
DOMINANT	29	0.675	-0.064	-0.281	0.036
MASCULIN	30	0.308	-0.105	-0.287	-0.032
STAND	31	0.593	0.244	0.043	0.162
HAPPY	32	0.122	0.069	0.641	0.106
SOFTSPOK	33	-0.290	0.129	0.388	0.160
WARM	34	0.149	0.483	0.594	-0.150
TRUTHFUL	35	0.139	0.320	0.139	0.166
TENDER	36	0.107	0.446	0.551	-0.142
GULLIBLE	37	-0.041	0.085	0.110	-0.448
LEADACT	38	0.727	-0.024	0.020	0.106
CHILDLIK	39	0.005	-0.068	-0.111	-0.418
INDIVID	40	0.435	0.094	0.069	0.182
FOULLANG	41	-0.017	0.032	0.150	0.030
LOVECHIL	42	0.024	0.201	0.282	-0.127
COMPETE	43	0.541	-0.083	0.163	-0.109
AMBITIOU	44	0.466	0.000	0.199	0.087
GENTLE	45	0.023	0.447	0.554	-0.084
	VP	6.014	4.002	3.400	2.227

THE VP FOR EACH FACTOR IS THE SUM OF THE SQUARES OF THE ELEMENTS OF THE COLUMN
OF THE FACTOR PATTERN MATRIX CORRESPONDING TO THAT FACTOR. WHEN THE ROTATION
IS ORTHOGONAL, THE VP IS THE VARIANCE EXPLAINED BY THE FACTOR.

TABLE 12.22   (Continued)

(b) Men

ROTATED FACTOR LOADINGS (PATTERN)

		FACTOR 1	FACTOR 2	FACTOR 3	FACTOR 4	FACTOR 5
HELPFUL	2	0.553	-0.006	0.364	0.134	0.147
RELIANT	3	0.129	0.218	0.335	-0.020	0.605
DEFBEL	4	0.181	0.459	0.302	-0.202	0.091
YIELDING	5	0.431	-0.434	-0.148	0.237	0.095
CHEERFUL	6	0.465	0.069	0.080	0.663	-0.050
INDPT	7	0.038	0.286	0.293	0.029	0.736
ATHLET	8	0.014	0.109	0.126	0.027	0.026
SHY	9	0.068	-0.463	-0.015	-0.045	-0.052
ASSERT	10	0.034	0.670	0.113	0.070	-0.165
STRPERS	11	0.068	0.763	0.060	-0.082	0.054
FORCEFUL	12	-0.151	0.719	0.078	-0.200	-0.159
AFFECT	13	0.618	0.154	-0.132	0.282	0.030
FLATTER	14	0.218	0.101	-0.111	0.003	0.015
LOYAL	15	0.408	-0.030	0.384	0.000	0.137
ANALYT	16	0.183	0.144	0.441	-0.013	-0.002
FEMININE	17	0.237	-0.094	-0.473	-0.053	-0.042
SYMPATHY	18	0.720	0.002	0.098	0.001	-0.157
MOODY	19	-0.125	0.047	-0.300	-0.516	-0.053
SENSITIV	20	0.731	0.021	0.106	-0.054	0.033
UNDSTAND	21	0.778	-0.018	0.219	0.135	0.102
COMPASS	22	0.767	0.121	0.067	0.052	0.053
LEADERAB	23	0.114	0.688	0.105	0.091	0.199
SOOTHE	24	0.733	0.083	-0.164	0.091	0.088
RISK	25	0.050	0.553	-0.078	-0.084	0.157
DECIDE	26	0.049	0.548	0.146	0.261	0.236
SELFSUFF	27	0.215	0.237	0.286	0.166	0.671
CONSCIEN	28	0.537	-0.068	0.341	0.120	-0.002
DOMINANT	29	-0.130	0.721	0.003	-0.078	0.027
MASCULIN	30	-0.110	0.197	0.511	0.103	0.115
STAND	31	0.156	0.651	0.346	-0.022	0.034
HAPPY	32	0.345	0.160	0.065	0.714	0.001
SOFTSPOK	33	0.223	-0.322	-0.076	0.212	0.285
WARM	34	0.720	0.083	-0.124	0.370	0.059
TRUTHFUL	35	0.335	0.081	0.481	-0.013	0.222
TENDER	36	0.772	0.036	-0.167	0.193	0.148
GULLIBLE	37	0.156	-0.236	-0.323	-0.093	-0.112
LEADACT	38	0.121	0.706	0.042	0.124	0.121
CHILDLIK	39	0.059	0.107	-0.356	-0.172	-0.165
INDIVID	40	0.179	0.498	-0.033	-0.078	0.283
FOULLANG	41	0.113	-0.099	0.048	0.303	0.056
LOVECHIL	42	0.384	0.002	0.066	0.283	0.044
COMPETE	43	0.008	0.431	0.078	-0.026	-0.021
AMBITIOU	44	0.154	0.412	0.010	0.100	0.139
GENTLE	45	0.713	-0.037	-0.094	0.253	0.152
VP		6.845	5.843	2.433	2.170	2.001

THE VP FOR EACH FACTOR IS THE SUM OF THE SQUARES OF THE ELEMENTS OF THE COLUMN
OF THE FACTOR PATTERN MATRIX CORRESPONDING TO THAT FACTOR. WHEN THE ROTATION
IS ORTHOGONAL, THE VP IS THE VARIANCE EXPLAINED BY THE FACTOR.

TABLE 12.23   PERCENTS OF VARIANCE AND COVARIANCE EXPLAINED BY EACH OF THE ROTATED ORTHOGONAL FACTORS FOR MEN AND WOMEN

	Factors				
	1	2	3	4	5
Men					
SSL	6.84	5.84	2.43	2.17	2.00
Percent of variance	15.55	13.27	5.52	4.93	4.55
Percent of covariance	35.48	30.29	12.60	11.26	10.37
Women					
SSL	6.01	4.00	3.40	2.23	
Percent of variance	13.66	9.09	7.73	5.07	
Percent of covariance	38.43	25.88	21.74	14.26	

TABLE 12.24 SMCs FOR FACTORS WITH VARIABLES AS IVs, AND CORRELATIONS AMONG FACTOR SCORES. SELECTED OUTPUT FROM BMDP4M PFA WITH ORTHOGONAL (VARIMAX) ROTATION. SETUP IN TABLE 12.19.

(a) Women

FACTOR SCORE COVARIANCE (COMPUTED FROM FACTOR
STRUCTURE AND FACTOR SCORE COEFFICIENTS)

THE DIAGONAL OF THE MATRIX BELOW CONTAINS THE SQUARED
MULTIPLE CORRELATIONS OF EACH FACTOR WITH THE VARIABLES.

		FACTOR	FACTOR	FACTOR	FACTOR
		1	2	3	4
FACTOR	1	0.903			
FACTOR	2	0.021	0.849		
FACTOR	3	0.001	0.087	0.805	
FACTOR	4	0.044	-0.006	0.010	0.779

(b) Men

FACTOR SCORE COVARIANCE (COMPUTED FROM FACTOR
STRUCTURE AND FACTOR SCORE COEFFICIENTS)

THE DIAGONAL OF THE MATRIX BELOW CONTAINS THE SQUARED
MULTIPLE CORRELATIONS OF EACH FACTOR WITH THE VARIABLES.

		FACTOR	FACTOR	FACTOR	FACTOR	FACTOR
		1	2	3	4	5
FACTOR	1	0.922				
FACTOR	2	0.004	0.912			
FACTOR	3	0.003	0.033	0.769		
FACTOR	4	0.061	-0.007	0.007	0.791	
FACTOR	5	0.020	0.022	0.070	0.010	0.798

in definition of a factor. With the use of the .45 cut, Table 12.26 is generated to further assist interpretation. In more informal presentations of factor analytic results, this table might be reported instead of Table 12.28. Factors are put in columns and variables with the largest loadings are put on top. In interpreting a factor, items near the top of the columns are given somewhat greater weight. Variable names are written out in full detail and labels for the factors (e.g., Dominance) are suggested at the top of each column.

Comparisons between factor structures for women and men are made using visual inspection, the salient similarity index s, and r. It is first noted that the optimal number of factors is different for the two sexes, precluding exact correspondence in structure. Visual inspection reveals the resemblance between factor 1 for women and factor 2 for men, both labeled Dominant. Nine of the 12 variables defining the factor for men are among the 12 variables for women. It is further noted that the variables that are different tend to have lower loadings.

Visual inspection also reveals the similarity between factor 1 for men and factor 2 for women, interpreted as Empathy. Six of 12 variables for men are the same as the 6 variables for women. Again, the variables that are different tend to have lower loadings. Factor 4 for women and factor 5 for men are near-identical. Only the order of loadings for the first 2 of the 3 variables differs between the sexes.

Factor 3 for men, interpreted directly as Sex Role, has no counterpart for women. The three variables that compose the factor (MASCULIN, TRUTHFUL, FEMININE) for men do not load on any of the factors for women.

Factor 3 for women bears some similarity to factor 4 for men, both tentatively interpreted as Positive Affect. Two of the 3 variables for men correspond to 2 of the 5 variables for women (HAPPY and CHEERFUL). However, the 3 remaining variables for women (WARM, GENTLE, TENDER) are part of Factor 1 for men. The third variable for men (MOODY) does not load on any factor for women. Taken at face value, it appears that women and men think differently about empathy and positive affect—for women, warmth, gentleness, and tenderness are part of positive affect, for men those three variables are part of empathy.

TABLE 12.25 SORTED FACTOR LOADINGS FROM BMDP4M PFA FOR WOMEN AND MEN. SETUP IN TABLE 12.19.

(a) Women

SORTED ROTATED FACTOR LOADINGS (PATTERN)

		FACTOR 1	FACTOR 2	FACTOR 3	FACTOR 4
LEADERAB	23	0.739	0.000	0.000	0.000
LEADACT	38	0.727	0.000	0.000	0.000
STRPERS	11	0.701	0.000	0.000	0.000
DOMINANT	29	0.675	0.000	0.000	0.000
FORCEFUL	12	0.645	0.000	0.000	0.000
ASSERT	10	0.643	0.000	0.000	0.000
STAND	31	0.593	0.000	0.000	0.000
COMPETE	43	0.541	0.000	0.000	0.000
COMPASS	22	0.000	0.811	0.000	0.000
UNDSTAND	21	0.000	0.731	0.000	0.000
SENSITIV	20	0.000	0.660	0.000	0.000
SYMPATHY	18	0.000	0.649	0.000	0.000
SOOTHE	24	0.000	0.540	0.000	0.000
HAPPY	32	0.000	0.000	0.641	0.000
WARM	34	0.000	0.483	0.594	0.000
CHEERFUL	6	0.000	0.000	0.559	0.000
GENTLE	45	0.000	0.000	0.554	0.000
TENDER	36	0.000	0.000	0.551	0.000
SELFSUFF	27	0.000	0.000	0.000	0.657
DEFBEL	4	0.000	0.000	0.000	0.000
INDPT	7	0.466	0.000	0.000	0.484
HELPFUL	2	0.000	0.000	0.000	0.000
FLATTER	14	0.000	0.000	0.000	0.000
RISK	25	0.496	0.000	0.000	0.000
DECIDE	26	0.483	0.000	0.000	0.000
FEMININE	17	0.000	0.000	0.000	0.000
CONSCIEN	28	0.000	0.000	0.000	0.000
YIELDING	5	0.000	0.000	0.000	0.000
MASCULIN	30	0.000	0.000	0.000	0.000
ATHLET	8	0.000	0.000	0.000	0.000
LOYAL	15	0.000	0.000	0.000	0.000
SOFTSPOK	33	0.000	0.000	0.000	0.000
ANALYT	16	0.000	0.000	0.000	0.000
TRUTHFUL	35	0.000	0.000	0.000	0.000
MOODY	19	0.000	0.000	0.000	0.000
GULLIBLE	37	0.000	0.000	0.000	0.000
RELIANT	3	0.000	0.000	0.000	0.480
CHILDLIK	39	0.000	0.000	0.000	0.000
INDIVID	40	0.000	0.000	0.000	0.000
FOULLANG	41	0.000	0.000	0.000	0.000
LOVECHIL	42	0.000	0.000	0.000	0.000
SHY	9	0.000	0.000	0.000	0.000
AMBITIOU	44	0.466	0.000	0.000	0.000
AFFECT	13	0.000	0.000	0.000	0.000
	VP	6.014	4.002	3.400	2.227

TABLE 12.25 (Continued)

(b) Men

SORTED ROTATED FACTOR LOADINGS (PATTERN)

		FACTOR 1	FACTOR 2	FACTOR 3	FACTOR 4	FACTOR 5
UNDSTAND	21	0.778	0.000	0.000	0.000	0.000
TENDER	36	0.772	0.000	0.000	0.000	0.000
COMPASS	22	0.767	0.000	0.000	0.000	0.000
SOOTHE	24	0.733	0.000	0.000	0.000	0.000
SENSITIV	20	0.731	0.000	0.000	0.000	0.000
SYMPATHY	18	0.720	0.000	0.000	0.000	0.000
WARM	34	0.720	0.000	0.000	0.000	0.000
GENTLE	45	0.713	0.000	0.000	0.000	0.000
AFFECT	13	0.618	0.000	0.000	0.000	0.000
HELPFUL	2	0.553	0.000	0.000	0.000	0.000
CONSCIEN	28	0.537	0.000	0.000	0.000	0.000
STRPERS	11	0.000	0.763	0.000	0.000	0.000
DOMINANT	29	0.000	0.721	0.000	0.000	0.000
FORCEFUL	12	0.000	0.719	0.000	0.000	0.000
LEADACT	38	0.000	0.706	0.000	0.000	0.000
LEADERAB	23	0.000	0.688	0.000	0.000	0.000
ASSERT	10	0.000	0.670	0.000	0.000	0.000
STAND	31	0.000	0.651	0.000	0.000	0.000
RISK	25	0.000	0.553	0.000	0.000	0.000
DECIDE	26	0.000	0.548	0.000	0.000	0.000
MASCULIN	30	0.000	0.000	0.511	0.000	0.000
HAPPY	32	0.000	0.000	0.000	0.714	0.000
CHEERFUL	6	0.465	0.000	0.000	0.663	0.000
MOODY	19	0.000	0.000	0.000	-0.516	0.000
INDPT	7	0.000	0.000	0.000	0.000	0.736
SELFSUFF	27	0.000	0.000	0.000	0.000	0.671
RELIANT	3	0.000	0.000	0.000	0.000	0.605
FLATTER	14	0.000	0.000	0.000	0.000	0.000
DEFBEL	4	0.000	0.459	0.000	0.000	0.000
YIELDING	5	0.000	0.000	0.000	0.000	0.000
FEMININE	17	0.000	0.000	-0.473	0.000	0.000
SOFTSPOK	33	0.000	0.000	0.000	0.000	0.000
ATHLET	8	0.000	0.000	0.000	0.000	0.000
TRUTHFUL	35	0.000	0.000	0.481	0.000	0.000
LOYAL	15	0.000	0.000	0.000	0.000	0.000
GULLIBLE	37	0.000	0.000	0.000	0.000	0.000
ANALYT	16	0.000	0.000	0.000	0.000	0.000
CHILDLIK	39	0.000	0.000	0.000	0.000	0.000
INDIVID	40	0.000	0.498	0.000	0.000	0.000
FOULLANG	41	0.000	0.000	0.000	0.000	0.000
LOVECHIL	42	0.000	0.000	0.000	0.000	0.000
COMPETE	43	0.000	0.000	0.000	0.000	0.000
AMBITIOU	44	0.000	0.000	0.000	0.000	0.000
SHY	9	0.000	-0.463	0.000	0.000	0.000
	VP	6.845	5.843	2.433	2.170	2.001

Because visual inspection is insufficient to remove doubts about the similarity between factor 3 for women and factor 4 for men, s is calculated to compare the pattern of loadings (taken from Table 12.22), as shown in Table 12.27. As evaluated against Table C.7 in Appendix C, with a hyperplane percentage of .77 and 44 variables, the probability of getting an s value of .67 by chance alone is less than .001. By this test, the factors are similar.

A correlation coefficient is calculated between the third factor loadings for men and the fourth factor loadings for women in Table 12.22 to compare the pattern and magnitude of the loadings. The correlation is .75.

Table 12.28 presents a summary appropriate for a journal article reporting the results of this factor analytic study. Table 12.29 provides a checklist for FA. A Results section in journal format follows for the data analyzed in this section.

TABLE 12.26 ORDER (BY SIZE OF LOADINGS) IN WHICH VARIABLES CONTRIBUTE TO FACTORS FOR WOMEN AND MEN

(a) Women

Factor 1: *Dominance*	Factor 2: *Empathy*	Factor 3: *Positive affect*	Factor 4: *Independence*
Has leadership abilities	Compassionate	Happy	Self-sufficient
Acts as a leader	Understanding	Warm	Independent
Strong personality	Sensitive to needs of others	Cheerful	Self-reliant
Dominant	Sympathetic	Gentle	
Forceful	Eager to soothe hurt feelings	Tender	
Assertive	Warm		
Willing to take a stand			
Competitive			
Willing to take risks			
Makes decisions easily			
Independent			
Ambitious			

(b) Men

Factor 1: *Empathy*	Factor 2: *Dominance*	Factor 3: *Sex role*	Factor 4: *Positive affect*	Factor 5: *Independence*
Understanding	Strong personality	Masculine	Happy	Independent
Tender	Dominant	Truthful	Cheerful	Self-sufficient
Compassionate	Forceful	Feminine	Moody	Self-reliant
Eager to soothe hurt feelings	Acts as a leader			
Sensitive to needs of others	Has leadership abilities			
Sympathetic	Assertive			
Warm	Willing to take a stand			
Gentle	Willing to take risks			
Affectionate	Makes decisions easily			
Helpful	Individualistic			
Conscientious	Shy			
Cheerful	Defends own beliefs			

Note: Most important variables are near top of columns. Proposed labels are in italics.

TABLE 12.27 COMPARISON OF FACTOR 3 FOR MEN WITH FACTOR 4 FOR WOMEN THROUGH THE SALIENT SIMILARITY INDEX, s

		Factor 3, Women		
		PS	HP	NS
Factor 4, Men	PS	4		
	HP	5	34	
	NS			1

$$s = \frac{4 + 1 - 0 - 0}{4 + 1 + 0 + 0 + .5(0 + 5 + 0 + 0)}$$

$$= .67$$

TABLE 12.28 FACTOR LOADINGS, COMMUNALITIES (h^2), PERCENTS OF VARIANCE AND COVARIANCE FOR PRINCIPAL FACTORS EXTRACTION AND VARIMAX ROTATION FOR WOMEN AND MEN ON BSRI ITEMS

Item	Women					Men					
	F_1[a]	F_2	F_3	F_4	h^2	F_1	F_2	F_3	F_4	F_5	h^2
Leadership ability	.74	.00	.00	.00	.58	.00	.69	.00	.00	.00	.55
Acts as leader	.73	.00	.00	.00	.54	.00	.71	.00	.00	.00	.54
Strong personality	.70	.00	.00	.00	.51	.00	.76	.00	.00	.00	.60
Dominant	.68	.00	.00	.00	.54	.00	.72	.00	.00	.00	.54
Forceful	.64	.00	.00	.00	.46	.00	.72	.00	.00	.00	.61
Assertive	.64	.00	.00	.00	.44	.00	.67	.00	.00	.00	.50
Takes stand	.59	.00	.00	.00	.44	.00	.65	.00	.00	.00	.57
Competitive	.54	.00	.00	.00	.34	.00	.00	.00	.00	.00	.19
Takes risks	.50	.00	.00	.00	.28	.00	.55	.00	.00	.00	.35
Makes decisions	.48	.00	.00	.00	.38	.00	.55	.00	.00	.00	.45
Independent	.47	.00	.00	.48	.25	.00	.00	.00	.00	.74	.71
Ambitious	.47	.00	.00	.00	.26	.00	.00	.00	.00	.00	.22
Compassionate	.00	.81	.00	.00	.68	.77	.00	.00	.00	.00	.61
Understanding	.00	.73	.00	.00	.58	.78	.00	.00	.00	.00	.68
Sensitive	.00	.66	.00	.00	.44	.73	.00	.00	.00	.00	.55
Sympathetic	.00	.65	.00	.00	.44	.72	.00	.00	.00	.00	.55
Eager to soothe	.00	.54	.00	.00	.39	.73	.00	.00	.00	.00	.59
Warm	.00	.48	.59	.00	.63	.72	.00	.00	.00	.00	.68
Happy	.00	.00	.64	.00	.44	.00	.00	.00	.71	.00	.66
Cheerful	.00	.00	.56	.00	.36	.47	.00	.00	.66	.00	.67
Gentle	.00	.00	.55	.00	.51	.71	.00	.00	.00	.00	.61
Tender	.00	.00	.55	.00	.53	.77	.00	.00	.00	.00	.68

TABLE 12.28 (Continued)

Item	Women F₁ᵃ	F₂	F₃	F₄	h²	Men F₁	F₂	F₃	F₄	F₅	h²
Self-sufficient	.00	.00	.00	.66	.64	.00	.00	.00	.00	.67	.66
Self-reliant	.00	.00	.00	.48	.40	.00	.00	.00	.00	.61	.54
Affectionate	.00	.00	.00	.00	.48	.62	.00	.00	.00	.00	.50
Conscientious	.00	.00	.00	.00	.35	.54	.00	.00	.00	.00	.42
Defends beliefs	.00	.00	.00	.00	.25	.00	.46	.00	.00	.00	.38
Masculine	.00	.00	.00	.00	.19	.00	.00	.51	.00	.00	.34
Truthful	.00	.00	.00	.00	.17	.00	.00	.48	.00	.00	.40
Feminine	.00	.00	.00	.00	.16	.00	.00	−.47	.00	.00	.29
Helpful	.00	.00	.00	.00	.28	.55	.00	.00	.00	.00	.48
Individualistic	.00	.00	.00	.00	.24	.00	.50	.00	.00	.00	.37
Shy	.00	.00	.00	.00	.16	.00	−.46	.00	.00	.00	.22
Moody	.00	.00	.00	.00	.27	.00	.00	.00	−.52	.00	.38
Percent of variance	13.66	9.09	7.73	5.07		15.55	13.27	5.52	4.93	4.55	
Percent of covariance	38.43	25.58	21.74	14.26		35.48	30.29	12.60	11.26	10.37	

ᵃ Factor labels:

	Women	Men
F_1	Dominance	Empathy
F_2	Empathy	Dominance
F_3	Positive affect	Sex role
F_4	Independence	Positive affect
F_5		Independence

TABLE 12.29 CHECKLIST FOR FACTOR ANALYSIS

1. Limitations
   a. Outliers among cases
   b. Sample size
   c. Factorability of **R**
   d. Normality and linearity of variables
   e. Multicollinearity and singularity
   f. Outliers among variables

2. Major analyses
   a. Number of factors
   b. Nature of factors
   c. Type of rotation
   d. Importance of factors

3. Additional analyses
   a. Factor scores
   b. Distinguishability and simplicity of factors
   c. Complexity of variables
   d. Internal consistency of factors
   e. Comparison of factors
      (1) Between groups
      (2) With theory
   f. Outlying cases among the factors

Results

Principal factors extraction with varimax rotation was performed through BMDP4M on 44 items from the BSRI separately for women and for men. Principal components extraction was used prior to principal factors extraction to estimate number of factors, presence of outliers, absence of multicollinearity, and factorability of the correlation matrices. With an α = .001 cutoff level, 25 of 369 women and 4 of 162 men were identified as outliers and deleted from principal factors extraction.[10]

Four factors were extracted for women and five for men. As indicated by SMCs, all factors for both groups were internally consistent and well defined by the variables; the lowest of the SMCs for factors from variables was .787. [Information on SMCs is from Table 12.24.] The reverse was not true, however; variables were, by and large, not well-defined by this factor solution. Communality values, as seen in Table 12.28, tended to be low. With a cut of .45 for inclusion of a variable in interpretation of a factor, 20 and 11 of 44 variables for women and men, respectively, did not load on any factor. Failure of numerous variables to load on a factor reflects heterogeneity of items on the BSRI. However, only two of the variables in the solution for women, ''warm'' and ''independent,'' and one in the solution for men, ''cheerful,'' were complex.

[10] For each gender, outliers were compared as a group to nonoutliers through discriminant function analysis. As a group, at $p < .01$, the 4 men who were outliers were less dominant, happy, and truthful, but more warm than the men who were not outliers. As a group, at $p < .01$, the 25 women were more feminine, less self-reliant, more easily flattered, less truthful, and less likely to act as leaders than women who were not outliers.

Insert Table 12.28 about here

    Orthogonal rotation was retained
because of conceptual simplicity and ease of
description. Some correlation was revealed
when oblique rotation was requested,
however. For women, factors interpreted as
Empathy and Positive Affect correlated .33,
with 10% overlap in variance. For men,
factors interpreted as Dominance and Sex
Role correlated .39, with 15% overlap in
variance. However, because correlations
were moderate and limited to one pair of
factors in each analysis, and because
remaining correlations were low, orthogonal
rotation was deemed adequate.

    Loadings of variables on factors,
communalities, and percents of variance and
covariance are shown in Table 12.28.
Variables are ordered and grouped by size of
loading to facilitate interpretation.
Loadings under .45 (20% of variance) are
replaced by zeros. Interpretive labels are
suggested for each factor in footnotes.

    There was a strong similarity in the
Dominance factors (factor 1 for women and
factor 2 for men) for the sexes. Factor 4 for
women and factor 5 for men, both interpreted
as Independence factors, were virtually
identical, Factor 3 for men, Sex Role, had no
counterpart for women but was correlated
with Dominance for men (those who described
themselves as masculine also saw themselves
as dominant).

    Factor 1 for men and factor 2 for women,
Empathy, were similar but for the 5
additional variables that loaded on that
factor for men. Some of the variables showed
up on Factor 3 for women, which corresponded
somewhat to Factor 4 for men, and was labeled
Positive Affect. Because the correspondence
between factors 3 (women) and 4 (men) was

```
            less obvious, Cattell's Salient Similarity
            Index, s (Cattell and Baggaley, 1960;
            Cattell, 1957), and r were calculated from
            the full set of factor loadings. The s value
            was .67, which exceeded the value expected
            by chance at p < .001. The correlation among
            the loadings was .75. Both results indicated
            similarity between factor 3 for women and
            factor 4 for men. However, it seems that
            warmth, gentleness, and tenderness were
            associated with empathy for men, but with
            positive affect for women.
```

12.9 SOME EXAMPLES FROM THE LITERATURE

McLeod, Brown, and Becker (1977) used principal components extraction with varimax rotation to collapse responses to 17 items into 4 factors for use in a study of the relationship between perceived locus of blame for the Watergate scandal and several different indicators of political orientation and behaviors. Perceived locus of blame for Watergate among the 617 respondents was composed of 4 factors accounting for a total of 48.7% of the variance and labeled the Media, the Regime (Republican and Democratic Parties, Congress), Nixon and Entourage, and the System (party politics, the political system, the economic system). Factor scores were generated for 181 respondents who were part of a panel study; scores on each factor were correlated with such indicators of political behavior as political trust, party affiliation, campaign participation, and voting direction. Results were somewhat different for those who had or had not supported Nixon in 1972 and indicated that deleterious effects of Watergate were short-term and should be considered in the context of a more general and longer-term decline in political trust and participation.

Menozzi, Piazza, and Cavalli-Sforza (1978) studied the relationship between distributions of genetic characteristics and the expansion of early farming in Europe. The overall goal of analysis was to determine, if possible, if early farmers themselves had migrated or if farming techniques had been adopted by widening circles of local hunters and gatherers. PCA was used to collapse data from 10 genetic loci and 38 alleles (for blood group substances and the like) for 400 map locations into three components. Gradients (or clines) were then formed of the densities of each component over the geographical area of Europe and the Mid-East. The first component showed greatest density in the southeast (the Mid-East) and least in the northwest (Northern Europe and Scandinavia). This map was in remarkable agreement with a map showing archaeological evidence of spread of early farming; and both argue, but not exclusively, for a spread of farmers rather than a spread of only their technology. The second and third components show changes in genetic densities corresponding to east-west gradients and northeast to southwest gradients, respectively.

In a study of willingness to donate human body parts (Pessemier, Bemmoar and Hanssens, 1977), FA was used to collapse 10 items measuring willingness to donate into three factors: donation of blood, skin, marrow; donations upon death; and donation of kidney. FA was performed after a Guttman scale analysis revealed that willingness to donate was not unidimensional. Responses to items composing factors were summed to yield three factor scores for each respondent that served as DVs in separate analyses of variance.[11] Middle-aged female respondents with above-average incomes were most likely to donate regenerative tissue or parts after death.

To investigate the relationship between certain attitudinal constructs and willingness to donate, an attempt was made to reduce 36 attitudinal questions to a smaller set of underlying variables using FA. When the determinant of the covariance matrix was near-zero (indicating extreme multicollinearity), FA was abandoned in favor of inspection of partial correlations between the 36 variables, guided by face validity of the items. Of the 36 items, 25 were retained, composing 9 attitudinal constructs for which responses were summed. A high-medium-low split was used on each attitudinal construct to generate groups for IVs in ANOVA, with the willingness-to-donate variables used as DVs in three separate analyses of each IV. Significant differences were found between most attitudinal variables (e.g., liberalism, interest in physical attractiveness) and one or more of the DVs. However, because of the number of analyses performed among both IVs and DVs that are likely to be correlated, it is difficult to have much faith in any one of the significant differences.

A study by Wond and Allen (1976) combined categories of drugs with three major cognitive dimensions in FAs of semantic differentials to study the relationships between 18 drugs and various attitudes. The three major dimensions were "pleasantness," "strength," and "dangerousness." Four hundred and fifty respondents evaluated each of 18 drugs on each dimension using a Likert-type scale with 10 gradations (from "very safe" to "very dangerous," for instance). For the set of 18 drugs, principal factors extraction of 4 factors followed by rotation were performed separately for each attitudinal dimension plus usage. Only drugs loading .60 or higher with the factors were retained. Usage and "pleasantness" tended to have similar factor structures, as did "dangerousness" and "strength." Other analyses revealed that usage is inversely related to both perceived dangerousness and strength, but directly related to pleasantness. Alcoholic beverages and marijuana tended to show deviant patterns from the rest.

The meaning of money was investigated in a study by Wernimont and Fitzpatrick (1972). Principal components extraction and varimax rotation were used to identify 7 components underlying patterns of responses of 533 subjects to 40 adjective pairs, each rated in a 7-point scale. The 7 components were labeled "shameful failure," "social acceptability," "pooh-pooh attitude," "moral evil," "comfortable security," "social unacceptability," and "conservative business attitudes." Following PCA for the group as a whole, subjects were subdivided into 11 groups on the basis of occupation, which served as IVs in one-way ANOVAs, with component scores as

---

[11] The authors might well have used MANOVA instead of ANOVA.

DVs. Separate ANOVAs were performed for each component. The groups differed significantly in their component scores, with a break usually found between the meaning of money to the unemployed (hard-core trainees, college persons) versus the employed (scientists, managers, salespeople, and the like). In general, it was concluded that money means very different things to different segments of the population.

Chapter 13

An Overview of the General Linear Model

13.1 LINEARITY AND THE GENERAL LINEAR MODEL

In order to facilitate choice of the most useful technique to answer your research question, the emphasis has been on differences among statistical methods up to this point. We have repeatedly hinted, however, that all these techniques are special applications of the general linear model (GLM). The goal of this chapter is to introduce the GLM and to fit the various techniques into the model. In addition to the aesthetic pleasure provided by insight into the GLM, an understanding of it provides a great deal of flexibility in data analysis by promoting use of more sophisticated statistical techniques and computer programs. Most data sets are fruitfully analyzed by one or more of several techniques. Section 13.3 presents an example of the use of alternative research strategies.

Linearity and additivity are important to the GLM. Pairs of variables are assumed to have a linear relationship with each other; that is, it is assumed that relationships between pairs of variables are adequately represented by a straight line. Additivity is also relevant, because if one set of variables is to be predicted by a set of other variables, the effects of the variables within the set are additive in the prediction equation. The second variable in the set adds predictability to the first one, the third adds to the first two, and so on. In all multivariate solutions, the equation relating sets of variables is composed of a series of weighted terms added together.

These assumptions, however, do *not* preclude the inclusion of variables with curvilinear or multiplicative relationships. As is discussed throughout this book, variables can be dichotomized, transformed, or recoded so that even complex relationships are evaluated within the GLM.

13.2 BIVARIATE TO MULTIVARIATE STATISTICS AND OVERVIEW OF TECHNIQUES

13.2.1 Bivariate Form

The GLM is based on prediction or, in jargon, regression. A regression equation represents the value of a DV, Y, as a combination of one or more IVs, X's, plus error. The simplest case of the GLM, then, is the familiar bivariate regression:

$$A + BX + e = Y \qquad (13.1)$$

where B is the change in Y associated with a one-unit change in X; A is a constant representing the value of Y when X is zero; and e is a random variable representing error of prediction.

If X and Y are converted to standard z scores, z_x and z_y, they are now measured on the same scale and cross at the point where both z scores equal zero. The constant A automatically becomes zero because z_y is zero when z_x is zero. Further, after standardization of variances to 1, slope is measured in equal units (rather than the possibly unequal units of X and Y raw scores) and now represents strength of the relationship between X and Y; in bivariate regression with standardized variables, β is equal to the Pearson product-moment correlation coefficient. The closer β is to 1.00 or -1.00, the better the prediction of Y from X (or X from Y). Equation 13.1 then simplifies to

$$\beta z_x + e = z_y \qquad (13.2)$$

As discussed in Chapters 1 and 2, one distinction that is sometimes important in statistics is whether data are continuous or discrete.[1] There are, then, three forms of bivariate regression for situations where X and Y are (1) both continuous, analyzed by Pearson product-moment correlation, (2) mixed, with X dichotomous and Y continuous, analyzed by point biserial correlation, and (3) both dichotomous, analyzed by phi coefficient. In fact, these three forms of correlation are identical. If the dichotomous variable is coded 0–1, all the correlations can be calculated using the equation for Pearson product-moment correlation. Table 13.1 compares the three bivariate forms of the GLM.

13.2.2 Simple Multivariate Form

The first generalization of the simple bivariate form of the GLM is to increase the number of IVs, X's, used to predict Y. It is here that the additivity of the model first becomes apparent. In standardized form:

---

[1] When discrete variables have more than two levels, they are dummy variable coded into $k - 1$ (df) dichotomous variables to eliminate the possibility of nonlinear relationships. In this section, when we speak of statistical techniques using discrete variables, we imply that recoding is unnecessary or is handled internally in computer programs designed for the particular analysis.

$$\sum_{i=1}^{k} \beta_i z_{x_i} + e = z_y \qquad (13.3)$$

That is, Y is predicted by a weighted *sum* of X's. The weights, β_i, no longer reflect the correlation between Y and each X because they are also affected by correlations among the X's. Here, again, as seen in Table 13.1, there are special statistical techniques associated with whether all X's are continuous; here also, with appropriate coding, the most general form of the equation can be used to solve all the special cases.

If Y and all X's are continuous, the special statistical technique is multiple regression. Indeed, as seen in Chapter 5, Equation 13.3 is used to describe the multiple regression problem. But if Y is continuous while all X's are discrete, we have the special case of regression known as analysis of variance. The values of X represent "groups" and the emphasis is on finding mean differences in Y between groups rather than on predicting Y, but the basic equation is the same. A significant difference between groups implies that knowledge of X can be used to predict performance on Y.

Analysis of variance problems can be solved through multiple regression computer programs. There are as many X's as there are degrees of freedom for the effects. For example, in a one-way design, three groups are recoded into two dichotomous X's, one representing the first group vs. the other two and the second representing the second group vs. the other two. The third group is those who are not in either of the other two groups. Inclusion of a third X would produce singularity because it is perfectly predictable from the combination of the other two.

If IVs are factorially combined, main effects and interactions are still coded into a series of dichotomous X variables. Consider an example of one IV, anxiety level, divided into three groups and a second IV, task difficulty, divided into two

TABLE 13.1 OVERVIEW OF TECHNIQUES IN THE GENERAL
                LINEAR MODEL

A. Bivariate form (Eq. 13.2)
   1. Pearson product-moment correlation: X continuous, Y continuous
   2. Point biserial correlation: X dichotomous, Y continuous
   3. Phi coefficient: X dichotomous, Y dichotomous
B. Simple multivariate form (Eq. 13.3)
   1. Multiple regression: all X's continuous, Y continuous
   2. ANOVA: all X's discrete, Y continuous
   3. ANCOVA: some X's continuous and some discrete, Y continuous
   4. Two-group discriminant function analysis: all X's continuous, Y dichotomous
   5. Multiway frequency analysis (logit): all X's discrete, Y discrete
C. Full multivariate form (Eq. 13.4)
   1. Canonical correlation: all X's continuous, all Y's continuous
   2. MANOVA: all X's discrete, all Y's continuous
   3. MANCOVA: some X's continuous and some discrete, all Y's continuous
   4. Profile analysis: all X's discrete, all Y's continuous and commensurate
   5. Discriminant function analysis: all X's continuous, all Y's discrete
   6. Factor analysis (FA)/principal component analysis (PCA): all Y's continuous, all X's latent
   7. Multiway frequency analysis: all X's discrete, Y is category frequency

groups. There are two X components for the 2 df associated with anxiety level and one X component for the 1 df associated with task difficulty. An additional two X components are needed for the 2 df associated with the interaction of anxiety level and task difficulty. The five X components are combined to test each of the two main effects and the interaction or are tested individually if the comparisons coded into each component are of interest. Because a number of excellent books (e.g., Cohen and Cohen, 1975) describe analysis of variance through multiple regression, that fascinating topic has not been pursued in this book.

If some X's are continuous and others are discrete, with Y continuous, we have analysis of covariance. The continuous X's are the covariates and the discrete ones are the IVs. The effects of IVs on Y are assessed after adjustments are made for the effects of the covariates on Y. Actually, the GLM can deal with combinations of continuous and discrete Y's in much more general ways than traditional analysis covariance, as alluded to in Chapters 5 and 8.

If Y is dichotomous (two groups), with X's continuous, we have the simple multivariate form of discriminant function analysis. The aim is to predict group membership on the basis of the X's. There is a reversal in terminology between ANOVA and discriminant function analysis; in ANOVA the groups are represented by X, while in discriminant function analysis the groups are represented by Y. The distinction, although confusing, is trivial within the GLM. As seen in forthcoming sections, *all* the special techniques are simply special cases of the full GLM.

Finally, if Y and all X's are discrete we have multiway frequency analysis. The loglinear, rather than simple linear, model is required to evaluate relationships among variables. Logarithmic transforms are applied to cell frequencies and the weighted sum of these cell frequencies is used to predict group membership. Because the equation eventually boils down to a weighted sum of terms, it is considered here part of the GLM.

13.2.3 Full Multivariate Form

The GLM takes a major leap when the Y side of the equation is expanded. More than one equation may be required to relate the X's to the Y's:

Root

$$1: \quad \sum_{i=1}^{k} \beta_{i1} z_{x_{i1}} = \sum_{j=1}^{p} \gamma_{j1} z_{y_{j1}}$$

$$2: \quad \sum_{i=1}^{k} \beta_{i2} z_{x_{i2}} = \sum_{j=1}^{p} \gamma_{j2} z_{y_{j2}}$$

$$\qquad\qquad\qquad\qquad\qquad\qquad (13.4)$$

$$\cdot$$
$$\cdot$$
$$\cdot$$

$$m: \quad \sum_{i=1}^{k} \beta_{im} z_{x_{im}} + e = \sum_{j=1}^{p} \gamma_{jm} z_{y_{jm}}$$

where m equals k or p, whichever is smaller, and γ are regression weights for the standardized Y variables.

In general, there are as many equations as the number of X or Y variables, whichever is smaller. When there is only one Y, X's are combined to produce one straight-line relationship with Y. Once there is more than one Y, however, combined Y's and combined X's may fit together in several different ways. Consider first the simplest example where Y is a dichotomous variable representing two groups. The X's are combined to form a single variable that best separates the mean of Y for the first group from the mean of Y for the second group, as illustrated in Figure 13.1(a). A line parallel to the imaginary line that connects the two means represents the linear combination of X's, or the first root of X. Once a third group is added, however, it may not fall along that line. To maximally separate the three groups, it may be necessary to add a second linear combination of X's, or second root. In the example in Figure 13.1(b), the first root separates the means of the first group from the means of the other two groups but does not distinguish the means for the second and third groups. The second root separates the mean of the third group from the other two but does not distinguish between the first two groups.

Roots are called by other names in the special statistical technique in which they are developed: discriminant functions, principal components, canonical variates, and so forth. Full multivariate techniques need multidimensional space to describe relationships among variables. With 2 df, two dimensions might be needed. With 3 df, up to three dimensions might be needed, and so on.

With three groups, however, means for all groups might fall along a single straight line [i.e., $\bar{Y}_3$ of Figure 13.1(b) could also fall along the line representing the first root of X]. Then only the first root is needed to describe the relationship. The number of roots *necessary* to describe the relationship between two sets of variables may be smaller than the number of roots maximally available. For this reason, the error term in Equation 13.4 is not necessarily associated with the mth root. It is

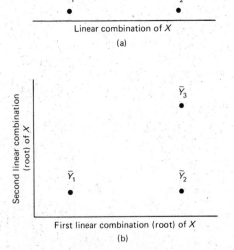

Figure 13.1   (a) Plot of two group means, $\bar{Y}_1$ and $\bar{Y}_2$, on a scale representing the linear combination of X. (b) Plot of two linear combinations of X required to distinguish among three group means: $\bar{Y}_1$, $\bar{Y}_2$, and $\bar{Y}_3$.

associated with the last necessary root, with "necessary" statistically or psychometrically defined.

As with simpler forms of the GLM, specialized statistical techniques are associated with whether variables are continuous, as summarized in Table 13.1. Canonical correlation is the most general form and the noble ancestor of the GLM where all X's and Y's are continuous. With appropriate recoding, all bivariate and multivariate problems (with the exception of PCA, FA, and MFA) could be solved through canonical correlation. Practically, however, the programs for canonical correlation tend not to give the kinds of information usually desired when one or more of the X or Y variables is discrete. Programs for the "multivariate general linear model" tend to be rich, but much more difficult to use.

With all X's discrete and all Y's continuous, we have multivariate analysis of variance. The discrete X variables represent groups, and combinations of Y variables are examined to see how their centroids differ as a function of group membership. If some X's are continuous, they can be analyzed as covariates, just as in ANCOVA; MANCOVA is used to discover how groups differ on Y's after adjustment for the effects of covariates.

If the Y's are all measured on the same scale and/or represent levels of a within-subjects IV, profile analysis is available—a form of MANOVA that is especially informative for these kinds of data. And if there are multiple DVs at each level of a within-subjects IV, doubly multivariate analysis of variance is used to discover the effects of the IVs on the Y's.

When Y is discrete (more than two groups) and X's are continuous, the full multivariate form of discriminant function analysis is used to predict membership in Y.

There is a family of procedures—FA and PCA—where the continuous Y's are measured empirically but the X's are latent. It is assumed that a set of roots underlies the Y's; the purpose of analysis is to uncover the set of roots, or factors, or X's.

Finally, MFA is appropriate when there is a set of discrete X's among which relationships are sought and none of the variables is designated Y. Application of a loglinear form of the GLM allows prediction of frequencies in the cells from factorially combined X's, so the implicit DV is cell frequency.

13.3 ALTERNATIVE RESEARCH STRATEGIES

For most data, there is more then one appropriate analytical strategy, depending on considerations such as how the variables are interrelated, your preference for interpreting statistics associated with certain techniques, and the audience you intend to address.

A data set for which alternative strategies are appropriate has groups of people who receive one of three types of treatment, behavior modification, short-term psychotherapy, and a waiting-list control group. Suppose a great many variables are measured: self-reports of symptoms and moods, reports of family members, therapist reports, and a host of personality and attitudinal tests. The major goal of analysis is probably to find out if, and on which variable(s), the groups differ after treatment.

The obvious strategy is MANOVA, but a likely problem is that the number of variables exceeds the number of clients in some group, leading to singularity. Further, with so many variables, some are likely to be highly related to combinations of others. You could choose among them or combine them on some rational basis, or you might choose first to look at empirical relationships among them.

A first step in reducing the number of variables might be examination of squared multiple correlations of each variable with all the others through regression or factor analysis programs. But the SMCs might or might not provide sufficient information for a judicious decision about which variables to delete and/or combine. If not, the next likely step is a principal component analysis on the pooled within-cells correlation matrix.

The usual procedures for deciding number of components and type of rotation are followed. Out of this analysis come scores for each client on each component and some idea of the meaning of each component. Depending on the outcome, subsequent strategies might differ. If the principal components are orthogonal, the component scores can serve as DVs in a series of univariate ANOVAs, with adjustment for experimentwise Type I error. If the components are correlated, then MANOVA is used with component scores as DVs. The stepdown hierarchy might well correspond to the order of components (the scores on the first component enter first, and so on).

Or, you might want to analyze the component scores through a discriminant function analysis to learn, for instance, that differences between behavior modification and short-term psychotherapy are most notable on components loaded heavily with attitudes and self-reports, while differences between the treated groups and the control group are associated with components loaded with therapist reports and personality measures.

You could, in fact, solve the entire problem through discriminant function analysis. Discriminant function programs protect against multicollinearity and singularity by setting a tolerance level so that variables that are highly predicted by the other variables do not participate in the solution.

These strategies are all "legitimate" and simply represent different ways of getting to the same goal. In the immortal words spoken one Tuesday night in the Jacuzzi by Sanford A. Fidell, "You mean you only know one thing, but you have a dozen different names for it?"

A Skimpy Introduction to Matrix Algebra

The purpose of this appendix is to provide readers with sufficient background to follow, and duplicate as desired, calculations illustrated in the fourth sections of Chapters 5 through 12. The purpose is not to provide a thorough review of matrix algebra or even to facilitate an in-depth understanding of it. The reader who is interested in more than calculational rules has several excellent discussions available, particularly those in Tatsuoka (1971) and Rummel (1970).

Most of the algebraic mnaipulations with which the reader is familiar—addition, subtraction, multiplication, and division—have counterparts in matrix algebra. In fact, the algebra that most of us learned is a special case of matrix algebra involving only a single number, a scalar, instead of an ordered array of numbers, a matrix. Some generalizations from scalar algebra to matrix algebra seem "natural" (i.e., matrix addition and subtraction) while others (multiplication and division) are convoluted. Nonetheless, matrix algebra provides an extremely powerful and compact method for manipulating sets of numbers to arrive at desirable statistical products.

The matrix calculations illustrated here are calculations performed on square matrices. Square matrices have the same number of rows as columns. Sums-of-squares and cross-products matrices, variance-covariance matrices, and correlation matrices are all square. In addition, these three very commonly encountered matrices are symmetrical, having the same value in row 1, column 2, as in column 1, row 2, and so forth. Symmetrical matrices are mirror images of themselves about the main diagonal (the diagonal going from top left to bottom right in the matrix).

There is a more complete matrix algebra that includes nonsquare matrices as well. However, once one proceeds from the data matrix, which has as many rows as research units (subjects) and as many columns as variables, to the sum-of-squares and cross-products matrix, as illustrated in Section 1.5, most calculations illustrated in this book involve square, symmetrical matrices. A further restriction on this appendix is to limit the discussion to only those manipulations used in the fourth sections of Chapters 5 through 12. For purposes of numerical illustration, two very simple matrices, square, but not symmetrical (to eliminate any uncertainty regarding which elements are involved in calculations), will be defined as follows:

$$\mathbf{A} = \begin{bmatrix} a & b & c \\ d & e & f \\ g & h & i \end{bmatrix} = \begin{bmatrix} 3 & 2 & 4 \\ 7 & 5 & 0 \\ 1 & 0 & 8 \end{bmatrix}$$

$$\mathbf{B} = \begin{bmatrix} r & s & t \\ u & v & w \\ x & y & z \end{bmatrix} = \begin{bmatrix} 6 & 1 & 0 \\ 2 & 8 & 7 \\ 3 & 4 & 5 \end{bmatrix}$$

A.1 ADDITION OR SUBTRACTION OF A CONSTANT TO A MATRIX

If one has a matrix, **A,** and wants to add or subtract a constant, k, to the elements of the matrix, one simply adds (or subtracts) the constant to every element in the matrix.

$$\mathbf{A} + k = \begin{bmatrix} a & b & c \\ d & e & f \\ g & h & i \end{bmatrix} + k$$

$$= \begin{bmatrix} a+k & b+k & c+k \\ d+k & e+k & f+k \\ g+k & h+k & i+k \end{bmatrix} \tag{A.1}$$

If $k = -3$, then

$$\mathbf{A} + k = \begin{bmatrix} 0 & -1 & 1 \\ 4 & 2 & -3 \\ -2 & -3 & 5 \end{bmatrix}$$

A.2 MULTIPLICATION OR DIVISION OF A MATRIX BY A CONSTANT

Multiplication or division of a matrix by a constant is a straightforward process.

$$k\mathbf{A} = k \begin{bmatrix} a & b & c \\ d & e & f \\ g & h & i \end{bmatrix}$$

$$k\mathbf{A} = \begin{bmatrix} ka & kb & kc \\ kd & ke & kf \\ kg & kh & ki \end{bmatrix} \tag{A.2}$$

and

$$\frac{1}{k}\mathbf{A} = \begin{bmatrix} \dfrac{a}{k} & \dfrac{b}{k} & \dfrac{c}{k} \\ \dfrac{d}{k} & \dfrac{e}{k} & \dfrac{f}{k} \\ \dfrac{g}{k} & \dfrac{h}{k} & \dfrac{i}{k} \end{bmatrix} \tag{A.3}$$

Numerically, if $k = 2$, then

$$kA = \begin{bmatrix} 6 & 4 & 8 \\ 14 & 10 & 0 \\ 2 & 0 & 16 \end{bmatrix}$$

A.3 ADDITION AND SUBTRACTION OF TWO MATRICES

These procedures are straightforward, as well as useful. If matrices **A** and **B** are as defined at the beginning of this appendix, one simply performs the addition or subtraction of corresponding elements.

$$\mathbf{A} + \mathbf{B} = \begin{bmatrix} a & b & c \\ d & e & f \\ g & h & i \end{bmatrix} + \begin{bmatrix} r & s & t \\ u & v & w \\ x & y & z \end{bmatrix}$$

$$= \begin{bmatrix} a+r & b+s & c+t \\ d+u & e+v & f+w \\ g+x & h+y & i+z \end{bmatrix} \tag{A.4}$$

and

$$\mathbf{A} - \mathbf{B} = \begin{bmatrix} a-r & b-s & c-t \\ d-u & e-v & f-w \\ g-x & h-y & i-z \end{bmatrix} \tag{A.5}$$

For the numerical example:

$$\mathbf{A} + \mathbf{B} = \begin{bmatrix} 3 & 2 & 4 \\ 7 & 5 & 0 \\ 1 & 0 & 8 \end{bmatrix} + \begin{bmatrix} 6 & 1 & 0 \\ 2 & 8 & 7 \\ 3 & 4 & 5 \end{bmatrix}$$

$$\mathbf{A} + \mathbf{B} = \begin{bmatrix} 9 & 3 & 4 \\ 9 & 13 & 7 \\ 4 & 4 & 13 \end{bmatrix}$$

Calculation of a difference between two matrices is required when, for instance, one desires a residuals matrix, the matrix obtained by subtracting a reproduced matrix from an obtained matrix (as in factor analysis, Chapter 12). Or, if the matrix that is subtracted happens to consist of columns with appropriate means of variables inserted in every slot, then the difference between it and a matrix of raw scores produces a deviation matrix.

A.4 MULTIPLICATION, TRANSPOSES, AND SQUARE ROOTS OF MATRICES

Matrix multiplication is both unreasonably complicated and undeniably useful. Note that the ijth element of the resulting matrix is a function of row i of the first matrix and column j of the second.

$$\mathbf{AB} = \begin{bmatrix} a & b & c \\ d & e & f \\ g & h & i \end{bmatrix} \begin{bmatrix} r & s & t \\ u & v & w \\ x & y & z \end{bmatrix}$$

$$= \begin{bmatrix} ar + bu + cx & as + bv + cy & at + bw + cz \\ dr + eu + fx & ds + ev + fy & dt + ew + fz \\ gr + hu + ix & gs + hv + iy & gt + hw + iz \end{bmatrix} \qquad \text{(A.6)}$$

Numerically,

$$\mathbf{AB} = \begin{bmatrix} 3 & 2 & 4 \\ 7 & 5 & 0 \\ 1 & 0 & 8 \end{bmatrix} \begin{bmatrix} 6 & 1 & 0 \\ 2 & 8 & 7 \\ 3 & 4 & 5 \end{bmatrix}$$

$$= \begin{bmatrix} 3 \cdot 6 + 2 \cdot 2 + 4 \cdot 3 & 3 \cdot 1 + 2 \cdot 8 + 4 \cdot 4 & 3 \cdot 0 + 2 \cdot 7 + 4 \cdot 5 \\ 7 \cdot 6 + 5 \cdot 2 + 0 \cdot 3 & 7 \cdot 1 + 5 \cdot 8 + 0 \cdot 4 & 7 \cdot 0 + 5 \cdot 7 + 0 \cdot 5 \\ 1 \cdot 6 + 0 \cdot 2 + 8 \cdot 3 & 1 \cdot 1 + 0 \cdot 8 + 8 \cdot 4 & 1 \cdot 0 + 0 \cdot 7 + 8 \cdot 5 \end{bmatrix}$$

$$= \begin{bmatrix} 34 & 35 & 34 \\ 52 & 47 & 35 \\ 30 & 33 & 40 \end{bmatrix}$$

Regrettably, $\mathbf{AB} \neq \mathbf{BA}$ in matrix algebra. Thus

$$\mathbf{BA} = \begin{bmatrix} 6 & 1 & 0 \\ 2 & 8 & 7 \\ 3 & 4 & 5 \end{bmatrix} \begin{bmatrix} 3 & 2 & 4 \\ 7 & 5 & 0 \\ 1 & 0 & 8 \end{bmatrix}$$

$$\mathbf{BA} = \begin{bmatrix} 25 & 17 & 24 \\ 69 & 44 & 64 \\ 42 & 26 & 52 \end{bmatrix}$$

If another concept of matrix algebra is introduced, some useful statistical properties of matrix algebra can be shown. The transpose of a matrix is indicated by a prime (′) and stands for a rearrangement of the elements of the matrix such that the first row becomes the first column, the second row the second column, and so forth. Thus

$$\mathbf{A}' = \begin{bmatrix} a & d & g \\ b & e & h \\ c & f & i \end{bmatrix}$$

$$= \begin{bmatrix} 3 & 7 & 1 \\ 2 & 5 & 0 \\ 4 & 0 & 8 \end{bmatrix} \qquad \text{(A.7)}$$

When transposition is used in conjunction with multiplication, then some advantages of matrix multiplication become clear, namely,

$$\mathbf{AA}' = \begin{bmatrix} a & b & c \\ d & e & f \\ g & h & i \end{bmatrix} \begin{bmatrix} a & d & g \\ b & e & h \\ c & f & i \end{bmatrix}$$

$$= \begin{bmatrix} a^2 + b^2 + c^2 & ad + be + cf & ag + bh + ci \\ ad + be + cf & d^2 + e^2 + f^2 & dg + eh + fi \\ ag + bh + ci & dg + eh + fi & g^2 + h^2 + i^2 \end{bmatrix} \quad \text{(A.8)}$$

The elements in the main diagonal are the sums of squares while those off the diagonal are cross products.

Had **A** been multiplied by itself, rather than by a transpose of itself, a different result would have been achieved.

$$\mathbf{AA} = \begin{bmatrix} a^2 + bd + cg & ab + be + ch & ac + bf + ci \\ da + ed + fg & db + e^2 + fh & dc + ef + fi \\ ga + hd + ig & gb + he + ih & gc + hf + i^2 \end{bmatrix}$$

If $\mathbf{AA} = \mathbf{C}$, then $\mathbf{C}^{1/2} = \mathbf{A}$. That is, there is a parallel in matrix algebra to squaring and taking the square root of a scalar, but it is a complicated business because of the complexity of matrix multiplication. If, however, one has a matrix **C** from which a square root is desired (as in canonical correlation, Chapter 6), one searches for a matrix, **A**, which, when multiplied by itself, produces **C**. If, for example,

$$\mathbf{C} = \begin{bmatrix} 27 & 16 & 44 \\ 56 & 39 & 28 \\ 11 & 2 & 68 \end{bmatrix}$$

then

$$\mathbf{C}^{1/2} = \begin{bmatrix} 3 & 2 & 4 \\ 7 & 5 & 0 \\ 1 & 0 & 8 \end{bmatrix}$$

A.5 MATRIX "DIVISION" (INVERSES AND DETERMINANTS)

If you liked matrix multiplication, you'll love matrix inversion. Logically, the process is analogous to performing division for single numbers by finding the reciprocal of the number and multiplying by the reciprocal: if $a^{-1} = 1/a$, then $(a)(a^{-1}) = a/a = 1$. That is, the reciprocal of a scalar is a number that, when multiplied by the number itself, equals 1. Both the concepts and the notation are similar in matrix algebra, but they are complicated by the fact that a matrix is an array of numbers.

To determine if the reciprocal of a matrix has been found, one needs the matrix equivalent of the 1 as employed in the preceding paragraph. The identity matrix, **I**, a matrix with 1's in the main diagonal and zeros elsewhere, is such a matrix. Thus

$$\mathbf{I} = \begin{bmatrix} 1 & 0 & 0 \\ 0 & 1 & 0 \\ 0 & 0 & 1 \end{bmatrix} \quad \text{(A.9)}$$

Matrix division, then, becomes a process of finding $\mathbf{A}^{-1}$ such that

$$\mathbf{A}^{-1}\mathbf{A} = \mathbf{AA}^{-1} = \mathbf{I} \quad \text{(A.10)}$$

One way of finding A^{-1} requires a two-stage process, the first of which consists of finding the determinant of A, noted $|A|$. The determinant of a matrix is sometimes said to represent the generalized variance of the matrix, as most readily seen in a 2×2 matrix. Thus we define a new matrix as follows:

$$D = \begin{bmatrix} a & b \\ c & d \end{bmatrix}$$

where

$$|D| = ad - bc \qquad (A.11)$$

If D is a variance-covariance matrix where a and d are variances while b and c are covariances, then $ad - bc$ represents variance minus covariance. It is this property of determinants that makes them useful for hypothesis testing (see, for example, Chapter 9, Section 9.4, where Wilks' Lambda is used in MANOVA).

Calculation of determinants becomes rapidly more complicated as the matrix gets larger. For example, in our 3 by 3 matrix,

$$|A| = a(ei - fh) + b(fg - di) + c(dh - eg) \qquad (A.12)$$

Should the determinant of A equal zero, then the matrix cannot be inverted because the next operation in inversion would involve division by zero. Multicollinear or singular matrices (those with variables that are linear combinations of one another, as discussed in Chapter 4) have zero determinants that prohibit inversion.

A full inversion of A is

$$A^{-1} = \begin{bmatrix} a & b & c \\ d & e & f \\ g & h & i \end{bmatrix}^{-1}$$

$$= \frac{1}{|A|} \begin{bmatrix} ei - fh & ch - bi & bf - ce \\ fg - di & ai - cg & cd - af \\ dh - eg & bg - ah & ae - bd \end{bmatrix} \qquad (A.13)$$

Please recall that because A is not a variance-covariance matrix, a negative determinant is possible, even somewhat likely. Thus, in the numerical example,

$$|A| = 3(5 \cdot 8 - 0 \cdot 0) + 2(0 \cdot 1 - 7 \cdot 8) + 4(7 \cdot 0 - 5 \cdot 1)$$
$$= -12$$

and

$$A^{-1} = \frac{1}{-12} \begin{bmatrix} 5 \cdot 8 - 0 \cdot 0 & 4 \cdot 0 - 2 \cdot 8 & 2 \cdot 0 - 4 \cdot 5 \\ 0 \cdot 1 - 7 \cdot 8 & 3 \cdot 8 - 4 \cdot 1 & 4 \cdot 7 - 3 \cdot 0 \\ 7 \cdot 0 - 5 \cdot 1 & 2 \cdot 1 - 3 \cdot 0 & 3 \cdot 5 - 2 \cdot 7 \end{bmatrix}$$

$$= \begin{bmatrix} {}^{40}\!/_{-12} & {}^{-16}\!/_{-12} & {}^{-20}\!/_{-12} \\ {}^{-56}\!/_{-12} & {}^{20}\!/_{-12} & {}^{28}\!/_{-12} \\ {}^{-5}\!/_{-12} & {}^{2}\!/_{-12} & {}^{1}\!/_{-12} \end{bmatrix}$$

$$= \begin{bmatrix} -3.33 & 1.33 & 1.67 \\ 4.67 & -1.67 & -2.33 \\ .42 & -.17 & -.08 \end{bmatrix}$$

Confirm that, within rounding error, Equation A.10 is true. Once the inverse of **A** is found, "division" by it is accomplished whenever required by using the inverse and performing matrix multiplication.

A.6 EIGENVALUES AND EIGENVECTORS: PROCEDURES FOR CONSOLIDATING VARIANCE FROM A MATRIX

We promised you a demonstration of computation of eigenvalues and eigenvectors for a matrix, so here it is. Like a Chinese dinner, however, you may well find that this discussion satisfies your appetite for only a couple of hours. During that time, round up Tatsuoka (1971), get the cat off your favorite chair, and prepare for an intelligible, if somewhat lengthy, description of the same subject.

Most of the multivariate procedures rely on eigenvalues and their corresponding eigenvectors (also called characteristic roots and vectors) in one way or another because they consolidate the variance in a matrix (the eigenvalue) while providing the linear combination of variables (the eigenvector) to do it. The coefficients applied to variables to form linear combinations of variables in all the multivariate procedures are rescaled elements from eigenvectors. The variance that the solution "accounts for" is associated with the eigenvalue, and is sometimes called so directly.

Calculation of eigenvalues and eigenvectors is best left up to a computer with any realistically sized matrix. For illustrative purposes, a 2 × 2 matrix will be used here. The logic of the process is also somewhat difficult, involving several of the more abstract notions and relations in matrix algebra, including the equivalence between matrices, systems of linear equations with several unknowns, and roots of polynomial equations.

Solutions of an eigenproblem involves solution of the following equation:

$$(\mathbf{D} - \lambda \mathbf{I})V = 0 \tag{A.14}$$

where λ is the eigenvalue and V the eigenvector to be sought. Expanded, this equation becomes

$$\left[\begin{bmatrix} a & b \\ c & d \end{bmatrix} - \lambda \begin{bmatrix} 1 & 0 \\ 0 & 1 \end{bmatrix} \right] \begin{bmatrix} v_1 \\ v_2 \end{bmatrix} = 0$$

or

$$\left[\begin{bmatrix} a & b \\ c & d \end{bmatrix} - \begin{bmatrix} \lambda & 0 \\ 0 & \lambda \end{bmatrix} \right] \begin{bmatrix} v_1 \\ v_2 \end{bmatrix} = 0$$

or, by applying Equation A.5,

$$\begin{bmatrix} a - \lambda & b \\ c & d - \lambda \end{bmatrix} \begin{bmatrix} v_1 \\ v_2 \end{bmatrix} = 0 \qquad (A.15)$$

If one considers the matrix **D**, whose eigenvalues are sought, a variance-covariance matrix, one can see that a solution is desired to "capture" the variance in **D** while rescaling the elements in **D** by v_1 and v_2 to do so.

It is obvious from Equation A.15 that a solution is always available when v_1 and v_2 are zero. A nontrivial solution may also be available when the determinant of the leftmost matrix in Equation A.15 is zero.[1] That is, if (following Equation A.11)

$$(a - \lambda)(d - \lambda) - bc = 0 \qquad (A.16)$$

then there may exist values of λ and values of v_1 and v_2 that satisfy the equation and are not zero. However, expansion of Equation A.16 gives a polynomial equation, in λ, of degree 2:

$$\lambda^2 - (a + d)\lambda + ad - bc = 0 \qquad (A.17)$$

Solving for the eigenvalues, λ, requires solving for the roots of this polynomial. If the matrix has certain properties (see footnote 1), there will be as many positive roots to the equation as there are rows (or columns) in the matrix.

If Equation A.17 is rewritten as $x\lambda^2 + y\lambda + z = 0$, the roots may be found by applying the following equation:

$$\lambda = \frac{-y \pm \sqrt{y^2 - 4xz}}{2x} \qquad (A.18)$$

For a numerical example, consider the following matrix.

$$\mathbf{D} = \begin{bmatrix} 5 & 1 \\ 4 & 2 \end{bmatrix}$$

Applying Equation A.17, we obtain

$$\lambda^2 - (5 + 2)\lambda + 5 \cdot 2 - 1 \cdot 4 = 0$$

or

$$\lambda^2 - 7\lambda + 6 = 0$$

The roots to this polynomial may be found by Equation A.18 as follows:

$$\lambda_1 = \frac{-(-7) + \sqrt{(-7)^2 - 4 \cdot 1 \cdot 6}}{2 \cdot 1}$$

$$= 6$$

---

[1] Read Tatsuoka (1971); a matrix is said to be positive definite when all $\lambda_i > 0$, positive semidefinite when all $\lambda_i \geq 0$, and ill-conditioned when some $\lambda_i < 0$.

and

$$\lambda_2 = \frac{-(-7) - \sqrt{(-7)^2 - 4 \cdot 1 \cdot 6}}{2 \cdot 1}$$

$$= 1$$

(The roots could also be found by factoring to get $[\lambda - 6][\lambda - 1]$.)

Once the roots are found, they may be used in Equation A.15 to find v_1 and v_2, the eigenvector. There will be one set of eigenvectors for the first root and a second set for the second root. Both solutions require solving sets of two simultaneous equations in two unknowns, to wit, for the first root, 6, and applying Equation A.15.

$$\begin{bmatrix} 5 - 6 & 1 \\ 4 & 2 - 6 \end{bmatrix} \begin{bmatrix} v_1 \\ v_2 \end{bmatrix} = 0$$

or

$$\begin{bmatrix} -1 & 1 \\ 4 & -4 \end{bmatrix} \begin{bmatrix} v_1 \\ v_2 \end{bmatrix} = 0$$

so that

$$-1v_1 + 1v_2 = 0$$

and

$$4v_1 - 4v_2 = 0$$

When $v_1 = 1$ and $v_2 = 1$, a solution is found.

For the second root, 1, the equations become

$$\begin{bmatrix} 5 - 1 & 1 \\ 4 & 2 - 1 \end{bmatrix} \begin{bmatrix} v_1 \\ v_2 \end{bmatrix} = 0$$

or

$$\begin{bmatrix} 4 & 1 \\ 4 & 1 \end{bmatrix} \begin{bmatrix} v_1 \\ v_2 \end{bmatrix} = 0$$

so that

$$4v_1 + 1v_2 = 0$$

and

$$4v_1 + 1v_2 = 0$$

When $v_1 = -1$ and $v_2 = 4$, a solution is found. Thus the first eigenvalue is 6, with [1, 1] as a corresponding eigenvector, while the second eigenvector is 1, with [-1, 4] as a corresponding eigenvector.

Because the matrix was 2×2, the polynomial for eigenvalues was quadratic and there were two equations in two unknowns to solve for eigenvectors. Imagine the joys of a matrix 15×15, a polynomial with terms to the 15th power for the first half of the solution and 15 equations in 15 unknowns for the second half. A little more appreciation for the computer center, please, next time you visit!

Research Designs for Complete Examples

B.1 WOMEN'S HEALTH AND DRUG STUDY

Data used in most of the large sample examples were collected with the aid of a grant from the National Institute on Drug Abuse (#DA 00847) to L. S. Fidell and J. E. Prather in 1974–1976. Methods of collecting the data and references to the measures included in the study are described here approximately as they have been previously reported (Hoffman and Fidell, 1979).

Method

A structured interview, containing a variety of health, demographic, and attitudinal measures, was given to a randomly selected group of 465 female, 20- to 59-year-old, English-speaking residents of the San Fernando Valley, a suburb of Los Angeles, in February 1975. A second interview, focusing primarily on health variables but also containing the Bem Sex Role Inventory (BSRI; Bem, 1974) and the Eysenck Personality Inventory (EPI; Eysenck, 1963), was conducted with 369 (79.4%) of the original respondents in February 1976.

The 1975 target sample of 703 names was approximately a .003 probability sample of appropriately aged female residents of the San Fernando Valley, and was randomly drawn from lists prepared by listers during the weeks immediately preceding the sample selection. Lists were prepared for census blocks that had been randomly drawn (proportional to population) from 217 census tracks, which were themselves randomly drawn after they were stratified by income and assigned probabilities proportional to their populations. Respondents were contacted after first receiving a letter soliciting their cooperation. Substitutions were not allowed. A minimum of four callbacks was required before the attempt to obtain an interview was terminated. The completion rate for the target sample was 66.1%, with a 26% refusal rate and a 7.9% "unobtainable" rate.

The demographic characteristics of the 465 respondents who cooperated in 1975 confirmed the essentially white, middle- and working-class composition of the San Fernando

TABLE B.1 DESCRIPTION OF SOME OF THE VARIABLES AVAILABLE FROM 1975–1976 INTERVIEWS

Variable	Abbreviation	Brief Description	Source
Demographic variables			
Socioeconomic level	SEL, SEL2	Measure of deference accorded employment categories (SEL2 from second interview)	Featherman (1973), update of Duncan Scale
Education	EDUC	Number of years completed	
	EDCODE	Categorical variable assessing whether or not education proceeded beyond high school	
Income	INCOME	Total family income before taxes	
	INCODE	Categorical variable assessing family income	
Age	AGE	Chronological age in 5-year categories	
Marital status	MARITAL	A categorical variable assessing current marital status	
	MSTATUS	A dichotomous variable assessing whether or not currently married.	
Parenthood	CHILDREN	A categorical variable assessing whether or not one has children	
Ethnic group membership	RACE	A categorical variable assessing ethnic affiliation	
Employment status	EMPLMNT	A categorical variable assessing whether or not one is currently employed	
	WORKSTAT	A categorical variable assessing current employment status *and*, if not, attitude toward unemployed status	
Religious affiliation	RELIGION	A categorical variable assessing religious affiliation	
Attitudinal variables			
Attitudes toward housework	ATTHOUSE	Frequency of experiencing various favorable and unfavorable attitudes toward homemaking	Derived from Johnson (1955)
Attitudes toward paid work	ATTWORK	Frequency of experiencing various favorable and unfavorable attitudes toward paid work	Johnson (1955)
Attitudes toward role of women	ATTROLE	Measure of conservative or liberal attitudes toward role of women	Spence and Helmreich (1972)

Locus of control	CONTROL	Measure of control ideology; internal or external	Rotter (1966)
Attitudes toward marital status	ATTMAR	Satisfaction with current marital status	From Burgess & Locke (1960); Locke & Wallace (1959)l and Rollins & Feldman (1970)
Personality variables			
Self-esteem	ESTEEM	Measures of self-esteem and confidence in various situations	Rosenberg (1965)
Neuroticism-stability index	NEUROTIC	A scale derived from factor analysis to measure neuroticism vs. stability	Eysenck & Eysenck (1963)
Introversion-extraversion index	INTEXT	A scale derived from factor analysis to measure introversion vs. extraversion	Eysenck & Eysenck (1963)
Androgny measure	ANDRM	A categorical variable based on femininity and masculinity	Derived from Bem (1974)
Health variables			
Mental health	MENHEAL	Frequency count of mental health problems (feeling somewhat apart, can't get along, etc.)	Langner (1962)
Physical health	PHYHEAL	Frequency count of problems with various body systems (circulation, digestion, etc.) general description of health	
Number of visits	TIMEDRS	Frequency count of visits to physical and mental health professionals	
Use of psychotropic drugs	DRUGUSE	A frequency, recency measure of involvement with prescription and nonprescription major and minor tranquilizers, sedatives-hypnotics, antidepressants, and stimulants	Balter & Levine (1971)
Use of psychotropic and over-the-counter drugs	PSYDRUG	DRUGUSE plus a frequency, recency measure of over-the-counter mood modifying drugs	
Attitudes toward medication	ATTDRUG	Items concerning attitudes toward use of medication	
Life change units	STRESS	Weighted items reflecting number and importance of change in life situation	Rahe (1974)

Valley, and agreed, for the most part, with the profile of characteristics of women in the valley that was calculated from 1970 Census Bureau data. The final sample was 91.2% white, with a median family income (before taxes) of $17,000 per year and an average Duncan scale (Featherman, 1973) socioeconomic level (SEL) rating of 51. Respondents were also well educated (13.2 years of school completed, on average), and predominantly Protestant (38%), with 26% Catholic, 20% Jewish, and the remainder "None" or "Other." A total of 52.9% worked (either full-time—33.5%—or part-time—19.4%). Seventy-eight percent were living with husbands at the time of the first interview, with 9% divorced, 6% single, 3% separated, 3% widowed, and fewer than 1% "living together." Altogether, 82.4% of the women had children; the average number of children was 2.7, with 2.1 children, on the average, still living in the same house as the respondent.

Of the original 465 respondents, 369 (79.4%) were reinterviewed a year later. Of the 96 respondents who were not reinterviewed, 51 refused, 36 had moved and could not be relocated, 8 were known to be in the Los Angeles area but were not contacted after a minimum of 5 attempts, and 1 was deceased. Those who were and were not reinterviewed were similar (by analyses of variance) on health and attitudinal variables. They differed, however, on some demographic measures. Those who were reinterviewed tended to be higher-SEL, higher-income white women who were better-educated, were older, and had experienced significantly fewer life change units (Rahe, 1974) in 1975.

The 1975 interview schedule was composed of items assessing a number of demographic, health, and attitudinal characteristics (see Table B.1). Insofar as possible, previously tested and validated items and measures were used, although time constraints prohibited including all items from some measures. Coding on most items was prearranged so that responses given large numbers reflected increasingly unfavorable attitudes, dissatisfaction, poorer health, lower income, increasing stress, increasing use of drugs, and so forth.

The 1976 interview schedule repeated many of the health items, with a shorter set of items assessing changes in marital status and satisfaction, changes in work status and satisfaction, and so forth. The BSRI and EPI were also included, as previously mentioned. The interview schedules for both 1975 and 1976 took 75 minutes on average to administer and were conducted in respondent's homes by experienced and trained interviewers.

To obtain median values for the masculine and feminine scores of the BSRI for a comparable sample of men, the BSRI was mailed to the 369 respondents who cooperated in 1976, with instructions to ask a man near to them (husband, friend, brother, etc.) to fill out and return it. The completed BSRI was received from 162 (46%) men, of whom 82% were husbands, 8.6% friends, 3.7% fiancés, 1.9% brothers, 1.2% sons, 1.2% ex-husbands, 0.6% brothers-in-law, and 0.6% fathers. Analyses of variance were used to compare the demographic characteristics of the men who returned the BSRI with those who did not (insofar as such characteristics could be determined by responses of the women to questions in the 1975 interview). The two groups differed in that, as with the reinterviewed women, the men who responded presented an advantaged socioeconomic picture relative to those who did not. Respondents had higher SEL2 ratings, were better educated, and enjoyed higher income. The unweighted averages of the men's and women's median masculine scores and median feminine scores were used to split the sample of women into those who were feminine, masculine, androgynous, and undifferentiated.

B.2 LEARNING DISABILITIES DATA BANK

Data for the large sample example in Chapter 10, Profile Analysis of Repeated Measures, were taken from a data bank developed at the California Center for Educational Therapy (CCET) in the San Fernando Valley.

Method

All children who are referred to the CCET are given an extensive battery of psychodiagnostic tests to measure current intellectual functioning, perceptual development, psycholinguistic abilities, visual and auditory functioning, and achievement in a number of academic subjects. In addition, an extensive Parent Information Outline queries parents about demographic variables, family health history, as well as child's developmental history, strengths, weaknesses, preferences, and the like. The entire data bank consists of 112 variables from the testing battery plus 155 variables from the Parent Information Outline.

Data collection began in July 1972 and continues to the present. The Chapter 10 sample includes children tested before February 1984 who were administered the Wechsler Intelligence Scale for Children, who were diagnosed as learning-disabled, whose parents agreed to be included in the data bank, and whose parents answered a question about the child's preference for playmates' age. Available answers to this question were: (1) older, (2) younger, (3) same age, and (4) no preference. The latter two categories were combined into a single one for the Chapter 10 analysis since either category by itself would have been too small. Of the 261 children tested between 1972 and 1984, 177 were eligible for inclusion in the Chapter 10 sample.

For the entire sample of 261 cases, average age is 10.58 years with a range of 5 to 61 years. (The Chapter 9 sample consists of school-age children only.) About 75% of the entire sample is male. At the time of testing, 63% of the entire sample attended public school; 33% were enrolled in various types of private schools. Of the 94% of parents who revealed their educational level, mothers had completed an average of 13.6 years of schooling and fathers had completed an average of 14.9 years.

B.3 SEXUAL ATTRACTION STUDY

Data used in the large sample Multiway Frequency Analysis example (Chapter 7) were collected in 1984 as part of a survey assessing issues surrounding the nature of sexual attraction to clients among clinical psychologists. Data-collection methods and demographic characteristics that follow are approximately as they appear in an *American Psychologist* paper (Pope, Keith-Speigel, and Tabachnick, 1986).

A cover letter, a brief 17-item questionnaire (15 structured questions and 2 open-ended questions), and a return envelope were sent to 1000 psychologists (500 men and 500 women) randomly selected from the 4356 members of Division 42 (Psychologists in Private Practice) as listed in the 1983 Membership Register of the American Psychological Association.

The questionnaire requested respondents to provide information about their gender, age group, and years of experience in the field. Information was elicited about the respondent's incidence of sexual attraction to male and female clients; clients' reactions to this experience of attraction; beliefs about the clients' awareness of and reciprocation of the attraction; the impact of the attraction on the therapy process; how such feelings were managed; the incidence of sexual fantasies about clients; why, if relevant, respondents chose to refrain from acting out their attraction through actual sexual intimacies with clients; what features determined which clients would be perceived as sexually attractive; incidence of actual sexual activity with clients; and the extent to which the respondents' graduate training and internship experiences had dealt with issues related to sexual attraction to clients.

Questionnaires were returned by 585 respondents. Of these 59.7% were men. Sixty-eight percent of the male respondents returned their questionnaires as compared with 49% of the female respondents. Approximately half of the respondents were 45 years of age and under. The sample's median age was approximately 46 years as compared with the median age of

40 years reported in a 1983 survey of mental health service providers (VandenBos and Stapp, 1983).

Respondents averaged 16.99 (SD = 8.43) years of professional experience with no significant differences between male and female psychologists. Younger therapists averaged 11.36 (SD = 8.43) years of experience and older therapists averaged 21.79 (SD = 8.13) years of experience. Only 77 of the 585 therapists reported never being attracted to any client.

Appendix C

Statistical Tables

TABLE C.1 NORMAL CURVE AREAS

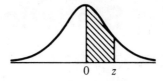

z	.00	.01	.02	.03	.04	.05	.06	.07	.08	.09
0.0	.0000	.0040	.0080	.0120	.0160	.0199	.0239	.0279	.0319	.0359
0.1	.0398	.0438	.0478	.0517	.0557	.0596	.0636	.0675	.0714	.0753
0.2	.0793	.0832	.0871	.0910	.0948	.0987	.1026	.1064	.1103	.1141
0.3	.1179	.1217	.1255	.1293	.1331	.1368	.1406	.1443	.1480	.1517
0.4	.1554	.1591	.1628	.1664	.1700	.1736	.1772	.1808	.1844	.1879
0.5	.1915	.1950	.1985	.2019	.2054	.2088	.2123	.2157	.2190	.2224
0.6	.2257	.2291	.2324	.2357	.2389	.2422	.2454	.2486	.2517	.2549
0.7	.2580	.2611	.2642	.2673	.2704	.2734	.2764	.2794	.2823	.2852
0.8	.2881	.2910	.2939	.2967	.2995	.3023	.3051	.3078	.3106	.3133
0.9	.3159	.3186	.3212	.3238	.3264	.3289	.3315	.3340	.3365	.3389
1.0	.3413	.3438	.3461	.3485	.3508	.3531	.3554	.3577	.3599	.3621
1.1	.3643	.3665	.3686	.3708	.3729	.3749	.3770	.3790	.3810	.3830
1.2	.3849	.3869	.3888	.3907	.3925	.3944	.3962	.3980	.3997	.4015
1.3	.4032	.4049	.4066	.4082	.4099	.4115	.4131	.4147	.4162	.4177
1.4	.4192	.4207	.4222	.4236	.4251	.4265	.4279	.4292	.4306	.4319
1.5	.4332	.4345	.4357	.4370	.4382	.4394	.4406	.4418	.4429	.4441
1.6	.4452	.4463	.4474	.4484	.4495	.4505	.4515	.4525	.4535	.4545
1.7	.4554	.4564	.4573	.4582	.4591	.4599	.4608	.4616	.4625	.4633
1.8	.4641	.4649	.4656	.4664	.4671	.4678	.4686	.4693	.4699	.4706
1.9	.4713	.4719	.4726	.4732	.4738	.4744	.4750	.4756	.4761	.4767
2.0	.4772	.4778	.4783	.4788	.4793	.4798	.4803	.4808	.4812	.4817
2.1	.4821	.4826	.4830	.4834	.4838	.4842	.4846	.4850	.4854	.4857
2.2	.4861	.4864	.4868	.4871	.4875	.4878	.4881	.4884	.4887	.4890
2.3	.4893	.4896	.4898	.4901	.4904	.4906	.4909	.4911	.4913	.4916
2.4	.4918	.4920	.4922	.4925	.4927	.4929	.4931	.4932	.4934	.4936
2.5	.4938	.4940	.4941	.4943	.4945	.4946	.4948	.4949	.4951	.4952
2.6	.4953	.4955	.4956	.4957	.4959	.4960	.4961	.4962	.4963	.4964
2.7	.4965	.4966	.4967	.4968	.4969	.4970	.4971	.4972	.4973	.4974
2.8	.4974	.4975	.4976	.4977	.4977	.4978	.4979	.4979	.4980	.4981
2.9	.4981	.4982	.4982	.4983	.4984	.4984	.4985	.4985	.4986	.4986
3.0	.4987	.4987	.4987	.4988	.4988	.4989	.4989	.4989	.4990	.4990

Source: Abridged from Table 1 of *Statistical Tables and Formulas,* by A. Hald. Copyright © 1952, John Wiley & Sons, Inc. Reprinted by permission of John Wiley & Sons, Inc.

TABLE C.2 CRITICAL VALUES OF THE t DISTRIBUTION FOR $\alpha = .05$ AND .01, TWO-TAILED TEST

Degrees of freedom	.05	.01
1	12.706	63.657
2	4.303	9.925
3	3.182	5.841
4	2.776	4.604
5	2.571	4.032
6	2.447	3.707
7	2.365	3.499
8	2.306	3.355
9	2.262	3.250
10	2.228	3.169
11	2.201	3.106
12	2.179	3.055
13	2.160	3.012
14	2.145	2.977
15	2.131	2.947
16	2.120	2.921
17	2.110	2.898
18	2.101	2.878
19	2.093	2.861
20	2.086	2.845
21	2.080	2.831
22	2.074	2.819
23	2.069	2.807
24	2.064	2.797
25	2.060	2.787
26	2.056	2.779
27	2.052	2.771
28	2.048	2.763
29	2.045	2.756
30	2.042	2.750
40	2.021	2.704
60	2.000	2.660
120	1.980	2.617
∞	1.960	2.576

Source: Abridged from Table 9 in *Biometrika Tables for Statisticians,* vol. 1, 2d ed., edited by E. S. Pearson and H. O. Hartley (New York: Cambridge University Press, 1958). Reproduced with the permission of the trustees of *Biometrika.*

TABLE C.3 CRITICAL VALUES OF THE F DISTRIBUTION

df_2		df_1 1	2	3	4	5	6	8	12	24	∞
1	0.1%	405284	500000	540379	562500	576405	585937	598144	610667	623497	636619
	0.5%	16211	20000	21615	22500	23056	23437	23925	24426	24940	25465
	1 %	4052	4999	5403	5625	5764	5859	5981	6106	6234	6366
	2.5%	647.79	799.50	864.16	899.58	921.85	937.11	956.66	976.71	997.25	1018.30
	5 %	161.45	199.50	215.71	224.58	230.16	233.99	238.88	243.91	249.05	254.32
	10 %	39.86	49.50	53.59	55.83	57.24	58.20	59.44	60.70	62.00	63.33
2	0.1	998.5	999.0	999.2	999.2	999.3	999.3	999.4	999.4	999.5	999.5
	0.5	198.50	199.00	199.17	199.25	199.30	199.33	199.37	199.42	199.46	199.51
	1	98.49	99.00	99.17	99.25	99.30	99.33	99.36	99.42	99.46	99.50
	2.5	38.51	39.00	39.17	39.25	39.30	39.33	39.37	39.42	39.46	39.50
	5	18.51	19.00	19.16	19.25	19.30	19.33	19.37	19.41	19.45	19.50
	10	8.53	9.00	9.16	9.24	9.29	9.33	9.37	9.41	9.45	9.49
3	0.1	167.5	148.5	141.1	137.1	134.6	132.8	130.6	128.3	125.9	123.5
	0.5	55.55	49.80	47.47	46.20	45.39	44.84	44.13	43.39	42.62	41.83
	1	34.12	30.81	29.46	28.71	28.24	27.91	27.49	27.05	26.60	26.12
	2.5	17.44	16.04	15.44	15.10	14.89	14.74	14.54	14.34	14.12	13.90
	5	10.13	9.55	9.28	9.12	9.01	8.94	8.84	8.74	8.64	8.53
	10	5.54	5.46	5.39	5.34	5.31	5.28	5.25	5.22	5.18	5.13
4	0.1	74.14	61.25	56.18	53.44	51.71	50.53	49.00	47.41	45.77	44.05
	0.5	31.33	26.28	24.26	23.16	22.46	21.98	21.35	20.71	20.03	19.33
	1	21.20	18.00	16.69	15.98	15.52	15.21	14.80	14.37	13.93	13.46
	2.5	12.22	10.65	9.98	9.60	9.36	9.20	8.98	8.75	8.51	8.26
	5	7.71	6.94	6.59	6.39	6.26	6.16	6.04	5.91	5.77	5.63
	10	4.54	4.32	4.19	4.11	4.05	4.01	3.95	3.90	3.83	3.76
5	0.1	47.04	36.61	33.20	31.09	29.75	28.84	27.64	26.42	25.14	23.78
	0.5	22.79	18.31	16.53	15.56	14.94	14.51	13.96	13.38	12.78	12.14
	1	16.26	13.27	12.06	11.39	10.97	10.67	10.29	9.89	9.47	9.02
	2.5	10.01	8.43	7.76	7.39	7.15	6.98	6.76	6.52	6.28	6.02

Group	%										
	5	4.36	4.53	4.68	4.82	4.95	5.05	5.19	5.41	5.79	6.61
	10	3.10	3.19	3.27	3.34	3.40	3.45	3.52	3.62	3.78	4.06
6	0.1%	15.75	16.89	17.99	19.03	20.03	20.81	21.90	23.70	27.00	35.51
	0.5%	8.88	9.47	10.03	10.57	11.07	11.46	12.03	12.92	14.54	18.64
	1%	6.88	7.31	7.72	8.10	8.47	8.75	9.15	9.78	10.92	13.74
	2.5%	4.85	5.12	5.37	5.60	5.82	5.99	6.23	6.60	7.26	8.81
	5%	3.67	3.84	4.00	4.15	4.28	4.39	4.53	4.76	5.14	5.99
	10%	2.72	2.82	2.90	2.98	3.05	3.11	3.18	3.29	3.46	3.78
7	0.1	11.69	12.73	13.71	14.63	15.52	16.21	17.19	18.77	21.69	29.22
	0.5	7.08	7.65	8.18	8.68	9.16	9.52	10.05	10.88	12.40	16.24
	1	5.65	6.07	6.47	6.84	7.19	7.46	7.85	8.45	9.55	12.25
	2.5	4.14	4.42	4.67	4.90	5.12	5.29	5.52	5.89	6.54	8.07
	5	3.23	3.41	3.57	3.73	3.87	3.97	4.12	4.35	4.74	5.59
	10	2.47	2.58	2.67	2.75	2.83	2.88	2.96	3.07	3.26	3.59
8	0.1	9.34	10.30	11.19	12.04	12.86	13.49	14.39	15.83	18.49	25.42
	0.5	5.95	6.50	7.01	7.50	7.95	8.30	8.81	9.60	11.04	14.69
	1	4.86	5.28	5.67	6.03	6.37	6.63	7.01	7.59	8.65	11.26
	2.5	3.67	3.95	4.20	4.43	4.65	4.82	5.05	5.42	6.06	7.57
	5	2.93	3.12	3.28	3.44	3.58	3.69	3.84	4.07	4.46	5.32
	10	2.29	2.40	2.50	2.59	2.67	2.73	2.81	2.92	3.11	3.46
9	0.1	7.81	8.72	9.57	10.37	11.13	11.71	12.56	13.90	16.39	22.86
	0.5	5.19	5.73	6.23	6.69	7.13	7.47	7.96	8.72	10.11	13.61
	1	4.31	4.73	5.11	5.47	5.80	6.06	6.42	6.99	8.02	10.56
	2.5	3.33	3.61	3.87	4.10	4.32	4.48	4.72	5.08	5.71	7.21
	5	2.71	2.90	3.07	3.23	3.37	3.48	3.63	3.86	4.26	5.12
	10	2.16	2.28	2.38	2.47	2.55	2.61	2.69	2.81	3.01	3.36
10	0.1	6.76	7.64	8.45	9.20	9.92	10.48	11.28	12.55	14.91	21.04
	0.5	4.64	5.17	5.66	6.12	6.54	6.87	7.34	8.08	9.43	12.83
	1	3.91	4.33	4.71	5.06	5.39	5.64	5.99	6.55	7.56	10.04
	2.5	3.08	3.37	3.62	3.85	4.07	4.24	4.47	4.83	5.46	6.94
	5	2.54	2.74	2.91	3.07	3.22	3.33	3.48	3.71	4.10	4.96
	10	2.06	2.18	2.28	2.38	2.46	2.52	2.61	2.73	2.92	3.28

TABLE C.3 (Continued)

df₂	df₁	1	2	3	4	5	6	8	12	24	∞
11	0.1	19.69	13.81	11.56	10.35	9.58	9.05	8.35	7.63	6.85	6.00
	0.5	12.23	8.91	7.60	6.88	6.42	6.10	5.68	5.24	4.76	4.23
	1	9.65	7.20	6.22	5.67	5.32	5.07	4.74	4.40	4.02	3.60
	2.5	6.72	5.26	4.63	4.28	4.04	3.88	3.66	3.43	3.17	2.88
	5	4.84	3.98	3.59	3.36	3.20	3.09	2.95	2.79	2.61	2.40
	10	3.23	2.86	2.66	2.54	2.45	2.39	2.30	2.21	2.10	1.97
12	0.1	18.64	12.97	10.80	9.63	8.89	8.38	7.71	7.00	6.25	5.42
	0.5	11.75	8.51	7.23	6.52	6.07	5.76	5.35	4.91	4.43	3.90
	1	9.33	6.93	5.95	5.41	5.06	4.82	4.50	4.16	3.78	3.36
	2.5	6.55	5.10	4.47	4.12	3.89	3.73	3.51	3.28	3.02	2.72
	5	4.75	3.88	3.49	3.26	3.11	3.00	2.85	2.69	2.50	2.30
	10	3.18	2.81	2.61	2.48	2.39	2.33	2.24	2.15	2.04	1.90
13	0.1%	17.81	12.31	10.21	9.07	8.35	7.86	7.21	6.52	5.78	4.97
	0.5%	11.37	8.19	6.93	6.23	5.79	5.48	5.08	4.64	4.17	3.65
	1 %	9.07	6.70	5.74	5.20	4.86	4.62	4.30	3.96	3.59	3.16
	2.5%	6.41	4.97	4.35	4.00	3.77	3.60	3.39	3.15	2.89	2.60
	5 %	4.67	3.80	3.41	3.18	3.02	2.92	2.77	2.60	2.42	2.21
	10 %	3.14	2.76	2.56	2.43	2.35	2.28	2.20	2.10	1.98	1.85
14	0.1	17.14	11.78	9.73	8.62	7.92	7.43	6.80	6.13	5.41	4.60
	0.5	11.06	7.92	6.68	6.00	5.56	5.26	4.86	4.43	3.96	3.44
	1	8.86	6.51	5.56	5.03	4.69	4.46	4.14	3.80	3.43	3.00
	2.5	6.30	4.86	4.24	3.89	3.66	3.50	3.29	3.05	2.79	2.49
	5	4.60	3.74	3.34	3.11	2.96	2.85	2.70	2.53	2.35	2.13
	10	3.10	2.73	2.52	2.39	2.31	2.24	2.15	2.05	1.94	1.80
15	0.1	16.59	11.34	9.34	8.25	7.57	7.09	6.47	5.81	5.10	4.31
	0.5	10.80	7.70	6.48	5.80	5.37	5.07	4.67	4.25	3.79	3.26
	1	8.68	6.36	5.42	4.89	4.56	4.32	4.00	3.67	3.29	2.87
	2.5	6.20	4.77	4.15	3.80	3.58	3.41	3.20	2.96	2.70	2.40

df	α										
	5	4.54	3.68	3.29	3.06	2.90	2.79	2.64	2.48	2.29	2.07
	10	3.07	2.70	2.49	2.36	2.27	2.21	2.12	2.02	1.90	1.76
16	0.1	16.12	10.97	9.00	7.94	7.27	6.81	6.19	5.55	4.85	4.06
	0.5	10.58	7.51	6.30	5.64	5.21	4.91	4.52	4.10	3.64	3.11
	1	8.53	6.23	5.29	4.77	4.44	4.20	3.89	3.55	3.18	2.75
	2.5	6.12	4.69	4.08	3.73	3.50	3.34	3.12	2.89	2.63	2.32
	5	4.49	3.63	3.24	3.01	2.85	2.74	2.59	2.42	2.24	2.01
	10	3.05	2.67	2.46	2.33	2.24	2.18	2.09	1.99	1.87	1.72
17	0.1	15.72	10.66	8.73	7.68	7.02	6.56	5.96	5.32	4.63	3.85
	0.5	10.38	7.35	6.16	5.50	5.07	4.78	4.39	3.97	3.51	2.98
	1	8.40	6.11	5.18	4.67	4.34	4.10	3.79	3.45	3.08	2.65
	2.5	6.04	4.62	4.01	3.66	3.44	3.28	3.06	2.82	2.56	2.25
	5	4.45	3.59	3.20	2.96	2.81	2.70	2.55	2.38	2.19	1.96
	10	3.03	2.64	2.44	2.31	2.22	2.15	2.06	1.96	1.84	1.69
18	0.1	15.38	10.39	8.49	7.46	6.81	6.35	5.76	5.13	4.45	3.67
	0.5	10.22	7.21	6.03	5.37	4.96	4.66	4.28	3.86	3.40	2.87
	1	8.28	6.01	5.09	4.58	4.25	4.01	3.71	3.37	3.00	2.57
	2.5	5.98	4.56	3.95	3.61	3.38	3.22	3.01	2.77	2.50	2.19
	5	4.41	3.55	3.16	2.93	2.77	2.66	2.51	2.34	2.15	1.92
	10	3.01	2.62	2.42	2.29	2.20	2.13	2.04	1.93	1.81	1.66
19	0.1%	15.08	10.16	8.28	7.26	6.61	6.18	5.59	4.97	4.29	3.52
	0.5%	10.07	7.09	5.92	5.27	4.85	4.56	4.18	3.76	3.31	2.78
	1 %	8.18	5.93	5.01	4.50	4.17	3.94	3.63	3.30	2.92	2.49
	2.5%	5.92	4.51	3.90	3.56	3.33	3.17	2.96	2.72	2.45	2.13
	5 %	4.38	3.52	3.13	2.90	2.74	2.63	2.48	2.31	2.11	1.88
	10 %	2.99	2.61	2.40	2.27	2.18	2.11	2.02	1.91	1.79	1.63
20	0.1	14.82	9.95	8.10	7.10	6.46	6.02	5.44	4.82	4.15	3.38
	0.5	9.94	6.99	5.82	5.17	4.76	4.47	4.09	3.68	3.22	2.69
	1	8.10	5.85	4.94	4.43	4.10	3.87	3.56	3.23	2.86	2.42
	2.5	5.87	4.46	3.86	3.51	3.29	3.13	2.91	2.68	2.41	2.09
	5	4.35	3.49	3.10	2.87	2.71	2.60	2.45	2.28	2.08	1.84
	10	2.97	2.59	2.38	2.25	2.16	2.09	2.00	1.89	1.77	1.61

TABLE C.3 (Continued)

df₂	df₁	1	2	3	4	5	6	8	12	24	∞
21	0.1	14.59	9.77	7.94	6.95	6.32	5.88	5.31	4.70	4.03	3.26
	0.5	9.83	6.89	5.73	5.09	4.68	4.39	4.01	3.60	3.15	2.61
	1	8.02	5.78	4.87	4.37	4.04	3.81	3.51	3.17	2.80	2.36
	2.5	5.83	4.42	3.82	3.48	3.25	3.09	2.87	2.64	2.37	2.04
	5	4.32	3.47	3.07	2.84	2.68	2.57	2.42	2.25	2.05	1.81
	10	2.96	2.57	2.36	2.23	2.14	2.08	1.98	1.88	1.75	1.59
22	0.1	14.38	9.61	7.80	6.81	6.19	5.76	5.19	4.58	3.92	3.15
	0.5	9.73	6.81	5.65	5.02	4.61	4.32	3.94	3.54	3.08	2.55
	1	7.94	5.72	4.82	4.31	3.99	3.76	3.45	3.12	2.75	2.31
	2.5	5.79	4.38	3.78	3.44	3.22	3.05	2.84	2.60	2.33	2.00
	5	4.30	3.44	3.05	2.82	2.66	2.55	2.40	2.23	2.03	1.78
	10	2.95	2.56	2.35	2.22	2.13	2.06	1.97	1.86	1.73	1.57
23	0.1	14.19	9.47	7.67	6.69	6.08	5.65	5.09	4.48	3.82	3.05
	0.5	9.63	6.73	5.58	4.95	4.54	4.26	3.88	3.47	3.02	2.48
	1	7.88	5.66	4.76	4.26	3.94	3.71	3.41	3.07	2.70	2.26
	2.5	5.75	4.35	3.75	3.41	3.18	3.02	2.81	2.57	2.30	1.97
	5	4.28	3.42	3.03	2.80	2.64	2.53	2.38	2.20	2.00	1.76
	10	2.94	2.55	2.34	2.21	2.11	2.05	1.95	1.84	1.72	1.55
24	0.1	14.03	9.34	7.55	6.59	5.98	5.55	4.99	4.39	3.74	2.97
	0.5	9.55	6.66	5.52	4.89	4.49	4.20	3.83	3.42	2.97	2.43
	1	7.82	5.61	4.72	4.22	3.90	3.67	3.36	3.03	2.66	2.21
	2.5	5.72	4.32	3.72	3.38	3.15	2.99	2.78	2.54	2.27	1.94
	5	4.26	3.40	3.01	2.78	2.62	2.51	2.36	2.18	1.98	1.73
	10	2.93	2.54	2.33	2.19	2.10	2.04	1.94	1.83	1.70	1.53
25	0.1%	13.88	9.22	7.45	6.49	5.88	5.46	4.91	4.31	3.66	2.89
	0.5%	9.48	6.60	5.46	4.84	4.43	4.15	3.78	3.37	2.92	2.38
	1 %	7.77	5.57	4.68	4.18	3.86	3.63	3.32	2.99	2.62	2.17

	2.5%	5.69	4.29	3.69	3.35	3.13	2.97	2.75	2.51	2.24	1.91
	5 %	4.24	3.38	2.99	2.76	2.60	2.49	2.34	2.16	1.96	1.71
	10 %	2.92	2.53	2.32	2.18	2.09	2.02	1.93	1.82	1.69	1.52
26	0.1	13.74	9.12	7.36	6.41	5.80	5.38	4.83	4.24	3.59	2.82
	0.5	9.41	6.54	5.41	4.79	4.38	4.10	3.73	3.33	2.87	2.33
	1	7.72	5.53	4.64	4.14	3.82	3.59	3.29	2.96	2.58	2.13
	2.5	5.66	4.27	3.67	3.33	3.10	2.94	2.73	2.49	2.22	1.88
	5	4.22	3.37	2.98	2.74	2.59	2.47	2.32	2.15	1.95	1.69
	10	2.91	2.52	2.31	2.17	2.08	2.01	1.92	1.81	1.68	1.50
27	0.1	13.61	9.02	7.27	6.33	5.73	5.31	4.76	4.17	3.52	2.75
	0.5	9.34	6.49	5.36	4.74	4.34	4.06	3.69	3.28	2.83	2.29
	1	7.68	5.49	4.60	4.11	3.78	3.56	3.26	2.93	2.55	2.10
	2.5	5.63	4.24	3.65	3.31	3.08	2.92	2.71	2.47	2.19	1.85
	5	4.21	3.35	2.96	2.73	2.57	2.46	2.30	2.13	1.93	1.67
	10	2.90	2.51	2.30	2.17	2.07	2.00	1.91	1.80	1.67	1.49
28	0.1	13.50	8.93	7.19	6.25	5.66	5.24	4.69	4.11	3.46	2.70
	0.5	9.28	6.44	5.32	4.70	4.30	4.02	3.65	3.25	2.79	2.25
	1	7.64	5.45	4.57	4.07	3.75	3.53	3.23	2.90	2.52	2.06
	2.5	5.61	4.22	3.63	3.29	3.06	2.90	2.69	2.45	2.17	1.83
	5	4.20	3.34	2.95	2.71	2.56	2.44	2.29	2.12	1.91	1.65
	10	2.89	2.50	2.29	2.16	2.06	2.00	1.90	1.79	1.66	1.48
29	0.1	13.39	8.85	7.12	6.19	5.59	5.18	4.64	4.05	3.41	2.64
	0.5	9.23	6.40	5.28	4.66	4.26	3.98	3.61	3.21	2.76	2.21
	1	7.60	5.42	4.54	4.04	3.73	3.50	3.20	2.87	2.49	2.03
	2.5	5.59	4.20	3.61	3.27	3.04	2.88	2.67	2.43	2.15	1.81
	5	4.18	3.33	2.93	2.70	2.54	2.43	2.28	2.10	1.90	1.64
	10	2.89	2.50	2.28	2.15	2.06	1.99	1.89	1.78	1.65	1.47
30	0.1	13.29	8.77	7.05	6.12	5.53	5.12	4.58	4.00	3.36	2.59
	0.5	9.18	6.35	5.24	4.62	4.23	3.95	3.58	3.18	2.73	2.18
	1	7.56	5.39	4.51	4.02	3.70	3.47	3.17	2.84	2.47	2.01
	2.5	5.57	4.18	3.59	3.25	3.03	2.87	2.65	2.41	2.14	1.79
	5	4.17	3.32	2.92	2.69	2.53	2.42	2.27	2.09	1.89	1.62
	10	2.88	2.49	2.28	2.14	2.05	1.98	1.88	1.77	1.64	1.46

TABLE C.3 (Continued)

df_2	df_1	1	2	3	4	5	6	8	12	24	∞
40	0.1%	12.61	8.25	6.60	5.70	5.13	4.73	4.21	3.64	3.01	2.23
	0.5%	8.83	6.07	4.98	4.37	3.99	3.71	3.35	2.95	2.50	1.93
	1 %	7.31	5.18	4.31	3.83	3.51	3.29	2.99	2.66	2.29	1.80
	2.5%	5.42	4.05	3.46	3.13	2.90	2.74	2.53	2.29	2.01	1.64
	5 %	4.08	3.23	2.84	2.61	2.45	2.34	2.18	2.00	1.79	1.51
	10 %	2.84	2.44	2.23	2.09	2.00	1.93	1.83	1.71	1.57	1.38
60	0.1	11.97	7.76	6.17	5.31	4.76	4.37	3.87	3.31	2.69	1.90
	0.5	8.49	5.80	4.73	4.14	3.76	3.49	3.13	2.74	2.29	1.69
	1	7.08	4.98	4.13	3.65	3.34	3.12	2.82	2.50	2.12	1.60
	2.5	5.29	3.93	3.34	3.01	2.79	2.63	2.41	2.17	1.88	1.48
	5	4.00	3.15	2.76	2.52	2.37	2.25	2.10	1.92	1.70	1.39
	10	2.79	2.39	2.18	2.04	1.95	1.87	1.77	1.66	1.51	1.29
120	0.1	11.38	7.31	5.79	4.95	4.42	4.04	3.55	3.02	2.40	1.56
	0.5	8.18	5.54	4.50	3.92	3.55	3.28	2.93	2.54	2.09	1.43
	1	6.85	4.79	3.95	3.48	3.17	2.96	2.66	2.34	1.95	1.38
	2.5	5.15	3.80	3.23	2.89	2.67	2.52	2.30	2.05	1.76	1.31
	5	3.92	3.07	2.68	2.45	2.29	2.17	2.02	1.83	1.61	1.25
	10	2.75	2.35	2.13	1.99	1.90	1.82	1.72	1.60	1.45	1.19
∞	0.1	10.83	6.91	5.42	4.62	4.10	3.74	3.27	2.74	2.13	1.00
	0.5	7.88	5.30	4.28	3.72	3.35	3.09	2.74	2.36	1.90	1.00
	1	6.64	4.60	3.78	3.32	3.02	2.80	2.51	2.18	1.79	1.00
	2.5	5.02	3.69	3.12	2.79	2.57	2.41	2.19	1.94	1.64	1.00
	5	3.84	2.99	2.60	2.37	2.21	2.09	1.94	1.75	1.52	1.00
	10	2.71	2.30	2.08	1.94	1.85	1.77	1.67	1.55	1.38	1.00

Source: Abridged from Table 18 in *Biometrika Tables for Statisticians*, vol. 1, 2d ed., edited by E. S. Pearson and H. O. Hartley (New York: Cambridge University Press, 1958). Reproduced with the permission of the trustees of *Biometrika*.

TABLE C.4 CRITICAL VALUES OF CHI SQUARE (χ^2)

df	0.250	0.100	0.050	0.025	0.010	0.005	0.001
1	1.32330	2.70554	3.84146	5.02389	6.63490	7.87944	10.828
2	2.77259	4.60517	5.99147	7.37776	9.21034	10.5966	13.816
3	4.10835	6.25139	7.81473	9.34840	11.3449	12.8381	16.266
4	5.38527	7.77944	9.48773	11.1433	13.2767	14.8602	18.467
5	6.62568	9.23635	11.0705	12.8325	15.0863	16.7496	20.515
6	7.84080	10.6446	12.5916	14.4494	16.8119	18.5476	22.458
7	9.03715	12.0170	14.0671	16.0128	18.4753	20.2777	24.322
8	10.2188	13.3616	15.5073	17.5346	20.0902	21.9550	26.125
9	11.3887	14.6837	16.9190	19.0228	21.6660	23.5893	27.877
10	12.5489	15.9871	18.3070	20.4831	23.2093	25.1882	29.588
11	13.7007	17.2750	19.6751	21.9200	24.7250	26.7569	31.264
12	14.8454	18.5494	21.0261	23.3367	26.2170	28.2995	32.909
13	15.9839	19.8119	22.3621	24.7356	27.6883	29.8194	34.528
14	17.1770	21.0642	23.6848	26.1190	29.1413	31.3193	36.123
15	18.2451	22.3072	24.9958	27.4884	30.5779	32.8013	37.697
16	19.3688	23.5418	26.2962	28.8454	31.9999	34.2672	39.252
17	20.4887	24.7690	27.5871	30.1910	33.4087	35.7185	40.790
18	21.6049	25.9894	28.8693	31.5264	34.8053	37.1564	42.312
19	22.7178	27.2036	30.1435	32.8523	36.1908	38.5822	43.820
20	23.8277	28.4120	31.4104	34.1696	37.5662	39.9968	45.315
21	24.9348	29.6151	32.6705	35.4789	38.9321	41.4010	46.797
22	26.0393	30.8133	33.9244	36.7807	40.2894	42.7956	48.268
23	27.1413	32.0069	35.1725	38.0757	41.6384	44.1813	49.728
24	28.2412	33.1963	36.4151	39.3641	42.9798	45.5585	51.179
25	29.3389	34.3816	37.6525	40.6465	44.3141	46.9278	52.620
26	30.4345	35.5631	38.8852	41.9232	45.6417	48.2899	54.052
27	31.5284	36.7412	40.1133	43.1944	46.9630	49.6449	55.476
28	32.6205	37.9159	41.3372	44.4607	48.2782	50.9933	56.892
29	33.7109	39.0875	42.5569	45.7222	49.5879	52.3356	58.302
30	34.7998	40.2560	43.7729	46.9792	50.8922	53.6720	59.703
40	45.6160	51.8050	55.7585	59.3417	63.6907	66.7659	73.402
50	56.3336	63.1671	67.5048	71.4202	76.1539	79.4900	86.661
60	66.9814	74.3970	79.0819	83.2976	88.3794	91.9517	99.607
70	77.5766	85.5271	90.5312	95.0231	100.425	104.215	112.317
80	88.1303	96.5782	101.879	106.629	112.329	116.321	124.839
90	98.6499	107.565	113.145	118.136	124.116	128.299	137.208
100	109.141	118.498	124.342	129.561	135.807	140.169	149.449

Source: Abridged from Table 8 in *Biometrika Tables for Statisticians*, vol. 1, 2d ed., edited by E. S. Pearson and H. O. Hartley (New York: Cambridge University Press, 1958). Reproduced with the permission of the trustees of *Biometrika*.

TABLE C.5 CRITICAL VALUES FOR SQUARED MULTIPLE CORRELATION
(R^2) IN FORWARD STEPWISE SELECTION
$\alpha = .05$

k	F	$N - k - 1$															
		10	12	14	16	18	20	25	30	35	40	50	60	80	100	150	200
2	2	43	38	33	30	27	24	20	16	14	13	10	8	6	5	3	2
2	3	40	36	31	27	24	22	18	15	13	11	9	7	5	4	2	2
2	4	38	33	29	26	23	21	17	14	12	10	8	7	5	4	3	2
3	2	49	43	39	35	32	29	24	21	18	16	12	10	8	7	4	2
3	3	45	40	36	32	29	26	22	19	17	15	11	9	7	6	4	3
3	4	42	36	33	29	27	25	20	17	15	13	11	9	7	5	4	3
4	2	54	48	44	39	35	33	27	23	20	18	15	12	10	8	5	4
4	3	49	43	39	36	33	30	25	22	19	17	14	11	8	7	5	4
4	4	45	39	35	32	29	27	22	19	17	15	12	10	8	6	5	3
5	2	58	52	47	43	39	36	31	26	23	21	17	14	11	9	6	5
5	3	52	46	42	38	35	32	27	24	21	19	16	13	9	8	5	4
5	4	46	41	38	35	52	29	24	21	18	16	13	11	9	7	5	4
6	2	60	54	50	46	41	39	33	29	25	23	19	16	12	10	7	5
6	3	54	48	44	40	37	34	29	25	22	20	17	14	10	8	6	5
6	4	48	43	39	36	33	30	26	23	20	17	14	12	9	7	5	4
7	2	61	56	51	48	44	41	35	30	27	24	20	17	13	11	7	5
7	3	59	50	46	42	39	36	31	26	23	21	18	15	11	9	7	5
7	4	50	45	41	38	35	32	27	24	21	18	15	13	10	8	6	4
8	2	62	58	53	49	46	43	37	31	28	26	21	18	14	11	8	6
8	3	57	52	47	43	40	37	32	28	24	22	19	16	12	10	7	5
8	4	51	46	42	39	36	33	28	25	22	19	16	14	11	9	7	5
9	2	63	59	54	51	47	44	38	33	30	27	22	19	15	12	9	6
9	3	58	53	49	44	41	38	33	29	25	23	20	16	12	10	7	6
9	4	52	46	43	40	37	34	29	25	23	20	17	14	11	10	7	6

k	F	10	12	14	16	18	20	25	30	35	40	50	60	80	100	150	200
10	2	64	60	55	52	49	46	39	34	31	28	23	20	16	13	10	7
10	3	59	54	50	45	42	39	34	30	26	24	20	17	13	11	8	6
10	4	52	47	44	41	38	35	30	26	24	21	18	15	12	10	8	6
12	2	66	62	57	54	51	48	42	37	33	30	25	22	17	14	10	8
12	3	60	55	52	47	44	41	36	31	28	25	22	19	14	12	9	7
12	4	53	48	45	41	39	36	31	27	25	22	19	16	13	11	9	7
14	2	68	64	60	56	53	50	44	39	35	32	27	24	18	15	11	8
14	3	61	57	53	49	46	43	37	32	29	27	23	20	15	13	10	8
14	4	53	49	46	42	40	37	32	29	26	23	20	17	13	11	9	7
16	2	69	65	61	58	55	53	46	41	37	34	29	25	20	17	12	9
16	3	61	58	54	50	47	44	38	34	31	28	24	21	17	14	11	8
16	4	53	50	46	43	40	38	33	30	27	24	21	18	14	12	10	8
18	2	70	67	63	60	57	55	49	44	40	36	31	27	21	18	13	9
18	3	62	59	55	51	49	46	40	35	32	30	26	23	18	15	12	9
18	4	54	50	46	44	41	38	34	31	28	25	22	19	15	13	11	8
20	2	72	68	64	62	59	56	50	46	42	38	33	28	22	19	14	10
20	3	62	60	56	52	50	47	42	37	34	31	27	24	19	16	12	9
20	4	54	50	46	44	41	39	35	32	29	26	23	20	16	14	11	8

α = .01

N − k − 1

k	F	10	12	14	16	18	20	25	30	35	40	50	60	80	100	150	200
2	2	59	53	48	43	40	36	30	26	23	20	17	14	11	9	7	5
2	3	58	52	46	42	38	35	30	25	22	19	16	13	10	8	6	4
2	4	57	49	44	39	36	32	26	22	19	16	13	11	8	7	5	4
3	2	67	60	55	50	46	42	35	30	27	24	20	17	13	11	7	5
3	3	63	58	52	47	43	40	34	29	25	22	19	16	12	10	7	5
3	4	61	54	48	44	40	37	31	26	23	20	16	14	11	9	6	5
4	2	70	64	58	53	49	46	39	34	30	27	23	19	15	12	8	6
4	3	67	62	56	51	47	44	37	32	28	25	21	18	14	11	8	6

TABLE C.5 (Continued)

											N − k − 1						
k	F	10	12	14	16	18	20	25	30	35	40	50	60	80	100	150	200
4	4	64	58	52	47	43	40	34	29	26	23	19	16	13	11	7	6
5	2	73	67	61	57	52	49	42	37	32	29	25	21	16	13	9	7
5	3	70	65	59	54	50	46	39	34	30	27	23	19	15	12	9	7
5	4	65	60	55	50	46	43	36	31	28	25	20	17	14	12	8	6
6	2	74	69	63	59	55	51	44	39	34	31	26	23	18	14	10	8
6	3	72	67	61	56	51	48	41	36	32	28	24	20	16	13	10	7
6	4	66	61	56	52	48	45	38	33	29	26	22	19	15	13	9	7
7	2	76	70	65	60	56	53	46	40	36	33	28	25	19	15	11	9
7	3	73	68	62	57	53	50	42	37	33	30	25	21	17	14	10	8
7	4	67	62	58	54	49	46	40	35	31	28	23	20	16	14	10	8
8	2	77	72	66	62	58	55	48	42	38	34	29	26	20	16	12	9
8	3	74	69	63	58	54	51	44	39	34	31	26	22	18	15	11	9
8	4	67	63	59	55	50	47	41	36	32	29	24	21	17	15	11	9
9	2	78	73	67	63	60	56	49	43	39	36	31	27	21	17	12	10
9	3	74	69	64	59	56	52	45	40	35	32	27	23	19	16	12	9
9	4	68	63	60	56	51	48	42	37	33	30	25	22	18	16	12	9
10	2	79	74	68	65	61	58	51	45	40	37	32	28	22	18	13	10
10	3	74	69	65	50	57	53	47	41	37	33	28	24	20	17	13	10
10	4	68	64	61	56	52	49	43	38	34	31	26	23	19	17	13	9
12	2	80	75	70	66	63	60	53	48	43	39	34	30	24	20	14	11
12	3	74	70	56	62	58	55	48	43	39	35	30	26	21	18	14	10
12	4	69	65	51	57	53	50	44	40	35	32	27	24	20	18	13	10
14	2	81	76	71	68	65	62	55	50	45	41	36	32	25	21	15	11
14	3	74	70	67	63	60	56	50	45	41	37	31	27	22	19	15	11
14	4	69	65	61	57	54	52	45	41	36	33	28	25	21	19	14	10
16	2	82	77	72	69	66	63	57	52	47	43	38	34	27	22	16	12

16	3	74	70	67	64	61	58	52	47	42	39	33	29	23	20	15	11
16	4	70	66	62	58	55	52	46	42	37	34	29	26	22	20	14	11
18	2	82	78	73	70	67	65	59	54	49	45	39	35	28	23	17	12
18	3	74	70	57	65	62	59	53	48	44	41	35	30	24	21	16	12
18	4	70	65	62	58	55	53	47	43	38	35	30	27	23	20	15	11
20	2	82	78	74	71	68	66	60	55	50	46	41	36	29	24	18	13
20	3	74	70	67	65	62	60	55	50	46	42	36	32	26	22	17	12
20	4	70	66	62	58	55	53	47	43	39	36	31	28	24	21	16	11

Note: Decimals are omitted; k = number of candidate predictors; n = sample size; F = criterion F-to-enter.

Source: Adapted from Tables 1 and 2 in "Tests of significance in forward selection regression," by L. Wilkinson and G. E. Dallal, *Technometrics*, 1981, 23(4), 377–380. Reprinted by permission.

TABLE C.6 CRITICAL VALUES FOR FMAX (S^2_{max}/S^2_{min}) DISTRIBUTION FOR $\alpha = .05$ AND $.01$

$\alpha = .05$

df \ k	2	3	4	5	6	7	8	9	10	11	12
4	9.60	15.5	20.6	25.2	29.5	33.6	37.5	41.1	44.6	48.0	51.4
5	7.15	10.8	13.7	16.3	18.7	20.8	22.9	24.7	26.5	28.2	29.9
6	5.82	8.38	10.4	12.1	13.7	15.0	16.3	17.5	18.6	19.7	20.7
7	4.99	6.94	8.44	9.70	10.8	11.8	12.7	13.5	14.3	15.1	15.8
8	4.43	6.00	7.18	8.12	9.03	9.78	10.5	11.1	11.7	12.2	12.7
9	4.03	5.34	6.31	7.11	7.80	8.41	8.95	9.45	9.91	10.3	10.7
10	3.72	4.85	5.67	6.34	6.92	7.42	7.87	8.28	8.66	9.01	9.34
12	3.28	4.16	4.79	5.30	5.72	6.09	6.42	6.72	7.00	7.25	7.48
15	2.86	3.54	4.01	4.37	4.68	4.95	5.19	5.40	5.59	5.77	5.93
20	2.46	2.95	3.29	3.54	3.76	3.94	4.10	4.24	4.37	4.49	4.59
30	2.07	2.40	2.61	2.78	2.91	3.02	3.12	3.21	3.29	3.36	3.39
60	1.67	1.85	1.96	2.04	2.11	2.17	2.22	2.26	2.30	2.33	2.36
∞	1.00	1.00	1.00	1.00	1.00	1.00	1.00	1.00	1.00	1.00	1.00

$\alpha = .01$

df \ k	2	3	4	5	6	7	8	9	10	11	12
4	23.2	37	49	59	69	79	89	97	106	113	120
5	14.9	22	28	33	38	42	46	50	54	57	60
6	11.1	15.5	19.1	22	25	27	30	32	34	36	37
7	8.89	12.1	14.5	16.5	18.4	20	22	23	24	26	27
8	7.50	9.9	11.7	13.2	14.5	15.8	16.9	17.9	18.9	19.8	21
9	6.54	8.5	9.9	11.1	12.1	13.1	13.9	14.7	15.3	16.0	16.6
10	5.85	7.4	8.6	9.6	10.4	11.1	11.8	12.4	12.9	13.4	13.9
12	4.91	6.1	6.9	7.6	8.2	8.7	9.1	9.5	9.9	10.2	10.6
15	4.07	4.9	5.5	6.0	6.4	6.7	7.1	7.3	7.5	7.8	8.0
20	3.32	3.8	4.3	4.6	4.9	5.1	5.3	5.5	5.6	5.8	5.9
30	2.63	3.0	3.3	3.4	3.6	3.7	3.8	3.9	4.0	4.1	4.2
60	1.96	2.2	2.3	2.4	2.4	2.5	2.5	2.6	2.6	2.7	2.7
∞	1.00	1.0	1.0	1.0	1.0	1.0	1.0	1.0	1.0	1.0	1.0

Note: S^2_{max} is the largest and S^2_{min} the smallest in a set of k independent mean squares, each based on degrees of freedom (df).

Source: Abridged from Table 31 in *Biometrika Tables for Statisticians*, vol. 1, 2d ed., edited by E. S. Pearson and H. O. Hartley (New York: Cambridge University Press, 1958). Reproduced with the permission of the trustees for *Biometrika*.

TABLE C.7 PROBABILITIES FOR THE SALIENT SIMILARITY INDEX, s

Probabilities $s \geq v_s$ for Hyperplane Count 60%

Number of variables, p

10	v_s:	.76	.51	.26	.01	.00						
	p:	.001	.020	.138	.364	.500						

20	v_s:	.63	.51	.26	.13	.01	.00					
	p:	.000	.004	.086	.207	.393	.500					

30	v_s:	.59	.51	.42	.34	.26	.17	.09	.01	.00		
	p:	.000	.001	.005	.019	.054	.131	.256	.411	.500		

40	v_s:	.51	.44	.38	.32	.26	.19	.13	.07	.01	.00	
	p:	.000	.001	.005	.016	.041	.090	.169	.282	.426	.500	

50	v_s:	.46	.41	.36	.31	.26	.21	.16	.11	.06	.01	.00
	p:	.000	.001	.004	.011	.026	.061	.115	.200	.301	.428	.500

60	v_s:	.42	.38	.34	.30	.26	.21	.17	.13	.09	.05	.01
	p:	.000	.001	.003	.008	.018	.039	.078	.132	.215	.318	.440

80	v_s:	.35	.32	.29	.26	.22	.19	.16	.13	.10	.07	.04
	p:	.000	.001	.003	.008	.018	.034	.063	.107	.170	.243	.338
	v_s:	.00										
	p:	.500										

100	v_s:	.36	.31	.28	.26	.23	.21	.18	.16	.13	.11	.08
	p:	.000	.001	.003	.006	.011	.019	.034	.058	.091	.136	.196
	v_s:	.03	.01	.00								
	p:	.353	.449	.500								

Probabilities $s \geq v_s$ for Hyperplane Count 70%

10	v_s:	.67	.34	.01	.00								
	p:	.002	.052	.316	.500								

20	v_s:	.67	.51	.34	.17	.01	.00						
	p:	.000	.002	.027	.135	.357	.500						

30	v_s:	.56	.45	.34	.23	.12	.01	.00					
	p:	.000	.002	.016	.064	.190	.383	.500					

40	v_s:	.51	.42	.34	.26	.17	.09	.01	.00				
	p:	.000	.002	.007	.034	.098	.222	.403	.500				

50	v_s:	.47	.41	.34	.27	.21	.14	.07	.01	.00			
	p:	.000	.001	.004	.018	.052	.123	.247	.407	.500			

60	v_s:	.39	.34	.28	.23	.17	.12	.06	.01	.00			
	p:	.000	.002	.007	.025	.066	.142	.262	.415	.500			

80	v_s:	.38	.34	.30	.26	.21	.17	.13	.09	.05	.01	.00	
	p:	.000	.001	.002	.007	.019	.046	.092	.174	.283	.425	.500	

100	v_s:	.34	.31	.27	.24	.21	.17	.14	.11	.07	.04	.01	.00
	p:	.000	.001	.002	.006	.016	.037	.069	.125	.207	.318	.438	.500

Source: Adapted from Tables 1–4 in "Factor matching procedures: An improvement of the s index; with tables," by R. B. Cattell, K. R. Balcar, J. L. Horn, and J. R. Nesselroade, *Educational and Psychological Measurement*, 1969, **29**, 781–792. Reproduced with permission of the publisher and authors.

TABLE C.7 (*Continued*)
Probabilities $s \geq v_s$ for Hyperplane Count 80%

Number of variables, p

10	v_s:	.51	.01	.00					
	p:	.012	.187	.500					

20	v_s:	.76	.51	.26	.01	.00			
	p:	.000	.003	.041	.279	.500			

30	v_s:	.67	.51	.34	.17	.01	.00		
	p:	.000	.001	.011	.083	.316	.500		

40	v_s:	.51	.38	.26	.13	.01	.00		
	p:	.000	.003	.024	.115	.347	.500		

50	v_s:	.51	.41	.31	.21	.11	.01	.00	
	p:	.000	.001	.006	.036	.142	.361	.500	

60	v_s:	.42	.34	.26	.17	.09	.01	.00	
	p:	.000	.002	.012	.052	.164	.369	.500	

80	v_s:	.38	.32	.26	.19	.13	.07	.01	.00
	p:	.000	.001	.004	.022	.076	.199	.301	.500

100	v_s:	.31	.26	.21	.16	.11	.06	.01	.00
	p:	.000	.002	.010	.036	.105	.220	.402	.500

Probabilities $s \geq v_s$ for Hyperplane Count 90%

Number of variables, p

10	v_s:	.01	.00			
	p:	.052	.500			

20	v_s:	.51	.01	.00		
	p:	.003	.099	.500		

30	v_s:	.67	.34	.01	.00	
	p:	.000	.007	.133	.500	

40	v_s:	.51	.26	.01	.00	
	p:	.000	.012	.167	.500	

50	v_s:	.41	.21	.01	.00	
	p:	.000	.018	.198	.500	

60	v_s:	.51	.34	.17	.01	.00
	p:	.000	.002	.029	.217	.500

80	v_s:	.38	.26	.13	.01	.00
	p:	.000	.004	.045	.251	.500

100	v_s:	.31	.21	.11	.01	.00
	p:	.000	.007	.061	.286	.500

References

Asher, H. B. (1976). *Causal Modeling*. Sage University Paper Series on Quantitative Applications in the Social Sciences, Series No. 07–003. Beverly Hills, Calif., and London: Sage.

Balter, M. D., and Levine, J. (1971). Character and extent of psychotropic drug usage in the United States. Paper presented at the Fifth World Congress on Psychiatry, Mexico City.

Bartlett, M. S. (1941). The statistical significance of canonical correlations. *Biometrika, 32*, 29–38.

———. (1954). A note on the multiplying factors for various chi square approximations. *Journal of the Royal Statistical Society, 16* (Series B), 296–298.

Beatty, J. R. (1977). Identifying decision-making policies in the diagnosis of learning disabilities. *Journal of Learning Disabilities, 10*(4), 201–209.

Bem, S. L. (1974). The measurement of psychological androgyny. *Journal of Consulting and Clinical Psychology, 42*, 155–162.

Benedetti, J. K., and Brown, M. B. (1978). Strategies for the selection of log-linear models. *Biometrics, 34*, 680–686.

Bentler, P. M. (1980). Multivariate analysis with latent variables: Causal modeling. *Annual Review of Psychology, 31*, 419–456.

———. (1985). Theory and implementation of E Q S: A structural equations program. Los Angeles: BMDP Statistical Software, Inc.

Bock, R. D. (1966). Contributions of multivariate experimental designs to educational research. In R. B. Cattell, ed., *Handbook of Multivariate Experimental Psychology*. Chicago: Rand McNally.

———. (1975). *Multivariate Statistical Methods in Behavioral Research*. New York: McGraw-Hill.

Bock, R. D., and Haggard, E. A. (1968). The use of multivariate analysis of variance in behavioral research. In D. K. Whitla, ed., *Handbook of Measurement and Assessment in Behavioral Sciences*. Reading, Mass.: Addison-Wesley.

Box, G. E. P., and Cox, D. R. (1964). An analysis of transformations. *Journal of the Royal Statistical Society, 26* (Series B), 211–243.

Bradley, J. V. (1982). The insidious L-shaped distribution. *Bulletin of the Psychonomic Society, 20*(2), 85–88.

———. (1984). The complexity of nonrobustness effects. *Bulletin of the Psychonomic Society, 22*(3), 250–253.

Bradley, R. H., and Gaa, J. P. (1977). Domain-specific aspects of locus of control: Implications for modifying locus of control orientation. *Journal of School Psychology, 15*(1), 18–24.

Brown, M. B. (1976). Screening effects in multidimensional contingency tables. *Applied Statistics, 25,* 37–46.

Brown, R. D., Braskamp, L. A., and Newman, D. L. (1978). Evaluator credibility as a function of report style. *Evaluation Quarterly, 2*(2), 331–334.

Browne, M. W. (1975). Predictive validity of a linear regression equation. *British Journal of Mathematical and Statistical Psychology, 28,* 79–87.

Burgess, E., and Locke, H. (1960). *The Family,* 2nd ed. New York: American Book.

Caffrey, B., and Lile, S. (1976). Similarly of attitudes toward science on the part of psychology and physics students. *Teaching of Psychology, 3*(1), 24–26.

Campbell, D. R., and Stanley, J. C. (1966). *Experimental and Quasi-experimental Designs for Research.* New York: Rand McNally.

Cattell, R. B. (1957). *Personality and Motivation Structures and Measurement.* Yonkers-on-Hudson, N.Y.: World Book.

———. (1966). The scree test for the number of factors. *Multivariate Behavioral Research, 1,* 245–276.

Cattell, R. B., and Baggaley, A. R. (1960). The salient variable similarity index for factor matching. *British Journal of Statistical Psychology, 13,* 33–46.

Cattell, R. B., Balcar, K. R., Horn, J. L., and Nesselroade, J. R. (1969). Factor matching procedures: An improvement of the *s* index; with tables. *Eductional and Psychological Measurement, 29,* 781–792.

Cattin, P. (1980). Note on the estimation of the squared cross-validated multiple correlation of a regression model. *Psychological Bulletin, 87*(1), 63–65.

Cohen, E., and Burns, P. (1977). SPSS-MANOVA. Document No. 413 (Rev. A). Evanston, Ill.: Northwestern University, Vogelback Computing Center.

Cohen, J., and Cohen, P. (1975). *Applied Multiple Regression/Correlation Analysis for the Behavioral Sciences.* New York: Erlbaum.

Cohen, P., Gaughran, E., and Cohen, J. (1979). Age patterns of childbearing: A canonical analysis. *Multivariate Behavioral Research, 14,* 75–89.

Comrey, A. L. (1962). The minimum residual method of factor analysis. *Psychological Reports, 11,* 15–18.

———. (1973). *A First Course in Factor Analysis.* New York: Academic Press.

Cooke, T. D., and Campbell, D. T. (1979). *Quasi-experimentation: Design and Analysis Issues for Field Settings.* Chicago: Rand McNally; Boston: Houghton Mifflin.

Cooley, W. W., and Lohnes, P. R. (1971). *Multivariate Data Analysis.* New York: Wiley.

Cornbleth, T. (1977). Effects of a protected hospital ward area on wandering and nonwandering geriatric patients. *Journal of Gerontology, 32*(5), 573–577.

Curtis, B., and Simpson, D. D. (1977). Differences in background and drug use history among three types of drug users entering drug therapy programs. *Journal of Drug Education, 7*(4), 369–379.

Dillon, W. R., and Goldstein, M. (1984). *Multivariate Analysis: Methods and Applications.* New York: Wiley.

Dixon, W. J., ed. (1985). *BMDP Statistical Software: 1985 Printing.* Berkeley: University of California Press.

Edwards, A. L. (1976). *An Introduction to Linear Regression and Correlation.* San Francisco: Freeman.

Eysenck, H. J., and Eysenck, S. B. G. (1963). *The Eysenck Personality Inventory.* San Diego, Calif.: Educational and Industrial Testing Service; London: University of London Press.

Featherman, D. (1973). Metrics of occupational status reconciled to the 1970 Bureau of Census Classification of Detailed Occupational Titles (based on Census Technical Paper No. 26, ''1970 Occupational and Industry Classification Systems in Terms of their 1960 Occupational and Industry Elements''). Washington, D.C.: U.S. Government Printing Office. (Update of Duncan's socioeconomic status metric described in J. Reiss, et al., *Occupational and Social Status.* New York: Free Press.)

Fidell, S. (1978). Nationwide urban noise survey. *Journal of the Acoustical Society of America, 16*(1), 198–206.

Fillenbaum, G. G., and Wallman, L. M. (1984). Change in household composition of the elderly: A preliminary investigation. *Journal of Gerontology, 39,* 342–349.

Fleming, J. S., and Pinneau, S. R. (1988). Measuring the stability of linear weighting systems via correlated scoring functions. In review.

Fornell, C. (1979). External single-set components analysis of multiple criterion/multiple predictor variables. *Multivariate Behavioral Research, 14,* 323–338.

Frane, J. W. (August 3, 1977). Personal communication. Health Sciences Computing Facility, University of California, Los Angeles.

———. (1980). The univariate approach to repeated measures-foundation, advantages, and caveats. BMD Technical Report No. 69. Health Sciences Computing Facility, University of California, Los Angeles.

Giarrusso, R. (1977). The effects of attitude similarity and physical attractiveness on romantic attraction and time perception. Masters thesis, California State University, Northridge.

Gini, C. (1912). *Variabilité Mutabilitá: Contributo allo Studio delle Distribuzioni e delle Relazioni Statistiche.* Bologna: Cuppini.

Glock, C., Ringer, B., and Babbie, E. (1967). *To Comfort and to Challenge.* Berkeley: University of California Press.

Goodman, L. A. (1978). *Analyzing Qualitative/Categorical Data.* Cambridge, Mass: Abt Books.

Goodman, M. J., Chung, C. S., and Gilbert, F. (1974). Racial variation in diabetes mellitus in Japanese and Caucasians living in Hawaii. *Journal of Medical Genetics, 11,* 328–338.

Goodman, M. J., Steward, C. J., and Gilbert, F., Jr. (1977). Patterns of menopause: A study of certain medical physiological variables among Caucasian and Japanese women living in Hawaii. *Journal of Gerontology, 32*(3), 291–298.

Gorsuch, R. L. (1983). *Factor Analysis.* Hillsdale, N.J.: Erlbaum.

Gray-Toft, P. (1980). Effectiveness of a counseling support program for hospice nurses. *Journal of Counseling Psychology, 27,* 346–354.

Haberman, S. J. (1982). Analysis of dispersion of multinominal responses. *Journal of the American Statistical Association, 77,* 568–580.

Hakstian, A. R., Roed, J. C., and Lind, J. C. (1979). Two-sample T^2 procedure and the assumption of homogeneity of covariance matrices. *Psychological Bulletin, 86,* 1255–1263.

Hall, S. M., Hall, R. G., DeBoer, G., and O'Kulitch, P. (1977). Self and external management compared with psychotherapy in the control of obesity. *Behavior Research and Therapy, 15*, 89–95.

Harman, H. H. (1967). *Modern Factor Analysis,* 2nd ed. Chicago: University of Chicago Press.

———. *Modern Factor Analysis,* 3rd ed. Chicago: University of Chicago Press.

Harman, H. H., and Jones, W. H. (1966). Factor analysis by minimizing residuals (Minres). *Psychometrika, 31,* 351–368.

Harris, R. J. (1975). *Primer of Multivariate Statistics.* New York: Academic Press.

Heise, D. R. (1975). *Causal Analysis.* New York: Wiley.

Hoffman, D., and Fidell, L. S. (1979). Characteristics of androgynous, undifferentiated, masculine and feminine middle-class women. *Sex Roles, 5*(6), 765–781.

Hull, C. H., and Nie, N. H. (1981). *SPSS Update 7–9.* New York: McGraw-Hill.

Jakubczak, L. F. (1977). Age differences in the effects of palatability of diet on regulation of caloric intake and body weight of rats. *Journal of Gerontology, 32*(1), 49–57.

Johnson, G. (1955). An instrument for the assessment of job satisfaction. *Personnel Psychology, 8,* 27–37.

Joreskog, K. G., and Sorbom, D. (1978). *LISREL IV Users Guide.* Chicago: National Education Research.

Kaiser, H. F. (1970). A second-generation Little Jiffy. *Psychometrika, 35,* 401–415.

———. (1974). An index of factorial simplicity. *Psychometrika, 39,* 31–36.

Keppel, G. (1973). *Design and Analysis: A Researcher's Handbook.* Englewood Cliffs, N.J.: Prentice-Hall.

———. (1982). *Design and Analysis: A Researcher's Handbook,* 2nd ed. Englewood Cliffs, N.J.: Prentice-Hall.

Kish, L. (1965). *Survey Sampling.* New York: Wiley.

Knoke, D., and Burke, P. J. (1980). *Log-linear Models.* Beverly Hills, Calif.: Sage.

Langer, T. (1962). A 22-item screening score of psychiatric symptoms indicating impairment. *Journal of Health and Human Behavior, 3,* 269–276.

Larzelere, R. E., and Mulaik, S. A. (1977). Single-sample test for many correlations. *Psychological Bulletin, 84*(3), 557–569.

Lawley, D. N., and Maxwell, A. E. (1963). *Factor Analysis as a Statistical Method.* London: Butterworth.

Lee, W. (1975). *Experimental Design and Analysis.* San Francisco: Freeman.

Levine, M. S. (1977). *Canonical Analysis and Factor Comparison.* Sage University Paper Series in Quantitative Applications in the Social Sciences, Series No. 07–006. Beverly Hills, Calif., and London: Sage.

Locke, H., and Wallace, K. (1959). Short marital-adjustment and prediction tests: Their reliability and validity. *Marriage and Family Living, 21,* 251–255.

McLeod, H. M., Brown, J. D., and Becker, L. B. (1977). Watergate and the 1974 congressional elections. *Public Opinion Quarterly, 41,* 181–195.

McNeil, K. S., Kelly, F. J., and McNeil, J. T. (1975). *Testing Research Hypotheses Using Multiple Linear Regression.* Carbondale: Southern Illinois University Press.

Maki, J. E., Hoffman, D. M., and Berk, R. A. (1978). A time series analysis of the impact of a water conservation campaign. *Evaluation Quarterly, 2*(1), 107–118.

Marascuilo, L. A., and Levin, J. R. (1983). *Multivariate Statistics in the Social Sciences: A Researcher's Guide*. Monterey, Calif.: Brooks/Cole.

Mardia, K. V. (1971). The effect of nonnormality on some multivariate tests and robustness to nonnormality in the linear model. *Biometrika, 58*(1), 105–121.

Marshall, S. P. (1983). Sex differences in mathematical errors: An analysis of distractor choices. *Journal for Research in Mathematics Education, 14*, 325–336.

Menozzi, P., Piazza, A., and Cavalli-Sforza, L. (1978). Synthetic maps of human gene frequencies in Europeans. *Science, 201*, 786–792.

Merrill, P. F., and Towle, J. J. (1976). The availability of objectives and performance in a computer-managed graduate course. *Journal of Experimental Education, 45*(1), 12–29.

Miller, J. K., and Farr, S. D. (1971). Bimultivariate redundancy: A comprehensive measure of interbattery relationship. *Multivariate Behavioral Research, 6*, 313–324.

Milligan, G. W. (1980). Factors that affect Type I and Type II error rates in the analysis of multidimensional contingency tables. *Psychological Bulletin, 87*, 238–244.

Mitchell, L. K., and Krumboltz, J. D. (1984, April). *The effect of training in cognitive restructuring on the inability to make career decisions*. Paper presented at the meeting of the Western Psychological Association, Los Angeles.

Moser, C. A., and Kalton, G. (1972). *Survey Methods in Social Investigation*. New York: Basic Books.

Mosteller, F., and Tukey, J. W. (1977). *Data Analysis and Regression*. Reading, Mass.: Addison-Wesley.

Mulaik, S. A. (1972). *The Foundation of Factor Analysis*. New York: McGraw-Hill.

Myers, J. L. (1979). *Fundamentals of Experimental Design*, 3rd ed. Boston: Allyn & Bacon.

Nicholson, W., and Wright, S. R. (1977). Participants' understanding of the treatment in policy experimentation. *Evaluation Quarterly, 1*(2), 245–268.

Nie, N. H., Hull, C. H., Jenkins, J. G., Steinbrenner, K., and Bent, D. H. (1975). *Statistical Package for the Social Sciences*, 2nd ed. New York: McGraw-Hill.

Norusis, M. J. (1985). *SPSS^x: Advanced Statistics Guide*. New York: McGraw-Hill.

O'Kane, J. M., Barenblatt, L., Jensen, P. K., and Cochran, L. T. (1977). Anticipatory socialization and male Catholic adolescent sociopolitical attitude. *Sociometry, 40*(1), 67–77.

Olson, C. L. (1976). On choosing a test statistic in multivariate analysis of variance. *Psychological Bulletin, 83*(4), 579–586.

———. (1979). Practical considerations in choosing a MANOVA test statistic: A rejoinder to Stevens. *Psychological Bulletin, 86*, 1350–1352.

Overall, J. E., and Spiegel, D. K. (1969). Concerning least squares analysis of experimental data. *Psychological Bulletin, 72*(5), 311–322.

Overall, J. E., and Woodward, J. A. (1977). Nonrandom assignment and the analysis of covariance. *Psychological Bulletin, 84*(3), 588–594.

Pessemier, E. A., Bemmoar, A. C., and Hanssens, D. M. (1977). Willingness to supply human body parts: Some empirical results. *Journal of Consumer Research, 4*(3), 131–140.

Pope, K. S., Keith-Spiegel, P., and Tabachnick, B. (1986). Sexual attraction to clients: The human therapist and the (sometimes) inhuman training system. *American Psychologist, 41*, 147–158.

Porter, J. R., and Albert, A. A. (1977). Subculture or assimilation? A cross-cultural analysis of religion and women's role. *Journal for the Scientific Study of Religion, 16*, 234–359.

Price, B. (1977). Ridge regression: Application to nonexperimental data. *Psychological Bulletin, 84*(4), 759–766.

Rahe, R. H. (1974). The pathway between subjects' recent life changes and their near-future illness reports: Representative results and methodological issues. In B. S. Dohrenwend and B. P. Dohrenwend, eds., *Stressful Life Events: Their Nature and Effects*. New York: Wiley.

Rao, C. R. (1952). *Advanced Statistical Methods in Biometric Research*. New York: Wiley.

Rock, D. L. (1982). Appointments for intake interviews: An analysis of appointment-keeping behavior at an urban community mental health center. *Psychological Reports, 58*, 863–868.

Rollins, B., and Feldman, H. (1970). Marital satisfaction over the family life cycle. *Journal of Marriage and Family, 32*, 29–38.

Rotter, J. B. (1966). Generalized expectancies for internal versus external control of reinforcement. *Psychological Monographs, 80*(1, Whole No. 609).

Rozeboom, W. W. (1979). Ridge regression: Bonanza or beguilement? *Psychological Bulletin, 82*(6), 242–249.

Rummel, R. J. (1970). *Applied Factor Analysis*. Evanston, Ill.: Northwestern University Press.

St. Pierre, R. G. (1978). Correcting covariables for unreliability. *Evaluation Quarterly, 2*(3), 401–420.

SAS Institute Inc. (1985). *SAS User's Guide: Statistics, Version 5 Edition*. Cary, NC: SAS Institute Inc.

Schall, J. J., and Pianka, E. R. (1978). Geographical trends in numbers of species. *Science, 201*, 679–686.

Scheffé, H. A. (1953). A method of judging all contrasts in the analysis of variance. *Biometrika, 40*, 87–104.

Shannon, C. E. (1948). A mathematical theory of communication. *Bell System Technical Journal, 50*, 379–423; 623–656.

Singh, B., Greer, P. R., and Hammond, R. (1977). An evaluation of the use of the law in a free society materials on "responsibility." *Evaluation Quarterly, 1*(4), 621–628.

Spence, J., and Helmreich, R. (1972). The attitudes toward women scale: An objective instrument to measure attitude toward rights and roles of women in contemporary society. *Journal Supplementary Abstract Service* (Catalogue of Selected Documents in Psychology), *2*, 66.

SPSS Inc. (1985). *SPSS^x Statistical Algorithms*. Chicago: SPSS Inc.

———. (1986). *SPSS^x User's Guide, Edition 2*. New York: McGraw-Hill.

Steiger, J. H. (1980). Tests for comparing elements of a correlation matrix. *Psychological Bulletin, 87*(2), 245–251.

Stewart, D., and Love, W. (1968). A general canonical index. *Psychological Bulletin, 70*, 160–163.

Strober, M. H., and Weinberg, C. B. (1977). Working wives and major family expenditures. *Journal of Consumer Research, 4*(3), 141–147.

Tabachnick, B. G., and Fidell, L. S. (1983). *Using Multivariate Statistics*. New York: Harper & Row.

Tatsuoka, M. M. (1971). *Multivariate Analysis: Techniques for Educational and Psychological Research*. New York: Wiley.

———. (1975). Classification procedures. In D. J. Amick and H. J. Walberg, eds., *Introductory Multivariate Analysis*. Berkeley, Calif.: McCutchan.

Theil, H. (1970). On the estimation of relationships involving qualitative variables. *American Journal of Sociology, 76,* 103–154.

Thurstone, L. L. (1947). *Multiple Factor Analysis.* Chicago: University of Chicago Press.

Timm, N. H. (1975). *Multivariate Statistics with Applications in Education and Psychology.* Belmont, Calif.: Brooks/Cole.

Vandenbos, G. R., and Stapp, J. (1983). Service providers in psychology: Results of the 1982 APA human resources survey. *American Psychologist, 38,* 1330–1352.

Vaughn, G. M., and Corballis, M. C. (1969). Beyond tests of significance: Estimating strength of effects in selected ANOVA designs. *Psychological Bulletin, 72*(3), 204–213.

Wade, T. C., and Baker, T. B. (1977). Opinions and use of psychological tests: A survey of clinical psychologists. *American Psychologist, 32*(10), 874–882.

Wernimont, P. F., and Fitzpatrick, S. (1972). The meaning of money. *Journal of Applied Psychology, 56*(3), 218–226.

Wesolowsky, G. O. (1976). *Multiple Regression and Analysis of Variance.* New York: Wiley-Interscience.

Wherry, R. J., Sr. (1931). A new formula for predicting the shrinkage of the coefficient of multiple correlation. *Annals of Mathematical Statistics, 2,* 440–457.

Wiener, Y., and Vaitenas, R. (1977). Personality correlates of voluntary midcareer change in enterprising occupation. *Journal of Applied Psychology, 62*(6), 706–712.

Wilkinson, L. (1979). Tests of significance in stepwise regression. *Psychological Bulletin, 86*(1), 168–174.

———. (1986). *SYSTAT: The System for Statistics.* Evanston, Ill.: SYSTAT, Inc.

Wilkinson, L., and Dallal, G. E. (1981). Tests of significance in forward selection regression with an *F*-to-enter stopping rule. *Technometrics, 23*(4), 377–380.

Willis, J. W., and Wortman, C. B. (1976). Some determinants of public acceptance of randomized control group experimental designs. *Sociometry, 39*(2), 91–96.

Winer, B. J. (1971). *Statistical Principles in Experimental Design,* 2nd ed. New York: McGraw-Hill.

Wingard, J. A., Huba, G. J., and Bentler, P. M. (1979). The relationship of personality structure to patterns of adolescent substance use. *Multivariate Behavioral Research, 14,* 131–143.

Wong, M. R., and Allen, T. A. (1976). A three dimensional structure of drug attitudes. *Journal of Drug Education, 6*(2), 181–191.

Woodward, J. A., and Overall, J. E. (1975). Multivariate analysis of variance by multiple regression methods. *Psychological Bulletin, 82*(1), 21–32.

Young, R. K., and Veldman, D. J. (1977). *Introductory Statistics for the Behavioral Sciences,* 3rd ed. New York: Holt, Rinehart and Winston.

Index